Communications in Computer and Information Science 1159

More information about this series at http://www.springer.com/series/7899

Linqiang Pan · Jing Liang · Boyang Qu (Eds.)

Bio-inspired Computing: Theories and Applications

14th International Conference, BIC-TA 2019
Zhengzhou, China, November 22–25, 2019
Revised Selected Papers, Part I

Springer

Editors
Linqiang Pan
Huazhong University of Science and Technology
Wuhan, China

Jing Liang
Zhengzhou University
Zhengzhou, China

Boyang Qu
Zhongyuan University of Technology
Zhengzhou, China

ISSN 1865-0929 ISSN 1865-0937 (electronic)
Communications in Computer and Information Science
ISBN 978-981-15-3424-9 ISBN 978-981-15-3425-6 (eBook)
https://doi.org/10.1007/978-981-15-3425-6

This Springer imprint is published by the registered company Springer Nature Singapore Pte Ltd.
The registered company address is: 152 Beach Road, #21-01/04 Gateway East, Singapore 189721, Singapore

Preface

Bio-inspired computing is a field of study that abstracts computing ideas (data structures, operations with data, ways to control operations, computing models, artificial intelligence, etc.) from biological systems or living phenomena such as cells, tissues, neural networks, the immune system, an ant colony, or evolution. The areas of bio-inspired computing include neural networks, brain-inspired computing, neuromorphic computing and architectures, cellular automata and cellular neural networks, evolutionary computing, swarm intelligence, fuzzy logic and systems, DNA and molecular computing, membrane computing, and artificial intelligence, as well as their application in other disciplines such as machine learning, image processing, computer science, and cybernetics. Bio-Inspired Computing: Theories and Applications (BIC-TA) is a series of conferences that aims to bring together researchers working in the main areas of bio-inspired computing to present their recent results, exchange ideas, and cooperate in a friendly framework.

Since 2006, the conference has taken place in Wuhan (2006), Zhengzhou (2007), Adelaide (2008), Beijing (2009), Liverpool and Changsha (2010), Penang (2011), Gwalior (2012), Anhui (2013), Wuhan (2014), Anhui (2015), Xi'an (2016), Harbin (2017), and Beijing (2018). Following the success of previous editions, the 14th International Conference on Bio-Inspired Computing: Theories and Applications (BIC-TA 2019) was held in Zhengzhou, China, during November 22–25, 2019, and was organized by Zhongyuan University of Technology with the support of Zhengzhou University, Zhengzhou University of Light Industry, Henan Normal University, Henan University of Technology, North China University of Water Resources and Electric Power, Pingdingshan University, Nanyang Institute of Technology, Peking University, Huazhong University of Science and Technology, Henan Electrotechnical Society, and Operations Research Society of Hubei.

We would like to thank the President of Zhongyuan University of Technology, Prof. Zongmin Wang, and Academician of the Chinese Academy of Engineering, Prof. Xiangke Liao, for commencing the opening ceremony.

Thanks are also given to the keynote speakers for their excellent presentations: Mitsuo Gen (Tokyo University of Science, Japan), Yaochu Jin (University of Surrey, UK), Derong Liu (Guangdong University of Technology, China), Ponnuthurai Nagaratnam Suganthan (Nanyang Technological University, Singapore), Kay Chen Tan (City University of Hong Kong, China), Mengjie Zhang (Victoria University of Wellington, New Zealand), and Ling Wang (Tsinghua University, China).

We gratefully acknowledge Zongmin Wang, Qingfu Zhang, Jin Xu, Haibin Duan, Zhoufeng Liu, Xiaowei Song, Jinfeng Gao, Yanfeng Wang, Yufeng Peng, Dexian Zhang, Hongtao Zhang, Xichang Xue, Qinghui Zhu, and Xiaoyu An for their contribution in organizing the conference.

A special thanks goes to Prof. Guangzhao Cui for his extensive guidance and assistance in the local affairs and financial support of the conference.

BIC-TA 2019 attracted a wide spectrum of interesting research papers on various aspects of bio-inspired computing with a diverse range of theories and applications. 121 papers were selected for inclusion in the BIC-TA 2019 proceedings, publish by Springer Nature in the series *Communications in Computer and Information Science* (CCIS).

We are grateful to the external referees for their careful and efficient work in the reviewing process, and in particular the Program Committee chairs Maoguo Gong, Rammohan Mallipeddi, Ponnuthurai Nagaratnam Suganthan, Zhihui Zhan, and the Program Committee members. The warmest thanks are given to all the authors for submitting their interesting research work.

We thank Lianghao Li, Wenting Xu, Taosheng Zhang, et al. for their help in collecting the final files of the papers and editing the volume. We thank Zheng Zhang and Lianlang Duan for their contribution in maintaining the website of BIC-TA 2019 (http://2019.bicta.org/). We also thank all the other volunteers, whose efforts ensured the smooth running of the conference.

Special thanks are due to Springer Nature for their skilled cooperation in the timely production of these volumes.

December 2019

Linqiang Pan
Jing Liang
Boyang Qu

Organization

Steering Committee

Atulya K. Nagar	Liverpool Hope University, UK
Gheorghe Paun	Romanian Academy, Romania
Giancarlo Mauri	Università di Milano-Bicocca, Italy
Guangzhao Cui	Zhengzhou University of Light Industry, China
Hao Yan	Arizona State University, USA
Jin Xu	Peking University, China
Jiuyong Li	University of South Australia, Australia
Joshua Knowles	The University of Manchester, UK
K. G. Subramanian	Liverpool Hope University, UK
Kalyanmoy Deb	Michigan State University, USA
Kenli Li	University of Hunan, China
Linqiang Pan (Chair)	Huazhong University of Science and Technology, China
Mario J. Perez-Jimenez	University of Sevilla, Spain
Miki Hirabayashi	National Institute of Information and Communications Technology, Japan
Robinson Thamburaj	Madras Christian College, India
Thom LaBean	North Carolina State University, USA
Yongli Mi	Hong Kong University of Science and Technology, Hong Kong

Honorable Chair

Zongmin Wang	Zhongyuan University of Technology, China

General Chairs

Qingfu Zhang	City University of Hong Kong, China
Jin Xu	Peking University, China
Haibin Duan	Beihang University, China
Zhoufeng Liu	Zhongyuan University of Technology, China
Jing Liang	Zhengzhou University, China

Program Committee Chairs

Boyang Qu	Zhongyuan University of Technology, China
Linqiang Pan	Huazhong University of Science and Technology, China
Dunwei Gong	China University of Mining and Technology, China

Maoguo Gong	Xidian University, China
Zhihui Zhan	South China University of Technology, China
Rammohan Mallipeddi	Kyungpook National University, South Korea
P. N. Suganthan	Nanyang Technological University, Singapore

Organizing Chairs

Xiaowei Song	Zhongyuan University of Technology, China
Jinfeng Gao	Zhengzhou University, China
Yanfeng Wang	Zhengzhou University of Light Industry, China
Yufeng Peng	Henan Normal University, China
Dexian Zhang	Henan University of Technology, China
Hongtao Zhang	North China University of Water Resources and Electric Power, China
Xichang Xue	Pingdingshan University, China
Qinghui Zhu	Nanyang Institute of Technology, China
Xiaoyu An	Henan Electrotechnical Society, China

Special Session Chairs

Yinan Guo	China University of Mining and Technology, China
Shi Cheng	Shaanxi Normal University, China

Tutorial Chairs

He Jiang	Dalian University of Technology, China
Wenyin Gong	China University of Geosciences, China

Publicity Chairs

Ling Wang	Tsinghua University, China
Aimin Zhou	East China Normal University, China
Hongwei Mo	Harbin Engineering University, China
Ke Tang	Southern University of Science and Technology, China
Weineng Chen	South China University of Technology, China
Han Huang	South China University of Technology, China
Zhihua Cui	Taiyuan University of Science and Technology, China
Chaoli Sun	Taiyuan University of Science and Technology, China
Handing Wang	Xidian University, China
Xingyi Zhang	Anhui University, China

Local Chairs

Kunjie Yu	Zhengzhou University, China
Chunlei Li	Zhongyuan University of Technology, China
Xiaodong Zhu	Zhengzhou University, China

Publication Chairs

Yuhui Shi	Southern University of Science and Technology, China
Zhihua Cui	Taiyuan University of Science and Technology, China
Boyang Qu	Zhongyuan University of Technology, China

Registration Chairs

Xuzhao Chai	Zhongyuan University of Technology, China
Li Yan	Zhongyuan University of Technology, China
Yuechao Jiao	Zhongyuan University of Technology, China

Program Committee

Muhammad Abulaish	South Asian University, India
Chang Wook Ahn	Gwangju Institute of Science and Technology, South Korea
Adel Al-Jumaily	University of Technology Sydney, Australia
Bin Cao	Hebei University of Technology, China
Junfeng Chen	Hoahi University, China
Wei-Neng Chen	Sun Yat-sen University, China
Shi Cheng	Shaanxi Normal University, China
Tsung-Che Chiang	National Taiwan Normal University, China
Kejie Dai	Pingdingshan University, China
Bei Dong	Shanxi Normal University, China
Xin Du	Fujian Normal University, China
Carlos Fernandez-Llatas	Universitat Politecnica de Valencia, Spain
Shangce Gao	University of Toyama, Japan
Wenyin Gong	China University of Geosciences, China
Shivaprasad Gundibail	MIT, Manipal Academy of Higher Education (MAHE), India
Ping Guo	Beijing Normal University, China
Yinan Guo	China University of Mining and Technology, China
Guosheng Hao	Jiangsu Normal University, China
Shan He	University of Birmingham, UK
Tzung-Pei Hong	National University of Kaohsiung, China
Florentin Ipate	University of Bucharest, Romania
Sunil Jha	Banaras Hindu University, India
He Jiang	Dalian University of Technology, China
Qiaoyong Jiang	Xi'an University of Technology, China
Liangjun Ke	Xian Jiaotong University, China
Ashwani Kush	Kurukshetra University, India
Hui L.	Xi'an Jiaotong University, China
Kenli Li	Hunan University, China
Yangyang Li	Xidian University, China
Zhihui Li	Zhengzhou University, China

Qunfeng Liu	Dongguan University of Technology, China
Xiaobo Liu	China University of Geosciences, China
Wenjian Luo	University of Science and Technology of China, China
Lianbo Ma	Northeastern University, China
Wanli Ma	University of Canberra, Australia
Xiaoliang Ma	Shenzhen University, China
Holger Morgenstern	Albstadt-Sigmaringen University, Germany
G. R. S. Murthy	Lendi Institute of Engineering and Technology, India
Akila Muthuramalingam	KPR Institute of Engineering and Technology, India
Yusuke Nojima	Osaka Prefecture University, Japan
Linqiang Pan (Chair)	Huazhong University of Science and Technology, China
Andrei Paun	University of Bucharest, Romania
Xingguang Peng	Northwestern Polytechnical University, China
Chao Qian	University of Science and Technology of China, China
Rawya Rizk	Port Said University, Egypt
Rajesh Sanghvi	G. H. Patel College of Engineering & Technology, India
Ronghua Shang	Xidian University, China
Zhigang Shang	Zhengzhou University, China
Ravi Shankar	Florida Atlantic University, USA
V. Ravi Sankar	GITAM University, India
Bosheng Song	Huazhong University of Science and Technology, China
Tao Song	China University of Petroleum, China
Jianyong Sun	University of Nottingham, UK
Yifei Sun	Shaanxi Normal University, China
Handing Wang	Xidian University, China
Yong Wang	Central South University, China
Hui Wang	Nanchang Institute of Technology, China
Hui Wang	South China Agricultural University, China
Gaige Wang	Ocean University of China, China
Sudhir Warier	IIT Bombay, China
Slawomir T. Wierzchon	Polish Academy of Sciences, Poland
Zhou Wu	Chongqing University, China
Xiuli Wu	University of Science and Technology Beijing, China
Bin Xin	Beijing Institute of Technology, China
Gang Xu	Nanchang University, China
Yingjie Yang	De Montfort University, UK
Zhile Yang	Shenzhen Institute of Advanced Technology, Chinese Academy of Sciences, China
Kunjie Yu	Zhengzhou University, China
Xiaowei Zhang	University of Science and Technology of China, China
Jie Zhang	Newcastle University, UK
Gexiang Zhang	Southwest Jiaotong University, China
Defu Zhang	Xiamen University, China

Contents – Part I

Contents – Part II

Bioinformatics and Systems Biology

Complex Networks

DNA and Molecular Computing

Neural Networks and Artificial Intelligence

Evolutionary Computation and Swarm Intelligence

Review on the Improvement and Application of Ant Colony Algorithm

Dongping Qiao, Wentong Bai(✉), Kanghong Wang, and Yajing Wang

Henan Key Laboratory of Intelligent Manufacturing of Mechanical Equipment, Zhengzhou University of Light Industry, Zhengzhou 450002, China
baiwentong2019@163.com

Abstract. The ant colony optimization algorithm is an approximation algorithm and it is also a probabilistic algorithm for finding optimized paths. Many combinatorial optimization problems have been solved by the ant colony optimization algorithm. Firstly, the basic principles of ant colony algorithm are first introduced by this reference. Secondly, it briefs several improvements method of ant colony algorithm and the application in solving practical problems, including the improvement of ant colony algorithm, parameter combination tuning and the application of ant colony algorithm in combination optimization problem. Finally, the problems existing in the ant colony algorithm are summarized and forecasted in this article.

Keywords: Ant colony algorithm · Optimization · Parameter factor · Hybrid algorithm

1 Introduction

In 1991, Italian scholar Dorigo et al. proposed the ant colony algorithm and successfully applied to solve the traveling salesman problem and the secondary distribution problem [1]. Since the introduction of the ant colony algorithm, many research institutions and researchers have paid close attention to it. In view of the simple operation of the ant colony algorithm, the strong positive feedback mechanism, and the ease of combining other heuristic algorithms, it is widely used in many fields. As a new optimization method, it can solve many combinatorial optimization problems [2], such as shop scheduling problems, vehicle routing problems, robot path planning, network routing and image processing, etc. Due to the lack of information in the early stage of ant colony algorithm, slow convergence, and susceptibility to stagnation, many domestic and foreign professional researchers have improved the ant colony algorithm to improve the stability of the algorithm.

This paper summarizes the development and research results of ant colony algorithm in recent years, introduces several improved forms of ant colony algorithm and its application in different fields, and summarizes the existing problems of ant colony algorithm. In this way, hope to provide the basis for the further development of ant colony algorithm.

L. Pan et al. (Eds.): BIC-TA 2019, CCIS 1159, pp. 3–14, 2020.
https://doi.org/10.1007/978-981-15-3425-6_1

2 Principle and Model of Ant Colony Algorithm

2.1 The Basic Principle of Ant Colony Algorithm

The ant colony algorithm mimics the process of ants looking for food in nature. When ants are looking for food, the path from the nest to the shorter food distance will accumulate more pheromones in the same period of time, so that the probability of being selected by the ants in the next cycle will become larger, and then the shortest path will be found [3,4].

The basic principle of the ant colony algorithm lies in the following three points: Firstly, the ants will continuously release pheromones in the process of searching for food. Secondly, the ant has a perceptual ability to perceive a certain range, and selects the position of the next node by comparing the strength of the pheromone concentration left by other ants in each path. Thirdly, the concentration of ant releasing pheromones will continue to evaporate over time [5,6].

2.2 Ant Colony Algorithm Model

The ant colony algorithm was originally used to solve the traveling salesman problem. The ant colony system is used to solve the TSP problem of N cities to describe the model of the ant colony algorithm [7]. The following markers are introduced: N represents the number of cities; M represents the number of ants; d_{ij} represents the distance between the city i and the city j; τ_{ij} represents the residual information on the city i and the city j connection; in a general way, $\eta_{ij}(t) = 1/d_{ij}(t)$ is used in tourist problem, where η_{ij} is called city visibility [8,9].

(1) State transition probability criterion. Each ant selects the location of the next city according to the size of the pheromone concentration on each path. $P_{ij}^{k}(t)$ represents the probability that ant k moves from city i to target city j at time T:

$$P_{ij}^{k} = \begin{cases} \frac{[\tau_{ij}(t)]^{\alpha}\,[\eta_{ij}]^{\beta}}{\sum\limits_{u \in allowed} [\tau_{ij}(t)]^{\alpha}\,[\eta_{ij}]^{\beta}} & if\ j \in allowed \\ 0 & otherwise \end{cases} \quad (1)$$

Among them, $allowed \in \{C - tabu_k\}$ represents that the ant can select a collection of cities in the next step. α represents the relative importance of pheromone concentration, $\tau_{ij}(t)$ and β represents the relative importance of urban visibility η_{ij}.

(2) Pheromone update rules. When all the ants have finished the path, they will update the pheromone of all the paths. This is called global pheromone update for short.

$$\tau_{ij}\,(t+n) = (1-\rho) * \tau_{ij}\,(t) + \Delta\tau_{ij}\,(t, t+n) \quad (2)$$

$$\Delta\tau_{ij}(t, t+n) = \sum_{k=1}^{m} \Delta\tau_{ij}{}^{k}(t, t+n) \tag{3}$$

Among them, ℓ represents pheromone volatilization factor, $\Delta\tau_{ij}(t, t+n)$ represents the amount of pheromone added by the path (i, j) over the n time, $\Delta\tau_{ij}^{k}(t, t+n)$ represents the amount of pheromone left by the kth ant in the path (i, j) in this cycle. The local pheromone update rule can also be used here.

$$\Delta\tau_{ij}^{k}(t, t+n) = \begin{cases} \frac{Q}{L_k} & \frac{Q}{L_k} \in (i, j) \\ 0 & \text{other}size \end{cases} \tag{4}$$

Among them, L_k represents the distance that the ant has traveled in this cycle, Q represents the sum of pheromones released by ants after completing a cycle.

3 Improvement of Ant Colony Algorithm

3.1 Improvement of Pheromone Updating Method

According to the characteristics of ant colony foraging, different pheromone updating methods have different effects on the final results of ant colony algorithm, such as the search ability and convergence rate of the algorithm. In reference [2], Dorigo and Maniezzo et al. proposed an elitist strategy. After each iteration, all optimal solutions are given additional pheromone enhancements to increase the probability of being selected in the next cycle.

The improved Ant Colony System was proposed in reference [10]. In the ACS algorithm, only the global pheromone update strategy is adopted for the optimal path, and other paths adopt the local pheromone update strategy, and the taboo table is used for conditional constraints. This not only speeds up the ant colony algorithm to find the optimal solution, but also effectively avoids the algorithm stagnation.

In reference [11,12], the max and min ant colony system (MMAS) was proposed. In the MMAS algorithm, the feedback of the optimal path information is strengthened. In order to avoid the local optimization or stagnation of the algorithm, the pheromone concentration of each path is limited to $[\tau_{\min}, \tau_{\max}]$ intervals, thus ensuring the stability of the ant colony algorithm. The MMAS algorithm not only solves the problem of excessive dependence of the ant colony algorithm on the initial solution, but also improves the global search ability of the algorithm.

In view of the problem of low accuracy of algorithm in reference [13], a sort-based ant colony algorithm (Rank Based Ant System, RAS) was proposed. The RAS algorithm releases pheromones in proportion according to the length of the path explored by ants, arranges the paths that ants pass in ascending order, and assigns different weights according to the length of the path, the shorter the path is, the larger the weight value is. This increases the probability that the shortest path is selected.

In view of the problem of insufficient information exchange among ants, in reference [14] mutual information diffusion ant colony algorithm was proposed. The mathematical model of frequency assignment problem in unequal shortwave network is established to enhance the information exchange among ants and improve the convergence speed of the algorithm.

In addition, many algorithms for pheromone improvement strategies have been proposed, such as local pheromone correction strategy [15], adaptive pheromone volatilization [16] and defining direction information element to strengthen the optimal solution [17] etc. To some extent, the stability and accuracy of the algorithm are improved.

3.2 Improvement of Path Search Strategy

Ant colony algorithm can also improve the performance of the algorithm by improving the path selection strategy to reduce the probability that the algorithm falls into the local optimal solution. The improved ant colony algorithm (ACS) is an improvement of the path selection strategy, It is the optimization of the previous solution through the state transition rule, which accelerates the convergence speed of the algorithm and accumulates the current information state of the algorithm [18], which not only solves the algorithm convergence. Slow speed and easy to fall into the local optimal solution problem, and also enhance the global search ability of ant colony algorithm.

For transmission line path planning problems, the reference [19] proposed an improved ant colony algorithm for transmission line path search. Firstly, through geographic information system technology (GIS) and the remote sensing image technology (RS), the geographic data collection and processing, integration and classification of complex geographic information for the required path planning area are completed as the initial population of the ant colony algorithm; It is introduced into the fuzzy hierarchical analysis model to obtain the corresponding grid attributes and positions, so as to improve the search efficiency of the ant colony algorithm.

In view of the problem of Shortcomings of premature convergence and stagnation for ant colony algorithm, in reference [20], a non-intersecting algorithm (NIAS) was proposed. The algorithm judges the intersection relation between the pending path and the selected path, changes the path selection strategy, carries on the differential update of the information element to the path of the existence of the intersection phenomenon, and sets a weight function to realize the update of the pheromone. This not only optimizes the quality of the initial solution of the population, but also enhances the global search ability of the algorithm.

An ant colony algorithm with random disturbance characteristic is proposed, in reference [21], the algorithm not only presents the disturbance factor described by the inverted exponential curve, but also gives the corresponding disturbance strategy and random selection strategy. This new path selection strategy not only solves the stagnation of ant colony algorithm, but also makes the algorithm have better global search ability.

In addition, there are also many improvements to the path selection strategy that have been proposed, for example, a weighted-valued polymorphic ant colony algorithm is proposed in reference [22] etc. Reference [23] defines the dynamic adjustment of dynamic search-induced operators at different stages. All of these are improvements to the path selection strategy, which not only avoids the premature stagnation of the algorithm, but also improves the global search ability of the algorithm.

3.3 Improvement of Parameter Combination Optimization

The selection of parameters in the ant colony algorithm has a great influence on the performance of the algorithm. The proper combination of parameters can optimize the performance of the algorithm, so that the global optimal solution can be obtained. In reference [24], an ant colony algorithm based on dynamic variation and optimal individual variation of parameters was proposed. The algorithm sets local parameters and global parameters and each parameter factors can be adjusted according to the distribution of pheromones in the search space. It not only increases the dynamic performance of the algorithm, but also improves the speed of the algorithm.

In view of the problem of the lack of information and the number of iterations in the early stage of ant colony algorithm, in reference [25], the adaptive adjustment of parameters α, β and ρ was proposed. Each parameter will change according to the change of the number of iterations of the algorithm, so that the convergence speed of the algorithm can be accelerated while avoiding the algorithm falling into local optimum.

In view of the unreasonable selection of the parameters in the algorithm, the problem of the long running period and slow convergence of the algorithm is caused, in reference [26], the optimal parameter combination method for α, β, ρ, m and σ was proposed. The performance of the ant colony algorithm is optimized by the combination of parameters, and the performance of the algorithm is more stable.

In addition, the reference [27] proposes the micro-particle swarm optimization (PSO) algorithm to optimize the combination parameters. The reference [28] proposes optimizing the parameters α, β, ρ, m and σ, etc and the random factor t using a uniform design method. In reference [29], according to the experimental object, the appropriate test level is selected, and then the orthogonal test method is used to obtain the optimal combination of parameters. These are reference has adopt an algorithm to obtain the optimal parameter combination to improve the global search ability and convergence rate of the algorithm.

3.4 Intelligent Fusion of Ant Colony Algorithms

Although the ant colony algorithm has strong positive feedback ability, the lack of information in the initial stage of the algorithm leads to a slower evolution. The combination of ant colony algorithm and other algorithms can not only make up for the shortcomings of ant colony algorithm, but also combine the advantages of the two algorithms [30]. In the reference [31,32], the ant colony algorithm and the genetic algorithm are combined, that is, using the randomness and the rapidity of the genetic algorithm, the parallelism of the ant colony algorithm and the capability of the positive feedback mechanism are also utilized, and the solution efficiency of the single algorithm is greatly improved.

In reference [33], an artificial colony improved ant colony path planning algorithm (AI-ACA) was proposed. The algorithm first uses Artificial Immune Algorithm to find out the best combination of parameters of Ant Colony Algorithm, and then uses pheromone accumulation and the positive feedback mechanism of ant colony algorithm to find out the best path. By combining ant colony algorithm with artificial immune algorithm, the efficiency of single ant colony algorithm is improved effectively.

In reference [34], the combination of ant colony algorithm and neural network algorithm was proposed. Firstly, the existing data of BP neural network are used as learning samples, and the weights of BP neural network are trained by ant colony algorithm to optimize the neural network. After training, the current optimal parameters are stored to find the global optimal solution. Ant swarm neural network can effectively improve the accuracy and efficiency of the algorithm.

In reference [35], combination of ant colony algorithm and particle swarm optimization was proposed. This algorithm is a pheromone update mode which combines global asynchronous and elite strategy, and reasonably determines the iterative times of the hybrid algorithm. The advantages of the hybrid algorithm in solving large-scale TSP are obvious, and the search time of the algorithm can be greatly shortened when there is little difference between the solutions in the set of solutions.

In addition, there are many other algorithms that can be integrated with ant colony algorithm, such as Tabu search algorithm [36], simulated annealing algorithm [37], 2-opt algorithm [38] and Fish Swarm algorithm [39] etc. With the continuous development of the algorithm, the combination of ant colony algorithm and other algorithms has a broad prospect.

4 Application of Ant Colony Algorithm

4.1 Job Shop Scheduling Problem

The problem of job-shop scheduling (JSP) is often encountered in the production of enterprises. It is widely concerned in the field of scheduling and CIMS. The problem of job-shop scheduling is a typical NP-hard problem. Ant colony algorithm (ACA) is used to solve the JSP problem, which is usually transformed into the problem of finding the best path.

In reference [40], a multi-information ant colony optimization algorithm for multi-objective job shop scheduling was proposed. The particle swarm optimization algorithm is used to adaptively adjust the parameters in the ant colony algorithm to balance the global search and localization of the ant colony algorithm, improve the quality of understanding and achieve an effective solution to the multi-target job shop scheduling problem.

A hybrid ant colony algorithm for solving JSP was proposed in [41]. The algorithm uses local pheromone updating strategy and global pheromone updating strategy to update pheromone in pheromone updating rule. By combining neighborhood search and ant colony algorithm, the efficiency and feasibility of the algorithm are improved. Thus, the optimal solution of JSP can be solved.

In reference [42], an asynchronous parallel ant colony algorithm based on Petri net is proposed to solve the JSP problem. Through roulette and asynchronous parallel search mechanism, the precocity and local stagnation of the algorithm are avoided to a certain extent, the complexity of the algorithm is reduced and the computational efficiency of the algorithm is improved.

In addition, ant colony algorithm is widely used in complex job shop scheduling problems such as flexible job shop scheduling [43], flow shop scheduling and dynamic shop scheduling [44,45] etc.

4.2 Vehicle Routing Problem

The problem of vehicle path planning is also known as VRP, which has important practical significance in logistics. With the development of ant colony algorithm, many scholars at home and abroad use ant colony algorithm to solve the VRP problem, and also put forward different kinds of ant colony algorithm solving methods.

An application of simulated annealing ant colony algorithm to VRP problem was proposed in [46]. In each search process, the hybrid algorithm performs a second search for the optimal solution of each ant, so as to change the quality of the solution. The combination algorithm greatly improves the convergence speed and accuracy, and greatly improves the practicability of VRP.

A hybrid ant colony algorithm was proposed to solve vehicle routing problem in reference [47]. The hybrid algorithm combines the adaptive ant colony algorithm and the maximum-minimum ant colony algorithm, and achieves the global search by changing the heuristic factor and adjusting the pheromone concentration, so that the problem in VRP can be solved effectively.

The application of ant colony algorithm in emergency VRP was proposed in reference [48]. The improved ant colony algorithm is added to the dynamic road condition factor to deal with the path damage caused by the sudden event, and the algorithm uses the ant colony algorithm MPDACO to process the pheromone concentration on the basis of the global pheromone update mechanism. The method not only solves the problem of logistics distribution under the emergency condition, but also the convergence speed, the calculation complexity and the efficiency of the algorithm are relatively high.

In addition, the ant colony algorithm is also applied to different fields such as image processing [49], network route [50], remote sensing image classification problem [51] and control parameter optimization [52], and also achieves good results.

4.3 Robot Path Planning

The robot planning path is to search for an optimal or sub-optimal path from the start position to the target position in the known or unknown environment with an obstacle (such as distance, time and so on) [53]. In recent years, with the in-depth study of robot path, the integration of the algorithm is undoubtedly the best choice, and ant colony algorithm is a common solution.

The reference [54], an adaptive search radius ant colony dynamic path planning algorithm was proposed. The algorithm can automatically change the optimization radius according to the complexity of the environment, search for the advantages in the local range, then, look for a new local region, repeat this operation until the global optimal path is found. The adaptive ant colony algorithm improves the computing ability of the mobile robot, thus accelerates the convergence speed of the algorithm and realizes the effective dynamic path planning for the complex unknown environment.

In reference [55], a path planning method based on fuzzy ant colony algorithm is proposed. The fuzzy controller is used to optimize the correction parameters on-line (optimize the obstacle avoidance performance of the robot), and the global pheromone updating mechanism is used to accelerate the accumulation of the information on the shorter path, which not only improves the search efficiency and search speed of the algorithm, but also improves the searching efficiency and speed of the algorithm. The mobile robot also has better obstacle avoidance function.

A method of creating robot working environment based on grid method was proposed in reference [56]. This method sets the parameters of ant colony algorithm in the initial stage, runs the program to get the trajectory of the robot, and then obtains the global optimal route. This algorithm not only avoids the local stagnation, but also improves the search efficiency of the algorithm.

In addition, the robot path planning based on dynamic search strategy was proposed in reference [57]. In reference [58], ant colony algorithm (ACA), which considers the path length and the rotation of the robot, is proposed to plan the path of the robot. In reference [59], a stochastic extended ant colony algorithm was proposed to improve the route and global optimization ability of diversity to solve the robot path planning problem. In reference [60], a heterogeneous ant colony algorithm was proposed to solve the shortcomings of robot in global path planning.

5 Conclusion and Future Works

The ant colony algorithm has been continuously researched and developed, and its theory and application have made great progress. It has evolved from the initial solution of the classic traveler problem to the optimization problem in various fields and the optimization problem with different boundary conditions. Because the development of ant colony algorithm is still immature, many shortcomings need to be further solved: Ant colony algorithm has slow convergence speed, large precision and stagnation when dealing with large-scale problems;

The initial stage of ant colony algorithm has a certain blindness, which reduces the global search ability of the algorithm; The algorithm is prone to local optimal solutions and the ant colony algorithm lacks a complete theoretical system to prove it.

Based on the summary and analysis of this paper, hope that the ant colony algorithm can be further researched and developed in the following aspects:

(1) Strengthen the theoretical analysis of the ant colony algorithm. The development of ant colony algorithm is still in its infancy, and there is no systematic theoretical analysis. It does not give a reasonable mathematical explanation for its effectiveness, and the model of ant colony algorithm needs to be further researched.
(2) Solve the problem of ant colony algorithm dependence on initial information. How to reduce the blindness and randomness of ant colony algorithm search in the initial stage.
(3) The application of ant colony algorithm in practical engineering. Although the ant colony algorithm has made great progress in theory and application, it still stays in the experiment under ideal conditions and needs to face large-scale combinatorial optimization problems in practical engineering, so that the ant colony algorithm will have low convergence efficiency and It is easy to fall into stagnation and cannot be applied to deal with practical engineering application problems. Therefore, the effectiveness of ant colony algorithm in solving practical problems needs further research.
(4) Control the convergence speed of the ant colony algorithm. When the ant colony algorithm solves the large-scale problem, it will take a long time, and it is easy to fall into the stagnation phenomenon. The convergence speed of the control ant colony algorithm needs to be further researched in the future.

References

1. Dorigo, M., Maniezzo, V., Colorni, A.: The ant system: optimization by a colony of cooperating agents. IEEE Trans. Syst. Man Cybern. Part B **26**(1), 1–13 (1996)
2. Dorigo, M., Ganbardella, L.: Ant colony system: a cooperative learning approach to the traveling salesman problem. IEEE Trans. Evol. Comput. **1**(1), 53–66 (1997)
3. Gong, Y.: Research and application of ant colony algorithm. Anhui University of Technology (2014)
4. Li, Z.: Research and application prospect of intelligent optimization algorithm. J. Wuhan Univ. Light Ind. 35(04), 1–9 131 (2016)
5. Liang, X., Huang, M.: Modern Intelligent Optimization Hybrid Algorithm and its Application. Electronic Industry Press, Beijing (2011)
6. Huang, M.: Ant colony optimization algorithm and its application. Nanchang University (2007)
7. Liang, X., Huang, M., Ning, T.: Modern Intelligent Optimization Hybrid Algorithm and Its Application, 2nd edn. Publishing House of Electronics Industry, Beijing (2014)

8. Li, S., Chen, Y., Li, Y.: Ant Colony Algorithm and its Application. Harbin Institute of Technology Press, Harbin (2004)
9. Chen, S., Ma, L.: Basic principles and overview of ant colony algorithm. Sci. Technol. Innov. Appl. (31), 41 (2016)
10. Zhao, X., Tian, E.: The Ant Colony System (ACS) and its convergence proof. Comput. Eng. Appl. **43**(5), 67–70 (2007)
11. Zhao, X.: MAX-MIN ant colony system and its convergence proof. Comput. Eng. Appl. **08**, 70–72 (2006)
12. Jia, R., Ma, W.: Improved maximum and minimum ant colony algorithm based on neighborhood search. Comput. Simul. **31**(12), 261–264 (2014)
13. Bullnheimer, B., Hartl, R.F., Strauss, C.: A new rank based version of the ant system-a computational study. Cent. Eur. J. Oper. Res. Econ. **7**, 25–38 (1999)
14. Li, X., He, Q., Li, Y., Zhu, Z.: Short-wave frequency optimization assignment based on mutual information diffusion ant colony algorithm. J. Huazhong Univ. Sci. Technol. (Nat. Sci. Ed.) **44**(04), 6–11 (2016)
15. Dorigo, M., Gambardella, L.M.: Ant colony system: a cooperative learning approach to the traveling salesman problem. IEEE Trans. Evol. Comput. **1**(1), 53–56 (1997)
16. Zhou, N., Ge, G., Su, S.: An pheromone-based adaptive continuous-domain hybrid mosquito swarm algorithm. Comput. Eng. Appl. **53**(6), 156–161 (2017)
17. Chen, Y., Han, W., Cui, H.: Improved fusion of genetic algorithm and ant colony algorithm. Chin. J. Agric. Mechanization **35**(04), 246–249 (2014)
18. Monarche, N., Venturini, G., Slimane, M.: On how pachycondyllan apicalis ants suggests a new algorithm. Future Gener. Comput. Syst. **16**(8), 937–946 (2000)
19. Xie, J., Su, D., Lu, S., Jia, W., Sun, M., Guo, J.: Key technology of transmission line path planning based on improved ant colony algorithm. Electr. Meas. Instrum., 1–7 (2019)
20. Wang, Y., Huang, L.: An ant colony algorithm based on disjoint search strategy. J. Chongqing Univ. Technol. (Nat. Sci.) **25**(04), 65–69 (2011)
21. Wang, L., Wang, L., Zheng, C.: Identification of tag SNPs using set coverage ant colony algorithm with random disturbance characteristics. J. Yibin Univ. **15**(06), 81–85 (2015)
22. Bao, W., Zhu, X., Zhao, J., Xu, H.: Weighted-valued polymorphic ant colony algorithm. Softw. Eng. **19**(04), 1–4 (2016)
23. You, X., Liu, S., Lv, J.: An ant colony algorithm for dynamic search strategy and its application in robot path planning. Control Decis. **32**(03), 552–556 (2017)
24. Mou, L.: Ant colony algorithm based on parameter dynamic change and variation. Comput. Eng. **36**(19), 185–187 (2010)
25. You, H., Lu, Z.: Fast ant colony algorithm for adaptive adjustment of parameters α, β and ρ. Manuf. Autom. 4040(06), 99–102–112(2018)
26. Wei, X., Li, Y.: Parameter optimization and simulation research in ant colony algorithm. Manuf. Autom. **37**(10), 33–35 (2015)
27. Yang, S., Zhang, S.: Research on improved ant colony algorithm and parameter optimization. Electron. Technol. Softw. Eng. **13**, 186–188 (2016)
28. Huang, P., Chen, Y.: Study on parameter optimization of improved ant colony algorithm. Comput. Age **06**, 53–55 (2014)
29. Gan, Y., Li, S.: Study on parameter optimization configuration of ant colony algorithm. Manuf. Autom. **33**(05), 66–69 (2011)
30. Wang, S., Jiang, H.: Application of genetic-ant colony algorithm in post-disaster emergency material path planning. Comput. Appl. Softw. 35(09), 99–103+131 (2018)

31. Lin, F., Chen, J., Ding, K., Li, Z.: Optimization of DV-Hop localization algorithm based on genetic algorithm and binary ant colony algorithm. Instr. Tech. Sens. (01), 86–90+96 (2019)
32. Meng, X., Pian, Z., Shen, Z., et al.: Ant colony algorithm based on force pheromone coordination. Control Decis. Making **5**, 782–786 (2013)
33. Zhang, Y., Hou, Y., Li, C.: Path planning for handling robot ant colony based on artificial immune improvement. Comput. Meas. Control **23**(12), 4124–4127 (2015)
34. Wu, D., Shao, J., Zhu, Y.: Research on fault diagnosis technology of CNC machine tools based on ant colony algorithm and neural network. Mech. Des. Manuf. **1**, 165–167 (2013)
35. Zhang, C., Li, Q., Chen, P., Yang, S., Yin, Y.: An improved ant colony algorithm based on particle swarm optimization and its application. J. Univ. Sci. Technol. Beijing **35**(07), 955–960 (2013)
36. Yin, J., Shuai, J., Wen, B.: Research on logistics distribution optimization based on ant colony algorithm and tabu search algorithm. Light Ind. Technol. **3434**(08), 96–99 (2018)
37. Liu, K., Zhang, M.: Path planning based on simulated annealing ant colony algorithm. In: International Symposium on Computational Intelligence and Design, pp. 461–466 (2016)
38. Qin, D., Wang, C.: A hybrid ant colony algorithm based on 2-opt algorithm. Ind. Control Comput. **31**(01), 98–100 (2018)
39. Lv, S., Ma, K.: Application of mosquito-fish group hybrid algorithm in batch scheduling of differential workpieces. Comput. Syst. Appl. **27**(01), 162–167 (2018)
40. Huang, R.H., Yu, T.H.: An effective ant colony optimization algorithm for multi-objective job-shop scheduling with equal-size lot-splitting. Appl. Soft Comput. **57**, 642–656 (2017)
41. Ji, Y., Dang, P., Guo, X.: Study on job shop scheduling problem based on ant colony algorithm. Comput. Digit. Eng. 39(01), 4–6+52 (2011)
42. Tian, S., Chen, D., Wang, T., Liu, X.: An asynchronous ant colony algorithm for solving flexible job shop scheduling problems. J. Tianjin Univ. (Nat. Sci. Eng. Technol.) **49**(09), 920–928 (2016)
43. Lu, H., Lu, Y.: Study on flexible job shop scheduling method based on distribution estimation and ant colony hybrid algorithm. Mech. Electr. Eng. **36**(06), 568–573 (2019)
44. Fan, H., Xiong, H., Jiang, G., et al.: A review of scheduling rules algorithms in dynamic job shop scheduling problems. Comput. Appl. Res. **33**(3), 648–653 (2016)
45. Khoukhi, F.E., Boukachour, J., Hilali Alaoui, A.E.: The "Dual-Ants Colony": a novel hybrid approach for the flexible job shop scheduling problem with preventive maintenance. Comput. Ind. Eng. **106**, 236–255 (2016)
46. Zhang, J., Zhang, J., Song, X.: Application of simulated annealing mosquito swarm algorithm in VRP problem. J. Xihua Univ. (Nat. Sci. Ed.) **36**(06), 6–12 (2017)
47. He, W., Ni, Y., Wang, T.: Vehicle routing problem based on mixed behavior ant colony algorithm. J. Hefei Univ. Technol. Nat. Sci. Ed. (7), 883–887 (2014)
48. Li, W., Dong, Y., Li, X., Zhang, W.: Application and convergence analysis of improved ant colony algorithm in emergency VRP. J. Comput. Appl. 31(12), 3557–3559+3567 (2014)
49. Zhu, H., He, H., Fang, Q., Dai, Y., Jiang, D.: Peak clustering of mosquito population density for medical image segmentation. J. Nanjing Norm. Univ. (Nat. Sci. Ed.) **42**(02), 1–8 (2019)
50. Dai, T., Li, W.: Optimization of wireless sensor network routing based on improved ant colony algorithm. Comput. Meas. Control **24**(02), 321–324 (2016)

51. Wang, M., Wan, Y., Ye, Z., et al.: Remote sensing image classification based on the optimal support vector machine and modified binary coded ant colony optimization algorithm. Inf. Sci. **402**, 50–68 (2017)
52. Shen, C.: Design of PID controller based on adaptive ant colony algorithm. Instrum. Tech. Sens. (12), 126–128+156 (2016)
53. Yu, Y.: Research on path planning of mobile robot based on improved ant colony algorithm. J. Mech. Trans. **7**, 58–61 (2016)
54. Zhao, F., Yang, C., Chen, F., Huang, L., Tan C.: Adaptive search radius ant colony dynamic path planning algorithm. Comput. Eng. Appl. 54(19), 56–61+87 (2018)
55. Zhao, H., Guo, J., Xu, W., Yan, S.: Research on trajectory planning of mobile robot based on fuzzy ant colony algorithm. Comput. Simul. **35**(05), 318–321 (2018)
56. Lin, W., Deng, S., et al.: Research on path planning of mobile robot based on ant colony algorithm. Mech. Res. Appl. 31(04), 144–155+148 (2018)
57. You, X.-M., Liu, S., Lv, J.-Q.: Ant colony algorithm based on dynamic search strategy and its application on path planning of robot. Juece/Control Decis. **32**(3), 552–556 (2017)
58. Wang, Y., Ma, J., Wang, Y.: Path planning of robot based on modified ant colony algorithm. Tec./Tech. Bull. 55(3), 1–6 (2017)
59. Bai, J., Chen, L., Jin, H., Chen, R., Mao, H.: Robot path planning based on random expansion of ant colony optimization. In: Qian, Z., Cao, L., Su, W., Wang, T., Yang, H. (eds.) Recent Advances in Computer Science and Information Engineering. Lecture Notes in Electrical Engineering, vol. 125, pp. 141–146. Springer, Heidelberg (2012). https://doi.org/10.1007/978-3-642-25789-6_21
60. Lee, J.: Heterogeneous-ants-based path planner for global path planning of mobile robot applications. Int. J. Control Autom. Syst. **15**(4), 1754–1756 (2017)

Experimental Analysis of Selective Imitation for Multifactorial Differential Evolution

Deming Peng[1,2], Yiqiao Cai[1,2(✉)], Shunkai Fu[1,2], and Wei Luo[1,2]

[1] College of Computer Science and Technology, Huaqiao University, Quanzhou, China
942798207@qq.com, yiqiao00@163.com, 957472905@qq.com, luowei@hqu.edu.cn
[2] Fujian Key Laboratory of Big Data Intelligence and Security, Huaqiao University, Xiamen 361000, China

Abstract. Recently, evolutionary multitasking optimization (EMTO) is proposed as a new emerging optimization paradigm to simultaneously solve multiple optimization tasks in a cooperative manner. In EMTO, the knowledge transfer between tasks is mainly carried out through the assortative mating and selective imitation operators. However, in the literature of EMTO, little study on the selective imitation operator has yet been done to provide a deeper insight in the knowledge transfer across different tasks. Based on this consideration, we firstly study the influence of the inheritance probability (IP) of the selective imitation on an EMTO algorithm, multifactorial differential evolution (MFDE), through the experimental analysis. Then, an adaptive inheritance mechanism (AIM) is introduced into the selective imitation operator of MFDE to automatically adjust the IP value for different tasks at different evolutionary stages. The experimental results on a suite of single-objective multitasking benchmark problems have demonstrated the effectiveness of AIM in enhancing the performance of MFDE.

Keywords: Evolutionary multitasking optimization · Differential evolution · Knowledge transfer · Selective imitation · Adaptive inheritance mechanism

1 Introduction

Recently, evolutionary multitasking optimization (EMTO), as a new emerging optimization paradigm in evolutionary computation, has been proposed with the goal of improving the convergence characteristics for multiple tasks by seamlessly transferring knowledge across them [6,7]. Distinguished from the traditional optimization paradigms (i.e., single-objective and multi-objective optimization), EMTO solves multiple different optimization problems (i.e., tasks) concurrently, inspired by the remarkable capacity of the human mind in performing multiple tasks with apparent simultaneity [7]. In EMTO, multiple search spaces corresponding to different tasks concurrently exist, and each of them possesses a

L. Pan et al. (Eds.): BIC-TA 2019, CCIS 1159, pp. 15–26, 2020.
https://doi.org/10.1007/978-981-15-3425-6_2

unique functions landscape [6]. In this manner, the useful knowledge found during the optimization of each task can be reused and transferred automatically across different related tasks in the multitasking environment. By utilizing the latent synergies among tasks, EMTO has shown the efficacy and great potential in solving the complex optimization problems from the field of science and engineering [6].

Due to its attractive characteristics, EMTO has drawn lots of attention from the researchers, resulting in many EMTO algorithms for solving different types of optimization problems. Multifactorial EA (MFEA), as one of the most representative EMTO algorithms, has been proposed to exploit the latent genetic complementarities between tasks by implicitly transferring the biological and cultural building blocks [6]. After that, many works have been devoted into further improving the search ability of MFEA with different techniques. In [1], a linear domain adaptation strategy was incorporated into MFEA to transform the search space of a simple task to the search space similar to its constitutive complex task. In [9], a resource allocation mechanism was proposed for MFEA (named MFEARR) based on the parting ways detection mechanism to reallocate the fitness evaluations on different types of offspring. In [8], a group-based MFEA was proposed by grouping the similar tasks and transferring the genetic information between tasks within the same group. In [2], MFEA with online transfer parameter estimation was proposed to online learn and exploit the similarities between different tasks with the optimal mixture modeling.

In addition, several researches have focused on the offspring generation operator of MFEA. In [4], a multifactorial DE (MFDE) and a multifactorial particle swarm optimization (MFPSO) were proposed to generate the offspring through the mutation strategy of DE and the velocity update operator of PSO, respectively. In [10], a multi-factorial brain storm optimization algorithm (MFBSA) was proposed by introducing the brain storm optimization (BSO) into MFEA, where the clustering technique is applied to gather the similar tasks.

As the above EMTO algorithms show, the knowledge transfer is mainly realized via two features of multifactorial inheritance in a synergistic way, i.e., assortative mating and selective imitation [6]. However, among these works, much research effort of EMTO has been devoted into studying the assortative mating operator. In contrast, there has been little study on the selective imitation operator of EMTO.

Motivated by the above observation, we embark a preliminary study to analyze the influence of the selective imitation operator on the performance of the EMTO algorithm. For this purpose, an inheritance probability (IP) is defined to control the inheritance of skill factor for the offspring generated by the cross-cultural parents, and MFDE is selected as an instance algorithm. Furthermore, an adaptive inheritance mechanism (AIM) is introduced into MFDE to automatically adjust the IP value for different tasks at different evolutionary stages. A suite of single-objective multitasking benchmark problems from the CEC 2017 evolutionary multi-task optimization competition [3] is employed to evaluate the effectiveness of AIM in the knowledge transfer of MFDE.

2 Background

2.1 EMTO

In EMTO, the multiple decision spaces of the tasks are searched simultaneously through the implicit parallelism of population-based optimization algorithm, aiming at obtaining the optimal solutions for multiple different task. In general, a multitasking optimization problem with K tasks can be defined as follows:

$$\{X_1, X_2, \ldots, X_K\} = \arg\min\{f_1(X), f_2(X), \ldots, f_K(X)\} \tag{1}$$

where X_k is a feasible solution for $f_k(X)$. In addition, a population of solutions encoded in a unified representation space is used, and the following definitions are employed to compare these solutions in the multitasking environment.

- *Factorial cost:* The factorial cost (ψ_j^i) of a solution X_i represents its objective value $f_j(X_i)$ on a particular task T_j.
- *Factorial rank:* The factorial rank (r_j^i) of X_i on task T_j is the index of X_i in the sorted population in ascending order based on its ψ_j^i.
- *Scalar fitness:* The scalar fitness (φ^i) of X_i is defined based on its best factorial rank over all the tasks, which is calculated as $\varphi^i = 1/\min_{j\in\{1,\ldots,K\}}\{r_j^i\}$.
- *Skill factor:* The skill factor (τ^i) of X_i is defined as the index of the task that it performs most effective, which is given by $\tau^i = \arg\min_{j\in\{1,\ldots,K\}}\{r_j^i\}$.

With the above definitions, the performance comparison between the solutions of population in EMTO is carried out based on their scalar fitness values (i.e., φ^i). To elaborate, if $\varphi^i > \varphi^j$, X_i is considered to dominate X_j. Thus, for the multitasking optimization problem in Eq. (1), the multifactorial optimality can be defined as:

- *Multifactorial Optimality:* X^* is considered as a multifactorial optimum if and only if it is the global optimum of all the K tasks, i.e., $X^* = \arg\min\{f_k\}$, $k = 1, \ldots, K$.

2.2 MFEA and MFDE

MFEA is implemented to conduct the multi-tasking optimization with implicit knowledge transfer among tasks [6]. In MFEA, two key components, i.e., assortative mating and selective imitation, are employed for multifactorial inheritance. In the former, the knowledge is transferred implicitly among tasks for offspring generation, and, in the latter, the objective value of an offspring is evaluated by considering its inherited skill factors from its parents. The basic structure of MFEA is shown in Algorithm 1. More details of MFEA can be found in [6].

In the assortative mating operator of MFEA, two solutions are selected randomly to generate two offspring based on their skill factors. Specifically, if they possess the same skill factors or they possess the different skill factors but satisfy a prescribed random mating probability (rmp), the offspring will be generated

Algorithm 1. MFEA

Randomly generate NP solutions to initialize the population P^0, $g = 0$;
Calculate the factorial cost and factorial rank of each solution on each task;
Assign the skill factor to each solution;
WHILE $g \leq MaxG$ **DO**
  Apply the *assortative mating* on P^g to generate an offspring population O^g;
  Apply the *selective imitation* on O^g to evaluate each offspring based on its inherited skill factor;
  Update the scalar fitness and skill factor of each solution in $P^g \cup O^g$;
  Select the fittest NP solutions from $P^g \cup O^g$ to form the next population P^{g+1};
  $g = g + 1$;
END WHILE
Return the best solution for each task.

by the crossover operator. Otherwise, each of them will undergo the mutation operator to generate an offspring respectively.

In the selective imitation operator, the skill factor of each offspring is directly inherited from its parent if its parent possess the same skill factor. Otherwise, its skill factor will be randomly inherited from either of its parent. After that, each offspring will be evaluated selectively on the task corresponding to its inherited skill factor.

As a variant of MFEA, MFDE is proposed by incorporating the search mechanisms of DE (i.e., mutation and crossover) into MFEA. Due to using the same algorithm structure, MFDE only differs from MFEA in assortative mating and selective imitation, which are described in Algorithms 2 and 3, respectively.

Algorithm 2. Assortative mating of MFDE

Randomly select X^g_{r1}, X^g_{r2} and X^g_{r3} that have the same skill factor with X_i and $r1 \neq r2 \neq r3 \neq i$;
Randomly select X'^g_{r2} and X'^g_{r3} that have the different skill factor from X^g_i;
IF $rand(0, 1) < rmp$ **THEN**//mutation operator
  $V^g_i = X^g_{r1} + F \times X'^g_{r2} + X'^g_{r3}$;
ELSE
  $V^g_i = X^g_{r1} + F \times X^g_{r2} + X^g_{r3}$;
END IF
FOR $j = 1, \ldots, D$ **THEN**//crossover operator
  IF $rand(0, 1) \leq Cr$ or $j == j_{rand}$ **THEN**
    $u^g_{i,j} = v^g_{i,j}$;
  ELSE
    $u^g_{i,j} = x^g_{i,j}$;
  END IF
END FOR

In the assortative mating of MFDE (Algorithm 2), the mutation and crossover operator of DE are used to generate the offspring for each target solution (i.e., X_i^g). To elaborate, a randomly value $rand(0, 1)$ in the range of $[0, 1]$ is firstly generated to select the parents for the mutation operator. If $rand(0, 1)$ is less than the predefined mating probability value (rmp), the parents with the skill factor that differs from X_i^g (i.e., $X_{r2}^{\prime g}$ and $X_{r3}^{\prime g}$) are used for generating the mutant vector V_i^g. Otherwise, the parents with the same skill factor as X_i^g (i.e., X_{r2}^g and X_{r3}^g) are selected for mutation. In addition, DE/rand/1 is used as the mutation strategy in MFDE (lines 4 and 5), where F is the scaling factor that typically lies in the interval [0.4, 1]. After that, the binomial crossover operator of DE is employed to generate a trial vector U_i^g for each pair of X_i^g and V_i^g (lines 8–13). In the crossover operator, D is the dimension of the optimization problem, $Cr \in [0, 1]$ is the crossover rate and $j_{rand} \in [1, D]$ is a randomly selected integer.

Algorithm 3. Selective imitation of MFDE

1: **IF** U_i^g is generated the parents with different skill factors **THEN**
2:     **IF** $rand(0, 1) \leq 0.5$ **THEN**
3:         U_i^g inherits the skill factor from X_i^g;
4:     **ELSE**
5:         U_i^g inherits the skill factor from $X_{r2}^{\prime g}$;
6:     **END IF**
7: **ELSE**
8:     U_i^g inherits the skill factor from X_i^g;
9: **END IF**
10: Evaluate the factorial cost of U_i^g for the task corresponding to its inherited skill factor;
11: Set the factorial cost of U_i^g for other unevaluated tasks to ∞.

In the selective imitation of MFDE (Algorithm 3), the skill factor of U_i^g is inherited from its parents according to different circumstances. To elaborate, if its parents have the same skill factor, U_i^g will directly inherit their skill factor (line 8). Otherwise, U_i^g will randomly imitate the skill factor of either of its parents with an equal probability (lines 2–6). Finally, U_i^g will be evaluated only for the task corresponding to its inherited skill factor.

3 Proposed Method

3.1 Motivations

As shown in most EMTO algorithms, the knowledge transfer across different tasks is realized through both the assortative mating and selective imitation operators. The former is used to execute the intra-cultural and cross-cultural mating in a controlled manner, while the latter plays the role in the vertical cultural transmission in a random way. In the existing studies of EMTO, compared with the assortative mating, the selective imitation gains very little attention.

Therefore, it is worthy of being studied and analyzed, which may provide a deeper insight in the knowledge transfer by vertical cultural transmission via selective imitation.

Along this line, the inheritance probability (IP) is firstly defined for the selective imitation of an instance algorithm, MFDE, as follows:

- *Inheritance probability:* The inheritance probability (IP) represents the likelihood for the offspring generated by the cross-cultural mating that it inherits the skill factor from its corresponding target parent.

Based on the definition of IP, the inheritance operator in the selective imitation of MFDE is modified and shown in Algorithm 4.

Algorithm 4. The modified selective imitation with IP for MFDE

IF U_i^g is generated the parents with different skill factors **THEN**
    IF $rand(0,1) \leq IP$ **THEN**
        U_i^g inherits the skill factor from X_i^g;
    ELSE
        U_i^g inherits the skill factor from $X_{r2}^{\prime g}$;
    END IF
ELSE
    U_i^g inherits the skill factor from X_i^g;
END IF
Evaluate the factorial cost of U_i^g for the task corresponding to its inherited skill factor;
Set the factorial cost of U_i^g for other unevaluated tasks to ∞.

In the original MFDE, IP is set to 0.5 for all the problems and its value does not change during the evolutionary process (see line 2 in Algorithm 3). In this light, Algorithm 3 is a special case of Algorithm 4. However, we have observed that the setting of IP in Algorithm 4 can greatly affect the performance of MFDE through our preliminary experimental studies[1]. Therefore, to alleviate the influence of IP on the performance of MFDE, an adaptive inheritance mechanism (AIM) is thus proposed in the next subsection to automatically adjust the IP value for MFDE.

3.2 Adaptive Inheritance Mechanism (AIM)

In this paper, AIM is proposed based on the historical successful and failure experience of knowledge transfer. The pseudocode of AIM is shown in Algorithm 5, where c is the learning rate, $suRate_j^s$ and $suRate_j^o$ represent the success rate of that the cross-cultural offspring for task T_j enter into the next generation by inheriting the skill factor from the target and transferred solutions, respectively. In MFDE with AIM, the update of IP_j for each task T_j is executed at the end of every generation. Here, the initial value of IP_j is set to 0.5, and c is set to 0.3.

[1] The results of the preliminary experimental studies on IP will be shown in Sect. 4.2.

Algorithm 5. Adaptive inheritance mechanism (AIM)

```
FOR each task T_j DO
    IF suRate_j^s ≥ suRate_j^o THEN
        IP_j = min{IP_j + c × suRate_j^s, rand(0.9, 1)};
    ELSE
        IP_j = max{IP_j − c × suRate_j^o, rand(0, 0.1)};
    END IF
END FOR
```

As shown in Algorithm 5, with AIM, each task has its own IP for the knowledge transfer based on its evolutionary state. Specifically, if $suRate_j^s \geq suRate_j^o$, the IP value will be increased to encourage the communications between solutions with the same skill factor. On the contrary, the complementarity between tasks will be promoted if $suRate_j^s < suRate_j^o$.

In addition, although a new parameter c is introduced into AIM, the setting of IP can be carried out in an adaptive way. On the one hand, c is used to control the learning rate of IP for each task online and does not directly affect the knowledge transfer among tasks. On the other hand, the introduction of c in AIM can provide a more convenient way to analyse the influence of IP on the performance of MFDE. Note that AIM in this paper is an attempt to study the influence of selective imitation for MFDE. The sensitivity analysis of c and other implementations of AIM will be studied in our future work.

4 Experimental Results and Discussions

4.1 Benchmark Functions and Parameter Setup

Nine single-objective multitasking benchmark problems from the CEC 2017 evolutionary multi-task optimization competition [3] are used here. In each benchmark problems, two distinct tasks with different global optima and search ranges are included. Moreover, based on the landscape similarity and the degree of intersection of the global optima, these problems can be classified into different categories, i.e., complete intersection and high similarity (F1, CI+HS), complete intersection and medium similarity (F2, CI+MS), complete intersection and low similarity (F3, CI+LS), partial intersection and high similarity (F4, PI+HS), partial intersection and medium similarity (F5, PI+MS), partial intersection and low similarity (F6, PI+LS), no intersection and high similarity (F7, NI+HS), no intersection and medium similarity (F8, NI+MS), and no intersection and low similarity (F9, NI+LS). The properties of these problems are shown in Table 1. More details of them can be found in [3].

The parameter setting of MFDE are specified as follows:

- Population size (NP): 100.
- Maximum number of generations ($MaxG$): 1000.
- Independent number of runs (NR): 20.

- Random mating probability (rmp): 0.3.
- Crossover rate (Cr): 0.5.
- Scale factor (F): 0.9.

In addition, the experimental studies in this paper are conducted via MATLAB 2017a on a Windows 10 PC with Intel Core i5 CPU at 3.3 GHz and 8GB RAM.

Table 1. Summary of the properties of nine single-objective multitasking benchmark problems.

Problem	Task	Landscape	Intersection	Similarity
F1 CI+HS	T1: Griewank T2: Rastrigin	Multimodal, nonseparable multimodal, nonseparable	Complete intersection	1.0000
F2 CI+MS	T1: Ackley T2: Rastrigin	Multimodal, nonseparable Multimodal, nonseparable	Complete intersection	0.2261
F3 CI+LS	T1: Ackley T2: Schwefel	Multimodal, nonseparable Multimodal, separable	Complete intersection	0.0002
F4 PI+HS	T1: Rastrigin T2: Sphere	Multimodal, nonseparable Unimodal, separable	Partial intersection	0.8670
F5 PI+MS	T1: Ackley T2: Weierstrass	Multimodal, nonseparable Multimodal, nonseparable	Partial intersection	0.2154
F6 PI+LS	T1: Ackley T2: Weierstrass	Multimodal, nonseparable Multimodal, nonseparable	Partial intersection	0.0725
F7 NI+HS	T1: Rosenbrock T2: Rastrigin	Multimodal, nonseparable Multimodal, nonseparable	No intersection	0.9434
F8 NI+MS	T1: Griewank T2: Weierstrass	Multimodal, nonseparable Multimodal, nonseparable	No intersection	0.3669
F9 NI+LS	T1: Rastrigin T2: Schwefel	Multimodal, nonseparable Multimodal, separable	No intersection	0.0016

4.2 Influence of IP on MFDE

To show the influence of IP on MFDE, the values of IP considered in this experiment are as follows: $IP \in \{0.1, 0.3, 0.5, 0.7, 0.9\}$. The results of the MFDE variants with different IP values on the multitasking benchmark problems are shown in Table 2, where the best mean values obtained by the corresponding algorithm are highlighted in bold. In addition, the average ranking values of MFDE with different IP values on different types of problems are presented in Table 3. From the results in Tables 2 and 3, some observations can be obtained:

- There are greatly differences in the performances of MFDE with different IP values. To elaborate, MFDE with $IP = 0.7$ can obtain the best results in 9 problems (i.e., T1 and T2 in F1, T1 and T2 in F3, T1 in F4, T2 in F6, T2 in F8, and T1 and T2 in F9) in terms of the mean of the best error values. Relatively, MFDE with $IP = 0.9$, $IP = 0.5$, $IP = 0.3$ and $IP = 0.1$ obtain the best results in 6, 4, 3, and 2 problems, respectively.

- In terms of the intersection degree for a pair of tasks, MFDE with a small IP value (e.g., $IP = 0.3$) performs better than that with a larger IP value (e.g., $IP = 0.9$) overall on the problems with complete intersection (CI). Inversely, MFDE with a larger IP value is better than that with a small IP value on the problems both with PI and NI.
- In terms of the similarity for a pair of tasks, MFDE with a larger IP value can achieve the better results than that with a smaller IP value on the problems with HS, MS and LS.
- In terms of the average ranking value for all the benchmark problems, MFDE with $IP = 0.7$ achieves the best performance, followed by MFDE with $IP = 0.9$ and MFDE with $IP = 0.5$. MFDE with $IP = 0.1$ obtains the worst results averagely.
- Overall, MFDE with a larger IP value (e.g., 0.7 and 0.9) performs significantly better than MFDE with a smaller IP value (e.g., 0.1 and 0.3).

According to the above analysis, the performance of MFDE is significantly affected by the setting of IP, especially for the problems with different degrees of intersection.

Table 2. Mean of the best error values obtained by MFDE with different IP values on different multitasking benchmark problems.

Problem	Task	$IP = 0.1$	$IP = 0.3$	$IP = 0.5$	$IP = 0.7$	$IP = 0.9$
F1	T1	5.46E−04	4.03E−04	1.05E−03	**5.94E-06**	1.23E−03
CI+HS	T2	2.03E+00	1.30E+00	4.31E+00	**8.17E-03**	1.94E+00
F2	T1	8.93E−02	6.49E−02	**5.55E−03**	1.55E−01	1.22E−01
CI+MS	T2	5.51E−01	1.00E+00	**1.99E−01**	2.74E+00	3.23E+00
F3	T1	**2.12E+01**	**2.12E+01**	**2.12E+01**	**2.12E+01**	**2.12E+01**
CI+LS	T2	**1.20E+04**	1.23E+04	1.24E+04	**1.20E+04**	1.24E+04
F4	T1	2.43E+02	7.50E+01	8.41E+01	**7.31E+01**	7.48E+01
PI+HS	T2	9.81E−03	1.10E−04	1.59E-05	3.98E-06	**1.09E−06**
F5	T1	4.00E−03	2.04E−03	1.47E−03	1.04E−03	**1.00E−03**
PI+MS	T2	8.40E+01	8.24E+01	7.15E+01	6.86E+01	**6.62E+01**
F6	T1	3.57E−01	**2.46E−01**	5.54E−01	4.63E−01	4.52E−01
PI+LS	T2	2.02E−01	1.55E−01	5.13E−02	**3.16E−02**	1.52EV01
F7	T1	1.05E+02	9.37E+01	**8.53E+01**	8.83E+01	1.57E+02
NI+HS	T2	2.19E+01	**1.76E+01**	2.46E+01	2.76E+01	4.11E+01
F8	T1	3.35E−03	2.61E−03	1.80E−03	2.38E−03	**7.61E−04**
NI+MS	T2	3.93E+00	3.46E+00	3.40E+00	**2.91E+00**	3.31E+00
F9	T1	4.20E+02	2.04E+02	9.55E+01	**9.45E+01**	**9.45E+01**
NI+LS	T2	4.56E+03	4.46E+03	4.21E+03	**3.70E+03**	3.97E+03

Table 3. Average ranking value of MFDE with different IP values on different types of multitasking benchmark problems.

Category	$IP = 0.1$	$IP = 0.3$	$IP = 0.5$	$IP = 0.7$	$IP = 0.9$
CI	2.333	**2.000**	2.667	2.167	3.667
PI	4.500	3.333	3.333	2.000	**1.833**
NI	4.333	3.333	2.500	**2.000**	2.667
HS	3.833	2.500	3.333	**1.833**	3.500
MS	4.167	3.500	**2.167**	2.833	2.333
LS	3.167	2.667	3.000	**1.500**	2.333
Avg Rank	3.72	2.89	2.83	**2.06**	2.72

4.3 Performance Enhancement of MFDE with AIM

To test the effectiveness of the proposed AIM, the comparison between the original MFDE (i.e., $IP = 0.5$) and its augmented algorithm with AIM is made on the nine multitasking benchmark problems in this experiment. The results are shown in Table 4, where "+", "=", and "−" obtained by the Wilcoxon test [5] mean that MFDE with AIM is significantly better than, equivalent to and worse than MFDE on the corresponding problem, respectively.

Table 4. Mean and standard deviation of the best error values obtained by MFDE and MFDE with AIM on different benchmark problems.

Problem	Task	MFDE		MFDE with AIM		
F1	T1	1.05E−03	3.15E−03	**6.22E−04**	2.77E−03	=
CI+HS	T2	4.31E+00	1.27E+01	**1.74E+00**	7.79E+00	+
F2	T1	**5.55E−03**	2.32E−02	1.77E−01	3.58E−01	−
CI+MS	T2	**1.99E−01**	8.90E−01	3.75E+00	8.34E+00	−
F3	T1	2.12E+01	3.50E−02	2.12E+01	3.89E−02	=
CI+LS	T2	1.24E+04	1.14E+03	**1.21E+04**	1.08E+03	=
F4	T1	8.41E+01	1.96E+01	**7.85E+01**	1.34E+01	=
PI+HS	T2	1.59E−05	1.13E−05	**1.36E−06**	7.78E−07	+
F5	T1	1.47E−03	7.34E−04	**9.83E−04**	4.88E−04	+
PI+MS	T2	**7.15E+01**	2.22E+01	7.97E+01	2.24E+01	=
F6	T1	5.54E−01	6.67E−01	**2.25E−01**	4.71E−01	=
PI+LS	T2	5.13E−02	1.02E−01	**3.49E−02**	5.95E−02	=
F7	T1	**8.53E+01**	3.82E+01	1.06E+02	6.75E+01	=
NI+HS	T2	**2.46E+01**	1.65E+01	3.25E+01	1.72E+01	=
F8	T1	1.80E−03	3.57E−03	**1.37E−03**	3.35E−03	+
NI+MS	T2	3.40E+00	1.13E+00	**2.67E+00**	9.42E−01	+
F9	T1	9.55E+01	2.67E+01	**9.45E+01**	2.30E+01	=
NI+LS	T2	4.21E+03	6.84E+02	**3.64E+03**	9.70E+02	+
SUM(+/=/−)		−		6/10/2		

As shown in Table 4, AIM is able to achieve the improvements for MFDE on most benchmark problems. To elaborate, MFDE with AIM can obtain the better results than the original MFDE on 12 out of 18 problems in terms of mean of the best error values, while is worse than it on 5 problems. Furthermore, based on the single-problem analysis by the Wilcoxon test, MFDE with AIM is significantly better than MFDE on 6 problems and is outperformed by it on 2 problems. In general, the effectiveness of AIM has been clearly verified through the above comparisons.

5 Conclusions

In evolutionary multitasking optimization (EMTO), the knowledge transfer is mainly carried out through the assortative mating and selective imitation strategy. Compared with the assortative mating, the selective imitation draws very little attention in the field of EMTO. To study the influence of the selective imitation on the performance of EMTO, an inheritance probability (IP) is defined to control the inheritance of skill factor for an offspring generated by the cross-cultural parents in an instance EMTO algorithm, multifactorial differential evolution (MFDE). An experimental analysis on a suite of single-objective multitasking benchmark problems has been performed and the results have shown that IP has the great effect on the performance of MFDE. Motivated by this observation, an adaptive inheritance mechanism (AIM) for adaptively adjusting the IP value has been put forward to alleviate its influence on MFDE. Though the comparison studies between MFDE and its augmented algorithm with AIM, the effectiveness of the proposed method has been confirmed on the multitasking benchmark problems. In the future, the effect of AIM on different EMTO algorithms will be studied comprehensively.

Acknowledgments. This work was supported in part by the Natural Science Foundation of Fujian Province of China (2018J01091), the National Natural Science Foundation of China (61572204, 61502184), the Postgraduate Scientific Research Innovation Ability Training Plan Funding Projects of Huaqiao University (18014083013), and the Opening Project of Guangdong Province Key Laboratory of Computational Science at the Sun Yat-sen University. The authors would like to express their deep gratitude to Prof. Y. S. Ong from School of computer science and engineering, Nanyang technological university (NTU), Singapore, for his patient guidance, and the Data Science and Artificial Intelligence Research Center at NTU for the support to our work.

References

1. Bali, K.K., Gupta, A., Feng, L., Ong, Y.S., Siew, T.P.: Linearized domain adaptation in evolutionary multitasking. In: 2017 IEEE Congress on Evolutionary Computation (CEC), pp. 1295–1302. IEEE (2017)
2. Bali, K.K., Ong, Y.S., Gupta, A., Tan, P.S.: Multifactorial evolutionary algorithm with online transfer parameter estimation: MFEA-II. IEEE Trans. Evol. Comput. **24**, 69–83 (2019)

3. Da, B., et al.: Evolutionary multitasking for single-objective continuous optimization: benchmark problems, performance metrics and baseline results. Technical report, Nanyang Technological University (2016)
4. Feng, L., et al.: An empirical study of multifactorial PSO and multifactorial DE. In: 2017 IEEE Congress on Evolutionary Computation (CEC), pp. 921–928. IEEE (2017)
5. García, S., Fernández, A., Luengo, J., Herrera, F.: A study of statistical techniques and performance measures for genetics-based machine learning: accuracy and interpretability. Soft Comput. **13**(10), 959–977 (2009)
6. Gupta, A., Ong, Y.S., Feng, L.: Multifactorial evolution: toward evolutionary multitasking. IEEE Trans. Evol. Comput. **20**(3), 343–357 (2015)
7. Ong, Y.-S.: Towards evolutionary multitasking: a new paradigm in evolutionary computation. In: Senthilkumar, M., Ramasamy, V., Sheen, S., Veeramani, C., Bonato, A., Batten, L. (eds.) Computational Intelligence, Cyber Security and Computational Models. AISC, vol. 412, pp. 25–26. Springer, Singapore (2016). https://doi.org/10.1007/978-981-10-0251-9_3
8. Tang, J., Chen, Y., Deng, Z., Xiang, Y., Joy, C.P.: A group-based approach to improve multifactorial evolutionary algorithm. In: IJCAI, pp. 3870–3876 (2018)
9. Wen, Y.W., Ting, C.K.: Parting ways and reallocating resources in evolutionary multitasking. In: 2017 IEEE Congress on Evolutionary Computation (CEC), pp. 2404–2411. IEEE (2017)
10. Zheng, X., Lei, Y., Gong, M., Tang, Z.: Multifactorial brain storm optimization algorithm. In: Gong, M., Pan, L., Song, T., Zhang, G. (eds.) BIC-TA 2016. CCIS, vol. 682, pp. 47–53. Springer, Singapore (2016). https://doi.org/10.1007/978-981-10-3614-9_6

Brain Storm Optimization Algorithm with Estimation of Distribution

Jia-hui Luo[1(✉)], Ren-ren Zhang[2], Jin-ta Weng[1], Jing Gao[2], and Ying Gao[1(✉)]

[1] Guangzhou University, Guangzhou 510006, China
gaoying_gzhu@outlook.com

[2] Guangdong Hengdian Information Technology Co., Ltd., Guangzhou 510640, China

Abstract. Brain Storm Optimization (BSO) algorithm is a new intelligence optimization algorithm, which is effective to solve the multi-modal, high-dimensional and large-scale optimization problems. However, when the BSO algorithm deals with the complex problems, there are still some disadvantages, such as the slow speed of the search algorithm for the late, premature convergence and easy to fall into local optimal solutions and so on. In order to solve these problems, a BSO algorithm with Estimation of Distribution (EDBSO) is proposed. Similarity as the DMBSO, the EDBSO algorithm is divided the discussion process into two parts, including intra-group discussion and inter-group discussion. The Estimation of Distribution algorithm in continuous domains, that is based on the variables subject to Gaussian distribution, is used to improve the process of inter-group discussion of DMBSO algorithm. In this paper, five benchmark functions are used to evaluate the search performance of EDBSO algorithm. In order to verify the convergence and accuracy of the EDBSO algorithm, the EDBSO algorithm is compared with 4 improved algorithm in different dimensions. The simulation results show that the EDBSO algorithm can effectively avoid to falling into the local optimum and prevent premature convergence of this algorithm, and it can find better optimal solutions stably. With the increase of the problem dimensions, the EDBSO algorithm which has better robustness is suitable for solving complex optimization problems.

Keywords: Swarm Intelligence · Brain storm optimization algorithm · Estimation of Distribution Algorithm · Extension continuous domains

1 Introduction

Inspired by the problem-solving of human beings, Brain Storm Optimization (BSO) algorithm proposed by Shi [1] is a new swarm intelligence algorithm, which is effective to solve the multi-modal high-dimensional and large-scale optimization problems. In the BSO algorithm, every individual is abstracted into a

Supported by Natural Science Foundation of Guangdong Province, China, under Grant No. 2014A030313524; by Science and Technology Projects of Guangdong Province, China, under Grant No. 2016B010127001, and Science and Technology Projects of Guangzhou under Grant Nos. 201607010191 and 201604016045; by 2018 Guangzhou University Graduate "Basic Innovation" Project under Grant Nos. 2018GDJC-M13.

L. Pan et al. (Eds.): BIC-TA 2019, CCIS 1159, pp. 27–41, 2020.
https://doi.org/10.1007/978-981-15-3425-6_3

feasible potential solution. These solutions are separated into several clusters and the best solutions of each cluster are kept to the next iteration. By comparing the local optimal solutions generated in each cluster, the global optimal solution of the whole population would be found eventually. Although the BSO algorithm has many excellent characteristics in solving the optimization problems, the research on the algorithm is either still in the preliminary stage. When the BSO algorithm deals with the complex problems, there are still some disadvantages, such as the slow speed of the search algorithm for the late, premature convergence and easy to fall into local optimal solutions and so on [2].

In order to solve the above problems, an improved BSO algorithm called MBSO was proposed by Zhang et al. [3]. This algorithm replaced the k-means cluster method by using a simple cluster method (SGM) and replaced the Gaussian mutation method of the original BSO algorithm by using DE algorithm. MBSO algorithm improves the search accuracy of the algorithm and reduces the running time of the algorithm. Wu et al. [4] used non-dominated sorting to update non-inferior solutions into archive sets for solving multi-objective optimization problems. The inter-group discussion and the intra-group discussion mechanism were into the original BSO algorithm by Yang et al. [5], and the BSO algorithm based on discussion mechanism (DMBSO) was proposed. Shi et al. [6] used k-medians cluster method instead of k-means cluster method to reduce the running time of the original BSO algorithm. Yang et al. [7] proposed a BSO algorithm based on differential step size by using the characteristics of differential evolution (DE) algorithm. Shi [8] replaced the complex decision space clustering by dividing the target space into common classes and elite classes, and proposed BSO algorithm based on the objective space (BSO-OS), which greatly reduced the complexity of the algorithm and improved the operating efficiency of this algorithm. Diao et al. [9] combined the DE algorithm with the BSO algorithm to improve the robustness and global convergence. A differential BSO Algorithm (DBSO) is proposed to improve the ability of the BSO algorithm to jump out of the local optimal solutions. Wu [10] proposed a DBSO for objective space clustering (DBSO-OS). Liang et al. [11] proposed a modified BSO (MBSO) algorithm, which effectively avoided the search being trapped in the local optimal solutions. This method effectively solves the problem of slow convergence in the late stage of the algorithm by adjusting the probability parameters of the population grouping strategy. Using the Cauchy random variable to optimize the BSO algorithm, Cheng et al. [12] proposed the BSO in objective space algorithm with Cauchy random variable (BSO-OS-Cauchy) algorithm to solve the multi-modal optimization problem.

It can be seen from the above research that the core of the search ability of the BSO algorithm is the implementation method of the aggregation and dispersion behaviors. Based on the previous research, this paper proposed BSO algorithm with Estimation of Distribution (EDBSO) algorithm. In order to verify the convergence and accuracy of the EDBSO algorithm, the paper compares the EDBSO algorithm with the BSO algorithm, DBSO algorithm, DMBSO algorithm from 10-dimensional to 100-dimensional and further analyzes of the results of the experiment.

2 The Related Work

2.1 Brain Storm Optimization Algorithm

In the original BSO algorithm, every individual represents a potential solution. The individual is updated through the aggregation and dispersion. The BSO algorithm involves five main operations, including population initialization, evaluation, clustering, adding perturbation and disturbance. The steps to implement the BSO algorithm are as follows [13]:

Step 1: The n individuals are randomly generated.
Step 2: This n individuals are clustered by the k-means algorithm, and the number of clusters is m.
Step 3: This n individuals are evaluated.
Step 4: Sort the individuals of each cluster separately. Select the best individuals in each class as the central individuals of these clusters. In order to maintain the diversity of the groups, the cluster centers are selected in a random manner. The number of general cluster centers is 3–5.
Step 5: Replace the cluster center with an arbitrary solution to a small probability.
Step 6: Generate new individuals according to certain rules. The BSO algorithm contains 4 ways to update individuals. The setting method of parameters refers to the literature [5].
Step 7: If n new individuals have been generated, go to step 8; otherwise, go to step 6.
Step 8: If the maximum number of iterations have been reached, the algorithm ends; otherwise, go to step 2.

From the above steps, each individual has the possibility of interaction in multiple directions for a specific optimization problem. Therefore, when setting an update strategy for a new individual, the generation mechanism of the new individual should be different depending on the probability of selection. The literature [5] contains the following two methods. The first method uses a single cluster center to generate individuals and the other method generates new individuals based on two existing individuals.

In this paper, We use the first method to generate new individuals. Original BSO algorithm uses k-means for clustering and it uses the best individuals in each cluster as the clustering center. When a new individual is generated, the individual is selected as a weighted sum of a cluster center, an individual in a cluster, two cluster centers, or an individual in a cluster with a certain probability. On this basis, the process of exploring the current solution is realized by adding mutation. The most common method of variation is Gaussian variation method. The Gaussian variation method is as following [13]:

$$X_{new}^{d} = X_{old}^{d} + \xi * n(\mu, \sigma) \tag{1}$$

In this formula, the parameter X_{old}^{d} represents the individual selected in the d-dimensional. The parameter X_{new}^{d} represents the new individual produced in the d-dimensional. $n(\mu,\sigma)$ is a Gaussian random function with μ as the mean and σ as the standard deviation. ξ is a coefficient that measures the weight of a Gaussian random value, which has an important influence on the degree of contribution of the individual. If the ξ is large, it means that it is beneficial to explore; otherwise, it means that it is beneficial to mining. This allows the algorithm to strike a balance between local search and global search. The update formula for ξ is as following [6]:

$$\xi(t) = \log sig(\frac{\frac{T}{2} - t}{K}) * rand(0, 1) \tag{2}$$

In this formula, $logsig(\cdot)$ is a logarithmic sigmoid transfer function. The parameter T is the maximum number of iterations. The parameter t is the current number of iterations. The parameter K is the slope of the $logsig(\cdot)$, and the speed of the algorithm can be adjusted by k from global search to local search. A random value between 0 to 1 is generated by *rand(0,1)*.

2.2 Estimation of Distribution Algorithms

The Estimation of Distribution Algorithms (EDAs) was proposed by Larranaga and Lozan [14] (1996). The basic steps of the EDAs algorithm are as following:

Step 1: Initialize the population.
Step 2: Choose the dominant group.
Step 3: Construct a probability model.
Step 4: Random sampling.
Step 5: Generate new groups. If the termination condition of the algorithm is met, the result is output; otherwise, it is turned to step 2.

At present, the Estimation of Distribution Algorithms for extension continuous domain are mainly divided into two kinds, including the Estimation of Distribution Algorithms based on Bayesian Network and Estimation of Gaussian Network Algorithm (EGNA) [15]. EGNA which use the Gaussian Network model to solve optimization problems is a kind of Estimation of Distribution algorithm, including $EGNA_{ee}$ algorithm, $EGNA_{BGe}$ algorithm and $EGNA_{BIC}$ algorithm. This paper mainly uses the $EGNA_{ee}$ algorithm to develop the current research. The basic steps of the $EGNA_{ee}$ algorithm is as following:

Step 1: The structure of the Gaussian network is learned by the method of edge exclusion test.

Step 2: The parameters of the Gaussian network are learned, and the Gaussian network model is constructed by using the network structure and parameters in learning.

Step 3: The Gaussian network in step 2 is used to sample and encode.

EDAs have a strong global search capability, which are suitable for solving the optimization problem of nonlinear and variable coupling. In order to enhance global search ability of DMBDO algorithm in the inter-group discussion and improve the convergence speed of DMBSO algorithm, it is feasible to use the EDA to improve the inter-group discussion process of DMBSO algorithm.

3 BSO Algorithm with Estimation of Distribution

DMBSO algorithm can avoid being stagnated in the local optima solutions, more effectively and steadily find better results than the original BSO algorithm, so the DMBSO algorithm is widely used. In order to further improve the performance of the algorithm, the EDBSO algorithm is improved on the basis of the DMBSO algorithm. EDBSO algorithm implementation process is shown in Fig. 1.

In the sampling process of the EDBSO algorithm, the edge distribution of each group is calculated. The calculated result is used to calculate the joint probability distribution of two clusters in the process of the inter-group discussion. The calculation process is as following:

$$R = R_1 * R_2^T \tag{3}$$

In this formula, the parameter R_1 represents the result of the joint probability distribution of two clusters. the individual diversity is enhanced by building edge distribution.

Similarity as the DMBSO algorithm, the EDBSO algorithm divides the discussion process of the BSO algorithm into two parts: intra-group discussion and inter-group discussion. The intra-group discussion mode is reserved as the intra-group processing method of the DMBSO algorithm, and the inter-group discussions were implemented using a Gaussian distribution model. In EDBSO algorithm, the joint probability distribution is obtained from the product of the edge distributions of their clusters, and the next generation of new populations is obtained by Gaussian distribution.

$$\hat{\mu}_i = \frac{1}{N_i}\sum_{k=1}^{N_i} x_{ik} \tag{4}$$

$$\hat{\sigma}_i^2 = \frac{1}{N_i}\sum_{k=1}^{N_i} (x_{ik} - \hat{\mu}_i)^2 \tag{5}$$

In these formulas, the parameter $\hat{\mu_i}$ represents the means of this Gaussian distribution and the parameter $\hat{\sigma_i}$ represents the means of this Gaussian ditribution. The parameter N_i represents the number of individuals in i-dimensional. The parameter X_{ik} represents the value of k-th individual in i-dimensional.

Similarity as the DMBSO algorithm, the number of discussions in intra-group and the inter-group are as following [5]:

$$N_{t\_i\ n} = N_{m_t}(\frac{N_{c\_i}}{N_{m\_i}}) \tag{6}$$

$$N_{t\_e\ x} = N_{m\_t}(1 - \frac{N_{c\_i}}{N_{m\_i}}) \tag{7}$$

In these formulas, the parameter N_{t_in} represents the upper limit of the number of discussions in the current group, while the parameter N_{t_ex} represents the upper limit of the number of discussions between the current groups. The parameter N_{c_i} represents the current generation of individuals, the parameter N_{m_i} represents the total number of iterations, and the parameter N_{m_t} represents the maximum of discussions in the group and between groups.

Two individual fusion processes involved in the algorithm step are performed as following [5]:

$$X_{new} = \nu X_1 + (1 - \nu) X_2 \tag{8}$$

In this formula, the parameter X_{new} represents a new individual produced by the fusion of two individuals. X_1 and X_2 represent two individuals undergoing fusion operations. The parameter v represents a random number from 0 to 1. Random disturbances should be added to the new individual generation process. For the specific method, please refer to formula (7). For the method of increasing the disturbance, please refer to formula (6).

In the inter-group discussion, compared the previously generated individual X_{old} to the newly generated individual X_{new}. The better individual X_{best} is retained by this compared result. This improvement improves the global search capability and effectively improves the efficiency of the DMBSO algorithm. The inter-group discussion process of EDBSO algorithm is shown in Algorithm 1.

Algorithm 1. An example for inter-group discussion process of EDBSO algorithm

```
while the upper time limit of inter-group discussion do
    for each cluster is combined with each other do
        Select an individual from n individuals randomly X_old
        Calculate the joint probability distribution of these two clusters
        Calculate the mean of the joint probability distribution Ex
        Calculate the variance of the joint probability distribution Dx
        Generate a new generation of individuals based on Gaussian distribution
        X_new
        if X_new is better than X_old then
            X_best = X_new
        else
            X_best = X_old
        end if
    end for
end while
```

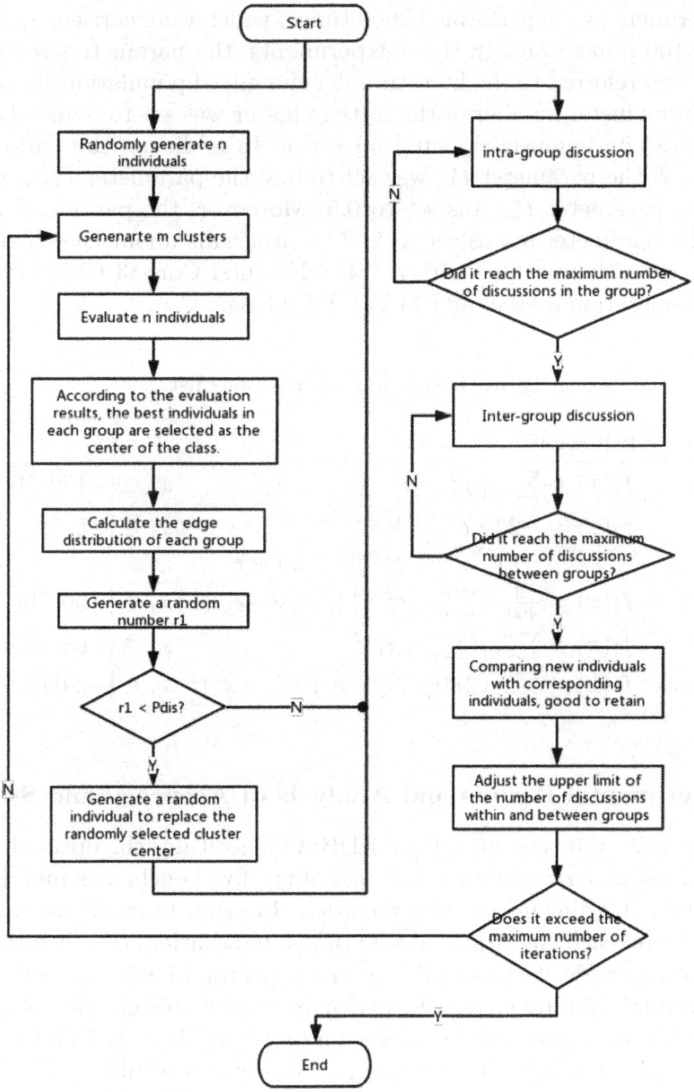

Fig. 1. Overall EDBSO algorithm.

4 Experiment and Performance Analysis

4.1 Parameter Setting and Selection

In this section, the performance of the DEBSO algorithm was compared with BSO algorithm, DBSO algorithm and DMBSO algorithm. In order to evaluate the performance of the DEBSO algorithm, totally five benchmark functions were used in the experiments. The benchmark functions are shown in Table 1.

These experiments were performed 2000 times, which were carried out on 10, 20, 30, 50 and 100 dimensions. In these experiments, the parameters setting of the algorithm were referred to the literature [5], the initial population size was set to 200, while the cluster number of the initial cluster was set to 5 and the optimal threshold was continuously updated 50 times. In addition, the parameter P_{5a} was set to 0.2, the parameter P_{6b} was set to 0.8, the parameter P_{6biii} was set to 0.4, and the parameter P_{6c} was set to 0.5. Moreover, the parameter μ was set to 0 and the parameter σ was set to 1. The programs about these experiments were running on the machine with a 2.40 GHz Intel Core i3 CPU, 4 GB RAM, Windows 7 operation system and MATLAB 2018a.

Table 1. Benchmark functions list.

Function	Equation	Range
Sphere	$f_1(x) = \sum_{i=1}^{n} x_i^2$	$x_i \in [-100, 100]$
Ackley	$f_2(x) = -20 * e^{-0.2*\sqrt{\frac{1}{\pi}*\sum_{i=1}^{n} x_i^2}} - e^{\frac{1}{n}\sum_{i=1}^{n}\cos(2\pi x_i)} + 20 + e$	$x_i \in [-5, 5]$
Griewank	$f_3(x) = \frac{1}{4000} * \sum_{i=1}^{n} x_i^2 - \prod_i^n \cos(\frac{x_i}{\sqrt{i}}) + 1$	$x_i \in [-50, 50]$
Schwefel	$f_4(x) = \sum_{i=1}^{n}(\sum_{j=1}^{i} x_i)^2$	$x_i \in [-65.532, 65.532]$
Rosenbrock	$f_5(x) = \sum_{i=1}^{n} 100(x_{i+1} - x_i^2) + (1 - x_i)^2$	$x_i \in [-2.048, 2.048]$

4.2 Experimental Results and Analysis of Accuracy and Stability

In order to verify the accuracy of the EDBSO algorithm, the optimal solutions, worst solutions, mean values and variances of the five benchmark functions in 10, 20, 30, 50 and 100 dimensions were recorded. In order to avoid the randomness of the algorithm, each algorithm runs 50 times. In addition, the allowable error is 0.0001. In other words, if the optimization result of the EDBSO algorithm and the function optimal solution error is less than 10^{-4}, the optimization is considered successful. The simulation results are shown in Tables 2, 3, 4, 5 and 6.

From the above results, we can see the following two rules:

First, from the aspect of stability, the EDBSO algorithm is generally worse than other algorithms when dealing with low-dimensional problems, and better than other algorithms when dealing with high-dimensional problems. However, When optimizing for the function f_5, the stability of the EDBSO algorithm generally performs poorly.

Because EDA has a good global optimization ability, it balances the local search and global search capabilities of the algorithm in the process of combining with the BSO algorithm in the high-dimensional. Therefore, the EDBSO algorithm is more stable than the DMBSO algorithm in general.

Secondly, from the aspect of accuracy, the EDBSO algorithm is similar to the optimal solutions obtained by the DMBSO algorithm in general. Especially in dealing with high-dimensional problems, the precision far exceeds the original

Table 2. Comparison of simulation results between EDBSO algorithm and other three algorithms with f_1 function.

Dimension	Method	Best	Worst	Mean	Std	Time
10	BSO	7.5691e−45	0.36063	0.075117	0.021549	770.175385
	DBSO	2.0506e−15	1.0285e−06	1.0792e-07	8.2943e-14	293.165853
	DMBSO	5.4163e−45	346.21	6.9865	4049.1	4299.456328
	EDBSO	5.4263e−45	0.9719	0.078402	0.078092	2560.939262
20	BSO	1.6887e−43	2.3559	0.13196	0.47268	832.313335
	DBSO	7.3772e−19	0.00038485	1.7995e-05	5.6424e-09	309.762115
	DMBSO	5.0108e−44	2.0619	0.23044	0.57639	4273.036396
	EDBSO	4.0897e−44	17.34	0.4644	10.404	2879.269162
30	BSO	3.2455e−43	6.8921	0.37916	3.913	836.985255
	DBSO	4.8418e−17	0.00016954	1.4382e−05	2.125e−09	324.934638
	DMBSO	1.9328e−43	3.4557e−43	2.572e−43	2.4972e−87	4426.811928
	EDBSO	1.4622e−43	9.443	0.29078	3.2774	3161.113494
50	BSO	2.7989e−42	5.4325e-07	1.0865e−08	9.9732e−15	894.578408
	DBSO	4.2554e−11	0.00021077	1.8249e−05	2.1293e−09	355.554967
	DMBSO	6.3141e−43	14.544	0.55271	12.677	4502.695085
	EDBSO	6.5306e−43	1.0188e−42	8.1285e−43	1.0423e−86	3558.646687
100	BSO	0.27906	2.0028	0.80144	0.24411	995.158772
	DBSO	4.0315e−20	0.00082609	5.588e−05	3.2104e−08	439.499676
	DMBSO	2.6301e−42	4.3119e−42	3.5717e−42	2.4358e−85	4918.480215
	EDBSO	2.6544e−42	4.5026e−42	3.5506e−42	2.3978e−85	5041.192595

Table 3. Comparison of simulation results between EDBSO algorithm and other three algorithms with f_2 function.

Dimension	Method	Best	Worst	Mean	Std	Time
10	BSO	4.4409e−15	1.6722	0.40589	0.74277	728.503127
	DBSO	1.2057e−08	0.00034917	2.0891e−05	4.6402e−09	311.621410
	DMBSO	8.8818e−16	2.1374	0.42474	0.81411	2460.411436
	EDBSO	1.8784	0.20404	0.40497	3165.178900	1.8784
20	BSO	4.4409e−15	2.7773	0.26599	1.0997	854.985797
	DBSO	2.0865e−09	0.00049461	0.00012496	2.3614e−08	346.537867
	DMBSO	4.4409e−15	2.8021	0.36682	1.4289	3379.315426
	EDBSO	4.4409e−15	2.9366	0.20459	0.87007	3597.044907
30	BSO	7.9936e−15	3.378	0.20359	1.1011	923.957796
	DBSO	4.8205e−08	0.00044112	9.9372e−05	1.8365e−08	387.234620
	DMBSO	4.4409e−15	2.9623	0.059247	0.29655	4837.391121
	EDBSO	4.4409e−15	3.0428	0.064729	0.31319	4055.648705
50	BSO	2.2204e−14	4.012	0.9882	0.77481	1505.443256
	DBSO	1.089e−09	0.00047706	0.0001047	2.1236e−08	423.867895
	DMBSO	4.4409e−15	3.3037	0.066074	0.36883	4746.394647
	EDBSO	4.4409e−15	7.9936e−15	5.1514e−15	3.4819e−30	6371.046706
100	BSO	1.345	2.5039	1.9773	0.075596	1505.443256
	DBSO	1.8993e−08	0.00042797	0.00012383	2.1577e−08	581.047732
	DMBSO	7.9936e−15	1.8652e−14	1.2825e−14	1.2848e−29	5304.769384
	EDBSO	7.9936e−15	2.2204e−14	1.3465e−14	1.0628e−29	7424.734967

Table 4. Comparison of simulation results between EDBSO algorithm and other three algorithms with f_3 function.

Dimension	Method	Best	Worst	Mean	Std	Time
10	BSO	0.036919	0.45891	0.17116	0.011544	803.929645
	DBSO	3.3307e−16	5.5278e-08	4.1035e-09	1.9574e−16	349.529557
	DMBSO	0.019697	0.81871	0.087912	0.022225	4025.457376
	EDBSO	0	0.18516	0.0075497	0.0012578	3769.519374
20	BSO	0	0.3285	0.054682	0.009047	897.480801
	DBSO	0	8.0417e−06	4.6717e−07	3.0153e−12	50.787879
	DMBSO	0	0.39761	0.034355	0.0068181	4214.867362
	EDBSO	0	0.35461	0.022364	0.0085674	3437.500599
30	BSO	0	0.30339	0.029291	0.0086693	971.577982
	DBSO	0	1.4594e−05	6.1898e−07	7.4953e−12	515.217462
	DMBSO	0	0.27573	0.028178	0.0067842	4391.354511
	EDBSO	0	0.2033	0.0058469	0.0015142	5837.440756
50	BSO	1.7671e−05	0.051549	0.0074397	0.00016719	1056.782318
	DBSO	6.3283e−15	4.1557e−06	3.9879e−07	1.524e−12	599.665957
	DMBSO	0	0.027037	0.007241	8.9556e−05	4712.325434
	EDBSO	0	0.019697	0.00083989	1.718e−05	5732.384206
100	BSO	0.041859	0.13226	0.084249	0.0008353	1219.363575
	DBSO	0	3.4345e−06	2.8825e−07	2.8825e−07	781.974582
	DMBSO	2.7018e−06	0.027079	0.0031711	7.1493e−05	5642.504073
	EDBSO	3.0268e-11	5.2219e−05	1.1129e−05	3.396e−10	7036.574346

Table 5. Comparison of simulation results between EDBSO algorithm and other three algorithms with f_4 function.

Dimension	Method	Best	Worst	Mean	Std	Time
10	BSO	1.7838e−44	0.38703	0.086087	0.02536	874.555300
	DBSO	2.7083e−15	3.6602e-06	1.6544e-07	5.7512e-13	379.768567
	DMBSO	5.989e−45	123.49	2.5128	514.95	3957.924598
	EDBSO	4.506e−45	0.31933	0.042819	0.0093859	2926.852289
20	BSO	1.1217e−43	2.5231	0.25065	0.81504	975.324421
	DBSO	6.8536e−19	2.5545e-05	3.1151e−06	6.6801e−11	475.574934
	DMBSO	5.2834e−44	1.9616	0.10169	0.28415	4237.019520
	EDBSO	5.0513e−44	1.1171	0.056071	0.086907	3518.125449
30	BSO	4.0226e−43	6.0813	0.12163	1.2497	1003.238078
	DBSO	5.8451e−17	0.00019091	1.3357e−05	1.7754e−09	544.732076
	DMBSO	1.5415e−43	3.2164e−43	2.4237e-43	2.2951e−87	4683.178679
	EDBSO	1.6449e−43	5.1815	0.16553	1.2088	3801.966827
50	BSO	2.929e−42	1.3257e−08	2.7481e-10	5.9381e−18	1279.731608
	DBSO	8.3551e−13	0.00013867	1.5265e−05	1.0953e−09	668.654830
	DMBSO	6.2084e−43	9.3132e−43	7.8301e-43	1.0248e−86	5804.566591
	EDBSO	5.9873e−43	9.4787e−43	8.115e-43	1.0639e−86	5006.061714
100	BSO	0.32672	2.1354	0.7434	0.20522	3741.186808
	DBSO	6.86e-16	0.00018819	1.9614e−05	2.6175e−09	1626.673663
	DMBSO	2.8508e−42	4.0873e−42	3.4774e-42	1.5864e−85	9888.189891
	EDBSO	2.7141e−42	4.3691e−42	3.5139e-42	2.459e−85	9981.639613

Table 6. Comparison of simulation results between EDBSO algorithm and other three algorithms with f_5 function.

Dimension	Method	Best	Worst	Mean	Std	Time
10	BSO	8.6099	658.99	54.594	28826	816.044947
	DBSO	8.6099	8.8097	8.6291	0.0021323	287.409354
	DMBSO	8.6099	5.2365e+05	10496	9.2654e+09	1877.469488
	EDBSO	8.6099	35.752	12.953	128.5	2357.664396
20	BSO	18.513	2856.9	160.77	2.9497e+05	847.495563
	DBSO	18.729	18.919	18.801	0.0015177	307.243327
	DMBSO	18.513	3113.3	153.28	4.1619e+05	1958.874346
	EDBSO	18.513	78.194	25.646	573	2748.065361
30	BSO	28.416	2373.2	212.09	3.7381e+05	892.798270
	DBSO	28.7	28.978	28.735	0.0040831	325.809832
	DMBSO	28.416	3.2164e-43	2.4237e-43	2.2951e−87	4683.178679
	EDBSO	28.416	116.07	31.915	506.32	3109.158825
50	BSO	48.245	4154.4	620.98	2.0728e+06	955.031143
	DBSO	48.518	48.865	48.578	0.011457	360.953601
	DMBSO	48.221	2628	195.77	3.688e+05	2212.602511
	EDBSO	48.221	194.12	58.844	2283.8	3650.214542
100	BSO	367.69	5173.3	1281.9	1.5198e+06	1089.589691
	DBSO	98.035	99.104	98.151	0.058087	447.651992
	DMBSO	97.736	2208.2	208.09	2.0404e+05	2432.138448
	EDBSO	97.736	619.88	181.23	29123	5273.746324

BSO algorithm and DBSO algorithm. When optimizing for the f_3 function, the EDBSO algorithm is better than the DMBSO algorithm in dealing with high-dimensional problems.

Because the edge distribution of each group is calculated during the sampling process and the calculated results are applied to the joint probability distribution in the discussion between the groups, the diversity of the EDBSO algorithm is enhanced. To a certain extent, this mechanism avoids the premature convergence of the algorithm.

In summary, the EDBSO algorithm has good precision and is suitable for optimization of high-dimensional data models, but the running time of the algorithm is longer than other three algorithms. In the process of using, the algorithm should be parallelized, and the computing power of the computer is effectively utilized to reduce the running time of the algorithm.

4.3 Experimental Results and Analysis of Convergence

In order to verify the convergence of the EDBSO algorithm, Figs. 2, 3, 4, 5 and 6 shows the convergence curves of the four algorithms for the simulation results of the five benchmark functions from 10-dimensional to 100-dimensional.

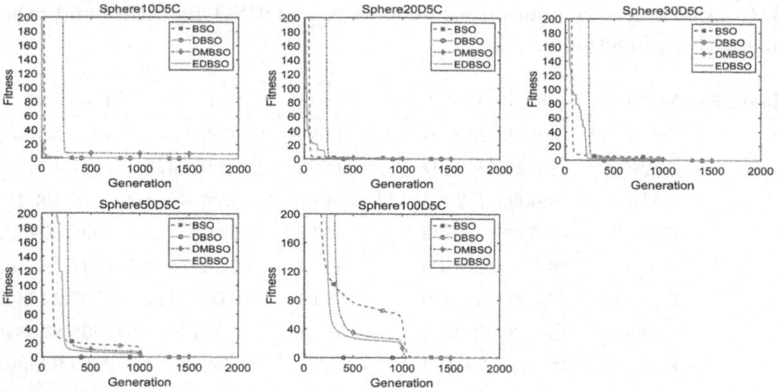

Fig. 2. Convergence curve of 4 algorithms with f_1 function.

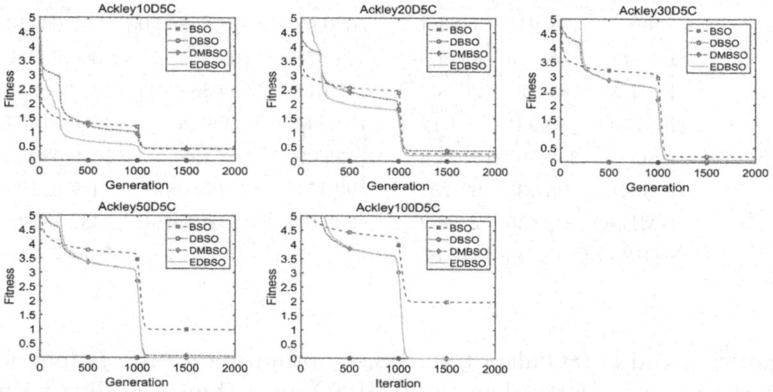

Fig. 3. Convergence curve of 4 algorithms with f_2 function.

It can be seen from the convergence curve that in the simulation experiments of benchmark functions f_1, f_2, f_3 and f_5, the convergence performance of EDBSO algorithm in each dimension is better than the original BSO algorithm and DMBSO algorithm. Especially when dealing with the two benchmark functions of f_2 and f_5, the convergence speed and convergence precision are much better than the original BSO algorithm and DMBSO algorithm. Compared with BSO algorithm and DMBSO algorithm, the EDBSO algorithm is more easier to fall into the local solutions than other algorithms. The convergence performance of this algorithm in high-dimensional is not as good as BSO algorithm and DMBSO algorithm, but the convergence performance in low-dimensional is better than original BSO algorithm and DMBSO algorithm. Therefore, from the overall perspective in these experiments, the convergence performance of the EDBSO algorithm is better than the original BSO and DMBSO, but the overall convergence performance is not as good as the DBSO algorithm.

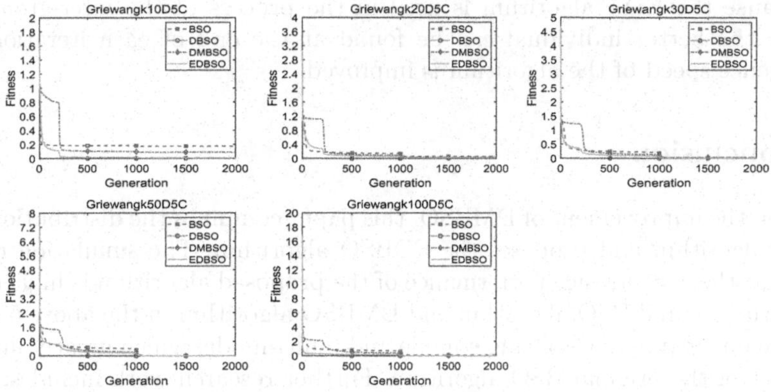

Fig. 4. Convergence curve of 4 algorithms with f_3 function.

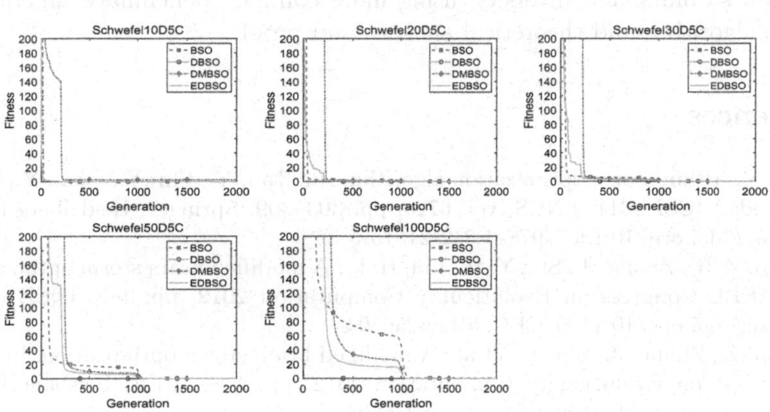

Fig. 5. Convergence curve of 4 algorithms with f_4 function.

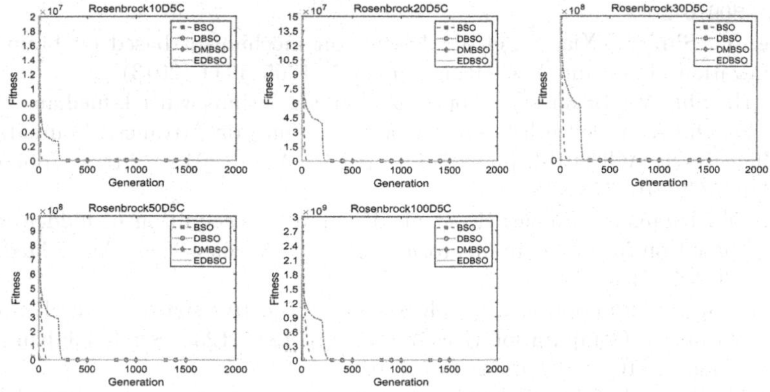

Fig. 6. Convergence curve of 4 algorithms with f_5 function.

Because the EDA algorithm is used in the process of the inter-group discussion, the better individual can be found at the end of each iteration, the convergence speed of the algorithm is improved.

5 Conclusion

Based on the improvement of DMBSO, this paper combines the distribution estimation algorithm and proposes the EDBSO algorithm. The simulation results show that the convergence performance of the proposed algorithm is higher than that of the original BSO algorithm and DMBSO algorithm in the above 5 different benchmark functions of single mode and multi-mode, which greatly improve the effect of the original BSO algorithm. Further research work includes: using different estimation distribution variability methods instead of Gaussian distribution estimation variability to the convergence of the algorithm, EDBSO algorithm stability and diversity, using more complex benchmark functions to test the algorithm and theoretical analysis and proof.

References

1. Shi, Y.: Brain storm optimization algorithm. In: Tan, Y., Shi, Y., Chai, Y., Wang, G. (eds.) ICSI 2011. LNCS, vol. 6728, pp. 303–309. Springer, Heidelberg (2011). https://doi.org/10.1007/978-3-642-21515-5_36
2. Zhan, Z. H., Zhang, J., Shi, Y. H., Liu, H. L.: A modified brain storm optimization. In: IEEE Congress on Evolutionary Computation 2012, pp. 1–8. IEEE (2012). https://doi.org/10.1109/CEC.2012.6256594
3. Zhan, Z., Zhang, J., Shi, Y., et al.: A modified brain storm optimization. In: IEEE Congress on Evolutionary Computation 2012, pp. 1–8. IEEE, Brisbane (2012). https://doi.org/10.1109/CEC.2012.6256594
4. Xue, J., Wu, Y., Shi, Y., Cheng, S.: Brain storm optimization algorithm for multi-objective optimization problems. In: Tan, Y., Shi, Y., Ji, Z. (eds.) ICSI 2012. LNCS, vol. 7331, pp. 513–519. Springer, Heidelberg (2012). https://doi.org/10.1007/978-3-642-30976-2_62
5. Yang, Y., Shi, Y., Xia, S., et al.: Discussion mechanism based on brain storm optimization algorithm. J. Zhejiang Univ. **47**, 1705–1711 (2013)
6. Zhu, H., Shi, Y.: Brain storm optimization algorithms with k-medians clustering algorithms. In: Seventh International Conference on Advanced Computational Intelligence (ICACI 2015), pp. 107–110. IEEE, Wuyi (2015) . https://doi.org/10.1109/ICACI.2015.7184758
7. Yang, Y., Duan, D., Zhang, H., et al.: Kinematic recognition of hidden markov model based on improved brain storm optimization algorithm. Space Med. Med. Eng. 403–407 (2015)
8. Shi, Y.: Brain storm optimization algorithm in objective space. In: IEEE Congress on Evolutionary Computation (CEC) 2015, pp. 1227–1234. Sendai, Japan (2015). https://doi.org/10.1109/CEC.2015.7257029
9. Diao, M., Wang, X., Gao, H., et al.: Differential brain storm optimization algorithm and its application to spectrum sensing. Appl. Sci. Technol. **43**, 14–19 (2016)

10. Wu, Y., Fu, Y., Wang, X., Liu, Q.: Difference brain storm optimization algorithm based on clustering in objective space. Control Theory Appl. **34**, 1583–1593 (2017)
11. Liang, Z., Gu, J., Hou, X.: A modified brainstorming optimization algorithm. J. Hebei Univ. Technol. 56–62 (2018)
12. Cheng, S., Chen, J., Lei, X., et al.: Locating multiple optima via brain storm optimization algorithms. IEEE Access **6**, 17039–17049 (2018)
13. Shi, Y.: An optimization algorithm based on brainstorming process. In: Emerging Research on Swarm Intelligence and Algorithm Optimization, pp. 1–35 (2015)
14. Larranaga, P., Lozano, J.: Estimation of Distribution Algorithms: A New Tool for Evolutionary Computation. Springer, Heidelberg (2001). https://doi.org/10.1007/978-1-4615-1539-5
15. Lozano, J., Bengoetxea, E.: Estimation distribution algorithms based on multivariate normal and Gaussian networks. Department Computer Science Artificial Intelligence University, Basque Country, Vizcaya, Spain, Technical report KZZA-1K-1-01 (2001)

Tentative Study on Solving Impulse Control Equations of Plant-pest-predator Model with Differential Evolution Algorithm

Huichao Liu(✉), Fengying Yang, Liuyong Pang, and Zhong Zhao

Huanghuai University, Zhumadian 363000, Henan, China
lhc@huanghuai.edu.cn

Abstract. In recent years, using ecological control method for pest management has become a hot topic, and some pest-predator models have been proposed. These models are expressed by impulse control equations, and which can be transformed into a global optimization problem. But it is easy to fall into local optimal when solving these equations by traditional method. On the other hand, differential evolution (DE) algorithm has been widely used to solve a variety of complex optimization problems. Therefore, attempting to solve the impulse control equations of plant-pest-predator model by DE algorithm is an important motivation of this paper. In order to further enhance the optimization capability, a rotation-based differential evolution (RDE) was introduced by embedding a rotation-based learning mechanism into DE. The simulation experiments show that the RDE algorithm can solve the impulse control equations effectively, and the results are more competitive than those obtained by the traditional algorithms. Meanwhile, the convergence speed of RDE algorithm is also very fast. This preliminary study may provide a new method for solving ecological control problem.

Keywords: Evolutionary computation · Differential evolution algorithm · Rotation-based learning · Plant-pest-predator model · Impulse control equation

1 Introduction

Pest control is very important for agricultural production. The outbreak of pests and diseases will lead to serious ecological and economic losses. Traditionally, pesticide spraying can easily lead to environmental pollution and increase disease resistance. Therefore, integrated pest management (IPM) methods were proposed in the 1950s [14]. IPM methods utilize biological, chemical and cultural strategies to reduce disease populations, and have proved to be more effective than traditional methods [5,16].

Recently, mathematical models of IPM strategies have been extensively studied. The authors in [13] consider a predator-prey model with disease in the prey.

L. Pan et al. (Eds.): BIC-TA 2019, CCIS 1159, pp. 42–52, 2020.
https://doi.org/10.1007/978-981-15-3425-6_4

Reference [4] studies a predator-pest model with periodic release of infected pest and predators. Gakkhar and Naji [3] study the dynamics of a three-species tri-trophic food web for a generalist-specialist-prey system with modified Holling type II functional response. Further, [11] studies the dynamics of a tri-trophic food web system containing a Leslie-Gower type generalist predator. Based on a two species predator–prey model [10], Liang et al. [7] proposes a three species model which includes a top predator (the enemy of the pest), the pest, and a basal producer (a renewable biotic resource, referred to as plant). The pest is the prey of the predator, and the generalist predator of the basal producer. These mathematical models mentioned above are mainly represented by impulse control equations, and which can be transformed into a global optimization problems with multiple decision variables. In the literatures, traditional methods are often used to solve these optimization problem, but which will fall into local optimal easily.

In the past decades, researchers have proposed many excellent intelligent algorithms (such as differential evolution algorithm (DE) [15], particle swarm optimization (PSO) [6], et al.), which have shown better performance compared to traditional methods when tackling the complex problems. However, there is no relevant research on solving the impulse equation of the predator–prey model by DE algorithm. Therefore, an important motivation of this paper is trying to solve impulse control equations with DE algorithm in order to obtain better results.

In [9], a novel rotation-based learning mechanism, named RBL, is proposed by extending the classical OBL mechanism [12,17]. In order to verify the effectiveness of RBL mechanism, the rotation-based differential evolution (RDE) algorithm was introduced by embedding the RBL mechanism into DE [8]. In this article, the RDE algorithm will be used to solve a three populations ecological control model which proposed in the literature [7]. Experiments show that this solution is feasible, and the RDE algorithm can solve the problem effectively and converge quickly.

2 Related Work

2.1 Differential Evolution Algorithm

Ever since differential evolution (DE) algorithm was proposed by Price and Storn in 1995 [15], vast DE variants have been introduced to optimize the various benchmark functions and real-world problems. A comprehensive explanation of DE and its variants can be found in the surveys [1,2]. Among many evolutionary schemes of DE, the *DE/rand/1/exp* scheme will be used in this paper, and its main frame can be found in [2].

2.2 Rotated-Based Learning (RBL)

The rotation-based learning (RBL) mechanism is extended by the OBL mechanism [18]. In OBL mechanism, an *opposite number* is mirrored to its original

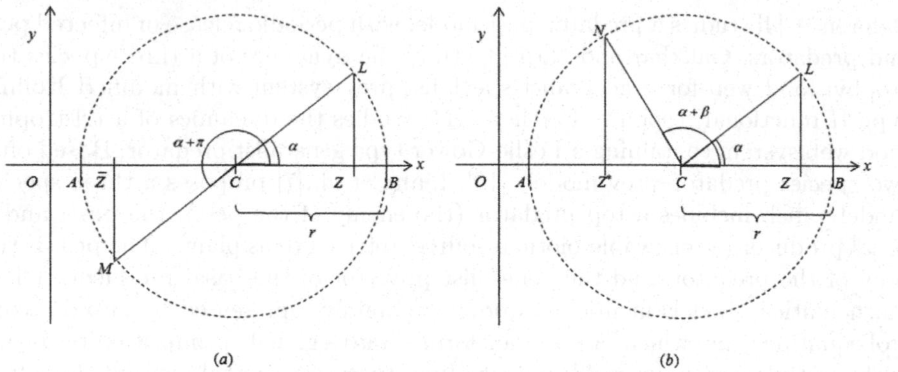

Fig. 1. The geometric interpretation of opposite number (*a*), and rotation number (*b*) in two-dimensional plane

number by the center of the defined boundary in one-dimensional space (i.e. a number axis). However, a one-dimensional number axis can be placed in a two-dimensional space as well, so that the *opposite number* can be explained in another way.

As shown in Fig. 1(a), for a given two-dimensional plane with x axis and y axis, a number z, and its lower boundary a and upper boundary b, can be marked by points Z, A and B on x axis. The center point of the interval $[a, b]$ is denoted by C, its coordinate is $\frac{a+b}{2}$. Then, a circle can be drawn on the plane with the center of C, and its radius is $r(=\frac{b-a}{2})$. Through the point Z, a straight line, which is perpendicular to x axis and intersects with the circle on point L, can be drawn. Obviously, the number z is the x position of point L, and the length of line segment $\overline{CL}$ is equal to the radius r of the circle. For the sake of convenience, two variables u and v are defined by:

$$\begin{aligned} u &= |\overrightarrow{CZ}| = z - \frac{a+b}{2} \\ v &= |\overline{LZ}| = \sqrt{r^2 - u^2} = \sqrt{(z-a)(b-z)}, \end{aligned} \tag{1}$$

where $|\overline{LZ}|$ indicates the length of line segment $\overline{LZ}$, and that $|\overrightarrow{CZ}|$ denotes the length of the directed line segment $\overrightarrow{CZ}$ together with the sign which indicates the direction of the line segment.

Let the point L rotate 180° counterclockwise around the circle, a new point M will be reached. The projection of M on x axis is denoted by $\overline{Z}$. Obviously, its coordinate $\overline{z}$ is just the *opposite number* of z. Therefore, an *opposite number* in OBL can be explained as rotating 180° of the original number in the two-dimensional plane.

In fact, the rotation mentioned above can be at any angle, this is the rotation-based learning mechanism (RBL). As shown in Fig. 1(b), for a given number z, the point Z, L, C, and the circle with radius r are all fixed. If rotating the L

point by β degrees counterclockwise around the circle, then a new point N will be reached. So, the angle of $\angle NCZ$ is $\alpha + \beta$. The projection of point N on x axis is point Z^*, its coordinate z^* on x axis is the **rotation number** of z. Let the quantity of directed line segment $\overrightarrow{CZ^*}$ be denoted by u^*, which is obtained by:

$$u^* = r \times (\cos(\alpha + \beta)) = u \times \cos\beta - v \times \sin\beta \tag{2}$$

Then, the **rotation number** z^* can be calculated by

$$z^* = (\frac{a+b}{2}) + u^*. \tag{3}$$

It is clearly that the rotation number can search any point in the rotation space by rotating different angles. Moreover, the concept of rotation number can be easily generalized to higher dimensions. Let $Z = (z_1, z_2, \ldots, z_D)$ be a vector with D variables. $a = [a_1, a_2, \ldots, a_D]$ and $b = [b_1, b_2, \ldots, b_D]$ are the lower boundary and upper boundary of Z, respectively, and $z_i \in [a_i, b_i]$, $i = 1, 2, \ldots, D$. Therefore, the center point of rotation space $[a, b]^D$ is indicated by $C = (c_1, c_2, \ldots, c_D)$, and $c_i = \frac{a_i+b_i}{2}$. The radius vector is $R = (r_1, r_2, \ldots, r_D)$, and $r_i = \frac{b_i-a_i}{2}$.

For every dimension, let the point Z rotate β degrees counterclockwise by the center point C. Similar to (2), then the **rotation point** $Z^* = (z_1^*, z_2^*, \ldots, z_D^*)$ can be achieved, where the z_i^* is defined by

$$z_i^* = (\frac{a_i + b_i}{2}) + (u_i \times \cos\beta - v_i \times \sin\beta). \tag{4}$$

Based on the above analysis, the proposed RBL mechanism can rotate any degrees between 0° and 360° (same as the interval (−180°, 180°)), and explore any point in the search space. Actually, RBL mechanism also can be adapted to different application modes by specifying different angles [9]. Therefore, the RBL mechanism is very flexible for finding the potential promising solutions. Nevertheless, the Gaussian distribution is simply used to determine the deflection angle, and it is defined as

$$\beta = \beta_0 \cdot N(1, \sigma), \tag{5}$$

where β and β_0 indicate the deflection angle and its basic number, respectively. σ is the standard deviation of the Gaussian function $N(\cdot,\cdot)$, and is used to adjust the variation of the basic number. Before calculating the parameter, an initial value β_0 should be set firstly. In the following experiments, β_0 and σ will be set to 180° and 0.25, respectively. Then, β will vary mainly within [90°, 270°], and fluctuate around the 180° randomly. This has been proved to be an effective parameter setting in literature [8].

In order to take RBL into practice, the RBL mechanism should be embedded into a population-based algorithm. The implementation pseudocode of RBL mechanism is shown in Algorithm 1. The rotation space should be obtained firstly by calculating the minimum and maximum values of each dimension of all individuals in the population. Moreover, the center point of the rotation

Algorithm 1. The RBL mechanism for a population-based algorithm

1: {Update the boundary vectors a, b and its center point C for current population $P = \{Z_1, Z_2, \ldots, Z_D\}$}
2: **for** i =1 to D **do**
3:     $a_i \leftarrow$ the minimum value of ith variable in P;
4:     $b_i \leftarrow$ the maximum value of ith variable in P;
5:     $c_i \leftarrow \frac{a_i+b_i}{2}$;
6: **end for**
7: {Produce the rotation-based population $RP = \{Z_1^*, Z_2^*, \ldots, Z_D^*\}$}
8: **for** $Z_i \in P$ **do**
9:     {Generate the new rotation-based individual $Z_i^* = \{z_{i1}^*, z_{i2}^*, \ldots, z_{iD}^*\}$}
10:     Set the rotation degree β;
11:     **for** j = 1 to D **do**
12:         Calculate the values of $u_{ij} \in u_i$ and $v_{ij} \in v_i$ by (1);
13:         Calculate z_{ij}^* by (4).
14:     **end for**
15: **end for**

space, the lower and upper bound vectors should be obtained as well. Then, the rotation-based population RP which consisting the rotation-based individuals generated by (1)–(4) will be produced. Note that, the rotation angle β in step 10 should be generated by (5) in this paper.

Algorithm 2. The RDE Algorithm

Initialize the parameters pr_0 and β_0;
Randomly initialize each individual in population P;
Conduct RBL mechanism to produce the rotation-based population RP;
Select NP fittest individuals from $\{P, RP\}$ as the initial population P;
while $FEs \leq MAX\_FEs$ **do**
    if $rand(0, 1) \leq pr$ **then**
        Conduct the RBL mechanism to produce the rotation-based population RP;
        Select NP fittest individuals from $\{P, RP\}$ as the new population P;
        Update the pr by (5);
    else
        Execute the $DE/rand/1/exp$ scheme;
    end if
    Update the *best* individual;
end while
Output the *best* individual;

2.3 Rotation-Based Differential Evolution

In order to verify the effectiveness of RBL mechanism, the rotation-based differential evolution (RDE) algorithm was introduced by embedding the RBL mechanism into DE algorithm [8]. The pseudocode of the RDE algorithm is shown in

Algorithm 2. In RDE algorithm, the RBL mechanism is used for both population initialization and generation jumping. In population initialization phase, a basic population P is firstly initialized by randomly generating individuals. And then, the RBL mechanism shown in Algorithm 1 is used to produce the rotation-based population RP. Finally, all individuals in $P \bigcup RP$ are evaluated, and the NP best individuals are selected as the initial population P.

A new *probability of rotation* (*pr*) parameter, which is a real number within (0, 1), is introduced to control the execution of RBL mechanism. In each generation, if a random number belongs to (0, 1) is less than *pr*, then RBL mechanism will be conducted to produce the rotation-based population. Then, the NP best individuals from $P \bigcup RP$ are selected as the new population for next generation. Otherwise, the classical $DE/rand/1/exp$ scheme will be executed to generate the next evolution-based population.

In RDE algorithm, the new probability of rotation (*pr*) parameter should be set ahead. For convenience, the self-adaptive mechanism defined in (5) is also used to generate the parameter *pr*. Using the self-adaptive mechanism, RDE algorithm only needs to predefine two initial values of parameters *pr* and β, which are indicated by pr_0 and β_0, respectively.

3 Plant–Pest–Predator Model

3.1 Impulse Control Equation

Pest control is very important in agriculture. In order to reduce environmental pollution and improve the effect of pest control, many scholars have studied the integrated pest management(IPM) method which integrates biological, chemical and ecological factors to control pests. On the basis of predator-prey model proposed by predecessors, a new three-species pest control model was proposed in literature [7], in which a new IPM model is formulated involving a top predator (the enemy of the pest), the pest, and a basal producer (a renewable biotic resource, referred to as plant below). The pest is the prey of the predator, and the generalist predator of the basal producer. At time $t \neq nT, n = 1, 2, ..., N$, the three species plant, pest and predator satisfy the following ordinary differential equations:

$$\begin{cases} \dot{S} = rS(1 - \dfrac{S+I}{K}) - \lambda IS - \dfrac{\alpha_2 SP}{1+a_2 S} + \dfrac{\beta_1 SX}{1+a_1 X} \\ \dot{I} = \lambda IS - \mu I \\ \dot{P} = \dfrac{\beta_2 SP}{1+a_2 S} - \gamma P \\ \dot{X} = bX(1 - \dfrac{X}{L}) - \dfrac{\alpha_1 SX}{1+a_1 X} \end{cases} \tag{6}$$

and at time $t = nT$, the impulsive controls are implemented

$$\begin{cases} \Delta S(t) = -\theta_1 S(t), \\ \Delta I(t) = -\theta_2 I(t) + u, \\ \Delta P(t) = -\theta_3 P(t) + v, \\ \Delta X(t) = 0. \end{cases} \tag{7}$$

The state $X(t)$ denotes a basal producer which is a renewable biotic resource such as plants or vegetation. The states $S(t)$ and $I(t)$ denote the densities of the susceptible and infective prey (pest), respectively. They serve as the food of the predator $P(t)$. At the same time, the pest is the generalist predator for $X(t)$. That is, the pest feeds not only on the $X(t)$, but also on some alternative food resource. Hence, the dynamics of $S(t)$ does not totally depend on $X(t)$. The amounts of released infective pest and predator population are positive constants u and v, respectively. Parameter T is the period of the impulse. Parameters θ_i (i = 1, 2, 3) denote the fraction of susceptible, infective pests and natural enemies killed due to the spraying of pesticides. In Eq. 6, r, K, b, L, λ, μ, γ, α_1, α_2, a_1, a_2, β_1 and β_2 are some constant parameters. Therefore, this model is also constructed with some assumptions, the details can be seen in literature [7].

3.2 Optimal Control Strategies

The stable susceptible pest-eradication periodic solution means that the plant can reach its carrying capacity under impulsive controls. However, Considering the cost of various controls which is an important problem in practice. Hence, this section will address how to maximizing the economic yield of plants at the terminal time with minimum control cost. Two commonly used tactics, chemical control and biology control, are applied to realize the goal. That is, the number of pest is controlled by means of selecting the kill fractions θ_i which are in direct proportion to the amount of associated pesticides, or the amounts of u and v released which match the cost of biological cultivation. These two impulsive controls are implemented at fixed periodic time points nT. These problems can be treated as optimal parameter selection problems, and can be stated as finding the optimal control parameters u , v, and θ_i, $i = 1, 2, 3$, to minimize the objective function

$$J(\theta_i, u, v, T) = -CX(NT) + \Phi_1(T, u, v, \theta_1, \theta_2, \theta_3), \tag{8}$$

where $C > 0$ is the profit of X per unit and Φ_1 represents the cost of the various controls.

Considering the practical application, the problem can be divided into three scenarios.

- 1. Optimization by biological control with fixed period
  In this situation, the maximization of $X(t)$ is obtained by use of biological controls. That is, the optimization is realized by choosing optimal parameters u and v. The corresponding objective function is defined as

$$J(u, v) = -CX^N(1) + \frac{1}{2}N(u^2 + v^2). \tag{9}$$

- 2. Optimization by chemical control with fixed period
  In this situation, the maximization of X(t) is obtained by the use of chemical control. That is, the optimization is realized by choosing optimal parameters $\theta_i, i = 1, 2, 3$. Hence, the corresponding objective function is defined as

$$J(\theta_1, \theta_2, \theta_3) = -CX^N(1) + \frac{1}{2}N(\theta_1^2 + \theta_2^2 + \theta_3^2). \tag{10}$$

- 3. Optimization by integrated control with fixed period
  In this situation, both biological controls and chemical controls are used in the optimization. The corresponding objective function is defined as

$$J(u, v, \theta_1, \theta_2, \theta_3) = -CX^N(1) + \frac{1}{2}N(\theta_1^2 + \theta_2^2 + \theta_3^2 + u^2 + v^2). \tag{11}$$

4 Experimental Verifications

4.1 Settings

In order to get the optimal parameter for each situations, the proposed RDE algorithm is used. In order to facilitate the comparison of experimental results, the constant parameters of the impulse equation are set according to the values in literature [7].

$r = 0.8; \lambda = 0.6; \mu = 0.35; a_2 = 0.5; a_1 = 0.5; \gamma = 0.35; K = 3;$
$b = 0.5; L = 2.8; \alpha_1 = 0.5; \beta_1 = 0.3; ; \alpha_2 = 0.35; \beta_2 = 0.25; C = 10.$

With these parameters and some control parameters, the states $S(t)$ and $I(t)$, the number of the pest, do not tend to zero. And the state $X(t)$, the basal producer, does not reach its maximal yield. We hope find the optimal control parameters to maximize the output of $X(t)$ with minimal cost.

For each situation, two pairs of pulse control periods are selected. One pair is $N = 4, T = 4$, another is $N = 8, T = 2$. Therefore, the experiment will consist of six groups.

For the parameters of RDE algorithm, we have not made too many choices, only selected some commonly used empirical values. i.e. $F = 0.5, Cr = 0.9, Populationsize = 20, br_0 = 0.1, \beta_0 = 180, Numberofgeneration = 100$. The stopping condition of the algorithm is to arrive at the predefined generation, or the optimal value is not updated for 20 successive generations.

4.2 Results of the Experiment

With the above settings, the optimization results are summarized in Tables 1, 2 and 3. In these tables, TSM denotes the time scaling method, which is adopted in literature [7], and RDE denotes the RDE algorithm proposed in this paper. As can be seen from Tables 1, 2 and 3, RDE can achieve better values in most cases than TSM method. In fact, finding the optimal control parameters of impulsive control equations is a two-objective optimization problem, which includes two objectives, $X^N(1)$ and *cost* (See Eqs. 9–11). But in the process of solving, the

Table 1. Optimization results for situation 1

	TSM	RDE	TSM	RDE
N	4	4	8	8
T	4	4	2	2
u	0.9000	0.9000	0.6301	0.6347
v	0.6413	0.3603	0.2277	0.2316
θ_1	0	0	0	0
θ_2	0	0	0	0
θ_3	0	0	0	0
J	−24.2006	−24.5358	−25.4923	−25.5307
$X^N(1)$	2.6559	2.6416	2.7288	2.7357
cost	2.3589	1.8798	1.7953	1.8259

Table 2. Optimization results for situation 2

	TSM	RDE	TSM	RDE
N	4	4	8	8
T	4	4	2	2
u	0	0	0	0
v	0	0	0	0
θ_1	0.1000	0.8999	0.9000	0.9000
θ_2	0.1000	0.1004	0.1000	0.0501
θ_3	0.1000	0.1005	0.1000	0.0519
J	−16.0142	−19.9062	−23.6087	−23.7405
$X^N(1)$	1.6074	2.1566	2.6929	2.7001
cost	0.006	1.6600	3.3200	3.2608

Table 3. Optimization results for situation 3

	TSM	RDE	TSM	RDE
N	4	4	8	8
T	4	4	2	2
u	0.7056	0.6417	0.6520	0.6235
v	0.2694	0.2889	0.2320	0.2285
θ_1	0.9000	0.8997	0.2586	0.2531
θ_2	0.1000	0.0517	0.1000	0.0500
θ_3	0.1000	0.0745	0.1000	0.0502
J	−24.4605	−24.6137	−24.9703	−25.2931
$X^N(1)$	2.7261	2.7240	2.7230	2.7333
cost	2.8009	2.6259	2.2599	2.0402

two objectives are combined into a single objective J. It can be seen that RDE algorithm is better not only on all single objective J, but also on two objective $X^N(1)$ and *cost* in some cases.

The Convergence curve of RDE algorithm in situation 3 when $N = 8$ and $T = 2$ is shown in Fig. 2. It can be seen that the algorithm can converge quickly and stop the optimization process before reaching the predefined generations. Similar results also can be found in other situations.

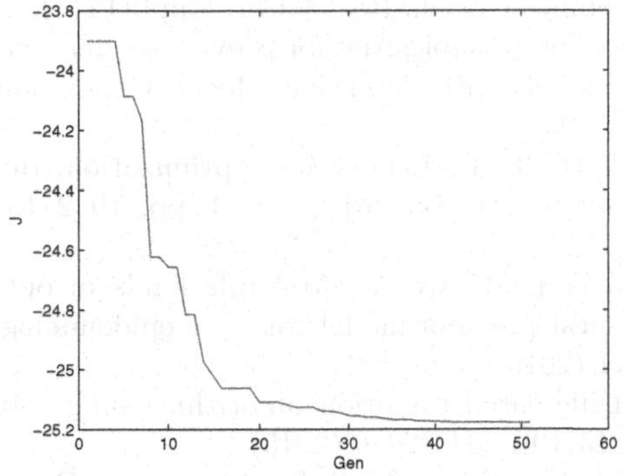

Fig. 2. The Convergence curve of RDE algorithm in situation 3 when $N = 8$ and $T = 2$

5 Conclusions

Ecological control of pests and diseases is a research hotspot in recent years, which is of great significance to agricultural production and environmental protection. Mathematical models of pest control are usually expressed by impulse control equations. Traditional method, such as time scale method, is easy to fall into local optimal. In order to further improve the solution quality, this paper attempts to use the RDE algorithm to solve the pulse control equation in three different situations. Experiment results demonstrate that RDE algorithm have competitive performance and fast convergence speed, which means it is feasible to solve IPM problem with evolutionary algorithm.

Obviously, this study is still relatively preliminary. There is also a fourth case in the plant-pest-predator model, that is, the impulse period and the number of impulses are changed. How to effectively solve this situation will be a research direction in the future.

Acknowledgement. This work was supported in part by Science and Technology Research Program in Henan Province of China (182102210411); Science and Technology Key Research Project of Henan Provincial Education Department of China (18A520040); and Young Backbone Teacher of Henan Province (2018GGJS148). The first two authors contributed equally to this work.

References

1. Das, S., Mullick, S.S., Suganthan, P.N.: Recent advances in differential evolution-an updated survey. Swarm Evol. Comput. **27**(1), 1–30 (2016)
2. Das, S., Suganthan, P.N.: Differential evolution: a survey of the state-of-the-art. IEEE Trans. Evol. Comput. **15**(1), 4–31 (2011)
3. Gakkhar, S., Naji, R.: On a food web consisting of a specialist and a generalist predator. J. Biol. Syst. **11**(4), 365–376 (2003)
4. Georgescu, P., Zhang, H.: An impulsively controlled predator-pest model with disease in the pest. Nonlinear Anal.: Real World Appl. **11**(1), 270–287 (2010)
5. Hui, J., Zhu, D.: Dynamic complexities for prey-dependent consumption integrated pest management models with impulsive effects. Chaos Solitons Fractals **29**(1), 233–251 (2006)
6. Kennedy, J., Eberhart, R.: Particle swarm optimization. In: 1995 IEEE International Conference on Neural Networks, vol. 4, pp. 1942–1948. IEEE, November 1995
7. Liang, X., Pei, Y., Zhu, M., Lv, Y.: Multiple kinds of optimal impulse control strategies on plant-pest-predator model with eco-epidemiology. Appl. Math. Comput. **287**(288), 1–11 (2016)
8. Liu, H., Wu, Z.: Differential evolution algorithm using rotation-based learning. Chin. J. Electron. **43**(10), 2040–2046 (2015)
9. Liu, H., Wu, Z., Li, H., Wang, H., Rahnamayan, S., Deng, C.: Rotation-based learning: a novel extension of opposition-based learning. In: Pham, D.-N., Park, S.-B. (eds.) PRICAI 2014. LNCS (LNAI), vol. 8862, pp. 511–522. Springer, Cham (2014). https://doi.org/10.1007/978-3-319-13560-1_41
10. Pei, Y., Ji, X., Li, C.: Pest regulation by means of continuous and impulsive nonlinear controls. Math. Comput. Model. **51**(5–6), 810–822 (2010)
11. Priyadarshi, A., Gakkhar, S.: Dynamics of Leslie–Gower type generalist predator in a tri-trophic food web system. Commun. Nonlinear Sci. Numer. Simul. **18**(11), 3202–3218 (2013)
12. Rahnamayan, S., Tizhoosh, H.R., Salama, M.M.: Opposition-based differential evolution. IEEE Trans. Evol. Comput. **12**(1), 64–79 (2008)
13. Shi, R., Jiang, X., Chen, L.: A predator-prey model with disease in the prey and two impulses for integrated pest management. Appl. Math. Model. **33**(5), 2248–2256 (2009)
14. Stern, V.M.: Economic thresholds. Annu. Rev. Entomol. **18**(1), 259–280 (1973)
15. Storn, R., Price, K.: Differential evolution-a simple and efficient adaptive scheme for global optimization over continuous spaces. Technical report TR-95-012, International Computer Science Institute, Berkeley, CA, March 1995
16. Tang, S., Cheke, R.: State-dependent impulsive models of integrated pest management (IPM) strategies and their dynamic consequences. J. Math. Biol. **50**(3), 257–292 (2005). https://doi.org/10.1007/s00285-004-0290-6
17. Tizhoosh, H.R.: Opposition-based learning: a new scheme for machine intelligence. In: International Conference Computational Intelligence for Modellling, Control and Automation, and International Conference Intelligent Agents, Web Technologies and Internet Commerce, vol. 1, pp. 695–701. IEEE, November 2005
18. Xu, Q., Wang, L., Wang, N., Hei, X., Zhao, L.: A review of opposition-based learning from 2005 to 2012. Eng. Appl. Artif. Intell. **29**(1), 1–12 (2014)

Microgrid Frequency Control Based on Genetic and Fuzzy Logic Hybrid Optimization

Haibin Su(✉), Haisong Chang, and Yixiao Cao

North China University of Water Resources and Electric Power, Zhengzhou, China
601423336@qq.com

Abstract. Power system frequency is an important performance index of AC microgrid operation. The power quality and user load, and even the safe and stable operation of microgrid, are affected directly by the frequency. Most modern microgrid contains a large number of intermittent renewable energy generation, this generation units have small capacity, many quantity, larger environmental and user load disturbance, strong nonlinear and unpredictable. The traditional control theory is difficult to effectively solve the problem of the microgrid frequency control. A new combination intelligent control method is presented. Two stage combinatorial optimization techniques of genetic algorithm and fuzzy logic is used. First, genetic algorithm is used to optimize the input and output fuzzy triangular membership function parameters. Second, fuzzy logic is used to optimize PI controller parameter to achieve the optimal control of the microgrid system frequency. The effectiveness of the two stage combination intelligent control algorithm is validated by the MATLAB simulation experiment.

Keywords: Microgrid · Frequency control · Intelligent control · Genetic algorithm optimization · Fuzzy logic

1 Introduction

With the increasing popularity of renewable energy, a large number of nontraditional power is introduced in power system. This promotes further rapid development of new energy as well as efficient use, but also brings a series of problems to power system stability, such as increasing the traditional power system complexity, uncertainty and power quality. The most new energy generation device is distributed at low voltage grid to form a local microgrid, the microgrid is integrated in the large grid by the distributed generation. Typical distributed generation microgrid consists of diesel generator (or micro gas turbine generator), photovoltaic generator, wind turbine, energy storage device. In microgrid system, because of the small inertia characteristics of the generation unit, the generator unit dynamic is impacted greatly by load disturbance. Energy storage unit is very important to improve the power quality and stability of the microgrid system, and the common energy storage device are storage battery and super capacitor.

The main performance index (such as voltage and frequency) of the microgrid system must be controlled by appropriate control strategies to maintain the system's good

L. Pan et al. (Eds.): BIC-TA 2019, CCIS 1159, pp. 53–64, 2020.
https://doi.org/10.1007/978-981-15-3425-6_5

performance and stability. The current control structures are centralized control, single agent control and decentralized control. In the centralized control structure, central control unit controls the load and system parameters, collects all information of the microgrid load and DGS (distribute generations) device, and determines the system load size as well the microsource power supply. In the single agent control structure, local load and system parameters of the microgrid system is controlled by a bigger controllable microsource device, the disadvantage of this method is that the system needs to be assembled with a high cost microsource device. In decentralized control structures, each microsource device is equipped with a local controller to realize the distributed control of the microgrid. For microgrid frequency control, the current method is mostly traditional PI controller. A modified PID control of diesel governor is used to deal with the slow and big net-load fluctuations to achieve the zero steady-state error frequency control of islanded micro-grid [1]. Using traditional proportion integration (PI) block to replace damping block in rotor motion equation, It eliminated frequency deviation in standalone mode in the microgrid system [2]. Because of the bandwidth limitation, the control dynamic performance can not meet the microgrid requirements. Some complex control algorithms are proposed, Such as model predictive control, robust H∞ control [3–6]. However, these control methods are based on exact mathematical models. When the microgrid structure or the nonlinearity is changed, the controller can hardly achieve the desired performance. Intelligent control method, such as H2/H∞ controller based on improved particle swarm optimization, combination of the neural network and fuzzy logic optimization, combination of the fuzzy logic and the particle swarm optimization, is used for the regulation frequency of the micro grid [6–8]. Also some paper present virtual synchronous generator control strategy [9–11], the frequency stability of microgrid is improved by using synchronous generator's rotor motion equation. With the aid of data communication networks, generation unit should be selected reasonably as control leader in secondary frequency control of microgrids [12–14]. For multi-microgrids system, distributed method is applied microgrid frequency control [15, 16].

In order to improve the reliability of microgrid power supply, microgrid can operate at the grid connected mode or island mode. Under the island mode, the voltage and frequency of microgrid need to be regulated independently in order to ensure the quality of power supply. Because the microgrid is placed the low voltage site and medium voltage distribution networks, so it is affected greatly by the load disturbance. What's more, the new energy generation device has intermittent characteristics, these factors will bring some difficulties to the microgrid voltage/frequency control. The traditional PI control method is difficult to achieve satisfactory results. For improving the power quality of microgrid, this paper proposes intelligent method based on Fuzzy logic and genetic algorithm hybrid to achieve microgrid frequency control. In the proposed control strategy, the PI parameters are automatically tuned using fuzzy rules, according to the online measurements. In order to obtain an optimal performance, the GA technique is used online to determine the membership functions parameters.

2 Microgrid Structure and Control Scheme

AC microgrid is a interconnection network system of the distributed load and low voltage distributed energy. It is composed of microsource unit and new energy power generation unit, such as micro diesel generator (DEG), wind turbine generator (WTG), photovoltaic (PV) and energy storage equipment. An example of a microgrid architecture is shown in Fig. 1, the distributed system is consisted of many radial feeders, its load can be divided into sensitive (L1, L2) and non sensitive (L3, L4) in the microgrid, sensitive load can be fed by one or more microsource unit, while the non sensitive load can be closed in case of emergency or severe disturbances. All microsource unit is connected by a circuit breaker and power flow controller which are controlled by the central controller or energy management system. The circuit breaker is disconnected from the feeder when serious disturbance occurs, so that other equipment is not affected by the serious fault. Microgrid is connected to the main grid by a fast switching (SS). The SS is capable to island the MG for maintenance purposes or when faults or a contingency occurs.

Microsource and energy storage devices are connected generally to microgrid via a power electronic circuit, the connection mode depends on the type of device, respectively DC/AC, AC/DC/AC, AC/AC power electronic converter interface, and microgrid control depends on the control of the inverter. The diesel generator and the energy storage system must be sufficient energy compensation to avoid power supply interruption. With the aid of advanced control technology, microgrid central controller can control microgrid operating in stable and economic. The microsource controller controls the microsource and energy storage system, while load controller controls the controllable load in user side. In order to increase the reliability of the traditional power system, the microgrid must maintain good performance both in main grid and island mode. In main grid mode, the main grid is responsible for controlling and maintaining power system in desired conditions, and the MG systems act as real/reactive power injectors. But in island mode, the microsource is responsible for maintaining the local loads and keeping the frequency and voltage indices at specified nominal values.

Control is one of the key problem for the safe and stable operation of AC microgrid. The microgrid has a hierarchical control structure with different layers, which needs advanced control technology to control effectively. As mentioned above, the microgrid can not only run at the island, but also connected with the main grid. In the island mode, microgrid should provide proper control loops to cope to the variations of the load disturbances, and perform active power/frequency and reactive power/voltage regulation. The AC MG operates according to the available standards, and the existing controls must properly work to supply the required active and reactive powers as well as to provide voltage and frequency stability. From Fig. 1, each of microsource has a local controller, each of load controller can control the controllable load, microgrid need central controller to achieve microgrid and the main grid connected control. Similar to the traditional power system, AC microgrid control is divided into different levels of control, which mainly includes the primary local control layer, the secondary level control layer, the central control layer and the emergency control layer. The primary local control layer achieves the current and voltage loop control of the microsource. The secondary level control layer realizes the control of microgrid's frequency and voltage deviation control. The central control layer realizes the microgrid economic optimum operation and scheduling control. The emergency control layer covers all possible emergency control schemes and special protection plans to maintain the system stability and availability in the face of contingencies. The emergency controls identify proper preventive and corrective measures that mitigate the effects of critical contingencies. In contrast to the local control, the primary local control layer does not need communication. The secondary level control layer, the central control layer and the emergency control layer need communication channels. Therefore, the primary local control layer can also be referred to as decentralized control, the secondary level control layer, the central control layer and the emergency control layer are called centralized control. Because of the variety of power generation and load, the AC microgrid has a high nonlinearity, dynamic and uncertainty. It needs an advanced intelligent control strategy to solve complex control problems.

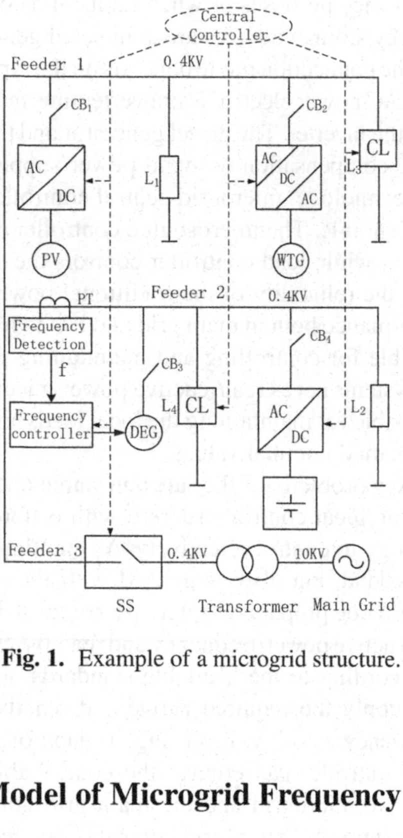

Fig. 1. Example of a microgrid structure.

3 Mathematical Model of Microgrid Frequency Control

Under the island mode, the control of the microgrid system is more important than the connected grid mode. Because the island mode microgrid has not the large grid support, the control is more difficult. Based on Fig. 1 example of microgrid structure, this paper establishes the AC microgrid control model of island mode, as shown in Fig. 2. The diesel generator is only as frequency control, photovoltaic and wind turbine keep always the maximum power output, storage battery is used for compensating diesel oil generator power shortage.

Wind generation and photovoltaic generation are intermittent power source, which its rated power are set respectively to 30 KW and 100 KW, and connected to feeder by power electronic control interface. The inverter control interface is modeled two series inertia model, the inverter switching time constant TIN equals 0.04, and filter time constant TLC equals 0.003. The diesel generator rated power is set to 120 KW, its model is equivalent to two series inertia model, the electromechanical time constant Tg equals 0.09, and inertia time constant Tt equals 0.5. Diesel generator frequency is controlled by the intelligent controller. Storage battery rated power is set to 80 KW, its model is equivalent to a first order inertia, time constant TB equals 0.15. L1 and L2 is the sensitive load, and its load power may not be controlled. L3 and L4 is non sensitive load, its load power can be controlled according to the energy balance need to increase or decrease the load at any time, even to remove the load.

In the microgrid, the new energy generation has the characteristics of large intermittent, fluctuation, nonlinear and non predictability. The advanced intelligent control method is needed to control the system frequency, so as to ensure the stability, reliability and power quality.

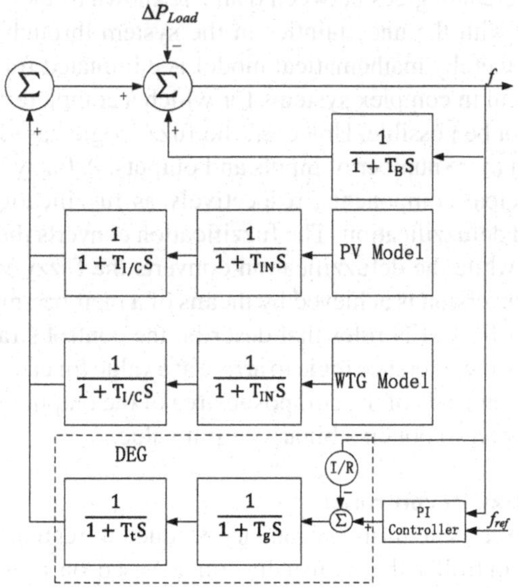

Fig. 2. Model of microgrid frequency control.

4 Genetic-Fuzzy Controller Algorithm Design

A. Theoretical Background

The GA is a popular heuristic optimization technique usually applied when the problem is highly nonlinear and high dimensional order. The idea behind it is based on biological genetics and the concept of "survival-of-the-fittest." Successive generations inherit features from their parents in a random fashion through the crossover of chromosomes. Also, certain changes, called mutations, may occur to the structure of the chromosomes at random. It is thought that through the process of evolution, successive generations will breed "better" populations, since only the fittest will survive. In the frame of optimization, the algorithm is built such that each chromosome is composed of a vector of possible values of the variables to be found. Initially, a number of these chromosomes are initialized at random, and designated as the initial population. Through successive iterations, the fitness of these chromosomes are evaluated, and a certain fraction of the chromosomes having the best fitness are retained in the next generation, which the rest discarded. These are replicated to make up the population number. Then, crossovers and mutations are applied on the chromosomes, and the process is repeated for a given number of iterations. The chromosomes are eventually converged to an optimum. As all other heuristic techniques, however, GA can be susceptible to being stuck in local minima. Fortunately, this effect is less so in GA than other heuristic techniques such as PSO with proper tuning of parameters, particularly the number of mutations, which introduce chromosomes randomly, hence enhance the capability of exiting local minima.

The fuzzy logic is a many-valued logic where the fuzzy logic variables may have truth values ranging in different degrees between 0 and 1, known as their membership grade. Fuzzy logic can deal with the uncertainties in the system through a simple IF-THEN rule based approach, thereby mathematical model is eliminated for the system control. This is especially useful in complex systems for which a complete mathematical model representation may not be possible. However, the fuzzy logic based system complexity increases rapidly with more number of inputs and outputs. A fuzzy logic control system consists of four principal components, respectively as fuzzification, fuzzy rule base, inference system, and defuzzification. The fuzzification converts the binary logic inputs into fuzzy variables, while the defuzzification converts the fuzzy variables into binary logic outputs. This conversion is achieved by means of a membership function. The rule base is a collection of IF-THEN rules that describe the control strategy. The output of each rule is deduced by the inference logic to arrive at a value for each output membership function. The "fuzzy centroid" of the composite area of the output membership function is then computed in order to obtain a binary output value.

B. **Conventional fuzzy PI controller**

In the traditional power system, the secondary frequency regulation is controlled by the conventional PI controller that is usually tuned based on preset operating points. When the operating condition is changed, the PI controller will not meet the desirable performance requirements. While, if the PI controller can be continuously able to track the changes occurred in the power system, the optimum performance will be always achieved. Fuzzy logic can be used as an suitable intelligent method for online tuning of PI controller parameters.

Fuzzy PI controller is composed of two parts, respectively a traditional PI controller and fuzzy logic unit. The input variables of the Fuzzy logic units are the Δf (microgrid frequency deviation) and ΔPL (load disturbance). The output variable of the fuzzy PI controller parameter are KP (the proportional gain) and Ki (the integral gain). Fuzzy rule base consists of a group of 18 rule, as shown in Table 1. The input variable Δf fuzzy subsets $T(\Delta f)$ equals {Negative Large (NL), Medium (NM), Negative Small (NS), Positive Small (PS), Positive Medium (PM), and Positive Large (PL)}. The input variable ΔPL fuzzy subsets $T(\Delta PL)$ equals {Positive Small (PS), Positive Medium (PM), and Positive

Large (PL)}. The output variable fuzzy subset T(KP, Ki) equals {Negative Large (NL), Medium (NM), Negative Small (NS), Positive Small (PS), Positive Medium (PM), and Positive Large (PL)}. They have been arranged based on triangular membership function which is the most popular one. The antecedent parts of each rule are composed by using AND function (with interpretation of minimum). Here, Mamdani fuzzy inference system is also used.

Fuzzy PI controller has better performance than the traditional control method, but its control performance is highly dependent on the membership function. If there is no precise information of the microgrid system, the membership function will not be accurately selected. And the designed fuzzy PI controller will not achieve the optimal performance under the large range of operating conditions.

Table 1. The fuzzy inference rules.

ΔPL	Δf					
	NL	NM	NS	PS	PM	PL
S	NL	NM	NS	PS	PS	PM
M	NL	NL	NM	PS	PM	PM
L	NL	NL	NL	PM	PM	PM

C. **Genetic algorithm to optimize membership function parameters**

Genetic algorithm is an efficient global optimization search algorithm, which simulates biological evolution process. The algorithm has simple structure, it can be processed in parallel, and get the global optimal solution or sub optimal solution without any initial information. The optimization design frame is shown in Fig. 3, the optimization target is the minimum frequency deviation Δf.

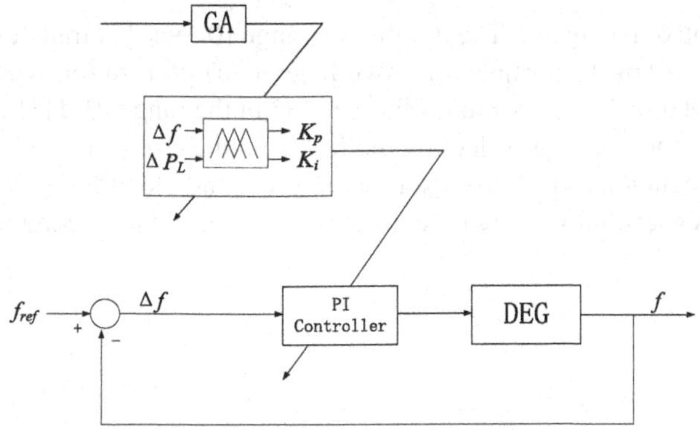

Fig. 3. GA-Fuzzy PI frequency controller

The membership function of the input and output variable is isosceles triangle function. The Parodi and Bonelli coding method is adopted. Real number coding parameters (c, w) are expressed respectively the position and width of the single membership degree function. R_{jk} is expressed as the *j-th* linguistic variables of the *k-th* membership degree function, C_{jk} is expressed as fuzzy domain position of the R_{jk}, and W_{jk} is expressed as fuzzy domain width of the R_{jk}. Thus membership function MF_j of the fuzzy linguistic variable set I_j can be expressed as string encoding $(c_{j1}, \mathrm{w_{j1}})$ … (c_{jk}, w_{jk}) (where k is fuzzy set number). The string encoding is expressed as a gene in the genetic algorithm. The individual's chromosome is composed of a number of such genes. The chromosome length depends on the number of input and output variable number.

Genetic operations include selection, crossover and mutation. The best individual preservation method is adopted for Genetic selection, that is, a number of individuals with the lowest fitness value of the population are copied. A special crossover and mutation method is introduced for real number encoding, that is the max-min-arithmetical crossover method and one-point mutation method. The mutation operation has been achieved to result in a new individual by adding random number e $(-w_{jk} \leq \mathrm{e} \leq w_{jk})$ to the any value (c_{jk}, w_{jk}). If the mutation operation aims at the position of the membership function. After the mutation, the C must be reordered to ensure that the membership function is ordered on the domain by this value. According to the mutation rate, many gene (c, w) of the same chromosome may be selected.

The calculation formula of the optimal algorithm index is as follows

$$J = \int_0^{\infty} \Big(\omega_1 |\Delta f(t)| + \omega_2 u^2(t) \Big) dt + \omega_3 t_u \tag{1}$$

where $\Delta f(t)$ is for microgrid frequency error, $u(t)$ is the controller output, t_u is rise time, ω is weights, and $\omega_1 + \omega_2 + \omega_3 = 1$. The smaller J value is, the better performance of the system is. In the Genetic algorithm, the bigger individual fitness value is, the better chromosome is. The fitness function is defined as follows.

$$F(x) = \frac{1}{1 + \alpha J} \tag{2}$$

where α is sensitive parameter. The J value of range $[0, +\infty]$ is transformed into $F(x)$ value of range [0, 1] by the formula, meanwhile genetic optimization is transformed into solving the problem of the maximum fitness value in the range [0, 1] (Fig. 4).

The wheel game is used as selection method, and the crossover probability $\mathrm{P_c}$ is set to 0.6, and the mutation probability $\mathrm{P_m}$ is set to 0.01, and the initial population is set to 50, and the largest evolutionary G is set to 100 as the termination condition.

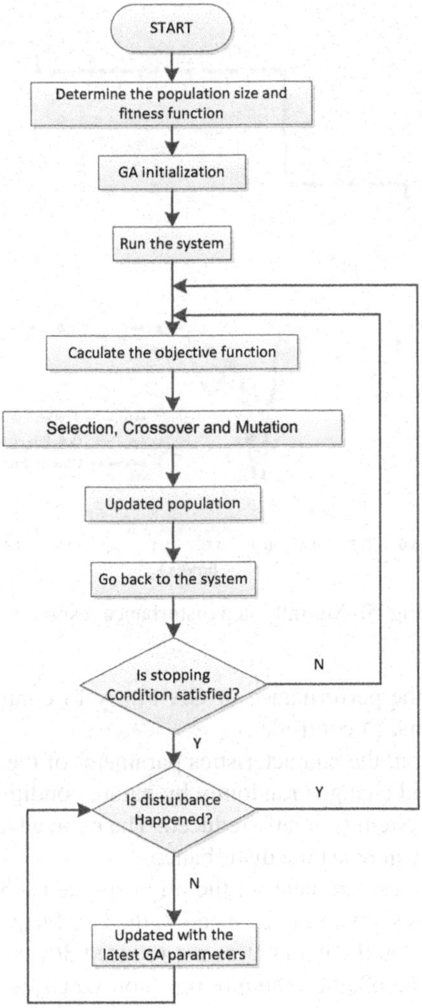

Fig. 4. Flowchat of GA-Fuzzy PI control

5 Simulation Analysis

In order to compare the performance of the traditional PI, the classical fuzzy PI and GA-Fuzzy PI controller, the three control schemes are modeled and simulated using MATLAB programming. The Fig. 5 shows the dynamic response of the microgrid frequency control when multiple step disturbance is inputted three controller respectively. From Fig. 5, the performance index of the GA-Fuzzy PI controller is superior to the classical fuzzy PI controller and the conventional PI controller.

Figure 5 shows that the proposed GA-Fuzzy PI control method has better performance, and the new control algorithm quickly eliminates the frequency deviation before the secondary load step disturbance starts. For more severe disturbances, such as 0.084 pu

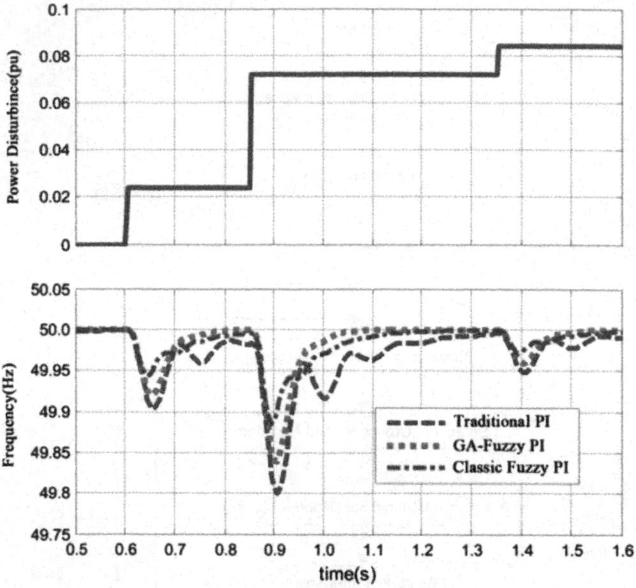

Fig. 5. Multiple step disturbance response

perturbance amplitude, the performance of GA-Fuzzy PI controller is better than the fuzzy PI and the traditional PI controller.

In practical application, the characteristics parameter of the new energy generation device can be affected and changed randomly by nature condition. The performance of the closed-loop control system is greatly reduced. The main advantage of the intelligent control method is that it can resist the disturbance.

When the characteristics parameter of the microsource is changed, the control performance of the system is shown in Figs. 6 and 7, the GA-Fuzzy PI controller has good adaptive performance. when the parameters are reduced 30% in the Fig. 6 (the turbine time constant T_g is set to 0.063, the generator time constant T_t is set to 0.35, and the time constant T_B of the energy storage battery is set to 0.105), the conventional PI controller is unable to adapt to the parameter perturbation. In the Fig. 7, when the parameters are increased 30% (the turbine time constant T_g is set to 0.117, the generator time constant T_t is set to 0.65, and the time constant T_B of the energy storage battery is set to 0.195), the GA-Fuzzy PI controller has more advantages than the other two controllers.

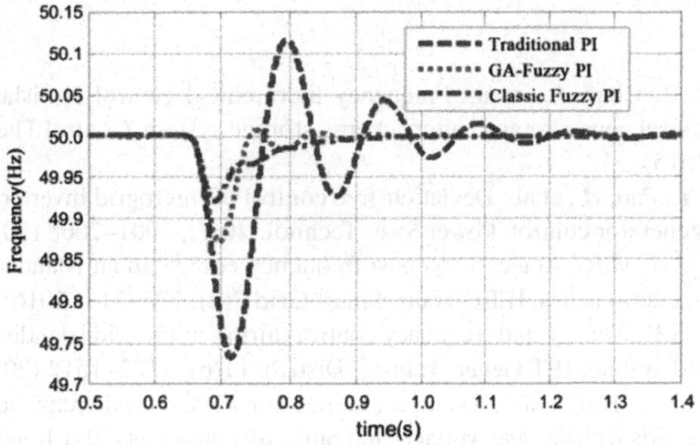

Fig. 6. Multiple step disturbance response

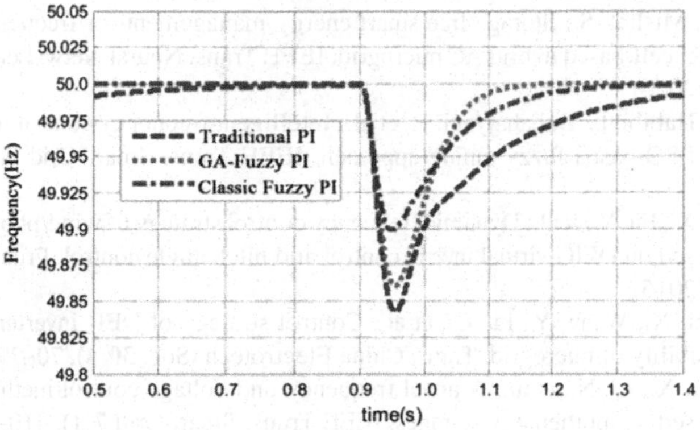

Fig. 7. Multiple step disturbance response

6 Conclusions

For AC microgrid island operation, the grid frequency is supported by the microsouce in the microgrid, the microgrid frequency control is more important, especially in the face of large disturbances, uncertainties and load changes. Traditional frequency control method is difficult to meet the required performance index. A new adaptive intelligent control method is proposed for the microgrid. Fuzzy membership function parameters are optimized by genetic algorithm. It has better effect than the classical fuzzy PI frequency controller. Based on GA-Fuzzy PI control method applied to the microgrid, the simulation result shows good response characteristics, and has fast response and small overshoot. Especially when the microsource parameters are changed largely, the controller still has strong robustness.

References

1. Ma, Y., Yang, P., Chen, S., et al.: Frequency hierarchical control for islanded micro-grid consisting of diesel generator and battery energy storage system. Control Theory Appl. **32**(8), 1098–1105 (2015)
2. Yang, J., Liu, Y., Pan, H., et al.: Deviation-free control of microgrid inverter based on virtual synchronous generator control. Power Syst. Technol. **40**(7), 2001–2008 (2016)
3. Bevrani, H., Feizi, M.R., Ataee, S.: Robust frequency control in an islanded microgrid H∞ and μ-synthesis approaches. IEEE Trans. Smart Grid **7**(2), 706–717 (2016)
4. John, T., Lam, S.P.: Voltage and frequency control during microgrid islanding in a multi-area multi-microgrid system. IET Gener. Transm. Distrib. **11**(6), 1502–1512 (2017)
5. Liang, L., Hou, Y., Hill, D.J.: Design guidelines for MPC-based frequency regulation for islanded microgrids with storage, voltage, and ramping constraints. IET Renew. Power Gener. **11**(8), 1200–1210 (2017)
6. Wu, Z., Zhao, X., Wang, X.: Frequency H2/H∞ optimization control for isolated micro-grid based on IPSO algorithm. Fuzzy Syst. Math. **30**(6), 1–10 (2016)
7. Sekhar, P.C., Mishra, S.: Storage free smart energy management for frequency control in a diesel-PV-fuel cell-based hybrid AC microgrid. IEEE Trans. Neural Netw. Learn. Syst. **27**(8), 1657–1671 (2016)
8. Bevrani, H., Habibi, F., Babahajyani, P., et al.: Intelligent frequency control in an AC micro-grid: online PSO-based fuzzy tuning approach. IEEE Trans. Smart Grid **3**(4), 1935–1944 (2012)
9. Zhao, J., Lü, X., Fu, Y., et al.: Dynamic frequency control strategy of wind/photovoltaic/diesel microgrid based on DFIG virtual inertia control and pitch angle control. Proc. CSEE **35**(15), 3815–3822 (2015)
10. Meng, J., Shi, X., Wang, Y., Fu, C., et al.: Control strategy of DER inverter for improving frequency stability of microgrid. Trans. China Electrotech. Soc. **30**(4), 70–79 (2015)
11. Tang, X., Hu, X., Li, N., et al.: A novel frequency and voltage control method for islanded microgrid based on multienergy storages. IEEE Trans. Smart Grid **7**(1), 410–419 (2016)
12. Malik, S.M., Ai, X., Sun, Y., et al.: Voltage and frequency control strategies of hybrid AC/DC microgrid: a review. IET Gener. Transm. Distrib. **11**(2), 303–313 (2017)
13. Amani, A.M., Gaeini, N., Jalili, M.: Which generation unit should be selected as control leader in secondary frequency control of microgrids? IEEE. J. Emerg. Sel. Top. Circ. Syst. **7**(3), 393–402 (2017)
14. Li, Y., Xie, Y., Cheng, Z., et al.: Research of non-steady state error control strategy of bus voltage and frequency in micro-grid system. Electric Mach. Control **20**(7), 49–57 (2016)
15. Wang, Z., Chen, L., Liu, F., et al.: Distributed frequency control of multi-microgrids with regulation capacity constraints of controllable loads. Autom. Electric Syst. **40**(15), 47–66 (2016)
16. Hajimiragha, A.H., Zadeh, M.R.D., Moazeni, S.: Microgrids frequency control considerations within the framework of the optimal generation scheduling problem. IEEE Trans. Smart Grid **6**(2), 534–547 (2015)

Using Multi-objective Particle Swarm Optimization to Solve Dynamic Economic Emission Dispatch Considering Wind Power and Electric Vehicles

Baihao Qiao and Jing Liu(✉)

School of Artificial Intelligence, Xidian University, Xi'an 710071, China
neouma@163.com

Abstract. As a kind of clean energy, wind power can reduce the fuel cost and pollution emission of tradition thermal power generators effectively. As a transportation, electric vehicle (EV) can not only save energy, but also protect the environment. However, the large-scale development of EVs will increase the load pressure on the power grid. Therefore, in order to deal with the rapid development of wind power and EVs, a multi-objective dynamic economic emission dispatch model with wind power and EVs is proposed considering both the total fuel cost and pollution emission objectives in this paper. The two-*lbests* based MOPSO (2LB-MOPSO) with constraint handling method is developed to optimize the proposed model. The 10-unit system, 100 wind turbines and 50000 EVs are employed as the test case to demonstrate the performance of the 2LB-MOPSO in the proposed model. In addition, other evolutionary algorithms are compared with 2LB-MOPSO. The simulation results show that 2LB-MOPSO is superior in solving the complex constrained DEED problem, and the proposed model can guide the charging and discharging behavior of EVs to serve the power grid.

Keywords: Particle swarm optimization · Dynamic power dispatch · Wind power · Electric vehicles

1 Introduction

The tradition dynamic economic emission dispatch (DEED) is a key problem of the power system operation that has the objective of optimally dispatching the thermal power generators' output power in a period, so that the total fuel cost and pollution emission are minimized while meeting some certain constraints [1]. Compared with the static economic emission dispatch (SEED) problem, because the DEED problem employs the dynamic dispatch for a load period of one day and considers ramp limits between different intervals, which is more in line with the real short-term dispatching requirements, it has gained more and more researchers' attention [2].

Wind power is considered to be one of the most promising renewable energy sources in the 21st century, and plays an important role in mitigating the energy crisis and avoiding environmental pollution [3]. Due to zero emissions energy saving and mitigating

L. Pan et al. (Eds.): BIC-TA 2019, CCIS 1159, pp. 65–76, 2020.
https://doi.org/10.1007/978-981-15-3425-6_6

noise, the development of electric vehicle (EV) is supported by car manufacturers and governments policy [4]. However, the extra load demand by the random charging behavior of EVs will further increase the system's peak valley load difference. But the development of vehicle to grid (V2G) technology for EVs is able to provide peak load shaving and load leveling services to the power system [5]. So, the rapid development of wind power and EV has brought new challenges to the DEED problem, and some researches have been published about the DEED problem with wind power and EVs [6–10].

Khodayar *et al.* in [6] proposed the security-constrained unit commitment (SCUC) model to study the coordination and integration of aggregated plug-in EV and wind energy in the power system. Shao *et al.* in [7] focused on EV aggregators constraints proposed the SCUC model. Zhao *et al.* in [8] took in to account the uncertainties of plug-in EVs and wind turbines and developed the economic dispatch (ED) model. Then the particle swarm optimization (PSO) and interior point method were employed to solve the ED model. Qu *et al.* in [9] considered the uncertain wind power and large-scale EVs and developed the DEED model. But the wind power is stationary in the whole dispatching period, which does not meet the actual dispatching requirements. Zhang *et al.* in [10] proposed a novel dynamic multi-objective dispatch framework for the hydro-thermal-wind and EVs coordination scheduling problem. The improved PSO algorithm with dual population evolution mechanism and a hierarchical elitism preserving strategy was employed to solve this dispatch problem.

Many evolutionary algorithms (EAs) [1, 4, 8–15] have been successfully applied to solve power dispatch problems. PSO is a population-based heuristic search technique proposed by Kennedy and Eberhart who were inspired by the social behavior of bird flocks [16]. The relative simplicity of the PSO has made it a natural candidate for solving multi-objective problems (MOPs). However, in most multi-objective particle swarm optimization (MOPSO), the position of *pbest* and *gbest* determines the flight direction of particles in the search space. However, these positions may be far apart from each other in the parameter space. Therefore, using two far apart *pbest* and *gbest* positions to guide the directions of acceleration of particles may not be very effective in solving hard problems, especially, high-dimensional complex DEED problem. The two-*lbests* based MOPSO (2LB-MOPSO) variant was proposed by Zhao *et al.* in [17]. In 2LB-MOPSO, each objective function in the external archive is divided into a number of bins. The two *lbests* are selected from the tops fronts in a non-domination sorted external archive located in two neighbouring bins, so that they are near each other in the parameter space. The detailed selection process and method of two *lbests* are shown Subsection 3.2 in [17]. 2LB-MOPSO is effectively focus the search around a small region in the parameter space in the vicinity of the best existing fronts. Therefore, 2LB-MOPSO does not cause potential chaotic search process by the random nature of the particle's *pbest* and *gbest* (or *lbest*) selection process. Thus, 2LB-MOPSO has the advantage of solving the complex DEED problem.

In this paper, the non-liner, lager-scale, high-dimension, non-convex and multi-stage multi-objective DEED with wind power and EVs model is developed considering both the total fuel cost and pollution emission objectives. Furthermore, the 2LB-MOPSO based on constraint handling method is proposed to obtain the optimal dispatching schemes. Moreover, to demonstrate the performance of the 2LB-MOPSO in the proposed model, the 10-unit system, 100 wind turbines and 50000 EVs are employed as the test case. Compared with other EAs, 2LB-MOPSO is superior in solving the complex constrained DEED problem. In addition, the proposed model can guide the charging and discharging behavior of EVs to serve the power grid.

The rest of this paper is organized as follows. Section 2 studies the modeling of DEED with wind power and EVs. The constraints handling method and 2LB-MOPSO implemented in the proposed model are introduced in Sect. 3. The experimental discussions are given in Sect. 4. Finally, conclusions are given in Sect. 5.

2 Modeling of DEED with Wind Power and EVs

The DEED problem with wind power and EVs can be formulated as a multi-objective optimization model. The two conflicting objectives, i.e., the total fuel cost and pollutants emission of the thermal power generators, should be minimized simultaneously while fulfilling power balance constraint, EVs remain power constraint, travel constraint of EVs owners, ramp rate limits and up and down spinning reserve constraints. The DEED considering wind power and EVs model is formulated in this section.

2.1 Objective Functions

Since the wind power and EVs are clean energy, and almost zero emission to the environment, the main emission is discharged by coal-fired units in the process of thermal power generation. In addition, the EV cost is assumed to be a one-time investment by the owners, i.e. the EVs are regarded as zero cost during the dispatching period. Therefore, the wind power and EVs are not considered in the objective functions.

Objective 1: Minimization of Total Fuel Cost. The total fuel cost of each thermal power generator considering pulsatile valve-point are represented by quadratic functions with sine components [12]. Therefore, the total fuel cost of N thermal power generators over T dispatching time intervals is expressed as

$$F_C = \sum_{t=1}^{T}\sum_{i=1}^{N}\{a_i + b_i P_{i.t} + c_i(P_{i.t})^2 + |d_i \sin[e_i(P_{i.\min} - P_{i.t})]|\} \tag{1}$$

where a_i, b_i, c_i, d_i and e_i are the fuel cost coefficients of the ith power generator. $P_{i.t}$ is the output power of ith unit in interval time t. $P_{i.\min}$ is the minimum output power of ith unit. F_C is the total fuel cost of the thermal generators in a dispatching period.

Objective 2: Minimization of Total Pollutants Emission. Generally, the atmospheric pollutants such as SO_x and NO_x caused by fossil-fueled thermal generators can be modeled as the sum of a quadratic and an exponential function [9, 12]. Thus, the total pollution emission of N thermal power generators over the whole dispatching period can be modelled as follows,

$$F_M = \sum_{t=1}^{T}\sum_{i=1}^{N}[(\alpha_i + \beta_i P_{i.t} + \gamma_i P_{i.t}^2 + \zeta_i \exp(\varphi_i P_{i.t})] \tag{2}$$

where α_i, β_i, γ_i, ζ_i and φ_i are the emission coefficients of the ith power generator, and F_M is the total pollution emission of the thermal generators in a dispatching period.

2.2 System Constraints

The DEED problem with wind power and EVs subjects to the following equality and inequality constraints.

Power Balance Constraint. Considering the wind power and EVs charging and discharging, in each period, the total output power of thermal generators, the EVs charging and discharging power, and the wind power should be balanced with the sum of transmission loss and load demand.

$$\sum_{t=1}^{N} P_{i.t} + \sum_{j=1}^{N_w} P_{\mathrm{w}.t} + P_{\mathrm{Dch}.t} = P_{\mathrm{D}.t} + P_{\mathrm{L}.t} + P_{\mathrm{ch}.t} \quad (3)$$

where $P_{\mathrm{w}.t}$ is the wind power in the time interval t. $P_{\mathrm{ch}.t}$ is the EVs charging load in the time interval t. $P_{\mathrm{Dch}.t}$ is the EVs discharging power in the time interval t. $P_{\mathrm{D}.t}$ is the load demand of system in the time interval t. $P_{\mathrm{L}.t}$ is the transmission loss in the time interval t, which can be calculated by using the B-coefficients method [9].

$$P_{\mathrm{L}.t} = \sum_{i=1}^{N}\sum_{j=1}^{N} P_{i.t} B_{ij} P_{j.t} + \sum_{i=1}^{N} P_{i.t} B_{i0} + B_{00} \quad (4)$$

where $B_{i,j}$, B_{i0} and B_{00} are the network loss coefficients.

EVs Remain Power Constraint. The remain power S_t of EVs at time interval t is defined as,

$$S_t = S_{t-1} + \lambda_{\mathrm{C}} P_{\mathrm{ch}.t} \Delta t - \frac{1}{\lambda_{\mathrm{D}}} P_{\mathrm{Dch}.t} \Delta t - S_{\mathrm{Trip}.t} \quad (5)$$

where λ_C and λ_D are the coefficients of the charging and discharging efficiencies, respectively. Δt is the dispatch interval, here, and $\Delta t = 1$. $S_{\mathrm{Trip}.t}$ is the amount of power consumed during the driving of EVs, which can be calculated as follows,

$$S_{\mathrm{Trip}.t} = \Delta SL \quad (6)$$

where ΔS is the average power consumption per unit time of EVs driving, and L is driving distance.

To ensure the safety of the operation and service lifespan of the EVs battery, remaining power S_t is constrained by lower and upper limits.

$$S_{\min} \leq S_t \leq S_{\max} \quad (7)$$

Travel Constraint of EVs Owners. To meet the EVs owners travel demand, in a charging and discharging cycle, $S_{\text{Trip}.t}$ should satisfy,

$$\sum_{t=1}^{T} S_{\text{Trip}.t} = \sum_{t=1}^{T} \lambda_C P_{\text{ch}.t} \Delta t - \sum_{t=1}^{T} \frac{1}{\lambda_D} P_{\text{Dch}.t} \Delta t \tag{8}$$

Power Limits. For a safe and stable operation, the charging and discharging power of the EVs and wind power should be less than the rate power, which can be defined as (9) and (10), the output power of thermal generators should be limited between its upper and lower, expressed as (11).

$$\begin{cases} P_{\text{ch}.t} \leq P_{Nch} \\ P_{\text{Dch}.t} \leq P_{NDch} \end{cases} \tag{9}$$

$$0 \leq P_{\text{w}.t} \leq P_{rate} \tag{10}$$

$$P_{i.\min} \leq P_{i.t} \leq P_{i.\max} \tag{11}$$

where P_{Nch} and P_{NDch} are the charging and discharging rated power of the EVs, respectively. P_{rate} is the rate power of wind turbine. $P_{i.\max}$ is the upper output limit of the ith thermal generator.

Ramp Rate Limits. The output power of thermal generator should be no significant change between adjacent intervals. Therefore, the generators' power ramp rates are limited.

$$\begin{cases} P_{i.t} - P_{i.t-1} \leq U_{Rt} \Delta t \\ P_{i.t-1} - P_{i.t} \leq D_{Rt} \Delta t \end{cases} \tag{12}$$

where U_{Rt} and D_{Rt} are the increase and decrease rate limits of the ith thermal generator, respectively.

Spinning Reserve Constraints. In the traditional power system, it is necessary to maintain a certain spinning reserve capacity to ensure the safe and reliable operation of the system. Due to the increasing scale of wind power and EVs into the power grid, the up and down spinning reserve is more needed to deal with the risks brought by the randomness and intermittentness of wind power and EVs. The up and down spinning reserve that power system can provide are calculated as the system's 10 min response reserve.

$$\sum_{i=1}^{n} \min(P_{i.\max} - P_{i.t}, \frac{U_{Rt}}{6}) + P_{\text{Dch}.t} \geq \omega_u P_{\text{w}.t} + P_{\text{ch}.t} + S_{\text{R}.t} \tag{13}$$

$$\sum_{i=1}^{n} \min(P_{i.t} - P_{i.\min}, \frac{D_{Rt}}{6}) + P_{\text{Dch}.t} \geq \omega_d (P_{rate} - P_{\text{w}.t}) + P_{\text{ch}.t} \tag{14}$$

(13) and (14) are the up down spinning reserve, respectively. ω_u and ω_d are the demand coefficients of wind power on up and down spinning reserve. $S_{\text{R}.t}$ is the spinning reserve capacity requirements in the time interval t, which is usually taken as 5%–10% of the demand power in the dispatch period and are the maximum and minimum output power of the ith unit in the time interval t.

3 Constraints Handling Method and 2LB-MOPSO Implemented in Proposed Model

3.1 Constraints Handling Method

The proposed DEED with wind power and EVs model is a non-linear, lager-scale, high-dimension, non-convex and multi-stage. Therefore, the key to optimize the proposed model is to require an efficient constraint handling method. So, the constraint handling method of dynamic adjustment of decision variables based on penalty function is adopted in this paper. The procedure of constraints handling method is shown in Algorithm 1.

Algorithm 1: Constraint Handling Method Procedure

Input: Decision variables of $P_{i.t}$, $P_{\mathrm{ch}.t}$ and $P_{\mathrm{Dch}.t}$; Wind power of $P_{\mathrm{w}.t}$; Decision variable boundary values of $P_{i.\min}$, $P_{i.\max}$, P_{Nch} and P_{NDch}; Demand power of $P_{\mathrm{D}.t}$; Network loss coefficients of $B_{i,j}$, B_{i0} and B_{00}; Increase and decrease rate limits of U_{Rt} and D_{Rt}; Threshold value of ε; Maximum adjustment of L;

Output: Adjusted $P_{i.t}$, $P_{\mathrm{ch}.t}$, $P_{\mathrm{Dch}.t}$ and constraint violation;

```
for l=1 to L do   % Charging and discharging dynamic adjustment of EVs
    Calculate θ according to (8);
    Add θ/T to P_ch.t and P_Dch.t;
    Cross boundary processing according to (9);
   if l<L or θ ≤ ε then
     Calculate the constraint violation;
    break;
   end if;
end for;
while (within a dispatch period) do
    Calculate P_L.t according to (4);
%Dynamic adjustment of thermal generator output power
   for l=1 to L do
        Calculate θ according to (3);
        Add θ/N to P_i.t;
        Cross boundary processing according to inequality equations (11) and (12);
        if l<L or θ ≤ ε then
             Calculate the constraint violation;
          break;
        end if;
    end for;
end while;
```

3.2 2LB-MOPSO Implemented in Proposed Model

The decision variables in proposed model is the output power and the EVs charging/discharging power of $P_{\mathrm{ch}.t}$ or $P_{\mathrm{Dch}.t}$. The one of particle $\boldsymbol{x}_i$, whose dimension is

$(N + 1) \times T$, can be expressed as (15).

$$x_i = \begin{bmatrix} P_{1,1} & P_{1,2} & \cdots & P_{1,T} \\ P_{2,1} & P_{2,2} & \cdots & P_{2,T} \\ \vdots & \vdots & \vdots & \vdots \\ P_{N,1} & P_{N,2} & \cdots & P_{N,T} \\ P_{\mathrm{ev},1} & P_{\mathrm{ev},2} & \cdots & P_{ev,T} \end{bmatrix} \tag{15}$$

The detailed description of 2LB-MOPSO is shown in [17], and the procedure of 2LB-MOPSO implemented in proposed model is shown in Algorithm 2.

Algorithm 2: 2LB-MOPSO implemented in proposed model

Input: Decision variable boundary values of $P_{i.\min}$, $P_{i.\max}$, P_{Nch} and P_{NDch}; Wind power of $P_{\mathrm{w}.t}$; Penalty coefficient of *s*; Maximum function evaluations of *Max_FEs*; Population size of *NP*; 2LB-MOPSO parameters.
Output: Dispatch schemes.

Initialize:
Initialize *NP* number of particles randomly and uniformly in the search space.
Calculate the objective functions of all particles and get the adjusted particles by Algorithm 1.
Select *lbests* from the external archive.
Optimization Loop:
while *(FEs ≤Max_FEs)* ***do***
FEs←1;
for *i*=1 ***to*** *NP*
For particle *i*, get a pair of *lbests* according to Subsection 3.2 in [17].
% updating velocity
$V(i) = \omega * (V(i) + c_1 * rand() * (lbest(i) - particle(i)$
$+c_2 * rand() * (lbest(i+NP) - particle(i)))$
% Limit the velocity
$V(i) = (\min(V_{\max}(i), V(i))) \,\&\, \max(V_{\min}(i), V(i)))$
%updating position of each particle
$X(i) = X(i) + V(i)$
% Limit the velocity
$X(i) = (\min(X_{\max}(i), X(i))) \,\&\, \max(X_{\min}(i), X(i)))$
Calculate the objective functions of *particle*(*i*) and get the adjusted *particle*(*i*) by Algorithm 1.
end for
Updating the external archive according to Subsection 3.2 in [17];
FEs ← *FEs*+1;
end while

4 Experiments and Discussion

In order to verity the feasibility of the proposed DEED with wind power and EVs model, the 10-unit system, 100 wind turbines and 50000 EVs are selected in this paper. Load demand and parameters of thermal power generator are obtained from [9]. In addition, to verify the effectiveness of 2LB-MOPSO, other algorithms including the MOPSO [18], NSGA-II [12] and SPEA2 [9] are compared with it in the same test case, and their parameters are selected in the corresponding literature. In each algorithm, *NP* is set to 100, and the *Max_FEs* is set to 500000. In the constant handling method, threshold value ε is set to 10^{-6}, and the maximum adjustment L is set to 50. The parameter settings of the algorithms can be found in the corresponding references.

The wind power is from the historical average wind power of 2017 in northwestern China's wind farms. The rated power of one wind turbine is 1.5 MW, and the average wind power of one wind turbine in a day is shown Fig. 1.

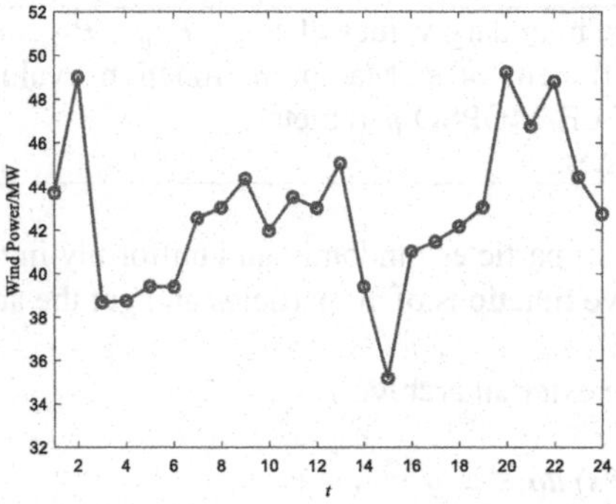

Fig. 1. The average wind power of one wind turbine in a day

The battery capacity is 24 kW·h (for example: Nissan Leaf) and driving consumption is 15 kW·h/100 km. Assuming that the state of charge (SOC) of an EV every morning is 100%, EVs have two-time stages (07:00–08:00 and 17:00–18:00) driving in the commute (total 50 km) every day, and the rest time can participate in power grid to be scheduled. In the dispatch period the minimum SOC and the rate power charging and discharging are limited to 20% of the battery capacity. The charging and discharging efficiency are 0.85, up and down spinning reserve coefficients are set to 0.3, the dispatch period is 24 h.

All the algorithms are implemented using MATLAB 2018b and executed on a personal computer with Core I5-6500 CPU, 8G RAM and windows 10 64-bits operating system. Each algorithm is run 30 times and the best value of the objective functions and corresponding solutions are recorded.

The optimal results of proposed model obtained by different algorithms are shown in Table 1. The fuzzy-based method is employed to get the best compromise solutions [9]. The minimum values are given in bold type. In the terms of best cost and best emission, the 2LB-MOPSO are significantly better than MOPSO, SPEA2 and NSGA-II. In the best compromise solution, the 2LB-MOPSO better than the other algorithms except MOPSO. The F_C of MOPSO is 2.4166E + 06$ which is 1800$ less than that of 2LB-MOPSO, but its F_M is 2.8702E + 05 lb which is 14110 lb more than that of 2LB-MOPSO. Obviously, the increase in F_M of MOPSO accounted for a greater proportion (5.17%) compared to the reduction in F_C (0.075%). Therefore, 2LB-MOPSO gets a better compromise result.

Table 1. The optimal results by different algorithms

Algorithms	2LB-MOPSO		MOPSO		SPEA2		NSGA-II	
	F_C	F_M	F_C	F_M	F_C	F_M	F_C	F_M
Best cost	**2.3641E+06**	2.9631E+05	**2.3939E+06**	3.0869E+05	**2.4820E+06**	2.8260E+05	**2.4161E+06**	2.7911E+05
Best emission	2.5071E+06	**2.6441E+05**	2.5040E+06	**2.7137E+05**	2.4923E+06	**2.7276E+05**	2.4496E+06	**2.7247E+05**
Best compromise	2.4184E+06	**2.7291E+05**	**2.4166E+06**	2.8702E+05	2.4844E+06	2.7781E+05	2.4280E+06	2.7561E+05

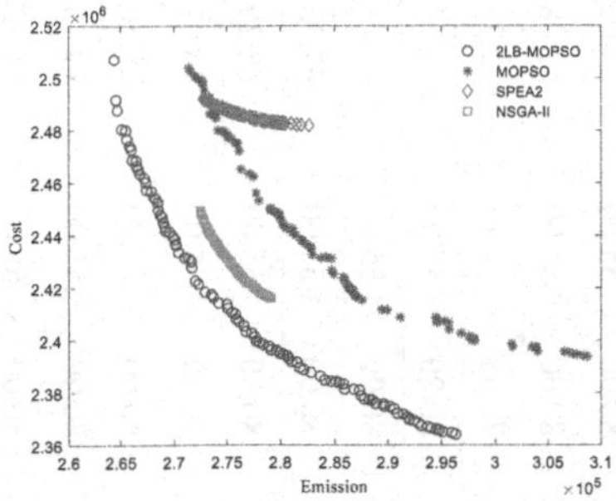

Fig. 2. The pareto front by different algorithms

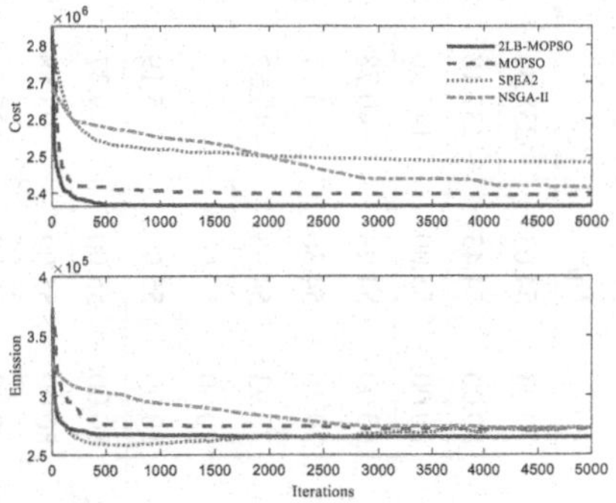

Fig. 3. The convergence curves of different algorithms

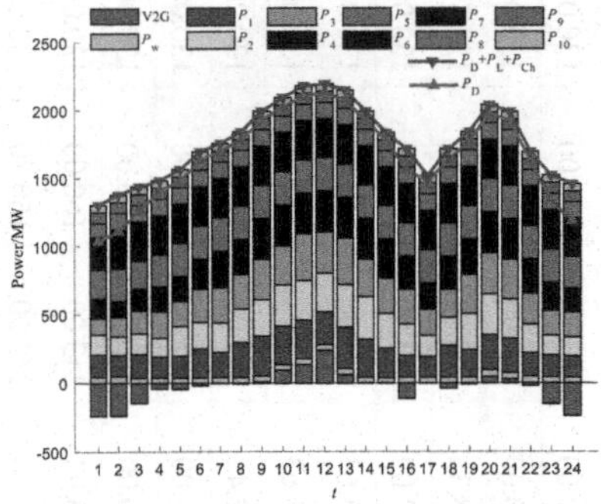

Fig. 4. Constraints check for the best compromise solution

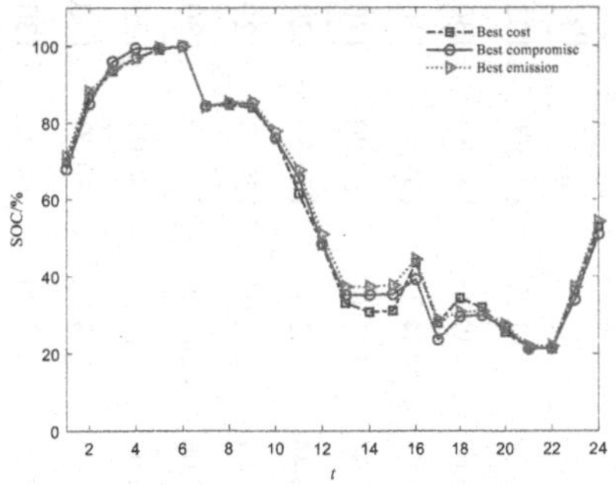

Fig. 5. SOC of EVs in optimal solutions

Figure 2 demonstrates the Pareto front with uniformly distributed Pareto solutions in the objective space by the different algorithms. Figure 3 shows the convergence curves of objective functions of different algorithms. It can be seen from Figs. 2 and 3, 2LB-MOPSO is significantly better than MOPSO, SPEA2 and NSGA-II in solving complex DEED problems with constraints.

The detailed dispatch schemes of the best compromise solution obtained by 2LB-MOPSO is shown in Table 2, and the power balanced constraints can be checked at each interval in Fig. 4. Figure 5 demonstrates the SOC of EVs in optimal solutions. From Table 2, Figs. 4 and 5, we can obverse that the EVs charging at 22:00–6:00, 8:00, 16:00

Table 2. The best compromise solution obtained by 2LB-MOPSO

t	P_1	P_2	P_3	P_4	P_5	P_6	P_7	P_8	P_9	P_{10}	P_{ev}	P_w	P_L	P_D
1	162.96	145.70	118.56	146.32	212.29	133.51	102.66	103.71	79.03	54.03	−238.85	43.71	27.66	1036
2	153.03	136.43	141.22	116.68	239.05	157.89	130.00	119.97	80.00	54.43	−237.18	49.02	30.54	1110
3	169.89	153.11	165.25	162.48	204.27	158.58	130.00	120.00	80.00	55.00	−145.50	38.68	33.77	1258
4	152.43	137.43	194.29	181.39	233.16	160.00	130.00	120.00	80.00	54.98	−40.78	38.75	35.65	1406
5	158.20	217.43	181.90	182.44	242.35	159.35	129.35	119.35	79.36	54.36	−43.17	39.41	40.34	1480
6	209.64	194.34	242.45	219.61	242.93	159.93	129.93	119.93	79.93	54.93	−17.52	39.39	47.52	1628
7	181.99	215.08	255.95	269.06	243.00	160.00	130.00	120.00	80.00	55.00	–	42.55	50.63	1702
8	252.67	244.26	253.01	256.92	243.00	160.00	130.00	120.00	80.00	55.00	−5.19	43.04	56.72	1776
9	288.12	266.84	293.80	300.00	243.00	160.00	130.00	120.00	80.00	55.00	9.37	44.36	66.49	1924
10	277.78	302.98	285.63	300.00	243.00	160.00	130.00	120.00	80.00	55.00	93.58	41.96	67.93	2022
11	281.50	290.32	340.00	300.00	243.00	160.00	130.00	120.00	80.00	55.00	133.66	43.51	70.99	2106
12	237.33	284.55	302.74	300.00	243.00	157.88	130.00	120.00	80.00	55.00	237.82	43.02	64.34	2127
13	299.90	312.34	340.00	300.00	243.00	160.00	130.00	120.00	80.00	55.00	61.02	45.06	74.31	2072
14	276.22	310.37	275.01	300.00	243.00	160.00	130.00	120.00	80.00	55.00	2.68	39.40	67.68	1924
15	215.28	253.48	257.54	288.33	242.69	159.36	125.59	119.46	78.58	54.15	2.73	35.20	56.38	1776
16	159.48	229.45	253.15	243.09	243.00	160.00	130.00	120.00	80.00	55.00	−111.72	41.01	48.46	1554
17	151.73	153.14	189.34	193.63	243.00	160.00	130.00	120.00	80.00	55.00	–	41.47	37.31	1480
18	231.73	204.37	204.21	243.63	243.00	160.00	130.00	120.00	80.00	55.00	−37.23	42.17	48.88	1628
19	198.78	264.37	284.21	262.42	243.00	160.00	130.00	120.00	80.00	55.00	−8.29	43.05	56.53	1776
20	254.80	297.42	303.32	300.00	243.00	160.00	130.00	120.00	80.00	55.00	46.07	49.24	66.86	1972
21	248.29	287.23	289.81	300.00	243.00	160.00	130.00	120.00	80.00	55.00	28.56	46.76	64.65	1924
22	169.95	208.89	224.32	253.94	243.00	160.00	130.00	120.00	80.00	55.00	−19.15	48.81	46.76	1628
23	157.59	145.42	177.93	207.06	243.00	160.00	130.00	120.00	80.00	55.00	−151.12	44.43	37.32	1332
24	151.76	139.68	182.52	175.26	229.01	152.66	129.31	119.45	78.19	55.00	−237.37	42.74	34.21	1184

and 18:00–19:00, respectively, total 13 h, and these hours are the moments of low load demand in a day. In other times except driving, the EVs discharging to the power grid to mitigate the peak load demand. Therefore, the charging and discharging behavior of EVs in proposed model has the function of shaving peaks and filling valleys.

5 Conclusions

In this paper, the multi-objective DEED with wind power and EVs model is established to deal with the conflicting fuel cost and pollution emissions objectives in all dispatching time intervals. And the 2LB-MOPSO based on constraint handling method is proposed to obtain the optimal dispatching schemes based on the 10-unit system, the test case is simulated to demonstrate the performance of the 2LB-MOPSO in proposed model. The simulation results show that 2LB-MOPSO is significantly better than MOPSO, SPEA2 and NSGA-II in solving complex DEED problems with constraints. In the proposed model, wind power and EVs can reduce the fuel cost and pollution emission of thermal power generators effectively. Moreover, the proposed model can guide the charging and discharging behavior of EVs to alleviate the load pressure on the power grid.

Acknowledgments. This work was supported in part by the General Program of NSFC under Grant 61773300, in part by the Key Program of Fundamental Research Project of Natural Science of Shaanxi Province, China under Grant 2017JZ017.

References

1. Basu, M.: Particle swarm optimization based goal-attainment method for dynamic economic emission dispatch. Electric Power Compon. Syst. **34**(9), 1015–1025 (2006)
2. Qu, B., Zhu, Y., Jiao, Y., Wu, M., Suganthan, P., Liang, J.: A survey on multi-objective evolutionary algorithms for the solution of the environmental/economic dispatch problems. Swarm Evol. Comput. **38**, 1–11 (2018)
3. De Vries, B.J.M., Van Vuuren, D.P., Hoogwijk, M.: Renewable energy sources: their global potential for the first-half of the 21st century at a global level: an integrated approach. Energy Policy **35**(4), 2590–2610 (2007)
4. Zakariazadeh, A., Jadid, S., Siano, P.: Multi-objective scheduling of electric vehicles in smart distribution system. Energy Convers Manage **79**(3), 43–53 (2014)
5. Kempton, W., Letendre, S.E.: Electric vehicles as a new power source for electric utilities. Transp. Res. Part D: Transp. Environ. **2**(3), 157–175 (1997)
6. Khodayar, M., Wu, L., Shahidehpour, M.: Hourly coordination of electric vehicle operation and volatile wind power generation in SCUC. IEEE Trans. Smart Grid **3**(3), 1271–1279 (2012)
7. Shao, C., Wang, X., Wang, X., Du, C., Dang, C., Liu, S.: Cooperative dispatch of wind generation and electric vehicles with battery storage capacity constraints in SCUC. IEEE Trans. Smart Grid **5**(5), 2219–2226 (2014)
8. Qiao, B.J., Liu, J.J., Liu, J.X., Yang, Z.B., Chen, X.F.: An enhanced sparse regularization method for impact force identification. Mech. Syst. Signal Process. **126**, 341–367 (2019)
9. Qu, B., Qiao, B., Zhu, Y., Liang, J., Wang, L.: Dynamic power dispatch considering electric vehicles and wind power using decomposition based multi-objective evolutionary algorithm. Energies **10**(12), 2017 (1991)

10. Zhang, Y., Le, J., Liao, X., Zheng, F., Liu, K., An, X.: Multi-objective hydro-thermal-wind coordination scheduling integrated with large-scale electric vehicles using IMOPSO. Renew. Energy **128**, 91–107 (2018)
11. Zhu, Y., Wang, J., Qu, B.: Multi-objective economic emission dispatch considering wind power using evolutionary algorithm based on decomposition. Int. J. Electr. Power Energy Syst. **63**(12), 434–445 (2014)
12. Basu, M.: Dynamic economic emission dispatch using nondominated sorting genetic algorithm-II. Int. J. Electr. Power Energy Syst. **30**(2), 140–149 (2008)
13. Pan, L.Q., He, C., Tian, Y., Wang, H.D., Zhang, X.Y., Jin, Y.C.: A classification-based surrogate-assisted evolutionary algorithm for expensive many-objective optimization. IEEE Trans. Evol. Comput. **23**(1), 74–88 (2019)
14. Qu, B., Qiao, B., Zhu, Y., Jiao, Y., Xiao, J., Wang, X.: Using multi-objective evolutionary algorithm to solve dynamic environment and economic dispatch with EVs. In: Tan, Y., Takagi, H., Shi, Y., Niu, B. (eds.) ICSI 2017. LNCS, vol. 10386, pp. 31–39. Springer, Cham (2017). https://doi.org/10.1007/978-3-319-61833-3_4
15. Zhile, Y., Kang, L.I., Qun, N.I.U., Xue, Y.S., Foley, A.: A self-learning TLBO based dynamic economic/environmental dispatch considering multiple plug-in electric vehicle loads. J. Mod. Power Syst. Clean Energy **2**(4), 298–307 (2014)
16. Eberhart, R., Kennedy, J.: A new optimizer using particle swarm theory. In: Proceedings of the Sixth International Symposium on Micro Machine and Human Science, MHS 1995, pp. 39–43. IEEE (1995)
17. Zhao, S.Z., Suganthan, P.N.: Two-lbests based multi-objective particle swarm optimizer. Eng. Optim. **43**(1), 1–17 (2011)
18. Coello, C.C.A., Pulido, G.T., Lechuga, M.S.: Handling multiple objectives with particle swarm optimization. IEEE Trans. Evol. Comput. **8**(3), 256–279 (2004)

Evolutionary Optimization of Three-Degree Influence Spread in Social Networks Based on Discrete Bacterial Foraging Optimization Algorithm

Tian Zhang, Lianbo Ma, and Mingli Shi(✉)

College of Software, Northeastern University, Shenyang 110819, China
3053376397@qq.com

Abstract. The influence maximization (IM) problem is an important issue in social network, which is to seek k nodes with maximal influence cascade such that the influence spread invoked by the k nodes in the network is maximized. The traditional approaches for resolving influence maximization, including Greedy, Distance, DegreeDiscount and PageRank, usually suffer from several drawbacks, such as high computational cost and unstable accuracy. In this paper, we propose a new optimization model, i.e., complete-three-layer-influence evaluation (CTLI), based on an improved three-degree model by considering the intra-layer and inter-layer's communication effect. A discrete bacterial foraging optimization algorithm is proposed to optimize CTLI model. In this algorithm, the update and mutation rules for the bacteria are redefined to improve the search ability. Finally, the proposed model and algorithm are tested on four real-world social network instances. Results demonstrate that the proposed method outperforms its compared algorithms in terms of solution accuracy and computation efficiency.

Keywords: Bacterial foraging optimization · Social networks · Influence maximization

1 Introduction

Over the years, with the great popularity of internet technology, a number of online social networks, significant examples being Facebook, Linkedin and Twitter [1,2], have boomed in modern society [3]. In these applications, they enable individuals to spread their information passively by the means of viral marketing. The studies have shown that people obtain more values on recommendations

Supported by the National Natural Science Foundation of China under Grant No. 61773103, the Fundamental Research Funds for the Central Universities No. N180408019 and N181713002, and the Program for Liaoning Innovative Research Team in University under Grant No. LT2016007.

L. Pan et al. (Eds.): BIC-TA 2019, CCIS 1159, pp. 77–87, 2020.
https://doi.org/10.1007/978-981-15-3425-6_7

form their friends than those form other ways, such as TV, letter, book and so on. Moreover, social networks not only communication channels but also a dominant virtual marketing tool for commercial companies and public services. These facts are a major challenge for decision makers to find a small number of influential nodes that seek to maximize influence spread of information (i.e., interests, opinions, and attitudes) in complex networks. The problem is also named as influence maximization (IM), which seeks k nodes that can acquire maximum influence spread in a random cascade model.

Domingos and Richardson [5] formulated the IM model as a near-optimal problem firstly, which can be solved by Evolutionary algorithms. Kempe et al. [4] articulated this problem that the optimization of IM is an NP-hard previously. And they prove that the one of the optimal solutions is a greedy algorithm (KK-greedy) which can be approximated to a ratio of (1−1/e) [4]. The greedy method determine an initial set with excellent performance on independent cascade (IC) model [3] and weighted concatenation (WC) model [4], which is surprisingly better in accuracy compared with degree based heuristics. In addition, with the KK-greedy algorithm, the influence spread is combined with Monte Carlo simulation on the IC and WC models. However, the KK-greedy algorithm needs so many simulations in computing that it has expensive computational overhead in each round, it performs deficiently in efficiency for large-scale social networks with thousands of nodes.

Recently, many greedy approaches focused on enhance methods to decrease computations in the targeted social network. Leskovec et al. [6] proposed a Cost-Effective Lazy Forward strategy (CELF) by exploiting the sub-modularity property into the greedy approach. On account of the sub-modularity function, the CELF algorithm is almost 700 times faster than the previous greedy algorithm. Based on CELF, a novel CELF++ algorithm is proposed by Goyal et al. [7], which demonstrated it is 35%–55% quicker than CELF. Subsequently, some heuristic methods find a subset of k-sized seeds that can affect other seeds in a particular propagation mode in the social networks. Moreover, the PageRank (PR) approach proposed in [8] is used to solve the IM problem. Nevertheless, those algorithms are still low efficient and very poor to approximate the influence spread. Evolutionary algorithms (EAs) are inspired by the evolutionary of biology in nature [9–13]. And they have been utilized on real world applications widely [3,14,15]. BFO, as one of EAs, has become a successful population-based numerical optimization algorithm [10] due to its complex social behaviors and strong robustness. For extensive experiments, our work concentrates on search for an efficient approach to seek k nodes to solve IM problem in the real-world social networks' instances.

In this paper, with the three-degree cascade model [16], we formulate a novel optimization model, and then propose a discrete bacterial foraging optimization algorithm to optimize the model. The contributions of our algorithm can be listed as follows:

- Based on the strategy of three-degree model [16], we propose a complete-three-layer-influence (CTLI) evaluation model. The main idea of the function is to integrate the cooperating influence of the three-hop influence of initial seeds for three-degree model. Next, the IM problem is transformed into the optimization of the CTLI model.

- To avoid the disadvantage of BFO, we develop a discrete bacterial foraging optimization algorithm (DBFO) for optimizing CTLI model. In DBFO, the representations, initialization definition, local-communication (LC) mechanism and mutation operation are redefined to evaluate the performance and promote the convergence.

The remainder of this paper is organized as follows: In Sect. 2, we describe the model and algorithms. Section 3 shows the proposed CTLI model and LCDBFO algorithm. Section 4 is the experiments explanation about parameter settings of our algorithm and other compared algorithms. Finally, the conclusion is drawn in Sect. 5.

2 Model and Algorithms

The online social network is formulated as a graph $G = (V; E; A)$. The node set V and the edge set E represent all users and the social relationship between two linked users in a social network. And A is an adjacent matrix. Specifically, $a_{vu} = a_{uv}$ for an unweighed graph, (v, u)(from v to u) $= (u, v)$ denotes a same relationship or edge and $d_v^{in} = d_v^{out}$ represents the in-degree is the same as out-degree for node v. Moreover, the undirected social network is a special case of directed networks.

2.1 Propagation Model

Based on IC [3] model, focus to simulate information propagation in a social network. The activation probability of the IC model is defined as follows:

$$p_{vu} = 1 - (1 - p)^{a_{vu}} \tag{1}$$

Where p_{vu} is the probability that node v actives u, $p \in (0, 1)$ is the initial probability, and a_{vu} is the weight of edge (v, u). Moreover, node v only has one chance to activate node u in the targeted social network.

2.2 Algorithms of Influence Maximization

Based on three-degree theory, Qin et al. [16] propose a TLAA model and can be computed as follows:

$$p(v) = \begin{cases} 1 & v \in D(S, 0) \\ 1 - \prod_{u \in D_{i-1}(u,v) \in E} (1 - p(u) \cdot p(u, v) \cdot \lambda_i) & v \in D(S, i), 0 < i < 4 \\ 0 & v \in D(S, i), i \geq 4 \end{cases} \tag{2}$$

Where λ_i is the propagation attenuation coefficient of the i-th layer (when $\lambda_1 \geq \lambda_2 \geq \lambda_3$ and $i \geq 4, \lambda_i = 0$), and $v \in D(S, i)$ represent v is owned by layer i.

In addition, DegreeDiscount [17], a heuristics algorithm, described more precisely in the exact degree of attenuation, which is defined as:

$$1 + (d_v - 2t_v - (d_v - t_v) \cdot t_v \cdot p + o(t_v)) \cdot p \quad (3)$$

Where d_v represents the degree of node v, and t_v is the number of nodes in all neighbors of node v have been seeded. And p is the probability of influence from node v to the neighborhood.

2.3 Bacterial Foraging Optimization Algorithm

Bacterial foraging behaviors, including chemotaxis, reproduction and elimination-dispersal, are shown in Fig. 1. In original BFO, each position of bacterium represents a possible optimization solution. Chemotaxis can also be considered as the optimal foraging decision making capabilities of bacteria [18]. Swimming is more frequent as the bacteria approaches a nutrient gradient and tumbling is more frequent as the bacteria moves away from some food source to search for more sources. After completing the chemotaxis steps, the fitness value of each bacterium is calculated and sorted in order according to fitness value. In the reproduction step, a half of bacteria with worse value die and the other bacteria splits into two identical ones. And the population of bacteria keeps constant. In classical BFO, the elimination-dispersal mechanism happens after a certain number of reproduction processes. According to a probability p_{ed}, some bacteria are killed and moved to a random position in the environment. Let N_{ed} and N_c represent elimination-dispersal steps and chemotactic steps, respectively.

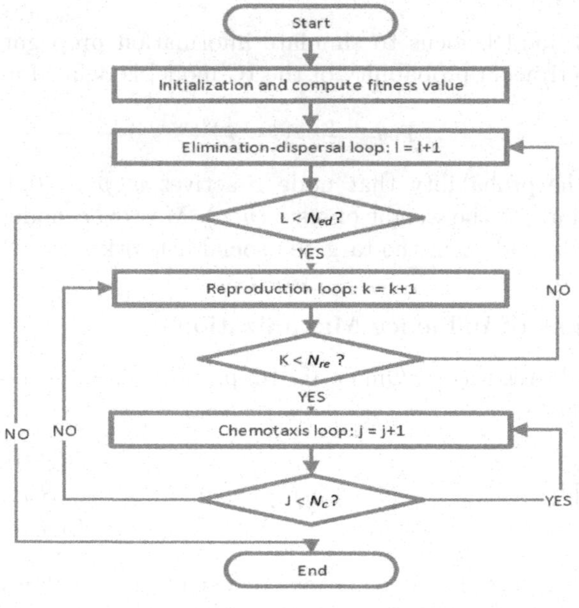

Fig. 1. The framework of bacterial foraging optimization algorithm.

3 Proposed Model and DLCBFO Algorithm

3.1 Proposed Model

The named complete-three-layer-influence (CTLI) evaluation model is to approximately compute the propagation range of the nodes in the IC model. Based on [16], the main idea of our model is to combine the intra-layer and backward inter-layer's propagation into the three-degree model.

Definition 1. *S is the initial node set, u, v and m are nodes.*

1. *If $u \in S$, $\exists(u, v)$, which $v \in u_{laver1}$, then v is a layer-1 neighbor of u.*
2. *$\exists(v, m)$, that $m \in v_{laver1}$, $m \in u_{laver1}$, then m is a layer-2 neighbor of u.*
3. *$\exists(m, t)$, that $v \in t_{laver1}$, $t \in u_{laver3}$, then t is a layer-3 neighbor of u.*

In the IC model, the number of diffusion nodes is defined and calculated by CTLI evaluation function as follows:

$$\begin{aligned}
\delta_{CTLI} &= \text{k} + \delta_1(S) + \delta_2(S) + \delta_3(S) \\
&= \text{k} + \delta_{0-1}(S) + \delta_{1-1}(S) + \delta_{1-2}(S) + \delta_{2-2}(S) + \delta_{2-1}(S) + \delta_{2-3}(S) + \delta_{3-3}(S) + \delta_{3-2}(S) \\
&= \text{k} + \sum_{v\in L(1)} \left[1 - \prod_{u\in L(0),(u,v)\in E} (1 - p_1(S,v)p\lambda_1)\right] + \sum_{v\in L(1)} \left[1 - \prod_{w\in L(1)/v,(w,v)\in E} (1 - p_1(S,w)p\lambda_2)\right] \\
&+ \sum_{m\in L(2)} \left[1 - \prod_{v\in L(1),(u,m)\in E} (1 - p_2(S,m)p\lambda_2)\right] + \sum_{m\in L(2)} \left[1 - \prod_{n\in L(2)/m,(m,n)\in E} (1 - p_2(S,n)p\lambda_3)\right] \\
&+ \sum_{w\in L(1)} \left[1 - \prod_{m\in L(2)/m,(m,w)\in E} (1 - p_2(S,w)p\lambda_2)\right] + \sum_{t\in L(3)} \left[1 - \prod_{m\in L(2),(m,t)\in E} (1 - p_3(S,t)p\lambda_3)\right] \\
&+ \sum_{t\in L(3)} \left[1 - \prod_{r\in L(3)/t,(r,t)\in E} (1 - p_3(S,r)p\lambda_4)\right] + \sum_{n\in L(2)} \left[1 - \prod_{t\in L(3),(t,n)\in E} (1 - p_3(S,n)p\lambda_3)\right]
\end{aligned} \tag{4}$$

where, u is a node in S, v and w are nodes in layer-1, m and n are nodes in layer-2, t and r are nodes in layer-3, $u \in L(0), v, w \in L(1), m, n \in L(2), t, r \in L(3)$, k is the number of initial nodes, λ_i is a propagation attenuation coefficient for layer-i, $\delta_i(S)$ is the total number of nodes propagated in layer-i, p_i is the final activation probability of layer-i neighbor.

3.2 Discrete Local Communication BFO (DLCBFO)

Firstly, the bacteria are initialized by the DegreeDiscount method, whose details is shown in Algorithm 1. Then, in the Chemotaxis operation, the local-communication (LC) mechanism is defined to filter excellent nodes and accelerate the convergence, as presented in Algorithm 2. Finally, after the dead probability of bacteria is calculated, in elimination-dispersal operation, the mutation strategy is to hold the excellent population and enhance the variety of population, as manifested Algorithm 3.

Initialization. In DLCBFO, the DegreeDiscount [17] initialization is used to select more excellent of the initial population. Each the local of bacteria represents a solution for influence maximization problem. As shown in Algorithm 1, the DegreeDiscount of each individual is calculated (Line 9). In this way, k-node is chosen by DegreeDiscount heuristics algorithm.

Algorithm 1. Initial_DegreeDiscount algorithm

Input: graph G = (V; E); the size of initial activation node set k.
Output: Initial set S and suboptimal nodes point.
1: Initialize S = $\emptyset$
2: **for** each $i = 1$ *to* k **do**
3: compute its degree d_v
4: initialize t_v to 0
5: **for** each $m = 1$ *to* u **do**
6: **u** = $\text{argmax}_v \{d_v | v \in v | S\}$
7: $S = S \cup \{u\}$
8: **for** each $n = 1$ *to* c **do**
9: calculate each node's DegreeDiscount based on Eq. (3)
10: **end for**
11: **end for**
12: **end for**
13: $r \leftarrow$ sort(G, descend)
14: **for** each $j = 1$ *to* i **do**
15: point $\leftarrow$ save the excellent nodes in G
16: **end for**

Local Communication Mechanism. The purposeful local communication strategy, which can learn excellent experience from other solutions and improve the convergence. Here the purposeful evaluation strategy is introduced in Algorithm 2. With the CTLI fitness of the population, a set of excellent solutions are obtained by using two operations *find* (Line 1 in Algorithm 2) and *choose* (Line 4 in Algorithm 2). Next, the population get_number is merged with current solution. Finally, a new solution for p is generated by operation *select* (Line 9 in Algorithm 2), which the object of operation include kk and *point* (suboptimal nodes).

Algorithm 2. Local communication operation

Input: graph G = (V; E); population P; the CTLI value *num*; population size *seedsize*; suboptimal nodes *point*.

Output: new_population P.

```
number ←find( num )
bb ← ceil(length( number ))
if bb > 0 then
    get_number ← Choose (number, bb)
    for each j = 1 to l e n g t h(get_number) do
        i ← Merge (get_number )
    end for
    for each k = 1 to seedsize do
        p ← Select (k, point)
    end for
end if
```

Algorithm 3. Mutation operation

Input: the excellent nodes in G *point*; the size of initial activation nodes *seedsize*; the old local of bacteria w.

Output: the new local of bacteria w.

```
a ←randperm(seedsize , {1,2,...,k})
b ←random_select (point , {1,2,...,k})
for i = [1,{1,2,...,k}] do
    while f = ismember(b,w) == 1 do
        b ←point (randperm (length (point), 1))
        f ←ismember(b,w)
    end while
    w ←Merge (b)
end for
```

Mutation Operation. In original BFO, the update equation to generate a positional change is rather similar to a blind mutation operator, and the probability that a completely randomly chosen solution becomes a better or worse solution. To solve this problem, the modified mutation operation is proposed and described in Algorithm 3, and the new candidate solution is better than the previous one.

4 Experiments

4.1 Experiment Settings

We evaluated the efficiency of the proposed algorithm(DLCBFO) and tested other four well known algorithms for influence maximization on four real-world social networks. All the experiments are conducted at a server with a 2.30 GHz Inter Core i5-8300 CPU and 16 GB host memory.

In our experiments, four real-world social network instances are used, including Football, NetScience, Polblogs and Power. The active probability p of the propagation model is set to 0.01 for Football and Polblogs. Due to the sparsity of NetScience and Power, p is set to 0.06 for NetScience and 0.08 for Power networks. The details of these instances on each network are shown in Table 1.

Table 1. Statistic characteristics of four real-world social network instances.

Network	Nodes	Edges	Ave-Degree
Football	115	613	5.33
Polblogs	1490	19090	12.81
NetScience	1589	2742	1.73
Power	4941	6594	1.33

4.2 Experiment Results

With optimizing the CTLI of Football, Polblogs, NetScience and Power network instances, the experimental results of DLCBFO with different population numbers are shown in Fig. 2. The number of the size k of initial node set is 40. The number of populations are selected from 2, 20, 40, 60, 80, 100, 120 and 140. As shown in Fig. 2, the curve of CTLI values on the four network instances looks like a straight line. Here, when *number* is 80, the value of CTLI will keep stable and powerful on most of the network instances. Therefore, we set the population number n to 80.

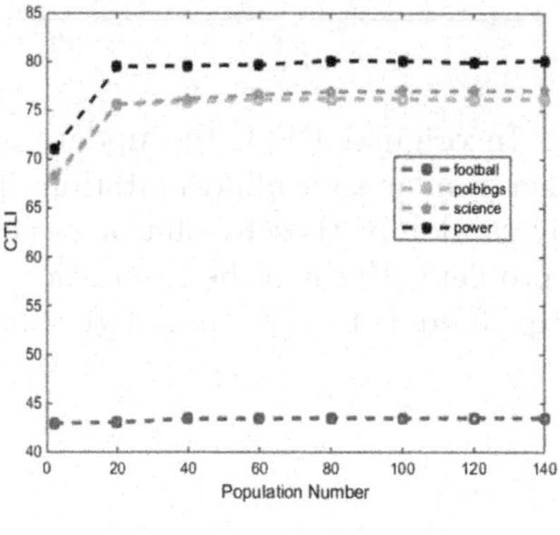

Fig. 2. The different value of CTLI on four social network instances with different population number n.

The Fig. 3 shows the three-hop influence spread on four real-world networks instances under IC model. The Fig. 3 shows the influence spread in Football, Polblogs, Netscience and Power respectively. The y-axis represents the three-hop influence spread evaluation, and the x-axis represents the set of the seed size. And the curve obtained by DLCBFO in terms of the best CTLI value over each network. Specially on Fig. 3(a), in order to better observe the difference of the influence spread evaluation (ISE), the seed size is set 20 to 40. And we can observe all algorithms are steadily rising up and the DLCBFO algorithm do better than others on all time. From Fig. 3(b), it is shown that DLCBFO, Degree and DegreeDiscount do better than other algorithms. In these performances, Distance is ranked at a medium level. However, PageRank has an extremely poor performance. From Fig. 3(c), it indicates that DLCBFO produces the largest influence spread. When the set of seed size is 40, the performance of PageRank is much better than that of DegreeDiscount. Meanwhile, Degree outperforms Distance. From the Fig. 3(d), Distance has the worst performance on the Power network. Finally, it is obvious that DLCBFO does best in terms of accuracy in IC model.

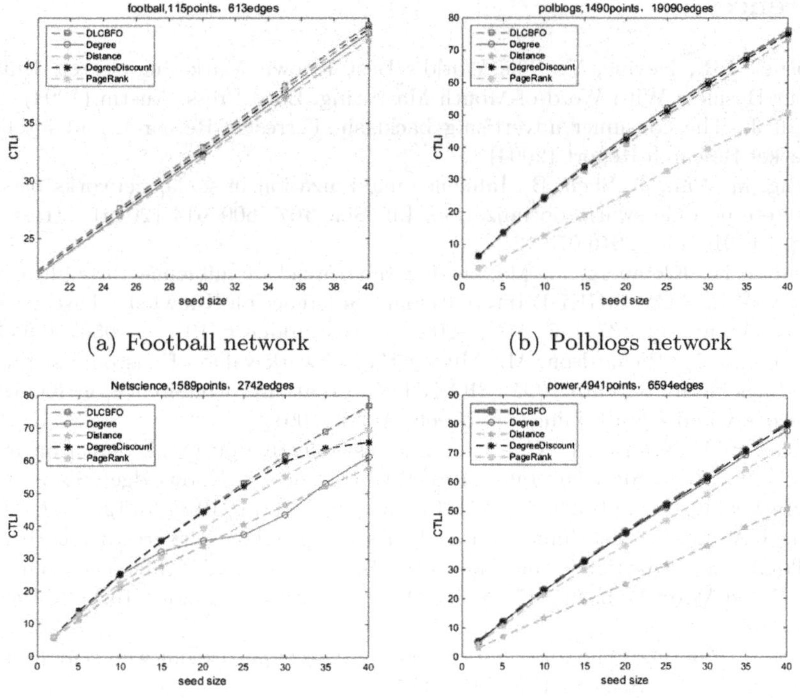

(a) Football network (b) Polblogs network

(c) Netscience network (d) Power network

Fig. 3. Influence spread on the four real-world social networks in the IC model

5 Conclusion

In this paper, we have proposed an intelligent optimization algorithm called DLCBFO, which is to consider the IM problem in social networks as a discrete optimization problem. And the optimization problem by developing a function called complete-three-layer-influence evaluation. Next, a discrete optimizer DLCBFO is developed to solve the CTLI. In DLCBFO, DegreeDiscount initialization, the improvement update and mutation rules are used to improve the search ability of the operators. We have conducted the experiments on four real-world social network instances. In the future, we will focus on proposing a set of objective functions and developing the influence maximization problem in other cascade models and social network instances.

Acknowledgments. This work is supported by the National Natural Science Foundation of China under Grant No. 61773103 and 61503373, the Fundamental Research Funds for the Central Universities No. N180408019 and N181713002, and the Program for Liaoning Innovative Research Team in University under Grant No. LT2016007.

References

1. Misner, I.R., Devine, V.: The World's Best Known Marketing Secret: Building Your Business With Word-of-Mouth Marketing. Bard Press, Austin (1994)
2. Nail, J.: The consumer advertising backlash. Forrester Research and intelliseek Market Research Report (2004)
3. Gong, M., Yan, J., Shen, B.: Influence maximization in social networks based on discrete particle swarm optimization. Inf. Sci. **367**, 600–614 (2016). https://doi.org/10.1016/j.ins.2016.07.012
4. Kempe, D., Kleinberg, J.: Maximizing the spread of influence through a social network. In: ACM SIGKDD International Conference on Knowledge Discovery and Data Mining, pp. 137–146. ACM (2003). https://doi.org/10.1145/956750.956769
5. Domingos, P., Richardson, M.: Mining the network value of customers. In: Proceedings of the Seventh ACM SIGKDD International Conference on Knowledge Discovery and Data Mining, pp. 57–66. ACM (2001)
6. Leskovec, J., Krause, A., Guestrin, C.: Cost-effective outbreak detection in networks. In: ACM SIGKDD International Conference on Knowledge Discovery and Data Mining, pp. 420–429. ACM (2007). https://doi.org/10.1145/1281192.1281239
7. Goyal, A., Lu, W., Lakshmanan, L.: CELF++: optimizing the greedy algorithm for influence maximization in social networks. In: International Conference Companion on World Wide Web, pp. 47–48. ACM (2011). https://doi.org/10.1145/1963192.1963217
8. Brin, S., Page, L.: The anatomy of a large-scale hypertextual web search engine. Comput. Netw. ISDN Syst. **30**, 107–117 (1998)
9. Ma, L., et al.: Cooperative two-engine multi-objective bee foraging algorithm with reinforcement learning. Knowl.-Based Syst. **133**, 278–293 (2017). https://doi.org/10.1016/j.knosys.2017.07.024
10. Passino, K.M.: Biomimicry of bacterial foraging for distributed optimization and control. IEEE Control Syst. Mag. **22**(3), 52–67 (2002)

11. Ma, L., Wang, R., Chen, M., Wang, X., Cheng, S., Shi, Y.: A novel many-objective evolutionary algorithm based on transfer learning with kriging model. Inf. Sci. **509**, 437–456 (2019). https://doi.org/10.1016/j.ins.2019.01.030
12. Marinakis, Y., Marinaki, M., Matsatsinis, N.: A hybrid discrete Artificial Bee Colony - GRASP algorithm for clustering. In: International Conference on Computers & Industrial Engineering, pp. 548–553. IEEE (2009). https://doi.org/10.1109/ICCIE.2009.5223810
13. Ma, L., Zhu, Y., Liu, Y., Tian, L.: A novel bionic algorithm inspired by plant root foraging behaviors. Appl. Soft Comput. **37**, 95–133 (2015). https://doi.org/10.1016/j.asoc.2015.08.014
14. Ma, L., Wang, X., Huang, M., Lin, Z., Tian, L., Chen, H.: Two-level master-slave RFID networks planning via hybrid multi-objective Artificial Bee Colony optimizer. IEEE Trans. Syst. Man Cybern. Syst. **49**(5), 861–880 (2019). https://doi.org/10.1109/TSMC.2017.2723483
15. Ma, L., Hu, K., Zhu, Y., Chen, H.: Cooperative Artificial Bee Colony algorithm for multi-objective RFID network planning. J. Netw. Comput. Appl. **42**, 143–162 (2014). https://doi.org/10.1016/j.jnca.2014.02.012
16. Qin, Y., Ma, J., Gao, S.: Efficient influence maximization under TSCM: a suitable diffusion model in online social networks. Soft. Comput. **21**(4), 827–838 (2016). https://doi.org/10.1007/s00500-016-2068-3
17. Chen, W., Wang, Y., Yang, S.: Efficient influence maximization in social networks. In: ACM SIGKDD International Conference on Knowledge Discovery and Data Mining, pp. 199–208. DBLP, Paris (2009). https://doi.org/10.1145/1557019.1557047
18. Crespi, B.J.: The evolution of social behavior in microorganisms. Trends Ecol. Evol. **16**(4), 178–183 (2001)

Ant Colony Algorithm Based on Upper Bound of Nodes for Robot Path Planning Problems

Sizhi Qiu, Bo Hu, Yongbin Quan, Xiaoxin Jian, and Haibin Ouyang(✉)

School of Mechanical and Electric Engineering, Guangzhou University, Guangzhou 510006, China
oyhb1987@163.com

Abstract. In order to solve the problems of ant colony algorithm such as low search efficiency, slow convergence speed and easy to fall into local optimum, an improved ant colony algorithm is proposed in this paper. In the proposed algorithm, the upper bound of path nodes is adjusted and the grid map mode is integrated. The upper bound of nodes are adjusted for driving the optimization process of algorithm. Experiment simulation is carried out based on two specific robot path-planning examples. Results show that the proposed ant colony algorithm can guarantees the global optimization capability, and it also can improve the efficiency of path planning.

Keywords: Ant colony algorithm · Robot path planning · Grid map · Global optimization

1 Introduction

Robots can help humans do dangerous, repetitive and difficult work, and it also can liberate our human from high-risk, or physically demanding labor. One most important parts of robot is path planing, which it has become a hot issue. A good planning route not only can save time, but it also can reduce a great much of consume energy. Therefore, the research of path planning has an important theory and practical significance. Whatever the environment that the robot in, simple or complex, static or dynamic, known or unknown, the primary task for robot is to perceive the environment and avoid the barrier, which promote robot can finishes its work with minimum loss include time and space. This process can be called path planning. Nowadays based on the environment information, path planning can be divided into two categories: global planning (static planning) and local planning (dynamic planning). According to obstacle avoidance strategies of mobile robots, the path planning can be broadly divided into two categories [3]: The first kind of obstacle avoidance strategy is based on the completely known environment, for example, the obstructions' position and shape have been known or given. This kind of obstacle avoidance belongs to static planning issues. The main methods include the grid method, visibility graph method, and so on. The second category is the obstacle avoidance strategy based on sensor information, such as the algorithm with fuzzy logic, neural network and artificial potential field method.

L. Pan et al. (Eds.): BIC-TA 2019, CCIS 1159, pp. 88–97, 2020.
https://doi.org/10.1007/978-981-15-3425-6_8

Traditional path planning methods, such as artificial potential field method [4], simulated annealing algorithm [5], and fuzzy logic algorithm [6], which generally have such defects as slow convergence speed and poor global search ability. With the development of intelligent control technology, some new methods are presented and developed. For example, neural network algorithm, genetic algorithm, ant colony algorithm and other intelligent algorithms are gradually applied to solve the problem of path planning. Among them, neural network algorithm has the ability of self-adaptation and self-learning, but poor generalization ability is its fatal shortcoming. Genetic algorithm is suitable for global path planning, but the search space is large, the operation efficiency is not high, the running speed is slow. Ant colony optimization (ACO) is widely concerned by researchers because of its heuristic, parallelism and robustness. Ant colony optimization (ACO) is proposed by Marco Dorigo in 1992. It showed a probability model which is used to find optimal path algorithm [7], and its inspiration from the path in the process of ants searching for food, this algorithm has distributed computing, information positive feedback and heuristic search, the characteristics of the nature is a heuristic global optimization algorithm.

The establishment of environment model is the basic premise and key of mobile robot autonomous navigation and path planning. The establishment of environment model is not only a reflection of the robot's perception ability and intelligence level, but also enhances the robot's ability to complete intelligent tasks, and improves the flexibility, stability, robustness and efficiency of robot navigation. The most commonly used environmental modeling methods are grid, geometry and topology [8–10].

Based on the grid map modeling, an improved ant colony algorithm is proposed in this article through in the process of iteration the limitation on the path to the node number, effectively blocked obviously unreasonable ants search path, through the modeling and simulation, compared with other algorithms under the same environment, we obtained a better effect, and to a certain extent, it improved the ant search accuracy, the convergence speed of the algorithm and the efficiency of path planning.

2 Environment Overview and Modeling

The actual working environment of robot is a real physical space, while the space processed by path planning algorithm is an abstract space of the environment. Environment modeling is a mapping of physical space to abstract space. The purpose of environment modeling is to build an effective environment model, which it is convenient for using computer to solve path planning problem.

In this paper, the grid method can be equivalent to two-dimensional finite area AS through the fractal algorithm of text [1]. AS can be changed into any shape. Therefore, when building the robot environment, it is necessary to supplement the obstacle grid at the AS boundary to take the robot working environment as a square, each square is called a grid, and the size of the grid area is generally determined by the size of the robot car body. If the robot traverses the whole site, we can record the movement of the robot at the center of each grid, and we record the obstacle information during the robot movement. In this way, the movement and obstacle information of the robot are recorded on a grid to define the grid properties. All of the grid properties constitute a map of the site, in which the grid occupied by obstacles is called the obstacle grid; and a grid without obstacles is called a free grid [11].

The rectangular coordinate system method was used to construct the grid plane, and the grid was divided into two color grids, black and white, as shown in Fig. 1. Black represents the obstacle grid, white represents the free grid, and the robot can move freely in the free grid in the plane. The obstacles are fixed, and the grid number is labeled 1, 2, 3, …, n. Each number represents a grid, and establish the relationship between the grid number and the coordinates in the plane of XOY.

$$\begin{cases} x = \text{floor}((C - 1)/N) \\ y = N - \text{mod}((C, 20) + 0.5) \end{cases} \tag{1}$$

In the formula (1): *N* is the number of grids for each row and column in the grid environment; *C* is the grid number, floor is the downward integer, mod is the remainder.

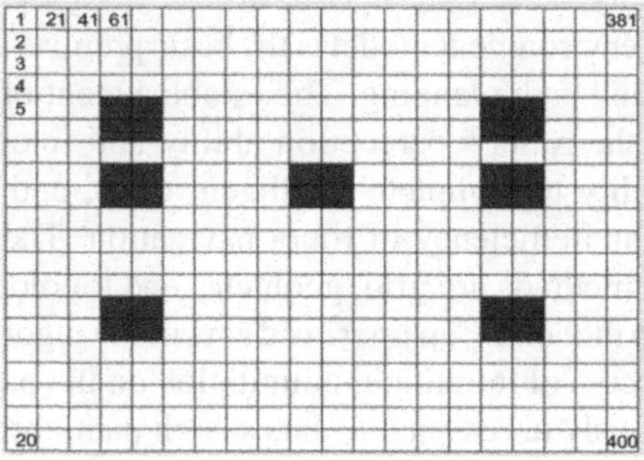

Fig. 1. Simple grid diagram

3 Improved Ant Colony Optimization Algorithm

3.1 Basic Principle of Ant Colony Algorithm

Biologists have found that ant's foraging is a group activity, and it is not a solitary search for food source. When an ant searches for a food source, it releases a pheromone in its path, and it senses pheromones released by other ants. The pheromone concentration indicates the distance of the path. The higher the pheromone concentration, the shorter the corresponding path distance.

Generally, ants prefer the path with higher pheromone concentration with higher probability, and release a certain amount of pheromone to enhance the pheromone concentration in that path, thus these actions forming a positive feedback.

The basic optimization process of ant colony algorithm can be described as: the feasible solution of the problem to be optimized is represented by the ant's walking path, and all the paths of the whole ant colony constitute the solution space of the problem to be optimized. Ants with shorter paths release more pheromones. With the increase of time, the pheromone concentration accumulated on the shorter paths gradually increases, and the number of ants choosing this path is increasing. In the end, the whole colony of ants will focus on the optimal path under the effect of positive feedback, and the corresponding path is the optimal path.

3.2 The Basic Model of Ant Colony Algorithm

Through pheromones, individuals of ant colony communicate with each other indirectly. A group composed of simple ants can accomplish some complex tasks that are difficult for individual ants to complete.

The mathematical model of basic ant colony algorithm is given by analyzing the classic TSP problem of n cities. The TSP problem is a combination optimization problem, in which the artificial ant has to choose the path to visit n cities, and the limit of the path is that each city can only visit once, and finally return to the original city. It is required to travel the nearest distance, visit each city only once, start from a certain city, and finally return to that city. In the optimization process, artificial ants calculate the state transition probability according to the information amount on different paths and heuristic information, as shown in Eq. (2):

$$p_{ij}^{k}(\mathrm{t}) = \begin{cases} \frac{[\tau_{ij}(t)]^{\alpha} \cdot [\eta_{ij}(t)]^{\beta}}{\sum_{s \in allow_k} [\tau_{is}(t)]^{\alpha} \cdot [\eta_{is}(t)]^{\beta}}, & s \in allow_k \\ 0, & s \notin allow_k \end{cases} \tag{2}$$

p_{ij}^{k} (t) represents the transfer probability of ant k from element (city) I to element (city) j at time t; α and β represent the relative importance of pheromone and heuristic information in ant colony search path; $\tau_{ij}(t)$ is the information quantity on the path between city I and city j at time t; $Tabu_k$ (k = 1, 2, …, m) is used here to record the set of elements (cities) that the KTH ant has walked through. The explanatory formula (2) will not select the elements (cities) in $Tabu_k$ again, ensuring the legitimacy of solving the optimal path selection. In order to make the information contained in the pheromones after the ant has traversed all the elements (cities) effective, it is necessary to update the remaining pheromones. The change process of pheromone is shown in formula (3):

$$\begin{cases} \tau_{ij}(t+1) = (1-\rho)\tau_{ij}(t) + \Delta\,\tau_{ij}, \\ \Delta\,\tau_{ij} = \sum_{k=1}^{n} \Delta\,\tau_{ij}^{k}, \end{cases} \quad 0 < \rho < 1 \tag{3}$$

$\Delta\,\tau_{ij}^{k}$ represents the pheromone concentration released by the kth ant on the connection path between the *i*th and the *j* grids; $\Delta\,\tau_{ij}$ represents the sum of pheromone concentration released by all ants in the connection path of grid I and grid *j*.

To solve the problem of pheromone release by ants, in order to make better use of the overall information, this paper adopts the ant cycle model proposed by Dorigo, $\Delta\,\tau_{ij}^{k}$ is calculated as a formula (4)

$$\Delta\,\tau_{ij}^{k} = \begin{cases} \frac{Q}{L_k}, \text{the } k^{th}\ ant\ from\ grid\ i\ to\ j \\ 0,\ others \end{cases} \tag{4}$$

Where Q is a constant, represents the amount of pheromone released by the ant in a cycle; L_k is the length of the path traveled by the kth ant.

3.3 Improved Ant Colony Optimization Algorithm

3.3.1 Improved Strategy

The concept of node upper limit is introduced and the coordinates traversed by ant are regarded as nodes, the sum of the distances between nodes is called the node upper limit. In the whole ant colony, when the number of nodes traversed by an individual is far more than that of the currently known optimal individual, the individual cannot plan towards the optimal solution, it results in the problem of long algorithm time and low efficiency. By setting the upper limit of nodes, this paper avoids the influence of too many individual nodes and effectively improves the execution efficiency of the algorithm.

3.3.2 Improved Method

Y is set as the upper limit of dynamic nodes. In the whole ant colony, when the number of path nodes of an ant in the traversal process is obviously larger than the number of the optimal path nodes at present, the ant is eliminated in advance and the next ant traverses. For example, the path of the first ant is

$$1 \to 2 \to 3 \to 4 \to 5 \to 6 \to 7 \to 8 \to 9$$

The path traversed by the ant is a node. If 1 is the starting point of the ant and 9 is the target point of the ant, the distance of the first ant is 8. The path of the other ant is

$$1 \to 3 \to 2 \to 4 \to 5 \to 4 \to 6 \to 7 \to 6 \to 8 \to 7$$

The second ant has not traversed all the nodes, and the ant's path distance is 14. The difference between the two ants is significant, and the upper limit of nodes is 1.5, so the ants that are within 12, and it can reach the target point are kept. Like the second ant, the distance is greater than 12, and it didn't go through all the nodes, so we eliminated it ahead of time and went through the next ant. In order to plan the path towards the optimal solution, the upper limit of dynamic nodes decreases with the increase of *i* of iteration number $\mathrm{iter_i}$, so that the number of nodes of the planned path is closer and closer to the optimal solution. Take the upper bound multiple of the initial node Yo = 2 as an example, set the dynamic upper limit of the ith iteration node as Y_i, the maximum number of iterations is $\mathrm{iter_{max}}$, the i times iteration is iter_i, Then the upper limit *Y* formula of dynamic node is calculated

$$Y_i = \frac{1}{3600}\mathrm{iter_i^2} - \frac{1}{30}iter_i + 2,\ \ \mathrm{iter} \in [0, iter_{\mathrm{max}}] \tag{5}$$

When the traditional ant colony algorithm is used to solve the robot path-planning problem, it is easy to fall into the local extreme value, and it is difficult to avoid the obstacles and find a reasonable path. After many experiments, it is found that in the middle and later stages of planning, because the path gradually converges, and in order to eliminate the errors caused by individual ants deviating from the optimal path greatly in the course of path transformation, we adjusted the upper limit of nodes dynamically. Through the dynamic adjustment in the evolutionary process of each generation, it not only guarantees the overall situation in the early stage, but also guarantees the operation efficiency in the later stage. The simulation experiment shows that this adjustment mode has obvious effect on the improvement of convergence speed.

3.3.3 Basic Steps of Solving Robot Path Planning

By combining the node upper limit ant colony algorithm with robot path planning, a reasonable and reliable path is finally found. The steps are as follows:

Step 1. Adopt the grip method to build the map, initialize the algorithm, empty the grid of pheromones, reset the number of iterations. Initialize the ant colony, put all the ants at the starting point S.
Step 2. Calculate the transition probability of the next grid by formula (2), the ant migrates between the grid points according to the probability and records the migration path. When the migration path is less than the upper limit of nodes and the ant does not reach the target point, repeat the migration path logging process; When the ant migration path is equal to or greater than the upper limit of nodes and does not reach the target point, reinitialize the ants, clear migration path and node access records, go back to our starting point S and the number of ants' re-initialization plus 1. If the re-initialization count value is less than the upper limit, the migration path recording process is performed; otherwise, the ant is discarded.
Step 3. For ants reaching the target point, calculate its path distance, and the upper limit of nodes is dynamically adjusted according to formula (5).
Step 4. Execute on all ants in the colony from **Step 2** to **Step 3**. Choose the smallest path of each ant.
Step 5. In the optimal path, the pheromone is updated by formula (3) and (4) according to the movement direction of the ant when it reaches each grid point. Number of iterations plus 1. When the number of algorithm iterations is less than the maximum iteration limit, go to **Step 2**; otherwise, the algorithm terminates.

3.3.4 The Search Block Diagram

Dispatch 50 ants every time, the maximum number of iterations $iter_{max} = 60$, Pheromone importance factor $\alpha = 1$, Pheromone volatility $\beta = 0.5$.

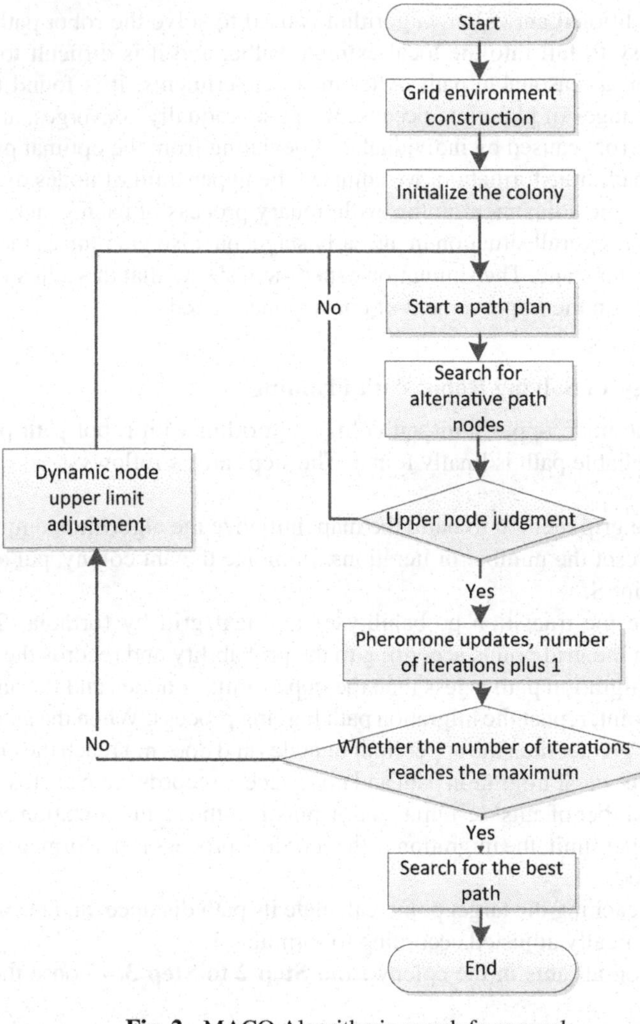

Fig. 2. MACO Algorithmic search framework

4 Discussion and Analysis of Experimental Simulation Results

Considering the comprehensiveness and fairness of the algorithm performance test, the general path problem in the literature is selected and it compared with the basic ACO algorithm and the improved ant colony algorithm. All algorithms are implemented by MATLAB 2016Ra programming, Intel (R) CORE(TM)2 Quad CPU Q9400 @2.66 GHz, 2.66 GHz, with the 3.50 GB of physical address expansion of memory, Microsoft Windows XP. It ran 30 times on the computer. In order to ensure the fairness of the comparison experiment, all the parameters in the comparison algorithm are set in strict accordance with the original literature. Parameter setting of the algorithm in this paper: the number of ants $N = 50$, iterations: 100, the relative importance of pheromones is 0.5; the pheromone volatile factor is 0.1.

4.1 Simulation and Results of Environment 1

According to the environment map given in literature [12], using the algorithm in this paper, A* algorithm, ACO and MACO apply to this path diagram, the simulation results are shown in Table 1. It can be seen from Table 1 that the algorithm in this paper is superior to A* in optimization efficiency, ACO and MACO, and the optimization time is the shortest. At the same time, the results of the optimization environment path of this algorithm in Fig. 1 are shown in Fig. 3.

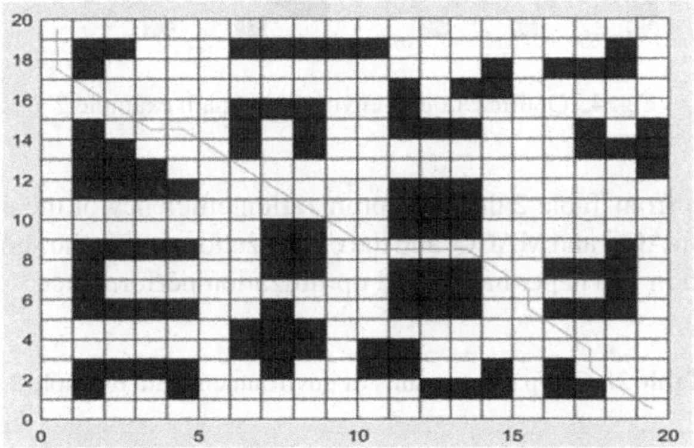

Fig. 3. Optimization of environment path example 1

Table 1. Comparison results of environment path example 1

Algorithm	Average shortest path length	Average running time
A*	31.30	40.15
ACO	30.10	39.06
Reference [12]	29.21	38.20
MACO	29.44	30.84

4.2 Simulation and Results of Environment 2

According to the environment map given in reference [11], the algorithm in this paper, ACO and MACO are applied to the path planning diagram the simulation results are shown in Table 2. Meanwhile, the results of the optimized environment path of the algorithm in Fig. 2 are shown in Fig. 4.

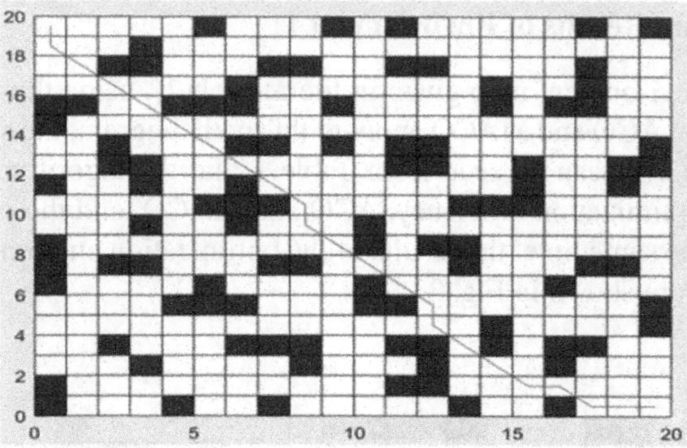

Fig. 4. Optimization of environment path example 2

It can be seen from Table 2 that the optimization efficiency of the algorithm in this paper is better than ACO and MACO, and the optimization time is the shortest, indicating that the algorithm in this paper has a good optimization performance.

Table 2. Comparison results of environment path example 2

Algorithm	Average shortest path length	Average running time
ACO	36.1	46.3
Reference [11]	35.8	32.4
MACO	29.7	27.1

5 Conclusion

In this paper, an improved ant colony algorithm (MACO) is proposed based on grid map modeling. By setting the upper limit of nodes, the number of path nodes is compared with the number of existing path nodes as the number of iterations increases, if the limit is exceeded, the ants are weeded out. The results show that the proposed ant colony algorithm can not only ensure the global optimization ability of the ant colony algorithm, but also improve the efficiency of path planning.

References

1. Ge, S., Cui, J.: New potential functions for mobile robot path planning. IEEE Trans. Robot. Autom. **16**(5), 615–620 (2000)
2. Mo, H., Xu, L.: Research of biogeography particle swarm optimization for robot path planning. Neurocomputing **148**, 91–99 (2015)
3. Mac, T., Copot, C., Tran, T., et al.: Heuristic approaches in robot path planning: a survey. Robot. Auton. Syst. **86**, 13–28 (2016)

4. Zhang, J., Zhao, Z., Liu, D.: A path planning method for mobile robot based on artificial potential field. J. Harbin Inst. Technol. **38**(8), 1306–1309 (2006)
5. van Laarhoven, P.J.M., Aarts, E.H.L.: Simulated annealing. In: van Laarhoven, P.J.M., Aarts, E.H.L. (eds.) Simulated Annealing: Theory and Applications. MAIA, vol. 37, pp. 7–15. Springer, Dordrecht (1987). https://doi.org/10.1007/978-94-015-7744-1_2
6. Qing, L., Chao, Z., Caiwei, H., et al.: Path planning based on fuzzy logic algorithm for mobile robots in dynamic environments. J. Central South Univ. (Sci. Technol.) **44**(s2), 104–108 (2013)
7. Dorigo, M., Gambardella, M.: Ant colony system: a cooperative learning approach to the traveling salesman problem. IEEE Trans. Evol. Comput. **1**(1), 53–66 (1997)
8. Delmerico, J., Mueggler, E., Nitsch, J., et al.: Active autonomous aerial exploration for ground robot path planning. IEEE Robot. Autom. Lett. **2**(2), 664–671 (2017)
9. Fu, B., Chen, L., Zhou, Y., et al.: An improved A* algorithm for the industrial robot path planning with high success rate and short length. Robot. Auton. Syst. **106**, 26–37 (2018)
10. Yu, J., Rus, D.: An effective algorithmic framework for near optimal multi-robot path planning. In: Bicchi, A., Burgard, W. (eds.) Robotics Research. SPAR, vol. 2, pp. 495–511. Springer, Cham (2018). https://doi.org/10.1007/978-3-319-51532-8_30
11. Deng, G., Zhang, X., Liu, Y.: Ant colony optimization and particle swarm optimization for robot-path planning in obstacle environment. Control Theory Appl. **26**(08), 879–883 (2009)
12. Xinlei, H., Lianzhi, Y.: Mobile robot path planning based on improved ant colony algorithm. Softw. Guide **12**, 162–164 (2017)

Adaptive Brain Storm Optimization Based on Learning Automata

Yan Xu, LianBo Ma(✉), and Mingli Shi

Software college, Northeastern University, Shenyang, China
malb@swc.neu.edu.cns

Abstract. Brain storm optimization algorithm is a new swarm intelligence algorithm based on the simulation of the human brainstorming process. However, the original BSO may suffer from getting trapped into local optima, due to its fixed search pattern and fixed step-size disturbance mode. Aiming to address this issue, this paper proposes Adaptive Brain Storm Optimization Based on Learning Automata called LABSO to resolve complex optimization problems. The main idea of LABSO is to use the learning automata mechanism to adaptively select optimal parameters and related search strategy for each idea at different search stages of BSO, which can improve the adaptability and flexibility of the algorithm. Especially, the learning automaton engine is apt to select a convergent operation with a large probability in the early search stage, and to choose a divergent operation in the later search stage. The update of the selection probabilities of candidate strategies is not fixed but learned from the evolution process by the learning automaton. Experimental results on a set of CEC2017 benchmark functions have verified the outstanding performance of LABSO in comparison with several representative BSO variants.

Keywords: Evolutionary algorithms · Brain storm optimization · Learning automata · Swarm intelligence

1 Introduction

Swarm intelligence (SI) is an artificial intelligence technique based on the study of behavior of simple individuals, which is derived from the simulation of intelligent behaviors of social insects or animals in the ecosystem. Swarm Intelligence and bio-inspired computation have become increasing popular in the last two decades. Bio-inspired algorithms such as ant colony system(ACS) [1], particle swarm optimization (PSO) [2,3], artificial bee colony algorithm (ABC) [4–7] and cuckoo search have been applied in almost every area of science. In most SI

Supported by the National Natural Science Foundation of China under Grant No. 61773103, the Fundamental Research Funds for the Central Universities No. N180408019 and N181713002, and the Program for Liaoning Innovative Research Team in University under Grant No. LT2016007.

L. Pan et al. (Eds.): BIC-TA 2019, CCIS 1159, pp. 98–108, 2020.
https://doi.org/10.1007/978-981-15-3425-6_9

algorithms, the learning strategies are usually applied for neighborhood information exchange between individuals in the population to emerge cooperative intelligence. Hence, the learning mechanism of the individuals in the population plays a significant role on the algorithm performance. Brain storm optimization (BSO) is a young and promising swarm intelligence algorithm originally proposed by Y. Shi [8,9]. Unlike the other swarm intelligence algorithms simulating collective behavior of simple insects. BSO is based on the collective behavior of human being, that is, the brainstorming process [10]. As for the optimization problem, each position within the searching space can be regarded as an idea. In each generation of the evolution, the ideas are gathered into separate groups by k-means clustering operation, while the superior one is the cluster center of each group. New individual can be generated based on one or two individuals in clusters. In addition, the updating of ideas is accomplished by adding the Gaussian factor or combining with ideas from other clusters. BSO has previously proven itself as a worthy competitor to its better known rivals.

Learning automata [10] is one of a machine learning algorithms [11]. A learning automata [12], is a machine which can perform finite number of actions. Each selected action is evaluated by a probability environment and evaluated result in the form of a positive or negative signal is given to learning automata and learning automata uses this response to select next action and thus approaches to selection of an action which gets the most reward from environment. In other words, automata learn the action which receives the most reward from the environment.

In an attempt to enhance the performance of BSO. We introduce learning automatas for every ideas; the main contributions of this paper including the follow:

- Using learning automatas mechanism in new individual generation strategy, Each particle adaptively chooses whether its offspring is generated from one or more clusters, depending on its situation, instead of a fixed probability.
- Three dynamic searching strategies are adaptively used in different search stages to control the path and velocity of the ideas.

The rest of this paper is organized as follows. Section 2 reviews the related work of BSO and Learning Automata. Section 3 describes LABSO in detail. Section 4 presents the experimental settings and comparison results on 14 benchmark functions with its peers. Conclusions are drawn in Sect. 5.

2 Related Work

2.1 Brain Storm Optimization Algorithm

It's in [13] and [14], the process of BSO can be described as follows.

In BSO algorithm, an individual is represented by an idea, denoted as Xi = [xi1,xi2,...,xiD], where i = 1,2,...,m, and D is the dimension of the optimization problem to be solved. Then each idea in turn represents a potential solution for the problem to be solved, and its fitness value f(Xi) is evaluated. During each generation, all the ideas are clustered into several clusters using k-means clustering method, and the best idea in each cluster is chosen as the cluster center. In the updating individuals operation [15], newly generated individuals are expected to move towards better and better search areas. BSO first randomly chooses one cluster or two. Then the cluster center, which has higher priority or another idea in the cluster, is selected. Afterward, the selected idea(s) is updated according to (1):

$$X_{new} = X_{old} + \xi N(\mu, \sigma) \tag{1}$$

$$\mathrm{X}_{old} = \begin{cases} X_{ij}, & one\ cluster \\ \omega_1 X_{i1,j} + \omega_2 X_{i2,j}, & two\ clusters \end{cases} \tag{2}$$

where is the Gaussian random value with the mean and variance. 1 and 2 are weight values of the two ideas. is an adjusting factor, as expressed by

$$\xi = rand() \times \log sig(\frac{N_{iter\,\max}/2 - N_{iter}}{K}) \tag{3}$$

where logsig() is a logarithmic sigmoid transfer function, Nitermax is the maximum number of iterations, and Niter is the current iteration number, K is for changing logsig() functions slope, and rand() returns a random value within (0, 1).

2.2 Learning Automata(LA)

The learning automata (LA) approach can be used to determine an optimal action from a set of actions [16]. It selects an action from its finite set of actions. As shown in Fig. 1, each selected action is evaluated by a probabilistic environment and the result of evaluation as a feedback, returns to the automaton in form of a positive or negative signal, and consequently the automaton takes this signal into account, in the next action selection phase. LA with a variable structure can be presented by a quadruple, P,T, where =1,2,...,r, is the set of actions; = 1,2,...,s is the set of inputs; P = P1,P2,...,Pr is the action probability vector and T is a scheme of a learning algorithm, P(t+1) = T((t),(t),P(t)).

Then, we present a typical linear learning algorithm as follows. Assume that the action i is chosen at time k as a sample realization from distribution (k). By considering 0,1, A linear schema for updating probability vector of LA with r actions is defined as (4) when =0 (favorable answer) and (5) when =1 (unfavorable answer).

$$\begin{array}{ll} Pj(k+1) = Pj(k) + a[1 - Pj(k)] & i == j \\ Pj(k+1) = (1-a)Pj(k) & \forall j \;\; i == j \end{array} \tag{4}$$

$$\begin{array}{ll} Pj(k+1) = (b/r - 1) + (1-b)Pj(k) & i == j \\ Pj(k+1) = (1-b)Pj(k) & \forall j \;\; i \neq j \end{array} \tag{5}$$

Where a,b are respectively the reward and penalty parameters. If a and b are equal, the learning scheme is called LR-P (Linear Reward-Penalty). If the learning parameter b is set to 0, then the learning scheme is named LR-l (Linear

Reward-Inaction). And finally if the learning parameter b is much smaller than a, the learning scheme is called LRεP (Linear Reward-epsilon-Penalty).

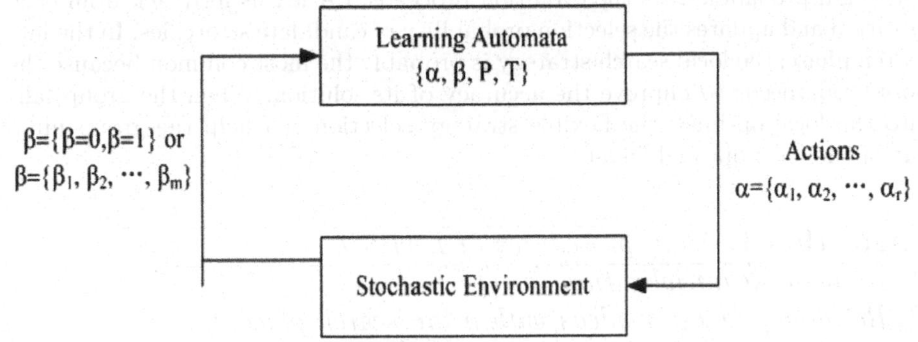

Fig. 1. Framework of LA

3 Adaptive Brain Storm Optimization Based on Learning Automata

3.1 Implementation of Learning Automata

In the proposed algorithms, we use two learning automatas to improve the performance of BSO. The first one is applied to set the probability parameters of the new individual generation strategy. Stochastic search techniques are usually problem dependent. Thus, an efficient parameter setting forms an important part of the algorithm. Generating an individual from one cluster could refine a search region, and it enhances the exploitation ability. On the contrast, an individual, which is generated from two or more clusters, may be far from these clusters.The exploration ability is enhanced under this scenario.Therefore, the selected strategy should be obtained through continuous training of the environment rather than fixed parameter setting. The second one is applied to choose step size strategy. Although the probability of choosing a cluster center is higher than other ideas, such mechanism emphasized the key role of the elite individuals in the swarm, but there seems to be some lack of direction in it. In other words, global information of the entire swarm is not fully utilized. In order to enhance the performance and prevent the individual from not updating after multiple iterations, this paper considered the use of global information and cluster center information base on the step size strategy of the original BSO, and trained by the learning automaton to select the most suitable actions for the current stage through the feedback of the environment.

3.2 Main Steps

Algorithm 1 shows the main procedures of LABSO, which uses dynamic probability to select clusters to produce individuals and splits the search method into 3 sub methods. The probability vector of LA is updated at the end of each evaluation. It can guarantee that the population search strategy is different for each iteration. In the early search phase, LA chooses the global search strategy with high probability. As the evolution proceeds, LA learns feedback from each iteration and updates the selection probability of candidate strategies. In the late search phase, the local search strategy is probably the most common because the population needs to improve the accuracy of its solution. When the group falls into the local optimal, the flexible strategy selection can help the group jump out of the local optimal faster.

Algorithm 1. *Main procedures of LABSO*

1: Population initialization:Init(P)
2: Randomly create n ideas within the search place and valuate the ideas
3: Cluster n ideas by k-means algorithm
4: for each ideas cluster=(one-cluster,two-cluster) do
  Select an action of LAcluster as C
  ;end
5:for each idea Xi, i in [1,N]
6:if C=one-cluster then
  Select a cluster
  choose cluster center or normal individual to generate new individual
else
  Select two cluster
  choose cluster center or normal individual to generate new individual
end-if
7: for each ideas stepSize=step1,step2,step3 do
  Select an action of LAsteSize as S
end
8:choose appropriate step size strategy
9: end
10: Evaluate the population
11: Update the probability of LAcluster and LAstepSize
12: end-for

In original BSO, the probability of selecting one cluster is fixed to Pone-cluster = 0.8, and the probability of two clusters is fixed to Ptwo-cluster = 0.2. In BSOLA, the set of cluster is used to store the probability of selecting one or two clusters to generate new individuals, Both of them are set to 0.5. With the LA learning constantly, each individual can choose the strategy flexibly in the process of iteration. The set of stepSize is used to store the search method. If only a single search strategy is used throughout the iteration, its effect may be reduced as the search proceeds. The reasons are as follows: First, the used auxiliary vector is apt to be similar to Sselect, which makes the information loss. Second, a single guidance vector may be biased and lead the individual to converge to a local-optimal region. We construct auxiliary transmission vectors, stepSize = (step1, step2, step3), and defined as below.

- Centralized global and center best search: step1=Sselect+rand()*(CR-Sselect) +rand()*(Sbest-Sselect), where Sbest is the best solution in the population, and CR is a cluster center randomly selected from the M clusters. This vector enables the individual to move quicker towards the global best region or a potentially good direction.
- Explore search: The search strategy in original BSO, Eq. 1.
- Centralized center best search: step3 = Sselect + rand()*(Pi-Pj), where Pi and Pj is an individual from the population. This can assist individuals to diverge and jump out of the local optima.

4 Experiments and Results

4.1 Parameter Settings and Test Functions

To evaluate the performance of LABSO, a set of 14 benchmarks from CEC2017 test beds are employed, including unimodal function F1, F3, multimodal functions F4, F5, F6, F7, hybrid functions F11, F12, F13, F14, and composition functions F21, F25, F27 and F30. In these test functions, F11, F12, F13 and F14 are hybrid functions, in which the variables are randomly divided into several subcomponents and then different functions are used for different subcomponents. The composition functions F21, F25, F27 and F30 are more difficult to be optimized, because they merge the properties of the sub-functions better and maintains continuity around the global and local optima. Parameters of test functions and settings of algorithm are shown in Tables 1 and 2.

The LABSO is compared against several state-of-the-art BSOs, including BSO [13], QBSO [17], PPBSO [18] and BSO-OS [19]. Their parameters refer to the default settings of their original references [14,18,19] [20].

In order to reduce the error, all algorithms use the same population size n and the same number of cluster m, where n is set to 50 and m is set to 5. All the algorithms run 30 times on each benchmark function with dimensions 50. The maximum number of function evaluations is set to 50,000.

Table 1. Parameters of CEC2017 test functions

f	Functions	Dimensions	Initial Range	X*	f(x*)
f1	Shifted and Rotated Bent Cigar Function	50	$[-100,100]^D$	O1	100
f3	Shifted and Rotated Zakharov Function	50	$[-100,100]^D$	O3	300
f4	Shifted and Rotated Rosenbrocks Function	50	$[-100,100]^D$	O4	400
f5	Shifted and Rotated Rastrigins Function	50	$[-100,100]^D$	O5	500
f6	Shifted and Rotated Expanded Scaffers F6?Function	50	$[-100,100]^D$	O6	600
f7	Shifted and Rotated Lunacek Bi_Rastrigin?Function	50	$[-100,100]^D$	O7	700
f11	Hybrid Function 1 (N = 3)	50	$[-100,100]^D$	O11	1100
f12	Hybrid Function 2 (N = 3)	50	$[-100,100]^D$	O12	1200
f13	Hybrid Function 3 (N = 3)	50	$[-100,100]^D$	O13	1300
f14	Hybrid Function 4 (N = 4)	50	$[-100,100]^D$	O14	1400
f21	Composition Function 1 (N = 3)	50	$[-100,100]^D$	O21	2100
f25	Composition Function 5 (N = 5)	50	$[-100,100]^D$	O25	2500
f27	Composition Function 7 (N = 6)	50	$[-100,100]^D$	O27	2700
f30	Composition Function 10 (N = 3)	50	$[-100,100]^D$	O30	3000
Search range: $[-100,100]^D$					

Table 2. Parameters of involved algorithm

Algorithm	Parameters settings	Reference
QBSO	N = 100, m = 5, Preplace = 0.2, Pone = 0.8, Pone-center = 0.4, Ptwo-center = 0.5, K = 25	[9]
PPBSO	N=100, m = 5, Preplace = 0.2, Pone = 0.8, Pone-center = 0.4, Ptwo-center = 0.5, Wpredator = 0.05, Pprey = 0.1	[10]
BSO-OS	N = 100, Perce = 0.1, Preplace = 0.2, Pone = 0.8	[11]
BSO	N = 100, m = 5, Preplace = 0.2, Pone = 0.8, Pone-center = 0.4, Ptwo-center = 0.5	[5]
LABSO	N = 100, m = 5, Preplace = 0.2, Pone-center = 0.5, Ptwo-center = 0.5	–

4.2 Result and Discussion

From Table 3, it is obvious that LABSO can obtain satisfactory results on most benchmark functions. The results show that LABSO does better than BSO on all the functions considerably on these test functions. Therefore, it also turns out that learning Automata mechanism can indeed improve the performance of BSO.As shown clearly in Fig. 2 the convergence rate of LABSO is the highest among those algorithms. For the two unimodal functions, LABSO obtains the best results among all the five algorithms on f1 and f3. This may be due to that the LA can match the search environment and provide suitable noise mode to create better ideas around the global optimal or cluster center optimal region, which refines the solution for high accuracy. In BSO, according to Eqs. (1) and (3), the new idea is disturbed based on the current idea by Gaussian noise.

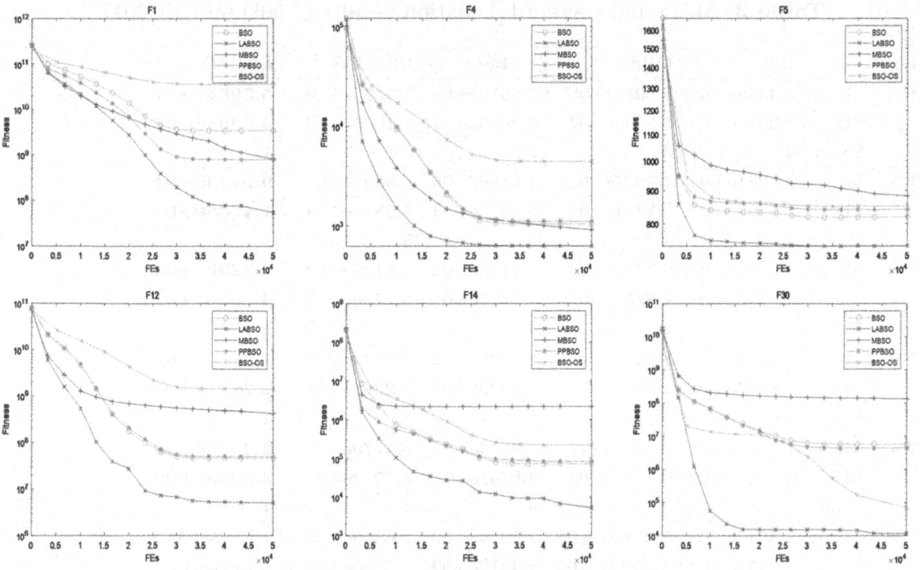

Fig. 2. Convergence curves of different BSO variants

However, this noise may be coarse and is not efficient enough to reach more accurate accuracy. In the contrast, LABSO uses three disturbed noise mode and makes full use of the globally optimal and cluster center optimal information. This may cause two advantages. One is that LABSO can use large disturbed noise in the early phase to accelerate the search speed, and the other is that LABSO can use small disturbed noise in the late phase to refine the finial solution.

For the four multimodal functions, Table 3 also shows that LABSO is the best algorithm for f4-f6 and the second best algorithm for f7. The good performance of LABSO on multimodal functions may be due to use LA to choose the strategy to generate new individuals. When the algorithm being trapped by the current ideas(trapped by the current local optimum). LA mechanism will increase the probability of choosing two clusters to generate new individuals, to increase the exploration ability. Moreover, once a new better search region appears, the three disturbed noise mode of LA mechanism can make the algorithm fast converge to this new region and to refine the solution again. Therefore, LABSO not only can avoid local optima of multimodal functions, but also can obtain very high accurate solution to these functions.

With regards to hybrid and composite functions, because of more number of local optimum solutions prevalent in these solutions, it is easy to get the algorithm to fall into local optimum. Due to the Learning Automata mechanism, LABSO has the ability of jumping out of local optima, and find a better position with higher efficiency. LABSO performs better in the environment of high dimension and high evaluation times, because with the feedback of the environment, LA mechanism can more clearly find the search strategy suitable for each idea.

Table 3. Mean and standard deviation results of 50D on CEC2017

Function		BSO	QBSO	PPBSO	BSO-OS	LABSO
F1	Mean	3.4358e+09	8.0015e+08	7.8612e+08	3.5675e+10	**5.3984e+07**
	SD	1.2016e+09	5.5400e+07	3.9680e+07	1.1635e+10	**5.3964e+07**
	Rank	4	3	2	5	1
F3	Mean	1.0231e+05	1.2937e+05	1.2136e+05	1.9628e+05	**9.2423e+04**
	SD	1.9509e+03	2.7831e+04	1.1917e+04	7.9878e+04	**1.4708e+03**
	Rank	2	4	3	5	1
F4	Mean	1.0517e+03	9.1988e+02	1.1159e+03	4.4126e+03	**6.3259e+02**
	SD	1.4163e+02	1.8757e+01	2.7127e+02	3.7730e+02	**4.4096e+01**
	Rank	3	2	4	5	1
F5	Mean	8.1839e+02	8.8247e+02	8.4128e+02	8.5172e+02	**7.3880e+02**
	SD	6.9645e+00	2.3735e+01	1.5922e+01	2.2388e+01	**2.2883e+01**
	Rank	2	5	3	4	1
F6	Mean	6.4515e+02	6.5382e+02	6.4914e+02	6.4770e+02	**6.4399e+02**
	SD	1.9249e+00	2.7177e+00	4.0741e-01	3.7211e-01	**3.8229e+00**
	Rank	2	5	4	3	1
F7	Mean	1.8912e+03	**1.4161e+03**	2.0984e+03	2.0340e+03	1.8219e+03
	SD	1.9172e+02	**1.0171e+02**	2.4312e+02	2.5900e+01	1.1995e+01
	Rank	3	1	5	4	2
F11	Mean	2.5388e+03	4.5389e+04	5.1929e+03	1.7690e+04	**1.7837e+03**
	SD	9.2422e+02	1.4797e+03	4.8311e+02	2.5614e+03	**8.5643e+00**
	Rank	2	5	3	4	1
F12	Mean	4.6583e+07	4.3471e+08	5.0801e+07	1.4417e+09	**5.0721e+06**
	SD	2.2396e+06	4.7242e+07	1.5457e+07	6.0230e+08	**2.0709e+06**
	Rank	2	4	3	5	1
F13	Mean	1.1802e+05	4.2751e+06	8.2001e+04	8.0027e+04	**6.1422e+04**
	SD	1.5927e+04	2.7339e+05	2.4894e+04	1.0536e+03	**2.3715e+04**
	Rank	4	5	3	2	1
F14	Mean	7.1286e+04	2.1585e+06	8.2569e+04	2.1906e+05	**5.2529e+03**
	SD	1.2111e+04	1.0953e+06	6.1791e+03	5.3911e+04	**4.2453e+02**
	Rank	2	5	3	4	1
F21	Mean	2.5130e+03	2.5941e+03	2.4802e+03	**2.3754e+03**	2.4736e+03
	SD	5.6901e+01	4.8502e+01	3.0428e+01	**7.5350e+01**	1.9767e+01
	Rank	4	5	3	1	2
F25	Mean	3.2767e+03	3.4890e+03	3.3295e+03	**2.7029e+03**	3.2417e+03
	SD	2.1521e+01	6.7793e+01	5.6767e+01	**1.0618e+00**	3.1600e+01
	Rank	3	5	4	1	2
F27	Mean	8.3614e+03	3.8266e+03	3.2000e+03	9.4115e+03	**3.2000e+03**
	SD	2.6135e+02	5.0567e+02	1.5566e-04	5.2368e+02	**1.0950e-04**
	Rank	4	3	2	5	1
F30	Mean	5.7510e+06	1.3340e+08	4.4038e+06	7.0314e+04	**1.1693e+04**
	SD	3.3206e+06	3.7443e+07	5.0380e+05	1.3078e+04	**1.8363e+03**
	Rank	4	5	3	2	1

4.3 Computational Complexity

LABSO has shown a superior ability for a majority of benchmark functions. In this subsection, we calculate its computational time complexity together with BSOs. The time complexity in each procedure of BSO is described as follows: In BSO, the time complexity for initializing is O(N) where N is the population size. Evaluating the fitness of population is O(N). Using K-means to divide the population into c clusters needs $O(cN^2)$. The process of individual selection and step length generation both cost $O(N^2)$. The generation of new individuals and the fitness calculation need $O(N^2)$, respectively. Thus, the overall time complexity of BSO is

$$O(N) + O(N) + O(cN^2) + O(N^2) + O(N^2) = 2O(N^2) + O(cN^2) + 2O(N)$$

To be simplified, its overall time complexity is $O(N^2)$. LABSO is modified based on BSO. The obvious difference is: Using learning automata to make decisions $O(3N^2)$. Thus, the overall time complexity of LABSO is

$$O(N) + O(N) + O(cN^2) + O(3N^2) + O(N^2) = O(cN^2) + 2O(N) + O(4N^2)$$

The overall time complexity of LABSO can be seen as $O(N^2)$. The main differences between LABSO and BSO are in search strategy. As LABSO uses learning automata to applies multiple step length strategies, it costs $O(3N^2)$ which is greater than $O(N^2)$ of BSO, Thus, LABSO and BSO have the same time complexity, which indicates that both are competitive in computational efficiency.

5 Conclusion

In this paper, a novel BSO variant called LABSO is proposed based on LA. LA is used to choose the best search strategy for the current stage and improve the search efficiency. In LABSO, the learning automatas are used in new individual generation process and step size disturbance process. The learning automata for new individual generation can maximize exploration and exploitation capabilities, and the learning automata for step size disturbance mode can utilize the information between global and cluster center individuals in an efficient manner. These two operations can alleviate the evolution stagnation of clusters.

Experiments have been conducted on a set of the CEC2017 benchmark functions where LABSO is compared with several BSO algorithms. Experimental results demonstrate that the LABSO generally performs better than other BSO algorithms in terms of the accuracy and convergence on multimodal and unimodal test problem, and LABSO also obtains satisfactory results on more complex hybrid and composition functions.

In the future, the LABSO will be compared with more state-of-the-art SI algorithms on different test functions. Multiple alternative strategies can help better jump out of local optima. We should add different strategies to LA to cope with the unknown environment and make the algorithm more flexible. Moreover, the proposed LABSO should be applied for resolving complex real-world problems.

References

1. Dorigo, M., Gambardella, L.M.: Ant colony system: a cooperative learning approach to the traveling salesman problem. IEEE Trans. Evol. Comput. **1**(1), 53–56 (1997)
2. Eberhart, Shi, Y.: Particle swarm optimization: developments, applications and resources. In: Congress on Evolutionary Computation (2002)
3. Ma, L., Wang, X., Huang, M., Lin, Z., Tian, L., Chen, H.: Two-level master-slave RFID networks planning via hybrid multi-objective artificial bee colony optimizer. IEEE Trans. Syst. Man Cybern.: Syst. **49**(5), 861–880 (2019)
4. Shi, Y.H., Eberhart, R.C.: Empirical study of particle swarm optimization. In: Congress on Evolutionary Computation (2002)
5. Ma, L., Wang, R., Chen, M., Wang, X., Cheng, S., Shi, Y.: A novel many-objective evolutionary algorithm based on transfer learning with Kriging model. Inf. Sci. (2019). https://doi.org/10.1016/j.ins.2019.01.030
6. Ma, L., Hu, K., Zhu, Y., Chen, H.: Cooperative artificial bee colony algorithm for multi-objective RFID network planning. J. Netw. Comput. Appl. **42**, 143–162 (2014)
7. Ma, L., Zhu, Y., Liu, Y., Tian, L., et al.: A novel bionic algorithm inspired by plant root foraging behaviors. Appl. Soft Comput. **37**, 95–133 (2015)
8. Li, B., Tang, K., Li, J., et al.: Stochastic ranking algorithm for many-objective optimization based on multiple indicators. IEEE Trans. Evol. Comput. **20**, 924–938 (2016)
9. Laumanns, M., Thiele, L., Deb, K., Zitzler, E.: Combining convergence and diversity in evolutionary multiobjective optimization. Evol. Comput. **10**(3), 263–282 (2002)
10. Narendra, K.S., Thathachar, M.A.: Learning automata - a survey. IEEE Trans. Syst. Man Cybern. **4**, 323–334 (1974)
11. Ma, L., et al.: Cooperative two-engine multi-objective bee foraging algorithm with reinforcement learning. Knowl.-Based Syst. **133**, 278–293 (2017)
12. Narendra, K.S., Thathachar, M.A.L.: Learning Automata: An Overview. Prentice Hall, Upper Saddle River (1989)
13. Shi, Y.: Brain storm optimization algorithm. In: Tan, Y., Shi, Y., Chai, Y., Wang, G. (eds.) ICSI 2011. LNCS, vol. 6728, pp. 303–309. Springer, Heidelberg (2011). https://doi.org/10.1007/978-3-642-21515-5_36
14. Shi, Y.: An optimization algorithm based on brainstorming process. Int. J. Swarm Intell. Res. **2**(4), 35–62 (2011)
15. Zhan, Z., Zhang, J., Shi, Y., Liu, H.: A modified brain storm optimization. In: Proceedings of2012 IEEE World Congress Computational Intelligence, Brisbane, Australia, 10–15 June, pp. 1–8 (2012)
16. Narendra, K.S., Thathachar, M.A.L.: Learning Automata: An Introduction. Prentice-Hall Inc., Upper Saddle River (1989)
17. Duan, H., Li, C.: Quantum-behaved brain storm optimization approach to solving loneys solenoid problem. IEEE Trans. Mag. **51**(1), 1–7 (2015)
18. Duan, H., Li, S., Shi, Y.: Predator-prey brain storm optimization for DC brushless motor. IEEE Trans. Mag. **49**(10), 5336–5340 (2013)
19. Shi, Y.: Brain storm optimization algorithm in objective space. In: Congress on Evolutionary Computation (2015)

A Reference Point-Based Evolutionary Algorithm for Many-Objective Fuzzy Portfolio Selection

Jian Chen[1,2], Xiaoliang Ma[1,2], Yiwen Sun[3], and Zexuan Zhu[1,2(✉)]

[1] College of Computer Science and Software Engineering, Shenzhen University, Shenzhen 518060, China

[2] Guangdong Laboratory of Artificial Intelligence and Digital Economy (SZ), Shenzhen University, Shenzhen 518060, China

zhuzx@szu.edu.cn

[3] School of Medicine, Shenzhen University, Shenzhen 518060, China

Abstract. Portfolio selection is an important problem in the practice and theory of finance. This paper uses a five-objective (including mean, variance, skewness, kurtosis, and entropy) model to replace the classical Markowitz mean-variance model for finding better portfolio selection. To obtain a more accurate estimation of risk asset returns, a fuzzy number variable, instead of a random variable, based on the acknowledge of experts is used to estimate the return of a risk asset. A new reference point-based evolutionary algorithm (NRPEA) is proposed to obtain well-convergence and well-distributed solutions for the many-objective optimization problems. In NRPEA, the auxiliary reference points are generated and selected to guide the population evolution. Experiment results on six well-known data sets demonstrate the effectiveness and efficiency of NRPEA in the comparison with other three state-of-the-art many-objective optimization algorithms.

Keywords: Fuzzy portfolio selection · Many-objective optimization · High-order moment · Reference point

---

This work was supported in part by the National Natural Science Foundation of China, under Grants 61976143, 61471246, 61603259, and 61871272, Guangdong Special Support Program of Top-notch Young Professionals, under Grant 2014TQ01X273, Shenzhen Fundamental Research Program, under Grant JCYJ20170302154328155, Scientific Research Foundation of Shenzhen University for Newly-introduced Teachers, under Grant 2019048, and Zhejiang Lab's International Talent Fund for Young Professionals. This work was supported by the National Engineering Laboratory for Big Data System Computing Technology, Shenzhen University, Shenzhen 518060, China and Guangdong Laboratory of Artificial Intelligence and Digital Economy(SZ), Shenzhen University, Shenzhen 518060, China.

L. Pan et al. (Eds.): BIC-TA 2019, CCIS 1159, pp. 109–123, 2020.
https://doi.org/10.1007/978-981-15-3425-6_10

1 Introduction

Portfolio selection problem is one hot topic in economic theory research. Modern portfolio selection problem stems from the Markowitz mean-variance model [1], which is a bi-objective optimization problem via maximizing the return and minimizing the risk simultaneously. The mean-variance portfolio model assumes an underlying normal probability distribution or quadratic utility function on the asset return [2–5]. However, many studies show that returns on assets tend to have a heavy-tailed and asymmetric leptokurtic distribution [2], i.e., not a normal distribution in statistics [2]. This implies that the higher-order moments should be considered. Recently, 3-order moment or 4-order moment of asset return have been considered in [3–5] and shown to be helpful for portfolio selection problem. Shannon's entropy can also be used to measure the dispersion degree of assets for effective risk diversification [6,7].

Besides high-order moments, uncertainty is another key factor in portfolio selection problem. The majority of the current portfolio selection problems over the decades are based on probability theory. However, the uncertain returns of risk assets have many influence factors, such as politics, social conditions, economic change, and the status of the related company. Compared with a random variable, a fuzzy number variable based on the acknowledge of experts is more suitable for estimating the return of a risk asset [4]. Thus, a few researchers have extended the probabilistic portfolio model to the fuzzy environment [4–6].

To find a better portfolio selection,this paper uses a five-objective fuzzy portfolio selection model [18]. The portfolio selection problem is formulated as a five-objective optimization problem. On top of the traditional 1-order and 2-order moments (mean and variance), the model also considers objectives including 3-order and 4-order moments (skewness and kurtosis) and the mean proportional entropy. The mean proportional entropy based on the sum of Minkowski distance between weights of the invested assets is used to acquire a well-diversified portfolio. The fuzzy variable based on the acknowledge of experts is introduced to handle the uncertain return of risk asset in the portfolio selection problem. Pareto dominance-based multi-objective evolutionary algorithms (MOEAs) have been widely used to solve multi-objective optimization problems. However, traditional MOEAs might have difficulty in dealing with many-objective optimization problems (with number of objectives greater than three) [8–14]. This paper introduces a new reference point-based evolutionary algorithm (NRPEA) to deal with the five-objective portfolio selection problem. The auxiliary reference points in NRPEA are generated and selected to guide the population evolution for five-objective optimization. Experimental results on real-world data sets show that the proposed NRPEA can obtain superior or comparable performance to the other state-of-the-art many-objective evolutionary algorithms.

In the rest of this paper, the related background acknowledge is introduced in Sect. 2. Section 3 presents the used five-objective fuzzy portfolio selection problem. The details of the proposed NRPEA are described in Sect. 4. Section 5 presents the experimental studies. Finally, Sect. 6 concludes this paper.

2 Background

This section introduces the background of fuzzy theory, portfolio selection model, and many-objective portfolio selection model.

2.1 Fuzzy Numbers and Weighted Possibilistic Moments

Definition 1. A fuzzy number $\tilde{A}$ is defined as any fuzzy subset of R, whose membership function $\mu_{\tilde{A}}(x)$ meets the following four conditions: (i) $\tilde{A}$ is normal, i.e., $\exists x \in R$, s.t. $\mu_{\tilde{A}}(x) = 1$; (ii) $\mu_{\tilde{A}}(x)$ is quasi-concave, i.e., $\mu_{\tilde{A}}(\lambda x + (1-\lambda)y) \leq \min\{\mu_{\tilde{A}}(x), \mu_{\tilde{A}}(y)\}$ for $\forall \lambda \in [0,1]$; (iii) $\mu_{\tilde{A}}(x)$ is upper semicontinuous, i.e., $\{x \in R | \mu_{\tilde{A}}(x) \leq \epsilon\}$ is a closed set for any $\epsilon \in [0,1]$; and (iv) the closure of set $\{x \in R | \mu_{\tilde{A}}(x) > 0\}$ is compact.

Definition 2. Given the left width θ, the right width δ and the core $[c, d]$, a fuzzy number $\tilde{A}$ is a trapezoidal fuzzy number if its membership function is of the following form:

$$\mu_{\tilde{A}}(x) = \begin{cases} 1 - \frac{c-x}{\delta}, & \text{if } c - \delta \leq x \leq c \\ 1, & \text{if } c \leq x \leq d \\ 1 - \frac{x-d}{\theta}, & \text{if } d \leq x \leq d + \theta \\ 0, & \text{if otherwise} \end{cases} \tag{1}$$

The trapezoidal fuzzy number $\tilde{A}$ can be denoted by $\tilde{A} = (c, d, \sigma, \theta)$.

Definition 3. γ-level set of the trapezoidal fuzzy number $\tilde{A}$ is defined as $[\underline{a}(\gamma), \overline{a}(\gamma)] = \{x \in R | \mu_{\tilde{A}}(x) \geq \gamma\}$, where $\gamma \in [0,1]$.

Definition 4. Suppose $\tilde{A}$ be a fuzzy number having $[\underline{a}(\gamma), \overline{a}(\gamma)], \gamma \in [0,1]$.

The weighted possibilistic moment (WPM) and the weighted possibilistic variance (WPV) of the fuzzy number $\tilde{A}$ are defined as follows:

$$E_f(\widetilde{A}) = \int_0^1 f(\gamma)((\underline{a}(\gamma) + \overline{a}(\gamma))/2) d\gamma, \tag{2}$$

$$Var_f(\widetilde{A}) = \frac{1}{2} \int_0^1 f(\gamma) \cdot \left[(\underline{a}(\gamma) - E)^2 + (\overline{a}(\gamma) - E)^2 \right] d\gamma, \tag{3}$$

where $f(\gamma) = (n+1)\gamma^n$ is a probability density function such that $\int_0^1 f(\gamma) d\gamma = 1$. Specifically, WPM (2) is the first-order moment of fuzzy number $\tilde{A}$.

Definition 5. Let $E_l(\tilde{A}), l = 1, ..., L$ be the l-order weighted possibilistic moments of fuzzy number $\tilde{A}$.

The weight possibilistic skewness (WPS) and weight possibilistic kurtosis (WPK) are defined as follows

$$Skew_f(\widetilde{A}) = \text{WPS}(\widetilde{A}) = E_3(\widetilde{A}) / E_2(\widetilde{A})^{1/3} \tag{4}$$

$$Kur_f(\widetilde{A}) = \text{WPK}(\widetilde{A}) = E_4(\widetilde{A}) / \sqrt{E_2(\widetilde{A})} \tag{5}$$

2.2 The Classical Portfolio Selection Model

The basic assumptions of Markowitz's portfolio theory include that (1) investors are risk-averse and pursue maximum expected utility, (2) investors choose their portfolios according to the expected return and variance of return, and (3) all investors are in the same investment period. Markowitz proposed to find the effective portfolios that maximize the return and minimizing the risk simultaneously.

The expected return E is used to measure the return of securities, and the variance σ^2 of return is used to represent the investment risk. The Markowitz optimization model is defined as follows:

$$\begin{aligned} &\max E\left(r_p\right)=\sum w_i r_i \\ &\min \sigma^2\left(r_p\right)=\sum\sum w_i w_j \operatorname{cov}\left(r_i, r_j\right), \end{aligned} \tag{6}$$

where r_p is the portfolio return, r_i, r_j denote the return of assets i and j, respectively, w_i, w_j are the weight of assets i and j in the portfolio, respectively, $\delta^2(r_p)$ denotes the total risk variance of portfolio return, and $cov(r_i, r_j)$ is the covariance between two assets.

2.3 Multi-objective Optimization Problem

Without loss of generality, a multi-objective optimization problem can be defined as follows:

$$\min_{\mathbf{x}\in\Omega} \mathbf{F}(\mathbf{x}) = \left[f_1(\mathbf{x}), \cdots, f_M(\mathbf{x})\right]^T,$$

where Ω is an n-dimensional feasible variable space, $\mathbf{x}$ is an n-dimensional decision vector, and $\mathbf{F}(\mathbf{x}) : \Omega \to R^m$ consists of m objective functions.

(1) ***Pareto dominance:*** Solution $x_1 \in \Omega$ is said to dominate solution $x_2 \in \Omega$ (denoted as $x_1 \succ x_2$) if $\forall m = 1, 2, \cdots, M, f_m(x_1) \leq f_m(x_2)$, and $\exists i = 1, 2, \cdots, M, f_i(x_1) < f_i(x_2)$.
(2) ***Pareto optimal set:*** For solution $x^\star \in S$, if there is no $x' \in S$ satisfying $x' \succ x^\star$, then $x^\star$ is called the Pareto-optimal solution. All Pareto-optimal solutions are called Pareto-optimal set, i.e., $PS = \{x^* | \neg\exists x \in \Omega : x \succ x^*\}$.
(3) ***Pareto front:*** The mapping of the PS on the objective space is known as the Pareto-optimal front (PF), i.e., $PF = \{\mathbf{F}(\mathbf{x}) | \mathbf{x} \in PS\}$.

3 The Five-Objective Portfolio Selection Model and Individual Encoding

For the convenience of introduction, this paper defines the following relevant variables for the portfolio selection problem with n risk assets:

x_i: The rate of the total investment assigned to the risk asset $i, i = 1, 2, \cdots, n$;
k_i: The rate of transaction cost on the risk asset $i, i = 1, 2, \cdots, n$;
$\tilde{r}_i$: The fuzzy return rate on the risk asset $i, i = 1, 2, \cdots, n$;
$\mathbf{x}^0 = (x_1^0, ..., x_n^0)$: The previous portfolio;

The fuzzy portfolio selection problem in this paper is defined as a five-objective optimization model [18].

$$\begin{cases} \max & f_1(\mathbf{x}) = E_f\left(\sum_{i=1}^{n} \tilde{r}_i x_i\right) - \sum_{i=1}^{n} k_i \left|x_i - x_i^0\right| \\ \min & f_2(\mathbf{x}) = Var_f\left(\sum_{i=1}^{n} \tilde{r}_i x_i\right) \\ \max & f_3(\mathbf{x}) = Skew_f\left(\sum_{i=1}^{n} \tilde{r}_i x_i\right) \\ \min & f_4(\mathbf{x}) = Kur_f\left(\sum_{i=1}^{n} \tilde{r}_i x_i\right) \\ \max & f_5(\mathbf{x}) = 2 - \sum_{i=1}^{n-1} |x_i - x_{i+1}| \\ s.t & K_{min} \leq ||\mathbf{x}||_0 \leq K_{\max}, \\ & \sum_{i=1}^{n} x_i = 1, \quad x_i \geq 0, i = 1, 2, \ldots, n \end{cases} \tag{7}$$

In problem (7), $E_f(\cdot), V_f(\cdot), Skewness_f(\cdot)$, and $Kur_f(\cdot)$ are defined in Eqs. (2–5). In $f_1(\mathbf{x})$, the total transaction cost from the previous portfolio $\mathbf{x}^0$ to the current portfolio $\mathbf{x}$ is $\sum_{i=1}^{n} k_i|x_i - x_i^0|$.

The fifth objective function $f_5(\mathbf{x})$ in Eq. (7), i.e., the mean proportional entropy $2 - \sum_{i=1}^{n-1} |x_i - x_{i+1}|$, is introduced in the model to diversify the investment. This entropy is defined based on the Minkowski distance. Compared with Shannon entropy, it has higher diversity and matches a better market asset allocation.

To decentralize the investment risk and control the asset management complexity, the amount of assets usually is subject to a lower limit and an upper limit, i.e., $K_{min} \leq ||\mathbf{x}||_0 \leq K_{max}$ in the constraint of the problem (7). Here, K_{min} and K_{max} are the minimum and maximum amount of the selected assets, respectively. In the optimization of NRPEA, all maximization objectives are transferred to minimization ones by multiplying -1.

4 A New Reference Point-Based Evolutionary Algorithm

This section introduces the proposed new reference point-based evolutionary algorithm (NRPEA) for portfolio selection problem, including the general framework, individual encoding scheme, evolutionary operators, reference point generation, Tchebychev selection, and environmental selection.

4.1 The General Framework

The general framework of the proposed NRPEA is outlined in Algorithm 1. The procedure is similar to most of MOEAs. Firstly, a population P_t is randomly initialized. The non-dominated individuals of P_t is used to set the reference points R_t and calculate the ideal point $\mathbf{z}^*$. Secondly, an offspring population O_t is generated by genetic operators, where the Tchebychev distance $d_{i,j}^{tch}$ of a population individual to a reference point is defined as follows:

$$d_{i,j}^{tch} = \max_{m=1,2,\ldots,M} \omega_m \left(\frac{f_m(x_i) - r_j^m}{f_m^{\max} - f_m^{\min}} \right), \tag{8}$$

where $i = 1, ..., N, j = 1, ..., |R_t|$. r_j is the m-th objective value of the j-th reference point. $f_j max$ and $f_j min$ are respectively the maximum and minimum values of the j-th objective in the population. Thirdly, a clone population P_c is generated by immune clone operator on non-dominated individuals of $P_t \cup O_t$ based on their normalized crowding distance. The number of clones q_i for the i-th individual is denoted as follows:

$$q_i = \lceil N \times \frac{CD(P_t(i))}{\sum_{j=1}^{N} CD(P_t(j))} \rceil.$$

Fourthly, a set of reference points R_t is selected from $P_t \cup P_c \cup O_t$ each generation introduced in Algorithm 2. Finally, N best solutions are chosen based on the reference points as the next parent population introduced in Algorithm 3.

4.2 Individual Encoding Scheme

To deal with the constraints of problem (7), we use an individual encoding scheme including asset weight vector and asset selection vector. Let n be the number of risk assets. The following example assumes $K_{max} = 10$, i.e., the maximum amount of the selected assets is set as 10.

A asset weight vector $\mathbf{w} = (w_1, ..., w_n)$, $w_i \in [0.01, 0.99), \forall i \in \{1, ..., n\}$. *For example,*

w_1	w_2	w_3	w_4	w_5	w_6	w_7	w_8	w_9	w_{10}

A asset selection vector $\mathbf{z} = (z_1, ..., z_n), z_i \in \{0, 1\}$

1	0	0	0	0	1	1	0	1	1

$\Downarrow$ *normalized*

A investment rate $\mathbf{x} = (x_1, ..., x_n), x_i \in [0.01, 0.99)$

x_1	x_2	x_3	x_4	x_5	x_6	x_7	x_8	x_9	x_{10}

The investment rate $\mathbf{x} = (x_1, ..., x_n)$ on each asset is a normalization of the asset weight vector, i.e.,

$$x_i = l_i z_i + \frac{w_i z_i}{\sum_{i=1}^{m} w_i z_i} \left(u_i z_i - \sum_{i=1}^{m} l_i z_i \right), \tag{9}$$

There is a special case $x_i = 0$ if $z_i = 0$. When $z_i = 1$, the investment rate x_i usually has the upper limit u_i and lower limit l_i due to the actual trading volume.

Algorithm 1. New reference point-based archiving evolutionary algorithm (NRPEA).

Input: Many-objective portfolio selection model, a stopping condition, the population size N, the number of evaluations.
Output: The final external archive A;
Set $t = 0$, initialize a population P_t of N individuals, set reference points R_t as non-dominated individuals of P_t, and calculate the ideal point $\mathbf{z}^*$ based on P_t;
While the stopping condition is not met **do**
    Compute the Tchebychev distance $d_{i,j}^{tch}$ of each population individual to each reference point by Eq. (8);
    Generate mating population P_m by tournament selection based on Tchebychev distance $d_i^{tch} = \min_{1 \le j \le |R|} d_{i,j}^{tch}$;
    Apply genetic operators on P_m to generate an offspring population O_t;
    Perform non-dominated sorting for $P_t \cup O_t$;
    Generate a clone population P_c by an immune clone operator on non-dominated individuals of $P_t \cup O_t$;
    Use $P_t \cup P_c \cup O_t$ to update reference points R based on Algorithm 2;
    Conduct Environment selection based on reference points R; //see Algorithm 3;
    $t = t + 1$;
End While
Return P_t;

4.3 Evolutionary Operator with a Heuristic Repair Strategy

The individual expression includes real-valued asset weight vector and binary asset selection vector. The proposed evolutionary operator consists of crossover and mutation operators. The simulated binary crossover (SBX) is used in this paper for the real-valued asset weight vector, while the single-point crossover is used for the binary asset selection vector. The detail is showed in the following example, where $K_{max} = 10$, i.e., the maximum amount of the selected assets is set as 10.

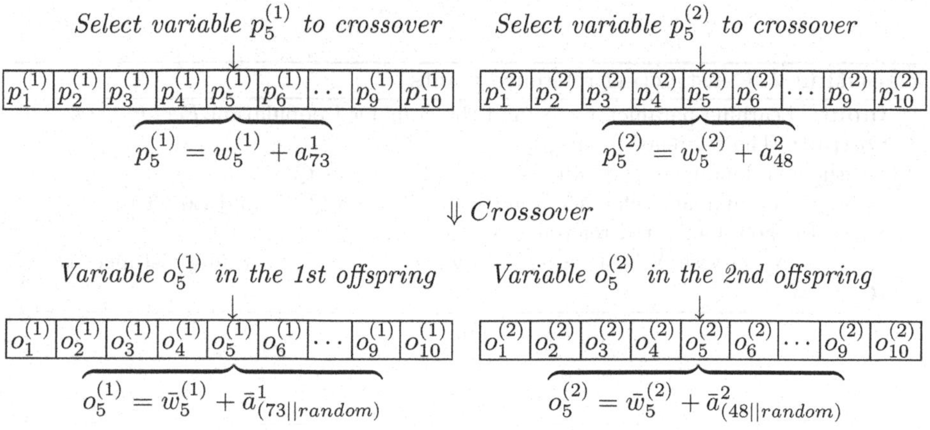

In the above crossover operator, $w_5^{(1)}$ is the 5th asset weight in the 1st parent individual, and a_{73}^1 is the 73-th asset in the 1st parent individual. $\bar{a}^1_{(73||random)}$ indicates that the index of the select asset is 73 mainly and a random integer partly.

The polynomial mutation is used for the real-valued asset weight vector, while the bit-flip mutation is used for the binary asset selection vector.

After crossover operator and/or mutation operator, the same asset may appear in different variables of offspring. Thus, the repeatedly selected asset need to be replaced. To deal with this issue, this paper proposed a heuristic repair strategy based on the occurrence frequency of each asset in the non-dominated archive A, i.e.,$c_i = \frac{\sum_{j=1}^{|A|} s_{i,j}}{|A|}, i = 1, ..., n$.

where $s_{i,j} \in \{0, 1\}$ is the number of the i-th asset appearing in the j-th archived individual. The higher the asset c_i score is, the more likely it will appear in the next generation. According to the occurrence frequency in the non-dominated archive, roulette wheel select is used to select an asset to replace the repeated selected asset in the offspring.

4.4 Constructing Reference Points

In the proposed NRPEA, a set of reference points with good convergence and diversity is constructed to guide the population evolution. Algorithm 2 presents the constructing process of reference points. First, we set Q_t as non-dominated individuals of $P_t \cup P_c \cup O_t$. Second, given a tolerant vector δ, each individual $F(\boldsymbol{x}^i) = [f_1(\boldsymbol{x}^i), ..., f_M(\boldsymbol{x}^i)]$ in set Q_t generates M auxiliary reference points as follows:

$$\mathbf{r}_m^{(i)} = [f_1(\boldsymbol{x}^i), \ldots, f_m(\boldsymbol{x}^i) - \varepsilon_m, f_{m+1}(\boldsymbol{x}^i), \ldots, f_M(\boldsymbol{x}^i)] \tag{10}$$

Thus, $M|Q_t|$ auxiliary reference points are generated as shown in Fig. 1. Finally, N best individuals are selected from the auxiliary reference points as the reference points based on non-dominated sorting for the subsequent environment selection.

Algorithm 2. Constructing reference points

1: **Input:** Learning parameter δ, α, and the combined population $P_t \cup P_c \cup O_t$.
2: **Output:** The reference points R_t;
3: Set the non-dominated individuals of $P_t \cup P_c \cup O_t$ as Q_t;
4: Generate an auxiliary reference point set R base on Q_t, δ, and Eq. (10);
5: Select the non-dominated reference points of R as R_t;
6: If $|R_t| > N$, choose N best individuals with the largest crowding distances as R_t;
7: **Return**R_t

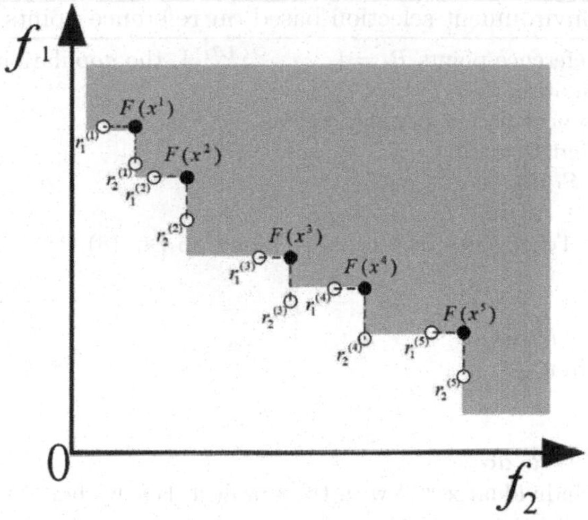

Fig. 1. Generate the auxiliary reference points.

4.5 Environment Selection Based on Reference Points

Algorithm 3 gives a detailed procedure of the environment selection based on reference points. More specifically, the fitness assignment of an individual is based on the distance to reference points R_t. First, the Tchebychev distances (8) between the individuals and the reference points are calculated to evaluate the individuals in lines 2–6. Second, for each reference point, the individual with the smallest Tchebychev distance to the reference point is chosen to constitute the next parent population in lines 9–11. It is noted that the number of reference points R_t may be less than the population size N. To deal with this issue, all reference points are available again to choose the rest individuals in line 8.

5 Experimental Study

In this section, the proposed NRPEA is compared with three state-of-the-art many-objective evolutionary algorithms. All compared algorithms are implemented in MATLAB. Thirty independent runs are performed for each algorithm on each test problem.

5.1 Dataset and Data Processing

Six datasets collected from well-known international stock markets are used to test the performance of NRPEA. They are authoritative test cases for portfolio selection optimization and are publicly available from Pubmed central (PMC) at: https://www.ncbi.nlm.nih.gov/pmc/articles/PMC4959918/. These datasets provide weekly return time series for assets and stock indexes of several major markets across the world. The details of datasets are shown as in Table 1.

Algorithm 3. Environment selection based on reference points

```
Input: The reference points R_t = {r^1, ..., r^{|R_t|}}, the population size N, and the
  combined population S_t = P_t ∪ P_c ∪ O_t;
Output: The next parent population P_{t+1};
Set P_{t+1} = ∅ and R' = R_t;
For each x^i ∈ S_t do
   For each r^j ∈ R_t do
      Calculate Tchebychev distance d_{ij}^{tch} based on Eq. (8).
   End For
End For
While |P_{t+1}| < N do
   If R_t = ∅ then
      R_t = R';
   End If
   For each r^j ∈ R_t do
      Find the individual x^{j_min} with the smallest Tchebychev distance to r^j;
   End For
   R_t = R_t \ {r^j}, P_{t+1} = P_{t+1} ∪ {x^{j_min}}, S_t = S_t \ {x^{j_min}};
End While
Return P_{t+1};
```

For the fuzzy return rates of the assets, we use the statistical indicator of sample percentage to approximate the trapezoidal fuzzy returns distribution [15]. In numerical statistics, the core $[a_i, b_i]$ of the fuzzy return $\tilde{r}_i$ is set to the interval $[40th, 60th]$, and the quantities $40th - 5th$ and $95th - 60th$ as the left α_i and right β_i spreads, respectively, where $k-th$ is the $k-th$ percentile of the sample. The corresponding membership function is defined in *Definition 1*, i.e.,

$$\mu_{\tilde{r}_i}(x) = \begin{cases} 1 - \frac{a_i - x}{\alpha_i}, & \text{if } a_i - \alpha_i \le x \le a_i \\ 1, & \text{if } a_i \le x \le b_i \\ 1 - \frac{x-d}{\beta_i}, & \text{if } b_i \le x \le b_i + \beta_i \\ 0, & \text{if otherwise} \end{cases} \tag{11}$$

At the same time, we calculate the correlation coefficient matrix $r_{i,j}$ among these assets, the average value of weekly return, and the expected return $\bar{r}$ on each asset. The relevant data is used in the learning mechanism to guide the generation of candidate good assets.

5.2 Performance Metric and the Experimental Settings

In this paper, hyper-volume (HV) metric is used as the performance measure to assess the convergence and diversity of the obtained solution set [16], Generally speaking, a larger value of HV metric indicates a better approximation to the true PF. The HV indicator is defined as follows:

$$HV(S) = \text{Volume}\left(\bigcup_{x^i \in S} \left[F(x^i), \mathbf{r} \right] \right), \tag{12}$$

Table 1. This benchmark instances datasets from PMC

	Dataset name	N	|T|	Time interval	Country	Description	# of rebalancing
1	DowJones	28	1363	Feb 1990–Apr 2016	USA	Dow Jones Industrial Average	110
2	NASDAQ100	82	596	Nov 2004–Apr 2016	USA	NASDAQ 100	46
3	FTSE100	83	717	Jul 2002–Apr 2016	UK	FTSE 100	56
4	SP500	442	595	Nov 2004–Apr 2016	USA	S& P 500	46
5	NASDAQComp	1203	685	Feb 2003–Apr 2016	USA	NASDAQ Composite	53
6	FF49Industries	49	2325	Jul 1969–Jul 2015	USA	Fama and French 49 Industry	190

where $[F(x^i), \mathbf{r}]$ indicates a hyper-rectangle formed by the reference point $\mathbf{r}$ and the i-th non-dominated individual $F(x^i)$. To compute the HV metric, the reference point $\mathbf{r}$ is calculated as $1.1 \times (f_1^{max}, f_2^{max}, \cdots, f_m^{max})$, where f_k^{max} $(k = 1, 2, \cdots, m)$ is the maximum value of the kth objective in the true PF.

NRPEA is compared with the other three state-of-the-art MOEAs, i.e., NSGA-III [17], RSEA [18], and SPEAR [12,19]. The population size of the four MOEAs is set to 100, and the maximal function evaluation times is set to 100,000. η_c and η_m indicate the distribution indexes of SBX and PM, respectively. p_c and p_m represent the probabilities of SBX and PM, respectively. More details of the parameter settings for the MOEAs are summarized in Table 2. where CR and F in DE operator use the recommended setting in the proposed paper.

Table 2. Parameter settings of the compared algorithms

Algorithm	Parameter settings
RSEA	$CR = 1.0, F = 0.5, T = 1/N$
	$p_c = 1, p_m = 1/n, \eta_c = 20, \eta_m = 20$
NSGA-III	$p_c = 1, p_m = 1/n, \eta_c = 20, \eta_m = 20$
SPEAR	$p_c = 1, p_m = 1/n, \eta_c = 20, \eta_m = 20$
NRPEA	$\alpha = 0.4, \beta = 0.1$
	$p_c = 1.0, p_m = 1/n, \eta_c = 20, \eta_m = 20$

5.3 Experimental Results

Table 3 shows the mean and standard deviation of the HV-metric values by each algorithm in solving the six portfolio problems. Wilcoxon's rank-sum test is performed to show the statistically significant differences between the results obtained by NRPEA and another algorithm with a 5% significance level. The results show that the proposed NRPEA outperforms the other algorithms in most of the benchmark problems in terms of the HV metric.

Figure 2 reports the evolution process of the mean HV-metric values. It can be seen that the proposed NRPEA obtains the best performance on most of the benchmark problems based on the HV metric.

Table 3. Mean and standard deviation values of the HV metric obtained by the compared algorithm

Problem	NSGAIII	SPEAR	RSEA	NRPEA
TR_port1	2.8517e-1 (4.28e-2) -	2.3560e-1 (3.29e-2) -	2.3790e-1 (4.30e-2) -	**3.3139e-1 (1.39e-2)**
TR_port2	2.0261e-1 (1.46e-2) -	1.6440e-1 (1.07e-2) -	1.5152e-1 (4.04e-2) -	**2.1360e-1 (1.51e-2)**
TR_port3	2.0583e-1 (1.84e-2) -	1.7636e-1 (1.89e-2) -	**2.4176e-1 (3.06e-2)** =	2.2059e-1 (2.31e-2)
TR_port4	1.8954e-1 (2.41e-2) -	1.8246e-1 (3.38e-2) -	1.8869e-1 (3.31e-2) -	**2.3850e-1 (1.58e-2)**
TR_port5	2.1240e-1 (2.62e-2) -	1.8481e-1 (2.15e-2) -	1.9716e-1 (3.92e-2) -	**2.3395e-1 (2.66e-2)**
TR_port6	2.1585e-1 (2.47e-2) -	1.8538e-1 (2.93e-2) -	2.1990e-1 (1.55e-2) =	**2.3297e-1 (2.34e-2)**
+/-/=	0/6/0	0/6/0	1/4/1	-

5.4 Portfolio Performance Evaluation Based on the Improved Sharpe Ratio

Sharpe rate (SR) is usually used to evaluate the performance of the portfolio. It measures the asset return under different risk situations and is calculated by the following form:

$$\mathrm{SR}(\mathrm{r}(\mathbf{x})) = E(r(\mathbf{x}))/\sqrt{Var(r(\mathbf{x}))}, \tag{13}$$

where $r_{\mathbf{x}}$ is the return of a portfolio $\mathbf{x}$. However, the Sharpe ratio based on the mean-variance theory is not accurate when the return distribution appears skew or kurtosis. To deal with this issue, we use an adjusted Sharpe ratio (ASR) [20], which takes into account portfolio skewness and kurtosis:

$$\begin{aligned} ASR(r(\mathbf{x})) = SR(r(\mathbf{x}))[1 + (Skew(r(\mathbf{x}))/6)SR(r(\mathbf{x})) \\ -(Kur(r(\mathbf{x}))/24)SR(r(\mathbf{x}))^2] \end{aligned} \tag{14}$$

Skewness $Skew(r(x))$ and kurtosis $Kur(r(x))$ are two important high-order moments. Different from SR (13), ASR (14) reflects the average expected utility of the portfolio $\mathbf{x}$. The bigger ASR is, the better the performance of the corresponding portfolio $\mathbf{x}$ is.

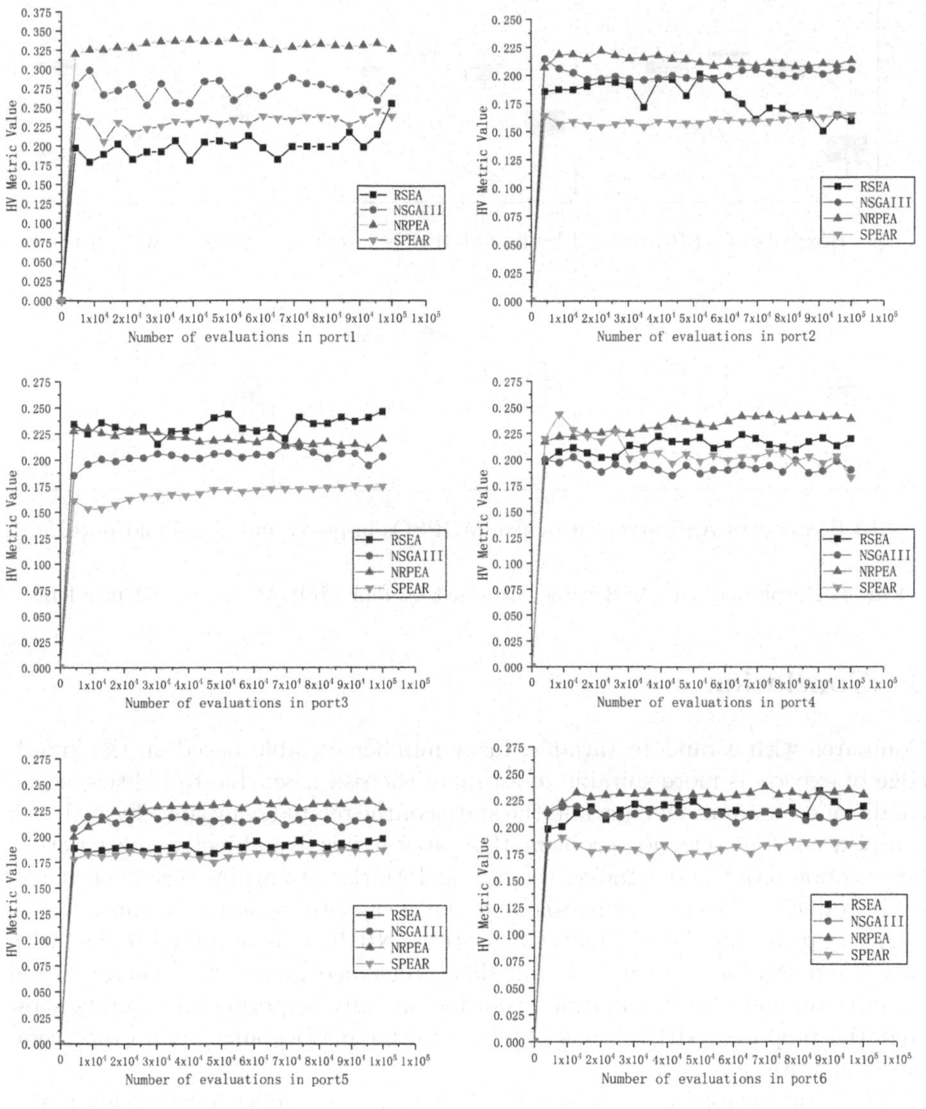

Fig. 2. The evolution process of the mean HV-metric values.

Figure 3 shows boxplots of the ASR values attained by four MOEAs on six benchmark problems. It can be seen that the proposed NRPEA obtains superior or comparable performance to other MOEAs on most of the benchmark problems based on the ASR values.

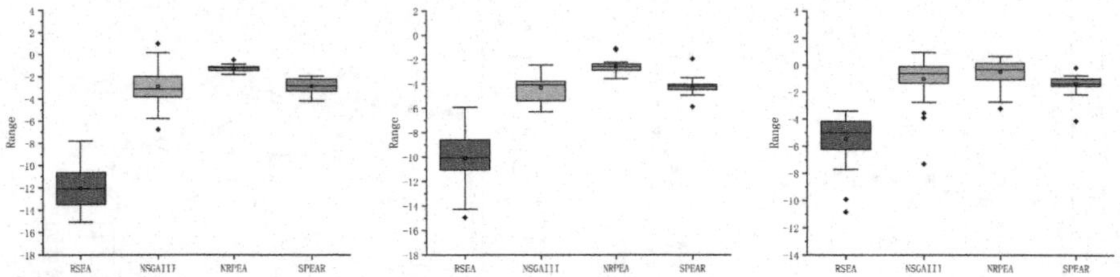

(a) Boxplots of ASR attained by five MOPSOs in port1 and port2 and port3

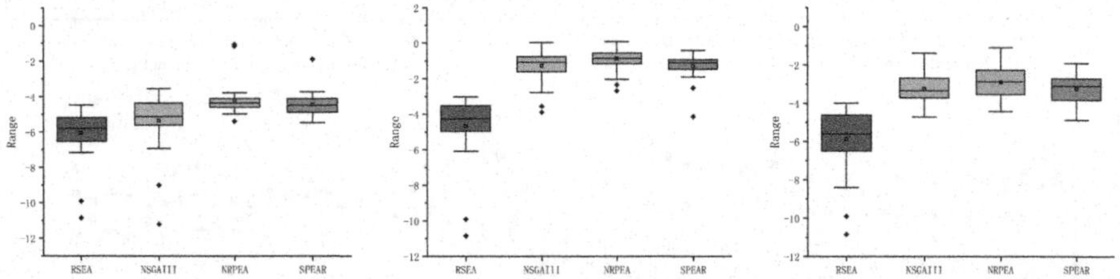

(b) Boxplots of ASR attained by five MOPSOs in port4 and port5 and port6

Fig. 3. Boxplots of the ASR values attained by four MOEAs on six test problems

6 Conclusion

Compared with a random variable, fuzzy number variable based on the knowledge of experts is more suitable to estimate the risk asset due to politics, social conditions, economic change, and the status of the related company. To deal with complex portfolio selection problem, this paper builds five-objective fuzzy portfolio selection problem enhancing the classical Markowitz mean-variance problem by introducing skewness, kurtosis, and the mean proportional entropy. A new reference point-based evolutionary algorithm (NRPEA) is proposed to deal with this five-objective problem. The auxiliary reference points are generated and selected to guide the population evolution. Finally, experimental results show that the proposed NRPEA has obtained better performance than other compared algorithm.

For future work, the proposed NRPEA can be extended considering multi-period portfolio and constraints, such as assets liquidity measurement and short selling.

References

1. Markowitz, H.: Portfolio selection. J. Finan. **7**(1), 77–91 (1952)
2. Narayan, P.K., Ahmed, H.A.: Importance of skewness in decision making: evidence from the Indian stock exchange. Glob. Finan. J. **25**(3), 260–269 (2014)
3. Brito, R.P., Sebastião, H., Godinho, P.: Efficient skewness/semivariance portfolios. J. Asset Manag. **17**(5), 331–346 (2016)

4. Saborido, R., Ruiz, A.B., Bermúdez, J.D., Vercher, E., Luque, M.: Evolutionary multi-objective optimization algorithms for fuzzy portfolio selection. Appl. Soft Comput. **39**, 48–63 (2016)
5. Mashayekhi, Z., Omrani, H.: An integrated multi-objective markowitz-dea cross-efficiency model with fuzzy returns for portfolio selection problem. Appl. Soft Comput. **38**, 1–9 (2016)
6. Huang, X.: Mean-entropy models for fuzzy portfolio selection. IEEE Trans. Fuzzy Syst. **16**(4), 1096–1101 (2008)
7. Qin, Z., Li, X., Ji, X.: Portfolio selection based on fuzzy cross-entropy. J. Comput. Appl. Math. **228**(1), 139–149 (2009)
8. Hisao Ishibuchi, Y., Setoguchi, H.M., Nojima, Y.: Performance of decomposition-based many-objective algorithms strongly depends on pareto front shapes. IEEE Trans. Evol. Comput. **21**(2), 169–190 (2016)
9. Yuan, Y., Hua, X., Wang, B., Zhang, B., Yao, X.: Balancing convergence and diversity in decomposition-based many-objective optimizers. IEEE Trans. Evol. Comput. **20**(2), 180–198 (2015)
10. Liu, Y., Gong, D., Sun, X., Zhang, Y.: Many-objective evolutionary optimization based on reference points. Appl. Soft Comput. **50**, 344–355 (2017)
11. Pan, L., He, C., Tian, Y., Su, Y., Zhang, X.: A region division based diversity maintaining approach for many-objective optimization. Integr. Comput.-Aided Eng. **24**(3), 279–296 (2017)
12. He, C., Tian, Y., Jin, Y., Zhang, X., Pan, L.: A radial space division based evolutionary algorithm for many-objective optimization. Appl. Soft Comput. **61**, 603–621 (2017)
13. Pan, L., He, C., Tian, Y., Wang, H., Zhang, X., Jin, Y.: A classification-based surrogate-assisted evolutionary algorithm for expensive many-objective optimization. IEEE Trans. Evol. Comput. **23**(1), 74–88 (2018)
14. Pan, L., Li, L., He, C., Tan, K.C.: A subregion division-based evolutionary algorithm with effective mating selection for many-objective optimization. IEEE Trans. Cybern. (2019). https://doi.org/10.1109/TCYB.2019.2906679
15. Yue, W., Wang, Y., Dai, C.: An evolutionary algorithm for multiobjective fuzzy portfolio selection models with transaction cost and liquidity. Math. Prob. Eng. **2015**, 1–15 (2015)
16. Deb, K., Sinha, A., Kukkonen, S.: Multi-objective test problems, linkages, and evolutionary methodologies. In: Proceedings of the 8th Annual Conference on Genetic and Evolutionary Computation, pp. 1141–1148. ACM (2006)
17. Mkaouer, W., Kessentini, M., Shaout, A., Koligheu, P., Bechikh, S., Deb, K., Ouni, A.: Many-objective software remodularization using NSGA-III. ACM Trans. Softw. Eng. Method. **24**(3), 17–27 (2015)
18. Yue, W., Wang, Y.: A new fuzzy multi-objective higher order moment portfolio selection model for diversified portfolios. Physica A **465**, 124–140 (2017)
19. Jiang, S., Yang, S.: A strength pareto evolutionary algorithm based on reference direction for multiobjective and many-objective optimization. IEEE Trans. Evol. Comput. **21**(3), 329–346 (2017)
20. Zakamouline, V., Koekebakker, S.: Portfolio performance evaluation with generalized sharpe ratios: beyond the mean and variance. J. Bank. Finan. **33**(7), 1242–1254 (2009)

Hybrid Bacterial Forging Optimization Based on Artificial Fish Swarm Algorithm and Gaussian Disturbance

Ruozhen Zheng[1], Zhiqin Feng[1], Jiaqi Shi[2](✉), Shukun Jiang[3](✉), and Lijing Tan[4]

[1] College of Management, Shenzhen University, Shenzhen, China
[2] College of Economy, Shenzhen University, Shenzhen, China
szdsjq@163.com
[3] College of Software, Shenzhen University, Shenzhen, China
15625562736@163.com
[4] College of Management, Shenzhen Institute of Information Technology, Shenzhen, China

Abstract. Traditional Bacterial Forging Optimization (BFO) has poor convergence speed and is easily trapped in the local optimum while dealing with some complex problems. Facing these disadvantages, a new hybrid algorithm for BFO based on Artificial Fish Swarm Algorithm (AFSA) and Gaussian disturbance is proposed, abbreviated as AF-GBFO. The algorithm combines following and swarming behaviors in AFSA with the chemotaxis part of BFO so that bacteria can update positions by evaluating the value of their own and others positions. The convergence speed can be improved in this way. The algorithm also combines Gaussian disturbance to change bacteria's positions by adding a number following Gaussian distribution. In that case, if all bacteria gather around the local optimum, they still have chance to get out of it. Meanwhile the elimination-dispersal way has been changed to have half of the bacteria eliminated and keep the positions with good values so that the convergence speed is increased. Compared with original BFO, GA, BFOLIW and BFONIW, AF-GBFO outperforms in most cases especially for the multimodal functions.

Keywords: Bacterial Forging Optimization · Artificial Fish Swarm Algorithm · Gaussian disturbance

1 Introduction

In 2002, Passino [9] proposed an optimization algorithm for bacterial foraging based on the behavior of E. coli in human intestinal tract to devour food. The algorithm simulates three kinds of behavior, the chemotaxis of E. coli, reproduction and elimination-dispersal. It holds great global searching ability and can search in parallel. Because of its efficiency, it is widely used in engineering and

L. Pan et al. (Eds.): BIC-TA 2019, CCIS 1159, pp. 124–134, 2020.
https://doi.org/10.1007/978-981-15-3425-6_11

science, such as PID controller design [7], power network [8] and others. However, it has poor convergence rate and is easily trapped in local minimum while dealing with some complex problems [10].

To address problem of poor convergence, researchers have developed many optimization algorithms. For example, Gupta [6] combined PSO algorithm with BFO algorithm to improve the convergence of mixed BF-PSO algorithm. Chen [2] proposed the BCF algorithm and added adaptive and cooperative modes to BFO. Wang [12] proposed the BBBFO algorithm, which adopted convergent strategy of gaussian distribution used the historical information of individuals and the shared information to promote the convergence, and enhanced the diversity of replicated groups. Feng [5] proposed ES-ABFO algorithm, which adopts adaptive chemotaxis. Dass [3] adopts adaptive step size and reproductive mechanism with different birth locations of clones. Dasgupta [4] proposed two schemes for adaptive chemotaxis step level. However, with convergence speed increasing, premature convergence occurs. Previous researches have failed to consider original global searching ability as they try to improve convergence performance. Therefore, the purpose of this paper is to enhance the convergence of BFO while keeping the global searching ability.

AFSA has a fast convergence speed while BFO has a strong global search ability. Inspired by BFO-AFSA proposed by Teng [11] that adopt the chemotaxis of BFO to AFSA to settle the problem of being caught in the local optimum in AFSA, AFSA clustering mechanism is used in new algorithm. The distinct features of AF-GBFO can be listed as follows: (1) Following and swarming behaviors as essential information mechanisms are introduced to the chemotaxis part. Bacteria change their current positions by the best position information in following behavior and central position information in swarming behavior so that they converge more quickly. (2) Gaussian disturbance is added to adjust former positions randomly to alleviate the premature convergence caused by convergence speed improving and balance the strategies.

This paper is organized as follows: The original BFO and AFSA are described in Sects. 2 and 3 specifies the improved algorithm. Following Sect. 4 shows the experimental results. Finally, Sect. 5 summarizes the whole paper.

2 Related Work

2.1 Bacterial Foraging Optimization

The bacterial foraging optimization algorithm was firstly proposed by Passino [9] in 2002. It simulates the behavior of E. coli searching for food in the human intestine. This process contains three main behavioral patterns: chemotaxis, reproduction, elimination and dispersal. The specific process is as follows:

Chemotaxis. E. coli has two behavioral patterns flagellum during chemotaxis: runs and tumble. First calculate the fitness value $Jhealth\,(j, k, l)$ of the bacteria, store the bacteria i as the best value of the current fitness, rotate and compare the fitness value before and after the rotation. If it is better, then continue to swim a unit in this direction, otherwise, jump out directly Chemotactic cycle.

$$\theta^i(j+1,k,l) = \theta^i(j,k,l) + C(i)\frac{\Delta(i)}{\sqrt{\Delta^T \Delta(i)}} \tag{1}$$

where: $\theta^i = [\theta^i_1, \theta^i_2, ..., \theta^i_d]$ is location information of bacteria. j is step of chemotaxis, k is step of reproduction and l is step of elimination and dispersal. C(i) is the unit length of each run, $\Delta(i)$ is the direction angle of jth step.

Reproduction. During the breeding process, the bacteria follow the principle of survival of the fittest, and the bacteria are arranged in ascending order according to J^i_{health}. Only top 50 of the bacteria can survive and the copying operation will keep the population size unchanged. A new generation of individuals enter the next cycle:

$$J^i_{health} = \sum_{j=1}^{Nc+1} J(i,j,k,l) \tag{2}$$

where: J^i_{health}, Nc and J are value, chemotaxis steps and fitness function.

Elimination and Dispersal. Replication will reduce the diversity of the population. In order to avoid premature convergence caused by reproduction, the BFO algorithm adds the dispelling behavior. Some individuals in are randomly regenerated in space, and a new generation of individuals enters a new chemotactic loop. Dispersal or not is determined by a given possibility Ped.

2.2 Artificial Fish Swarm Algorithm

Based on imitations of natural fishes' foraging behaviors, Artificial Fish Swarm Algorithm (AFSA) was proposed by Xiaolei et al. [13] in 2002. In water, each fish in the swarm is swimming through in search of their group's food. Similarly, each fish in AFSA with stochastic initial setting such as X = $(x_1, x_2, x_3, ..., x_n)$ is a potential solution to the problem to be solved and refers to the position of fish in certain dimensions; $Y = f(X)$ indicates the food concentration of local position; Y is the final optimal value. There are other parameters, such as d_{ij} (the distance between two fishes), step (the biggest movement of fish swimming), T (the numbers of attempts) and δ (a crowd factor). In AFSA, fishes search for optimization goal with three major behaviors preying, swimming and swarming. The description is as follows:

Preying Behavior. It is one of behaviors of fishes looking for food. Firstly, randomly initialize a fish's position X_i and choose another position X_j as well within the fish's visual scope with the fitness function. Then compute the food concentration of these two positions (Y_j and Y_i). If Y_j is larger than Y_i, fish will move one step toward X_j with formula (4). Otherwise, repeat selecting and comparing until T times. If Y_i is still larger than Y_j, X_i will be replaced by a random position within the visual scope using formula (3)

$$X_j = X_i + Visual * Rand. \tag{3}$$

$$X_i^{t+1} = X_i^t + \frac{X_j - X_i^t}{||X_j - X_i^t||} * Step * Rand \tag{4}$$

where: $Visual$ represents the field of view. $Rand$ is a random number between -1 to 1. $Step$ is the size of fish's step.

Swarming Behavior. In the process of swarming, fish searches other fishes and get the central position X_c. If $Y_c/n_f > \delta Y_i$ (n_f refers to the numbers of fishes inside the visual scope of bacteria (i) is satisfied, the fish will take a step to X_c with formula as follows. Otherwise, still preying.

$$X_i^{t+1} = X_i^t + \frac{X_c - X_i^t}{||X_c - X_i^t||} * Step * Rand \tag{5}$$

where: $Step$ is the size of step. $Rand$ is a random number between -1 to 1.

Following Behavior. Fish looks for the fish j with best fitness value in the its sight. If the condition $Y_c/n_f > \delta Y_i$ (n_f is met, it will move one step to the position of the best fish X_i or else it continues preying behavior.

Artificial Fish Swarm Algorithm performs well in solving optimization problems with fast convergence speed and easy-adjusted parameters [1,13]. And it is applied to machine learning [14] and other fields.

3 Based on AFSA and Gaussian Disturbance

This study incorporates swarming and following behaviors of AFSA in BFO to improve BFO searching performance with Gaussian disturbance and new elimination-dispersal way. The modified BFO is named AF-GBFO. Pseudo codes for AF-GBFO are shown in Algorithm 1.

3.1 Following and Swarming Behaviors

Chemotaxis is the most important part of BFO. In original BFO, there is no information sharing or learning during chemotaxis. In that case, each bacterium just searches for food positions randomly or by its own experience, which limits bacteria's performance. In order to improve the original BFO, AF-GBFO follows the principle of cohesion in following behavior: finds the position of the best bacteria within its visual scope. As shown in Fig. 1(a), in following part, according to the fitness value, the bacteria finds the position of best bacteria within its visual scope and takes one step tumble toward it along with uncertain steps of swimming. The formula can be described as:

$$\theta^i(j+1,k,l) = \theta^i(j,k,l) + \frac{\theta^p(j,k,l) - \theta^i(j,k,l)}{||\theta^p(j,k,l) - \theta^i(j,k,l)||} * C(i) \tag{6}$$

where: $\theta^i(j,k,l)$ refers to the position of bacteria i after j^{th} chemotaxis, k^{th} reproduction and l^{th} dispersal. And $\theta^i(j,k,l)$ is the position of best bacteria in the visual scope. C^i refers to the length of step.

Meanwhile, theory of swarming behavior is added to chemotaxis. As shown in Fig. 1(b), after tumbling and swimming, the bacteria calculates the distances again to find the positions of bacteria whose fitness is better and then get the central position of this colony. In this method, another parameter is introduced to ensure the central position not too crowded. The equation is as follows.

$$\theta^i(j+1,k,l) = \theta^i(j+1,k,l) + \frac{\theta^c(j,k,l) - \theta^c(j+1,k,l)}{||\theta^c(j,k,l) - \theta^i(j+1,k,l)||} * C(i) \tag{7}$$

where: $\theta^c(j,k,l)$ is the central position of good bacteria colony.

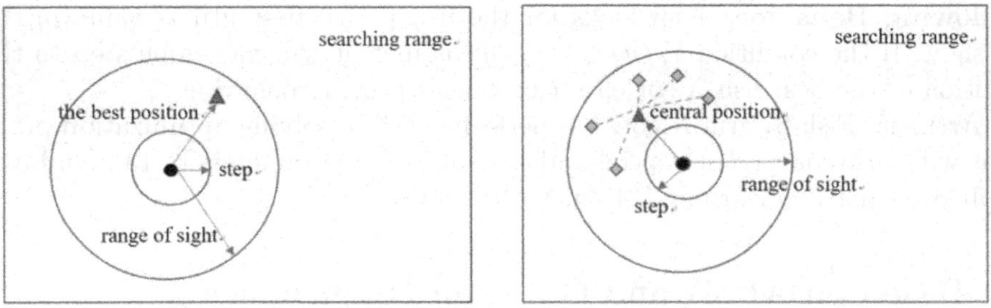

(a) Behavior of following (b) Behavior of swarming

Fig. 1. Following and Swarming Behaviors

3.2 Gaussian Disturbance

Swarming and following improve the convergence speed greatly but the problem of bacteria trapped in local optima remains. In original BFO, if there is a relatively good bacteria in the colony, it will attract the rest of bacteria to approach to it, which lowers the diversity of population. In this paper, we proposed a random number following Gaussian distribution to change the former position using $\theta^{ii} = \theta^{i} * (1 + N(0.1))$. If the new position is better than the old one, the position will be updated by new one. The equation can be defined as:

$$\theta^{best} = \begin{cases} \theta^{ii} & f(\theta^{ii}) < f(\theta^{i}) \\ \theta^{i} & otherwise \end{cases} \tag{8}$$

where: θ refers to the position of bacteria. f represents the fitness value.

In this way, diversity gets improved and bacteria have stronger searching ability.

3.3 Elimination and Dispersal

In original method, elimination-dispersal is determined by a probability p, meaning both the good individuals and bad ones hold the same opportunity to be eliminated. However, it is hard to keep the useful position information if good bacteria are eliminated. In that case, a new elimination dispersal step is adopted. The colony of health bacteria are split into two equal groups. And one of them are eliminated according the fitness value. If the fitness in new position is better than the fitness before eliminating, bacteria locate in the new positions, otherwise back to the former positions.

In the process of modifying algorithm, the algorithm with following and swarming behaviors outperforms the algorithms with only Gaussian disturbance and only elimination-dispersal change, proving that following and swarming mechanism is the most effective one among the three individual strategies. But the combination of strategies is much inferior. It is concluded that with the following and swarming behaviors, each bacterium is able to self-adaptively adjust its positions according to others' information in the most important chemotaxis mechanism and Gaussian disturbance strategy helps to escape local optima while following and swarming improve the convergence speed.

Algorithm 1. Pseudo code for AF-GBFO

```
Initial positions and parameters: S,Nc,Nre,Ned...(S is the number of bacteria,, Nre
is the max number of reproduction, Ned is the chances of elimination and dispersal.)
for l = 1 → Ned do
    for k = 1 → Nre do
        for j = 1 → Nc do
            Compute all the fitness function
            for i = 1 → S do
                1.Compute the distance between the bacteria and others and find the
                best one within visual scope.2.update the position using Eq. (6).
                3.Swim for N steps and Compute the central position
                if Y_c/n_f > δY_i then
                    Update the position using Eq. (7)
                end if
            end for
        end for
        1.Change the current position with a random number following the Gaussian
        distribution. 2.Set the position of the particle equal to the boundary value
        if it crosses the search range 3.Update the position if the new one is better.
        4.Compute the health values, sort the bacteria with health values and then
        copy half of population with better health values
    end for
    Eliminate bacteria by randomly regenerated
end for
```

4 Experiments and Results

4.1 Benchmark Functions

To measure the effectiveness of the novel AF-GBFO algorithm, we adopted 12 benchmark functions to test it in different situations. The search range and the global optimum position X* are listed in Table 1, g is dimension.

4.2 Experimental Settings

The performance of standard BFO, BFOLIW, BFONIW and GA are compared with AF-GBFO. In BFO, BFOLIW, BFONIW and AF-GBFO, the parameter we set for all the functions include: S = 50, Dimension = 15 or 2, Nc = 1000, Nre = 5, Ned = 2, and Ns = 4, which ensures a fair comparison. The chemotaxis step length in BFO and AF-GBFO is set 0.01 for all functions. In BFOLIW and BFONIW, a linearly decreasing and nonlinear chemotaxis step length are adopted respectively. The coefficient is 0.2 and initial step length is 0.01 in BFOLIW while the coefficient of BFONIW is 0.6 and step length is 0.001. In GA, the parametric setup is recommended in former researches. The number of group population, generations that max value remains constant, mutation children (Gaussian), mutation children (random) and elitism children are set 50, 20, 20, 20, 2 respectively.

Table 1. Benchmark functions

Function		Search	X*
f1	Rosenbrock	$[-5, 10]$	$[1, 1, \ldots, 1]$
f2	Dixon-Price	$[-10, 10]$	$2\hat{}((2-2^g)/2^g), g = 1, \ldots, 15$
f3	Styblinski-Tang	$[-5, 5]$	$[-2.90, -2.90, \ldots, -2.90]$
f4	Ackley	$[-32.768, 32.768]$	$[0, 0, \ldots, 0]$
f5	Levy	$[-10, 10]$	$[1, 1, \ldots, 1]$
f6	Rastrigin	$[-5.12, 5.12]$	$[0, 0, \ldots, 0]$
f7	Eggholder	$[-512, 512]$	$[512, 404.23]$
f8	Michalewicz10	$[-10, 10]$	$[2.20, 1.57]$
f9	Michalewicz20	$[-10, 10]$	$[2.20, 1.57]$
f10	Bealc	$[-4.5, 4.5]$	$[3, 0.5]$
f11	Levy N.13	$[-10, 10]$	$[1, 1]$
f12	Drop-Wave	$[-5.12, 5.12]$	$[0, 0]$

Table 2. Functions results

Function		BFOLIW	BFO	GA	BFONIW	AFGBFO	p-value
f1	Mean	8.02E+00	1.87E+01	5.73E+01	6.49E+00	**4.40E+00**	1.27E−01
	Std.	8.99E+00	2.05E+01	2.67E+02	8.28E+00	**1.32E−02**	
f2	Mean	6.76E−01	1.38E+00	2.09E+00	**6.56E−01**	6.67E−01	1.80E−02
	Std.	2.64E−06	2.13E−02	1.10E+00	4.97E−03	**1.16E−12**	
f3	Mean	−5.59E+02	−5.44E+02	−5.44E+02	−4.79E+02	**−5.78E+02**	0.00E+00
	Std.	**2.00E+02**	5.16E−02	3.96E+02	7.27E+02	2.66E+02	
f4	Mean	1.57E+01	1.56E+01	1.97E+00	4.47E−02	**8.88E−16**	1.40E−02
	Std.	9.98E−01	1.78E+00	4.43E−01	5.52E−05	**0.00E+00**	
f5	Mean	8.84E+00	1.13E+01	**3.50E−01**	4.01E−01	4.83E−01	1.49E−01
	Std.	2.49E+00	5.92E+00	1.68E−02	5.24E−01	**9.26E−03**	
f6	Mean	2.56E+01	4.14E+01	1.21E+01	5.34E+01	**0.00E+00**	4.10E−02
	Std.	2.28E+01	1.62E+01	1.62E+01	1.36E+02	**0.00E+00**	
f7	Mean	−9.35E+02	−9.35E+02	−9.59E+02	−9.67E+02	**−5.15E+04**	3.30E−02
	Std.	**5.57E−16**	2.12E−13	1.48E+00	4.92E+03	1.13E+09	
f8	Mean	−5.58E+00	−4.69E+00	−7.74E+00	−5.19E+00	**−9.62E+00**	2.00E−03
	Std.	2.15E−01	8.91E−02	1.91E+00	**8.64E−02**	7.48E−01	
f9	Mean	−5.08E+00	−3.95E+00	−6.96E+00	−4.88E+00	**−8.87E+00**	2.00E−03
	Std.	1.12E−01	**3.33E−02**	7.83E−01	3.40E−01	7.07E−01	
f10	Mean	6.11E−07	9.09E−06	4.99E−09	1.32E−06	**7.56E−12**	2.70E−02
	Std.	6.21E−13	1.03E−10	2.49E−16	1.60E−12	**4.31E−23**	
f11	Mean	3.93E−06	2.28E−04	1.35E−31	3.13E−06	**1.85E−10**	3.58E−01
	Std.	1.28E−11	3.89E−08	0.00E+00	5.23E−12	**3.26E−20**	
f12	Mean	1.78E+03	**−1.00E+00**	1.92E+03	−9.94E−01	**−1.00E+00**	1.79E−01
	Std.	1.53E−12	**0.00E+00**	1.62E+04	1.13E−05	**0.00E+00**	

After adopting the AFSA into BFO, its essential to value the visual scope affected by the search range and dimensions. We take different rate of searching range as the visual scope. According to experiments, C = 0.3 should be adopted.

4.3 Experimental Results

In this part, 12 functions are adopted to measure the global searching ability and the convergence rate of algorithms. Table 2 shows the best averaged results of ten independent runs. From the Table 2, we can see that AF-GBFO excels other algorithm at most cases according to mean value and standard deviation.

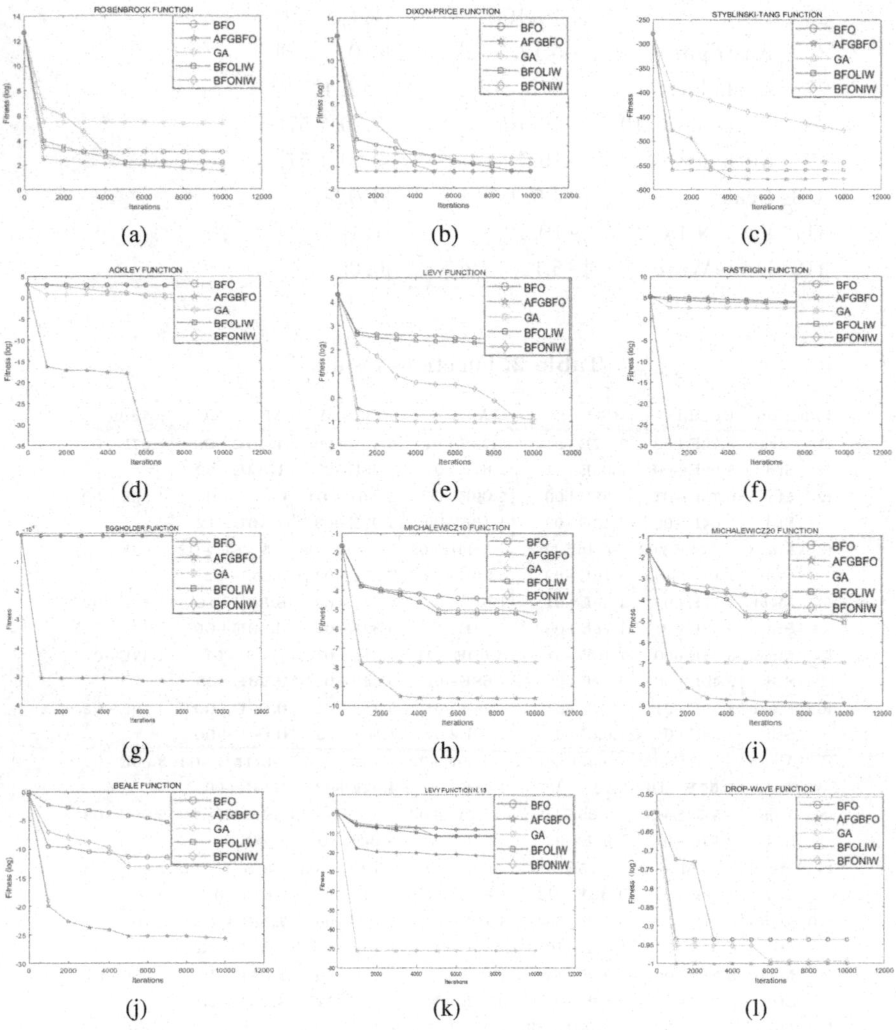

Fig. 2. Convergence curve of function

Figure 2 shows the comparison results of 5 algorithms. The functions shown in Fig. 2(a)–(b) are unimodal functions, Rosenbrock function and Dixon-price function are referred to as the valley. In such shape, convergence to the minimum is difficult. According to experiments, AF-GBFO finds relatively good solutions and shows fast convergence speed.

The functions shown in Fig. 2(c)–(l) are multimodal functions which exist many local optimum. For functions (f) and (l), AF-GBFO finds the global optimum directly. In function (g), we can see that bacteria in AG-BFO finds the fitness value compared with others. According to the figure of function (j), AF-GBFO doesn't return the best solution but its convergence is still outstanding. When it comes to the function (h) and (i), the Michalewicz function has steep valleys and ridges leading to a more difficult search. According to the figure of (h) and (i), AF-GBFO locates the best fitness value, showing the strong ability of jumping local optimum. But when it comes to functions (e) and (k), GA algorithm excels AF-GBFO in results and in function (k) the disparity becomes obvious. Function (k) is a two-dimensional function. The result may be that in multi-dimensional functions AF-GBFO can exert its searching ability. In order to distinguish whether the results achieved by AF-GBFO were statistically significantly better than other algorithms', T-tests were introduced and the confidence level was set to 95%. From the Table 2 p-value is usually less than 5%, which means that the experimental results from AF-GBFO were significantly different from other results. In conclusion, the AF-GBFO is an effective and competitive algorithm especially in solving multimodal and multi-dimensional functions.

5 Conclusion

Based on BFO and AFSA, a new algorithm AF-GBFO is proposed. This algorithm proposes a new chemotaxis mode that gives the bacteria virtual vision, approaching to the bacteria with the highest concentration of nutrients in the sight, and the convergence speed is greatly accelerated. A number based on Gaussian perturbation is also added to the learning process to avoid bacteria trapping into local optimum. To test the new modified algorithms' effectiveness, AF-GBFO and other four algorithms BFO, BFO-LIW, BFO-NIW, GA, are tested over 12 benchmark functions. Experiment results show that AF-GBFO is usually superior to other algorithms especially in multimodal problems, which greatly proves its fast convergence and strong global searching ability. Now, AF-GBFO is just tested to the benchmark functions. In the future, we will do more to improve this algorithm and apply it to real problems.

Acknowledgment. This work is partially supported by the Natural Science Foundation of Guangdong Province (2018A030310575), Natural Science Foundation of Shenzhen University (8530 3/00000155), Project supported by Innovation and Entrepreneurship Research Center of Guangdong University Student (2018A073825), Research Cultivation Project from Shenzhen Institute of Information Technology (ZY201717) and Innovating and Upgrading Institute Project from Department of Education of Guangdong Province (2017GWT SCX038). Ruozhen Zheng and Zhiqin Feng are first authors. They contributed equally to this paper.

References

1. Azizi, R.: Empirical study of artificial fish swarm algorithm. Comput. Sci. **17**(6), 626–641 (2014)
2. Chen, H., Niu, B., Ma, L., Su, W., Zhu, Y.: Bacterial colony foraging optimization. Neurocomputing **137**, 268–284 (2014)
3. Daas, M.S., Chikhi, S., Batouche, M.: Bacterial foraging optimization with double role of reproduction and step adaptation. In: Proceedings of the International Conference on Intelligent Information Processing, Security and Advanced Communication, p. 71. ACM (2015)
4. Dasgupta, S., Das, S., Abraham, A., Biswas, A.: Adaptive computational chemotaxis in bacterial foraging optimization: an analysis. IEEE Trans. Evol. Comput. **13**(4), 919–941 (2009)
5. Feng, X.H., He, Y.Y., Yu, J.: Economic load dispatch using bacterial foraging optimization algorithm based on evolution strategies. In: Advanced Materials Research, vol. 860, pp. 2040–2045. Trans Tech Publ. (2014)
6. Gupta, N., Saxena, J., Bhatia, K.S.: Optimized metamaterial-loaded fractal antenna using modified hybrid BF-PSO algorithm. Neural Comput. Appl., 1–17 (2019). https://doi.org/10.1007/s00521-019-04202-z
7. Kou, P.G., Zhou, J.Z., Yao-Yao, H.E., Xiang, X.Q., Chao-Shun, L.I.: Optimal PID governor tuning of hydraulic turbine generators with bacterial foraging particle swarm optimization algorithm. Proc. CSEE **29**(26), 101–106 (2009)
8. Mishra, S.: Bacteria foraging based solution to optimize both real power loss and voltage stability limit. In: Power Engineering Society General Meeting (2007)
9. Passino, K.M.: Biomimicry of bacterial foraging for distributed optimization and control. IEEE Control Syst. Mag. **22**(3), 52–67 (2002)
10. Tan, L., Lin, F., Hong, W.: Adaptive comprehensive learning bacterial foraging optimization and its application on vehicle routing problem with time windows. Neurocomputing **151**(3), 1208–1215 (2015)
11. Teng, F., Zhang, L.: Application of BFO-AFSA to location of distribution centre. Cluster Comput. **20**(3), 3459–3474 (2017). https://doi.org/10.1007/s10586-017-1144-5
12. Wang, L., Zhao, W., Tian, Y., Pan, G.: A bare bones bacterial foraging optimization algorithm. Cogn. Syst. Res. **52**, 301–311 (2018)
13. Xiaolei, L.I., Shao, Z., Qian, J.: An optimizing method based on autonomous animats: fish-swarm algorithm. Syst. Eng.-Theory Pract. **22**, 32–38 (2002)
14. Yazdani, D., Golyari, S., Meybodi, M.R.: A new hybrid algorithm for optimization based on artificial fish swarm algorithm and cellular learning automata. In: International Symposium on Telecommunications (2010)

Research on Multiobjective Optimization Strategy of Economic/Environmental Energy Management for Multi-energy Ship Based on MOEA/D

Xi Chen[1], Qinqi Wei[1], and Xin Li[2](✉)

[1] China Ship Development and Design Centre, Wuhan 430064, China
[2] School of Energy and Power Engineering, Wuhan University of Technology, Wuhan 430063, China
xinli0503@hotmail.com

Abstract. The economic/environmental energy management (EEEM) for multi-energy ship is to optimize the output power of the generation devices in the power system to meet the load demand and navigation speed need as well as satisfy the practical constraints, while decreasing both operation cost and pollutants simultaneously. Aiming at the frequently changing problem of the optimization model of EEEM for multi-energy ship, this paper proposed an optimization method based on improved multiobjective evolutionary algorithm based on decomposition (MOEA/D) framework. To deal with various types of constraints in the practical EEEM problem, a constraints handling approach is suggested to replace the method utilized in original Non-dominated Sorting Genetic Algorithm II (NSGAII). Thereafter, the improved algorithm is applied to three typical EEEM for multi-energy ship with different combinations of power generation devices. The efficiency of the algorithm is verified and the simulation results obtained can meet the navigation speed requirements as well as reduce the operation cost and emissions.

Keywords: Economic/environmental energy management · Multi-energy ship · Multiobjective optimization · MOEA/D · Constraints handling

1 Introduction

In recent years, the global shipping market is facing two major challenges: the shortage of fossil resources and the deterioration of environmental pollution. Thus, it is an effective way to replace traditional fossil fuel energy by multi-energy in ship energy system [1,2]. With the increasingly stringent environmental protection requirements of International Maritime Organization (IMO), "green ship" and "green shipping" have become the main topics of the future development of shipbuilding and shipping industry, and the study of the multiobjective optimization strategies of multi-energy ship energy management system (EMS) will improve the efficiency of energy utilizing as well as reducing the air pollutions [3].

L. Pan et al. (Eds.): BIC-TA 2019, CCIS 1159, pp. 135–146, 2020.
https://doi.org/10.1007/978-981-15-3425-6_12

EMS is not only to manage the power supply of generator units, but also to control various types of loads, like electricity and heat. On the basis of ensuring the reliability and stability of ship power supply, the optimal fuel consumption can be achieved, the economy and environmental protection of ship can be improved, and the optimal allocation management of energy can be realized from power generation to power consumption [4,5].

With the increase of ship power grid capacity, the focus of research on how to improve economy and environmental protection has been shifted to multi-energy ships while satisfying reliability and safety [6]. In other words, how to improve the ship's EMS strategy has become a growing concern. In the existing research, the optimization strategy of EMS aiming at improving ship safety, reliability and fuel economy mainly considers the following aspects: generator set limitation, load limitation, power balance limitation, power loss prevention and global working condition limitation, etc. [7].

In order to achieve the global optimal control effect, more scholars began to study the energy allocation strategy based on optimization theory. The control method based on optimization can be divided into global optimum and instantaneous optimum. The most representative energy management strategies based on global optimum are dynamic programming, Pontryagin minimum principle and energy management control method combined with intelligent algorithm [8–11]. Hou proposed a bi-objective optimization problem of reducing power fluctuation and energy consumption for the marine hybrid energy storage system composed of batteries and flywheels, and the dynamic programming method was utilized to obtain the optimal solution set, which can restrain load fluctuation to a certain extent and reduce fuel consumption cost [12]. Abkenar et al. used genetic algorithm to adjust the fuel and air flow rate for marine power plants including fuel cells and storage batteries, so that the fuel cells could operate safely under different power requirements, which not only maintained the performance of batteries, but also reduced fuel consumption [13]. Tang et al. proposed an adaptive co-evolutionary strategy to solve the dimension disaster problem of the economic dispatching model of diesel-fired storage ship power system, which effectively solved the energy optimal allocation problem under different operation modes [14]. In addition, some scholars put forward the idea of using multi-objective optimization method to study energy allocation [15,16]. Considering the factors of total cost, system reliability and environmental benefits, Liao and Yu used multi-objective optimization method to obtain the optimal capacity allocation scheme for the micro-grid of wind, solar and firewood storage ships. Based on the forecasting load, the impact of the cost realization value, load loss rate and pollutant emission control of different energy allocation operation strategies were studied [17,18].

The current studies mainly focus on the improvement of the optimization methods for some specific practical scenarios, but it is a more adaptive and efficient algorithm that is necessary to solve all kinds of economic/environmental energy management (EEEM) for multi-energy ship, since the optimization model is changing frequently during the voyage [19]. The multiple objectives and constraints may make the search space complicated for the algorithms to archive feasible and Pareto optimal solutions [20]. Therefore, this paper proposes an efficient multiobjective optimization algorithm, which can adapt to the change of EEEM problems for multi-energy ships, and the constraints can be handled efficiently. In reminder of this paper, the typical optimization model of EEEM for multi-energy ship is described in Sect. 2. The improved multiobjective evolutionary algorithm based on decomposition (MOEA/D) is proposed in Sect. 3. Thereafter, the improved MOEA/D is applied to several practical cases to study the effects of the algorithm performance in Sect. 4. Finally, the conclusions of the study are presented.

2 Optimization Model of EEEM for Multi-energy Ship

In this section, based on the analysis of the specific structure and technical requirements of the ship power system, a mathematical model for the economic and environmental protection of the ship power system is established, which mainly includes three parts: traditional diesel generators (DG), wind turbine (WT), and energy storage devices (ESD). The optimization model contains two objective functions considering minimizing the cost as well as the emission, and several typical constraints.

2.1 Mathematical Models of the Devices

DG is the traditional power generators in ships, of which the fuel consumption model can be described as:

$$FC = \sum_{i=1}^{nG} \sum_{t=1}^{nT} \left(FCg_i^t \Delta t + C_{su,i}^t \right), \tag{1}$$

$$FCg^t = a_2 {r_g^t}^2 + a_1 r_g^t + a_0, \tag{2}$$

where FCg_i^t is the fuel consumption of generator unit i in t period, $C_{su,i}^t$ is the fuel consumption of starting-up of generator unit i in t period, a is the coal consumption coefficient, and r_g^t is the active power standard unitary value.

The output power of WT can be calculated by the formula below:

$$P_{WT}(t) = \begin{cases} 0 v(t) < v_{ci} \\ a v(t)^3 - b P_r v_{ci} < v(t) < v_r \\ P_r v_r < v(t) < v_\infty \\ 0 v(t) > v_\infty \end{cases}, \tag{3}$$

where v_{ci},v_r and v_∞ are the Cut-in wind speed, the rated wind speed and the cut-off wind speed of WT, respectively. P_r is the rated power output. a and b are the coefficients.

In addition, the ESD is needed in a multi-energy ship, of which the capacity model is shown below:

$$Ebat(t) = ESTC[1 + B(Tbat(t)Tbat, ST\ C)], \tag{4}$$

where $ESTC$ is the rated capacity of batteries under standard test conditions, δ_B is the capacity temperature coefficient, T_{bat} is the actual working temperature of the battery and $T_{bat,STC}$ is the temperature of under the standard test conditions, which is 25°C. The discharge/charge process can be described as:

$$S_{SOC}(t) = (1 - \sigma) S_{SOC}(t-1) + \frac{P_{bat}(t) \times \Delta t \times \eta_c}{E_{bat}(t)}, \tag{5}$$

$$S_{SOC}(t) = (1 - \sigma) S_{SOC}(t-1) + \frac{P_{bat}(t) \times \Delta t}{E_{bat}(t) \times \eta_d}, \tag{6}$$

where σ is the self-discharge rate per hour of the battery, P_{bat} is the charge/discharge power per unit time of the battery, η_c and η_d are the charge efficiency and discharge efficiency, respectively, and S_{SOC} is the state of charge of the storage battery.

2.2 Objective Functions

The overall cost function of the multi-energy ship in this paper mainly contains the operation cost by the three types of devices, which is shown below:

$$TC = FC + SC + WC, \tag{7}$$

where FC, SC and WC are the operation costs of DG, ESD and WT, respectively.

Besides, the objective function for minimum emission considers the energy efficiency operation index (EEOI) by IMO, which can be described as follows:

$$EEOI = \frac{\sum_{t=1}^{nT} \sum_{i=1}^{Ng} (GE_i^t \Delta t)}{\sum_{t_p \in T_p} \left(m_{AES}^{t_p} Dist^{t_p} \right)}, \tag{8}$$

$$GE^t = b_2 {r_g^t}^2 + b_1 r_g^t + b_0, \tag{9}$$

where GE is the pollutant emission by the DGs, t_p is the time of berthing, $m_{AES}^{t_p}$ is the cargo load of ships after berthing and unloading cargo at ports, $Dist^{t_p}$ is the distance of ships before berthing, b is the pollutant discharge coefficient, and r_g^t is the standard unitary value of the active contribution.

2.3 Typical Constraints

(a) **Load Balance Constraints**
The power generated by the devices in the power system should meet the load demand in every time interval, which can be described as:

$$\sum_{i=1}^{nG} P^t_{DG,i}\eta_g\eta_{tr} = \frac{P^t_{lp}}{\eta_p} + P^t_{ls}, \quad (10)$$

where $P^t_{DG,i}$ is the output power of the devices. η_g, η_{tr} and η_p are the efficiency of generation, transmission and propulsion, respectively. P^t_{lp} and P^t_{ls} are the load of propulsion and service, respectively.

(b) **Power Output Limits**
All the devices have power generation limits, which can be expressed as:

$$r_{g\min} < r^t_g \leq 1.0, r_{g\min} = \frac{P_{DG\min}}{P_{DGn}}, \quad (11)$$

where $P_{DG\min}$ and $r_{g\min}$ are the minimum allowed output and per unit values.

(c) **Navigation Speed Constraints**
The navigation speed of the ship cannot change dramatically, as shown below:

$$(1-\eta_v)v^t_n \leq v^t_p \leq (1+\eta_v)v^t_n, \quad (12)$$

$$v^{t_h}_n = \eta_h v^{t_f}_n t_h \in T_h t_f \in T_c - T_h, \quad (13)$$

where v^t_p is the actual speed of the ship, v^t_n is the preset rated speed, and η_v is the range of speed. T_c is the cruise time, T_h is the local speed time. v_n keeps constant at full cruise interval ($T_c - T_h$), while keeping low ratio T_h at local cruise interval η_h and being zero at berthing period.

(d) **State of Charge Constraints**
The ESD cannot be overcharged or overused, so the limits of the state of charge (SOC) are as follows.

$$SOC_{\min} \leq SOC_t \leq SOC_{\max}, \quad (14)$$

where $SOC_{\min}$ and $SOC_{\max}$ are the minimum and maximum SOC.

3 Optimization Method

3.1 MOEA/D-M2M Framework

In this paper, the algorithm framework of MOEA/D-M2M is suggested to solve the practical EEEM problems for multi-energy ships. MOEA/D-M2M is a variant of MOEA based on decomposition, which decomposes a multiobjective optimization problem (MOP) into a set of sub-MOPs and solves them simultaneously, as is called "multiple to multiple (M2M)" [21]. This algorithm framework can well adapt to the complexity of the search space and achieve a good diversity of population. The procedure of the algorithm can be found in [21].

Comparing with the original MOEA/D, MOEA/D-M2M is more qualified for EEEM problems for multi-energy ships for the following reasons:

(a) The MOEA/D decomposition procedure is an approximate transformation. However, MOEA/D-M2M transforms a MOP into several sub-MOPs by an equivalent optimization process.
(b) The EEEM problems for multi-energy ships may have large numbers of variables and constraints, which makes the feasible search spaces distribute uniformly. But for MOEA/D-M2M, the decomposition procedure can "protect" every subpopulation in the sub-MOPs, so that some of the potential solutions can be obtained and would not be replaced by some super individuals in other subpopulations. In this way, the diversity of the whole population is ensured.
(c) The parameter settings of MOEA/D-M2M framework are simple which means it does not need much manual operation during the optimization process. Hence MOEA/D-M2M can solve the practical MOPs which always changes in objective functions, parameters or constraints.

In conclusion, the algorithm framework of MOEA/D-M2M can well adapt to the change of the EEEM problems for multi-energy ships. In this paper, Non-dominated Sorting Genetic Algorithm II (NSGAII) is utilized in the MOEA/D-M2M framework as the optimization tool to solve each sub-MOPs.

3.2 Improvement in Constraints Handling

The original MOEA/D-M2M focuses on non-constrained MOPs [21], which may not solve the practical EEEM problems for multi-energy ships [22]. Thus, the constraints handling process proposed in [23] is introduced, and the method of calculating overall violations is modified to deal with the multi-constrained MOPs in this paper, which is presented in the following steps:

Step 1: Calculate the violations of the solution to every constraint which belongs to one type, and sum them up. For example, if Individual x_1 violates some constraints which belong to Type A, then the overall violation for the constraints of Type A is expressed as:

$$sumC_{A,\mathrm{x}_1} = \sum_{n=1}^{N_{A,\mathrm{x}_1}} C_{A,\mathrm{x}_1}, \tag{15}$$

where C_{A,x_1} is the violation of one constraint which belongs to Type A, and N_{A,x_1} is the number of the violated constraints.

Step 2: Normalize the overall violation for the constraints of every type as follows:

$$\overline{C_{A,\mathrm{x}_1}} = \frac{sumC_{A,\mathrm{x}_n} - sumC_{A,\min}}{sumC_{A,\max} - sumC_{A,\min}}, \tag{16}$$

where $sumC_{A,\max}$ and $sumC_{A,\min}$ are the minimum overall violation for the constraints of Type A in the population.

Step 3: Sum up all the normalized overall violation of every type of constraints:

$$C_{x_1} = \sum_{m=1}^{M} \overline{C_{m,x_1}} w_m, \tag{17}$$

where M is the number of the constraints types in the optimization model, and w_m is the weight for the m-th constraint.

4 Simulation Results and Discussion

4.1 Parameters Settings

In this paper, the power system of the multi-energy ship contains four DGs, one ESD and one WT. Three cases are studied during the operation. In Case One, only DGs are utilized, and In Case Two, ESD is added to the power system. In Case Three, all of the three types of devices are applied. The parameters needed are shown below in Tables 1, 2 and 3.

Table 1. Navigation parameters of ships.

$\Delta t = 1\,\text{h},\ nT = 24\ intervalsi.e.24\,\text{h}$
$\eta_v = 10\%, \eta_h = 65\%$
$v_n = 19, Dist_n^{tp} = \{59.85, 119.7, 179.55\}$
$c_{p1} = 0.00235, c_{p2} = 3, m_{AES}^{tp} = \{60, 50, 40\}$

Table 2. Parameters of DG.

P_{DG_n} (MW)	$P_{DG_{min}}$ (MW)	$F_{g0,1,2}$ (m.u.)	$G_{e_{0,1,2}}$ (uCO_2/p.u)	C_{su} (m.u.)
13.5	4.5	300.0, 2187.2, 9.2	386.1, −2070.4, 8383.5	1350
13.5	4.5	291.0, 2330.1, 18.8	386.1, −2070.4, 8383.5	1500
3.75	0.6	210.0, 622.5, 9.4	862.43, −613.12, 956.25	165
3.75	0.6	204.4, 618.8, 9.9	362.43, −613.12, 956.25	168

Table 3. Parameters of WT.

Parameter	Value
v_{ci}	3
v_r	12
v_{∞}	24

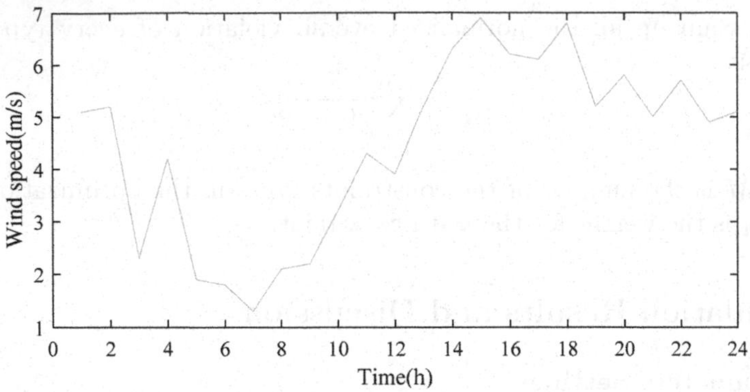

Fig. 1. Wind speed.

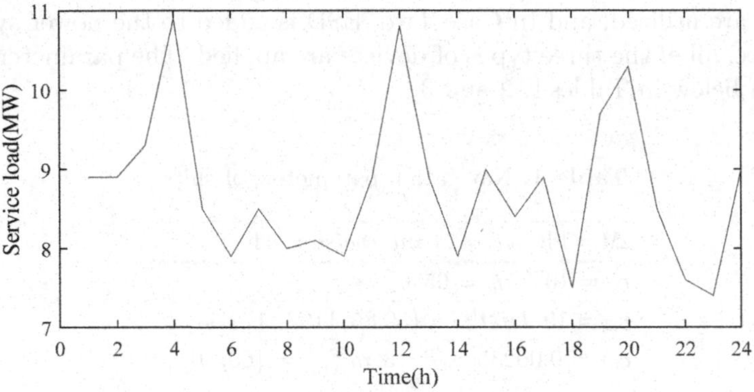

Fig. 2. Service load demand.

In addition, the wind speed data on May 20, 2018 in Shenzhen harbor, China is selected, as shown in Fig. 1. And the service load demand curve is shown in Fig. 2.

4.2 Simulation Experiments

In this section, the proposed I-MOEA/D is applied in the three cases and the results are obtained. For I-MOEA/D, K = S = 10. The population is 50 and the maximum generation number is 5000. The algorithm runs for 50 times and the best results are recorded. The navigation speed and the output power of different devices are presented.

Figure 1 shows the optimum navigation speed curves in three cases. It can be seen that in Case 2, the navigation speed is a little higher than that in Case 1, which indicates that the power system of multi-energy ship may supply more power with less cost and emission when a ESD is utilized. However, the speed is more volatile in Case 2, and even over the rated speed in 14:00 and 22:00.

On the other hand, in Case 3, the speed didn't change much comparing with that in Case 1, which is within the limitations of navigation speed of ship. Moreover, it is more stable than that in Case 2, which means the navigation speed meets the practical requirements of safety and stability when the ESD and WT are applied in the power system of multi-energy ship.

Since the result of multiobjective optimization is a set of Pareto solutions, it is difficult to compare the results in three cases directly. Therefore, in this paper, the solutions with minimum cost are selected, of which the output power of the devices are shown in Figs. 3, 4, 5 and 6.

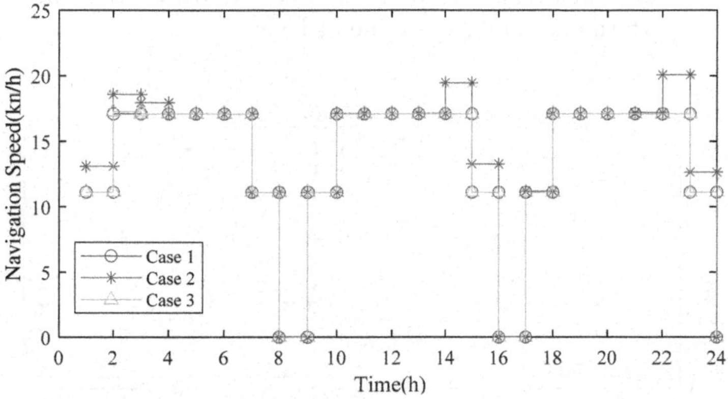

Fig. 3. Optimum navigation speed curves in three cases (Rated speed = 19 kn/h).

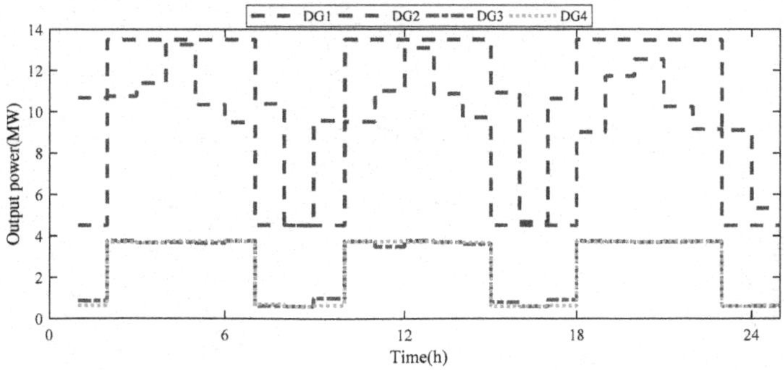

Fig. 4. Output power of the devices in multi-energy ship in Case 1.

As can be seen from Fig. 4, the peak load periods mainly concentrates on three parts: 2:00–7:00, 10:00–15:00 and 18:00–23:00. During these periods, four DGs will run, of which 1 # and 2 # DGs bear the main load of the whole power system, and 3 # and 4 # DGs output less power. In other periods of time, the power output of 3 # and 4 # DGs to power the grid is very low, while mainly 1 # and 2 # DGs provide the whole ship's power load demand.

In Case 2, when the ESD is added, the system dispatches the power generation devices more frequently. But compared with Case 1, the output power of 1# and 2# DGs also decreases, avoiding the long-term high-load and high-intensity operation, and making the system dispatch the power generation equipment in a balanced and reasonable manner. Therefore, the economic performance of DGs is improved.

Comparing with Case 1 and Case 2, the most obvious improvement of Case 3 is that when two additional power generation devices are added, the power supply of the whole energy system tends to be more stable. From Fig. 6, it can be found that the output power of 2 #, 3 # and 4 # diesel generators is more stable and fluctuates in a very short period (14:00–17:00), which greatly improves the economic performance and service life of DGs.

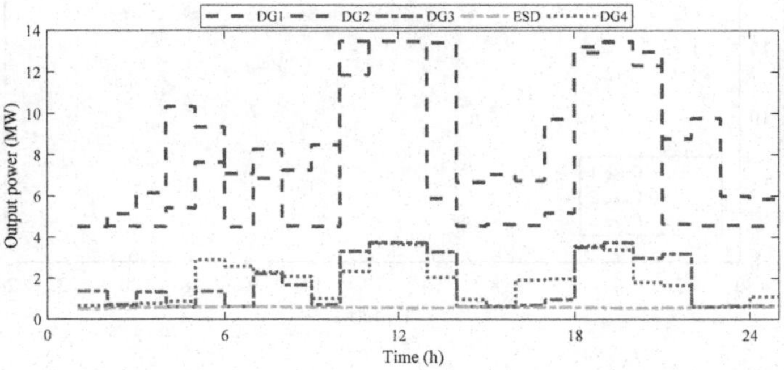

Fig. 5. Output power of the devices in multi-energy ship in Case 2.

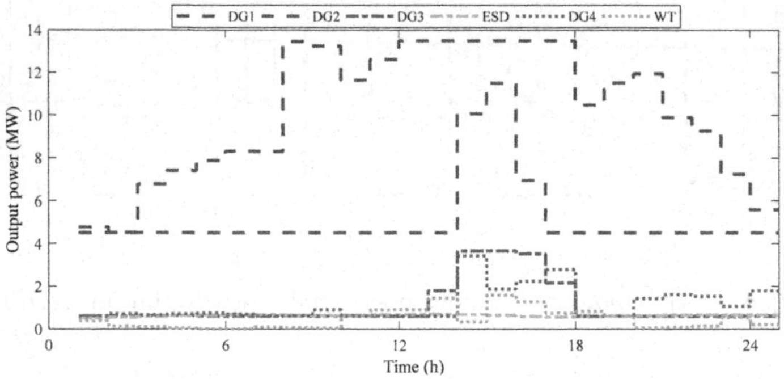

Fig. 6. Output power of the devices in multi-energy ship in Case 3.

5 Conclusion

This paper proposes an improved MOEA/D framework to solve the practical EEEM problems for multi-energy ship, of which the optimization model is always changing to meet the navigation requirements. The MOEA/D-M2M framework is introduced to adapt to the change of parameters and constraints in the optimization models. Besides, a constraints handling method based on classification and normalization is suggested to deal with various types of constraints. Case studies show that by utilizing the proposed optimization algorithm, the navigation speed is optimized and meets the practical requirements of safety and stability. In addition, the output power by different power generation devices can meet the load demand and other constraints, and the operation cost and pollutants are decreased. Further studies are needed to analyze the obtained Pareto fronts and compare the results with those achieved by other efficient MOEAs.

References

1. Wen, S., et al.: Optimal sizing of hybrid energy storage sub-systems in PV/diesel ship power system using frequency analysis. Energy **140**, 198–208 (2017)
2. Chong, L.W., Wong, Y.W., Rajkumar, R.K., Isa, D.: An optimal control strategy for standalone PV system with battery-supercapacitor hybrid energy storage system. J. Power Sources **331**, 553–565 (2016)
3. Niu, W., et al.: Sizing of energy system of a hybrid lithium battery RTG crane. IEEE Trans. Power Electron. **32**(10), 7837–7844 (2016)
4. Yuan, Y., Wang, J., Yan, X., Li, Q., Long, T.: A design and experimental investigation of a large-scale solar energy/diesel generator powered hybrid ship. Energy **165**, 965–978 (2018)
5. Hou, J., Sun, J., Hofmann, H.F.: Mitigating power fluctuations in electric ship propulsion with hybrid energy storage system: design and analysis. IEEE J. Oceanic Eng. **43**(1), 93–107 (2017)
6. Fang, S., Xu, Y., Li, Z., Zhao, T., Wang, H.: Two-step multi-objective management of hybrid energy storage system in all-electric ship microgrids. IEEE Trans. Veh. Technol. **68**(4), 3361–3373 (2019)
7. Dolatabadi, A., Mohammadi-Ivatloo, B.: Stochastic risk-constrained optimal sizing for hybrid power system of merchant marine vessels. IEEE Trans. Industr. Inf. **14**(12), 5509–5517 (2018)
8. Luo, Y., Chen, T., Zhang, S., Li, K.: Intelligent hybrid electric vehicle ACC with coordinated control of tracking ability, fuel economy, and ride comfort. IEEE Trans. Intell. Transp. Syst. **16**(4), 2303–2308 (2015)
9. Wang, R., Zhan, Y., Zhou, H.: Application of artificial bee colony in model parameter identification of solar cells. Energies **8**(8), 7563–7581 (2015)
10. Faddel, S., El Hariri, M., Mohammed, O.: Intelligent control framework for energy storage management on MVDC ship power system. In: 2018 IEEE International Conference on Environment and Electrical Engineering and 2018 IEEE Industrial and Commercial Power Systems Europe (EEEIC/I&CPS Europe), pp. 1–6. IEEE (2018)
11. Hou, J., Sun, J., Hofmann, H.: Adaptive model predictive control with propulsion load estimation and prediction for all-electric ship energy management. Energy **150**, 877–889 (2018)

12. Hou, J., Sun, J., Hofmann, H.: Control development and performance evaluation for battery/flywheel hybrid energy storage solutions to mitigate load fluctuations in all-electric ship propulsion systems. Appl. Energy **212**, 919–930 (2018)
13. Abkenar, A.T., Nazari, A., Jayasinghe, S.D.G., Kapoor, A., Negnevitsky, M.: Fuel cell power management using genetic expression programming in all-electric ships. IEEE Trans. Energy Convers. **32**(2), 779–787 (2017)
14. Tang, R., Li, X., Lai, J.: A novel optimal energy-management strategy for a maritime hybrid energy system based on large-scale global optimization. Appl. Energy **228**, 254–264 (2018)
15. Bi, X., Wang, C.: An improved NSGA-III algorithm based on objective space decomposition for many-objective optimization. Soft. Comput. **21**(15), 4269–4296 (2017)
16. Bi, X., Wang, C.: A niche-elimination operation based NSGA-III algorithm for many-objective optimization. Appl. Intell. **48**(1), 118–141 (2018)
17. Dextreit, C., Kolmanovsky, I.V.: Game theory controller for hybrid electric vehicles. IEEE Trans. Control Syst. Technol. **22**(2), 652–663 (2013)
18. Wang, K., Yan, X., Yuan, Y., Jiang, X., Lin, X., Negenborn, R.R.: Dynamic optimization of ship energy efficiency considering time-varying environmental factors. Transp. Res. Part D: Transp. Environ. **62**, 685–698 (2018)
19. Alasali, F., Haben, S., Holderbaum, W.: Energy management systems for a network of electrified cranes with energy storage. Int. J. Electr. Power Energy Syst. **106**, 210–222 (2019)
20. Li, X., Lai, J., Tang, R.: A hybrid constraints handling strategy for multiconstrained multiobjective optimization problem of microgrideconomical/environmental dispatch. Complexity **2017**, 12 (2017)
21. Liu, H.L., Gu, F., Zhang, Q.: Decomposition of a multiobjective optimization problem into a number of simple multiobjective subproblems. IEEE Trans. Evol. Comput. **18**(3), 450–455 (2013)
22. Deb, K., Rao N., U.B., Karthik, S.: Dynamic multi-objective optimization and decision-making using modified NSGA-II: a case study on hydro-thermal power scheduling. In: Obayashi, S., Deb, K., Poloni, C., Hiroyasu, T., Murata, T. (eds.) EMO 2007. LNCS, vol. 4403, pp. 803–817. Springer, Heidelberg (2007). https://doi.org/10.1007/978-3-540-70928-2_60
23. Deb, K., Pratap, A., Agarwal, S., Meyarivan, T.: A fast and elitist multiobjective genetic algorithm: NSGA-II. IEEE Trans. Evol. Comput. **6**(2), 182–197 (2002)

Many-Objective Evolutionary Optimization Based Economic Dispatch of Integrated Energy System with Multi-microgrid and CHP

Jinlei Wang[1], Xiaoyan Sun[1(✉)], Dunwei Gong[1], Lin Zhao[1], Yongli Wang[2], and Changhe Du[3]

[1] School of Information and Control Engineering, China University of Mining and Technology, Xuzhou 221116, China
xysun78@126.com
[2] School of Economic and Management, North China Electric Power University, Beijing 102206, China
[3] Qingdao University Information Industry Co., Ltd., Qingdao 266001, China

Abstract. Integrated energy system (IES) containing a variety of heterogeneous energy supplies has been widely focused on energy conversion and power dispatching for effective utilization on energy. However, existing studies are most directed to single micro-grid based IES without considering energy exchange among several micro-grids and the corresponding high-dimensional dispatching models. Motivated by these, we here consider an IES with many micro-grids supplies for combinations of cooling, heating and power (CCHP). The structure of such a system is first presented, and then the corresponding model of many-objective based power dispatching is given in detail. In our model, the operational economy and the environment pollution of each micro-grid are taken as optimized objectives. Then, NSGA-III, a powerful evolutionary algorithm for many-objective optimization is used to solve the dispatching model. The effectiveness of the proposed algorithm is experimentally demonstrated by applying it to a practical problem.

Keywords: Integrated energy system · CCHP · Many micro-grids · NSGA-III

1 Introduction

In recent years, global environmental pollution has become increasingly serious, and the trend of primary energy reservations has declined. However, the energy demands from science and technology continue to rise. At this time, it is necessary to transform the traditional single energy supply mode to improve the utilization of renewable energy, and thus to form an environment-friendly energy supplying and consuming. It is an effective way to improve the energy utilization by incorporating renewable energy, grid, gas and heat together in a unified way according to the diversity of energy demands and the complementarity between energy sources. Hence the concept of integrated energy system

L. Pan et al. (Eds.): BIC-TA 2019, CCIS 1159, pp. 147–163, 2020.
https://doi.org/10.1007/978-981-15-3425-6_13

(IES) is proposed. Among them, the application of combined cooling, heating and power (CCHP) is most popular [1–3].

Researches on the optimizations of IES, especially the economic dispatching and system planning, have been carried out in recent years. Geidl [4] firstly proposed the concept of energy center, which addressed the transforming and interactive among different energy sources in the IES and used intelligent optimization to obtain the optimal energy dispatching strategy. Rees [5] established a thermoelectric energy system model based on traditional power demand response, and dispatched thermal and electric energy through thermoelectric relationship. Chang [6] used matrix modeling to optimize the system operation and built a model of CCHP system including wind power generation. Chen [7] considered the uncertain factors in IES, analyzed the electric current and gas trends of the system through the coupling of the gas turbine, electric drive pressure station and energy hub. In the last, the influence of uncertainty factors on the system's probabilistic energy flow was further addressed. Wei [8] introduced a carbon trading mechanism by constructing a low carbon scheduling model. It also considered the impact of price factors and power constraints on system operation.

For different optimization objectives, the optimal method on solving the CCHP model is also an important part in the study of IES. In [9], a Tchebycheff decomposition with l_p-norm constraints on direction vector in which the sub-problem objective functions have intuitive geometric property was proposed. Yuan [10] proposed an improved NGSA-III procedure, called theta-NSGA-III, aiming to better tradeoff the convergence and diversity in many-objective optimization. In theta-NSGA-III, the non-dominated sorting scheme based on the proposed theta-dominance is employed to rank solutions in the environmental selection phase. Liu [11] proposed an improved chaotic particle swarm optimization algorithm based on particle dimension entropy and the greedy mutation strategy to solve the comprehensive energy dispatching cost model. Song [12] applied the taboo search to solve the model of micro-grid energy planning. In [13], the comprehensive energy dispatching model was constructed with the minimum network loss, and the genetic algorithm was used to search the optimal location and configurations of two types of IESs (radiation and mesh). Rezvan [14] designed a genetic algorithm to solve the optimal power supply when considering the load fluctuation. The refined power to gas model proposed in [15] adopted robust and stochastic optimization to deal with the uncertain factors in the system.

The existing researches have proposed a variety of dispatching models and were effectively solved by different intelligent algorithms for an IES. But the supply and demand coupling problems between different energy sources in the IES, the randomness of new energy generation and uncertainties in various types of loads have made it difficult [16,17]. Besides, the energy supply is often supposed to be an integrity, i.e., only one micro-grid with renewable energy is considered. In fact, for an IES, there can be many micro-grid systems, and energy exchange or dispatching also exist among these multiple energy supplies, which has not been considered. In such a case, the dispatching is not only for the CCHP, but also for the multiple micro-grids, therefore, the optimization objectives and constraints can be greatly increased compared with those existing studies. Accordingly, the model construction and optimization should be specially focused.

Motivated by the aforementioned, we here consider the combination of CCHP system supplied with many micro-grid system (MCHP). The corresponding structure of the MCHP is first defined, and then the dispatching model with many objectives is given in detail. The NSGA-III is used to optimize the model by relaxing the constraints. The experiment with three micro-grids is designed to demonstrate the effectiveness of the proposed algorithm.

2 Multi-microgrid IES with CCHP (MCHP)

The structure of the MCHP considered here is shown in Fig. 1. It consists of n micro-grids distributed in different regions. Each micro-grid includes renewable energy sources, e.g., wind and photovoltaic, diesel generator, storages and loads. The CCHP has micro-turbines, lithium bromide refrigerators, ground source heat pumps and cold-heat-electric load.

In the MCHP system, the thermal and cooling load are supplied by the ground source heat pump and the lithium bromide refrigerator. The power exchange can be performed between the micro-grid and the CCHP. Among them, the micro-gas turbine is a rotary heat engine powered by gas and air and it is the main equipment for supplying electricity to CCHP. The waste heat generated by the micro-gas turbine during power generation can be absorbed by the lithium bromide refrigerator. Ground source heat pump has two working modes, i.e., cooling and heating, which can provide both heat and cooling to the corresponding loads.

According to the conversion relationship between electric energy and thermal energy in the IES, its operation mainly includes three modes: determining power based on heat, determining heat based on power and mixing operation [18]. We here perform the energy dispatching on MCHP in the mode of determining power based on heat due to its easy operation [19]. The micro-grid absorbs excessive power generated from CCHP. As can be seen from Fig. 1, energy exchanges will occur among the micro-grids and the cogeneration system.

3 Economic Dispatch Modeling of MCHP

3.1 Modeling of Energy Conversion Equipment

(1) Micro-gas turbine
The residual heat Q_{MT} generated by a micro-gas turbine (MT) is given in Eq. (1) with η_{MT1} being the heat loss coefficient of MT [20].

$$Q_{\mathrm{MT}} = P_{\mathrm{MT}} \frac{1 - \eta_{\mathrm{MT}} - \eta_{\mathrm{NTI}}}{\eta_{\mathrm{MT}}} \tag{1}$$

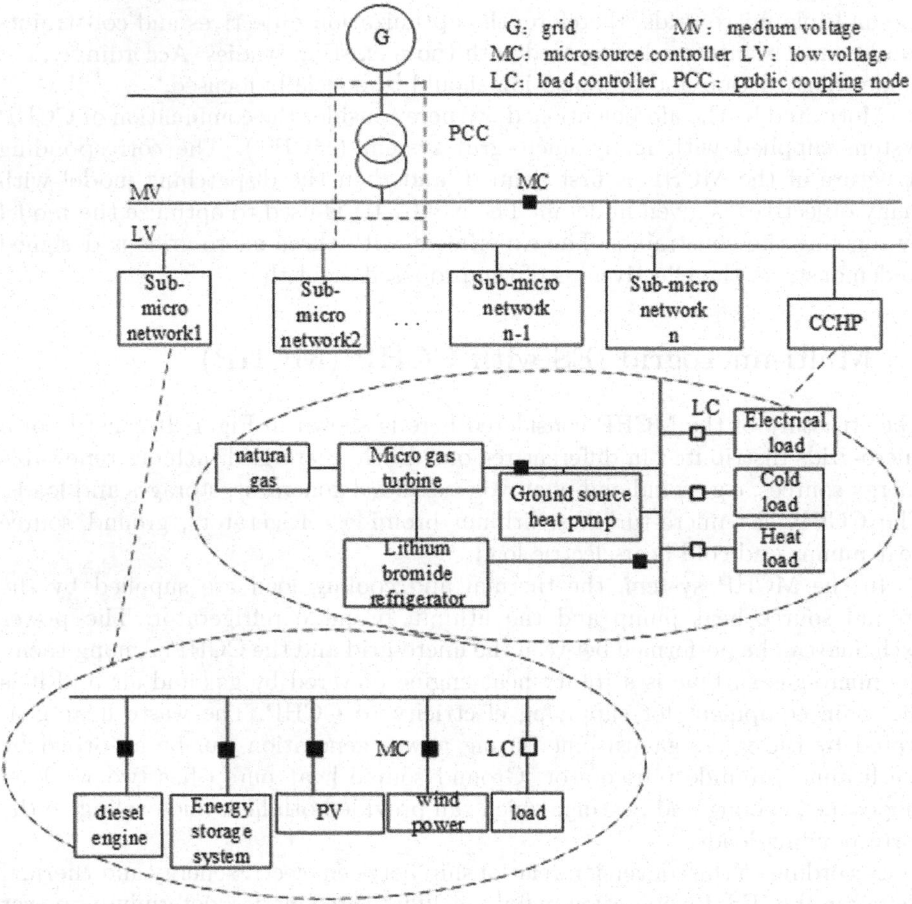

Fig. 1. The structure diagram of MCHP

where, η_{MT} is the conversion rate of a MT and defined as in Eq. (2):

$$\eta_{\mathrm{MT}} = a\left(\frac{P_{\mathrm{MT}}}{P_{\mathrm{MTmax}}}\right)^3 - b\left(\frac{P_{\mathrm{MT}}}{P_{\mathrm{MTmax}}}\right)^2 + c\left(\frac{P_{\mathrm{MT}}}{P_{\mathrm{MTmax}}}\right) + d \tag{2}$$

where P_{MT} and P_{MTmax} are the real and maximum output power respectively; parameters a, b, c and d are the generation coefficients.

(2) Lithium bromide refrigerator
The lithium bromide refrigerator can produce cooling and heat by absorbing the heat generated by a MT, and the mathematical models of its refrigeration and heat production are:

$$Q_{\mathrm{AM,c}} = Q_{\mathrm{MT}}\eta_{\mathrm{AM,c}}\eta_{\mathrm{rec}} \tag{3}$$

$$Q_{\mathrm{AM,h}} = Q_{\mathrm{MT}}\eta_{\mathrm{AM,h}}\eta_{\mathrm{rec}} \tag{4}$$

where $Q_{\text{ec,h}}$ and $Q_{\text{ec,c}}$ are the capacities of the cooling and heating provided by the lithium bromide refrigerator; P_{ec} is the electrical energy required by the ground source heat pump to produce cooling or heat; $\eta_{\text{ec,h}}$ and $\eta_{\text{ec,c}}$ are the converting efficiency of the ground source heat pump.

(3) Ground source heat pump

$$Q_{\text{ec,h}} = \eta_{\text{ec,h}} \times P_{\text{ec}} \tag{5}$$

$$Q_{\text{ec,c}} = \eta_{\text{ec,c}} \times P_{\text{ec}} \tag{6}$$

where $Q_{\text{ec,h}}$ and $Q_{\text{ec,c}}$ are the heating and cooling capacity provided by the heat pump system; P_{ec} is the energy required by the ground source heat pump; $\eta_{\text{ec,h}}$ and $\eta_{\text{ec,c}}$ are the efficiency when converting electrical energy into cooling and thermal energy, which are determined by the performance of the ground source heat pump equipment.

3.2 Many Dispatching Objectives

The stable operation of the CCHP with many micro-grids is the essential condition. The operation economy and environmental protection of the system are comprehensively considered under the condition of effectively and possibly using the renewable energy of each micro-grid.

(1) Economic objective
One day dispatching is considered and divided into 24 periods with time step being 1 h. Then the operating cost of the i-th micro-grid is shown in Eq. (7):

$$\min \quad C_{i\text{MG}} = C_{i\text{DG}} + C_{i\text{ESS}} - \sum_{j=1, j\neq i}^{n} C_{ij} + C_{i,\text{grid}} - C_{i.\text{CCHP}} \tag{7}$$

where, $C_{i.\text{DG}} = a_{i.\text{DG}} P_{i\text{DG}} + b_{i\text{DG}}$ and $C_{i.\text{ESS}} = c_{i.\text{ESS}} P_{i.\text{ESS}}$; parameter $C_{ij} = \kappa_{ij} P_{ij}$ is the transaction cost between the i-th and j-th micro-grids; $C_{i,\text{grid}}$ is the trading function between the i-th and j-th micro-grids. $P_{i\text{DG}}$ and $P_{i.\text{ESS}}$ are the output of diesel engine and energy storage, respectively; $P_{i}j$ is the trading power between the $i-$ th and $j-$ th micro-grids. $P_{ij} = -P_{ji}, P_{ij} > 0$ indicates microgrid i sells electricity to microgrid j. $P_{i.\text{grid}}$ is the amount of electricity traded between the micro-grid i and the power grid. $P_{i.\text{grid}} > 0$ indicates that micro-grid i purchases electricity from the power grid. $a_{i.\text{DG}}$ and $b_{i.\text{DG}}$ are the cost coefficients of diesel engine power generation in micro-grid i; $C_{i.\text{ESS}}$ is the energy cost coefficient of the energy storage system; $\mathcal{K}_{ij}$ is the transaction price; $\mathcal{K}_{\text{grids}}$ and K_{gridb} are the prices that the micro-grid selling electricity to the grid and the micro-grid buying electricity from the grid. $P_{n.\text{CCHP}}$ and $K_{n.\text{CCHP}}$ are the amount and the price of electricity that the micro-grid n purchases form CCHP, respectively. When $P_{n\text{cchp}} > 0$, micro-grid i sells electricity to CCHP.

The cost function C_{CCHP} of CCHP consists of the gas cost C_{fuel}, the electricity purchasing cost $C_{i.\text{buy}}$ from the i-th micro-grid and the electricity purchasing cost C_{gird} from the power grid:

$$\min \quad C_{\text{CCHP}} = C_{\text{fuel}} + \sum_{i=1}^{n} C_{i.\text{buy}} \tag{8}$$

where $C_{\text{fuel}} = \kappa_{\text{ng}} V_{\text{MT}} = \kappa_{\text{ng}} \frac{P_{\text{MT}}}{\eta_{\text{MT}} \text{LHV}_{\text{ng}}}$, $C_{i.buy} = -C_{i.ccup}$, K_{ng} is the natural gas price, V_{MT} is the amount of natural gas consumed by the gas turbine, P_{MT} is the power generation of gas turbine, η_{MT} is the natural gas conversion rate of gas turbine, LHV_{ng} is the low calorific value of natural gas.

Accordingly, the economic operation dispatching models of our system can be described with the following two expressions:

$$\min \quad C_{i,\text{MG}} = C_{i\text{DG}} + C_{i.\text{ESS}} - \sum_{j=1, j\neq i}^{n} C_{ij} + C_{i,\text{grid}} - C_{i.\text{CCHP}} \tag{9}$$

$$\min \quad C_{\text{CCHP}} = C_{\text{ftel}} + \sum_{i=1}^{n} C_{i,\text{buy}} \tag{10}$$

(2) Environmental cost
The environmental cost is to minimize the pollutant emissions F during the system operation period. The micro-gas turbine in the system burns natural gas and mainly produces carbon oxides C_xO_y. The pollutants generated by diesel engine power generation and grid power supply include CO_2, SO_2 and NO_x. Therefore, the optimal objective function of environmental protection is as follows:

$$\min \quad F = \sum_{i=1}^{n} \sum_{r=1}^{R} \sigma_r \left(P_{i\text{DG}} + P_{i,\text{grid}}\right) + vP_{\text{MT}} \tag{11}$$

where R is the number of types of pollutants, $r = 1, 2, \ldots R$ represents the r-th pollutant, σ_r is the emission factor of the r-th pollutant; v is the pollutant production coefficient of natural gas burning when the micro gas turbine generates unit electricity.

3.3 Constraints

In order to ensure the stable operation of the MCHP system, that is, supply and demand balance, the following constraints must be met: electric load balance of CCHP, electric load balance of micro-grid, heat and cold load balance. The following constraints are given:

$$P_{i.\text{DG}} + P_{i.\text{ESS}} + P_{i.\text{PV}} + P_{i.\text{WP}} + P_{i.\text{grid}} - \sum_{m=1, m\neq i}^{N} P_{im} - P_{i.\text{CCHP}} = P_{i.\text{L}} \tag{12}$$

$$P_{\mathrm{MT}} + \sum_{i=1}^{n} P_{i,\mathrm{buy}} = P_{\mathrm{L.CCHP}} \tag{13}$$

$$Q_{\mathrm{AM.h}} + Q_{\mathrm{ec.h}} = Q_h \tag{14}$$

$$Q_{\mathrm{AM.c}} + Q_{\mathrm{ec.c}} = Q_c \tag{15}$$

Where Q_c and Q_h are the cold and heat loads in the integrated energy system respectively; $P_{\mathrm{L.CCHP}}$ is the electrical load of the integrated energy system.

The micro-gas turbine, diesel engine and energy storage need to meet the output constraint and climbing constraints when working: $P_{\mathrm{MT\,min}} \leq P_{\mathrm{MT}} \leq P_{\mathrm{MT\,max}}$, $-R_{\mathrm{MTdown}} \leq P^{t}_{\mathrm{MTDG}} - P^{t-1}_{\mathrm{MTDG}} < R_{\mathrm{MTup}}$, $P_{i.\mathrm{DGmin}} \leq P_{i.\mathrm{DG}} \leq P_{i.\mathrm{DGmax}}$, $-R_{i,\mathrm{down}} \leq P^{t}_{i\mathrm{DG}} - P^{t-1}_{i\mathrm{DG}} < R_{i.\mathrm{up}}$, $P_{i.\mathrm{WP}} \in \left[P^{\min}_{i.\mathrm{WP}}, P^{\max}_{i.\mathrm{WP}}\right]$, $P_{i.\mathrm{PV}} \in \left[P^{\min}_{i.\mathrm{PV}}, P^{\max}_{i.\mathrm{PV}}\right]$.

4 Optimization of Dispatching Mode Based on NSGA-III

4.1 Cool and Heat Constraints Conversion Based on Dispatching Rules

The specific dispatching rules in the condition of "determining heat based on power" are R1: Determine the amount of heat and cool load required by CCHP and judge whether it is greater than the maximum heat capacity that the system itself can provide. R2: If the heat and cold load is greater than the system's own heat supply, the system will purchase electricity from other micro-grids to start the refrigerant and heat pump. R3: If the self-heating capacity meets the load demand, it will be provided by the system's own gas turbine. R4: Calculate the total power load demand of the whole system and start the operation dispatching optimization strategy to make sure the optimal energy supply combination of each energy supply end. The dispatching rules are shown in Fig. 2.

Under the above-mentioned rules, the electric energy in the CCHP system is determined by the cool and heat load. Therefore, the cool and heat loads in the CCHP are first converted into corresponding electric loads through the energy converter. Accordingly, the total electrical load $P_{\mathrm{L.CCHP}}$ required by the CCHP system is:

$$P_{\mathrm{L.CCHP}} = P_{\mathrm{Lh}} + P_{\mathrm{L.c}} + P_{\mathrm{Load}} \tag{16}$$

P_{Load} is the pure electric load of CCHP system, P_{Lh} is the electric load required to meet the thermal load and $P_{\mathrm{L.c}}$ is the electric load required to meet the cooling load of the system. When the number of sub-grids is large, the number of constraints also increases sharply.

4.2 Algorithm Implementation

It can be seen that if the system studied in this paper contains $n (n \geq 3)$ micro-grids, the dispatching optimization model of the system contains $n+2$ nonlinear high-dimensional optimization objectives, and also contains a large amount of uncertainty and multiple constraints. It is extremely difficult to find the optimal

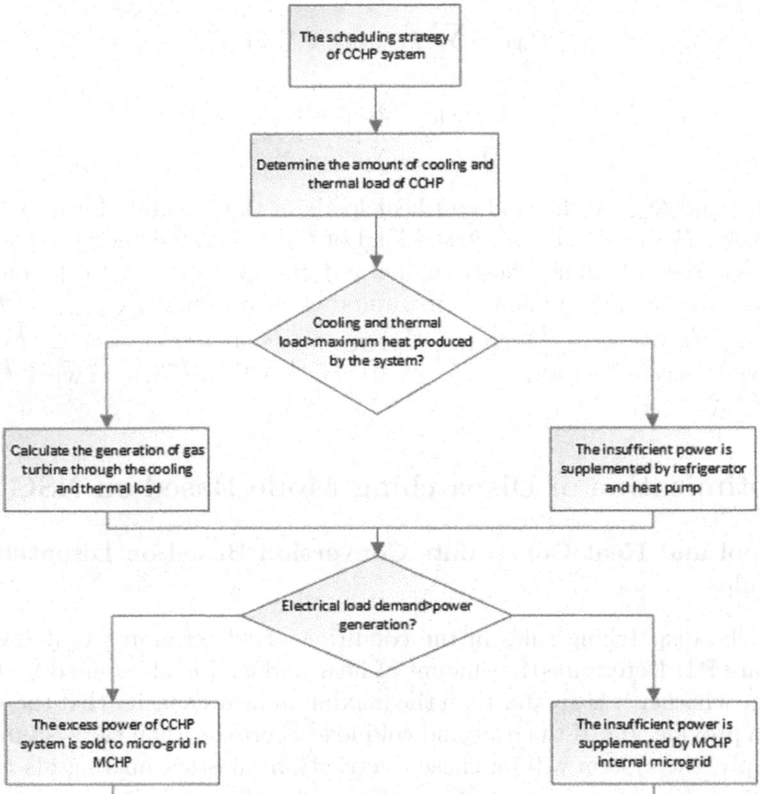

Fig. 2. Dispatching rules based on "Determining heat based on power"

solution using traditional optimization methods. Many evolutionary algorithms have been proposed to solve such problems [21,22]. Here, the optimization problem with high-dimensional objectives can be solved with the NSGA-III algorithm [23]. The flow chart of NSGA-III can been seen in Fig. 3.

Both wind and photovoltaic power generation are operated in the maximum power tracking mode, then the decision variables of the i-th micro-grid are $[P_{i.\text{DG}}, P_{i.\text{ESS}}, P_{ij}, P_{i.\text{grid}}, P_{i.\text{WP}}, P_{i.\text{PV}}, \kappa_{ij}]$. The optimization variables of CCHP are P_{MT}, P_{ec}. For convenient description, the decision variable of the i-th micro-grid is defined as $\mathbf{X}_n$ and the objective function is $F(\mathbf{X}_n)$, then the total decision variable X of the system is $X = [X_1, X_2, \ldots, X_n, P_{MT}, P_{ec}]$. That is, for a CCHP system with n microgrids, the total decision variable dimension is $7n + 2$.

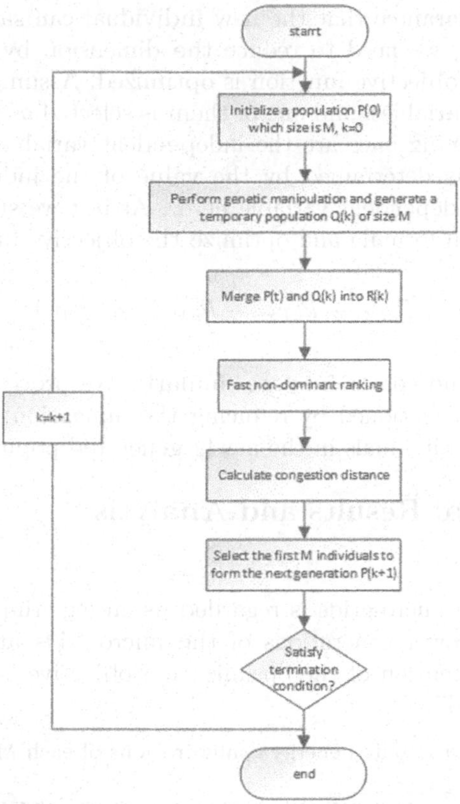

Fig. 3. The flow chart of NSGA-III

Obviously, in addition to the characteristics of high-dimensional multi-objectives, a large number of constraints further increase the difficulty in solving the model. This paper presents a penalty function method based on constrained relaxation to increase the probability of finding a feasible optimal solution. It can be known from the constraint expression that except for the decision bounds of the decision variables, most of the constraints are equality constraints. Therefore, the constraints are handled by adding a penalty function to all high-dimensional functions of multiple objectives for relaxation:

$$\min F_i(\mathbf{x}) = \mathbf{f_i}(\mathbf{x}) + \mathbf{c}\sum_{\mathbf{j}}^{\mathbf{l}} (\mathbf{h_j}(\mathbf{X}))^{\mathbf{2}} \tag{17}$$

Where c is the penalty coefficient, $h_j(X)$ is the j-th constraint, and l is the total number of constraints.

However, the convergence becomes worse as the number of objectives increases when using the penalty function method, although the probability that the inappropriate solutions eliminated can be increased by increasing the fitness value, there is no guarantee that the new individual can satisfy the constraint condition. Therefore, we need to reduce the dimension by using the equality constraint when the objective function is optimized. Assume that the equation constraint contains variables, and one of them is selected as the dependent variable, then the remaining ones are the independent variables, and the value of dependent variable is determined by the value of the independent variables. $P_{i\text{grid}}$ is used as the dependent variable here . At last we substitute $P_{i\text{grid}}$ into the objective function formula and optimize the objective function.

$$P_{i.\text{grid}} = P_{i.\text{L}} - P_{i.\text{DG}} - P_{i.\text{ESS}} - P_{i.\text{PV}} - P_{i.\text{WP}} + \sum_{m=1, m \neq i}^{N} P_{im} \tag{18}$$

The remaining load constraints are similarly. We process the equality constraint of the above-mentioned by reducing the dimension, it can increase the number of satisfied individuals in the newly generated population.

5 Experimental Results and Analysis

5.1 Background

A MCHP with three micro-grids is regarded as an experimental platform, and the corresponding power generations of the micro-grids are given in Table 1. Accordingly, the dimension of the optimization objective is 5, and that of the decision variables is 37.

Table 1. Distributed energy configurations of each Micro-grid

	WP/kW	PV/kW	DG/kW	ESS/kWh
MG1	120	–	200	50
MG2	–	90	200	50
MG3	80	100	200	50

The upper and lower limits of the diesel engine and energy storage system in the above three micro-grids, the climbing rate constraint, the parameters related to the output power, the initial energy values of the energy storage system, the energy storage cost and the charging and discharging efficiency are referred from [24]. The upper and lower limits of the micro gas turbine power in CCHP are 15 kW and 65 kW, and the climbing rate ranges from 5 to 10 kW/min. The related parameters of output efficiency and combustion parameters please refer to [25]. The energy efficiency coefficients of lithium bromide refrigerator are set as: $\eta_{\text{AM,c}} = 1.36$, $\eta_{\text{AM,h}} = 1.2$, its flue gas recovery rate $\eta_{\text{rec}} = 0.85$ [26]; The energy efficiency coefficients of ground source heat pump are $\eta_{\text{ec,h}} = 3$ and $\eta_{\text{ec,c}} = 3$ [27]. When considering environmental targets, it involves pollutants generated by diesel engines and power grids. The emission factor of each pollutant when outputting unit energy is referred to [28].

The prediction of renewable energy output and various loads on a typical day of a CCHP, corresponding predictions of an micro-grid are shown in Fig. 4.

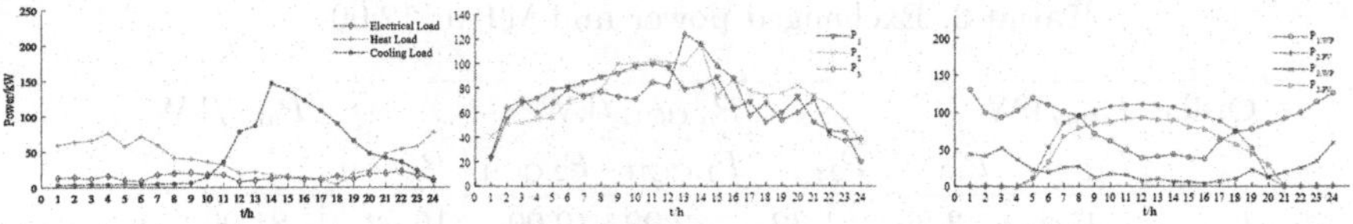

(a) Load of CCHP (b) Load of Micro-grid in MCHP (c) Renewable energy power

Fig. 4. Forecast on loads and renewable power

5.2 Results and Analysis

As we have addressed before, five objectives with 37 variables need to be optimized with NSGA-III under lots of constraints. For clear demonstration, we here take the detailed dispatching results of P.M. 12 o'clock of the typical day given in Fig. 5 as an example. The renewable energy generated in the MCHP system of the typical day can be seen from Fig. 5. Then, we randomly select 10 solutions from the Pareto set of NSGA-III, which are listed in Table 2.

Table 2. Randomly selected optimal solutions from Pareto set based on NSGA-III at 12:00

Order	MG1			MG2			MG3		
	$P_{1.\mathrm{DG}}$	$P_{1.\mathrm{ESS}}$	$P_{1.\mathrm{grid}}$	$P_{2.\mathrm{DG}}$	$P_{2.\mathrm{ESS}}$	$P_{2.\mathrm{grid}}$	$P_{3.\mathrm{DG}}$	$P_{3.\mathrm{ESS}}$	$P_{3.\mathrm{grid}}$
1	0.00	9.92	2.92	1.17	11.98	32.93	0.49	9.94	−14.44
2	0.02	10.00	0.11	1.22	3.50	41.10	0.57	6.64	−13.32
3	2.45	−7.12	9.65	0.12	−1.71	21.55	0.24	19.91	−38.90
4	9.00	−7.25	41.94	0.12	27.96	−26.86	0.05	16.48	−28.25
5	0.07	9.89	4.07	1.15	8.69	36.69	0.71	12.79	−16.94
6	11.96	−9.28	53.88	0.31	28.53	−31.32	1.15	18.18	−36.42
7	0.00	10.00	−7.10	1.56	−5.66	46.59	2.03	−22.28	18.31
8	0.20	−7.59	−1.20	0.73	28.58	−1.22	1.78	−17.41	−0.67
9	8.27	−9.77	48.52	0.16	24.59	−23.71	0.17	19.85	−32.30
10	0.13	10.00	−7.56	1.64	−5.75	47.79	2.04	−22.55	18.12

Table 3 shows the corresponding 10 optimization results of the interactions between the micro-grids in the MCHP system and the CCHP system. It also shows the gas turbine output at 12:00.

From these two Tables, we can clearly observe the energy exchanges among the micro-grids and CCHP, indicating that multi-microgrids in an IES is necessary for obtaining an optimized dispatching.

As is well known, the optimized solutions of NSGA-III are a nondominated set, and few solutions must be selected to make decisions. First, from the viewpoint of minimizing the entire cost of the MCHP system, the results of the diesel

Table 3. Exchanged power and MT at 12:00

Order	P_{nm}/kW			$P_{n.\mathrm{CCH}}$/kW			P_{MT}/kW
	P_{12}	P_{13}	P_{23}	$P_{1.\mathrm{CCH}}$	$P_{2.\mathrm{CCH}}$	$P_{2.\mathrm{CCH}}$	
1	−47.41	3.26	−1.32	−2.02	10.00	16.92	81.09
2	−47.73	3.49	−1.40	−4.63	9.49	14.98	86.16
3	−30.46	6.44	−0.50	−30.00	0.00	6.19	129.81
4	−3.74	9.80	1.65	−21.36	5.82	18.74	102.81
5	−47.97	4.33	−1.37	−1.34	9.93	18.53	78.88
6	−0.26	10.00	1.14	−12.18	6.12	13.05	99.01
7	−49.03	−5.00	−2.00	−2.07	5.46	10.06	92.55
8	−37.70	−2.65	0.14	−27.25	0.26	0.19	132.80
9	−3.21	9.69	2.00	−18.46	5.83	18.42	100.21
10	−50.00	−4.62	−1.96	−1.81	5.63	10.03	92.14

generator output, energy storage output and energy interactions among the grids within 24 h of the system are shown in Fig. 5. The exchanged power of the three micro-grids is shown in Fig. 6.

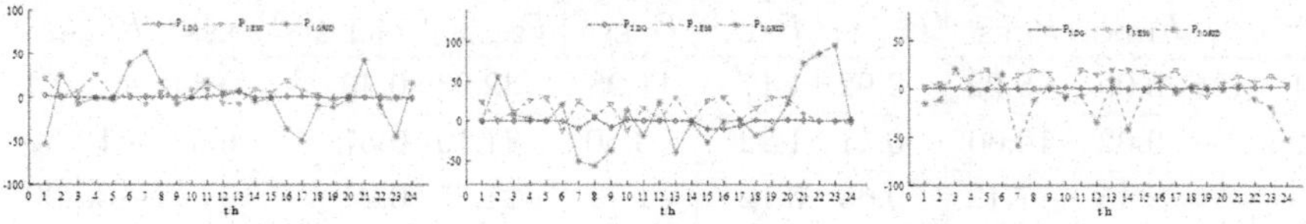

(a) Results of MG1 (b) Results of MG2 (c) Results of MG3

Fig. 5. Optimized results of 3 micro-sources

According to the variation of the energy sources between the adjacent time periods from Fig. 5, it can be seen that the optimization results of the proposed model satisfy the output constraints and climbing constraints of each energy source, so it is feasible. As for the power exchange among the three micro-grids shown in Fig. 6, it can be concluded that the energy interactions of the system in different time periods are different. For example, the energy interactions between the micro-grids 1, 3 and the micro-grid 2 are concentrated from 1 to 8 o'clock, 19 to 24 o'clock, while the power exchange between micro-grid 1 and 3 is more obvious during 9 to 18 o'clock.

Figure 7 demonstrates the 24-h optimal dispatching for CCHP in a multi-microgrid system with cogeneration, including the power generated by the micro-turbine, the exchanged power among the CCHP system and the threes micro-grid. Combining the CCHP system load forecasting situation in Fig. 4, it can be seen from Fig. 7 that the CCHP system in the MCHP system can interact

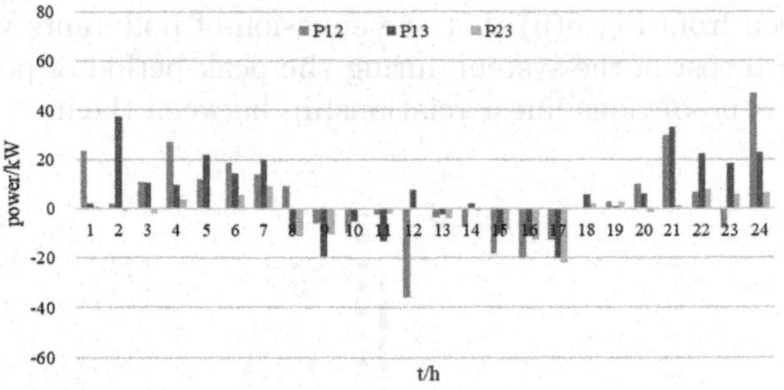

Fig. 6. Interactive power between Micro-grids

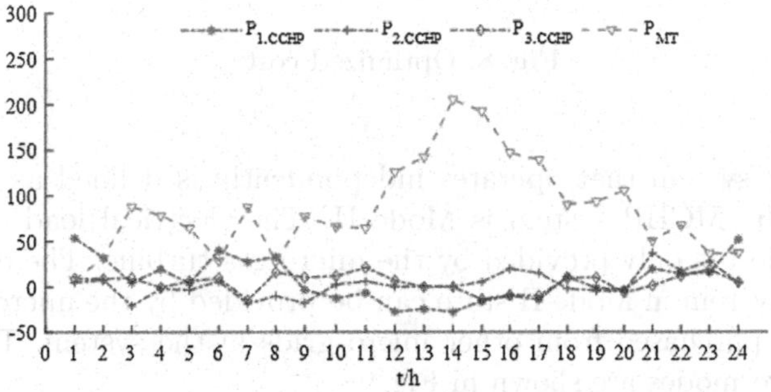

Fig. 7. CCHP System optimal dispatching results

with the micro-grid to significantly reduce the operating pressure of the micro gas turbine. In addition, it is not difficult to find that the power exchange between the micro-grid 1 and the CCHP system is more frequent during the valley electricity price. We can see that the renewable energy of micro-grid 1 is greater than the load required in Fig. 4(b) and (c), so CCHP can absorb the such power in the micro-grid and improve the utilization of renewable energy.

The corresponding cost of each micro-grid and CCHP, together with the total system cost and pollutant emissions of the optimized dispatching are given in Fig. 8 (the histogram and line chart represent the total system cost and pollutant emissions respectively). It can be seen from Fig. 8(a) that micro-grid 1 is mostly in the state of profit, because the predicted power of the renewable energy of micro-grid 1 during the valley electricity price is greater than its own load requirement, some of the remaining electricity can be sent to other micro-grids or sold to CCHP. Other micro-grids have different levels of cost reduction and even revenue generation according to their own renewable energy generation and load forecasting. The CCHP system has a slightly higher cost than the other micro-grids because of the large thermal load and the absence of renewable energy supply.

It can be seen from Fig. 8(b) that the emission of pollutants varies with the trend of the total cost of the system during the peak period of power consumption, but there is no obvious linear relationship between them.

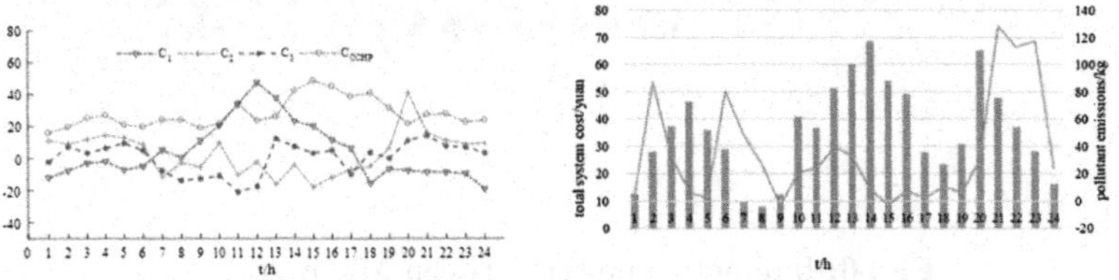

(a) Optimized cost of MGs and CCHP

(b) Total system cost and emissions of pollutants

Fig. 8. Optimized cost

The CCHP system that operates independently is defined as Mode I. The operation in the MCHP system is Mode II. The electrical load of the CCHP system in Mode I is only provided by the micro gas turbine. The electrical load of the CCHP system in Mode II state can be provided by the micro gas turbine. It can also be purchased from other micro-grids in the system. The operating costs in the two modes are shown in Fig. 9.

The blue histogram and the orange histogram represent Mode 1 and Mode II respectively. It can be seen from Fig. 9 that during the peak period of power consumption such as 8:00, 16:00, 18:00 and 19:00, the operating cost of Mode I is lower than that of Mode II. Mode II is more economical than Mode I in most of the time because the load of each micro-grid is large, the grid electricity price is high during the peak period. The micro-grid purchases electricity from CCHP, which leads to an increase in the operating cost of the CCHP system. It is calculated that the total 24-h cost of the CCHP system operating in Mode I is 731.19 yuan and the total cost of Mode II is 669.46 yuan, so the economy of the CCHP system operating in Mode II is slightly better than Mode I.

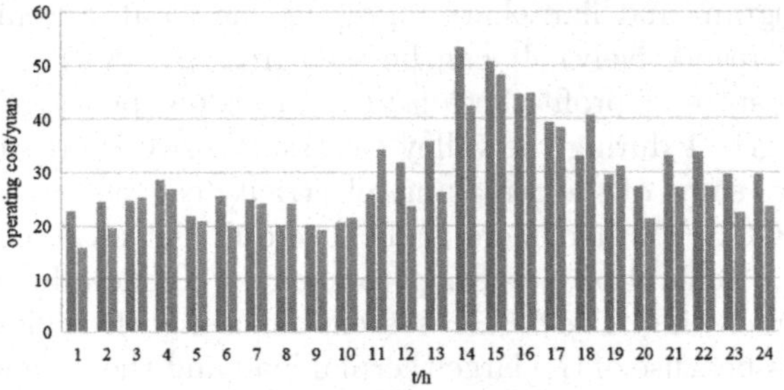

Fig. 9. Cost of CCHP in different modes

Table 4 gives the 24-h cumulative operating costs of the dispatching results on the micro-grid, CCHP, total system costs and pollutant emissions in the MCHP. It can be seen that the total operating cost of our MCHP system is reduced by 7.22%, and the pollutant emissions are reduced by 55.38%, so the model and the obtained solutions are more economical and environment-friendly.

Table 4. 24-h cumulative dispatching results for two modes

	C_1/yuan	C_2/yuan	C_3/yuan	C_{CCHP}/yuan	C_{all}/yuan	F/kg
MCHP	104.03	74.69	8.51	669.46	856.69	467.97
CCHP	67.95	120.58	3.64	731.19	923.36	621.13

In summary, the IES with multi-microgrid system and CCHP optimized with NSGA-III can effectively absorb the power of the renewable energy of the micro-grid and meet the needs of its own load consumption. The micro-grid can also purchase the surplus electric energy generated by the cogeneration system when generating heat energy at a lower price, thereby improving energy utilization, reducing the electrical energy interaction between the micro-grid, CCHP and the grid, and reducing the pollutant emissions.

6 Conclusions

Dispatching of an IES with many micro-grids and CCHP is considered here. We first present the structure of such a system by containing renewable suppliers in each micro-grid. Then, we further give the optimization model by using the economic and environment indices as the objectives, and derive a many-objective optimization under lots of constraints. The constraints are relaxed based on the dispatching rules of CCHP and used in NSGA-III for intelligently obtaining the optimized dispatching results. The proposed method is applied to an IES with three micro-grids, and the results show its feasibility and effectiveness.

References

1. Jin, M., Feng, W., Marnay, C., Spanos, C.: Microgrid to enable optimal distributed energy retail and end-user demand response. Appl. Energy **210**, 1321–1335 (2018)
2. Chen, J., Liu, J., Wang, Q., Zeng, J., Yang, N.: A multi-energy microgrid modelling and optimization method based on exergy theory. In: 2018 Chinese Automation Congress (CAC), pp. 483–488. IEEE (2018)
3. Long, T., Zheng, J., Zhao, W.: Optimization strategy of CCHP integrated energy system based on source-load coordination. In: 2018 International Conference on Power System Technology (POWERCON), pp. 1781–1788. IEEE (2018)
4. Geidl, M., Andersson, G.: Optimal power flow of multiple energy carriers. IEEE Trans. Power Syst. **22**(1), 145–155 (2007)

5. Rees, M., Wu, J., Awad, B.: Steady state flow analysis for integrated urban heat and power distribution networks. In: 2009 44th International Universities Power Engineering Conference (UPEC), pp. 1–5. IEEE (2009)
6. Chang, Y., Wang, H., Xie, H., Zhang, L.: Operation optimization of cogeneration system using matrix modeling. J. Huaqiao Univ. (Nat. Sci.) **39**(2), 233–239 (2018)
7. Chen, S., Wei, Z., Sun, G.: Probabilistic energy flow analysis of electric-gas hybrid integrated energy system. Proc. CSEE **35**(24), 6331–6340 (2015)
8. Wei, Z., Zhang, S., Sun, G., et al.: Low-carbon economy operation of electric-gas interconnected integrated energy system based on carbon trading mechanism. Autom. Electr. Power Syst. **40**(15), 9–16 (2016)
9. Ma, X., Zhang, Q., Tian, G., et al.: On Tchebycheff decomposition approaches for multiobjective evolutionary optimization. IEEE Trans. Evol. Comput. **22**(2), 226–244 (2018)
10. Yuan, Y., Xu, H., Wang, B., et al.: An improved NSGA-III procedure for evolutionary many-objective optimization. In: Christian (eds.) GECCO 2014, pp. 661–668. ACM, Vancouver (2014). http://doi.acm.org/10.1145/2576768.2598342
11. Liu, H., Chen, X., Li, J., et al.: Economic dispatch of regional electrothermal integrated energy system based on improved CPSO algorithm. Electr. Power Autom. Equip. **37**(6), 193–200 (2017)
12. Song, Y., Wang, Y., Yi, J.: Microgrid energy optimization planning considering demand side response and thermal/electrical coupling. Power Syst. Technol. **42**(11), 3469–3476 (2018)
13. Singh, R.K., Goswami, S.K.: Optimum siting and sizing of distributed generations in radial and networked systems. Electr. Mach. Power Syst. **37**(2), 127–145 (2009)
14. Rezvan, A.T., Gharneh, N.S., et al.: Optimization of distributed generation capacities in buildings under uncertainty in load demand. Energy Build. **57**(1), 58–64 (2013)
15. Zhu, L., Wang, J., Tang, L., et al.: Robust stochastic optimal scheduling for integrated energy systems with refinement model. Power Syst. Technol. **43**(1), 116–126 (2019)
16. Wang, S., Wu, Z., Zhuang, J.: Optimal scheduling model for multi-microgrid in combined cooling and cogeneration mode considering power interaction between microgrids and micro-source output coordination. Proc. CSEE **37**(24), 7185–7194 (2017)
17. Wang, J., Zhong, H., Xia, Q., et al.: Optimal joint-dispatch of energy and reserve for CCHP-based microgrids. IET Gener. Transm. Distrib. **11**(3), 785–794 (2017)
18. Hu, R., Ma, J., Li, Z., et al.: Optimization configuration and applicability analysis of distributed cogeneration system. Power Syst. Technol. **41**(2), 83–90 (2017)
19. Ren, H., Xu, P., Wu, Q.: Analysis on the concept and method of planning and design of distributed cogeneration system. Power Syst. Technol. **42**(3), 722–730 (2018)
20. Shi, Q.: Research on microgrid capacity optimization configuration and energy optimization management. Zhejiang University (2012)
21. Pan, L., He, C., Tian, Y., Wang, H., Zhang, X., et al.: A classification-based surrogate-assisted evolutionary algorithm for expensive many-objective optimization. IEEE Trans. Evol. Comput. **23**(1), 74–88 (2018)
22. Pan, L., Li, L., He, C., et al.: A subregion division-based evolutionary algorithm with effective mating selection for many-objective optimization. IEEE Trans. Cybern. **PP**(99), 1–14 (2019)

23. Deb, K., Jain, H.: An evolutionary many-objective optimization algorithm using reference-point-based nondominated sorting approach, part i: solving problems with box constraints. IEEE Trans. Evol. Comput. **18**(4), 577–601 (2014)
24. Zhao, T., Chen, Q., Zhang, J., et al.: Interconnected microgrid energy/standby shared scheduling based on adaptive robust optimization. Power Syst. Technol. **40**(12), 3783–3789 (2016)
25. Luo, Y., Liu, M.: Environmental economic dispatching of isolated network system considering risk reserve constraints. Power Syst. Technol. **37**(10), 2705–2711 (2013)
26. Yang, Z., Zhang, F., Liang, J., et al.: Economic operation of cogeneration microgrid with heat pump and energy storage. Power Syst. Technol. **42**(6), 1735–1743 (2018)
27. Guo, Y., Hu, B., Wan, L., et al.: Short-term optimal economic operation of cogeneration microgrid with heat pump. Autom. Electr. Power Syst. **39**(14), 16–22 (2015)
28. Lin, W., Le, D., Mu, Y., et al.: Multi-objective optimal hybrid power flow algorithm for regional integrated energy system. Proc. CSEE **37**(20), 5829–5839 (2017)

Multiobjective Particle Swarm Optimization with Directional Search for Distributed Permutation Flow Shop Scheduling Problem

Wenqiang Zhang[1(✉)], Wenlin Hou[1], Diji Yang[1], Zheng Xing[1], and Mitsuo Gen[2]

[1] Henan University of Technology, Zhengzhou, China
zhangwq@haut.edu.cn
[2] Fuzzy Logic Systems Institute, Tokyo University of Science, Tokyo, Japan

Abstract. The distributed permutation flow shop scheduling problem (DPFSP) is a variant of the permutation flow shop scheduling problem (PFSP). DPFSP is closer to the actual situation of industrial production and has important research significance. In this paper, a multiobjective particle swarm optimization with directional search (MoPSO-DS) is proposed to solve DPFSP. Directional search strategy are inspired by decomposition. Firstly, MoPSO-DS divides the particle swarm into three subgroups, and three subgroups are biased in different regions of the Pareto front. Then, particles are updated in the direction of the partiality. Finally, combine the particles of the three subgroups to find the best solution. MoPSO-DS updates particles in different directions which speed up the convergence of the particles while ensuring good distribution performance. In this paper, MoPSO-DS is compared with the NAGA-II, SPEA2, MoPSO, MOEA/D, and MOHEA algorithms. Experimental results show that the performance of MoPSO-DS is better.

Keywords: Distributed permutation flow shop scheduling problem · Particle swarm optimization · Directional search

1 Introduction

The distributed permutation flow shop scheduling problem (DPFSP) is a practical problem in industrial production, DPFSP is a variant of the permutation flow shop scheduling problem (PFSP). DPFSP adds the concept of a multi-factory to the PFSP. The machines of each factory are identical. Solving the DPFSP need to assign the jobs to factories, then arrange the order of jobs in each factory [12]. The job ordering problem in each factory is a PFSP. In this paper, the goal is to minimize the makespan and maximum flow time.

The PFSP proved to be a NP-hard problem [6], and DPFSP is a more complicated problem than PFSP. So, the traditional mathematical method is difficult

L. Pan et al. (Eds.): BIC-TA 2019, CCIS 1159, pp. 164–176, 2020.
https://doi.org/10.1007/978-981-15-3425-6_14

to solve DPFSP in an effective time, and intelligent optimization algorithm is needed to solve DPFSP. Li et al. proposed an improved artificial bee colony algorithm for DPFSP [9]. Chen et al. proposed a collaborative optimization algorithm to solve the problem effectively [2]. Jiang et al. proposed a modified multiobjective evolutionary algorithm based on decomposition for DPFSP [8].

Particle Swarm Optimization (PSO) is a kind of meta-heuristic algorithm, which is particle-based stochastic optimization technique proposed by Eberhart and Kennedy in 1995 [4]. PSO has obvious advantages in solving NP-hard problems. Zhao et al. proposed a factorial based PSO with a population adaptation mechanism for the no-wait flow shop scheduling problem [21]. Eddaly et al. proposed a hybrid combinatorial PSO as a resolution technique for solving blocking PFSP [5]. Wang et al. proposed a particle swarm optimization based clustering algorithm with mobile sink for wireless sensor network [16].

In actual factory manufacturing, DPFSP requires multiple objectives to be considered simultaneously, and these objectives may be in conflict. Therefore, the traditional PSO algorithm can not solve such problems well. Coello et al. proposed a multiobjective PSO, which allows the PSO algorithm to be able to deal with multiobjective optimization problems [3]. Han et al. proposed an adaptive multiobjective PSO algorithm based on a hybrid framework of the solution distribution entropy and population spacing information to improve the search performance in terms of convergent speed and precision [7]. However, using population spacing may result in a large amount of computation. Al Moubayed et al. proposed DMOPSO: MOPSO based on decomposition and dominance with archiving using crowding distance in objective and solution spaces [1], which incorporates dominance with decomposition used in the context of multi-objective optimisation. Adding a decomposition strategy can enhance the distribution of the algorithm, but too much decomposition will prematurely converge.

In this paper, a multiobjective particle swarm optimization with directional search (MoPSO-DS) is proposed for the DPFSP, MoPSO-DS algorithm is inspired by the idea of decomposition. The MoPSO-DS algorithm divides the particle swarm into three subgroups, which are biased to different regions of the Pareto front, and the particle update direction is determined by the direction of the preferred region. Finally, the particles are combined to find the optimal solution. This form of update speeds up the convergence of the particles towards the Pareto front while ensuring a good solution distribution. The proposed MoPSO-DS algorithm is compared with the MoPSO, NSGA-II, SPEA2, MOEA/D, and Fast Multi-objective Hybrid Evolutionary Algorithm (MOHEA) [19]. Because MoPSO, NSGA-II, SPEA2 and MOEA/D are classical evolutionary algorithms, MOHEA are similar to this algorithm.

This paper is organized as follows: Sect. 2 introduces the DPFSP formulation. The proposed MoPSO-DS algorithm is described in Sect. 3. The experiment data of the MoPSO-DS algorithm will be discussed in Sect. 4. Finally, the conclusion is shown in Sect. 5.

2 Problem Formulation

The DPFSP can be briefly described as follows: There are f identical factories, each of which is a flow shop with m machines. There are n jobs to process, and each job can be assigned to any factory. After the assignment is determined, the job cannot be transferred to another factory. Each job will be processed sequentially on machines $1, \ldots, m$, and the job processing orders on the m machines are the same. An illustration of the DPFSP is shown in Fig. 1. The processing time $d_{(i,j,k)}$ of the job i on the machine j in factory k is known and does not change when the factory changes. In addition, at any time, each machine can process up to one job, and each job can be processed on up to one machine. The goal is to reasonably assign jobs to the factories and determine the order of jobs for each factory. In this paper, the scheduling goal is to minimize the makespan and maximum flow time.

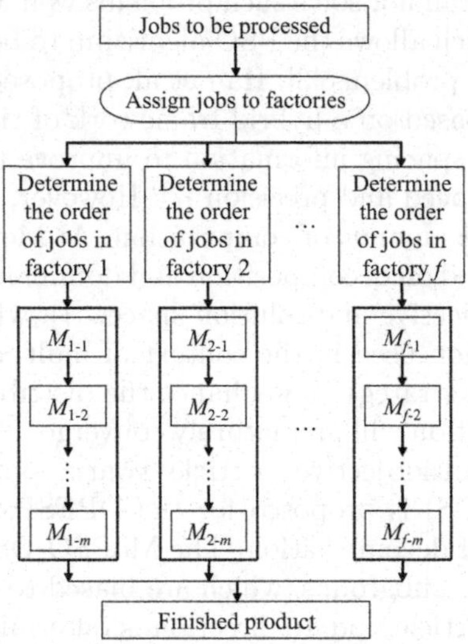

Fig. 1. An illustration of the DPFSP.

Let $X_k = \{X_k(1), X_k(2), \ldots, X_k(n_k)\}$ is defined as the job processing sequence of factory k, n_k is the number of jobs in factory k, $C_{(i,j,k)}$ is the completion time of job i on machine j in factory k. Then, the makespan C_{max} and the maximum flow time FT can be calculated as follows:

$$C_{(1,1,k)} = d_{(1,1,k)}, k = 1, 2, \ldots, f \tag{1}$$

$$C_{(i+1,1,k)} = C_{(i,1,k)} + d_{(i+1,1,k)}, i = 1, 2, \ldots, n-1; k = 1, 2, \ldots, f \tag{2}$$

$$C_{(1,j+1,k)} = C_{(1,j,k)} + d_{(1,j+1,k)}, j = 1, 2, \ldots, m-1; k = 1, 2, \ldots, f \tag{3}$$

$$C_{(i,j,k)} = \max\{C_{(i,j-1,k)}, C_{(i-1,j,k)}\} + d_{(i,j,k)}, i = 2, \ldots, n; \quad j = 2, \ldots, m; k = 1, 2, \ldots, f \tag{4}$$

$$C_{\max} = \max\{C_{(n_k,m,k)}\}, k = 1, 2, \ldots, f \tag{5}$$

$$FT = \sum_{k=1}^{f} \sum_{i=1}^{n_k} C_{(i,m,k)} \tag{6}$$

3 Multiobjective Particle Swarm Optimization with Directional Search

In this section, the details of proposed MoPSO-DS algorithm are introduced. The obvious feature of this algorithm is to update the particles from the concept of Pareto front. According to the characteristics of the particles in different regions of Pareto front, the particles are divided into multiple regions, and the particles in each region are updated in a specific direction.

3.1 Overview of the MoPSO-DS

The algorithmic framework of MoPSO-DS is shown in Fig. 2. First initialize the population, the individual historical optimal group (*pbestset*) and global optimal group (*gbestset*). Then, the following process is iteratively run in MoPSO-DS: divide subgroups, update particles in subgroups, merge subgroups into a new particle swarm, and update *pbestset* and *gbestset* according to the new particle swarm. Note that the direction of the three subgroups to update the particles is different. Once the termination condition is met, particles in the *gbestset* will be output as the final result. The pseudo code of the MoPSO-DS is shown in Algorithm 1.

3.2 Directional Search

Directional search is the core of this algorithm. The algorithm searches the solution space in three directions. The three directions are the upper part of the Pareto front, the middle of the Pareto front, and the lower part of the Pareto front. The tools used to select particles which biased toward these three regions are the vector evaluation genetic algorithm (VEGA) sampling strategy and the PDDR-FF fitness function [18].

VEGA Sampling Strategy. The VEGA sampling strategy comes from the vector evaluation genetic algorithm [13] proposed by Schaffer et al. Its main operation is to sort the particles in a single objective. VEGA sampling strategy can find the good particles on objective 1 and the good particles on objective 2. The good particles on the objective 1 are the particles in the upper part of the Pareto front, the good particles on the objective 2 are the particles in the lower part of the Pareto front.

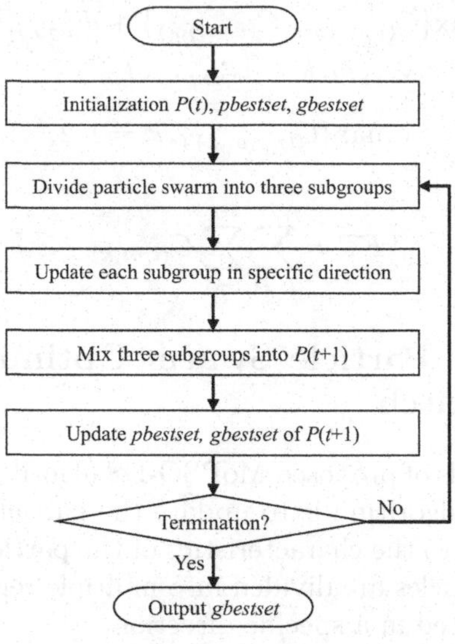

Fig. 2. Algorithmic framework of MoPSO-DS.

PDDR-FF Fitness Function. PDDR-FF is a fitness function based on the dominant relationship between individuals [18]. PDDR-FF is used to evaluate individuals. The first Pareto front particles and particles in the central region of the Pareto front will be rated excellent. Particles in the central region of the Pareto front can be selected by the excellent PDDR-FF value. The specific formula of PDDR-FF is shown below:

$$eval\left(S\right) = q\left(S\right) + \frac{1}{p\left(S\right) + 1} \tag{7}$$

Where S is the evaluated particle, $q(S)$ is the number of particles that dominate S, $p(S)$ is the number of the particles which are dominated by S. When the integer part of *eval* are equal, the fractional part can distinguish the quality of different particles. Particle who dominates more particles has larger domination area, and the fractional part of *eval* is smaller. Obviously, the first Pareto front particles and particles in the central region of the Pareto front will be small. Therefore, the smaller the *eval*, means the particle is better.

MoPSO-DS algorithm uses the VEGA sampling strategy and the PDDR-FF fitness function to divide the particle swarm into three subgroups, and then the three subgroups update the particles according to their dominant direction. As shown in Fig. 3, sub1 performs well on objective 1, sub2 performs well on objective 2, and sub3 performs well on objective 1 and objective 2. Sub1 updates the particle in the direction of the objective 1, sub2 updates the particle in the direction of the objective 2, and sub3 updates the particle in the direction of the center of the Pareto front. In the iteration, in order to ensure that the size of the particle swarm does not change, the sum of the sizes of the three subgroups needs to equal the size of the particle swarm.

Algorithm 1. The pseudo code of the MoPSO-DS

Input: problem data, PSO parameters, *maxgeneration*;
Output: the best Pareto solution;
t=0;
initialize $P(t)$ (velocity $\mathbf{v}_k(t)$ and position $\mathbf{x}_k(t)$) by encoding routine;
set *pbestset* for every particles, add each particle in their *pbest*;
set *gbestset*, keep the best Pareto solution by $gbestset = \text{argmin}\ \{eval(pbestset)\}$;
while ($t < maxgeneration$) **do**
    calculate each objective $f_i(\mathbf{x}_k)$ of $\mathbf{x}_k$, i=1,2 by decoding routine;
    calculate fitness function $eval(\mathbf{x}_k)$ by decoding routine;
    divide the $P(t)$ to three subgroups by VEGA and PDDR-FF;
    for each particle $(\mathbf{v}_k, \mathbf{x}_k)$ in each subgroup **do**
        get velocity $\mathbf{v}_k(t+1)$;
        update position $\mathbf{x}_k(t+1)$;
    end for
    mix particles from three subgroups into a new particle swarm $P(t+1)$;
    update $pbestset = \text{argmin}\ \{eval(pbestset), eval(P(t+1))\}$;
    update $gbestset = \text{argmin}\ \{eval(pbestset), eval(gbestset)\}$;
    $t=t+1$;
end while
output the best Pareto solution *gbestset*;

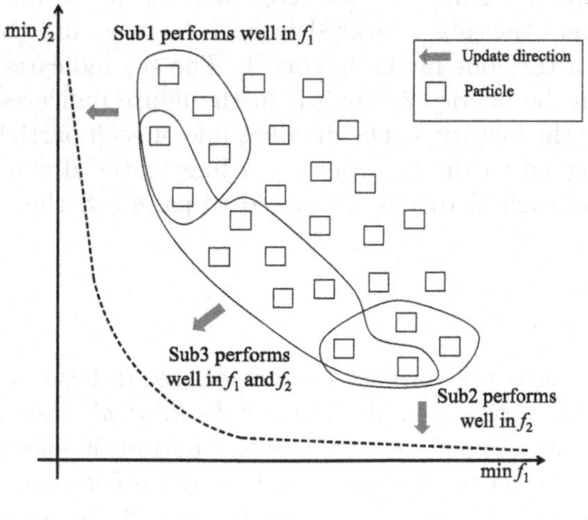

Fig. 3. Directional search.

3.3 Encoding and Decoding

For DPFSP, the problem of assigning jobs to factories is first solved, and then the order of the jobs of each factory is solved. Therefore, DPFSP has many coding methods, which are described in the [2,9]. In this paper, the encoding and decoding method is described below.

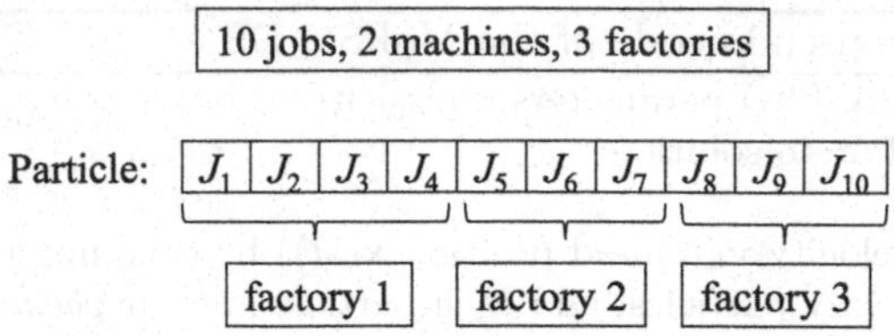

Fig. 4. Encode method of MoPSO-DS.

Encoding Method. As shown in Fig. 4, each particle contains all the jobs. In addition, each particle records two dividing lines, which are used to allocate jobs to three factories. The position of the dividing line is random, but it does not change once the particle is determined. When a new particle is generated, the dividing line of the new particle uses the dividing line of the original particle. In order to ensure the diversity of the dividing line, the dividing line is mutated once after each generation of the update. The mutation method is to randomly generate a new dividing line by random numbers. When changing the order of the particles' jobs, it can regulate the assignment of jobs to the factory and the order of the jobs in each factory.

Decoding Method. Figure 5 is decoding method of the above encoding, in which (a) represents the jobs allocated to each factory, and (b) represents the production flow of the jobs in the factory 1. The (c) indicates the production flow of the jobs in the factory 2. The (d) in the figure indicates the production flow of the jobs in the factory 3. The dividing line of each particle is determined. The jobs are assigned to the factories according to the dividing line, and the scheduling table of each factory is calculated to judge whether the particles are good or bad.

3.4 Particle Update

This section introduces the relevant content of the MoPSO algorithm and the variant of the update formula. In MoPSO, *pbestset* and *gbestset* are collections of particle. When updating particles, the reference particle is selected from *pbestset* and *gbestset*, and the particles within *pbestset* and *gbestset* are non-dominated particles. The method to update the *pbestset* is to add the new particles to their corresponding *pbestset*, and then check the non-dominated condition for each particle in *pbestset*. If the particle is dominated, the particle is deleted. The way to update *gbestset* is to add the last particle in each *pbestset* to *gbestset*, then check the particles in *gbestset*, if the particle is dominated, the particle is deleted.

The selection method used in this paper is the binary tournament method, which is also used in [11]. The particles in sub1 select the good particle on objective 1, the particles in sub2 select the good particle on objective 2, and the particles in sub3 select the particle with the small value of *eval*.

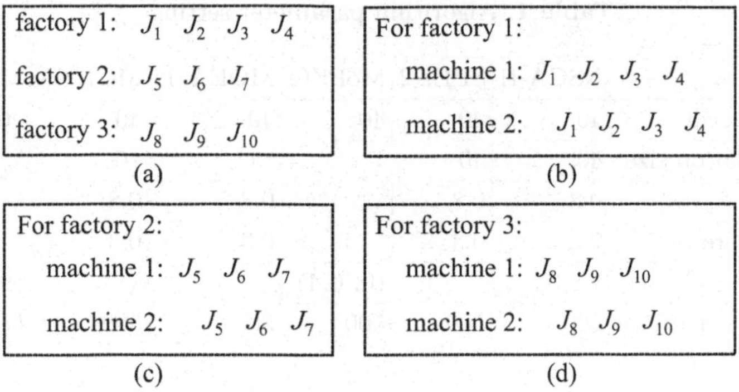

Fig. 5. Decode method of MoPSO-DS.

In this paper, the sequence coding method is used, and the difference between the two particles is expressed by the exchange order, which is also used in [10,15]. And, we made some improvements to the update formula based on sequence coding method [17]. The formulas which used to update the velocity and position of the particles are as follows:

$$\mathbf{v}_k(t+1) = \mathbf{rs} + c_1 r_1(pbest - \mathbf{x}_k(t)) + c_2 r_2(gbest - \mathbf{x}_k(t)) \tag{8}$$

$$\mathbf{x}_k(t+1) = \mathbf{x}_k(t) + \mathbf{v}_k(t+1) \tag{9}$$

Formula 8 is the particle velocity update formula, **rs** is a pair of random exchange orders, which replaces the initial velocity in the original formula to enhance the particle diversity. The *pbest* is selected from *pbestset*, and *gbest* is selected from *gbestset*. The r_1, r_2 is a random number between (0, 0.4). Formula 9 is the particle position update formula.

4 Experiments and Results

This section is the experimental analysis of the algorithm. The experimental data used is the benchmark data sets proposed by Taillard in 1993 [14]. Select 7 problems from the benchmark data sets. In this experiment, indicators C, GD, IGD, HV, $Spacing$ and $Spread$ are used to evaluate the performance of the algorithm [20]. The C, GD, IGD, and HV are mainly used to verify convergence performance, while IGD, $Spacing$ and $Spread$ are used to check the distribution performance.

In order to assess the performance of the algorithm, the MoPSO-DS algorithm is compared with NSGA-II, SPEA2, MoPSO, MOEA/D and MOHEA. In each experiment, the algorithm was run 30 times, the maximum generation was 500 generations, and the obtained data was processed by the above indicators. Finally, the wilcoxon rank sum test is performed on the index results, where '+' represents significantly better, '−' represents significantly worse, and '∗' represents no significant. The algorithm parameters are shown in the Table 1.

Table 1. Algorithm parameter setting

	NSGA-II	SPEA2	MoPSO	MOEA/D	MOHEA	MoPSO-DS
Population size	40	40	40	40	40	40
Elite population size	40	40	\	\	10	\
Cross rate	0.8	0.8	\	0.8	0.8	\
Mutation rate	0.3	0.3	\	0.3	0.3	\
$r_1 \backslash r_2$ range	\	\	(0, 0.4)	\	\	(0, 0.4)
Maximum generation	500	500	500	500	500	500

It can be seen from the results of the *GD* index in Table 2: When using the MoPSO-DS to solve the FSP problem, 5 of the 7 questions showed the best performance. 20_5 shows that MOHEA works best, 200_10 shows that MoPSO works best. From the results of the saliency analysis, the MoPSO-DS algorithm has significant advantages over other algorithms. This indicates that the solution set obtained by MoPSO-DS is closer to the Pareto front. This means that MoPSO-DS has better convergence performance.

Table 2. The results of *GD* index

problem	NSGA-II		SPEA2		MoPSO		MOEA/D		MOHEA		MoPSO-DS
20_5	3.19E-02	-	2.47E-02	+	4.41E-02	-	2.74E-02	+	2.46E-02	+	3.92E-02
50_5	4.71E-02	-	4.44E-02	-	4.74E-02	-	3.41E-02	-	3.16E-02	-	2.43E-02
50_20	4.41E-02	-	4.07E-02	-	2.71E-02	*	2.81E-02	*	2.68E-02	*	2.54E-02
100_5	5.92E-02	-	5.50E-02	-	3.38E-02	*	3.76E-02	-	4.22E-02	-	2.45E-02
100_10	6.88E-02	-	6.61E-02	-	3.40E-02	*	4.80E-02	-	4.46E-02	-	2.26E-02
100_20	5.12E-02	-	4.65E-02	-	2.44E-02	+	3.37E-02	-	3.18E-02	-	1.46E-02
200_10	8.88E-02	-	8.69E-02	-	2.51E-02	+	8.39E-02	-	6.89E-02	-	2.74E-02
+/-/*	0/7/0		1/6/0		2/2/3		1/5/1		1/5/1		\

From the results of *IGD* index in Table 3, MoPSO-DS performed best on 5 problems. SPEA2 performs best on 20_5, and MOHEA performs best on 50_20. From the results of the saliency analysis, the contrast algorithms have no good result than MoPSO-DS, so the MoPSO-DS algorithm is superior to all other algorithms. *IGD* index cannot only judge the convergence performance of the algorithm, but also considers the distribution performance of the solution. Results of the *IGD* show that the final solution set obtained by MoPSO-DS is not only closer to Pareto front, but also more consistent with the distribution of Pareto front.

Table 3. The results of *IGD* index

problem	NSGA-II		SPEA2		MoPSO		MOEA/D		MOHEA		MoPSO-DS
20_5	3.78E-02	+	2.84E-02	+	4.80E-02	*	3.36E-02	+	3.55E-02	+	4.61E-02
50_5	5.42E-02	-	4.83E-02	-	6.61E-02	-	5.54E-02	-	4.51E-02	*	4.35E-02
50_20	4.84E-02	-	4.28E-02	-	4.55E-02	-	3.90E-02	*	3.49E-02	*	3.63E-02
100_5	6.05E-02	-	5.77E-02	-	4.89E-02	-	6.42E-02	-	4.72E-02	-	3.56E-02
100_10	6.52E-02	-	6.36E-02	-	3.94E-02	*	6.04E-02	-	4.44E-02	-	2.75E-02
100_20	5.59E-02	-	5.01E-02	-	3.82E-02	-	5.22E-02	-	3.67E-02	-	2.67E-02
200_10	8.30E-02	-	8.07E-02	-	2.94E-02	*	7.90E-02	-	6.29E-02	-	2.89E-02
+/-/*	1/6/0		1/6/0		0/4/3		1/5/1		1/4/2		\

From the results of *HV* index in Table 4, MoPSO-DS ranks first on 6 problems, SPEA2 performs well on 20_5. From the results of the saliency analysis, the MoPSO-DS algorithm has significant advantages over other algorithms. The larger the *HV* value, the larger the dominant area. So, the final solution set obtained by MoPSO-DS has a larger area of advantage than the solution set obtained by the contrast algorithm. In other words, MoPSO-DS performs better in terms of convergence and distribution performance.

Table 4. The results of *HV* index

problem	NSGA-II		SPEA2		MoPSO		MOEA/D		MOHEA		MoPSO-DS
20_5	2.59E+05	+	2.78E+05	+	2.47E+05	*	2.65E+05	+	2.66E+05	+	2.48E+05
50_5	2.53E+06	-	2.68E+06	-	2.45E+06	-	2.52E+06	-	2.64E+06	-	2.77E+06
50_20	1.95E+06	-	2.15E+06	-	2.29E+06	*	2.04E+06	-	2.31E+06	*	2.42E+06
100_5	5.32E+06	-	5.64E+06	-	6.21E+06	-	5.31E+06	-	5.83E+06	-	6.73E+06
100_10	6.43E+06	-	6.88E+06	-	8.22E+06	*	6.73E+06	-	7.50E+06	-	8.86E+06
100_20	6.64E+06	-	7.29E+06	-	8.87E+06	+	7.07E+06	-	8.19E+06	-	9.87E+06
200_10	1.85E+07	-	2.02E+07	-	2.98E+07	*	1.87E+07	-	2.20E+07	-	2.98E+07
+/-/*	1/6/0		1/6/0		1/2/4		1/6/0		1/5/1		\

The results of the *Spacing* index are shown in Table 5. From the results of saliency analysis, the MoPSO-DS algorithm is better than NSGA-II, SPEA2, MOEA/D, and MOHEA. MoPSO is better than MoPSO-DS on 2 problems. *Spacing* index is mainly used to judge the distribution performance of the solution set and whether the solution is evenly distributed. Therefore, the uniformity of solution distribution obtained by MoPSO-DS is not as good as MoPSO. MoPSO-DS ranks second among the six algorithms.

Results of *Spread* index are shown in Table 6. From the results of the saliency analysis, MoPSO-DS has a better indicator value than NSGA-II, SPEA2, MOEA/D and MOHEA. The significant difference between the MoPSO-DS and the MoPSO is not obvious, so it is impossible to judge the pros and cons of these two algorithms. In general, MoPSO-DS has a good effect on the distribution of solutions.

Table 5. The results of *Spacing* index

problem	NSGA-II	SPEA2	MoPSO	MOEA/D	MOHEA	MoPSO-DS
20_5	8.41E+01 *	7.10E+01 *	7.20E+01 *	9.47E+01 *	8.20E+01 *	8.52E+01
50_5	2.62E+02 *	2.60E+02 *	2.07E+02 *	3.99E+02 *	2.76E+02 *	2.31E+02
50_20	3.46E+02 *	3.33E+02 *	1.75E+02 *	5.91E+02 -	3.62E+02 *	3.18E+02
100_5	8.62E+02 -	6.19E+02 -	3.95E+02 *	1.21E+03 -	6.85E+02 -	4.54E+02
100_10	7.58E+02 -	6.26E+02 -	4.40E+02 *	1.31E+03 -	8.39E+02 -	4.71E+02
100_20	8.87E+02 -	9.07E+02 -	3.51E+02 +	1.58E+03 -	1.13E+03 -	5.47E+02
200_10	1.88E+03 -	1.44E+03 -	4.99E+02 +	3.61E+03 -	1.85E+03 -	9.13E+02
+/-/*	0/4/3	0/4/3	2/0/5	0/5/2	0/4/3	\

Table 6. The results of *Spread* index

problem	NSGA-II	SPEA2	MoPSO	MOEA/D	MOHEA	MoPSO-DS
20_5	8.45E-01 *	8.19E-01 *	7.80E-01 *	9.03E-01 -	8.68E-01 -	8.03E-01
50_5	9.27E-01 *	9.12E-01 *	9.02E-01 *	9.80E-01 -	9.65E-01 -	9.08E-01
50_20	9.43E-01 *	9.18E-01 *	9.17E-01 *	9.76E-01 *	9.79E-01 *	9.14E-01
100_5	9.58E-01 *	9.27E-01 *	9.09E-01 *	1.00E+00 -	9.74E-01 -	8.83E-01
100_10	9.43E-01 *	9.30E-01 *	8.99E-01 *	9.79E-01 -	9.93E-01 -	9.20E-01
100_20	9.61E-01 *	9.42E-01 *	9.39E-01 *	9.77E-01 *	1.03E+00 -	9.53E-01
200_10	9.74E-01 -	9.62E-01 -	9.40E-01 *	1.03E+00 -	9.90E-01 -	9.04E-01
+/-/*	0/1/6	0/1/6	0/0/7	0/5/2	0/6/1	\

Many results of the index *Spacing* and *Spread* are not significant, the performance of the algorithm cannot be accurately determined. In order to more accurately judge the performance of the improved algorithm, the C index is used to evaluate performance of the algorithm. Results of C index are shown in Table 7. Most of the 7 questions show that the results of MoPSO-DS are good. MoPSO-DS has a big advantage compared to the contrast algorithms.

Through the analysis of the convergence indicators GD, IGD, HV results, MoPSO-DS algorithm has more obvious convergence performance. Through the analysis of the distribution indicators *Spacing*, *Spread* results, MoPSO-DS algorithm has good distribution performance. Through the analysis of the

Table 7. The results of C index

A:MoPSO-DS, B:MoPSO, C:NSGA-II, D:SPEA2, E:MOEA/D, F:MOHEA										
problem	C(A,B)	C(B,A)	C(A,C)	C(C,A)	C(A,D)	C(D,A)	C(A,E)	C(E,A)	C(A,F)	C(F,A)
20_5	6.67E-02	0.00E+00	0.00E+00	1.33E-01	0.00E+00	2.67E-01	0.00E+00	3.33E-02	3.33E-02	2.00E-01
50_5	5.67E-01	0.00E+00	5.00E-01	0.00E+00	1.33E-01	0.00E+00	0.00E+00	0.00E+00	2.67E-01	6.67E-02
50_20	2.33E-01	1.00E-01	7.67E-01	3.33E-02	4.00E-01	6.67E-02	3.33E-02	0.00E+00	6.67E-02	1.00E-01
100_5	2.67E-01	1.00E-01	8.00E-01	3.33E-02	6.33E-01	3.33E-02	2.00E-01	0.00E+00	5.67E-01	3.33E-02
100_10	2.33E-01	1.67E-01	8.67E-01	0.00E+00	6.67E-01	0.00E+00	4.67E-01	0.00E+00	6.33E-01	3.33E-02
100_20	3.67E-01	1.00E-01	9.33E-01	0.00E+00	8.33E-01	0.00E+00	4.33E-01	0.00E+00	7.33E-01	0.00E+00
200_10	2.67E-01	3.33E-01	1.00E+00	0.00E+00	9.67E-01	0.00E+00	8.67E-01	0.00E+00	1.00E+00	0.00E+00

comprehensive performance indicator C, the performance of the algorithm is excellent than contrast algorithms. Searching the solution space by direction can enhance the convergence performance of the algorithm, and multiple direction ensures the distribution performance of the algorithm. Therefore, the proposed MoPSO-DS algorithm is effective.

5 Conclusion

In this paper, MoPSO-DS is proposed to solve the DPFSP. The algorithm divides the particle swarm into three subgroups according to the characteristics of the particles in different regions of the Pareto front. Each subgroup updates the particles in the direction of their own advantages, which enhances the convergence performance. The three subgroups have different update directions, which indirectly guarantee the distribution performance. According to the experimental result, MoPSO-DS performs better than MoPSO, NSGA-II, SPEA2, MOEA/D and MOHEA. The algorithm framework has greatly improved the performance of the algorithm. In the next step, we will make efforts on the details, such as how many subgroups will maximize the performance of the algorithm.

Acknowledgements. This research work is supported by the Sub-Project of National Key R&D Program of China (2017YFD0401001-02), Science & Technology Research Project of Henan Province (162102210044), Program for Science & Technology Innovation Talents in Universities of Henan Province (19HASTIT027), Key Research Project in Universities of Henan Province (17A520030), Ministry of Education and the Grant-in-Aid for Scientific Research (C) of Japan Society of Promotion of Science (JSPS) (19K12148).

References

1. Al Moubayed, N., Petrovski, A., Mccall, J.: *D2MOPSO*: MOPSO based on decomposition and dominance with archiving using crowding distance in objective and solution spaces. Evol. Comput. **22**(1), 47–77 (2014)
2. Chen, J.F., Wang, L., Peng, Z.P.: A collaborative optimization algorithm for energy-efficient multi-objective distributed no-idle flow-shop scheduling. Swarm Evol. Comput. **50** (2019). https://doi.org/10.1016/j.swevo.2019.100557
3. Coello, C.A.C., Pulido, G.T., Lechuga, M.S.: Handling multiple objectives with particle swarm optimization. IEEE Trans. Evol. Comput. **8**(3), 256–279 (2004)
4. Eberhart, R., Kennedy, J.: A new optimizer using particle swarm theory. In: Proceedings of the Sixth International Symposium on Micro Machine and Human Science, MHS 1995, pp. 39–43. IEEE (1995)
5. Eddaly, M., Jarboui, B., Siarry, P.: Combinatorial particle swarm optimization for solving blocking flowshop scheduling problem. J. Comput. Des. Eng. **3**(4), 295–311 (2016)
6. Garey, M.R., Johnson, D.S., Sethi, R.: The complexity of flowshop and jobshop scheduling. Math. Oper. Res. **1**(2), 117–129 (1976)
7. Han, H., Lu, W., Qiao, J.: An adaptive multiobjective particle swarm optimization based on multiple adaptive methods. IEEE Trans. Cybern. **47**(9), 2754–2767 (2017)

8. Jiang, E., Wang, L., Lu, J.: Modified multiobjective evolutionary algorithm based on decomposition for low-carbon scheduling of distributed permutation flow-shop. In: 2017 IEEE Symposium Series on Computational Intelligence (SSCI), pp. 1–7. IEEE (2017)
9. Li, J.Q., Bai, S.C., Duan, P.Y., Sang, H.Y., Han, Y.Y., Zheng, Z.X.: An improved artificial bee colony algorithm for addressing distributed flow shop with distance coefficient in a prefabricated system. Int. J. Prod. Res. **57**(22), 6922–6942 (2019)
10. Li, J.Q., Sang, H.Y., Han, Y.Y., Wang, C.G., Gao, K.Z.: Efficient multi-objective optimization algorithm for hybrid flow shop scheduling problems with setup energy consumptions. J. Clean. Prod. **181**, 584–598 (2018)
11. Martínez-Cagigal, V., Santamaría-Vázquez, E., Hornero, R.: A novel hybrid swarm algorithm for P300-based BCI channel selection. In: Lhotska, L., Sukupova, L., Lacković, I., Ibbott, G.S. (eds.) World Congress on Medical Physics and Biomedical Engineering 2018. IP, vol. 68/3, pp. 41–45. Springer, Singapore (2019). https://doi.org/10.1007/978-981-10-9023-3_8
12. Naderi, B., Ruiz, R.: The distributed permutation flowshop scheduling problem. Comput. Oper. Res. **37**(4), 754–768 (2010)
13. Schaffer, J.D.: Multiple objective optimization with vector evaluated genetic algorithms. In: 1985 Proceedings of the First International Conference on Genetic Algorithms and Their Applications, pp. 93–100. Lawrence Erlbaum Associates. Inc. (1985)
14. Taillard, E.: Benchmarks for basic scheduling problems. Eur. J. Oper. Res. **64**(2), 278–285 (1993)
15. Tang, D., Dai, M., Salido, M.A., Giret, A.: Energy-efficient dynamic scheduling for a flexible flow shop using an improved particle swarm optimization. Comput. Ind. **81**, 82–95 (2016)
16. Wang, J., Cao, Y., Li, B., Kim, H.J., Lee, S.: Particle swarm optimization based clustering algorithm with mobile sink for WSNs. Futur. Gener. Comput. Syst. **76**, 452–457 (2017)
17. Yu, X., Gen, M.: Introduction to Evolutionary Algorithms. Springer, Cham (2010). https://doi.org/10.1007/978-1-84996-129-5
18. Zhang, W., Gen, M., Jo, J.: Hybrid sampling strategy-based multiobjective evolutionary algorithm for process planning and scheduling problem. J. Intell. Manuf. **25**(5), 881–897 (2014). https://doi.org/10.1007/s10845-013-0814-2
19. Zhang, W., Lu, J., Zhang, H., Wang, C., Gen, M.: Fast multi-objective hybrid evolutionary algorithm for flow shop scheduling problem. In: Xu, J., Hajiyev, A., Nickel, S., Gen, M. (eds.) Proceedings of the Tenth International Conference on Management Science and Engineering Management. AISC, vol. 502, pp. 383–392. Springer, Singapore (2017). https://doi.org/10.1007/978-981-10-1837-4_33
20. Zhang, W., Wang, Y., Yang, Y., Gen, M.: Hybrid multiobjective evolutionary algorithm based on differential evolution for flow shop scheduling problems. Comput. Ind. Eng. **130**, 661–670 (2019)
21. Zhao, F., Qin, S., Yang, G., Ma, W., Zhang, C., Song, H.: A factorial based particle swarm optimization with a population adaptation mechanism for the no-wait flow shop scheduling problem with the makespan objective. Expert Syst. Appl. **126**, 41–53 (2019)

An Improved Pigeon-Inspired Optimization Combining Adaptive Inertia Weight with a One-Dimension Modification Mechanism

Meiwen Chen[1,2](✉), Yiwen Zhong[1,2], and Lijin Wang[1,2]

[1] College of Computer and Information Science, Fujian Agriculture and Forestry University, Fuzhou 350002, China
[2] Key Laboratory of Smart Agriculture and Forestry, Fujian Agriculture and Forestry University, Fuzhou 350002, China
cmw97@163.com

Abstract. In recent years, many population-based swarm intelligence (SI) algorithms have been developed for solving optimization problems. The pigeon-inspired optimization (PIO) algorithm is one new method that is considered to be a balanced combination of global and local search by the map and compass operator and the landmark operator. In this paper, we propose a novel method of adaptive nonlinear inertia weight along with the velocity in order to improve the convergence speed. Additionally, the one-dimension modification mechanism is introduced during the iterative process, which aims to avoid the loss of good partial solutions because of interference phenomena among the dimensions. For separable functions, the one-dimension modification mechanism is more effective for the search performance. Our approach effectively combines the method of adaptive inertia weight with the strategy of one-dimension modification, enhancing the ability to explore the search space. Comprehensive experimental results indicates that the proposed PIO outperforms the basic PIO, the other improved PIO and other improved SI methods in terms of the quality of the solution.

Keywords: Swarm intelligence · Pigeon-inspired optimization · Adaptive inertia weight · One-dimension modification

1 Introduction

In recent years, population-based swarm intelligence algorithms have been successfully applied to solve optimization problems, mostly simulating natural behavior of biological systems in nature, such as the artificial bee colony (ABC) [1], cuckoo search (CS) [2], bat algorithm (BA) [3], and firefly algorithm

Supported by the Special Fund for Scientific and Technological Innovation of Fujian Agriculture and Forestry University of P.R. China (No. CXZX2016026, No. CXZX2016031).

L. Pan et al. (Eds.): BIC-TA 2019, CCIS 1159, pp. 177–192, 2020.
https://doi.org/10.1007/978-981-15-3425-6_15

(FA) [4]. Unlike traditional single-point-based algorithms such as hill-climbing algorithms, population-based swarm intelligence algorithms consist of a set of points (population) that solve the problem through information sharing to cooperate and/or compete among themselves [5]. These techniques attempt to mimic natural phenomena or social behavior in order to generate better solutions for the optimization problem by using iterations and stochasticity [6].

Pigeon-inspired optimization (PIO) was first put forward by Duan [7] and is derived from the simulation of the collective behavior of homing pigeons searching for the path to home by using navigation tools such as the magnetic field, sun, and landmarks. The original PIO consists of two operators. In early iterations, the speed of the pigeon mainly depends on its inertia weight according to the map and compass operator, while in later iterations, every pigeon flies straight toward the center of all of the pigeons according to the landmark operator. The PIO algorithm has been proven to converge more quickly and is more stable than the standard DE algorithm [7]. Recently, PIO has been used successfully in many applications including UAV three-dimensional path planning [8], PID parameter optimization [9], target detection [10], image restoration [11] and other applications. Additionally, a novel hybrid pigeon-inspired optimization and quantum theory (BQPIO) algorithm, which can avoid the premature convergence problem [12], was proposed for solving the continuous optimization problem. The above studies exemplify the contributions to PIO research.

In the original PIO method, the velocity of the pigeon mainly depends on the map and compass operator, which play the same role as the inertia weight (w) in PSO. The inertia weight (w) was originally proposed by Shi and Eberhart [13] to achieve a balance between the exploration and exploitation characteristics of particle swarm optimization (PSO). Since the introduction of this parameter, many different inertia weight strategies have been proposed, which can be easily categorized into two classes: linear strategy [14] and nonlinear strategy [15,16]. In this paper, an adaptive nonlinear inertia weight approach is presented that adjusts the inertia weight dynamically with the increase in the number of iterations.

PIO is a new nature-inspired intelligence algorithm that uses the updated velocity and position of all dimensions to generate the candidate solutions in the early iterations. For multidimensional function optimization problems, this strategy degrades the intensification ability of the algorithm because different dimensions may interfere with each other [17]. To address this problem, the strategy of dimension-by-dimension improvement was introduced in the improved CS algorithm (DDICS) [18], which effectively improves the convergence speed and the quality of the solutions. The mechanism of one-dimensional learning is employed in the local search of the standard bat algorithm, which effectively avoids the trapping of the whole swarm at the local optimum, thereby preventing premature convergence [19]. Peio *et al.* [20] introduced a new approach of selecting a promising dimension to shift by using the OAT (one-at-a-time) method in the ABC algorithm. In their work, a one-dimension modification strategy is proposed. The individual is updated with the velocity and position of a single dimension to generate the candidate solution. This approach is similar to that of the ABC algorithm, which produces a modification of its associated solution at the employed bee stage. Here, we propose a new improved PIO combining an adaptive inertia weight with one-dimension modification strategy, abbreviated as ODaPIO. Extensive experiments have been carried out to test the ODaPIO method, and the results reveal that the proposed approach can improve the quality of the solutions effectively, making the method competitive with other improved PIO and SI methods in terms of global numerical optimization.

The remainder of this paper is organized as follows. The original PIO is presented in Sect. 2. Section 3 describes the ODaPIO algorithm in detail. Comprehensive experiments are conducted to verify the performance of ODaPIO and investigate the influence of the parameter control, inertia weight, and dimension-selection strategy in Sect. 4. Section 5 draws the conclusions and points out directions for future work.

2 Original PIO

PIO is a novel type of swarm intelligence-based algorithm derived from the simulation of the collective behavior of homing pigeons searching for the path home using navigation tools such as the magnetic field, sun, and landmarks [7]. To idealize some of the homing characteristics of pigeons, two operators are designed by using several rules as follows:

1. *Map and compass operator.* At early iterations, the pigeons are far away from the destination, and the map and compass operator is used, which is based on the magnetic field and sun. The rules are defined with the position X_i and the velocity V_i of pigeon i, and the positions and velocities in a D-dimensional search space are updated in each iteration. The new position X_i and velocity V_i of pigeon i at the t-th iteration can be calculated with the following equations:

$$V_i^t = V_i^{t-1} \cdot e^{-Rt} + rand \cdot (X_g - X_i^{t-1}) \tag{1}$$

$$X_i^t = X_i^{t-1} + V_i^t \tag{2}$$

where R is the map and compass factor, $rand$ is a random number, and X_g is the current global best position. As seen from the equations, the pigeons are guided by the current best position with decreasing influence from its inertial velocity.

2. *Landmark operator.* In the later iterations, for the pigeons to fly close to their destination, every pigeon can fly straight to the destination, which is at the center of all of the pigeons, while the number of the pigeons decreases to half of the last generation. In other words, the pigeons that are far away from their destination will be discarded.

$$SN^t = SN^{t-1}/2 \tag{3}$$

where SN^t denotes the size of the population at the t-th generation.
The center of all of the pigeons is their destination in the t-th iteration, which can be described as:

$$X_C^t = \frac{\sum_{i=1}^{SN^t} X_i^t \cdot f(X_i^t)}{SN^t \cdot \sum_{i=1}^{SN^t} f(X_i^t)} \tag{4}$$

where $f(X_i^t)$ represents the objective function value corresponding to the position of pigeon i at the t-th generation. The new positions of all of the pigeons can be updated according to the following equation:

$$X_i^t = X_i^{t-1} + rand \cdot (X_C^t - X_i^{t-1}) \tag{5}$$

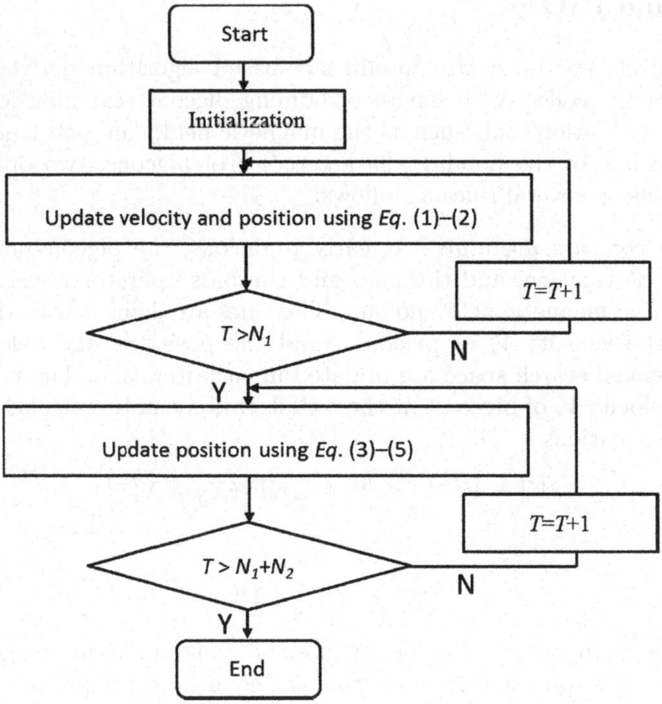

Fig. 1. Structure of the original PIO.

Based on these idealized rules, the flow chart of PIO is shown in Fig. 1. As mentioned above, the original PIO algorithm can be considered as a balanced combination of global and local search by controlling the map and compass operator and the landmark operator.

3 ODaPIO

3.1 Adaptive Inertia Weight Strategy

In the original PIO, according to Eqs. (1) and (2), we can see that the formulas for updating the velocity and the position are similar to those of PSO. The parameter e^{-Rt} plays the same role of the inertia weight w in PSO. In the literature [7], R is set to 0.3. For further clarification, the changes in the inertia weight e^{-Rt} are shown in Fig. 2. It is clear that the curve rapidly decreases to 0 when the iteration t is approximately the 20-th generation. According to Eq. (1), the new velocities are primarily determined by the currently global optimal position after 20-th generations, possibly leading to the loss of population diversity and the loss of strong global search ability in the later iterations. Shi and Eberhart [14] have observed that for most problems, the optimal solution can be improved by varying the value of the linear inertia weight with increasing iteration number. Inspired by the neural network sigmoid (activation) function, the novel nonlinear adaptive inertia weight strategy is proposed, which is represented mathematically in Eq. (6), and the graph is shown in Fig. 3. Here, T_{max} and t represent the maximum number of iterations and the current iteration number, respectively.

$$w = \frac{1}{1 + e^{(-(T_{max}-t)/T_{max})}} \tag{6}$$

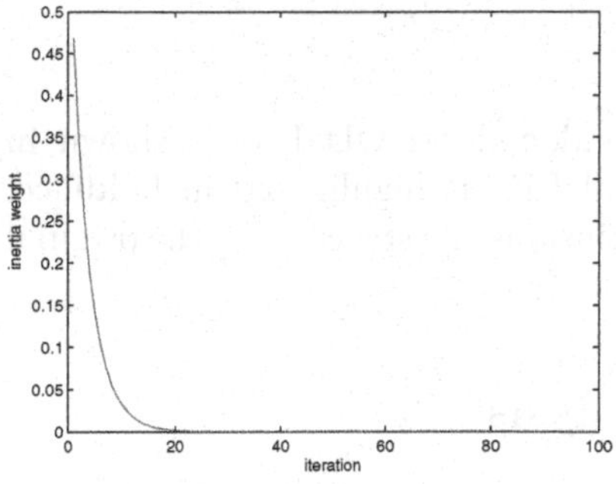

Fig. 2. The curve graph of e^{-Rt}

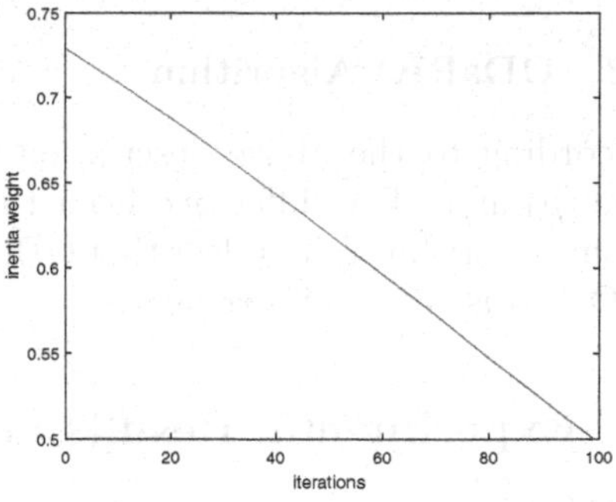

Fig. 3. The curve graph of w

Additionally, to increase the population diversity, the new velocity is updated by the guidance of the random individual instead of the current best solution. According to the above description, the new velocity V_i of pigeon i at the t-th iteration can be expressed as:

$$V_i^t = w \cdot V_i^{t-1} + rand \cdot (X_{rand}^{t-1} - X_i^{t-1}) \tag{7}$$

where X_{rand}^{t-1} represents the random individual.

3.2 One-Dimension Modification Mechanism

As shown in Eq. (2), when a new candidate solution is generated, the velocity and position are updated for all dimensions in the global search space. When solving multidimensional function optimization problems, the strategy of multidimensional modification may degrade the convergence speed and the quality of the algorithm solution due to interference phenomena among the dimensions [18]. For a multidimensional optimization problem, such as the minimization of the objective function $f(x) = 2x_1^2 + 3x_2^2 + x_3^2$, it is assumed that $x = (5, 0, 1)$ and $x' = (2, 1, 2)$ are the current solution and the new solution with multidimensional modifications, respectively. However, since $f(x') < f(x)$, x' is accepted, a part of good solution in the 2-*nd* dimension and 3-*rd* dimension is lost. The loss of a part of the good solution can be avoided if the individual is updated only for one dimension during the iterations, which is called the one-dimension modification strategy. Based on Eq. (2), the one-dimension modification mechanism is represented as follows:

$$X_{i,j}^t = \begin{cases} X_{i,j}^{t-1} + V_{i,j}^t, & \text{if} \quad j = j_1 \\ X_{i,j}^{t-1}, & \text{otherwise} \end{cases} \tag{8}$$

where $j_1 \epsilon [1, D]$ is a random integer. The advantage of this approach is that the original structure of the method has not changed and therefore is easy to implement.

3.3 ODaPIO Algorithm

According to the above description, the pseudocode of ODaPIO is shown in Algorithm 1. The difference from the original PIO is highlighted in boldface. From Algorithm 1, it is clear that ODaPIO maintains the structure of the original PIO and is easy to implement.

4 Experimental Results and Analysis

4.1 Benchmark and Experimental Settings

In this section, we perform comprehensive experiments to test the performance of the proposed ODaPIO for a suite of 18 benchmark functions that are shown in Table 1, consisting of three classes: (i) unimodal functions; (ii) multimodal functions; and (iii) shifted and rotated functions (F_{15}–F_{18}) introduced by Suganthan *et al.* [21]. The Rosenbrock function F_2 is a multimodal function when $D > 3$ [22]. F_3 is a step function that has one minimum and is discontinuous. These functions are more complex and can be used to compare the performance of different algorithms in a systematic manner. Each algorithm is executed 30 times for each function with the dimension $D = 30$. The population size $NP = 30$ and the maximum iteration is 1500. Improved results are marked in boldface.

Algorithm 1. ODaPIO

Initialize the parameters, the population size SN, map and compass factor R, space dimension D, the position X_i and velocity V_i for each pigeon, and the maximum iteration N_1 and the maximum iteration N_2 for two operators
Evaluate the fitness $f(X_i)$
while $t \leq N_1$ **do**
  for $i = 1$ to SN **do**
    Update the velocity and position using *Eqs.* (6)–(8)
    Check whether pigeons are located outside the search space
  end for
  Find the current best X_g.
end while
while $t \leq N_1 + N_2$ **do**
  Sort the pigeons and calculate the center using *Eqs.* (3)–(4)
  for $i = 1$ to SN **do**
    Update the new positions using *Eq.* (5)
    Check whether pigeons are located outside the search space
  end for
  Find the current best X_g.
end while

For a fair comparison, we use the same parameters unless mentioned otherwise. In Table 1, the column "domain" defines the lower and upper bounds of the definition domain in all dimensions. In addition, the average and standard deviation of the best error values, presented as "$AVG_{Er} \pm STD_{Er}$", are provided in the different tables.

Moreover, the effects of the parameter control, dimension selection and inertia weight on the performance of ODaPIO to are analyzed in detail.

4.2 Analysis on Parameter Control of N_1 and N_2

As mentioned above, the original PIO consists of the map and compass operator and landmark operator with the control parameters N_1 and N_2, respectively. In the early iterations, the pigeons are guided by the current best position using the map and compass operator which are far away from the destination. In the later iterations, the pigeons fly straight to the destination, which is the center of all of the pigeons, using the landmark operator. ODaPIO maintains the structure of the original PIO and has no additional parameters. ODaPIO has 3 parameters, namely, the numbers of iterations N_1 and N_2 for the map and compass operator and the landmark operator, respectively, and the population size SN. In this subsection, the control of the parameters N_1 and N_2 is discussed in detail. ODaPIO is considered to be a balanced combination of the global and local search by using the N_1 and N_2 control parameters. It is supposed that the number of iterations N_2 should be less than that of iterations N_1 during the search, because the value of N_2 controls the local search and accelerates the convergence. To further analyze the influence of the N_1 and N_2 parameters and

Table 1. Test functions

Test function	Domain	Name
F_1	[−600, 600]	Sphere
F_2	[−15, 15]	Rosenbrock
F_3	[−100, 100]	Step
F_4	[−10, 10]	Schwefel'sP2. 22
F_5	[−32. 768, 32. 768]	Ackley
F_6	[−600, 600]	Griewank
F_7	[−15, 15]	Rastrigin
F_8	[−500, 500]	Schwefel
F_9	[−100, 100]	Salomon
F_{10}	[−2π, 2π]	Easom
F_{11}	[0, π]	Michalewicz
F_{12}	[−2π, 2π]	X-She Yang
F_{13}	[−50, 50]	GeneralizedPenalized1
F_{14}	[−50, 50]	GeneralizedPenalized2
F_{15}	[−100, 100]	ShiftedSphere
F_{16}	[0, 600]	ShiftedRotatedGriewank
F_{17}	[−32, 32]	ShiftedRotatedAckley
F_{18}	[−5, 5]	ShiftedRastrigin

verify the assumption, several experiments are conducted on the 18 benchmark functions on the condition that the total number of iterations is 1500 so that the N_1/N_2 values are {0/1500, 750/750, 1000/500, 1400/100, 1420/80, 1450/50, 1490/10, 1500/0}. The results for increasing values of N_1 are shown in Table 2.

From Table 2, the results clearly demonstrate that the values of N_1 and N_2 significantly affect the performance of most functions. For example, for the Rosenbrock function F_2, which is a multimodal function when $D > 3$ and is difficult to converge, the global optimum is obtained when N_2 is varied from 50 to 750; however, a worse result in terms of the mean error value is obtained when N_2 is 10 or 0. For a clear illustration of this result, Fig. 4 shows the curves of the mean error value for some functions with the increase in N_1. As seen in Fig. 4, for functions F_2, F_8, F_{13}, and F_{14}, better results are obtained in terms of the mean error values with the cooperation of N_1 and N_2. The experimental results show that better performance is obtained for ODaPIO if N_2 is less than N_1, which is consistent with our assumption. Overall, ODaPIO achieves better performance when N_1 is 1450. However, the landmark operator (N_2) does not contribute to the performance for some functions (F_1, F_4, F_5, F_7) that are separable functions. This result indicates that the one-dimension modification mechanism improves the performance of the algorithm effectively, especially for separable functions.

Table 2. Comparison of the experimental results for different N_1/N_2

N_1/N_2	0/1500	750/750	1000/500	1400/100
Fun	$AVG_{Er} \pm STD_{Er}$	$AVG_{Er} \pm STD_{Er}$	$AVG_{Er} \pm STD_{Er}$	$AVG_{Er} \pm STD_{Er}$
F_1	1.89e + 01 ± 3.62e + 00	7.51e − 05 ± 4.37e − 05	2.55e − 08 ± 1.30e − 08	7.27e − 14 ± 4.77e − 14
F_2	7.48e + 02 ± 3.27e + 02	0.00e + 00 ± 0.00e + 00	0.00e + 00 ± 0.00e + 00	0.00e + 00 ± 0.00e + 00
F_3	5.11e + 01 ± 7.82e + 00	0.00e + 00 ± 0.00e + 00	0.00e + 00 ± 0.00e + 00	0.00e + 00 ± 0.00e + 00
F_4	1.73e + 01 ± 2.45e + 00	4.62e − 04 ± 9.15e − 05	8.99e − 06 ± 2.68e − 06	1.41e − 08 ± 3.66e − 09
F_5	3.04e + 00 ± 2.82e − 01	1.27e − 03 ± 3.72e − 04	2.57e − 05 ± 7.86e − 06	5.32e − 08 ± 1.62e − 08
F_6	7.22e − 01 ± 8.48e − 02	1.16e − 03 ± 1.15e − 03	7.32e − 04 ± 1.65e − 03	3.80e − 05 ± 6.19e − 05
F_7	2.36e + 01 ± 3.72e + 00	9.77e − 01 ± 5.50e − 01	7.11e − 03 ± 5.85e − 03	3.11e − 06 ± 5.15e − 06
F_8	1.01e + 04 ± 5.90e + 02	8.52e + 00 ± 2.43e + 01	1.84e + 01 ± 4.13e + 01	1.97e + 01 ± 5.46e + 01
F_9	5.21e − 01 ± 8.41e − 02	4.63e − 01 ± 8.85e − 02	4.46e − 01 ± 6.45e − 02	4.29e − 01 ± 5.59e − 02
F_{10}	−3.78e − 77 ± 6.83e − 77	−1.00e + 00 ± 4.97e − 09	−1.00e + 00 ± 2.22e − 12	−1.00e + 00 ± 0.00e + 00
F_{11}	−8.03e + 00 ± 7.47e − 01	−2.80e + 01 ± 2.05e − 01	−2.87e + 01 ± 1.67e − 01	−2.91e + 01 ± 9.56e − 02
F_{12}	5.32e − 07 ± 1.85e − 06	4.29e − 12 ± 1.63e − 13	3.88e − 12 ± 9.89e − 14	3.67e − 12 ± 4.40e − 14
F_{13}	4.01e + 00 ± 8.87e − 01	9.33e − 08 ± 2.23e − 07	3.52e − 11 ± 6.30e − 11	2.13e − 14 ± 8.85e − 14
F_{14}	5.40e − 01 ± 2.43e − 01	1.35e − 32 ± 5.57e − 48	1.35e − 32 ± 5.57e − 48	1.35e − 32 ± 5.57e − 48
F_{15}	7.93e + 04 ± 6.40e + 03	1.47e − 06 ± 8.69e − 07	6.22e − 10 ± 3.91e − 10	1.57e − 15 ± 2.78e − 15
F_{16}	4.89e + 03 ± 9.70e − 13	1.92e + 01 ± 1.33e + 01	8.38e + 00 ± 4.50e + 00	4.94e + 00 ± 2.62e + 00
F_{17}	2.13e + 01 ± 6.68e − 02	2.10e + 01 ± 7.10e − 02	2.10e + 01 ± 4.23e − 02	2.10e + 01 ± 6.32e − 02
F_{18}	4.54e + 02 ± 2.57e + 01	4.96e − 01 ± 4.86e − 01	3.27e − 03 ± 4.58e − 03	6.62e − 07 ± 9.63e − 07
N_1/N_2	1420/80	1450/50	1490/10	1500/0
Fun	$AVG_{Er} \pm STD_{Er}$	$AVG_{Er} \pm STD_{Er}$	$AVG_{Er} \pm STD_{Er}$	$AVG_{Er} \pm STD_{Er}$
F_1	4.01e − 14 ± 2.53e − 14	1.92e − 14 ± 1.74e − 14	4.78e − 15 ± 3.33e − 15	3.02e − 15 ± 1.94e − 15
F_2	0.00e + 00 ± 0.00e + 00	0.00e + 00 ± 0.00e + 00	4.38e − 07 ± 7.54e − 06	3.21e + 01 ± 1.16e + 01
F_3	0.00e + 00 ± 0.00e + 00	0.00e + 00 ± 0.00e + 00	0.00e + 00 ± 0.00e + 00	0.00e + 00 ± 0.00e + 00
F_4	1.10e − 08 ± 3.14e − 09	6.32e − 09 ± 2.02e − 09	3.28e − 09 ± 1.29e − 09	2.87e − 09 ± 1.15e − 09
F_5	3.29e − 08 ± 1.25e − 08	1.85e − 08 ± 6.73e − 09	1.15e − 08 ± 5.10e − 09	9.90e − 09 ± 3.04e − 09
F_6	7.22e − 05 ± 1.25e − 04	3.74e − 05 ± 1.43e − 04	1.45e − 04 ± 2.35e − 04	7.22e − 05 ± 1.38e − 04
F_7	2.10e − 06 ± 3.08e − 06	1.20e − 06 ± 9.84e − 07	4.24e − 07 ± 4.95e − 07	3.87e − 07 ± 4.24e − 07
F_8	4.97e + 00 ± 2.22e + 01	8.92e + 00 ± 3.03e + 01	3.98e + 00 ± 2.16e + 01	1.38e + 01 ± 3.67e + 01
F_9	4.28e − 01 ± 4.65e − 02	4.14e − 01 ± 5.30e − 02	4.17e − 01 ± 5.50e − 02	1.00e + 00 ± 7.94e − 02
F_{10}	−1.00e + 00 ± 0.00e + 00	−1.00e + 00 ± 1.03e − 16	−1.00e + 00 ± 7.71e − 17	−1.00e + 00 ± 8.50e − 17
F_{11}	−2.92e + 01 ± 1.03e − 01	−2.92e + 01 ± 8.76e − 02	−2.92e + 01 ± 6.86e − 02	−2.92e + 01 ± 9.37e − 02
F_{12}	3.67e − 12 ± 3.79e − 14	3.64e − 12 ± 3.50e − 14	3.63e − 12 ± 3.20e − 14	3.64e − 12 ± 3.28e − 14
F_{13}	3.00e − 16 ± 8.22e − 16	3.84e − 16 ± 1.43e − 15	1.76e − 17 ± 4.10e − 17	1.11e − 15 ± 5.98e − 15
F_{14}	1.35e − 32 ± 5.57e − 48	1.35e − 32 ± 5.57e − 48	9.81e − 23 ± 4.13e − 22	3.22e − 15 ± 1.69e − 14
F_{15}	7.11e − 16 ± 4.71e − 16	3.45e − 16 ± 4.50e − 16	8.57e − 17 ± 9.95e − 17	5.18e − 17 ± 3.52e − 17
F_{16}	3.70e + 00 ± 1.81e + 00	4.72e + 00 ± 3.10e + 00	4.39e + 00 ± 3.48e + 00	3.98e + 00 ± 1.92e + 00
F_{17}	2.10e + 01 ± 5.98e − 02	2.10e + 01 ± 5.41e − 02	2.10e + 01 ± 4.71e − 02	2.10e + 01 ± 5.67e − 02
F_{18}	3.91e − 07 ± 5.86e − 07	1.24e − 07 ± 1.45e − 07	1.14e − 07 ± 2.50e − 07	7.54e − 08 ± 9.46e − 08

4.3 Analysis on Inertia Weight

As seen in Fig. 3, the parameter e^{-Rt} considered as the inertia weight normally shows a trend of a fast reduction with increasing iteration number, which may lead to the loss of population diversity and loss of strong global search ability in the early iterations. R is set to 0.3 in the original PIO. Inspired by the sigmoid (activation) function of neural networks, the novel nonlinear adaptive inertia weight strategy is proposed, which is mathematically represented by Eq. (6).

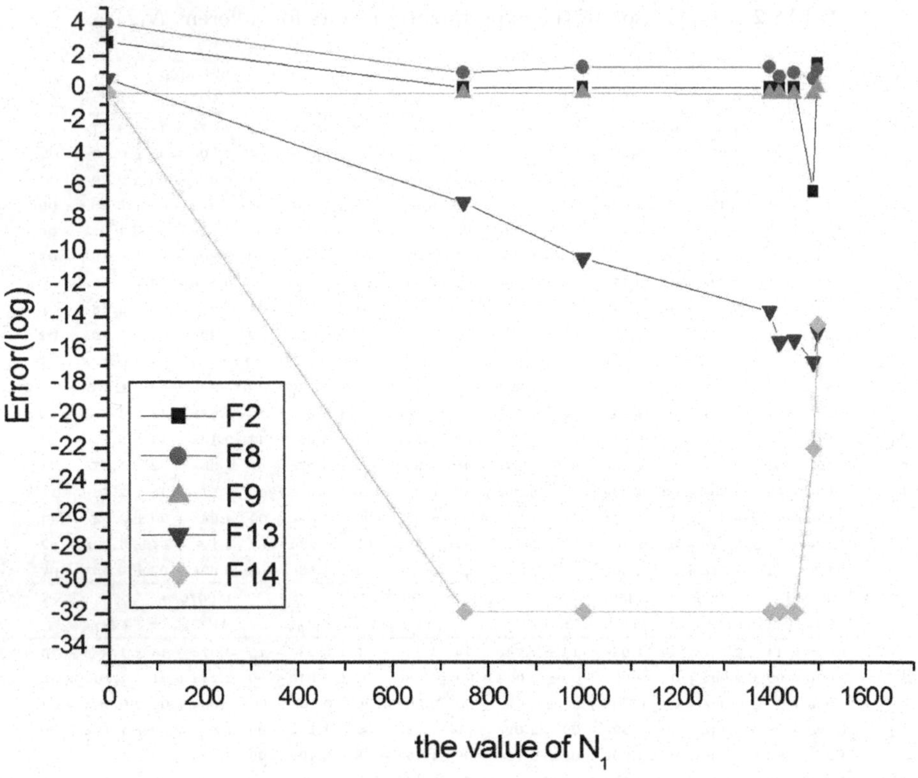

Fig. 4. Evolution of mean error values with increasing N_1 on some functions

Shi and Eberhart [14] proposed the classic linear inertia weight, which is mathematically represented as follows:

$$w_{max} - \frac{w_{max} - w_{min}}{T_{max}} \cdot t \tag{9}$$

where, in general, $w_{max} = 0.9$ and $w_{min} = 0.4$. Comparative experiments for the different inertia weight strategies are performed, with the results listed in Table 3, and the best results are marked in boldface. The classic linear inertia weight strategy and the parameter e^{-Rt} inertia weight are employed in the algorithms denoted as "$ODaPIO_l$" and "$ODaPIO_R$", respectively. For a fair comparison, we use the same parameters. Moreover, a Wilcoxon signed-rank test at the 5% significance level ($\alpha = 0.05$) is used to show the significant differences between the compared algorithms and ODaPIO. The "+" symbol indicates that the null hypothesis is rejected at the 5% significance level and that ODaPIO outperforms $ODaPIO_l$ and $ODaPIO_R$; the "−" symbol indicates that the null hypothesis is rejected at the 5% significance level and that the compared algorithm exceeds ODaPIO; and the "=" symbol reveals that the null hypothesis is accepted at the 5% significance level and that ODaPIO is equal to the compared algorithm. Additionally, we give the total number of statistically significant cases at the bottom of each table.

Table 3. Comparison of the experimental results for different inertia weight strategies

Fun	$ODaPIO_R$	$ODaPIO_l$	$ODaPIO$
	$AVG_{Er} \pm STD_{Er}$	$AVG_{Er} \pm STD_{Er}$	$AVG_{Er} \pm STD_{Er}$
F_1	$1.13e+00 \pm 2.63e+00$	$9.69e-09 \pm 6.89e-09$	$\mathbf{1.92e-14 \pm 1.74e-14}$
F_2	$\mathbf{0.00e+00 \pm 0.00e+00}$	$\mathbf{0.00e+00 \pm 0.00e+00}$	$\mathbf{0.00e+00 \pm 0.00e+00}$
F_3	$1.33e-01 \pm 7.30e-01$	$\mathbf{0.00e+00 \pm 0.00e+00}$	$\mathbf{0.00e+00 \pm 0.00e+00}$
F_4	$2.50e-03 \pm 4.69e-03$	$4.75e-06 \pm 1.29e-06$	$\mathbf{6.32e-09 \pm 2.02e-09}$
F_5	$5.32e-02 \pm 1.83e-01$	$1.47e-05 \pm 5.32e-06$	$\mathbf{1.85e-08 \pm 6.73e-09}$
F_6	$1.47e-02 \pm 3.09e-02$	$5.89e-04 \pm 7.88e-04$	$\mathbf{3.74e-05 \pm 1.43e-04}$
F_7	$3.16e-01 \pm 4.98e-01$	$6.64e-03 \pm 9.05e-03$	$\mathbf{1.20e-06 \pm 9.84e-07}$
F_8	$8.10e+02 \pm 2.52e+02$	$\mathbf{2.37e-05 \pm 3.34e-05}$	$8.92e+00 \pm 3.03e+01$
F_9	$4.76e-01 \pm 8.99e-02$	$\mathbf{4.21e-01 \pm 5.73e-02}$	$\mathbf{4.14e-01 \pm 5.30e-02}$
F_{10}	$-9.58e-01 \pm 7.83e-02$	$\mathbf{-1.00e+00 \pm 2.00e-12}$	$\mathbf{-1.00e+00 \pm 1.03e-16}$
F_{11}	$\mathbf{-2.93e+01 \pm 8.56e-02}$	$-2.88e+01 \pm 1.45e-01$	$-2.92e+01 \pm 8.76e-02$
F_{12}	$\mathbf{3.51e-12 \pm 8.03e-17}$	$3.91e-12 \pm 9.93e-14$	$3.64e-12 \pm 3.50e-14$
F_{13}	$5.48e-05 \pm 2.27e-04$	$9.94e-11 \pm 4.46e-10$	$\mathbf{3.84e-16 \pm 1.43e-15}$
F_{14}	$1.35e-32 \pm 5.57e-48$	$\mathbf{1.35e-32 \pm 5.57e-48}$	$\mathbf{1.35e-32 \pm 5.57e-48}$
F_{15}	$5.43e+01 \pm 1.52e+02$	$2.15e-10 \pm 2.14e-10$	$\mathbf{3.45e-16 \pm 4.50e-16}$
F_{16}	$4.88e+03 \pm 9.69e+00$	$\mathbf{5.34e+00 \pm 1.83e+00}$	$\mathbf{4.72e+00 \pm 3.10e+00}$
F_{17}	$\mathbf{2.09e+01 \pm 7.53e-02}$	$2.10e+01 \pm 6.12e-02$	$2.10e+01 \pm 5.41e-02$
F_{18}	$5.53e+00 \pm 2.31e+00$	$9.63e-04 \pm 6.61e-04$	$\mathbf{1.24e-07 \pm 1.45e-07}$
+/=/−	12/3/3	10/7/1	

From Table 3, the statistical results clearly show that ODaPIO outperforms $ODaPIO_R$ and $ODaPIO_l$ for most of the functions, demonstrating that the use of the nonlinear adaptive inertia weight can effectively enhance the performance of the algorithm. However, the three inertia weighting strategies have no significant effect for the Rosenbrock function F_2. $ODaPIO_l$ achieves the solution with the highest accuracy for F_8. As indicated by the "+/=/−" statistical results, ODaPIO outperforms $ODaPIO_R$ and $ODaPIO_l$ for 12 and 10 out of 18 functions, respectively, is equal to $ODaPIO_R$ and $ODaPIO_l$ for 3 and 7 out of 18 functions, and is inferior to $ODaPIO_R$ and $ODaPIO_l$ for 3 and 1 out of 18 functions.

4.4 Analysis on Dimension Selection

In the one-dimension modification mechanism, the candidate solution is updated for only a random dimension at each iteration, and because of the randomness, this approach may be wasteful with regard to the number of times that the solution is evaluated. Inspired by the definition of variance in statistics, we treat the data of each dimension in the population as a unit sample. A larger variance of a dimension indicates that the stability of this dimension is relatively poor and needs to be improved. Therefore, the approach of selecting the dimension with larger variance is proposed. In [23], the authors propose the methods of selecting the dimension based on difference from the best individual. To investigate the effects of the different strategies of dimension selection, we perform experiments for which "$ODaPIO_s$" denotes selecting the dimension based on the difference from the best individual and "$ODaPIO_v$" denotes selecting the dimension with a larger variance. As mentioned above, the random dimension-selected strategy is employed in ODaPIO. For a fair comparison, we use the same parameters. Moreover, a Wilcoxon signed-rank test at the 5% significance level ($\alpha = 0.05$) is used to show the significant differences between the compared algorithms and ODaPIO. The "+" symbol indicates that the null hypothesis is rejected at the 5% significance level and that ODaPIO outperforms $ODaPIO_s$ and $ODaPIO_v$; the "−" symbol indicates that the null hypothesis is rejected at the 5% significance level and that the compared algorithm outperforms ODaPIO; and the "=" symbol indicates that the null hypothesis is accepted at the 5% significance level and that ODaPIO is equal to the compared algorithm. The results are listed in Table 4, and better results are marked in boldface. Additionally, we specify the total number of statistically significant cases at the bottom of each table.

Table 4. Comparison of the experimental results for different dimension selection methods

Fun	$ODaPIO_s$	$ODaPIO_v$	$ODaPIO$
	$AVG_{Er} \pm STD_{Er}$	$AVG_{Er} \pm STD_{Er}$	$AVG_{Er} \pm STD_{Er}$
F_1	**8.28e − 25 ± 6.13e − 25**	3.53e − 19 ± 1.80e − 19	1.92e − 14 ± 1.74e − 14
F_2	2.61e + 01 ± 3.97e + 01	**0.00e + 00 ± 0.00e + 00**	**0.00e + 00 ± 0.00e + 00**
F_3	**0.00e + 00 ± 0.00e + 00**	**0.00e + 00 ± 0.00e + 00**	**0.00e + 00 ± 0.00e + 00**
F_4	**5.28e − 14 ± 1.91e − 14**	3.63e − 11 ± 8.82e − 12	6.32e − 09 ± 2.02e − 09
F_5	1.71e − 02 ± 9.38e − 02	**8.68e − 11 ± 3.64e − 11**	1.85e − 08 ± 6.73e − 09
F_6	3.19e − 04 ± 1.38e − 03	**3.87e − 05 ± 1.46e − 04**	**3.74e − 05 ± 1.43e − 04**
F_7	4.64e − 01 ± 7.27e − 01	**3.19e − 09 ± 4.84e − 09**	1.20e − 06 ± 9.84e − 07
F_8	2.01e + 02 ± 1.29e + 02	**4.20e + 00 ± 2.16e + 01**	8.92e + 00 ± 3.03e + 01
F_9	**4.47e − 01 ± 6.83e − 02**	**4.30e − 01 ± 4.07e − 02**	**4.14e − 01 ± 5.30e − 02**
F_{10}	**−1.00e + 00 ± 2.92e − 17**	**−1.00e + 00 ± 2.06e − 17**	**−1.00e + 00 ± 1.03e − 16**
F_{11}	−2.88e + 01 ± 3.23e − 01	**−2.92e + 01 ± 7.84e − 02**	**−2.92e + 01 ± 8.76e − 02**
F_{12}	**3.54e − 12 ± 7.10e − 14**	3.67e − 12 ± 2.08e − 14	3.64e − 12 ± 3.50e − 14
F_{13}	2.25e − 10 ± 1.23e − 09	**2.31e − 22 ± 2.69e − 22**	3.84e − 16 ± 1.43e − 15
F_{14}	**1.35e − 32 ± 5.57e − 48**	**1.35e − 32 ± 5.57e − 48**	**1.35e − 32 ± 5.57e − 48**
F_{15}	5.81e − 09 ± 3.18e − 08	**4.06e − 21 ± 2.13e − 21**	3.45e − 16 ± 4.50e − 16
F_{16}	2.68e + 03 ± 3.47e + 02	2.75e + 02 ± 1.15e + 02	**4.72e + 00 ± 3.10e + 00**
F_{17}	**2.10e + 01 ± 5.88e − 02**	**2.10e + 01 ± 6.57e − 02**	**2.10e + 01 ± 5.41e − 02**
F_{18}	2.06e + 00 ± 1.41e + 00	**1.85e − 10 ± 3.36e − 10**	1.24e − 07 ± 1.45e − 07
+/ = /−	10/5/3	2/8/8	

From Table 4, in terms of the results of the Wilcoxon signed-rank test, $ODaPIO_v$ is superior to ODaPIO and $ODaPIO_s$ for most functions, demonstrating that the method of selecting the dimension with a larger variance significantly improves the effectiveness of the algorithm for some functions. However, $ODaPIO_s$ obtains the best result for F_1 and F_4.

4.5 Comparison with SI Algorithms

To investigate the competitiveness of our approach, ODaPIO is compared with the original PIO and four improved SI algorithms, namely, BQPIO [12], *w*PSO [14], OXDE [22] and DDICS [14]. *w*PSO is the algorithm using the classic linearly decreasing inertia weight in the standard PSO. Each algorithm is executed 30 times for each function with the dimension $D = 30$. The population size $NP = 30$ and the maximum number of iterations = 1500. The parameters of the improved algorithms used in the comparison are set according to their

Table 5. Comparison of the experimental results for different SI algorithms

Fun	PIO	BQPIO	wPSO
	$AVG_{Er} \pm STD_{Er}$	$AVG_{Er} \pm STD_{Er}$	$AVG_{Er} \pm STD_{Er}$
F_1	1.46e + 01 ± 2.90e + 00	1.44e + 01 ± 2.44e + 00	5.51E − 05 ± 7.38E − 05
F_2	1.36e + 03 ± 2.55e + 02	1.49e + 03 ± 2.10e + 02	5.79E + 01 ± 4.39E + 01
F_3	5.62e + 01 ± 1.02e + 01	5.44e + 01 ± 7.98e + 00	2.67E − 01 ± 8.28E − 01
F_4	1.75e + 01 ± 3.17e + 00	1.50e + 01 ± 2.97e + 00	1.94E − 05 ± 1.30E − 05
F_5	2.68e + 00 ± 1.99e − 01	2.68e + 00 ± 2.12e − 01	1.27E − 01 ± 3.91E − 01
F_6	6.19e − 01 ± 1.31e − 01	5.97e − 01 ± 1.36e − 01	1.27E − 02 ± 1.15E − 02
F_7	1.68e + 01 ± 2.65e + 00	1.55e + 01 ± 2.37e + 00	4.52E + 01 ± 1.34E + 01
F_8	8.62e + 03 ± 6.72e + 02	5.95e + 03 ± 1.44e + 03	1.80E + 03 ± 4.23E + 02
F_9	4.87e − 01 ± 1.36e − 01	5.08e − 01 ± 1.55e − 01	6.84E − 01 ± 1.08E − 01
F_{10}	−3.80e − 45 ± 2.08e − 44	−6.22e − 17 ± 3.18e − 16	**−1.00E + 00 ± 1.72E − 08**
F_{11}	−1.03e + 01 ± 1.07e + 00	−1.07e + 01 ± 1.16e + 00	−2.47E + 01 ± 1.06E + 00
F_{12}	1.64e − 06 ± 2.72e − 06	1.25e − 06 ± 2.25e − 06	7.16E − 12 ± 6.79E − 13
F_{13}	4.36e + 00 ± 1.29e + 00	3.90e + 00 ± 1.07e + 00	3.02E − 01 ± 5.72E − 01
F_{14}	1.44e + 00 ± 3.10e − 01	1.51e + 00 ± 2.59e − 01	6.78E − 03 ± 1.13E − 02
F_{15}	6.04e + 04 ± 1.07e + 04	5.44e + 04 ± 9.41e + 03	2.91E − 06 ± 6.58E − 06
F_{16}	4.88e + 03 ± 3.27e + 00	1.72e + 03 ± 4.90e + 02	5.34E + 03 ± 2.06E + 02
F_{17}	2.11e + 01 ± 6.38e − 02	**2.09e + 01 ± 1.09e − 01**	2.10E + 01 ± 5.20E − 02
F_{18}	3.98e + 02 ± 3.63e + 01	3.74e + 02 ± 4.53e + 01	3.21E + 01 ± 7.64E + 00
Fun	OXDE	DDICS	ODaPIO
	$AVG_{Er} \pm STD_{Er}$	$AVG_{Er} \pm STD_{Er}$	$AVG_{Er} \pm STD_{Er}$
F_1	3.28e − 04 ± 2.54e − 04	2.52e − 06 ± 2.73e − 06	**1.92e − 14 ± 1.74e − 14**
F_2	2.41e + 01 ± 2.07e + 00	7.46e + 00 ± 4.88e + 00	**0.00e + 00 ± 0.00e + 00**
F_3	**0.00e + 00 ± 0.00e + 00**	**0.00e + 00 ± 0.00e + 00**	**0.00e + 00 ± 0.00e + 00**
F_4	7.57e − 04 ± 2.96e − 04	5.62e − 05 ± 1.59e − 05	**6.32e − 09 ± 2.02e − 09**
F_5	3.20e − 02 ± 1.70e − 01	9.96e − 04 ± 4.86e − 04	**1.85e − 08 ± 6.73e − 09**
F_6	1.54e − 03 ± 3.40e − 03	**1.51e − 05 ± 3.83e − 05**	3.74e − 05 ± 1.43e − 04
F_7	7.12e + 01 ± 1.43e + 01	3.26e + 00 ± 1.00e + 00	**1.20e − 06 ± 9.84e − 07**
F_8	**3.11e − 04 ± 4.22e − 04**	4.66e + 02 ± 1.43e + 02	8.92e + 00 ± 3.03e + 01
F_9	**3.58e − 01 ± 8.64e − 02**	2.51e + 00 ± 2.44e − 01	4.14e − 01 ± 5.30e − 02
F_{10}	**−1.00e + 00 ± 4.36e − 08**	**−1.00e + 00 ± 7.86e − 07**	**−1.00e + 00 ± 1.03e − 16**
F_{11}	−1.75e + 01 ± 1.44e + 00	−2.85e + 01 ± 1.87e − 01	**−2.92e + 01 ± 8.76e − 02**
F_{12}	2.73e − 09 ± 2.47e − 09	3.68e − 12 ± 1.24e − 13	**3.64e − 12 ± 3.50e − 14**
F_{13}	1.73e − 02 ± 7.74e − 02	3.27e − 09 ± 6.20e − 09	**3.84e − 16 ± 1.43e − 15**
F_{14}	3.90e − 03 ± 2.05e − 02	3.88e − 08 ± 4.93e − 08	**1.35e − 32 ± 5.57e − 48**
F_{15}	1.18e − 05 ± 7.31e − 06	2.52e − 07 ± 2.46e − 07	**3.45e − 16 ± 4.50e − 16**
F_{16}	1.04e − 01 ± 6.02e − 02	3.45e + 00 ± 1.08e + 00	4.72e + 00 ± 3.10e + 00
F_{17}	2.10e + 01 ± 5.36e − 02	**2.09e + 01 ± 5.63e − 02**	2.10e + 01 ± 5.41e − 02
F_{18}	6.43e + 01 ± 1.47e + 01	3.68e + 00 ± 1.41e + 00	**1.24e − 07 ± 1.45e − 07**

Table 6. Average ranking of six algorithms by the Friedman test

Algorithm	PIO	BQPIO	wPSO	OXDE	DDICS	ODaPIO
Ranking	4. 67	3. 83	3. 06	2. 76	2. 08	**1. 36**

original references. In the original PIO, $R = 0.3$, $N_1 = 1000$, and $N_2 = 500$. In BQPIO, $R = 0.2$, $N_1 = 1300$, and $N_2 = 200$. The compared results are listed in Table 5. Moreover, Table 6 gives the results of the Friedman test that have similarly been obtained in the literature [23] for the six algorithms.

From Table 5, the results suggest that some algorithms exhibit their advantage over parts of functions. OXDE does well in F_3, F_8, F_9 and F_{10}. DDICS obtains the best results in F_3, F_6, F_{10} and F_{17}. According to the statistical results of Friedman test, Table 6 shows that ODaPIO is the best, followed by DDICS, OXDE, *w*PSO, BQPIO and PIO. In summary, the proposed algorithm of ODaPIO is not only feasible but also competitive.

5 Conclusion and Future Work

Inspired by the homing behaviors of pigeons, a novel bio-inspired swarm intelligence optimizer called PIO was invented in 2014. In early iterations, the individual's velocity mainly depends on its fast decreasing inertia weight according to the map and compass operator, which may reduce the population diversity rapidly and lead to easy trapping in local optima. In this paper, we propose an approach that effectively combines the method of nonlinearly adaptive inertia weight with the strategy of one-dimension modification, called ODaPIO in short. Inspired by the sigmoid (activate) function of neural networks, the nonlinear inertia weight strategy is introduced, which aims to balance the exploration and exploitation abilities by dynamically adaptive inertia weight during the iterative optimization process. Meanwhile, the one-dimension modification mechanism aims to greatly enhance the quality of the solution in order to avoid interference phenomena among the dimensions. Extensive experimental results demonstrate that compared to other improved SI algorithms, the proposed approach can more feasibly and effectively solve complex continuous optimization problems. Additionally, the effects of the parameter control, different inertia weight strategies and dimension selection are analyzed in detail. Comprehensive experiments demonstrate that control of the N_1 and N_2 parameters significantly affects the performance of ODaPIO, that priority selection on the dimension with the larger variance is more effective, and that the proposed adaptive inertia weight strategy is more competitive than that using the classic linear inertia weight.

In future research, we will study further improvement of our approach to enable it to solve rotated problems more effectively. Additionally, we plan to apply the proposed algorithm to some real-world optimization problems for further examination.

References

1. Karaboga, D.: An idea based on honey bee swarm for numerical optimization. Technical report TR-06, Erciyes University, Engineering Faculty, Computer Engineering Department (2005)
2. Yang, X.S., Deb, S.: Cuckoo search via levy flights. In: Mathematics, pp. 210–214 (2010)
3. Yang, X.S.: A new metaheuristic bat-inspired algorithm. Comput. Knowl. Technol. **284**, 65–74 (2010)
4. Yang, X.S.: Nature-Inspired Metaheuristic Algorithms. Luniver Press, UK (2008)
5. Shi, Y.: Brain storm optimization algorithm. In: Proceedings of the 2nd International Conference on Swarm Intelligence, Chongqing, pp. 303–309 (2011)
6. Talbi, E.: Metaheuristics. From Design to Implementation. Wiley, Hoboken (2009)
7. Duan, H.B., Qiao, P.X.: Pigeon-inspired optimization: a new swarm intelligence optimizer for air robot path planning. Int. J. Intell. Comput. Cybern. **7**, 24–37 (2014)
8. Zhang, B., Duan, H.B.: Predator-prey pigeon-inspired optimization for UAV three-dimensional path planning. In: International Conference in Swarm Intelligence, pp. 96–105 (2014)
9. Sun, H., Duan, H.B.: PID controller design based on prey-predator pigeon-inspired optimization algorithm. In: IEEE International Conference on Mechatronics and Automation, pp. 1416–1421 (2014)
10. Li, C., Duan, H.B.: Target detection approach for UAVs via improved pigeon-inspired optimization and edge potential function. Aerosp. Sci. Technol. **39**, 352–360 (2014)
11. Duan, H., Wang, X.: Echo state networks with orthogonal pigeon-inspired optimization for image restoration. IEEE Trans. Neural Netw. Learn. Syst. **27**(11), 2413–2425 (2017)
12. Li, H.H., Duan, H.B.: Bloch quantum-behaved pigeon-inspired optimization for continuous optimization problems. In: Guidance, Navigation and Control Conference, pp. 2634–2638 (2014)
13. Shi, Y.H., Eberhart, R.: A modified particle swarm optimizer. In: Proceedings of the IEEE International Conference on Evolutionary Computation and IEEE World Congress on Computational Intelligence, Anchorage, Ala, USA, pp. 69–73 (1998)
14. Shi, Y., Eberhart, R.C.: Empirical study of particle swarm optimization. In: Proceedings of the 1999 Congress on Evolutionary Computation, pp. 320–324 (1999)
15. Shi, Y., Eberhart, R.C.: Fuzzy adaptive particle swarm optimization. In: Proceedings of the IEEE Conference on Evolutionary Computation, vol. 1, pp. 101–106 (2001)
16. Xu, S.H., Li, X.X.: An adaptive changed inertia weight particle swarm algorithm. Sci. Technol. Eng. **9**, 1671–1815 (2012)
17. Zhong, Y.W., Liu, X., Wang, L.J., et al.: Particle swarm optimization algorithm with iterative improvement strategy for multi-dimensional function optimization problems. Int. J. Innov. Comput. Appl. **4**, 223–232 (2012)
18. Wang, L.J., Yin, Y.L., Zhong, Y.W.: Cuckoo search algorithm with dimension by dimension improvement. J. Softw. **24**, 2687–2698 (2013). (in Chinese)
19. Chen, M.W., Zhong, Y.W., Wang, L.J.: Bat algorithm with one-dimension learning. J. Chin. Comput. Syst. **36**, 23–26 (2015)
20. Peio, L., Astrid, J., Patrick, S., et al.: A sensitivity analysis method for driving the artificial bee colony algorithm's search process. Appl. Soft Comput. **41**, 515–531 (2016)

21. Suganthan, P.N., Hansen, N., Liang, J.J., et al.: Problem definitions and evaluation criteria for the CEC2005 special session on real-parameter optimization. Nanyang Technology University, Singapore (2005)
22. Chen, G.M., Huang, X.B., Jia, J.Y., et al.: Natural exponential inertia weight strategy in particle swarm optimization. In: Proceedings of the 6th World Congress on Intelligent Control and Automation (2006)
23. Jin, X., Liang, Y.Q., Tian, D.P., et al.: Particle swarm optimization using dimension selection methods. Appl. Math. Comput. **219**, 5185–5197 (2013)

ESAE: Evolutionary Strategy-Based Architecture Evolution

Xue Gu[1,2], Ziyao Meng[1,2], Yanchun Liang[1,2], Dong Xu[3], Han Huang[1,4], Xiaosong Han[1], and Chunguo Wu[1](✉)

[1] Key Laboratory of Symbolic Computation and Knowledge Engineering of Ministry of Education, College of Computer Science and Technology, Jilin University, Changchun 130012, People's Republic of China
wucg@jlu.edu.cn

[2] Zhuhai Laboratory of Key Laboratory of Symbol Computation and Knowledge Engineering of Ministry of Education, School of Computer, Zhuhai College of Jilin University, Zhuhai 519041, People's Republic of China

[3] Department of Electrical Engineering and Computer Science, Informatics Institute, and Christopher S. Bond Life Sciences Center, University of Missouri, Columbia, MO 65211, USA

[4] School of Software Engineering, South China University of Technology, GuangZhou 510006, People's Republic of China

Abstract. Although deep neural networks (DNNs) play important roles in many fields, the architecture design of DNNs can be challenging due to the difficulty of input data representation, the huge number of parameters and the complex layer relationships. To overcome the obstacles of architecture design, we developed a new method to generate the optimal structure of DNNs, named Evolutionary Strategy-based Architecture Evolution (ESAE), consisting of a bi-level representation and a probability distribution learning approach. The bi-level representation encodes architectures in the gene and parameter levels. The probability distribution learning approach ensures the efficient convergence of the architecture searching process. By using Fashion-MNIST and CIFAR-10, the effectiveness of the proposed ESAS is verified. The evolved DNNs, starting from a trivial initial architecture with one single convolutional layer, achieved the accuracies of 94.48% and 93.49% on Fashion-MNIST and CIFAR-10, respectively, and require remarkably less hardware costs in terms of GPUs and running time, compared with the existing state-of-the-art manual screwed architectures.

Keywords: Deep neural networks (DNNs) · Evolutionary method · Automatic design · Probability distribution learning

Supported by organizations of the National Natural Science Foundation of China (61972174, 61876069 and 61876207), the Key Development Project of Jilin Province (20180201045GX and 20180201067GX), the Guangdong Key-Project for Applied Fundamental Research (2018KZDXM076), and the Guangdong Premier Key-Discipline Enhancement Scheme (2016GDYSZDXK036).

L. Pan et al. (Eds.): BIC-TA 2019, CCIS 1159, pp. 193–208, 2020.
https://doi.org/10.1007/978-981-15-3425-6_16

1 Introduction

The deep neural networks (DNNs) are typical implementation paradigms in the fields of data-driven artificial intelligence, and have been successfully applied to many fields including image recognition [1], speech processing [2], machine translation [3], etc. Specifically, after AlexNet [1] outperformed the traditional manual methods in the 2012 ImageNet Challenge, more and more complex DNNs were proposed, e.g. GoogLeNet [4], VGGNet [5], ResNet [6] and DenseNet [7]. As an example, VGG-16 has more than 130 million parameters, and needs 15.3 billion floating-point operations to complete an image recognition task. However, it should be noticed that these models were all manually designed by experts with an arduous trial-and-error process. Moreover, even with considerable expertise, we still have to spend a large amount of resources and time to design such well-performed models. Hence, it is necessary to study the automatic methods to design DNNs.

There are many search strategies that can be used to explore the neural architectures, including random search (RS) [8], bayesian optimization (BO) [9], evolutionary algorithms (EA) [10–18], reinforcement learning (RL) [19], and gradient-based methods [20,21]. BO is one of the most popular methods for hyper-parameter optimization, but the typical BO toolboxes are based on Gaussian processes and focus on lowdimensional continuous optimization problems. RS, BO, EA and RL, are the fact that architecture search is treated as a black-box optimization problem over a discrete domain. Instead of searching over a discrete set of candidate architectures, the gradient-based methods improve the efficiency by relaxing the search space to be continuous.

Inspired by the powerful search ability of evolutionary algorithms and the potentials of excellent architectures, this paper is devoted to design a bi-level encoding scheme with corresponding evolutionary operators, to evolve the feed forward DNNs. Especially, in order to accelerate the convergence and improve the accuracy, we proposed a learning strategy of probability distribution, which is used to dynamically adjust the frequency of various evolution operators and parameter values. In the proposed bi-level encoding scheme, the first level (GENE Level), indicates the general structure by the sequence of layers; the second level parameter level, shorted as PARA Level, specifies the parameters associated with each layer. The proposed encoding method can represent the network structures flexibly with variable depth.

The main contributions of this paper are:

- ESAE: a framework based on evolutionary strategy that automatically searches for the architectures and parameters of large scale DNNs;
- We design a learning approach of probability distribution, which can be speed up its search.

2 Related Work

There have been many works on evolutionary algorithms for optimizing the DNNs architecture [10–18]. The first work might be presented by Miller et al. [10] use genetic algorithm (GA) to generate architectures and use back propagation to optimize connecting weights. Based on evolutionary programming (EP), Yao et al. [11] proposed EPNet for evolving neural networks by five mutation operators. EPNet realizes the co-evolution of architectures and weights, and could generate compact neural networks with good generalization ability. Stanley et al. [12] proposed neuroevolution of augmenting topologies (NEAT). NEAT could optimize efficiently the weights and architectures at the same time. NEAT provides a novel way for evolving large-scale neural networks. Stanley et al. [13] designed HyperNEAT, which could evolve the large-scale networks with millions of connections. In 2010, the emergence of GPU for auxiliary neural network training mode greatly promoted the development of DNNs as well as the evolutionary search of architectures of neural networks. Real et al. [17] applied NEAT (shorted for neuroevolution of augmenting topologies) to implement a large-scale evolution of image classifiers, which explored architectures from basic initial conditions to large models with the performance comparable to the manual design models. Suganuma et al. [14] used cartesian genetic programming (CGP) to design an automatic image classification. Miikkulainen et al. [15] proposed CoDeepNEAT for automatically design DNNs, which used NEAT to design the architectures and hyper-parameters of CNNs, and then extended to evolution of components as well. Compared with the large-scale evolution of image classifiers, Real et al. [16] employed evolutionary algorithms (EA) to evolve an architecture from relatively complex initial conditions by introducing stacked modules, enhancing the probability to obtain high-performance networks. Assuncao et al. [18] proposed DENSER, a novel combination of GA with dynamic structured grammatical evolution (DSGE), which not only searches for the best network topology, but also tunes hyper-parameters, such as learning rate.

Some researchers try to apply swarm intelligence algorithms, such as particle swarm optimization (PSO) [22], monarch butterfly optimization (MBO) [23], earthworm optimization (EWA) [24], elephant herding optimization (EHO) [25], moth search algorithm (MS) [26], to search neural network architectures. Zhang et al. [22] proposed PSO-BP algorithm for shallow feedforward network search. Kiranyaz et al. [27] proposed a multi-dimensional PSO algorithm, which reforms the native structure of swarm particles in such a way that they can make interdimensional passes with a dedicated dimensional PSO process. Das et al. [28] proposed another PSO algorithm to train artificial neural networks (ANNs), where the numbers of layers and neurons are also regarded as parameters encoded into particles. However, this strategy also makes the search space extremely large and thus degrades its performance. Salama et al. [29] proposed an ant colony optimization (CACO) algorithm to learn ANNs structure, where the number of hidden neurons is determined by pruning the maximally connected network structure. However, these methods are only applicable to shallow networks. Zhou et al. [30] adapted water wave optimization (WWO) algorithm to optimize both the parameters and structures of both classical shallow ANNs and DNNs.

3 Methods

3.1 Bi-level Representation

The deep feed-forward neural networks consist of multiple functional layers, including convolution, pooling and so on. Each functional layer needs specific parameters. For example, the convolution layer needs kernel size, kernel numbers, padding and stride. Each candidate architecture is an ordered sequence consisting of these functional layers, which forms a variety of deep network instances under the characterization of different parameters. Inspired by the attribute description of DNNs with different granularities, this paper proposes a bi-level encoding scheme. GENE Level is used to describe the functional layer type of DNNs, which is called genotype coding. PARA Level is used to describe attributes of each genotype, which is called parameter coding. As shown in Fig. 1, Column (*a*) shows the genotype coding; Column(*b*) shows the parameter coding; and Column (*c*) shows the value set of the corresponding parameters.

Genotype coding describes the macro architecture of networks. The advantage of this representation is flexibility. It represents variable-length network structures and efficiently explore the global design space of networks. Parameter coding describes the specific attributes of genes, which enables to explore the local design space. When the network is to be expanded, a gene of certain genotype is added into its genotype coding string first and then, its parameters are generated according to some probability distribution. We will discuss the learning strategy of probability distribution in Sect. 3.4.

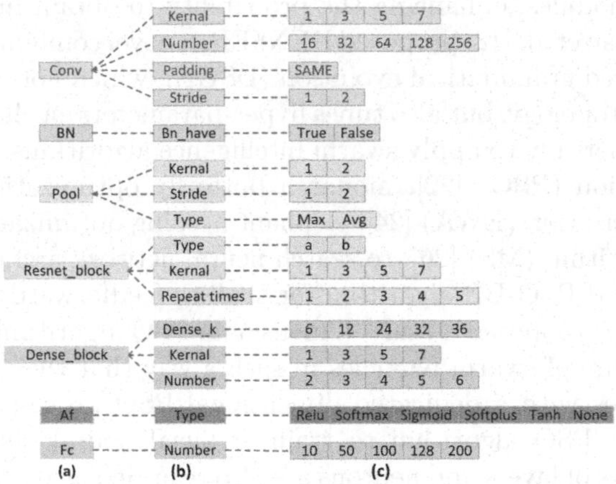

Fig. 1. The bi-level encoding scheme for the architecture of DNN ((a) genotype, (b) parameters of each genotype, (c) value set of the corresponding parameters).

Genotype Coding. Genotype coding encodes the overall structure of a network to indicate the functional layers of the candidate network. In paper, we introduce the feature modules of the ResNet and DenseNet. As shown in Fig. 1(a), 7 genotypes are designed, and the set of genotypes is denoted as G:

$$G = \{Conv, Bn, Pool, Resnet\_block, Dense\_block, Af, Fc\}$$

where *Conv*, *Bn*, *Pool*, *Resnet_block*, *Dense_block*, *Af*, and *Fc* are respectively called convolution, batch normalization, pooling, resnet block, dense block, activation function and full connection layer genes. The *Resnet_block* [6] and *Dense_block* [7] are shown in Fig. 2.

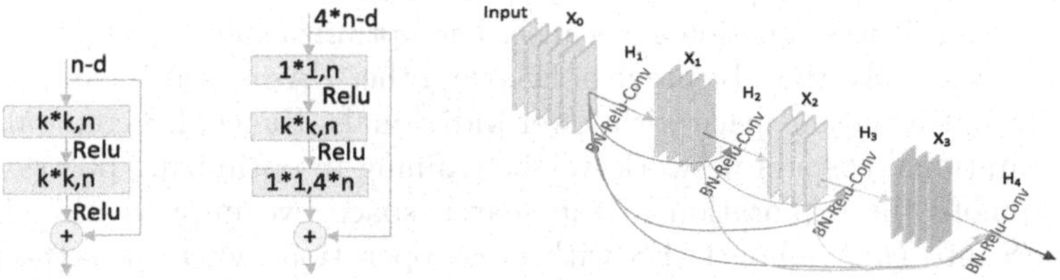

Fig. 2. The network blocks of *Resnet_block* and *Dense_block* (left: the structure diagram of the *Resnet_block* without $1 * 1$ convolution; middle: *Resnet_block* module structure diagram with $1 * 1$ convolution; right: the structure diagram of the 5-layer *Dense_block* module, the number of the modules' floors is variable).

Parameter Coding. Parameter coding encodes the parameters to specify the concrete gene properties. As shown in Fig. 1(b), the genetic parameters of the same colors jointly describe the characteristics of the corresponding genotype. During evolution, the genotypes and the parameters are evolved. The candidate parameters for a given gene are taken from a finite discrete set. With this bi-level encoding scheme, we can conveniently expand the candidate value sets of gene parameters. However, in order to verify the effectiveness and reliability of this algorithm, we only use the values reported in the existing literatures to construct the candidate value sets. The candidate value sets are shown in Fig. 1(c).

As an example, the well-known classical LeNet5 [31] (Fig. 3(a)) can be encoded by the proposed bi-level representation as follows in Fig. 3(b). Figure 3(c) presents the string form of the bi-level representation of LeNet5.

3.2 Evolutionary Strategy-Based Architecture Evolution

The architecture optimization problem belongs to dual optimization problems, because its performance needs to be shown by adjusting the network weights. The optimal architecture is strongly coupled with the optimal weights. However, the approach of optimizing both architectures and weights are time and memory consuming. Therefore, the popular way is to optimize only the architectures and hyper-parameters, and then use deep learning platforms, e.g., Keras, to optimize

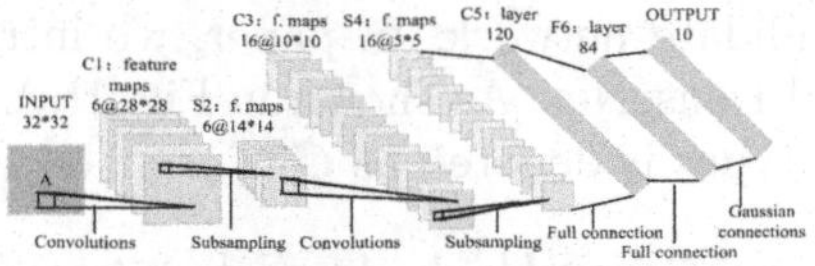

(a) The well-known LeNet5 [31]

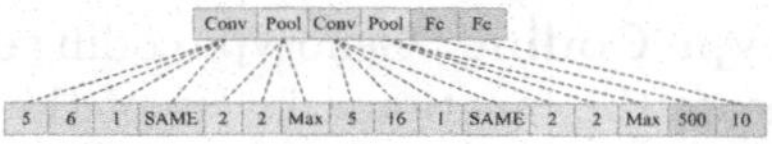

(b) The bi-level representation of LeNet5

(c) The string form of the bi-level representation of LeNet5

Fig. 3. A bi-level representation of the well-known LeNet5.

the weights. After several epochs, the task performance index is taken as the basis of constructing fitness functions. Because, the optimization of weights is time-consuming extremely, the advantage of ES over other EAs is that it can efficiently explore the optimization space with fewer individuals. Hence, ES could alleviate the computational costs of network weight training in optimizing process.

To promote the exploration of the search space, we improve the classical $(\mu + \lambda)$-ES into $(\mu, \lambda + \nu + 1)$-ES with cross operation, where μ is the size of the parental population, λ the number of temporary offsprings generated by mutation operations and ν the number of temporary offsprings generated by crossover operations, and 1 for the best one selected from parental individuals. Denote $T(S, k)$ as the operation of selecting k individuals from the individual set S through tournament selections. The specific ESAE steps are shown in Algorithm 1.

Algorithm 1. ESAE

Input: P: the initial population; μ: the size of the population;
Output: optimal *individual*
1: **repeat**
2: Each parent mutates 2 times, forming 2μ temporary individuals, denoted as O_m^1;
3: Homologous temporary offsprings compete, screening out μ offsprings;
4: The μ winning offsprings crossover mutually population, forming an offspring, denoted as O_m^2;
5: The offspring group is selected, $O = T(O_m^1 \cup O_m^2, \mu - 1)$;
6: Parental competition, $Q = T(P, 1)$;
7: Form a new parentsl population with the outperforming parents and offsprings, $P = Q \cup O$;
8: **until** (Iteration reach up to the upper bound or Accuracy = 1)

In this paper, we use $(\mu, \lambda + \nu + 1)$-ES with the parental population size as $\mu, \lambda = 2\,\mu$, and $\nu = C_\mu^2$. The proposed $(\mu, \lambda + \nu + 1)$-ES operates as follows:

Step (1): First, each parent mutates to produce 2 offsprings, called primary offspring. The size of primary offspring is 2 μ.
Step (2): The offspring generated by mutation from the same parent is called homologous offspring. Each group of homologous offsprings retains one individual through competition, and then the winning individuals are crossed mutually. The offsprings generated by crossover are called secondary offsprings. The population size of secondary offsprings is $\nu = C_{\mu}^{2}$, where C_{μ}^{2} is the combinatorial number.
Step (3) : $\mu - 1$ offsprings are selected through competition from the union of primary and secondary offsprings.
Step (4): Finally, the new parental population consists of the $\mu - 1$ offsprings selected in *Step* (3) and the best parent of the last generation.

We allow the best parental individual to enter into the next cycle, relieving the competitive pressure of the parental generation so that the best parental individual can be fully evolved. Figure 4 is a schematic diagram of the proposed $(3,\ 6+6+1)$-ES.

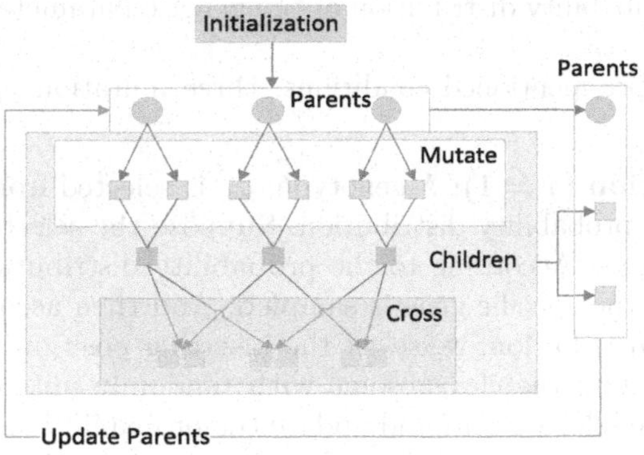

Fig. 4. Evolution flow chart of the proposed ESAE.

3.3 Variation Operators

To promote the exploration of DNNs architectures, we design variation operators that act on the bi-level encoding scheme, including mutation and crossover operators. The mutation operator can evolve in the levels of both genotype and gene parameter. At the genotype level, we design the operations that can insert or delete functional layers. And at the parameter level, we only evolve the value of parameters by selecting from a given candidate set. While the crossover operator can only evolve architectures by acting in genotype level.

Mutation. Let G be a set of genotypes:

$$G = \{Conv, Bn, Pool, Resnet\_block, Dense\_block, Af, Fc\}$$

$P(X_g = i)$ $(i = 1, 2, \ldots, 7)$ constitutes the probability distribution of X_g, denoted as $X_g \sim P_G$, where X_g is the random variable of genotypes. In Sect. 3.4, we will introduce the probability distribution learning strategy.

Similarly, the set of gene operation mutation operators is denoted as:

$$M = \{inserting, deleting, updating\},$$

where M contains 3 types of operators. $P(X_m = j)$ $(j = 1, 2, 3)$ constitutes the probability distribution of the genotype mutation operator, denoted as $X_m \sim P_M$.

First, we select a genotype i $(i = 1, 2, \ldots, 7)$ according to $X_g \sim P_G$, and then a specific mutation operator j $(j = 1, 2, 3)$ according to $X_m \sim P_M$. Here, denote the location set of gene i on the given chromosome as S_i, and gene i has n_i parameters. Then $P(X_{pt} = \alpha)$ $(\alpha = 1, 2, \ldots, n_i)$ constitutes the probability distribution of the genetic parameter type, denoted as $X_{pt} \sim P_{PT}$. Similarly, suppose the number of values of each parameter types is n_α, then $P(X_{pv} = \beta)$ $(\beta = 1, 2, \ldots, n_\alpha)$ constitutes the probability distribution of the genetic parameter values, denoted as $X_{pv} \sim P_{PV}$.

Based on the pre-mentioned conditions, three mutation operators are proposed as follows.

Inserting Mutation ($j = 1$): A genotype, g_i, is selected from G according to the corresponding probability distribution. Suppose the selected genotype contained n_i parameters. According to the probability distribution of the genetic parameter values, the specific gene is sampled. And then according to the uniform probability distribution, we select the inserting position where genotype i would be inserted into the chromosome with randomly initialized parameters. Then, we insert the selected gene and update the probability distribution of both genotype and genetic parameter value. The inserting mutation operator leads to the chromosome length to increase 1. The inserting mutation should satisfy the condition that Fc gene appears only in the tails of chromosomes.

Deleting Mutation ($j = 2$): Suppose a genotype, g_i, has been selected from G according to the corresponding probability distribution. Then, select randomly the location of the gene from S_i on the chromosome and delete it. Update the probability distribution of genotypes. The delete mutation operator leads to the chromosome length to reduce 1. The deleting mutation operator should satisfy the condition that the last gene Fc in the chromosome cannot be deleted.

Updating Mutation ($j = 3$): Similar with the deleting mutation process, a gene is selected on chromosome, denoted as s_k^i $(1 \leq k \leq |S_i|)$, where $|S_i|$ is the number of elements in S_i. Firstly, the gene parameter type to be mutated is selected according to $X_{pt} \sim P_{PT}$, and then the value of the parameter type is determined according to $P(X_{pv} = \beta)$. This produces offspring with different parameters from the parents. For a specific problem, in order to match the output dimensions, the cell number of the last layer, Fc, is fixed. For example, the cell number should only be 10 for CIFAR-10 to match with the recognition targets of 10 classes.

Crossover. To accelerate the convergence, the crossover operator is designed. Since the structure of the network layers is relatively stable for a DNN, the proposed crossover operation only works on GENE level, but does not change the gene parameters. Hence, the architecture of neural networks can be taken as a linear sequence of genotypes. Consequently, the crossover operation randomly selects the cut-point of two parents, and then exchanges the partial sequences behind the cut-point.

3.4 Probability Distribution of Evolutionary Operation and Its Learning Approach

The previous evolutionary algorithms generally use uniform probability distribution to perform evolutionary operations. The classical evolutionary algorithm only drives the population evolution based on the inherent adaptability of the algorithm, but not based on the historical evolution information. With the help of historical evolution information, the algorithm can evolve towards a more promising direction to find the optimal solution. Therefore, in order to realize the evolutionary mechanism of survival of the fittest more efficiently, we design a probability distribution learning approach, which enables the proposed algorithm to dynamically adjust the probability distribution of evolution operations and parameter values based on historical evolution information. The sliding window strategy is adopted here to make the probability distribution fully reflect the optimal evolution direction in the most recent steps.

The dynamic probability distribution used in Sect. 3.3 mainly includes:

1. $X_g \sim P_G$: the probability distribution of genotype X_g;
2. $X_m \sim P_M$: the probability distribution of genotype mutation operators X_m;
3. $X_{pt} \sim P_{PT}$: the probability distribution of genetic parameter types X_{pt};
4. $X_{pv} \sim P_{PV}$: the probability distribution of genetic parameter values X_{pv}.

For the generality, the proposed learning approach is stated with a discrete random variable, X, taking values within the set $\{1, 2, \ldots, m\}$. And the distribution is denoted as $X \sim P$. The probability distribution learning approach, named incremental learning approach (ILS), is proposed as follows.

Given the current probability distribution $X \sim P$, if the offspring generated with $X = i$ has a fitness higher than its parents, then the probability $p(X = i)$ should be increased with a small step; and vice versa, if the offspring generated with $X = i$ has a fitness lower than its parents, then the probability $p(X = i)$ should be reduced with a small step. The probability distribution updates using the following formulas:

$$p(X = i) = \begin{cases} (1+\gamma)p(X = i) & f(offspring) > f(parent) \\ (1-\gamma)p(X = i) & f(offspring) < f(parent) \end{cases} \tag{1}$$

where γ ($\gamma = 0.1$) is a predetermined small constant representing the incremental size of the probability distribution learning. Then to maintain the probability properties, $p(X = i)$ $(1 \leq i \leq m)$ are normalized to sum up to 1.

The proposed ILS is universally applicable to any dynamic adjustment of probability distributions, including genotype $X_g \sim P_G$, mutation operator $X_m \sim P_M$, parameter type $X_{pt} \sim P_{PT}$ and parameter value $X_{pv} \sim P_{PV}$.

4 Experiments

To verify the performance of the proposed ESAE, experiments are conducted using datasets, including Fashion-MNIST [32], CIFAR-10 [33], which are commonly used in the field of deep learning. For each problem, the dataset is partitioned into the training, validation and test dataset, where the train dataset is used to optimize the architectures and parameters, the validation dataset is used to evaluate the performance of the network during evolution, finally, the test dataset is used in a one-shot way to provide an unbiased evaluation of the trained models.

In order to reduce the evolving time, we only evolved the architectures and hyper-parameters including the kernel sizes, channel numbers, padding, stride, pooling type, resnet_block repeat time, dense_block k, bn, af, fc. The architectures were submitted to Keras for weights training. In the evolutionary process, each candidate network was trained 10 epochs. After 100 generations of evolution, the individual with the highest fitness was trained 400 epochs on Keras. And the accuracy on the test dataset is regarded as the final performance measure. Similar with Ref. [18], the policy of linearly varying learning rate is used. The learning rate starts at 0.01 and is increased linearly up to 0.1 on the 5th epoch; then decrease to 0.01 on the 250th epoch and finally, reduce to 0.001 on the 375th epoch and keep the fixed value till the 400th epochs.

Fashion-MNIST is similar to the well-known MNIST (i.e., 28×28 grayscale images), but the digits are replaced by fashion clothing items from Zalando's: t-shirt/top, trouser, pullover, dress, coat, sandal, shirt, sneaker, bag, or ankle boot [32]. The CIFAR-10 is composed by 32×32 RGB color images [33]. Each CIFAR-10 image contains one of the following objects: airplane, automobile, bird, cat, deer, dog, frog, horse, ship, or truck. In order to improve the network performance, the datasets are augmented with the method proposed by Snoek et al. [9].

In the following experiments, we used the Stochastic Gradient Descent (SGD) to train the weights on Keras with 32 as the mini-batch size. The cross entropy is taken as the loss function, as shown in Eq. (2):

$$L = -[ylog\hat{y} + (1 - y)log(1 - \hat{y})] \tag{2}$$

where y and $\hat{y}$ are the target and the predicted values, respectively. The Adam optimizer with $\beta_1 = 0.9$, $\beta_2 = 0.999$ and $\varepsilon = 1.0 \times 10^{-8}$ is used. The experimental parameters are shown in Table 1. And in the experiment, we use only one NVIDIA GTX 1080Ti.

Table 1. Experimental parameters

	Parameter name	Value
Evolutionary engine	Population size	3
	Number of generations	100
	Number of full training	400
	Number of runs	5
Fashion-MNIST dataset	Train set size	54000
	Validation set size	6000
	Test set size	10000
Cifar-10 dataset	Train set size	45000
	Validation set size	5000
	Test set size	10000
Training	Number of epochs	10
	Loss function	Categorical cross-entropy
	Batch size	32
	Initial learning rate	0.01
	learning rate range	[0.001, 0.1]
	Momentum	0.9
Data augmentation	Padding	4
	Random crop	4
	Horizontal flipping (%)	50

Evolutionary Deep Architecture for Fashion-MNIST Dataset. ESAE is used to explore the best architecture on Fashion-MNIST in this section. After 100 generations of evolution, the classification accuracy on validation set is 92.45%. After full train on the entire train set, the highest accuracy on test set is 94.48%. Figure 5 shows the best network architecture among 5 separate runs.

For the architecture of the best evolved networks (Fig. 5), the most interesting characteristics is that there are two adjacent pooling layers and the normalization layer is followed by an activation function layer. These characteristic haven't appeared in man-designed architectures.

Moreover, we conducte preliminary experiments with removing one of the pooling layers and removing the activation function layer after the normalization operation. The corresponding test accuracies are 93.78% and 93.85%, respectively, which are lower than the evolved architectures. Figure 6 shows the training curve of the modified networks. When one pooling layer is removed, the training curve tends to vibrate drastically; when the activation function layer behind the normalization layer is removed, the train accuracy is always below that of the evolved network. The two manually modified architectures result in the accuracy loss of 0.70% and 0.63%, respectively.

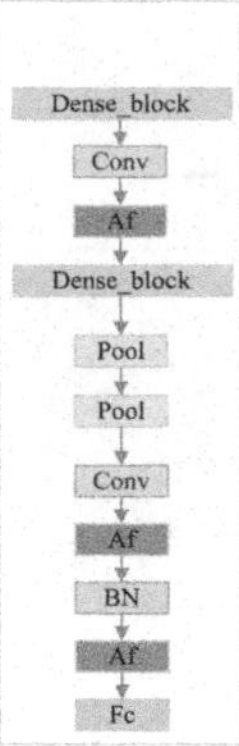

Layers	Net
Dense_block	$\begin{bmatrix}1\times1\ conv\\3\times3\ conv\end{bmatrix}\times3$
Conv	3×3 conv, stride 2
AF	Relu
Dense_block	$\begin{bmatrix}1\times1\ conv\\5\times5\ conv\end{bmatrix}\times4$
Pool	2×2 max pool, stride 2
Pool	2×2 max pool, stride 2
Conv	3×3 conv, stride 2
AF	Relu
BN	True
AF	Relu
FC	10 D fully-connected, softmax

Fig. 5. The evolved architecture for the Fashion-MNIST.

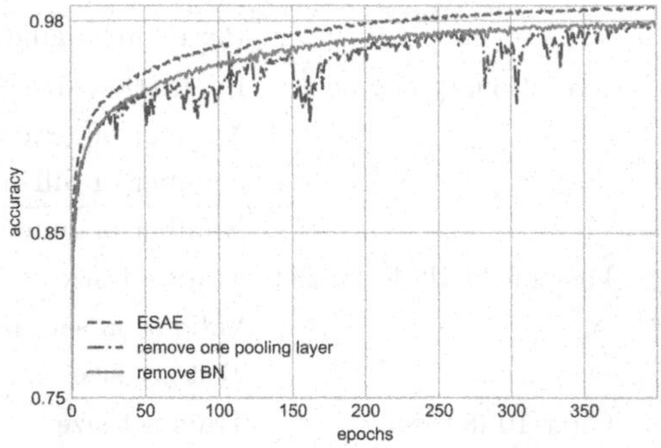

Fig. 6. Results of the optimal and the modified networks at Fashion-MNIST.

Evolutionary Deep Architecture for CIFAR-10 Dataset. In this section, ESAE is used to search the network architecture for CIFAR-10, with the same data enhancement and learning rate as those for Fashion-MNIST. After each run of architecture evolution, the best network is re-trained 5 times with 400 epochs with the same learning rate policy. The average accuracy is 93.49%.

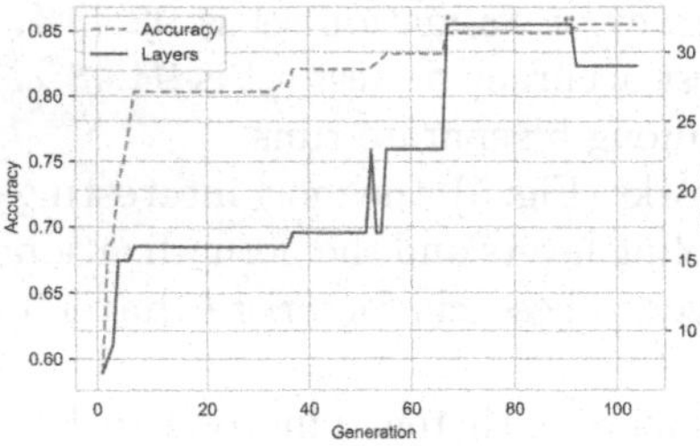

Fig. 7. Evolution of the fitness and number of layers of the best individuals across generations.

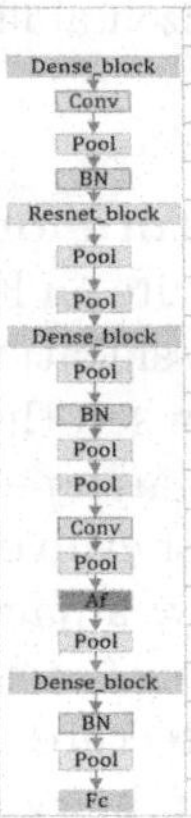

Layers	Net
Dense_block	$\begin{bmatrix}1\times1\ conv\\3\times3\ conv\end{bmatrix}\times4$
Conv	3×3 conv, stride 2
Pool	2×2 max pool, stride 2
BN	True
Resnet_block	$\begin{bmatrix}3\times3\ conv\\3\times3\ conv\end{bmatrix}\times2$
Pool	2×2 max pool, stride 2
Pool	2×2 max pool, stride 1
Dense_block	$\begin{bmatrix}1\times1\ conv\\5\times5\ conv\end{bmatrix}\times2$
Pool	2×2 max pool, stride 1
BN	True
Pool	2×2 max pool, stride 2
Pool	2×2 max pool, stride 1
Conv	3×3 conv, stride 1
Pool	2×2 max pool, stride 1
AF	Relu
Pool	2×2 max pool, stride 1
Dense_block	$\begin{bmatrix}1\times1\ conv\\5\times5\ conv\end{bmatrix}\times2$
BN	True
Pool	2×2 max pool, stride 1
FC	10 D fully-connected, softmax

Fig. 8. The evolved architecture for CIFAR-10.

Figure 7 shows the fitness curve of the optimal architecture for each generation in evolutionary process. The result shows that the performance of the evolved networks is steadily increasing, and the evolution converges around the 90th generation. There are two significant emergent behaviors in evolutionary process. The first one occurs in 65th generation, corresponding to the position marked "*" in Fig. 7, where the number of network layers increases from 7 to 23 with a *Dense_block* and the accuracy goes up from 58.94% to 83.26%. The second one occurs in 90th generation, corresponding to the position marked with "**" in Fig. 7, where the number of network layers decreases from 32 to 29 with removing a *Resnet_block* and the accuracy goes up from 84.84% to 85.52%.

It can be seen from Fig. 7 that the evolutionary behavior pattern alters around the 65th generation. Before the 65th generation, the increase of the fitness is accompanied by increasing network layers. And after the 65th generation, the increase of the fitness is accompanied by decreasing network layers.

At the early evolving stage, the inserting operators are used frequently. By increasing the number of layers in the network, inserting operator continuously explore the architecture space, increasing the fitness and the layers rapidly. At the later stage, deleting and updating operators contribute more to improve network fitness and their using frequency is also gradually increased, making the network reduce the complexity on the premise of maintaining the original performance. In this process, the algorithm constantly adjusts network parameters through the updating operator, so that the local development ability is gradually strengthened, and the complexity is gradually reduced.

Figure 8 shows the evolved optimal architecture for CIFAR-10. Different from manual design, the convolution kernel size of the evolutionary optimal architecture is 5×5, and contains the characteristics of the optimal network evolved by Fashion-MNIST, that is, two successive pooling layers are wrapped. In VGG network, the authors proved that two successive 3×3 convolutions can be replaced by one 5×5 convolution [5]. We replace the *Dense_block'* 5×5 convolution with two 3×3 convolutions. After the same full training, the accuracy of the replaced network is 91.5%. After removing the third pooling layer, the accuracy decreases to 90.2%. The validity of ESAE is demonstrated again by CIFAR-10.

Table 2. Classification accuracy comparisons with CNNs on Fashion-MNIST and CIFAR-10

Approach	Dataset	Accuracy (%)	GPU days	Resource (GPUs)	Parameters (10^6)
DENSER*	Fashion-MNIST	***95.26***	–	-	-
VGG		93.50	–	–	4
ESAE		94.48	18	1 2080Ti	**0.56**
DENSER*	CIFAR-10	94.13	–	–	10.81
AmoebaNet*		***97.45***	3150	450 K40	2.8
CoDeepNEAT*		92.70	–	–	–
CGP*		93.25	30.4	2	5.7
Large-Scale*		94.60	2500	250	5.4
ESAE		93.49	20	1 2080Ti	12.56

Note: Automatic approaches are marked with a symbol *

Table 2 shows the accuracy, GPU days, network parameters and resource in different EAs for Fashion-MNIST and CIFAR-10. Our algorithm obtains the best balance between resource effectiveness and architecture simplicity compared with other algorithms. On Fashion-MNIST, the accuracy of ESAE is 0.98 higher than VGG, but the parameters is 1/7 of VGG. For CIFAR-10, although the highest accuracy achieved by the current automatic method is 97.45%, 3.96% higher than ESAE, the GPU days is 157.5 times more than our ESAE (3150 vs 20) and the required GPU resource is 450 times more than our ESAE. Generally, ESAE obtains obvious improvements of computing resource and memory loading efficiency with the cost of 4.41% accuracy loss.

5 Conclusion

This paper proposes ESAE, i.e., a method for Evolutionary Strategy-based Architecture Evolution. The representation of candidate networks is built on the bi-level encoding scheme, enabling the searching space to include typical DNN modules, such as Resnet_block, Dense_block. This encoding scheme is beneficial to full exploring the architecture space of feed forward DNNs. To accelerate the searching speed and make the population evolve efficiently towards to the most promising directions, the learning approach of probability distributions is proposed, which significantly improves the convergent speed and enhance the architecture quality. Experiments on Fashion-MNIST and CIFAR-10 demonstrate that the proposed ESAE reduces drastically the consumption of computing resources and has better robustness and generalization ability, compared with the most exiting state-of-the-art methods, while keeping the competitive accuracy.

The weight training is a time-consuming task. Therefore, constructing the efficient surrogate model is an important branch of future works. The use of surrogate model would accelerate the exploration of DNNs architectures. In addition, many basic units of structure can be generated and evaluated by credit assembly mechanism. Consequently, the excellent modules can be recognized and then the most promising network is assembled with existing modules. A novel paradigm is forming that the architecture would evolve gradually from units, modules to network architectures. Moreover, according to the latest work, the architecture evolution may benefit from the weight agnostic idea remarkably [34], since the weight training is the most time consuming part on architecture search.

References

1. Krizhevsky, A., Sutskever, I., Hinton, G.-E.: ImageNet classification with deep convolutional neural networks. Commun. ACM **60**(6), 84–90 (2012)
2. Hinton, G., Deng, L., Yu, D., et al.: Deep neural networks for acoustic modeling in speech recognition: the shared views of four research groups. IEEE Signal Process. Mag. **29**(6), 82–97 (2012)
3. Wu, Y., et al.: Google's neural machine translation system: bridging the gap between human and machine translation. arXiv preprint arXiv:1609.08144 (2016)

4. Szegedy, C., et al.: Going deeper with convolutions. In: Proceedings of the IEEE Conference on Computer Vision and Pattern Recognition, CVPR, Boston, pp. 1–9 (2015)
5. Simonyan, K., Zisserman, A.: Very deep convolutional networks for large-scale image recognition. arXiv preprint arXiv:1409.1556 (2014)
6. He, K., Zhang, X., Ren, S., et al.: Deep residual learning for image recognition. In: Proceedings of the IEEE Conference on Computer Vision and Pattern Recognition, CVPR, Las Vegas, pp. 770–778 (2016)
7. Huang, G., Liu, Z., Van, D.M.L., et al.: Densely connected convolutional networks. In: IEEE Conference on Computer Vision and Pattern Recognition, CVPR, Puerto Rico, pp. 2261–2269 (2017)
8. Bergstra, J., Bengio, Y.: Random search for hyper-parameter optimization. J. Mach. Learn. Res. **13**(1), 281–305 (2012)
9. Snoek, J., Rippel, O., Swersky, K., et al.: Scalable Bayesian optimization using deep neural networks. Statistics **37**, 1861–1869 (2015)
10. Miller, G.F., Todd, P.M., Hegde, S.U.: Designing neural networks using genetic algorithms. In: Proceedings of the Third International Conference on Genetic Algorithms, ICGA, San Francisco, pp. 379–384 (1989)
11. Yao, X., Liu, Y.: A new evolutionary system for evolving artificial neural networks. IEEE Trans. Neural Netw. **8**(3), 694–713 (1997)
12. Stanley, K.O., Miikkulainen, R.: Evolving neural networks through augmenting topologies. Evol. Comput. **10**(2), 99–127 (2002)
13. Stanley, K.O., D'Ambrosio, D.B., Gauci, J.-A.: Hypercube-based encoding for evolving large-scale neural networks. Artif. Life **15**(2), 185–212 (2009)
14. Suganuma, M., Shirakawa, S., Nagao, T.A.: Genetic programming approach to designing convolutional neural network architectures. In: Proceedings of the Genetic and Evolutionary Computation Conference, pp. 497–504. ACM, Berlin (2017)
15. Miikkulainen, R., et al.: Evolving deep neural networks. In: Artificial Intelligence in the Age of Neural Networks and Brain Computing, pp. 293–312. Academic Press (2017)
16. Real, E., Aggarwal, A., Huang, Y., et al.: Aging evolution for image classifier architecture search. In: Thirty-Third AAAI Conference on Artificial Intelligence, pp. 5048–5056. AAAI, Hawaii (2019)
17. Real, E., Moore, S., Selle, A., et al.: Large-scale evolution of image classifiers. In: 34th International Conference on Machine Learning, ICML, Sydney, pp. 2902–2911 (2017)
18. Assunção, F., et al.: DENSER: deep evolutionary network structured representation. Genet. Program. Evolvable Mach. **20**(1), 5–35 (2019)
19. Zoph, B., Vasudevan, V., Shlens, J., et al.: Learning transferable architectures for scalable image recognition. In: The IEEE Conference on Computer Vision and Pattern Recognition, CVPR, Salt Lake, pp. 8697–8710 (2018)
20. Liu, H., Simonyan, K., Yang, Y.: DARTS: differentiable architecture search. In: International Conference on Learning Representations, ICLR, New Orleans (2019)
21. Hundt, A., Jain, V., Hager, G.-D.: sharpDARTS: faster and more accurate differentiable architecture search. arXiv preprint arXiv:1903.09900 (2019)
22. Zhang, J.-R., Zhang, J., Lok, T.-M., et al.: A hybrid particle swarm optimization-back-propagation algorithm for feedforward neural network training. Appl. Math. Comput. **185**(2), 1026–1037 (2007)
23. Wang, G.G., Deb, S., et al.: Monarch butterfly optimization. Neural Comput. Appl. **31**(7), 1–20 (2015)

24. Ge, W.G., Suash, D., Santos, C.L.D.: Earthworm optimisation algorithm: a bio-inspired metaheuristic algorithm for global optimisation problems. Int. J. Bio-Inspir. Comput. **12**(1), 1–17 (2018)
25. Wang, G.G., et al.: Elephant herding optimization. In: 2015 3rd International Symposium on Computational and Business Intelligence. IEEE, Bali Indonesia (2015)
26. Wang, G.-G., et al.: Moth search algorithm: a bio-inspired metaheuristic algorithm for global optimization problems. Mem. Comput. **10**(2), 151–164 (2016)
27. Kiranyaz, S., Ince, T., et al.: Evolutionary artificial neural networks by multi-dimensional particle swarm optimization. Neural Netw. **22**(10), 1448–1462 (2009)
28. Das, G., Pattnaik, P.-K., Padhy, S.-K.: Artificial neural network trained by particle swarm optimization for non-linear channel equalization. Expert Syst. Appl. **41**(7), 3491–3496 (2014)
29. Salama, K.-M., Abdelbar, A.-M.: Learning neural network structures with ant colony algorithms. Swarm Intell. **9**(4), 229–265 (2015)
30. Zhou, X.-H., Zhang, M.X., et al.: Shallow and deep neural network training by water wave optimization. Swarm Evol. Comput. **50**, 100561 (2019)
31. LeCun, Y., Boser, B.E., et al.: Handwritten digit recognition with a back-propagation network. In: Advances in neural Information Processing Systems, pp. 396–404 (1990)
32. Xiao, H., Rasul, K., Vollgraf, R.: Fashion-MNIST: a novel image dataset for benchmarking machine learning algorithms. arXiv preprint arXiv:1708.07747 (2017)
33. Krizhevsky, A., Hinton, G.: Learning multiple layers of features from tiny images. Technical report, University of Toronto (2009)
34. Gaier, A., et al.: Weight agnostic neural networks. arXiv preprint arXiv:1906.04358 (2019)

Species-Based Differential Evolution with Migration for Multimodal Optimization

Wei Li[1,2(✉)], Yaochi Fan[1], and Qiaoyong Jiang[1]

[1] School of Computer Science and Engineering, Xi'an University of Technology, Xi'an 710048, China
liwei@xaut.edu.cn

[2] Shaanxi Key Laboratory for Network Computing and Security Technology, Xi'an 710048, China

Abstract. Due to the fact that many realistic problems are multimodal optimization problems (MMOPs), multimodal optimization has attracted a growing interest. In this paper, a Species-based differential evolution with migration algorithm is proposed for MMOPs. First, a migration strategy is developed to improve the exploration ability of species with less individuals. Then, an improved operation of creating local individuals is introduced to help the algorithm to find high-quality solutions. Finally, an archive mechanism is applied to preserve the best solutions found during previous generations and avoid the loss of the peaks. Experimental results on CEC2013 test problems confirm the effectiveness of the proposed algorithm compared to several well-known multimodal optimization algorithms.

Keywords: Multimodal optimization problems · Species-based · Migration · Niching method

1 Introduction

Multimodal optimization, which requires finding one or more global optima in the search landscape, has attracted much attention. Many real-world problems such as the design of electric machines [1], multi-objective optimization [2], large-scale optimization [3], clustering problems [4] and simulation-based design of power systems [5] have the multimodality property. Evolutionary algorithms (EAs) are known for solving complex optimization problems, however, without specific mechanism, traditional EAs which are designed to search for a single optimal solution are difficult to locate all global optima simultaneously in a single run.

To tackle multimodal optimization problems (MMOPs), the famous niching techniques such as crowing [6], speciation [7], classification [8], clearing [9] and fitness sharing [10] are proposed to motivate multiple convergence behavior.

L. Pan et al. (Eds.): BIC-TA 2019, CCIS 1159, pp. 209–222, 2020.
https://doi.org/10.1007/978-981-15-3425-6_17

These methods have been integrated into various types of EAs, such as differential evolution (DE) [11], particle swarm optimization (PSO) [12], ant colony optimization (ACO) [13], to find the peak in multimodal landscape. Among these EAs, the DE variants have shown promising performance, such as CDE [6], crowding (NCDE) [22], fitness sharing (NSHrDE) [22], and speciation DEs (NSDE) [22], locally informative niching DE (LoINDE) [14], ensemble and arithmetic recombination-based speciation DE (EAESDE) [15], parent-centric normalized mutation with proximity-based crowding DE (PNPCDE) [16].

The species-based DE (SDE) proposed in [17] is one of the most effective methods to locate multiple global optima simultaneously with adaptive formation of multiple species. In SDE, two mechanisms are introduced to maintain a balance between exploitation and exploration in the search process. First, the exploitation ability of each species can be improved by creating local random individuals. Second, the exploration in the search space can be enhanced by removing redundant individuals and replacing them with randomly generated individuals. Finally, the sensitivity of SDE to different species radius values can be effectively alleviated with these two mechanisms. However, the determination of species is largely dependent on the species radius which is difficult to set without a priori knowledge. If the number of individuals in a species is too small, the species may not be able to find multiple global optima simultaneously because of the lack of diversity. To address the issue, a species-based differential evolution with migration (SDEM) is proposed in this paper. The main contributions of this paper are summarized as follows:

(1) To alleviate the problem that different species have distinct disequilibrium individuals, the migration method will be introduced in the proposed algorithm. The species with less individuals will increase the number of individuals, then the search efficiency can be improved.
(2) In order to further improve the exploration ability of the proposed algorithm, a new strategy is introduced to generate local individuals instead of a random way.
(3) The archive is introduced to save better individuals which are helpful to ensure the convergence of the algorithm. In addition, the loss of the peaks can be effectively alleviated.
(4) Systematic experiments conducted to compare the algorithms including SDE, FERPSO, r2PSO, r2PSO-*lhc* and NSDE on CEC2013 multimodal benchmark problems are described. The experimental results show that the proposed method is promising for solving multimodal optimization problems.

The remainder of this paper is organized as follows. Section 2 reviews species-based DE (SDE). Section 3 introduced the proposed species-based differential evolution with migration (SDEM) algorithm. Section 4 reports and discusses the experimental results while Sect. 5 concludes this paper.

2 Species-Based DE

SDE introduced the idea of speciation in a basic DE to solve multimodal optimization problems. The key steps in SDE are identifying species and determining species seeds. The population is classified into groups according to their similarity measured by Euclidean distance.

$$dist(x_i, x_j) = \sqrt{\sum_{k=1}^{D} (x_{i,k} - x_{j,k})^2} \tag{1}$$

where $\mathbf{x}_i = (x_{i,1}, x_{i,2}, \ldots, x_{i,D})$ is D-dimensional vector representing individual i from the population. The smaller the Euclidean distance between individual i and individual j, the more similar they are.

The fittest individual in the species is called species seed. All individuals that fall within the given distance from the species seed are considered as the same species. Here, the determination of the species seeds is briefly reviewed [17]. First, all individuals are sorted in decreasing order of fitness. The species seed set S is initially set to φ. Second, the fittest individual will be set as the first species seed. Then, the Euclidean distance between each individual with its species seed is computed. If an individual does not fall within the given radius of all the seeds of S, this individual will be identified as another species seed and be added to S. Finally, all species seeds of the current population will be identified. In the process of identifying species, if an identified species has less than m individuals, the missing individuals will be generated randomly within the radius of the species seed and added to that species. If the fitness of each DE child individual is the same as the fitness of its species seed, the child is replaced by a randomly generated new individual. The details of SDE are presented in Table 1 [22].

Table 1. Species-based DE (SDE).

Step 1	Generate an initial population of *NP* individuals randomly
Step 2	Evaluate the population
Step 3	Sort all individuals in descending order of their fitness values
Step 4	Determine the species seeds for the current population
Step 5	For each species as identified via its species seed, run a global DE variant
	5.1 If a species has less than m individuals, then randomly generate m new individuals within the radius of the species seed
	5.2 If a child's fitness is the same as that of its species seed, replace this child with a randomly generated new individual
Step 6	Keep only the *NP* fitter individuals from the combined population
Step 7	Stop if the termination criteria are met. Otherwise go back to Step 2

3 Species-Based Differential Evolution with Migration

In order to deal with MMOPs efficiently, the proposed SDEM consists of six main steps, namely initialization, construction of the species, migration, creation of local individuals, local search procedure and archive management.

Step-1 *Initialization*
Similar to traditional DE algorithm, *NP* individuals are randomly generated to form an initial population. All the individuals are uniformly distributed in the search space.

Step-2 *Construction of the species*
This paper adopts speciation which is an effective technique for MMOPs [18,19]. Like SDE [17], the best individual in each species is called a seed. At each generation, all the individuals are sorted in the descending order of their fitness values. The best individual in the whole population is selected as the seed at first. Then, the other individuals will be checked one by one. If the Euclidean distance between an individual and a seed is smaller than species radius r_s, the individual is assigned to the same species with the seed. On the contrary, if the Euclidean distances between an individual and all seeds are greater than r_s, the individual will be considered as a new seed. After checking all the individuals, the population is divided into several species. The pseudocode of species construction is given in Table 2.

Table 2. The pseudocode of species construction (Algorithm 1).

```
L_sorted: all the individuals sorted in decreasing order of fitness value
Species seeds S ← φ
while L_sorted ≠ φ
  p ← the current fittest individual from L_sorted
  seed_p ← φ
  for all s ∈ S
    if dist(s,p) ≤ r_s
      seed_p ← s
      break
    endif
  endfor
  if seed_p == φ
    S ← S ∪ {p}
    seed_p ← p
  endif
  L_sorted ← L_sorted − {p}
end while
```

Step-3 *Migration*
In general, numbers of individuals in different species are different during the search procedure. Some species have fewer individuals, while others have more. Species with many individuals will take over the population's resources and flood the next generation with its offspring. To address this problem, the migration method is introduced in the proposed algorithm. Briefly speaking, the number of individuals in larger species is decreased, while the number of individuals in smaller species is increased. The key idea is to distribute the individuals more evenly in the species. Then, the worst individuals in the larger species beyond the given size will migrate to some other species which have few individuals. The number of migrated individuals is calculated as follows.

$$MS_i = \frac{L_i - S_i}{2} \qquad i = 1, 2, ..., k \tag{2}$$

where k denotes the number of species beyond the specified size. L_i and S_i are the numbers of individuals in the ith larger species and the ith smaller one respectively.

The migration operation on ith smaller species is as follows.

$$\mathbf{x}_j = seed_i + seedL_i - \widetilde{x}_j \qquad j = 1, 2, ..., MS_i \tag{3}$$

where x_j denotes the jth individual that immigrates to ith smaller species. $seed_i$ and $seedL_i$ denote the seed of ith smaller species and ith larger one. $\widetilde{x}_j$ denotes the jth worst individual in the ith larger species.

Step-4 *Creation of local individuals*
In order to ensure the effective operation of DE, a species should have at least m individuals in the proposed algorithm. m is set to 10 in the following experiments. If ith species has less than m individuals, the missing individuals will be generated by:

$$\mathbf{newx} = \begin{cases} seed_i + \sigma \times \tan(\pi \times (rand - 0.5)) & rand > \delta \\ \mathbf{x}_{r1} + F \times (\mathbf{x}_{r2} - \mathbf{x}_{r3}) & otherwise \end{cases} \tag{4}$$

where σ is set to 0.1. $\delta = MaxFES/FES$. $MaxFES$ and FES denote the maximum number of function evaluations and the number of function evaluations, respectively. $rand$ denotes a uniformly selected random number from (0,1). $seed_i$ denotes the seed of ith species. F is the mutation factor. $\mathbf{x}_{r1}$, $\mathbf{x}_{r2}$ and $\mathbf{x}_{r3}$ are randomly selected from the population $newP$. $newP$ is defined by:

$$newP = \begin{cases} Pop & if\, i > len(Sp) \times \tau \\ \{Sp_{i-1}\} \cup \{Sp_i\} \cup \{Sp_{i+1}\} & otherwise \end{cases} \tag{5}$$

where τ is set to 0.7. i denotes the index of the current species. $len(Sp)$ denotes the number of species. Sp_i denotes the individuals of ith species. Pop denotes the current population. This process is repeated until the size of the species is equal to m.

Step-5 *Local search procedure*
In this step, all the individuals are updated with DE algorithm. For each individual, a mutant vector is produced by the mutation operation. Then, crossover operation is applied to each pair of the target vector and its corresponding mutant vector to generate a trial vector. Finally, the offspring which will enter the population of the next generation will be selected based on the selection operation.

Step-6 *Archive management*
In order to avoid losing the found peaks during the selection procedure, the archive is employed to save the individuals which have better performance. Specifically speaking, if the offspring performs worse, the individuals in the archive will be used as the parents for the next generation.

The details of SDEM are presented in Table 3.

Table 3. Species-based DE with migration (SDEM).

Step 1	Generate an initial population of *NP* individuals randomly
Step 2	Evaluate the population
Step 3	Species Construction according to Algorithm 1
Step 4	Migration operation according to Step-3
Step 5	Creating local individuals according to Step-4
Step 6	For each species as identified via its species seed, run a global DE variant
	6.1 Generate a child with DE and Evaluate the child
	6.2 If a child's fitness is the same as that of its species seed, replace this child with a randomly generated new individual
Step 7	Keep only the *NP* fitter individuals from the combined population
Step 8	Update the archive
Step 9	Stop if the termination criteria are met. Otherwise go back to Step 3

4 Experiments and Discussions

To verify the effectiveness of the SDEM proposed in this paper, 10 test problems from CEC2013 [20] multimodal benchmarks are used. The algorithms for testing include SDE, FERPSO [21], *r*2PSO [12], *r*2PSO-*lhc* [12] and NSDE [22]. To obtain an unbiased comparison, all the experiments are run on a PC with an Intel Core i7-3770 3.40 GHz CPU and 4 GB memory. All experiments are run 25 times, and the codes are implemented in Matlab R2013a.

4.1 Parameter Settings and Performance Criteria

Table 4 shows a brief description of testing problems. For each problem, the setting of number of function evaluations for each algorithm is given in Table 4.

The population size for SDEM, SDE and NSDE is given in Table 4. The mutation factor F and the crossover rate CR which are used in SDEM, SDE and NSDE are set to 0.9 and 0.1, respectively. The parameters of FERPSO, r2PSO and r2PSO-*lhc* are the same as those used in the corresponding references. F_1–F_{10} are simple and low dimensional multimodal problems. F_1–F_5 have a small number of global optima, and F_6–F_{10} have a large number of global optima. Some problems from the CEC2013 are drawn in the following. As shown in Fig. 1, Equal Maxima (F_2) has 5 global optima. There are no local optima. Himmelblau function (F_4) has 4 global optima. An example of the Shubert 2D function (F_6)shows that there are 18 global optima in 9 pairs. Moreover, Fig. 1 shows the 2D version of Modified Rastrigin (F_{10}), where D = 2, k1 = 3, k2 = 4, thus the total number of optima is 12. In addition, two well-known criteria [6] called average number of peaks (ANP) and success rate (SR) are employed to measure the performance of different multimodal optimization algorithms on each problem. Average number of peaks (ANP) denotes the average number of peaks found by an algorithm over all runs, while success rate (SR) denotes the percentage of successfully detecting all global optima out of all runs for each problem. In view of statistics, the Wilcoxon signed-rank test [23] at the 0.05 significance level is used to compare SDEM with other compared algorithms. "$\approx$", "+" and "−" are applied to express the performance of SDEM is similar to ($\approx$), worse (−) than, and better (+) than that of the compared algorithm, respectively.

Table 4. Parameter setting for test problems.

Fun	r	D	Number of global optima	Number of function evaluations	Population size (NP)
F_1	0.01	1	2	5E+04	80
F_2	0.01	1	5	5E+04	80
F_3	0.01	1	1	5E+04	80
F_4	0.01	2	4	5E+04	80
F_5	0.5	2	2	5E+04	80
F_6	0.5	2	18	2E+05	100
F_7	0.2	2	36	2E+05	300
F_8	0.5	3	81	4E+05	300
F_9	0.2	3	216	4E+05	300
F_{10}	0.01	2	12	2E+05	100

4.2 Experimental Results of Six Optimization Algorithms

Three accuracy thresholds, $\epsilon = 1.0\text{E}{-3}$, $\epsilon = 1.0\text{E}{-4}$ and $\epsilon = 1.0\text{E}{-5}$ are selected in the experiments. The experimental results and analyses are shown in the following. The results of test function from F_1 to F_{10} on $\epsilon = 1.0\text{E}{-3}$ are shown in Table 5. As can be seen, SDEM can find all the global optimal solutions on F_1, F_2,

F_3, F_5 and F_{10}. In addition, SDEM is significantly better than SDE, FERPSO, r2PSO, r2PSO-*lhc* and NSDE on F_6, F_7, F_8, F_9 and F_{10}. FERPSO performs the best on F_4. SDEM outperforms SDE, FERPSO, r2PSO, r2PSO-*lhc* and NSDE on seven, six, six, six, five test problems, respectively. The reason that SDEM has the outstanding performance may be because of migration mechanism.

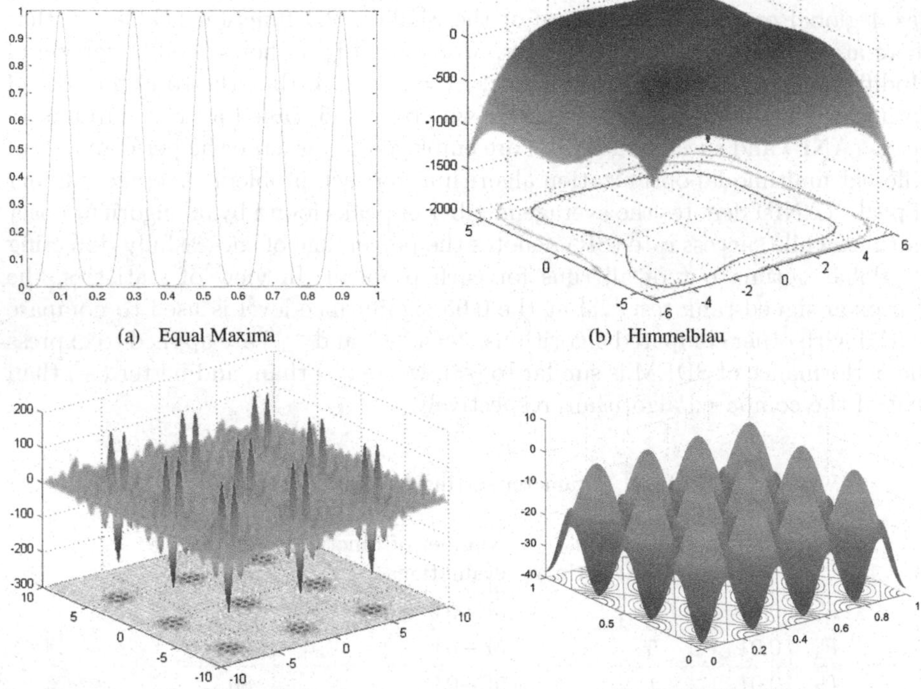

(a) Equal Maxima (b) Himmelblau

(c) Shubert 2D function (d) Modified Rastrigin

Fig. 1. Four problems from the CEC2013

From Table 6, we observe from the statistical results that SDEM can find all the global optimal solutions on F_1, F_2, F_3, F_5 and F_{10}. In addition, SDEM performs better than SDE, FERPSO, r2PSO, r2PSO-*lhc* and NSDE on F_6, F_7, F_8, F_9 and F_{10}. FERPSO performs the best on F_4. SDEM outperforms SDE, FERPSO, r2PSO, r2PSO-*lhc* and NSDE on seven, six, seven, six, six test problems, respectively. The reason is that the archive is introduced in SDEM, which is helpful to ensure the convergence of the algorithm.

In order to test the statistical significance of the six compared algorithms, the Wilcoxon's test at the 0.05 significance level, which is implemented by using KEEL software [24], is employed based on the PR values. Table 7 summarizes the statistical test results. It can be seen from Table 7 that SDEM provides higher R+ values than R− values compared with SDE, FERPSO, r2PSO, r2PSO-*lhc*

Table 5. Experimental results in ANP and SR on problems F_1–F_{10} at accuracy level $\epsilon = 1.0\text{E}{-3}$.

Func	SDEM		SDE		FERPSO		r2PSO		r2PSO-lhc		NSDE	
	ANP	SR	ANP	SR	ANP	SR	ANP	SR	ANP	SR	ANP	SR
F_1	2	1.00	1.56+	0.56	0+	0.00	0+	0.00	0+	0.00	2≈	1.00
F_2	5	1.00	5≈	1.00	5≈	1.00	5≈	1.00	5≈	1.00	5≈	1.00
F_3	1	1.00	1≈	1.00	1≈	1.00	1≈	1.00	1≈	1.00	1≈	1.00
F_4	3.92	0.92	1+	0.00	4≈	1.00	3.88≈	0.88	3.96≈	0.96	3.84≈	0.88
F_5	2	1.00	2≈	1.00	2≈	1.00	2≈	1.00	2≈	1.00	2≈	1.00
F_6	17.16	0.32	15.32+	0.08	15.92+	0.16	9.56+	0.00	11.48+	0.00	0.20+	0.00
F_7	25.72	0.00	22.08+	0.00	15.08+	0.00	19.16+	0.00	19.88+	0.00	23.24+	0.00
F_8	48.40	0.00	13.52+	0.00	2.08+	0.00	0.04+	0.00	0.04+	0.00	0.00+	0.00
F_9	141.80	0.00	50.20+	0.00	45.28+	0.00	38.6+	0.00	41.68+	0.00	57.16+	0.00
F_{10}	12	1.00	5.24+	0.00	10.68+	0.24	10.68+	0.28	11.20+	0.40	11.76+	0.80
+(SDEM is better)			7		6		6		6		5	
−(SDEM is worse)			0		0		0		0		0	
≈			3		4		4		4		5	

Table 6. Experimental results in ANP and SR on problems F_1-F_{10} at accuracy level $\epsilon = 1.0\text{E}{-4}$.

Func	SDEM		SDE		FERPSO		r2PSO		r2PSO-lhc		NSDE	
	ANP	SR	ANP	SR	ANP	SR	ANP	SR	ANP	SR	ANP	SR
F_1	2	1.00	1.56+	0.56	0+	0.00	0+	0.00	0+	0.00	2≈	1.00
F_2	5	1.00	5≈	1.00	5≈	1.00	5≈	1.00	5≈	1.00	5≈	1.00
F_3	1	1.00	1≈	1.00	1≈	1.00	1≈	1.00	1≈	1.00	1≈	1.00
F_4	3.92	0.92	1+	0.00	4≈	1.00	3.56≈	0.60	3.76≈	0.80	3.64≈	0.68
F_5	2	1.00	2≈	1.00	2≈	1.00	2≈	1.00	2≈	1.00	2≈	1.00
F_6	17.16	0.32	15.08+	0.04	15.32+	0.04	6.24+	0.00	9.04+	0.00	0.04+	0.00
F_7	25.68	0.00	22.08+	0.00	14.60+	0.00	17.64+	0.00	18.36+	0.00	17.92+	0.00
F_8	48.28	0.00	13.24+	0.00	0.88+	0.00	0.04+	0.00	0.00+	0.00	0.00+	0.00
F_9	141.68	0.00	50.20+	0.00	39.36+	0.00	17.64+	0.00	20.44+	0.00	28.20+	0.00
F_{10}	12	1.00	5.24+	0.00	10.56+	0.16	9.60+	0.08	10.12+	0.08	10.72+	0.28
+(SDEM is better)			7		6		7		6		6	
−(SDEM is worse)			0		0		0		0		0	
≈			3		4		3		4		4	

and NSDE. Furthermore, the p values of SDE, FERPSO, r2PSO, r2PSO-lhc and NSDE are less than 0.05, which means that SDEM is significantly better than compared algorithms. To further determine the ranking of the six compared algorithms, the Friedman's test, which is also implemented by using KEEL software, is conducted. As shown in Table 8, the overall ranking sequences of the test problems are SDEM, FERPSO, SDE, NSDE, r2PSO-lhc and r2PSO. The experimental results show that SDEM performs better than other compared algorithms. Therefore, it can be concluded that the improvement strategies on SDE are effective.

Figure 2 displays the average number of peaks in terms of the mean value achieved by each of six algorithms on $\epsilon = 1\text{E}{-04}$ for CEC2013 multimodal problems versus the number of *FES*. It can be seen that SDEM performs better than other algorithms, which suggest the migration mechanism and the improved operation of creating local individuals can help the algorithm to find multiple global optimal solutions. Figure 3 shows the results of ANP obtained in 25 independent runs by each algorithm for function F_1–F_{10} on $\epsilon = 1.0\text{E}{-5}$.

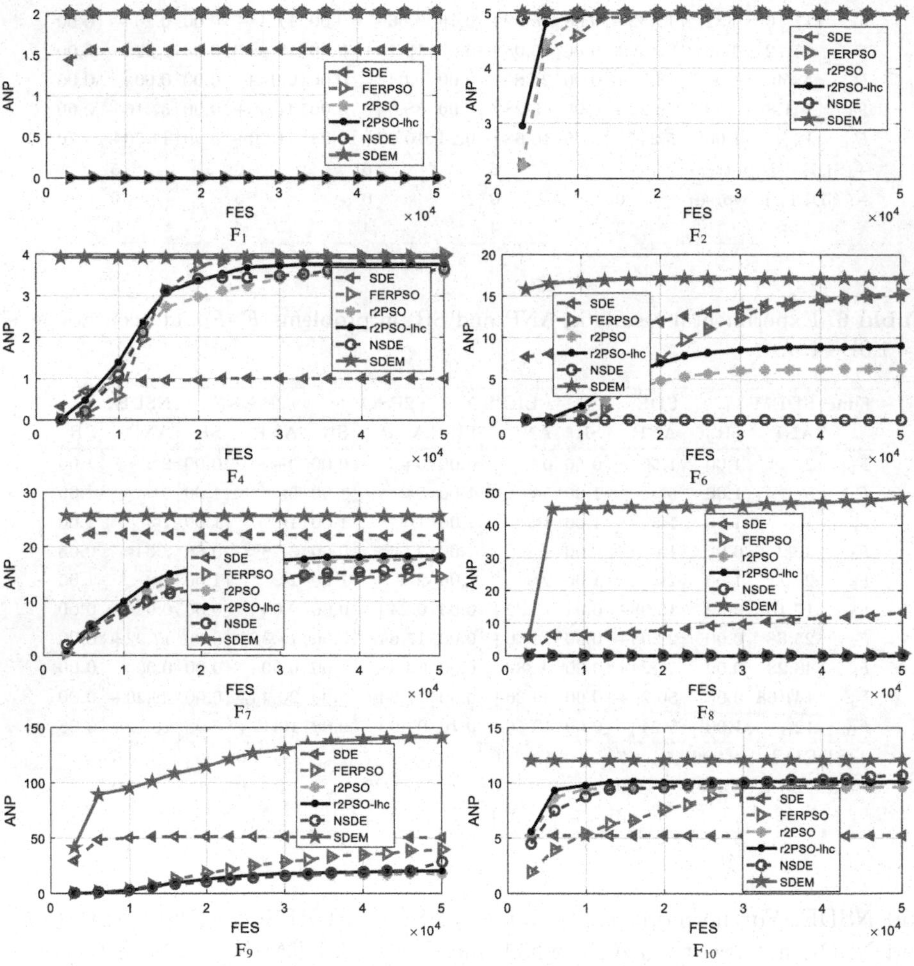

Fig. 2. Average number of peaks found by SDEM, SDE, FERPSO, *r*2PSO, *r*2PSO-*lhc* and NSDE versus the number of *FES* on eight test problems at accuracy level $\epsilon = 1.0\text{E}{-4}$.

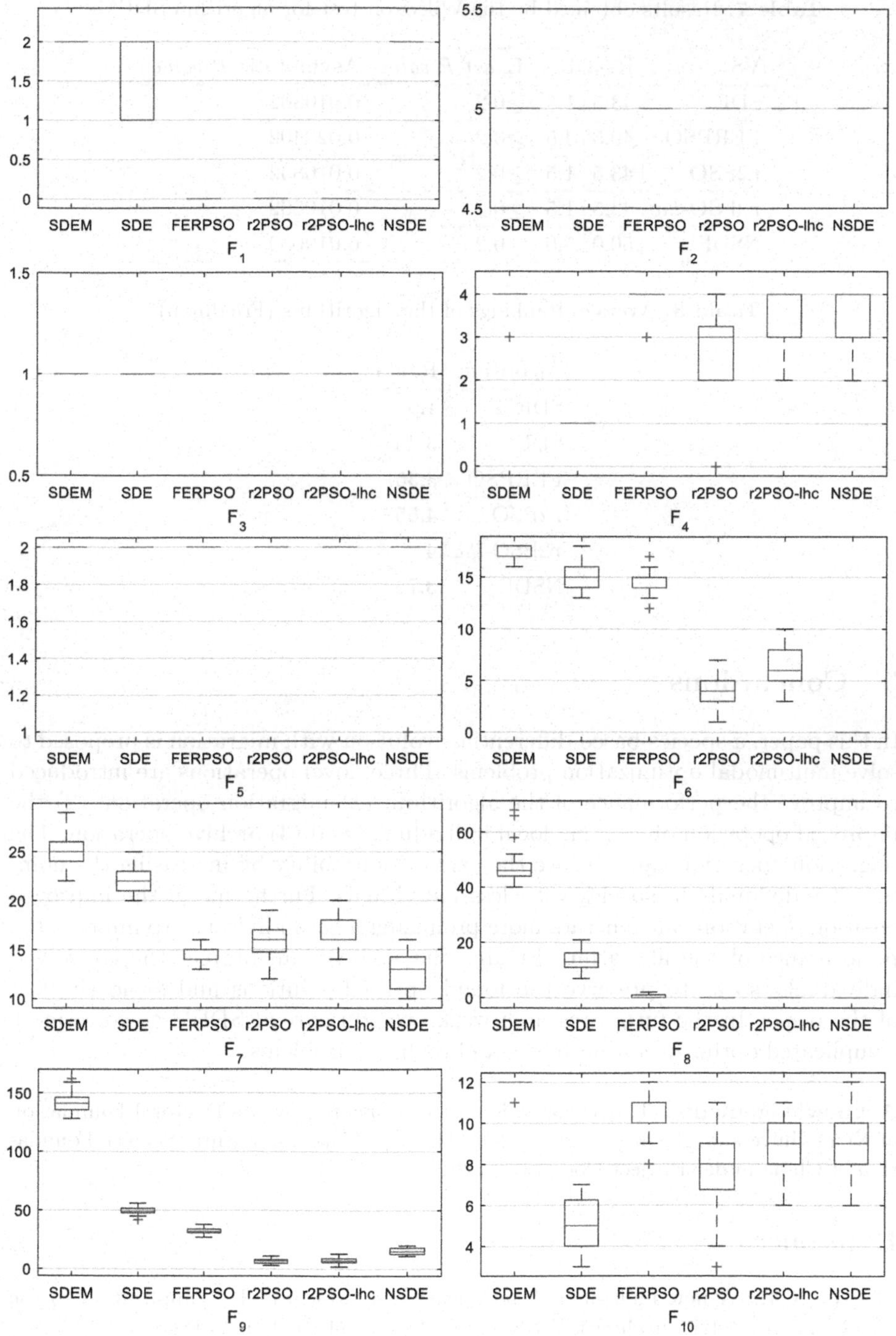

Fig. 3. Box-plot of Peaks found by SDEM, SDE, FERPSO, *r*2PSO, *r*2PSO-*lhc* and NSDE on ten test problems at accuracy level $\epsilon = 1.0\text{E}{-5}$.

Table 7. Results obtained by the Wilcoxon test for algorithm SDEM.

VS	R+	R−	Exact *P-value*	Asymptotic *P-value*
SDE	43.5	1.5	≥0.2	0.010862
FERPSO	40.5	4.5	≥0.2	0.028402
*r*2PSO	43.5	1.5	≥0.2	0.010862
*r*2PSO-*lhc*	43.5	1.5	≥0.2	0.010862
NSDE	50.0	5.0	≥0.2	0.019059

Table 8. Average Rankings of the algorithms (Friedman).

Algorithm	Ranking
SDEM	1.9
SDE	3.45
FERPSO	3.35
*r*2PSO	4.55
*r*2PSO-*lhc*	4
NSDE	3.75

5 Conclusions

In this paper, a species-based differential evolution with migration is proposed to solve multimodal optimization problems. Three novel operations are introduced to improve the performance of the algorithm: (1) migration operation; (2) the improved operation of creating local individuals; and (3) archive operation. The migration operation can improve the exploration ability by increasing the number of individuals in species with less individuals. Furthermore, the improved creation operation can generate more promising local individuals to improve the performance of the algorithm. Finally, the archive can protect the converged individuals, so as to preserve the found optimal solutions and avoid the loss of the peaks. In the future research work, we will extend SDEM to solve more complicated optimization problems and realistic problems.

Acknowledgements. This research is partly supported by the Doctoral Foundation of Xi'an University of Technology (112-451116017), National Natural Science Foundation of China under Project Code (61803301, 61773314).

References

1. Yoo, C.H.: A new multi-modal optimization approach and its application to the design of electric machines. IEEE Trans. Magn. **54**(3), 1–4 (2018)
2. Tanabe, R., Ishibuchi., H.: A review of evolutionary multi-modal multi-objective optimization. IEEE Trans. Evol. Comput., 1–9 (2019). https://doi.org/10.1109/TEVC.2019.2909744

3. Peng, X.G., Jin, Y.C., Wang, H.: Multimodal optimization enhanced cooperative co-evolution for large-scale optimization. IEEE Trans. Cybern. **49**(9), 3507–3520 (2019)
4. Naik, A., Satapathy, S.C., Ashour, A.S., Dey, N.: Social group optimization for global optimization of multimodal functions and data clustering problems. Neural Comput. Appl. **30**, 271–287 (2016). https://doi.org/10.1007/s00521-016-2686-9
5. Yazdanpanah, A., Singh, R., Gole, A., Filizadeh, S., Muller, J.C., Jayasinghe, R.P.: A parallel multi-modal optimization algorithm for simulation-based design of power systems. IEEE Trans. Power Delivery **30**(5), 2128–2137 (2015)
6. Thomsen, R.: Multimodal optimization using crowding-based differential evolution. In: IEEE Congress on Evolutionary Computation, 19–23 June, Portland, OR, USA, pp. 1382–1389 (2004)
7. Li, J.-P., Balazs, M.E., Parks, G.T., Clarkson, P.J.: A species conserving genetic algorithm for multimodal function optimization. Evol. Comput. **10**(3), 207–234 (2002)
8. Dong, W., Zhou, M.: Gaussian classifier-based evolutionary strategy for multimodal opti-mization. IEEE Trans. Neural Netw. Learn. Syst. **25**(6), 1200–1216 (2014)
9. Petrowski, A.: A clearing procedure as a niching method for genetic algorithms. In: IEEE Congress on Evolutionary Computation, 20–22 May, Nagoya, Japan, pp. 798–803 (1996)
10. Goldberg, D.-E., Richardson, J.: Genetic algorithms with sharing for multimodal function optimization. In: International Conference on Genetic Algorithms, pp. 41–49 (1987)
11. Storn, R., Price, K.V.: Differential evolution–a simple and efficient heuristic for global optimization over continuous spaces. J. Global Optim. **11**(4), 341–359 (1997). https://doi.org/10.1023/A:1008202821328
12. Li, X.: Niching without niching parameters: particle swarm optimization using a ring topology. IEEE Trans. Evol. Comput. **14**(1), 150–169 (2010)
13. Yang, Q., Chen, W.-N., Yu, Z., Li, Y., Zhang, H., Zhang, J.: Adaptive multimodal continuous ant colony optimization. IEEE Trans. Evol. Comput. **21**(2), 191–205 (2017)
14. Biswas, S., Kundu, S., Das, S.: Inducing niching behavior in differential evolution through local information sharing. IEEE Trans. Evol. Comput. **19**(2), 246–263 (2015)
15. Hui, S., Suganthan, P.-N.: Ensemble and arithmetic recombination based speciation differen-tial evolution for multimodal optimization. IEEE Trans. Cybern. **46**(1), 64–74 (2016)
16. Biswas, S., Kundu, S., Das, S.: An improved parent-centric mutation with normalized neighborhoods for inducing niching behavior in differential evolution. IEEE Trans. Cybern. **44**(10), 1726–1737 (2014)
17. Li, X.: Efficient differential evolution using speciation for multimodal function optimization. In: Conference on Genetic & Evolutionary Computation, pp. 873–880. ACM (2005)
18. Li, X.: Adaptively choosing neighbourhood bests using species in a particle swarm optimizer for multimodal function optimization. In: Deb, K. (ed.) GECCO 2004. LNCS, vol. 3102, pp. 105–116. Springer, Heidelberg (2004). https://doi.org/10.1007/978-3-540-24854-5_10
19. Li, J.P., Balazs, M.E., Parks, G.T., Clarkson, P.-J.: Erratum: a species conserving genetic algorithm for multimodal function optimization. Evol. Comput. **11**(1), 107–109 (2003)

20. Li, X., Engelbrecht, A., Epitropakis, M.: Benchmark functions for CEC 2013 special session and competition on niching methods for multimodal function optimization. Technical Report, Royal Melbourne Institute of Technology (2013)
21. Li, X.: A multimodal particle swarm optimizer based on fitness Euclidean-distance ratio. In: Genetic and Evolutionary Computation Conference (GECCO 2007), 7–11 July, pp. 78–85. ACM, London (2007)
22. Qu, B.Y., Suganthan, P.N., Liang, J.J.: Differential evolution with neighborhood mutation for multimodal optimization. IEEE Trans. Evol. Comput. **16**(5), 601–614 (2012)
23. Wang, Y., Cai, Z.X., Zhang, Q.F.: Differential evolution with composite trial vector generation strategies and control parameters. IEEE Trans. Evol. Comput. **15**, 55–66 (2011)
24. Alcal-Fdez, J., et al.: KEEL: a software tool to as-sess evolutionary algorithms to data mining problems. Soft. Comput. **13**(3), 307–318 (2009)

An Enhanced Bacterial Foraging Optimization Based on Levy Flight and Improved Roulette Wheel Selection

Xinzheng Wu[1], Aiqing Gao[2], Minyuan Lian[2(✉)], and Hong Wang[2]

[1] College of Economics, Shenzhen University, Shenzhen, China
[2] College of Management, Shenzhen University, Shenzhen, China
Lianmy_Carry@163.com

Abstract. For the purpose of promoting the convergence and global search capability of the original Bacterial Foraging Optimization (BFO), this paper proposes a novel BFO combined with Levy Flight and an improved Roulette Wheel Selection (LIRBFO). Due to the chemotaxis step length of the original BFO is set as a constant value. There is no balance between global search and local search. This restricts the application of BFO to tackle complex optimization problems. First of all, a random determining chemotaxis step size using Levy distribution is utilized to substitute fixed chemotaxis step size, which makes bacteria search for food through frequent short-distance search and occasional long-distance search and then improves the optimization ability and efficiency of the algorithm. Moreover, to make up errors caused by randomness strategies and maintain good population diversity, we present an improved roulette selection strategy based on evolutionary stability principle and apply it to the process of bacterial reproduction. The experiments are performed on eight benchmark functions to compare with the performance of LIRBFO, Bacterial Foraging Optimization with Levy Flight chemotaxis step size (LBFO) and improved BFO based on Improved Roulette Wheel Selection (IRBFO). The results obtained indicate this proposed algorithm greatly improves the convergence performance and search accuracy of original BFO in most cases.

Keywords: Bacterial Foraging Optimization · Levy Flight · Roulette wheel selection

1 Introduction

Over the past decades, numerous nature-inspired algorithms have been developed and widely used in many fields to solve real-world problems, such as genetic algorithm (GA) [6], ant colony optimization (ACO) [3] and particle swarm optimization (PSO) [7].

Bacterial Foraging Optimization (BFO), a swarm intelligence algorithm via imitating the foraging search strategy of E.coli bacteria was exploited by

L. Pan et al. (Eds.): BIC-TA 2019, CCIS 1159, pp. 223–232, 2020.
https://doi.org/10.1007/978-981-15-3425-6_18

Passino [11] in 2002. Over the years, BFO has been already used in various fields such as the PID controller [2], medical image [1], power systems [8], etc. However, due to the search directions are random and chemotaxis step size is fixed, the original BFO algorithm has a weak convergence performance and delaying in reaching global solutions may occur. In addition, with the dimensions of search space and the complexity of given problems increasing, the search ability will heavily decrease.

To solve these problems, various improvements have been made to enhance its optimization capability. The most common improvement method is improving the adjustment strategy of chemotaxis step size. There is an important step size optimization strategy being discussed below: Bacterial Foraging Optimization with linear chemotaxis step (BFO-LDC) [10]. A chemotaxis step length which begins with a large value C_{max} and decreases to C_{min} at the maximal number of iterations linearly is used over iterations. In the beginning, the chemotaxis step size is large so that it can't attend to exploration meticulously. With $C_{min} = C_{min}$, the system becomes the same as the original proposed BFO algorithm with fixed chemotaxis step length. It can't promote the search ability in the end.

Levy flight is widely used in PSO [5], bee colony algorithm (BCA) [12] to adjust the chemotaxis step and increase the ability of exploration. The core of the Levy flight strategy is frequent short-distance search and occasional long-distance search, which improves the global search ability of the BFO algorithm and helps to find the global optimal solution quickly and accurately.

Roulette wheel selection [9] is the case that the probability of each individual being selected equals to the proportional to its fitness. It was used in GA [4], PSO [16], BCA [15] and so on. The replication of the BFO algorithm adopts a relatively conservative elite preservation strategy, which has a strong local search ability, but the diversity of population is limited. The application of roulette wheel selection can maintain population diversity, but it has the disadvantage of leading to population degradation easily.

In this paper, an improved roulette wheel selection is used in the reproduction that will increase the diversity of the population while maintaining the superiority of the bacteria, thus improving the performance of BFO.

The rest of the paper is structured as follows: Sect. 2 provides a brief review of classical BFO. The proposed LIRBFO is presented in Sect. 3. In Sect. 4, we provided experimental result and discussion. Finally, the paper is concluded with a summary in Sect. 5.

2 Bacterial Foraging Optimization

Generally, the BFO models the bacterial foraging optimization process such as chemotaxis, reproduction, elimination and dispersal.

2.1 Chemotaxis

Chemotaxis simulates the movement of E. coli bacteria through "swimming" and "tumbling" via flagella. Biologically an E. coli cell can move in two different ways. It can swim in the same direction for a period of time, or it may tumble, and it will switch between these two modes for a lifetime.

Suppose $\theta^i(j,k,l)$ represents ith bacterium at jth chemotaxis, kth reproduction and lth elimination dispersal. $C(i)$ is the length of the unit walk and $\Delta(i)$ is a vector in the random direction whose elements are located in $[-1,1]$. In the chemotactic step, the movement of the bacterium can be represented by:

$$\theta^{\mathrm{i}}\left(j+1,k,l\right)=\theta^{i}\left(j,k,l\right)+C\left(i\right)\frac{\Delta\left(i\right)}{\sqrt{\Delta\mathrm{T}\left(i\right)\Delta\left(i\right)}} \tag{1}$$

2.2 Reproduction

In this process, relatively unhealthy bacteria die while each of the healthier bacteria asexually split into two bacteria. The swarm size constants. $J_{health}(i,j,k,l)$ is the fitness value (a standard for measuring bacterial health) of ith bacterium at the jth chemotaxis, kth reproduction, lth dispersal. It should be calculated after chemotaxis. S is the population of bacteria. Sorting J_{health} the in ascending order, the first $Sr(=S/2)$ bacteria survive and come into reproduction, the others are eliminated. The J_{health} is assessed as:

$$J_{\mathrm{health}}^{\mathrm{i}}=\sum\nolimits_{\mathrm{j}=1}^{N_C+1}J(i,j,k,l) \tag{2}$$

$$\theta^{(i+Sr)}(j,k,l)=\theta^{i}(j,k,l),i=1,\ldots,Sr \tag{3}$$

where N_c is the maximum number of steps for chemotaxis.

2.3 Elimination and Dispersal

After several replication operations, the colonies will present several clusters, and population diversity will be degraded. To avoid this phenomenon, the BFO algorithm introduces a mutation operation dispersion behavior. This operation simulates a biological phenomenon in which bacteria migrate with water or other organisms into a new environment. The mode of operation is that some bacteria are liquidated randomly with a small probability while the new replacements are initialized at random. After the dispersal operation, a new generation of individuals enters a new trend cycle.

3 The Proposed LIFBFO Algorithm

This part mainly explains the application of the random walk model of the Levy flight strategy and the roulette wheel selection in algorithm improvement. The Levy flight is used to improve the chemotaxis step and an improved roulette wheel selection is proposed to be used in the reproduction.

3.1 Levy Flights in Chemotaxis

The Levy flight was originally introduced by Paul Levy (1886–1971) in 1937. It is a statistical description of motion that extends beyond the more traditional Brownian motion discovered over one hundred years earlier. The special walk strategy of levy flight, whose random step is subject to a heavy-tailed probability distribution, which is called the levy distribution, obeys the power-law distribution and is expressed as where s is a random step size. The variance of Levy flight shows exponential relation with time.

$$\sigma^2(t)\, t^{3-\beta}, 1 \leq \beta \leq 3 \tag{4}$$

In an algorithm proposed by Mantegna [13], the method of generating levy random step size is as follows:

$$S = \frac{u}{|v|^{1/\beta}} \tag{5}$$

where u and v are drawn from normal distributions. That is

$$u \sim N(0, \sigma_u^2), v \sim N(0, \sigma_v^2) \tag{6}$$

$$\sigma_u = \left\{ \frac{r(1+\beta)\sin(\pi\beta/2)}{r\left[(1+\beta)/2\right]\beta 2^{(\beta-1)/2}} \right\}^{1/\beta}, \sigma_v^2 = 1 \tag{7}$$

The improved chemotaxis step size formula is as follows:

$$C(i) = scale * Levy(\lambda) \tag{8}$$

$$scale = \frac{1}{10p}\sum\nolimits_{d=1}^{\eta} \left|\theta_d^i(j,k,l) - \theta_d^*(j,k,l)\right| \tag{9}$$

In the formula, Levy(λ) is the Levy step size with λ as the parameter, the scale is the scale factor, $\theta_d^i(j,k,l)$ is the D-dimensional coordinate value of the position of the ith bacteria, and $\theta_d^*(j,k,l)$ is the D-dimensional coordinate value of the current optimal bacterial position. The scale is designed to take advantage of the positional relationship between the optimal bacteria and the current bacteria, with the aim of allowing the bacterial Levy chemotaxis step to be in the appropriate order of magnitude.

3.2 Improved Roulette Wheel Selection in Reproduction

The basic principle of the roulette wheel selection is that the fitness is converted into the selection probability proportionally. Suppose the group size is n. The standard roulette wheel selection is as follows:

Calculate the probability that each individual is selected to inherit to the next generation group $P(i)$ and $P(a_i)$ which is the accumulation of $P(i)$:

$$P(i) = \frac{F_i}{\sum_{i=1}^{n} F_i} \quad (10)$$

$$P(a_i) = \sum_{j=1}^{i} P(i) \quad (11)$$

where F_i is the fitness of each individual.

Use the simulated market operation to generate a random number r between 0 and 1, If $P(a_i) < r$, select individual i to enter the progeny population until meeting the quantity requirements of a new generation of the bacterial population.

Classical algorithms are prone to a "degenerate" phenomenon. Because of its strong randomness, the search ability of the algorithm decreases, and some particles with good fitness are eliminated. Evolutionarily stable strategy [14] is applied to improve the roulette selection so that it can reduce randomness meanwhile preserve the diversity.

Table 1. The pseudo-code 1.

```
Pseudo-code 1  Improved roulette wheel selection
Begin
  Sortind each bacterium based on Jhealth
  Select the top 20% of bacteria as a group of A
  For Jhealth1=Jhealth(the other 80% of the bacteria)
    PP=Jhealth1(i1)/sum(Jhealth1)
    Select 20% of bacteria that meet the request PP<rand as a group of B
  Reproduction: reproduce group A twice and group B once using Eq. (2).
  End
```

The process of the improved roulette wheel selection is shown in Table 1. Based on the sorted J_{health} in the pattern of reproduction, the top 20% of bacteria will be reserved into the next generation and reproduction twice. And select the other 20% bacteria using roulette wheel selection to reproduction only once.

3.3 The Enhanced Bacterial Foraging Optimization Algorithm Based on Levy Flight and Improved Roulette Wheel Selection (LIRBFO)

The whole process of the LIRBFO operation is illustrated in Fig. 1 and the pseudo-code of LIRBFO Algorithm is shown in Table 2. First, initialization is carried out, then chemotaxis based on Levy flight, swarming, a reproduction based on the improved roulette wheel selection, elimination and dispersal are carried out. Finally, the output of global optimum, average, minimum and standard deviation is achieved.

Table 2. The pseudo-code 2.

Pseudo-code 2 LIRBFO Algorithm

```
Input: dataset for training and testing: Tr and Te; number of features to be
   selected: D;
Initialization : Dim, initialization of population: S, fitness calculation: J,
   chemotaxis step: C, etc.
Optimization process:
For k=1,2,..,Ned
For k=1,2,..,Ned
For k=1,2,..,Ned
      Form parameter u and v based on normal distribution using Eq. (5)
      Define C based on Levy flight using Eqs. (7-8)
      Running: according to Eq. (1)
      Tumbling: according to Eq. (1)
    End chemotaxis iteration
    Reproduction : Select bacteria based on improved Roulette Wheel Selection
     using Eqs. (9-10)
    BFO reproduction using Eq. (2)
  End reproduction iteration
  Elimination: BFO elimination
End elimination iteration
Output: classification accuracy and its corresponding selected feature vector
```

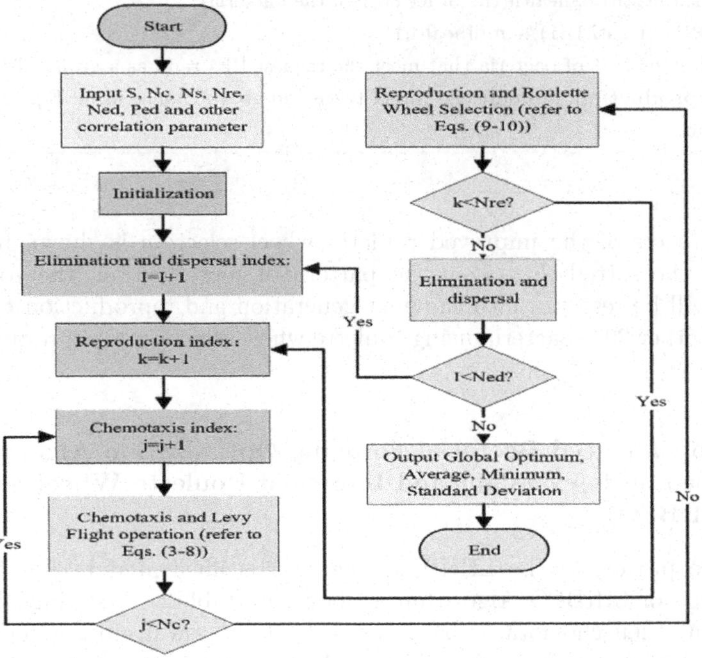

Fig. 1. The overall framework of the proposed LIRBFO.

4 Experiment and Result

In this section, the exhaustive evaluation of the proposed LIRBFO algorithm is presented. For comparison, three other algorithms are used. They are the classical BFO, an enhanced BFO algorithm with levy flight (LBFO) and an enhanced BFO algorithm with Improved Roulette Wheel Selection (IRBFO). As in [11], for each function, S = 50, Ns = 4, Ned = 2, Nre = 5, Nc = 1000, Ped = 0.25, the dimension d = 15. All experiments were repeated 30 runs.

Eight benchmark test functions shown in Table 3 are used to evaluate the performance of the proposed LIRBFO algorithm. The results of the five algorithms are shown in Table 4, and the best results are bolded. As given in Table 4, the LIRBFO algorithm gets optional results for all benchmark functions among 8 functions. As for the stability of the five algorithms, LIRBFO has a more stable performance than other algorithms because the standard deviation of LIRBFO is smaller. Figure 2 graphically presents the comparison of convergence characteristics for the five algorithms. On the eight functions, LIRBFO converges very fast to achieve the optimal result. LBFO converge very fast except f7. BFO and IRBFO do not perform well and get stuck in local minima. The BFO-LDC gets worse performances even than BFO in f1, f2, f5, and f8. Its performances haven‘t changed much compared with BFO in f6 and f7. The performances are much better than BFO and IRBFO but worse than LBFO and LIRBFO in f3 and f4. In general, LIBFO performance better than the other four algorithms because Levy flight improves the length of the chemotaxis step and the improved Roulette Wheel Selection promotes the quality of the flora.

Table 3. Descriptions of 8 benchmark function selected in the experiments.

	Function name	Mathematical formula	Domain	Minimum
$f1$	Sphere	$f_1 = \sum_{i=1}^{n} x_i^2$	(−100, 100)	0
$f2$	Moved-WeightedSphere	$f_2 = \sum_{i=1}^{d} (i * x^2)$	(−100, 100)	0
$f3$	Quartic	$f_3 = \sum_{i=1}^{n} ix_i^2$	(−50, 50)	0
$f4$	SumPowers	$f_4 = \sum_{i=1}^{d} \lvert x_i\rvert^{i+1}$	(−10, 10)	0
$f5$	Griewank	$f_5 = \sum_{i=1}^{n} x_i^2/4000 - \Pi_{i=1}^{n} cos(x_i/\sqrt{i}) + 1$	(−600, 600)	0
$f6$	Apline	$f_6 = \sum_{i=1}^{n} \lvert x_i sin(x_i) + 0.1x_i\rvert$	(−10, 10)	−78.33
$f7$	Step	$f_7 = \sum_{i=1}^{n} \lvert x_i + 0.5\rvert^2$	(−5, 5)	0
$f8$	Penalized1	$f_8 = \pi/n(10sin^2(\pi y_1) + \sum_{i=1}^{n-1} (y_1 - 1)^2[1 + 10sin^2(\pi y_{(i+1)})] + (y_n - 1)^2) + \sum_{i=1}^{n} u_i$	(−10, 10)	0

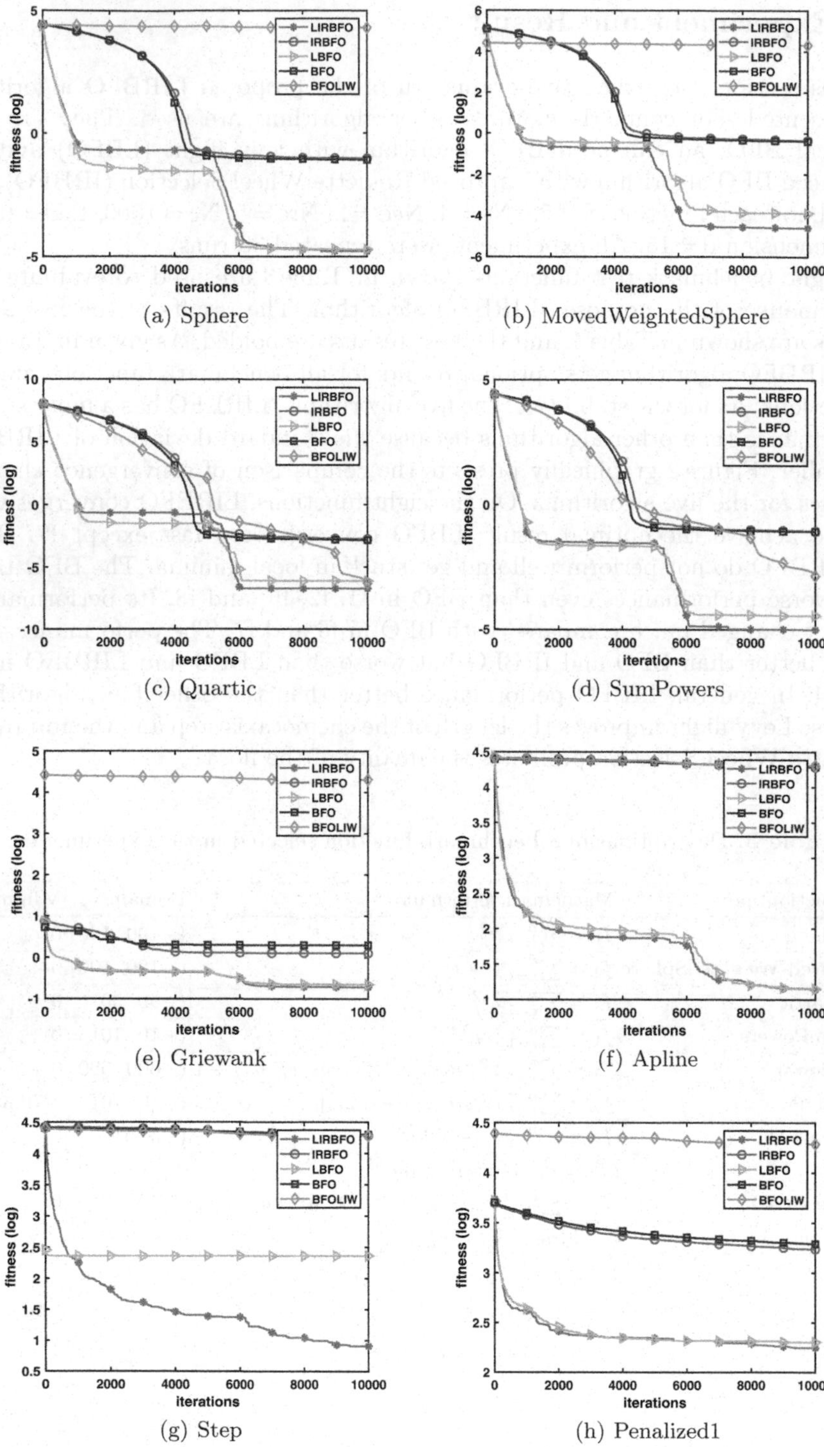

(a) Sphere (b) MovedWeightedSphere (c) Quartic (d) SumPowers (e) Griewank (f) Apline (g) Step (h) Penalized1

Fig. 2. Example of placing a figure with experimental results.

Table 4. Numerical results of eight benchmark functions for 15D.

Dim	Index	BFO	LDC-BFO	LBFO	IRBFO	LIRBFO
f1	Best	0.0824	11625	3.57E−05	0.0774	**4.48E−10**
	Mean	0.0814	2.02E+04	1.91E−05	0.0941	**1.62E−05**
	Std	0.0136	4.03E+03	3.35E−05	**2.49E−09**	5.49E−05
f2	Best	0.497	8360	3.34E−06	0.5036	**1.57E−07**
	Mean	0.4209	2.02E+04	1.29E−04	0.4384	**2.52E−05**
	Std	0.0761	3.53E+03	4.40E−04	0.0804	**7.90E−05**
f3	Best	9.69E−04	9.04 E−08	2.49E−09	0.0015	**7.27E−10**
	Mean	0.0014	2.78E−07	7.50E−07	0.0017	**1.85E−07**
	Std	6.00E−04	1.05E−07	3.99E−06	6.43E−04	**5.75E−07**
f4	Best	0.0716	6.77E−04	9.90E−06	0.0937	**6.21E−07**
	Mean	0.0866	0.0011	3.49E−05	0.0892	**9.51E−06**
	Std	0.0126	1.96E−04	9.48E−05	0.016	**1.74E−05**
f5	Best	2.2462	13101	0.0615	0.3391	**3.48E−07**
	Mean	1.3607	2.03E+04	0.2215	1.2658	**0.1903**
	Std	0.7426	3.41E+03	0.1866	0.8015	**0.1665**
f6	Best	25827	11717	**5**	13890	39
	Mean	1.96E+04	2.12E+04	14.2	1.92E+04	**13.8**
	Std	4.92 E+03	3.50E+03	12.891	3.05E+03	**9.773**
f7	Best	21909	10046	47	15618	**3**
	Mean	2.00E+04	1.92E+04	24.8	1.92E+04	**8.1**
	Std	4.34E+03	3.75E+03	16.9365	5.09E+03	**3.6953**
f8	Best	1.94E+03	13700	138.4434	2.01E+03	**176.5436**
	Mean	1.96E+03	1.93E+04	200.9101	1.73E+03	**176.7151**
	Std	130.1082	2.51E+03	83.9584	154.8545	**57.4789**

5 Conclusions

In this paper, the original BFO is combined with Levy flight and improved Roulette Wheel Selection to enhance the global search capability and convergence efficiency. The LIRBFO algorithm is a global search algorithm with several bene-fits. The advantages of the algorithm are: The Levy flight balanced local and global search capabilities and achieve the purpose of improving the algorithm's solution accuracy and convergence speed. And the improved Roulette Wheel Selection improved the quality of bacteria while ensuring the diversity of the flora. A series of benchmark functions are used to evaluate the proposed algorithm and the experimental results obtained from LIRBFO are compared with BFO, LBFO, IRBFO, BFO-LDC and the results show that the proposed LIRBFO improves the solution and has high convergence rate. Therefore, The LIRBFO algorithm is feasible and effective.

Future works should focus on optimizing the performance of LIRBFO algorithm, specifically, finding the optimal solution should be more time-saving and have stronger global search performance. In the near future, LIRBFO algorithm will be applied to solve practical problems.

Acknowledgments. This work is partially supported by the Natural Science Foundation of Guangdong Province (2018A030310575), Natural Science Foundation of Shenzhen University (85303/00000155). Xinzheng Wu and Aiqing Gao are first authors. They contributed equally to this paper.

References

1. Costin, H., Bejinariu, S.: Medical image registration by means of a bio-inspired optimization strategy. Comput. Sci. J. Moldova **20**(2), 178–202 (2012)
2. Kim, D.H.: Robust tuning of embedded intelligent PID controller for induction motor using bacterial foraging based optimization. In: Wu, Z., Chen, C., Guo, M., Bu, J. (eds.) ICESS 2004. LNCS, vol. 3605, pp. 137–142. Springer, Heidelberg (2005). https://doi.org/10.1007/11535409_19
3. Dorigo, M., Gambardella, L.M.: Ant colony system: a cooperative learning approach to the traveling salesman problem. IEEE Trans. EC **1**(1), 53–66 (1997)
4. Goldberg, D.E., Deb, K.: A comparative analysis of selection schemes used in genetic algorithms. Found. Genet. Algorithms **1**, 69–93 (1991)
5. Hakli, H., Uğuz, H.: A novel particle swarm optimization algorithm with levy flight. Appl. Soft Comput. J. **23**(5), 333–345 (2014)
6. Holland, J.: Adaptation in natural and artificial systems: an introductory analysis with applications to biology, control and artificial intelligence. Ann Arbor **6**(2), 126–137 (1992)
7. Kennedy, J., Eberhart, R.: Particle swarm optimization. In: Proceedings of 1995 IEEE International Conference on Neural Networks, Perth, Australia, 27 November–1 December, vol. 4, no. 8, pp. 1942–1948 (2011)
8. Kumar, K.S., Jayabarathi, T.: Power system reconfiguration and loss minimization for an distribution systems using bacterial foraging optimization algorithm. Int. J. Electr. Power Energy Syst. **36**(1), 13–17 (2012)
9. Lipowski, A., Lipowska, D.: Roulette-wheel selection via stochastic acceptance. Phys. A Stat. Mech. Appl. **391**(6), 2193–2196 (2012)
10. Niu, B., Yan, F., Pei, Z., Bing, X., Li, L., Chai, Y.: A novel bacterial foraging optimizer with linear decreasing chemotaxis step. In: International Workshop on Intelligent Systems & Applications (2010)
11. Passino, K.M.: Biomimicry of bacterial foraging for distributed optimization and control. IEEE Control Syst. **22**(3), 52–67 (2002)
12. Sharma, H., Bansal, J.C., Arya, K.V., Yang, X.S.: Lévy flight artificial bee colony algorithm. Int. J. Syst. Sci. **47**(11), 2652–2670 (2016)
13. Shlesinger, M.F.: Comment on "stochastic process with ultraslow convergence to a Gaussian: the truncated lévy flight". Phys. Rev. Lett. **74**(24), 4959 (1995)
14. Wang, Y., Wineberg, M.: Estimation of evolvability genetic algorithm and dynamic environments. Genet. Program Evolvable Mach. **7**(4), 355–382 (2006). https://doi.org/10.1007/s10710-006-9015-5
15. Xiang, W.L., An, M.Q.: An efficient and robust artificial bee colony algorithm for numerical optimization. Comput. Oper. Res. **40**(5), 1256–1265 (2013)
16. Xue, B., Qin, A.K., Zhang, M.: An archive based particle swarm optimisation for feature selection in classification. In: Evolutionary Computation (2014)

Three-Dimensional Packing Algorithm of Single Container Based on Genetic Algorithm

Shuting Jia(✉) and Li Wang(✉)

School of Computer Science and Software Engineering, University of Science and Technology Liaoning, Anshan 114051, China
jst_66666@163.com, Wangli9966@163.com

Abstract. In this paper, a mathematical model is established for the three-dimensional packing problem. This model is a multi-objective optimization problem which considers two factors: volume utilization ratio and load utilization ratio. This paper uses heuristic algorithm to design on the basis of the actual experience of packing. At the same time, combining genetic algorithm and heuristic algorithm, a three-dimensional hybrid genetic heuristic algorithm (3D-HGH) is proposed. This algorithm improves the performance of heuristic algorithm and improves the utilization ratio. The results of three-dimensional visualization of loading are obtained by simulation experiments.

Keywords: Three-dimensional packing · Heuristic algorithm · Genetic algorithm

1 Introduction

Modern logistics industry has developed rapidly in the world, and its development has become one of the standards to measure the degree of modernization of the country. At the present stage, in the face of a variety of logistics demand, the high logistics cost is a major problem to be solved in the logistics industry. The low efficiency of the packing link is one of the main factors. The packing problem is a typical NP problem. At present, container loading mainly depends on the actual experience of loading workers, resulting in low utilization of containers. Therefore, it is urgent to improve the loading efficiency of transport equipment and reduce transport costs.

The packing problem refers to placing several groups of rectangles or cuboids into a specified space to make the maximum space filling rate. As a typical combinatorial optimization problem, there have been a lot of related researches. On one-dimensional and two-dimensional packing problem, scholars at home and abroad have studied various strategies and methods to solve the problem. Zhang and Yao consider the FFD algorithm in several bin packing problems and tested its performance [1]. Alvim et al. propose a hybrid improvement procedure for the bin packing problem [2]. This heuristic use lower bounding strategies and the generation of initial solutions by reference to the dual min-max problem. Loh et al. use the concept of weight annealing to solve the one-dimensional bin packing problem [3]. They apply it to 1587 instances taken from benchmark problem

L. Pan et al. (Eds.): BIC-TA 2019, CCIS 1159, pp. 233–243, 2020.
https://doi.org/10.1007/978-981-15-3425-6_19

sets and find that the algorithm can produce very high-quality solutions very quickly and generates several new optimal solutions. Arbib et al. propose a time-indexed mixed integer linear programming method to solve the one-dimensional packing problem [4]. Some scholars put forward heuristic methods for two-dimensional packing problems, such as greedy method, local search method, neighborhood search algorithm and local enumeration algorithm [5–7]. And they applied them to some practical applications and obtained better results based on examples. A. Martinez-Sykora et al. present a number of variants of a constructive algorithm able to solve a wide variety of variants of the two-dimensional irregular bin packing problem (2DIBPP) [8].

Compared with the one-dimensional and two-dimensional packing problem, the research on three-dimensional packing problem is relatively less and immature. Most scholars use heuristic algorithm to solve three-dimensional packing problem. They propose heuristic algorithms such as tabu search, greedy search, and integer linear programming [9–11]. Kang et al. consider the variable sized bin packing problem where the objective is to minimize the total cost of used bins and use greedy algorithms to solve this problem [12]. Wu et al. present a three-stage heuristic algorithm for three-dimensional irregular packing problem [13]. Paquay et al. propose constructive heuristics for the three dimensional bin packing problem with transportation constraints [14]. And this paper considers additional constraints encountered in real world air transportation situations, such as cargo stability and particular shape of containers.

Above all, there is still little research on multi-objective three-dimensional packing problem with relatively complex constraints. Through analysis, this paper establishes a multi-objective mathematical model with the maximum load utilization ratio and the largest volume utilization ratio. And considers many constraints, such as center of gravity constraints, volume constraints and so on. It combines genetic algorithm to optimize on the basis of heuristic algorithm to get the optimal solution quickly.

This paper is organized as follows: Sect. 2 proposes problems and establishes models. Section 3 introduces the heuristic algorithm. Section 4 introduces the genetic algorithm. Section 5 presents the 3D-HGH algorithm to solve the proposed model. Section 6 designs the experimental scheme and analyzes the experimental results. Section 7 conclusion and future works.

2 Models of Three-Dimensional Packing

2.1 The Description of Problem

Solving single container loads different types of cargos and allows surplus cargos. Here, we define the problem as: Given that the length, width and height of a container are C_l, C_w, C_h, the volume is V, the maximum load is G, and the length, width and height of i kinds of cargos are l_i, w_i and h_i respectively, the volume is v_i. The shapes of containers and cargos are cuboids. The objective of the problem is to determine a feasible packing scheme to maximize the load and volume utilization of containers under certain constraints. And make the following assumptions:

1. The shape of the container is a cuboid;
2. The shape of the cargo is a cuboid with uniform mass distribution, and the length, width, and height do not exceed the length, width, and height of the container;
3. The cargos can keep their shape and size unchanged;
4. The cargos can be load-bearing and can be loaded in multiple layers;
5. The cargos are placed in an orthogonal layout parallel to the container, allowing rotation in any direction;
6. No loading or unloading of cargos during transportation.

2.2 Establish Model

This paper researches on three-dimensional packing problem. The spatial rectangular coordinate system definition is shown in Fig. 1.

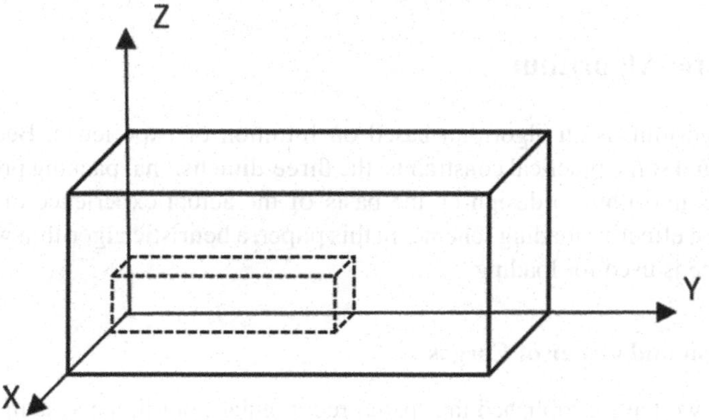

Fig. 1. Cargo packing diagram

And put forward the following constraints:

1. Volume constraint: the sum of the volumes of container-loaded cargos is not greater than the maximum volume of the container, i.e., $\sum_{i=1}^{n} v_i \leq V$;
2. Weight constraint: the sum of the weight of container-loaded cargos is not greater than the nominal weight of the container, i.e., $\sum_{i=1}^{n} m_i \leq G$;
3. Center of gravity constraint: in order to ensure the safety of the vehicle, the overall center of gravity of all cargos in the container is within the reasonable range of the center of gravity specified by the vehicle, i.e., $\frac{\sum_{i=1}^{n} m_i X_i}{\sum_{i=1}^{n} m_i} \in \left(\alpha_x, \alpha_x^{'}\right)$, $\frac{\sum_{i=1}^{n} m_i Y_i}{\sum_{i=1}^{n} m_i} \in \left(\alpha_y, \alpha_y^{'}\right)$, $\frac{\sum_{i=1}^{n} m_i Z_i}{\sum_{i=1}^{n} m_i} \in \left(\alpha_z, \alpha_z^{'}\right)$;
4. Dimension constraint: the loaded cargos shall be in the container. The sum of the coordinates of the loaded cargos and their length, width and height shall not exceed the length, width and height of the container, i.e., $x_i + l_i \in [0, C_l]$, $y_i + w_i \in [0, C_w]$, $z_i + h_i \in [0, C_h]$.

In conclusion, the model is established as follows:

$$maxf = \frac{\sum_{i=1}^{n} v_i}{V} + \frac{\sum_{i=1}^{n} m_i}{G} \quad (1)$$

According to the model in this paper, there are the following symbols:
C_l, C_w, C_h are the length, width and height of the container;
V is the volume of the container;
G is the maximum load;
i is the type of cargos, i = 1, 2, 3, …, n;
l_i, w_i, h_i are the length, width and height of the cargos respectively;
v_i is the volume of the cargos;
m_i is the quality of the cargos;
X_i, Y_i, Z_i are the center of gravity of the cargos in the container;
x_i, y_i, z_i are the range of center of gravity of the container, according to the actual determination.

3 Heuristic Algorithm

Heuristic algorithm is an algorithm based on intuition or experience. Because of its complexity and some practical constraints, the three-dimensional packing problem must use heuristic algorithm to design on the basis of the actual experience of packing in order to get an effective loading scheme. In this paper, a heuristic algorithm with optimal residual space is used for loading.

3.1 Location and Order of Cargos

In Sect. 2.2, we have established the spatial rectangular coordinate system. According to experience, the cargos to be put into containers must be close to the container or close to the cargos that have been put into the containers in front. The long side, wide side and high side of the cargos should be parallel to the long side, wide side and high side of the container. And the base point of the cargos should coincide with the base point of the loading space, that is, the cargos should be loaded in the lower left corner of the rear of the loadable space, which is the angle occupation strategy.

There are no special requirements for the loading order of the cargos, which is randomly generated and handled by the following optimization algorithm.

3.2 Division of Remaining Space

The cargo structure studied in this paper is regular cuboids. The initial remaining space when the cargo is not loaded is the entire container. As the cargos are continuously loaded, the remaining space will become more and more irregular. Therefore, this paper dismantles the irregular spaces and integrates them into regular space to form a large available space to improve the utilization rate of space. Figure 2 is a schematic diagram of the division of the remaining space.

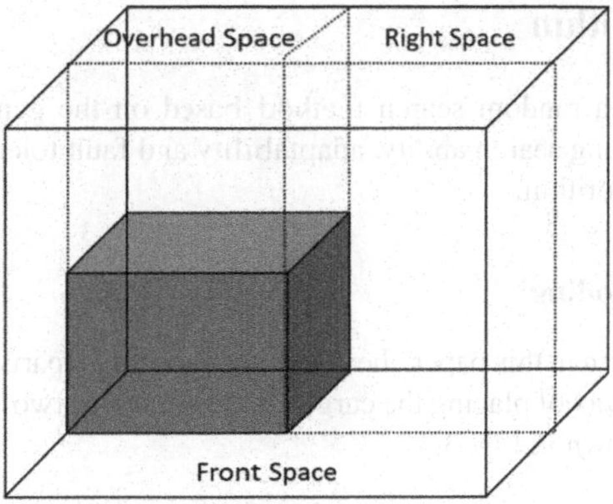

Fig. 2. Division of the remaining space diagram

3.3 Algorithm Description

According to the contents of the above algorithm and the model of three-dimensional packing problem established in this paper, the steps of solving the heuristic algorithm in this paper are summarized. Here, the set of cargo to be loaded is defined as CL, the set of remaining space is RS, and the set of goods that cannot be loaded into the container is RCL. The pseudo-code of the algorithm is shown in Table 1.

Table 1. Description of heuristic algorithm

Input	Data on container (L, W, H) and cargos (l_i, w_i, h_i)
Output	Loading results
	Initial RS to L×W×H
	repeat
	if CL is not an empty set **then**
	Select the cargos in the CL as current cargos c_i
	Select the space r_k in the RS
	if c_i can be put in r_k **then**
	Update RS by space remaining
	Select next cargo
	else
	Add c_i to RCL
	Select next cargo
	until CL is an empty set
	return Loading scheme and RCL

4 Genetic Algorithm

Genetic algorithm is a random search method based on the genetic law of biology. The algorithm has strong search ability, adaptability and fault tolerance ability, and is a relatively efficient algorithm.

4.1 Chromosome Coding

For the packing problem in this paper, the code consists of two parts: the order of loading of the cargos and the way of placing the cargos, that is, using a two-stage natural number coding method, as shown in Fig. 3.

Fig. 3. Coding method

Genotype of Loading Order. We define the genotype as G(11, 4, 7,…, 2, 6) and the length is N. Each gene bit in this gene represents the order in which the cargos are placed, as shown in Fig. 4.

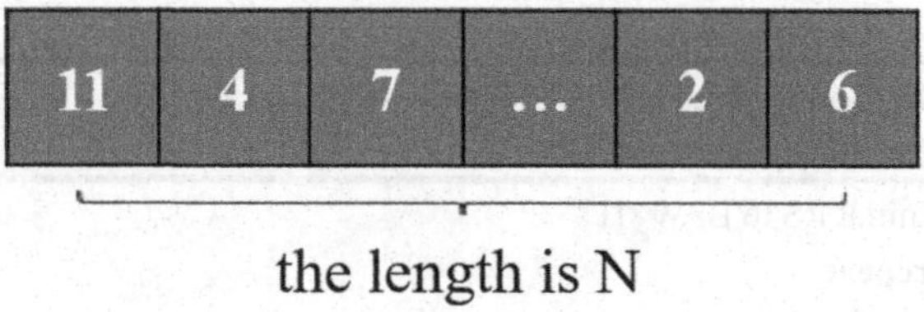

Fig. 4. Genotype of loading order

Genotype of Placing Way. The shape of the cargo is a cuboid with uniform mass distribution and the cargos can be allowed rotation in any direction. So there are six kinds of placement methods. The length, width and height can be (l, w, h), (w, l, h), (w, h, l), (h, w, l), (l, h, w), (h, l, w), (h, l, w), corresponding to natural numbers 1, 2, 3, 4, 5, 6, as shown in Fig. 5. We define the genotype G (1, 2, 3, 4, 5, 6) with a length of N. Each gene location in the gene represents the way cargos are placed.

The final gene length is 2N, as shown in Fig. 6.

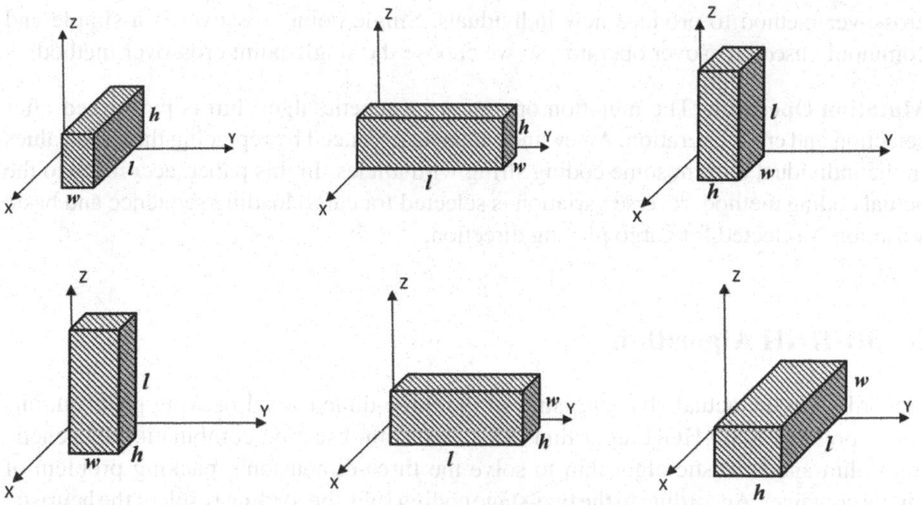

Fig. 5. Six ways of placing

Fig. 6. Chromosome

4.2 Population Initialization and Fitness Functions

The initial solution of the population is a random sequence of natural numbers. According to the coding scheme designed in this paper, a feasible sequence is randomly generated until the initial population that satisfies the population number is generated.

After determining the gene, in order to directly relate the fitness function with the fitness of individuals in the population, the fitness function used in this paper is the objective function of the model. It can be seen that the greater the function value, the greater the adaptability.

4.3 Genetic Operation

Selection Operator. The selection operation in genetic operation is to determine how to select the best individual to inherit to the next generation. According to the actual situation of this paper, we choose roulette strategy. Roulette strategy is suitable for the problem of fitness maximization, that is, the greater the fitness value, the greater the probability of individuals being selected.

Crossover Operator. The crossover operation in genetic operation refers to the process in which two paired chromosomes exchange some genes with each other by some crossover method to produce new individuals. Single point crossover is a simple and commonly used crossover operator, so we choose the single point crossover method.

Mutation Operator. The mutation operation in genetic algorithm is performed after selection and cross operation. A new individual is produced by replacing the gene values in the individual chromosome coding string with alleles. In this paper, according to the actual coding method, reverse variation is selected for cargo loading sequence and basic variation is selected for cargo placing direction.

5 3D-HGH Algorithm

According to the actual characteristics of the three-dimensional packing problem, this paper presents a 3D-HGH algorithm. This algorithm uses the combination of genetic algorithm and heuristic algorithm to solve the three-dimensional packing problem of single container. According to the two-stage coding rule, the loading result of the heuristic algorithm is optimized by randomly generating the initial population. Taking objective function as fitness function of genetic algorithm. Then carry on selection operation, crossover operation, and mutation operation of genetic algorithm. The optimal solution is

Table 2. Algorithm description

Input	Data on container (L, W, H) and cargos (l_i, w_i, h_i) Relevant parameters of genetic algorithm
Output	Optimal solution of loading
	Initialization produces the first generation of populations P(0)
	Evolutionary algebra t=0
	repeat
	for 1 to number of individuals in the population **do**
	Calculate the adaptability of each individual in P(T)
	Apply the selection operator to the population
	Apply the crossover operator to the population
	Apply the mutation operator to the population
	end for
	for 1 to number of individuals in the population **do**
	Get the next generation P(t + 1)
	end for
	t=t+1
	until t>algebra K
	return Final loading result

obtained by repeated iteration without satisfying the termination condition. The pseudo-code of the algorithm is shown in Table 2.

6 Simulation Experiment

In this paper, according to the proposed model and algorithm, simulation experiments are carried out by using C++. Computer configured as Intel(R) Core(TM) i5-8259U CPU @2.30 Hz, memory 8 GB, and the operating system is Windows 10 professional. Three groups of data are randomly generated, as shown in Table 3. Among them, the first group contains fewer different types of cargos and the number of cargos is moderate; the second group contains fewer different types of cargos, but the number of each kind of cargos is more; the third group contains more different kinds of cargos, but the number of each kind of cargos is less.

Table 3. Experimental data

	Container		Cargo					
	Size	Maximum load	Name	Length	Width	Height	Count	Weight
Group I	65 * 75 * 200	1000	A	25	26	20	40	6
			B	19	29	17	40	7
Group II	90 * 80 * 170	1200	C	20	4	15	120	2
			D	17	17	17	120	1
			E	12	14	5	120	2
			F	16	5	8	120	1
			G	17	10	11	120	3
Group III	40 * 40 * 15	500	H	2	7	11	2	4
			I	15	30	3	2	1
			J	4	5	10	2	1
			K	5	5	25	2	3
			L	30	40	4	2	2
			M	20	28	2	2	2
			N	10	10	10	2	2
			O	7	5	8	2	3
			P	5	10	8	2	3
			Q	9	10	10	2	3

The input parameters and control parameters of the genetic algorithm are determined before the genetic algorithm is solved. Based on past experience and combined with operational experiments, the parameters are determined as follows: population size is 200, crossover rate is 0.9, mutation rate is 0.1, iteration number is 200.

The utilization rates of heuristic algorithm and 3D-HGH algorithm are calculated, and the running time of heuristic algorithm is compared with that of 3D-HGH algorithm, as shown in Table 4. The experimental results show that the heuristic algorithm can obtain an effective loading scheme. Based on the heuristic algorithm, the 3d-hgh algorithm combined with the genetic algorithm improves the utilization in three groups of different types and quantities of cargo loading. Finally, the three-assembly scheme is obtained by 3D-HGH algorithm, as shown in Fig. 7.

Table 4. Simulation results

	Heuristic algorithm			3D-HGH algorithm		
	Utilization	Average time (s)	Remaining cargos	Utilization	Average time (s)	Remaining cargos
Group I	89.47%	5.507	A:1 B:1	90.80%	4.921	B:1
Group II	91.97%	26.329	D:2	92.77%	24.248	None
Group III	86.07%	1.136	N:1	90.24%	0.867	None

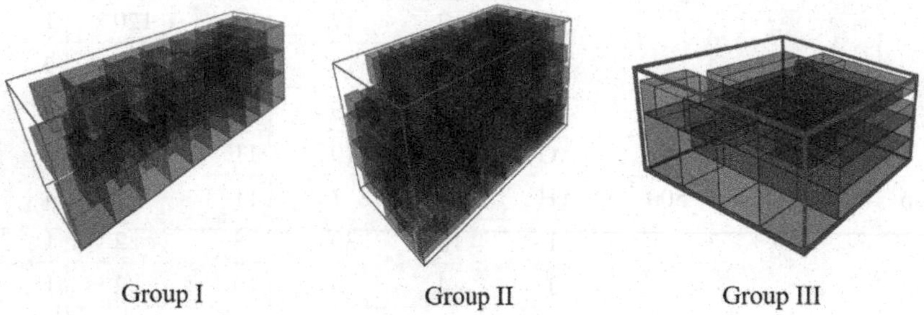

Fig. 7. Packing diagram

7 Conclusion and Future Works

The packing problem is a typical NP problem, which exists in the field of logistics and even in industrial production. According to the actual situation, this paper establishes a multi-objective mathematical model with complex constraints for the three-dimensional packing problem of single container, and combines genetic algorithm with heuristic packing algorithm to design a feasible loading scheme. The algorithm is tested by some examples and satisfactory results are obtained.

At present, it is only better to solve the packing problem of single container, and allow the cargo to have surplus when establishing the model. The next step is to solve more complex packing problems, such as multi-container loading problem, that is, loading all cargo into containers and reducing the number of containers as much as possible.

Acknowledgements. This work was supported by the National Natural Science Foundation of China (Project Number: 71472081).

References

1. Zhang, G., Yao, E.: The FFD algorithm for the bin packing problem with kernel items. Appl. Math.- J. Chin. Univ. Ser. B **13**(3), 335–340 (1998)
2. Alvim, A., Ribeiro, C., Glover, F., Aloise, D.: A hybrid improvement heuristic for the one-dimensional bin packing problem. J. Heuristics **10**(2), 205–229 (2004)
3. Loh, K., Golden, B., Wasil, E.: Solving the one-dimensional bin packing problem with a weight annealing heuristic. Comput. Oper. Res. **35**(7), 2283–2291 (2006)
4. Claudio, A., Fabrizio, M.: Maximum lateness minimization in one-dimensional bin packing. Omega-Int. J. Manag. Sci. **68**, 76–84 (2017)
5. Alvelos, F., Chan, T., Vilaca, P., Gomes, T., Silva, E., De Carvalho, J.: Sequence based heuristics for two-dimensional bin packing problems. Eng. Optim. **41**(8), 773–791 (2009)
6. Aringhieri, R., Duma, D., Grosso, A., Hosteins, P.: Simple but effective heuristics for the 2-constraint bin packing problem. J. Heuristics **24**(3), 345–357 (2018)
7. Lodi, A., Monaci, M., Pietrobuoni, E.: Partial enumeration algorithms for two-dimensional bin packing problem with guillotine constraints. Discrete Appl. Math. **217**, 40–47 (2017)
8. Martinezsykora, A., Alvarezvaldes, R., Bennell, J., Ruiz, R., Tamarit, J.: Matheuristics for the irregular bin packing problem with free rotations. Eur. J. Oper. Res. **258**(2), 440–455 (2017)
9. Lodi, A., Martello, S., Vigo, D.: Heuristic algorithms for the three-dimensional bin packing problem. Eur. J. Oper. Res. **141**(2), 410–420 (2002)
10. Silva, J., Soma, N., Maculan, N.: A greedy search for the three-dimensional bin packing problem: the packing static stability case. Int. Trans. Oper. Res. **10**(2), 141–153 (2003)
11. Hifi, M., Negre, S., Wu, L.: Hybrid greedy heuristics based on linear programming for the three-dimensional single bin-size bin packing problem. Int. Trans. Oper. Res. **21**(1), 59–79 (2014)
12. Kang, J., Park, S.: Algorithms for the variable sized bin packing problem. Eur. J. Oper. Res. **147**(2), 365–372 (2003)
13. Wu, H., Leung, S., Si, Y., Zhang, D., Lin, A.: Three-stage heuristic algorithm for three-dimensional irregular packing problem. Appl. Math. Modell. **41**, 431–444 (2017)
14. Paquay, C., Limbourg, S., Schyns, M., Oliveira, J.: MIP-based constructive heuristics for the three-dimensional bin packing problem with transportation constraints. Int. J. Prod. Res. **56**(4), 1581–1592 (2018)

A Hybrid Ant Colony Optimization Algorithm for the Fleet Size and Mix Vehicle Routing Problem with Time Windows

Xiaodong Zhu(✉) and Ding Wang(✉)

School of Electrical Engineering, Zhengzhou University, Zhengzhou 450001, China
zhu_xd@zzu.edu.com, wangding94@foxmail.com

Abstract. The fleet size and mix vehicle routing problem with time window (FSMVRPTW) is a combinatorial optimization and decision making problem. This problem requires the use of a fleet to provide services to customers at a minimum cost. In this paper, a hybrid ant colony algorithm for FSMVRPTW is proposed, which is composed of an insertion heuristic algorithm and an ant colony system. The ant colony optimization algorithm is used to generate an initial solution with the constraints of maximum capacity and time windows, and then the routing in the initial solution is divided into partial routing and individual customers. The insertion heuristic algorithm is used to reconstruct the solution by taking into acco unt the factors of time windows, distance, utilization and vehicle cost. Experiments on benchmark problems prove the feasibility of the algorithm. The hybrid algorithm expands the application of ant colony optimization algorithm and provides a new idea for solving FSMVRPTW.

Keywords: Ant colony optimization algorithm · Vehicle routing problem · Time windows · Heterogeneous fleet

1 Introduction

Vehicle routing problem has been extensively studied since it was proposed. Fleet size and mix vehicle routing problem with time windows (FSMVRPTW) is an important branch of vehicle routing problem. It refers to the use of a group of different types of fleets to deliver services to fixed customers, and to solve the minimum cost required to complete tasks under the conditions of meeting time window constraints and vehicle capacity constraints. Because the heterogeneity of fleet and the constraints of time windows are closer to real life, the research of FSMVRPTW is of great significance. FSMVRPTW has many application scenarios in real life, such as logistics distribution, school bus routing planning and so on. Good routing can not only save transportation costs and improve distribution efficiency, but also improve service quality, bringing lucrative profits to the market.

L. Pan et al. (Eds.): BIC-TA 2019, CCIS 1159, pp. 244–254, 2020.
https://doi.org/10.1007/978-981-15-3425-6_20

The vehicle routing problem is a NP-hard problem, and it is difficult to find the exact solution, so many heuristic algorithms are produced for solving it. Golden [1] described several effective methods to solve FSMVRP, and provides some benchmark problems for future research. Liu and Shen [2] summarized the previous algorithms of FSMVRP and VRPTW. They drew a conclusion that the traditional sequential routing construction method for VRPTW is not suitable for heterogeneous fleet problems, while using parallel routing construction method to solve FSMVRPTW itself is a difficult task. Several insertion-based saving heuristic algorithms are also proposed. Mauro [3] presented a constructive heuristic algorithm and a meta-heuristic algorithm to solve FSMVRPTW. A constructive heuristic algorithm is implemented by first generates an incomplete solution as the initial solution, and then uses parallel insertion algorithm to insert the remaining nodes. Based on the construction of heuristic algorithm, meta-heuristic algorithm uses ruin strategy to eliminate part of the routing and rebuild procedure to reconstruct the routing. Olli [4] presented a deterministic annealing heuristic algorithm. The algorithm consists of three stages. In the first stage, the initial solution is generated by combining diversification strategy with learning mechanism. In the second stage, local search process is used to reduce the number of routes in the initial solution. In the third stage, the solution is further improved by a set of four local search operations embedded in the deterministic annealing framework. Repoussis and Tarantilis [5] obtained a new adaptive memory programming (AMP) method for solving FSMVRPTW. This method consists of a probabilistic semi-parallel construction heuristic algorithm, a new solution reconstruction mechanism, an innovative iterative tabu search algorithm for enhancing local search and a frequency-based long-term storage structure. Vidal [6] proposed a hybrid genetic search algorithm with advanced diversity control for a class of vehicle routing problems with large time constraints. The algorithm achieves satisfactory results in dealing with various complex vehicle routing problems including time windows.

Ant colony optimization algorithm [7] is a meta-heuristic algorithm which simulates ants'foraging behavior in nature. It has unique advantages in solving routing optimization problems. Ant colony algorithm has been widely studied and applied in solving TSP [8,9], VRP [10,11], VRPTW [12,13], etc. But it has less research in FSMVRPTW. The main reason is that ants can choose the next node according to pheromone concentration, but there is no better solution to the choice of vehicle type. If the vehicle type is taken as a variable associated with the node coordinates, the search dimension will be improved, the search space will be greatly expanded, and the search efficiency will be reduced. If the vehicle type is chosen randomly when constructing feasible routing, the algorithm has some shortcomings such as inadequate searching ability and inability to find the optimal solution for cases with more vehicle types. This paper proposes a hybrid ant colony algorithm, which combines ant colony system and constructive heuristic algorithm. It can not only effectively help ants choose the right vehicle type in multi-vehicle environment, but also give full play to the advantages of ant colony algorithm with strong search ability.

This paper mainly includes the following sections: In Sect. 2, the FSMVRPTW is formulated; The Sect. 3 introduces an insertion heuristic algorithm; The Sect. 4 introduces ant colony system and hybrid algorithm combining the insertion heuristic algorithm; The Sect. 5 is numerical experiment, and the last part is the conclusion.

2 Mathematical Model of FSMVRPTW

2.1 Symbolic Description

The relevant symbols in the mathematical model are illustrated as follows:

$V = N \bigcup 0$: The set of all customers. Where $N = 1, 2, \cdots, n$ is the set of customers and 0 is the distribution center.
$A = \{(i, j)|i, j \in V\}$: The arc set.
d_{ij}: The distance of arc (i, j).
q_i: Quantity demanded of i.
s_i: Service time required by customer i.
$[a_i, b_i]$: Time window for customer i to receive services.
H: The set of different vehicle types for service.
$K = \{1, \cdots, k, \cdots\}$: The set of distinct vehicles obtained by defining n vehicles of type h for each $h \in H$.
Q_k: The capacity of vehicle k.
F_k: The fix-cost of vehicle k.
t_{ik}: The time when vehicle k arrives at customer i.
τ_i: The time when vehicle starts serving customer i.
π_k: The return time of vehicle k to the distribution center.

$$x_{ijk} = \begin{cases} 1, & if\ vehicle\ k\ passed\ through\ arc\ (i, j) \\ 0, & otherwise \end{cases}$$

$$y_{ik} = \begin{cases} 1, & if\ i\ is\ served\ by\ vehicle\ k \\ 0, & otherwise \end{cases}$$

$$z_k = \begin{cases} 1, & if\ vehicle\ k\ is\ used \\ 0, & otherwise \end{cases}$$

2.2 Mathematical Formulation

The mathematical model of FSMVRPTW can be expressed as follows:

$$\min Cost = \sum_k F_k + \sum_{i,j \in A} \sum_k d_{ij} x_{ijk} \tag{1}$$

Subject to :

$$\sum_{i \in V} d_i y_{ki} \le Q_k,\ \forall k \in K \tag{2}$$

$$\sum_{i \in V} y_{ki} = 1,\ \forall k \in K \tag{3}$$

$$\sum_{i \in V \setminus \{j\}} x_{ijk} = y_{kj},\ \forall arc(i,j) \in A, \forall k \in K \tag{4}$$

$$\sum_{j \in V \setminus \{j\}} x_{ijk} = y_{kj},\ \forall arc(i,j) \in A, \forall k \in K \tag{5}$$

$$\sum_{i \in N} x_{i0k} = \sum_{j \in N} x_{0jk} = 1,\ \forall k \in K \tag{6}$$

$$\tau_i = \max\{a_i, t_{ik}\},\ \forall i \in V,\ \forall k \in K \tag{7}$$

$$t_{jk} = \tau_i + s_i + d_{ij},\ \forall arc(i,j) \in A \tag{8}$$

$$t_{ik} < b_i,\ \forall i \in V \tag{9}$$

The objective function is to minimize the sum of vehicle fix-costs and routes distances. Constraint 2 means that the vehicle is not allowed to exceed its maximum capacity. Constraint 3 means that each customer can only be accessed once. Constraint 4 and 5 ensure that no loops are generated within the routing. Constraint 6 indicates that the starting and ending position of the vehicle must be the distribution center, and Constraints 7 to 9 are time window constraints.

3 Insertion Heuristic Algorithm

The purpose of insertion heuristic algorithm is to insert some unvisited customers into the current routes to form feasible solutions. The algorithm inserts each customer into the current routing and scores it, and inserts the customer with the highest score into the corresponding location to form a new routing until all customers are inserted.

3.1 The Cost Value of Insertion Customers

Let r be the current partial solution and $\bar{n}$ is the number of customer in it. U denotes the set of not visited customers. $g(u, r, p)$ is the time when the vehicle arrives at u, where $u \in U$ denote an insert customer and p is a insertion position of r. If r still satisfies the time window constraint after u insertion, then the insertion point p is called a feasible insertion position for u. Set $\Phi(u, r)$ as a set of feasible insertion positions when customer u is inserted into route r. Then we can calculate the cost value of each feasible insertion position for each insertion customer as formula 10.

$$\begin{aligned}\xi(u,r) = & \min_{p \in \Phi(u,r)} \{(d_{i_p u} + d_{u i_{p+1}} - d_{i_p i_{p+1}}) \\ & + \frac{\bar{n}}{n} Chg\_V(u,r) + \frac{\bar{n}}{n} \max(0, a_u - g(u,r,p))\}\end{aligned} \tag{10}$$

In formula 10, $\xi(u, r)$ is the weighted sum of three elements: (i) Regardless of vehicle capacity limitation, the increment of routing distance when customer u is inserted; (ii) The additional cost denoted as $Chg\_V(u, r)$ (see below) to be paid when the vehicle serving route r must be changed to the route which also visit customer u; (iii) The waiting time of the vehicle before the service customer u. The latter two terms are taken as coefficients by the ratio of $\bar{n}$ to n in order to increase the weight of the two terms as the insertion heuristic algorithm proceeds. The first item has no weight coefficient because the distance relation of three points satisfies the triangular inequality.

To formally describe cost $Chg\_V(u, r)$, let h be the type of vehicle serving route r and let h' be the cheapest type of vehicle we must use when u is added to the route. The cost is,

$$Chg\_V(u, r) = \left(1 - \frac{\bar{n}}{n}\right) C_R + \frac{\bar{n}}{n} C_A \tag{11}$$

Where C_R is the relative fixed cost increase.

$$C_R = F^{h'} \sum_{i \in r \cup \{u\}} \frac{q_i}{Q^{h'}} - F^h \sum_{i \in r} \frac{q_i}{Q^h} \tag{12}$$

And C_A is the absolute fixed cost increase.

$$C_A = F^{h'} - F^h \tag{13}$$

When calculating Chg_V, the ratio of $\bar{n}$ to n is used to adjust the weight of C_R and C_A. In most cases, the evaluation value of Chg_V is mainly determined by C_A, because insertion customers will increase the vehicle tolerance, so it is necessary to reselect the vehicle type. In a few cases, the insertion of customers will not change the vehicle type, and the Chg_V is mainly determined by C_R.

3.2 The Fitness Value of Inserted Customers

When the cost score of all insertion positions of each customer is obtained, then we use the fitness function to calculate the score of each customer, and select the next customer according to the fitness score and insert it into the corresponding position of the route. The fitness function can be shown as

$$\theta(u) = (\text{s}\min \xi(u, r) - \min \xi(u, r)) + \gamma_d d_{0u} - \gamma_w (b_u - a_u) \tag{14}$$

Where smin denotes the second minimum value; γ_d and γ_w are two weighing parameters, and this paper set $\gamma_d = 1.0$, $\gamma_w = 0.2$ according to Mauro [3]. $\theta(u)$ consists of three parts: (i) the added cost value when the customer u is not assigned to its preferred route; (ii) the distance between u and the distribution center; (iii) the time window width of u.

3.3 The Process of Insertion Heuristic Algorithm

The insertion heuristic algorithm can be described as follows:

Step 1: Input: partial solution r; unvisited customer U; parameters;
Step 2: Choose an unvisited customer u from U;
Step 3: Compute the score $\xi(u, r)$ [see 10] define $r(u)$ and $r(u)$ as the routes corresponding to the first and second minimum of $\xi(u, r)$;
Step 4: Compute the score $\theta(u)$ [see 14]; determine the customer $\bar{u}$ with maximum $\theta(u)$ value;
Step 5: If u can not form a feasible solution by inserting into the route, or the fixed cost of the fleet can be reduced if u is added to a new empty route, initialize a new route with customer $\bar{u}$. Else insert $\bar{u}$ into the position with the minimum $\xi(u, r)$.
Step 6: Update the solution r and set $U = U\backslash\{u\}$;
Step 7: Return to step 2 if U is not empty. Otherwise, the route r containing all the customers is output and the algorithm is terminated.

4 Hybrid Ant Colony Algorithm

In this paper, a hybrid ant colony algorithm is proposed. Firstly, the feasible solution is constructed using the same vehicle type (similar to VRPTW), and then the feasible solution is reconstructed using the insertion heuristic algorithm. The hybrid ant colony algorithm will be described in detail as follows.

4.1 The Preliminary Construction of Feasible Solutions

This paper uses the strategy of peripheral search to improve the search efficiency of ants in large-scale customers. Firstly, feasible customers are selected according to time window constraints and capacity constraints, and then several customers nearest to the current customers are selected as candidate customers set (Ω) from the set of feasible customers. As shown in the Fig. 1.

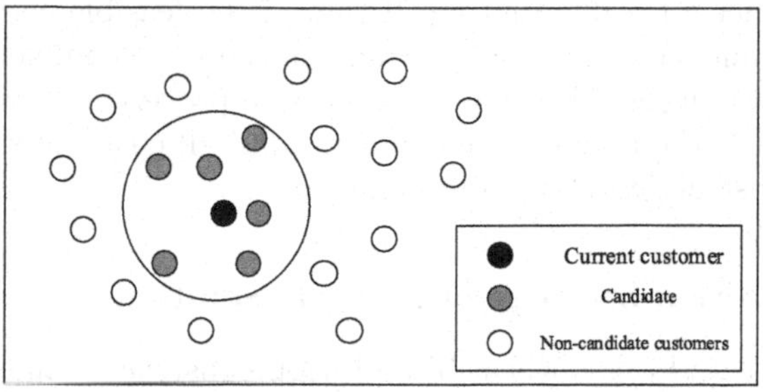

Fig. 1. Selection of candidate customers.

Assuming that the current customer is i, the transition probability of ant from customer i to customer j can be expressed as

$$p_{ij} = \begin{cases} \frac{S_{ij} \times \tau_{ij}^{\alpha} \times \eta_{ij}^{\beta}}{\sum\limits_{j \in \Omega} S_{ij} \times \tau_{ij}^{\alpha} \times \eta_{ij}^{\beta}}, j \in \Omega \\ 0, otherwise \end{cases} \tag{15}$$

Where p_{ij} is transition probability of ant from customer i to j; $S_{ij} = d_{0i} + d_{0j} - d_{ij}$ is the saving value of i and j; τ_{ij} is pheromone concentration between i and j; $\eta_{ij} = \frac{1}{d_{ij}}$ is visibility of pheromone concentration; α, β are parameters.

The next customer j is chosen according to the following formula.

$$j = \begin{cases} \arg\max\limits_{j \in \Omega} (p_{ij}), \ if \ q > p_t \\ randomj \in \Omega, \ otherwise \end{cases} \tag{16}$$

Where q is a value chosen randomly with uniform probability in the interval [0,1]. p_t $(0 < p_t \leq 1)$ interpreted as the definite selection probabilities is initiated as $p_0 = 1$ and is dynamically adjusted with the evolutionary process. The value of p_t is as follows

$$p_t = \begin{cases} 0.95 p_{t-1}, \ if \ 0.95 p_{t-1} > p_{\min} \\ p_{\min}, \ otherwise \end{cases} \tag{17}$$

Where $p_{(min)}$ is the minimum defined in evolutionary process. It is used to ensure that certain selection opportunities can be obtained even if the value of p_t is too small. Formula 15 shows that in the early stage of iteration, j is selected randomly from the candidate customers according to the transition probability. With the iteration, j has a certain probability to be directly the customer with the largest transfer probability among the candidate customers. This selection strategy can accelerate the convergence of the algorithm and expand the search space at the same time.

When the feasible solution is preliminarily constructed by using ant colony algorithm, the vehicle capacity value is first set to the maximum type of capacity. Each ant starts from the distribution center and randomly chooses the next customer from the candidate customer set according to the state transition probability to construct a feasible routing. If there is no feasible customer added to the routing, the ant returns to the distribution center to complete the construction of a feasible routing. Then repeat the above process and continue to build a viable route from the remaining customers until all customers are accessed to complete the construction of a viable solution.

4.2 Decomposition Strategy of Feasible Solutions

Let $R = \{R_1, R_2, ..., R_m\}$ is a feasible solution constructed initially, where R_a $(1 \leq a \leq m)$ is a feasible route. Since the solutions are all constructed according to the maximum vehicle load, they are not an appropriate solution for FSMVRPTW, so they need to be disassembled and reconstructed.

The purpose of the decomposition strategy is to decompose R into a routing set r containing partial customers and an unvisited customer set U. For each feasible route R_a, firstly, the actual load of the current feasible route l_a is calculated. And then the customers exceeding the maximum vehicle capacity of less than l_a in R_a are removed. The remaining route is r_a which turn into a subset of r, and the removed customers are added to the set U. In particular, when l_a is smaller than the minimum vehicle type capacity, all customers in R_a will be removed and added to U.

4.3 The Process of Hybrid Ant Colony Optimization

The steps of the hybrid ant colony optimization algorithm for solving the FSMVRPTW model can be described as follows:

Step 1: Input the number of ant m, the maximum iteration number $maxit$ and algorithmic parameters; initialize the current iteration number $it = 1$;
Step 2: Initialize the ant number $n = 1$;
Step 3: Preliminary construct a feasible solution of ant n using 4.1 algorithm, and the initial solution R^n is obtained.
Step 4: R^n is decomposed into r^n and U^n according to 4.2 algorithm. Where r^n is a partial solution of ant n and U^n is insert customer of ant n.
Step 5: The insertion heuristic algorithm in part 3 is used to insert the customers in U^n into r^n. After the reconstructed solution is obtained, the *cost* n of the solution is calculated according to formula 1.
Step 6: $n = n + 1$. If $n > m$, that is all ants have completed the solution construction, turn to step 7. Otherwise, return to step 2.
Step 7: Update the global pheromone as follows

$$\tau_{ij}^{new} = (1 - \rho) * \tau_{ij}^{old} + \Delta\tau_{ij} \tag{18}$$

$$\Delta\tau_{ij} = \begin{cases} \sum\limits_{n=1}^{m} \frac{Q}{\cos t(n)}, & if\ arc(i,j) \in r^n \\ 0, & otherwise \end{cases} \tag{19}$$

Where $\rho\,(0 < \rho < 1)$ is a constant representing the volatilization rate of pherom-ones; $\Delta\tau_{ij}$ is pheromone accumulation between customer i and j; Q is a constant, indicating the total amount of pheromones released by each ant. Because the pheromone superposition in ant colony optimization algorithm is a positive feedback process, in order to avoid stagnation of search, the maximum and minimum ant colony algorithm is adopted to limit the concentration of pheromone to $[\tau_{\min}, \tau_{\max}]$, as shown in formula 20.

$$\tau_{ij} = \begin{cases} \tau_{\max}, & if \tau_{ij} > \tau_{\max} \\ \tau_{\min}, & if \tau_{ij} < \tau_{\min} \end{cases} \tag{20}$$

Step 8: $it = it + 1$. If it is bigger than $maxit$, then the flow of the algorithm finishes. Otherwise, go back to Step 2, and repeat the steps.

5 Numerical Analysis

Solomon benchmark problems [14] is often used in the experiment of VRPTW algorithm. It has 56 problems in 6 categories: R1, R2, C1, C2, RC1 and RC2, each of which contains 100 customers. These 6 categories of problems can be divided into random distribution (R), cluster distribution (C) and mixed distribution (RC) according to customer distribution. The time windows of customer in R1, C1 and RC1 problems are narrower, while those in R2, C2 and RC2 problems are wider. For each instance of Solomon, Liu and Shen [2] introduced three different subclasses of instances, which are characterized by type a, type b and type c three different vehicle costs. In which type a vehicle costs are high, type b vehicle costs are medium, and type c vehicle costs are low. Combining each VRPTW instance with the cost of three types of vehicles, a total of 168 examples are obtained.

Table 1. Comparison of ACO and HACO in solving benchmark problems.

Problems	Type a			Type b			Type c		
	HACO	ACO	TS	HACO	ACO	TS	HACO	ACO	TS
R102	**4656.2**	5002.3	4762.4	**2190.3**	2421.4	2202.5	**1812.5**	2030.1	1989.4
R204	**3964.3**	4716.4	4283.3	**2211.8**	2876.1	2921.2	**1751.9**	2218.7	1903.7
C101	**8332.7**	9930.1	8549.6	**2649.2**	3248.3	2849.3	**1910.9**	2413.8	2371.5
C205	**7158.6**	8603.4	7934.6	**2084.2**	2513.4	2310.8	**1583.7**	2301.5	1863.4
RC103	**5673.4**	6057.1	5764.3	**2423.5**	2714.2	2692.1	**2003.2**	2551.3	2657.8
RC206	5831.6	6786.1	**5591.4**	**2438.2**	2846.1	2489.4	**2306.6**	2801.4	2841.6

In this paper, hybrid ant colony optimization algorithm (HACO), ant colony optimization algorithm (ACO) and tabu search (TS) are used to solve and compare the benchmark problems. The algorithm is programmed with MATLAB 2016a and runs on the personal computer of Intel Core i5-5200 CPU, 4G memory. The experimental parameters are as follows: the number of ants $m = 15$, the maximum number of iterations $maxit = 50$, $\rho = 0.08$, $\alpha = 1.7$, $\beta = 2$ and $Q = 80$. One instances are selected from each type of benchmark problems and running five times to and calculating average value as the result. The experimental results of HACO, ACO and TS are shown in Table 1.

Table 2. Average and comparison of problem results.

Problem type	ACO	HACO	$\Delta(\%)$	TS	$\Delta(\%)$
Type a	6849.2	5936.1	15.3	6147.6	3.6
Type b	2769.9	2332.8	18.7	2577.6	10.5
Tyoe c	2386.1	1894.8	25.9	2271.2	19.8

To illustrate the comparison of the results, we calculate the average values of various types of problems and average percentage error (label "$\Delta\%$"). As shown in Table 2.

From Tables 1 and 2, we can see that all the results of HACO are better than ACO algorithm. The average value of HACO in type a is better than ACO 15.3%, better than TS 3.6%, in type b is better than ACO 18.7%, better than TS 10.5% and in type c is better than ACO 25.9%, better than TS 19.8%, which shows that the proposed HACO is feasible and improves the quality of solving multi-vehicle problem.

Figure 2 shows the convergence of ACO, HACO and TS in solving C101b. Both results are run five times and averaged for each generation.

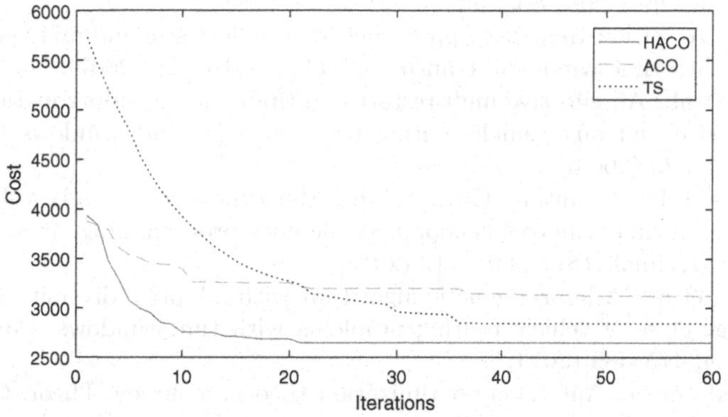

Fig. 2. The convergence of ACO, HACO and TS in solving C101b

From Fig. 2, we can see that with the progress of iteration, HACO converges faster and costs less, which shows that the hybrid algorithm proposed in this paper effectively improves the convergence speed of traditional ant colony optimization algorithm. The above results also show that the hybrid strategy proposed in this paper has the advantages of fast convergence and high solution quality. HACO effectively improves the ability of ant colony optimization algorithm to solve heterogeneous fleet problems, and provides a new idea for using ant colony algorithm to solve FSMVRPTW.

6 Conclusion

FSMVRPTW is closely related to real life. Because it is difficult to obtain exact solutions, a hybrid optimization algorithm combining ant colony algorithm and insertion heuristic algorithm is proposed in this paper. Firstly, a feasible solution is constructed by using a single type of vehicle, and then a new solution is generated by reconstructing the feasible solution through decomposition strategy and insertion strategy. Ant colony system has the advantages of strong global search ability and robustness in solving routing problems. The insertion heuristic algorithm solves the problem that ants can not select the appropriate vehicle type before constructing feasible routing. Hybrid ant colony optimization algorithm combines the advantages of ant colony algorithm and insertion heuristic algorithm, expands the application scope of ant colony algorithm, and provides a new way to solve FSMVRPTW.

References

1. Bruce, G.: The fleet size and mix vehicle routing problem. Comput. Oper. Res. **11**(1), 49–66 (1984)
2. Liu, F.H., Shen, S.Y.: The fleet size and mix vehicle routing problem with time windows. J. Oper. Res. Soc. **50**(7), 721–732 (1999)
3. Mauro, D.A., et al.: Heuristic approaches for the fleet size and mix vehicle routing problem with time windows. Transp. Sci. **41**(4), 516–526 (2007)
4. Olli, B., et al.: An effective multirestart deterministic annealing metaheuristic for the fleet size and mix vehicle-routing problem with time windows. Transp. Sci. **42**(3), 371–386 (2008)
5. Repoussis, P.P., Tarantilis, C.D.: Solving the fleet size and mix vehicle routing problem with time windows via adaptive memory programming. Transp. Res. Part C Emerg. Technol. **18**(5), 695–712 (2010)
6. Vidal, T., et al.: A hybrid genetic algorithm with adaptive diversity management for a large class of vehicle routing problems with time-windows. Comput. Oper. Res. **40**(1), 475–489 (2013)
7. Dorigo, M., et al.: Ant colony optimization theory: a survey. Theor. Comput. Sci. **344**(2–3), 243–278 (2005)
8. Yun, B., Li, T.Q., Zhang, Q.: A weighted max-min ant colony algorithm for TSP instances. IEICE Trans. Fund. **98**(3), 894–897 (2015)
9. Zhou, Y.: Runtime analysis of an ant colony optimization algorithm for TSP instances. IEEE Trans. Evol. Comput. **13**(5), 1083–1092 (2009)
10. Zhang, X.X., Tang, L.X.: A new hybrid ant colony optimization algorithm for the vehicle routing problem. Pattern Recogn. Lett. **30**(9), 848–855 (2009)
11. Mazzeo, S., Loiseau, I.: An ant colony algorithm for the capacitated vehicle routing. Electr. Notes Discrete Math. **18**, 181–186 (2004)
12. Deng, Y., et al.: Multi-type ant system algorithm for the time dependent vehicle routing problem with time windows. J. Syst. Eng. Electron. **29**(3), 625–638 (2018)
13. Zhang, H.Z., Zhang, Q.W.: A hybrid ant colony optimization algorithm for a multi-objective vehicle routing problem with flexible time windows. Inf. Sci. **490**, 166–190 (2019)
14. Solomon, M.M.: Algorithms for the vehicle routing and scheduling problems with time window constraints. Oper. Res. **35**(2), 254–265 (1987)

Multi-subpopulation Algorithm with Ensemble Mutation Strategies for Protein Structure Prediction

Chunxiang Peng[1], Xiaogen Zhou[2], and Guijun Zhang[1(✉)]

[1] College of Information Engineering, Zhejiang University of Technology, Hangzhou 310023, China
{pengcx,zgj}@zjut.edu.cn

[2] Department of Computational Medicine and Bioinformatics, University of Michigan, Ann Arbor, MI 48109, USA
xiaogenz@umich.edu

Abstract. One of the most challenging problems in protein structure prediction (PSP) is the ability of the sampling low energy conformations. In this paper, a multi-subpopulation algorithm with ensemble mutation strategies (MSEMS) is proposed, based on the framework of differential evolution (DE). In proposed MSEMS, multiple subpopulations with different mutation strategy pools are applied to balance the exploitation and exploration capability. One of the subpopulations is utilized to reward the one with the best performance in the other three subpopulations. Meanwhile, a distance-based fitness score is designed to alleviate inaccuracy of low-resolution energy model. The experimental results indicate that MSEMS has the potential to improve the accuracy of protein structure prediction.

Keywords: Protein structure prediction · Multi-subpopulation · Distance profile · Differential evolution

1 Introduction

Proteins are one of the essential components of cells, and their three-dimensional structure determines specific biological function, which make possible most of the key processes associated with life functions. Therefore, known the protein structure is the first step in understanding the biological functions of these proteins [1]. For instance, understanding the relationship between protein structure and function can help design novel proteins with distinctive function and help design drugs and vaccines. Some experimental approaches, such as X-ray and NMR (nuclear magnetic resonance), have been used to determine protein structures in the laboratory, but these methods are extremely time-consuming and expensive [2]. In many cases, different experimental protocols need to be

Supported by the National Nature Science Foundation of China (No. 61773346).

L. Pan et al. (Eds.): BIC-TA 2019, CCIS 1159, pp. 255–268, 2020.
https://doi.org/10.1007/978-981-15-3425-6_21

designed according to different proteins. The relative ease of determining the sequence information has led a growing gap between the known sequence and the determining structure of the protein [3]. However, computational approaches have become the cornerstone of protein structure analysis.

Predicting the three-dimensional structure of protein from the amino acid sequence of protein, is one of the most interesting challenges of modern computational biology, which is often referred to as a protein structure prediction problem [4]. According to the thermodynamic hypothesis [5], native structures are usually in the lowest energy state. PSP can be decomposed into two subproblems: (1) to define an appropriate energy function which guides the conformation to fold into native state, and (2) to develop an efficient and robust search strategy. For the second subproblem, it can be considered as a problem with energy minimization [6]. In order to find the unique native structure of protein in the huge sampling space by computer, efficient conformation space optimization algorithm must be designed and converted into practical computational problems. In recent years, different computational methods have been proposed with a view to tackling this problem, such as Molecular Dynamics (MD), Monte Carlo (MC), Evolution Algorithm (EA) and so on.

However, these approaches have limitations. MD is based on Newtonian mechanics, where the motion trajectory of all atoms is obtained by solving motion equations to minimize their energy [7]. However, the MD method requires a large amount of computation, which is obviously a great challenge at the current level of computer development. Therefore, at the present stage, MD method can only be used to simulate the folding process of small proteins and to refine the structure of low-resolution protein models [8]. MC is widely used in the field of de novo protein structure prediction, but the protein conformation energy hypersurface is extremely complex. MC sampling is often trapped in a local energy minimum state during conformational search [9]. Evolution Algorithms (EAs), inspired by the natural evolution of species, have been successfully applied to numerous optimization problems in various fields [10,11]. In EA, the population is first initialized. During each iteration, mutation and crossover operations are used to generate offspring individuals. The quality of the offspring individual is evaluated according to the fitness function, and the selection process is used to select a better individual to update the population [12].

Differential evolution (DE), as one of the most powerful EAs for global optimization over the continuous search space, is proposed by Storn and Price [13]. DE has been proven to have many advantages, such as ease of use, speed, simple structure, and robustness. These advantages make DE attractive to applications of diverse real-world optimization problems [14], such as communication, power systems [15], chemical [16], bioinformatics [17], and engineering design [18]. DE uses cooperation and competition among individuals to evolve the population to the global optimal solution. Amongst these operators, the mutation operator, which helps explore the search space by perturbing individuals, substantially influences the performance of DE [19]. Different mutation strategies have been proposed to enhance DE search performance [20], such as ranking-based [21], centroid-based [22] and so on. However, different mutation strategies are suitable for dealing with different problems. For example, ranking-based mutation strategies are suitable for exploring search spaces and locating the region of global optimum, but they are slow at exploitation of the solutions, whereas other strategies are the opposite. Thus unsuitable mutation strategy may lead to high computational costs and premature convergence. In order to balance the performance of exploration and exploitation, reachers proposed different approaches which cooperated different mutation strategies, such as the self-adaptive DE (SaDE) [23], DE with strategy adaptation mechanism (SaM) [24], a composite DE (CoDE) [25], DE with local abstract convex underestimate strategy (DELU) [26] and so on. DE algorithm has been widely used in PSP, such as DPDE [27], which used distance profile guided differential evolution algorithm to improve accuracy of protein structure prediction, SCDE [3], which utilized secondary structure and contact guided differential evolution algorithm to predict protein structure.

As the amino acid sequence length of proteins increases, the conformational space of proteins becomes extremely complex, which makes prediction of protein structure based on the full-atom level a challenging problem. In order to sufficiently sample the vast conformational space, low-resolution energy models are widely used in PSP, such as UNRES [28] which is utilized to simplify the conformational space. However, the low-resolution energy model can not sufficiently restrain the long-range interactions between residue pairs, but these interactions provide critical information for PSP. Because of the inherent inaccuracy of low-resolution energy models, it is possible that the low-energy conformation is not a near-native protein structure. For example, some near native conformations with high energy are missed.

In order to alleviate the impact of the inaccuracy of the energy models, Baker's research group [29], used residue-residue distance, contact, and other structure data of proteins to guide conformation to the native structure. Experimental results show that the structure data can improve prediction accuracy. Zhang research group [30] proposed a distance profile consisting of a histogram distribution of residue pairs distances extracted from unrelated experimental structure based on the number of occurrence of fragments at different positions but from same templates.

In this paper, we proposed a multi-subpopulation algorithm with ensemble mutation strategies for de novo protein structure prediction, which can effectively alleviate the shortcomings of existing methods of protein structure prediction, such as low sampling efficiency, poor population diversity and low prediction accuracy. In MSEMS, population is divided into multiple subpopulations, each of which performs different mutation strategies to enhance the ability of algorithm exploration and exploitation. In the sampling process, in order to reward the subpopulation with high acceptance rate, the mutation strategy of the last subpopulation is replaced by the mutation strategy of the subpopulation with high acceptance rate. This can be adaptive to enhance the performance of exploitation and exploration in different period. In the first subpopulation, the mutation strategy, which is good at locating the region of global optimum, is utilized to enhance the performance of exploration. In the second subpopulation, the mutation strategy, which can enhance the convergence speed to find the optimal solution by executing the local search in the promising solution regions, is used to enhance the performance of exploitation. In the third subpopulation, the mutation strategy combines the characteristics of parts the first subpopulation and the second subpopulation. The last subpopulation is a reward subpopulation, which enhances the ability of sampling according to the other three subpopulation. Meanwhile, in order to alleviate the inaccuracy of low-precision energy models, distance-based fitness score is design to improve prediction accuracy. If the energy of trial conformation is higher than that of target, the distance-based fitness score is used to decide whether the trial conformation can replace the target conformation. By using the distance-based fitness score in guiding sampling, these conformations which is reasonable structure with high, have an opportunity to be preserved. By using the multi-subpopulation algorithm with ensemble mutation strategies, the ability of search sampling efficiency is improved. Experimental results indicate that the proposed MSEMS can improve accuracy of PSP.

2 Background Information

2.1 Representation of Protein

It seems natural that each atom of the protein is expressed in Cartesian coordinates. However, with the increase of protein sequence length, the conformational space of protein structure becomes extremely complex. Structural changes maybe produce clash between atoms. To remove these clashes, it would be necessary to use a repair mechanism, which would be computationally very expensive [31]. To reduce the computational cost, an effective coarse-grained model is very important.

Here, we use coarse-grained model represented by heavy backbone atoms, while the side chain is represented by a pseudoatom at its centroid. As shown in Fig. 1, N, C, C_a, and O are heavy backbone atoms. This approach relies on the accurate and fast side-chain packing techniques [32] that are able to pack side chains on promising coarse-grained conformations before refinement. The structure of the peptide chain can be defined by a series of torsion angles, called backbone dihedral angles ϕ, ψ, ω. The dihedral angle formed by the plane of $C - N - C_a$ and the plane of $N - C_a - C$ is called ϕ. The dihedral angle formed by the plane of $N - C_a - C$ and the plane of $C_a - C - N$ is called ψ. The dihedral angle formed by the plane of $C_a - C - N$ and the plane of $C - N - C_a$ is called ω. Each backbone dihedral angle has a range of $[-180°, 180°]$. The protein structure can be represented by a series of backbone torsion angles (ϕ, ψ, ω). The backbone torsion angles (ϕ, ψ, ω) are the sole parameters that are varied during the optimization procedure. It is an efficient representation that reduces the dimensionality of conformational space to $3L$ dimensions for a chain of L amino acids [33].

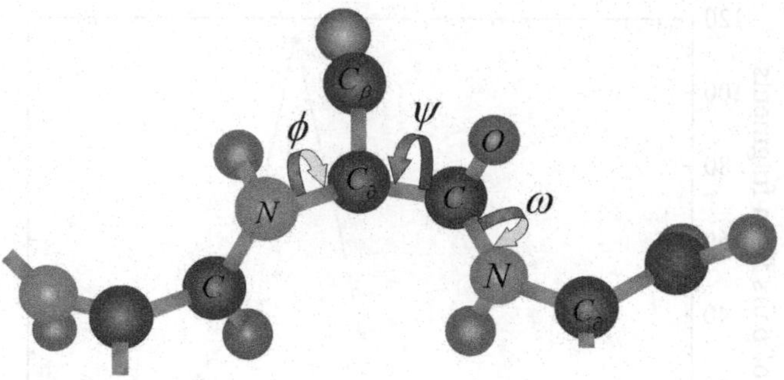

Fig. 1. Coarse-grained representation of a protein chain.

2.2 Energy Function

Low-resolution energy function is used here, corresponding to the *score*3 setting in the suite of energy functions used in the Rosetta de novo protocol and package [34]. The *score*3 is a linear combination of 10 terms that measure repulsion, amino-acid propensities, residue environment, residue pair interactions, interactions between secondary structure elements, density, and compactness. The *score*3 is defined as follows:

$$\begin{aligned} score3 = & W_{\text{repulsion}} E_{\text{repulsion}} + W_{\text{rama}} E_{\text{rama}} + W_{\text{slovation}} E_{\text{slovation}} \\ & + W_{\text{bb-schb}} E_{\text{bb-schb}} + W_{\text{bb-bbhb}} E_{\text{bb-bbhb}} + W_{\text{sc-schb}} E_{\text{sc-schb}} \\ & + W_{\text{pair}} E_{\text{pair}} + W_{\text{dunbrack}} E_{\text{dunbrack}} + W_{\text{reference}} E_{\text{reference}} \\ & + W_{\text{attraction}} E_{\text{attraction}} \end{aligned} \quad (1)$$

where the detailed description of each parameter and energy term can be found in [34]. In PSP, the low-energy conformations are usually considered to be near native structure. Thus, PSP can be considered as an energy minimization problem.

2.3 Distance Profile

The distance profile, proposed by Zhang's research group [30], which aims to derive long-range pair-wise distance and contact restraints from multiple fragments. A distance profile is the histogram of the distance between a residue pair, which is constructed for every residue pair in the query from the fragment pairs aligned with the target residue pair. The distance bin of the histograms is set to 0.5 Å. If the distance of residue pairs in the same template falls in a bin, the total number will add one in the corresponding bin. Distance profile can provide distance distribution between residue pairs, which is important for restraining the structure of proteins. A distance profile of a residue pair shown in Fig. 2, the peak distance is 6 Å, which indicates the distance of the residue pair has a high probability of being around 6 Å.

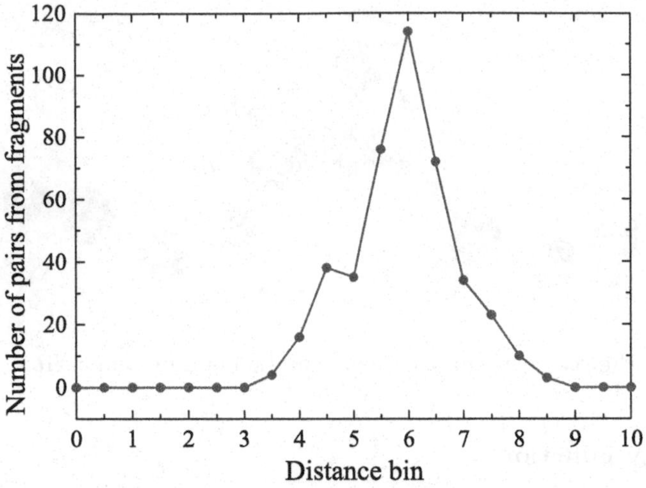

Fig. 2. A distance profile of a residue pair.

2.4 Fragment Assembly

The fragment assembly technique is one of the most successful methods for de novo protein structure prediction. In fragment assembly technique, each fragment insertion window of residues in the target sequence corresponds to a specific set of fragments from a known structural protein. Fragment assembly usually requires three processes: first, a fragment insertion window is randomly selected; second, a fragment for this window is randomly selected from the fragment library and replacing the residues backbone torsion angles of the fragment insertion window with the backbone torsion angles from the selected fragment; third, energy of the new conformation is calculated using an energy scoring function, which is preserved according to the Metropolis criterion [35,36]. The main advantage of fragment assembly is that it can use accurate atomic position in favor of rapidly and coarsely sampling the large conformational space.

3 Method

3.1 Distance-Based Fitness Score

The distance information between protein residue pairs is vital for PSP, but low-resolution energy model Rosetta *score*3 does not reflect this information. In this paper, a distance-based fitness score is designed to restrain the distance between residue pairs, because these residue pairs are assumed to have similar local interaction environment on different protein. In the distance-based fitness score, the peak distance of distance profile is utilized, because the query residue pair has a high probability of being near the peak distance. Meanwhile, the number of residue pairs from fragments are used to further reflect the peak distance information. The distance-based fitness score is given as follows:

$$S(x) = \sum_{(i,j)\in\Omega} \frac{N_{i,j}^{peak}}{\left|x_{i,j} - DP_{i,j}^{peak}\right| + \epsilon} \tag{2}$$

where Ω is the set of residue pairs which have distance profile. $x_{i,j}$ is the distance between the ith and jth C_α atoms in the conformation structure. $DP_{i,j}^{peak}$ is the peak distance between the ith and jth C_α atoms in distance profile. $N_{i,j}^{peak}$ is the number of fragment in residue pair (i,j) peak distance. $\epsilon = 0.001$ is a constant.

There could be structures very similar to the native structure, yet with very high energies, thus distance-based fitness score is designed to assist *score*3. The larger $S(x)$ is, the more reasonable the distance between the conformational residue pairs.

3.2 Algorithm Description

The proposed MSEMS searches low energy conformations based on the framework of DE, and the pipeline of MSEMS is shown in Fig. 3. In MSEMS, for a target amino acid sequence, the fragment library with homologous fragments (sequence identity >30%) removed is firstly generated by the ROBETTA server (http://robetta.bakerlab.org).

The initial population with NP conformations C_p, $p = \{1, 2, ..., NP\}$, is generated by executing random fragment assembly for each position from the corresponding fragment library. Then, the initial population is equally divided into four subpopulations C_m^1, C_n^2, C_k^3, C_t^4. Where $m = \{1, 2, ..., NP/4\}$, $n = \{NP/4+1, ..., NP/2\}$, $k = \{NP/2+1, ..., 3NP/4\}$, $n = \{3NP/4+1, ..., NP\}$.

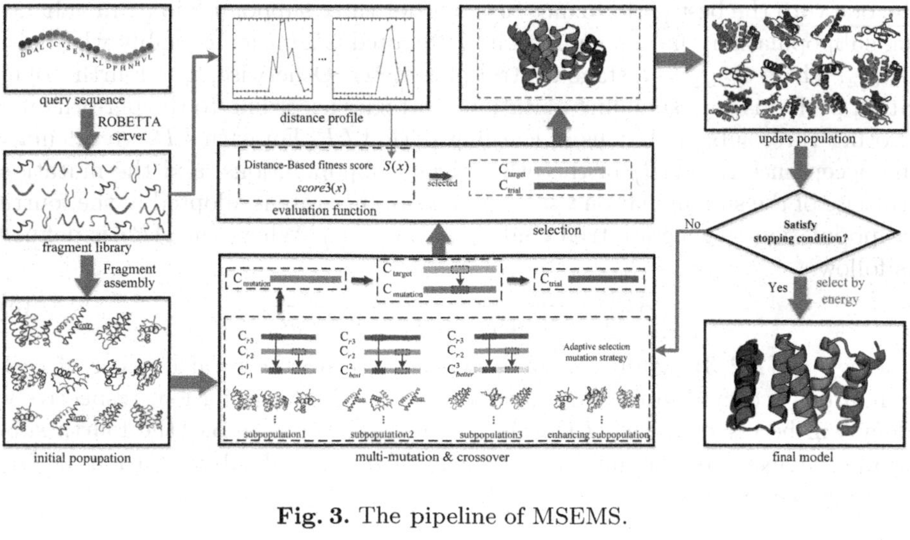

Fig. 3. The pipeline of MSEMS.

In the first subpopulation, the target conformation $C_m^{1,\text{target}}$ is executed the following mutation operation. We randomly select one conformation C_{r1}^1 from the first subpopulation, where $C_{r1}^1 \neq C_m^{1,\text{target}}$, and then randomly select two different conformations C_{r2}, C_{r3} from other three subpopulations, respectively. Two mutually different fragments (length = 9) are selected from C_{r2}, C_{r3}, and then the corresponding positions of C_{r1}^1 are replaced by the two fragments to generate a mutation conformation of the first subpopulation $C_{mutation}^1$. A fragment (length = 3) from $C_m^{1,\text{target}}$ is randomly selected to replace the corresponding position of $C_{mutation}^1$ to generate the trial conformation C_{trial}^1.

In the second subpopulation, the target conformation $C_n^{2,\text{target}}$ is executed the following mutation operation. We select the lowest energy conformation C_{best}^2 from the second subpopulation, and then randomly select two different conformations C_{r2}, C_{r3} from other three subpopulations, respectively. Two mutually different fragments (length = 3) are selected from C_{r2}, C_{r3}, and then the corresponding positions of C_{best}^2 are replaced by the two fragments to generate a mutation conformation of the second subpopulation $C_{mutation}^2$. A fragment (length = 3) from $C_n^{2,\text{target}}$ is randomly selected to replace the corresponding position of $C_{mutation}^2$ to generate the trial conformation C_{trial}^2.

In the third subpopulation, the target conformation $C_k^{3,\text{target}}$ is executed the following mutation operation. We randomly select one conformation C_{better}^3 from half the low energy conformations in the third subpopulation, and then randomly select two different conformations C_{r2}, C_{r3} from other three subpopulations, respectively. Two mutually different fragments (length = 9) are selected from C_{r2}, C_{r3}, and then the corresponding positions of C_{better}^3 are replaced by the two fragments to generate a mutation conformation of the second subpopulation $C_{mutation}^3$. A fragment (length = 3) from $C_k^{3,\text{target}}$ is randomly selected to replace the corresponding positions of $C_{mutation}^3$ to generate the trial conformation C_{trial}^3.

The fourth subpopulation, named as enhancing subpopulation, is used reward the one with the best performance in the other three subpopulations. In enhancing subpopulation, the subpopulation is executed fragments assembly when g < G_{stage}, where G_{stage} is a stage control parameter. Otherwise, The fourth population performs corresponding mutation strategies according to the performance of other three subpopulations in learning period LP. For each LP, we calculate the acceptance rate AR_f of the other three subpopulations, and the mutation strategy of the subpopulation with high acceptance rate is adopted in the fourth subpopulation to generate trial conformation C_{trial}^4. Where the AR_f is defined as follow:

$$AR_f = \frac{count_f}{count_1 + count_2 + count_3} \tag{3}$$

where $f = \{1, 2, 3\}$. $count_1$, $count_2$, and $count_3$ are the total number of conformations accepted in the first, second, and third subpopulation, respectively. They will be reset for each LP after g > G_{stage}. For example, the fourth subpopulation executed the mutation strategy of the second subpopulation, if AR_2 is higher than others.

In the selection stage, if $score3(C_{trial}) < score3(C_{target})$, C_{target} is replaced by corresponding C_{trial}. Otherwise, if $S(C_{trial}) > S(C_{target})$, C_{target} is replaced by corresponding C_{trial}.

Iterative mutation, crossover and selection operation, until evolutionary generation g reaches the maximum generation G_{MAX}, and then output final result and stop iterative.

4 Results

4.1 Test Set and Parameters Settings

In order to assess the performance of MSEMS, 6 proteins with size from 47 to 117 is used, which include 2 α folding proteins, 2 β folding proteins, and 2 α/β folding proteins. The test set contains all classes of proteins.

The parameters for the MSEMS used during the tests are: population size $NP = 100$, the max evolutionary generation $G_{MAX} = 3000$, stage control parameters $G_{stage} = 500$, learning period $LP = 100$, the initial values of $count_1$, $count_2$, and $count_3$ are set to zero.

Table 1. RMSD and energy of all test proteins final results obtained by Rosetta and MSEMS.

No.	PDB ID	Size	Type	RMSD(Å)		Energy	
				Rosetta	MSEMS	Rosetta	MSEMS
1	1B72	49	α	7.94	**2.23**	**0.31**	10.86
2	1ELW	117	α	10.18	**3.23**	−12.96	**−30.52**
3	1TEN	87	β	**11.72**	12.00	−82.86	**−104.52**
4	1TFI	47	β	4.52	**4.28**	−19.05	**−23.72**
5	1FO5	85	α/β	4.49	**4.05**	−41.91	**−72.89**
6	1GNU	117	α/β	16.15	**12.16**	**−39.48**	−17.24
Average				9.17	**6.32**	−32.66	**−39.67**

4.2 Predicted Results Analysis

In this part, the proposed MSEMS is compared with Rosetta [34]. It is because that Rosetta has shown a good performance for de novo PSP. Moreover, MSEMS uses stage 1 and stage 2 of Rosetta (ClassicAbinitio package) to generate initial population and employs Rosetta energy model *score*3. Therefore, a comparison to Rosetta is essential for highlighting the suitability of MSEMS.

The final generated models are compared with experimentally determined structure. Root Mean Square Deviation (RMSD) of C_α atoms is used to evaluated the accuracy of models. The smaller the RMSD, the more similar the two

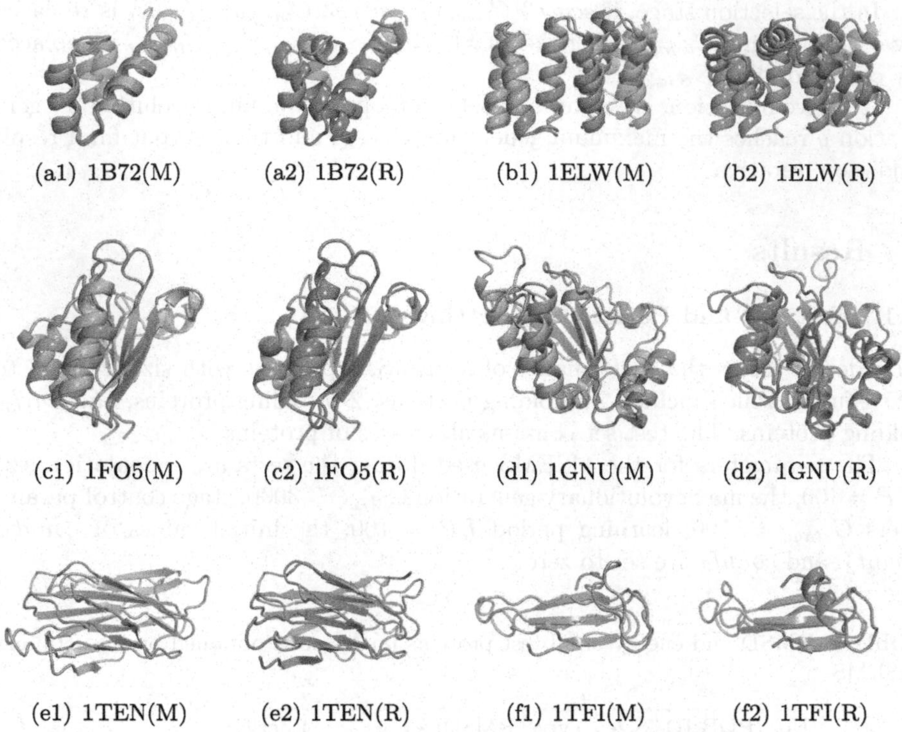

Fig. 4. Comparison of the structure obtained by Rosetta (orange) and MSEMS (blue) with the native structure (green), marked as R and M, respectively. (Color figure online)

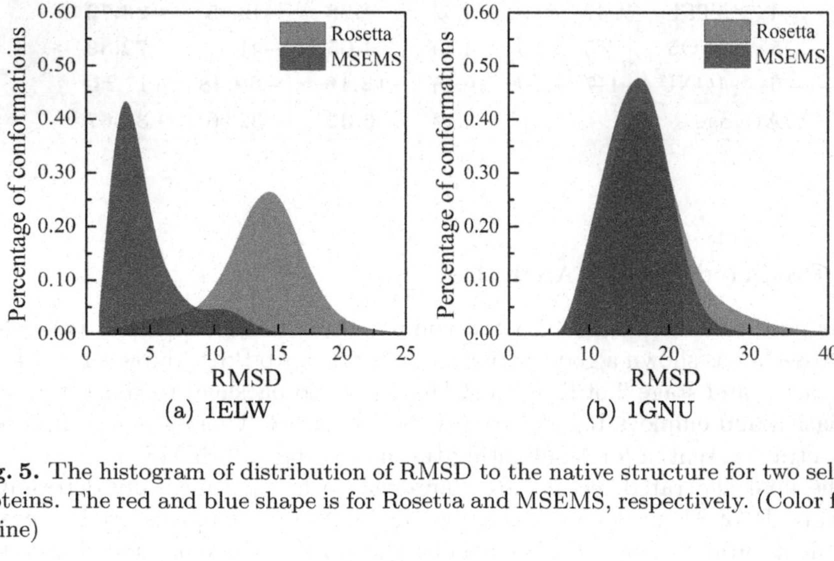

Fig. 5. The histogram of distribution of RMSD to the native structure for two selected proteins. The red and blue shape is for Rosetta and MSEMS, respectively. (Color figure online)

structures are. Statistics of MSEMS are calculated from the lowest energy conformation individual among the populations at the end of the runs. The results of Rosetta are obtained by running the Rosetta 3.10 version and statistics of Rosetta are calculated from the lowest energy model among 100 times standard Rosetta de novo protocol at the end of the runs.

The comparison of the final 3D structures obtained by Rosetta and MSEMS with the corresponding native structures for the six proteins is shown in Fig. 4. The results are shown in Table 1. The proposed MSEMS achieved better results than Rosetta in 5 out of 6 proteins (except with PDB ID 1TEN) in the aspect of RMSD. Specifically, MSEMS outperformed Rosetta by 3 Å or more in 3 proteins, including 2 α folding proteins (with PDB IDs 1B72 and 1ELW), and 1 α/β folding proteins (with PDB ID 1GNU). The average RMSD of MSEMS is 6.32 Å, while that of Rosetta is 9.17 Å. Meanwhile, MSEMS achieved lower energy than Rosetta in 4 out of 6 proteins (except with PDB IDs 1B72 and 1GNU). The average energy of MSEMS is −39.67 (lower is better), while that of Rosetta is −32.66.

For PDB ID 1B72, the energy of MSEMS is 10.86, and that of Rosetta is 0.31 Kcal/mol. However, the RMSD of MSEMS is 2.23 Å, which decreased by 71.91 % compared with that of Rosetta (7.74 Å). From Fig. 4 (a1), the result of MSEMS is more close to the native structure, while a α helix region misfolding occurred in the model obtained by Rosetta. It is probably because low-resolution model of Rosetta lack of distance information restraint. For PDB ID 1GNU, the energy of MSEMS is −17.24, and that of Rosetta is −39.48. However, the RMSD of MSEMS is 12.16 Å, which decreased by 24.71 % compared with that of Rosetta (16.15Å). These results indicate that low-resolution energy model is inaccuracy, which means the lowest energy conformation may not be the best one near the native structure. On the other hand, the results also indicate that distance-based fitness score assists energy to select conformation is effective.

For β folding proteins (with PDB IDs 1TEN and 1TFI), the proposed MSEMS does not show significant performance compared to Rosetta. For PDB IDs 1TEN and 1TFI, several loop regions are not predicted correctly, resulting in the β sheet structures not being in the correct position of these proteins.

To analyze the sampling ability of MSEMS to the near native structure conformations during the whole process. The histogram of distribution of predicted conformations with MSEMS and Rosetta as a function of RMSD to native for the two selected proteins is shown in Fig. 5. From Fig. 5 (a), we can clearly see that MSEMS can sample more near native conformations during the whole sampling process, and the distribution of MSEMS is significantly better than that of Rosetta. From Fig. 5 (b), their histograms of distribution are not significantly different, but MSEMS can sample more near native conformations than Rosetta. Moreover, the peak value of MSEMS histogram of distribution is better than that of Rosetta.

5 Conclusions

In this paper, we proposed a multi-subpopulation algorithm with ensemble mutation strategies for de novo protein structure prediction (MSEMS). First and foremost, the multi-subpopulation algorithm with ensemble mutation strategies is designed to enhance the sampling ability of the protein in low-energy region of conformation. MSEMS appropriately balances the exploration and exploitation of conformational space by the first three subpopulations, the fourth subpopulation is utilized to reward the one with the best performance in the other three subpopulations. Second, distance-based fitness score is designed to alleviate inaccuracy of low-resolution energy model. Therefore, MSEMS is able to generate the high quality models.

Experimental results indicated that proposed MSEMS deserves further investigation. Meanwhile, the experimental results also indicate that MSEMS does not perform very well in the prediction of loop region. In future work we will consider to use secondary structure information to restrain the protein structure, and design a sampling method for loop region.

References

1. Zhou, X., Hu, J., Zhang, C., Zhang, G., Zhang, Y.: Assembling multi-domain protein structures through analogous global structural alignments. Proc. Nat. Acad. Sci. U.S.A. **116**(32), 15930–15938 (2019)
2. Gntert, P.: Automated NMR protein structure calculation with CYANA. Prog. Nucl. Magn. Reson. Spectrosc. **43**(3–4), 105–125 (2004)
3. Zhang, G., Ma, L., Wang, X., Zhou, X.: Secondary structure and contact guided differential evolution for protein structure prediction. IEEE/ACM Trans. Comput. Biol. Bioinform. (2018)
4. Ben-David, M., Noivirt-Brik, O., Paz, A., Prilusky, J., Sussman, J.L., Levy, Y.: Assessment of CASP8 structure predictions for template free targets. Proteins: Struct. Funct. Bioinform. **77**(S9), 50–65 (2009)
5. Anfinsen, C.B.: Principles that govern the folding of protein chains. Science **181**(4096), 223–230 (1973)
6. Garza-Fabre, M., Kandathil, S.M., Handl, J., Knowles, J., Lovell, S.C.: Generating, maintaining, and exploiting diversity in a memetic algorithm for protein structure prediction. Evol. Comput. **24**(4), 1 (2016)
7. Gelin, B.R., Mccammon, J.A., Karplus, M.: Dynamics of folded proteins. Nature **267**(5612), 585–590 (1977)
8. Ovchinnikov, S., Park, H., David, E.K., Dimaio, F., Baker, D.: Protein structure prediction using Rosetta in CASP12. Proteins: Struct. Funct. Bioinform. **86**(10), 113–121 (2018)
9. Lee, J., Lee, J., Sasaki, T.N., Sasai, M., Seok, C., Lee, J.: De novo protein structure prediction by dynamic fragment assembly and conformational space annealing. Proteins: Struct. Funct. Bioinform. **79**(8), 2403–2417 (2011)
10. Yuan, B., Li, B., Chen, H., Yao, X.: A new evolutionary algorithm with structure mutation for the maximum balanced biclique problem. IEEE Trans. Cybern. **45**(5), 1054–1067 (2015)

11. Lee, K.B., Kim, J.H.: Multiobjective particle swarm optimization with preference-based sort and its application to path following footstep optimization for humanoid robots. IEEE Trans. Evol. Comput. **17**(6), 755–766 (2013)
12. Zhou, X., Zhang, G., Hao, X., Yu, L., Xu, D.: Differential evolution with multi-stage strategies for global optimization. In: IEEE Congress on Evolutionary Computation, pp. 2550–2557. IEEE, Vancouver (2016)
13. Storn, R., Price, K.: Differential evolution-a simple and efficient heuristic for global optimization over continuous spaces. J. Global Optim. **11**(4), 341–359 (1997)
14. Das, S., Mullick, S.S., Suganthan, P.N.: Recent advances in differential evolution-an updated survey. Swarm Evol. Comput. **27**, 1–30 (2016)
15. Glotić, A., Glotić, A., Kitak, P., Pihler, J., Tičar, I.: Parallel self-adaptive differential evolution algorithm for solving short-term hydro scheduling problem. IEEE Trans. Power Syst. **29**(5), 2347–2358 (2014)
16. Sharma, S., Rangaiah, G.P.: An improved multi-objective differential evolution with a termination criterion for optimizing chemical processes. Comput. Chem. Eng. **56**, 155–173 (2013)
17. Sudha, S., Baskar, S., Amali, S.M.J., Krishnaswamy, S.: Protein structure prediction using diversity controlled self-adaptive differential evolution with local search. Soft. Comput. **19**(6), 1635–1646 (2014). https://doi.org/10.1007/s00500-014-1353-2
18. De Melo, V.V., Carosio, G.L.: Investigating multi-view differential evolution for solving constrained engineering design problems. Expert Syst. Appl. **40**(9), 3370–3377 (2013)
19. Epitropakis, M.G., Tasoulis, D.K., Pavlidis, N.G., Plagianakos, V.P., Vrahatis, M.N.: Enhancing differential evolution utilizing proximity-based mutation operators. IEEE Trans. Evol. Comput. **15**(1), 99–119 (2011)
20. Zhou, X., Zhang, G.: Differential evolution with underestimation-based multimutation strategy. IEEE Trans. Cybern. **49**(4), 1353–1364 (2019)
21. Gong, W., Cai, Z.: Differential evolution with ranking-based mutation operators. IEEE Trans. Cybern. **43**(6), 2066–2081 (2013)
22. Zhou, X., Zhang, G.: Abstract convex underestimation assisted multistage differential evolution. IEEE Trans. Cybern. **47**(9), 2730–2741 (2017)
23. Qin, A.K., Huang, V.L., Suganthan, P.N.: Differential evolution algorithm with strategy adaptation for global numerical optimization. IEEE Trans. Evol. Comput. **13**(2), 398–417 (2009)
24. Gong, W., Cai, Z., Ling, C.X., Li, H.: Enhanced differential evolution with adaptive strategies for numerical optimization. IEEE Trans. Syst. Man Cybern. Part B (Cybern.) **41**(2), 397–413 (2010)
25. Wang, Y., Cai, Z., Zhang, Q.: Differential evolution with composite trial vector generation strategies and control parameters. IEEE Trans. Evol. Comput. **15**(1), 55–66 (2011)
26. Zhou, X., Zhang, G., Hao, X., Yu, L.: A novel differential evolution algorithm using local abstract convex underestimate strategy for global optimization. Comput. Oper. Res. **75**, 132–149 (2016)
27. Zhang, G., Zhou, X., Yu, X., Hao, X., Yu, L.: Enhancing protein conformational space sampling using distance profile-guided differential evolution. IEEE/ACM Trans. Comput. Biol. Bioinform. **14**(6), 1288–1301 (2017)
28. Liwo, A., Arłukowicz, P., Czaplewski, C., Ołdziej, S., Pillardy, J., Scheraga, H.A.: A method for optimizing potential-energy functions by a hierarchical design of the potential-energy landscape: application to the UNRES force field. Proc. Nat. Acad. Sci. **99**(4), 1937–1942 (2002)

29. Bowers, P.M., Strauss, C.E.M., Baker, D.: De novo protein structure determination using sparse NMR data. J. Biomol. NMR **18**(4), 311–318 (2000). https://doi.org/10.1023/A:1026744431105
30. Xu, D., Zhang, Y.: Toward optimal fragment generations for ab initio protein structure assembly. Proteins: Struct. Funct. Bioinform. **81**(2), 229–239 (2013)
31. Custdio, F.L., Barbosa, H.J.C., Dardenne, L.E.: Full-atom ab initio protein structure prediction with a Genetic Algorithm using a similarity-based surrogate model. In: IEEE Congress on Evolutionary Computation, pp. 1–8. IEEE (2010)
32. Xu, J.: Rapid protein side-chain packing via tree decomposition. In: Miyano, S., Mesirov, J., Kasif, S., Istrail, S., Pevzner, P.A., Waterman, M. (eds.) RECOMB 2005. LNCS, vol. 3500, pp. 423–439. Springer, Heidelberg (2005). https://doi.org/10.1007/11415770_32
33. Saleh, S., Olson, B., Shehu, A.: A population-based evolutionary search approach to the multiple minima problem in de novo protein structure prediction. BMC Struct. Biol.: Struct. Funct. Bioinform. **13**(S1), 1–19 (2013). https://doi.org/10.1186/1472-6807-13-S1-S4
34. Rohl, C.A., Strauss, C.E.M., Misura, K.M.S., Baker, D.: Protein structure prediction using Rosetta. Methods Enzymol. **383**(383), 66 (2003)
35. Simons, K.T., Bonneau, R., Ruczinski, I., Baker, D.: Ab initio protein structure prediction of CASP III targets using ROSETTA. Proteins: Struct. Funct. Bioinform. **37**(S3), 171–176 (1999)
36. Metropolis, N., Rosenbluth, A.W., Rosenbluth, M.N., Teller, A.H., Teller, E.: Equation of state calculations by fast computing machines. J. Chem. Phys. **21**(6), 1087–1092 (1953)

A Multi-objective Bat Algorithm for Software Defect Prediction

Di Wu, Jiangjiang Zhang, Shaojin Geng, Xingjuan Cai(✉), and Guoyou Zhang

Complex System and Computational Intelligent Laboratory, Taiyuan University of Science and Technology, Taiyuan 030024, China
xingjuancai@163.com

Abstract. Both the class imbalance of datasets and parameter selection of support vector machine (SVM) play an important role in the process of software defect prediction. To solve these two problems synchronously, the false positive rate (pf) and the probability of detection (pd) are considered as two objective functions to construct the multi-objective software defect prediction model in this paper. Meanwhile, a multi-objective bat algorithm (MOBA) is designed to solve this model. The individual update strategy in the population is performed using the individual update method in the fast triangle flip bat algorithm, and the non-dominated solution set is used to save the better individuals of the non-defective module and the support vector machine parameters. The simulation results show that MOBA can effectively save resource consumption and improve the quality of software compared with other commonly used algorithms.

Keywords: Software defect prediction · Multi-objective bat algorithm · False positive rate · Probability of detection

1 Introduction

With the popularization and dependence of computers, the computer applications have covered our lives in different areas. Meanwhile, the quality of software products has attracted more and more attention. In general, software quality mainly includes software reliability, comprehensibility, usability, maintainability and effectiveness [1], in which the reliability of software is particularly important. One of the main factors leading to software unreliability is software defect [2]. Software defects refer to the errors introduced in the process of software development, which will lead to errors, failures, collapses and even endanger the safety of human life and property in the process of software operation. Software defects are

---

National Natural Science Foundation of China under Grant No. 61806138, No. U1636220 and No. 61663028, Natural Science Foundation of Shanxi Province under Grant No. 201801D121127, PhD Research Startup Foundation of Taiyuan University of Science and Technology under Grant No. 20182002.

L. Pan et al. (Eds.): BIC-TA 2019, CCIS 1159, pp. 269–283, 2020.
https://doi.org/10.1007/978-981-15-3425-6_22

mainly caused by incorrect analysis of requirements in the software development process, insufficient experience of programmers or unreasonable arrangement of software administrators. Software testing can help developers find defects, but it is easy to increase the development time of the team when testing the software, which leads to excessive cost. Therefore, how to find defects as much as possible is especially important in ensuring software quality. The purpose of software defect prediction [3] is to use a specific method to discover which modules of the software system are more likely to be defective, or the number of defects and their distribution of possible software systems, thus providing meaningful guidance for software testing.

For practical application scenario, the enough training data need to be met for the precision requirements of traditional classifiers in a new project or the project under the circumstances of less historical data, which often requires performing the software defect prediction operations. Therefore, how to perform the prediction operate is very difficult for the staff. It is commonly known that datasets for software defect prediction have an inherent feature class imbalance of datasets. The main performance is that 80% of the defects in a software system are concentrated in 20% of the modules [4], which means that the number of defective modules is much smaller than the number of non-defective modules. The traditional two-classification algorithm is proposed based on relatively balanced datasets. When training the software defect prediction model, the defective module and the non-defective module give the same weight, which will cause the training model to pay more attention to the non-defective module. Insufficient training for defective modules ultimately leads to the significant reduction in predictive performance for defective modules. At the same time, in the commonly used classification algorithms, Vapnik and Elish proved that the performance of support vector machine (SVM) in software defect prediction is better than other classifiers [5]. Therefore, SVM is used as our base classifier. But with further research, scholars have found that SVM parameter selection is also an important factor affecting predictive performance.

Table 1 lists the study of class imbalances and SVM parameter selection for datasets in software defect prediction. It is used to explain which question the literature is studying. It is worth noting that the existing research solves the problem of the class imbalance of the datasets and the parameter selection of the SVM separately. However, when there are multiple factors that affect a problem, the final solution will be affected that one of the factors is optimized. The same is true for software defect prediction. Table 1 shows that the class imbalance of the datasets and the parameter selection of the SVM affect the prediction performance of the software defect. At present, the research is based on one of the factors. Although it can improve the ability of defect prediction, when considering another factor to solve another factor, a better software defect prediction model can be obtained. Thus, the above two problems are simultaneously optimized in this paper.

For the SVM parameter selection problem [17], most researchers use heuristic intelligent optimization algorithms to optimize parameters, and the performance

Table 1. The result of exiting researches.

Existing algorithms	Class imbalance in datasets	Parameters of SVM
GA-SVM [6]		√
ACO-SVM [7]		√
PSO-SVM [8]		√
MOCS-SVM [9]		√
LLE-SVM [10]		√
COIM [11]	√	
RUS [12]	√	
NearMiss [13]	√	
Evolutionary Sampling [14]	√	
SMOTE [15]	√	
SMOTE+RUS [16]	√	

of the original model is greatly improved in software defect prediction. To solve the imbalance of datasets problem, the method of under-sampling is adopted. Study existing under-sampling methods such as random under-sampling, neighborhood cleanup, NearMiss-2, etc. Most of them delete non-defective modules deterministically according to certain rules based on distance measurement. It will produce the problem about premature screening of the module and loss of valid information that combining the under-sampling method with SVM parameter selection based on intelligent optimization. To solve the problem, Drown et al. [14] proposed an evolutionary sampling method by using genetic algorithm to select the appropriate non-defective modules, but the parameters are not optimized.

Swarm intelligent optimization algorithm has been developed rapidly in recent years. In addition, the bat algorithm, as a typical swarm intelligence optimization algorithm, has the characteristics of fast convergence speed and strong searching ability. And it has showed good performance effects in parameter optimization area. What's more, many versions of improving bat algorithms have been proposed in recent years. Therefore, the bat algorithm will be employed to optimize the parameters of software defect prediction in this paper. Inspired by Drown, a multi-objective software defect prediction model for under-sampling strategy is proposed to synchronously solve the problem of selection of non-defective model and optimization of SVM parameters. Meanwhile, a multi-objective bat algorithm (MOBA), an extended bat algorithm based on the triangle flip strategy, is designed to solve the model in this paper.

The rest of this paper is organized as follows: Sect. 2 introduces the related works of software defect prediction problem. The multi-objective bat algorithm for under-sampling software defect prediction model is given in Sect. 3. In Sect. 4, MOEA was tested in the public database of promise and analyzed the experimental results in detail. A summary for this paper is described in Sect. 5.

2 Related Work

This section first introduces some basic principles of support vector machines. Then, the specific construction methods of the under-sampling software defect prediction model are expounded one by one.

2.1 Support Vector Machine (SVM)

SVM [18] is a new machine learning method based on statistics. It shows unique advantages in solving small sample, nonlinear and many-dimensional problems [19]. The main idea is to solve the kernel function and quadratic programming problem. The kernel function is used to map the data to the many-dimensional feature space to solve the nonlinear separable problem. Finally, an optimal classification hyperplane is found, which makes the two types of samples separate.

Assume that the training sample set of the SVM to $(x_i, y_i), i = 1, 2, ..., n, x_i \in R^n, y_i \in \{+1, -1\}$. Where x_i is the input vector and y_i is the output vector. Then the discriminant function is $f(x) = \omega^T x + b$, and the classification equation of the SVM is:

$$\omega^T x + b = 0 \tag{1}$$

where ω^T is the normal vector representing the optimal classification plane, and b is the bias. Normalize $f(x)$ so that all samples satisfy $|f(x) \geq 1|$ and the nearest sample from the classification face satisfies $f(x) = 1$. If the classification plane is such that all samples are accurately classified, then Eq. (2) is satisfied.

$$y_i(\omega^T x + b) - 1 \geq 0, i = 1, 2, ..., n \tag{2}$$

Finally, the obtained classification function is shown in Eq. (3).

$$f(x) = \operatorname{sgn}\left\{\sum_{i=1}^{L} y_i \cdot a_i \cdot k(x_i, x_j) + \mathrm{b}\right\} \tag{3}$$

where a_i denotes the Lagrangian multiplier ($a_i \in [0, C]$) and C is the penalty factor. Its choice has a greater influence on the SVM. And b is the bias term of the decision function, specifically $b = y_i - \sum_{i=1}^{l} y_i a_i k(x, x_i)$.

The choice of kernel function has a great influence on the classification algorithm. There are four main types of kernel functions of SVM, as shown in Table 2. In this paper, Radial Basis Function (RBF) is chosen as the kernel function. It is an ideal classification function with the advantage of a wide range of convergence.

2.2 Under-Sampling Software Defect Prediction Model

In software defect prediction, there are only two types of modules, defective and non-defective. Therefore, software defect prediction is essentially a two-category

Table 2. Kernel function of SVM.

SVM kernel function	Kernel function expression
Linear kernel	$KER(x_b, x_l) = x_b^T \cdot x_l$
Polynomial kernel	$KER(x_b, x_l) = [\varepsilon x_b^T \cdot x_l + m]^n, \varepsilon > 0$, n is the kernel dimension
Radial basis function	$KER(x_b, x_l) = exp\{-\frac{\lvert x_b - x_l \rvert^2}{\sigma^2}\}$, σ is the bandwidth of RBF
Sigmoid kernel function	$KER(x_b, x_l) = \tanh(\varepsilon x_b^T, x_l + m)$

Table 3. Parameter value.

Actural value	Predicted value	
	Defective module	Non-defective module
Defective module	TP	FN
Non-defective module	FP	TN

problem. In such problems, confusion matrices are often used to evaluate the advantages and disadvantages of the model, as shown in Table 3.

Where TP represents the number of modules that are actually predicted and are correctly predicted, FN represents the number of modules that are incorrectly predicted; FP represents the number of modules that are actually predicted to be non-defective, and TN represents the number of modules that are predicted correctly. Therefore, according to the description of the confusion matrix, some problems can be formulated.

The probability of detection (*pd*) indicates the correct prediction of the proportion of defective modules in the total defective module. The higher the number of defective modules correctly detected, the better the quality of the software.

$$pd = \frac{TP}{TP + FN} \tag{4}$$

The false positive rate (pf) is the ratio of the number of non-defective modules predicted to be defective modules to the total number of actual defective modules. In practical applications, modules that are predicted to be defective need to allocate certain test resources. Therefore, the larger the ratio, the more resources will be wasted.

$$pf = \frac{FP}{FP + TN} \tag{5}$$

However, in the practical problems of software defect prediction, it is undoubtedly the best to detect as many defects as possible. Only in this way, can we minimize the defects and reduce the cost of software maintenance. But many existing literatures [14,15] have shown that the limited testing resources will be inevitably wasted when non-defective modules are also tested. Therefore, it can be concluded that there is a conflict relationship between pd and pf.

Here we treat the under-sampling software defect prediction problem as a multi-objective optimization problem [20]. The under-sampling software defect prediction model can be described as the largest pd and the smallest pf.

$$\begin{cases} \max pd = \frac{TP}{TP+FN} \\ \min pf = \frac{FP}{FP+TN} \end{cases} \tag{6}$$

3 Multi-objective Bat Algorithm

In order to better describe the specific implementation of multi-objective bat algorithm (MOBA) in the under-sampling software defect prediction of unbalanced datasets, this section is divided into three parts. The first part introduces the basic theory and related principles of multi-objective optimization problems (MOPs). The second part will introduce multi-objective bat algorithm in detail. And the last part will explain the under-sampling software defect prediction process in the unbalanced datasets.

3.1 Multi-objective Optimization Problems (MOPs)

In practical engineering applications, there are often multiple conflicting objectives [21] that require simultaneous optimization. Such problems are often defined as multi-objective optimization problem [22], described as follows:

$$\min f(x) = (f_1(x), f_2(x), ..., f_n(x)) \tag{7}$$

where $f_i(x)(i = 1, 2, ..., n)$ represents the $i - th$ objective function [23]. For example, for a problem with two objective functions $f_1(x)$ and $f_2(x)$, when the value of $f_1(x)$ decreases, the value of $f_2(x)$ increases; when the value of $f_1(x)$ increases, the value of $f_2(x)$ decreases. At this time, which solution should be chosen to optimize the two objectives at the same time to achieve the best results, this is the problem that multi-objective optimization algorithm [24] needs to solve.

3.2 The Description of Multi-objective Bat Algorithm

To solve the multi-objective optimization problem by extending the bat algorithm, the multi-objective bat algorithm is proposed on the basis of the fast triangle flip bat algorithm. In addition, the principle of MOBA, the selection mechanisms and the specific updating strategies will be described in turn.

(1) Non-dominant solution set (pareto solution set)
Set p and q are two random individuals in the population, consider the following two cases:

(1) When there is $f_k(p) \leq f_k(q)(k = 1, 2, ..., n)$ for all objective functions
(2) There is at least one l that makes $f_l(p) < f_l(q)$, where $l \in \{1, 2, ..., n\}$

It represents p dominate q and is recorded as $p \prec q$. In other words, the objective function value of the individual p is better than q, and the composed set is called the non-dominated solution set. If the individual cannot be compared, it means that the individuals do non-dominated each other.

(2) Fast non-dominated sorting
Fast non-dominated sorting is the initial ranking of individuals in a population. When an individual is not dominated by any individual in the population, the individual is placed in the first layer of the pareto front (PF), and the individual is removed from the next sorted population.
Then, in the remaining individuals, the same way is used to find non-dominated individuals in the remaining layers. Until all individuals are distributed in each layer according to the non-dominated relationship, the PF is obtained. Where $Fr_1, Fr_2, .., Fr_u$ represents each layer. Because of the different characteristics of each generation of population, u is unknown and each generation is different.

(3) Crowding distance
After non-dominated sorting, multiple individuals in the same layer belong to a non-dominated relationship, and it is difficult to select a relatively superior individual. To solve this problem, a crowding distance strategy is proposed. As shown below:

$$Dist_i = \sum_{k=1}^{n} (Dist_{i+1} \cdot f_k - Dist_{i-1} \cdot f_k) \tag{8}$$

Where $Dist_i$ is the crowding distance of the $i-th$ population and $Dist_i \cdot f_k$ is the function value of the individual i at f_k.

(4) The description of fast triangle flip strategy

(1) Global search strategy
The global search strategy is an important part role in jumping out local optimal for single objective optimization problem, which will be help to find the global optimal solution. However, the search strategy will no longer be suitable for solving the multi-objective problems. To solve the problem, an improved global search strategy is adopted by using the external individuals in a population instead of global optimal individuals to increase the diversity of the population in this paper. In addition, the new individual update methods are showed as follows.
The random triangle flip strategy in the early stage:

$$v_{ik}(t+1) = v_{ik}(t) + (x_{mk}(t) - x_{uk}(t)) \cdot fr_i(t) \tag{9}$$

$$x_{ik}(t+1) = x_{ik}(t) + v_{ik}(t+1) \tag{10}$$

where, $x_{mk}(t)$and $x_{uk}(t)$ are two randomly selected non-dominant individuals from the population. The triangle flip with the optimal solution strategy in the later stage:

$$v_{ik}(t+1) = (p_k(t) - x_{ik}(t)) \cdot fr_i(t) \tag{11}$$

$$x_{ik}(t+1) = x_{ik}(t) + v_{ik}(t+1) \tag{12}$$

where $p_k(t)$ is a non-dominant individual randomly selected from the first frontier of the external population. And $x_{ik}(t)$ is a randomly selected individual in the population.
Therefore, it can be concluded from the above description that the improved global search method will be conducive to the individuals searching in the diversity direction, will enhance the diversity of the population.

(2) Local search strategy
To guide the evolution of individuals in a better direction, the improved local search model will be showed as follows.

$$x_{ik}(t+1) = p_{ik}(t) + \varepsilon_{ik} \cdot \overline{A}(t) \tag{13}$$

Where $p_{ik}(t)$ is the individual randomly selected form the first non-dominant front of the population; ε_{ik} is an uniformly distributed random number within the $(-1, 1)$; $\overline{A}(t) = \frac{\sum_{i}^{n} A_i(t)}{n}$ represents the average loudness at tth time.
Therefore, the specific process of MOBA is described as follows:

Step1: Initialize the frequency, pulse transmission frequency and loudness of the bat individual, and randomly generate the initial velocity and position of the bat individual within the defined domain interval.
Step2: Calculate the fitness value of population.
Step3: The current population is layered by using fast non-dominated sorting, and the individuals of the first layer are treated as non-dominated solutions front and saved.
Step4: Update the speed and position of individual bats.
Step5: Local perturbation: generate a random number $rand1$, if $rand1 > r_i(t)$, update the population.
Step6: Calculate a set of fitness values for the new location.
Step7: Update populations by using fast non-dominated sorting and crowding distance.
Step8: Update the optimal non-dominated solution set: combine with the non-dominated solution set front with the new population, and use the fast non-dominated sorting to update the PF using the individuals sorted at the first level;
Step9: When the maximum number of iterations are satisfied, outputs the non-dominated solution set front. On the contrast, returns to step4.

3.3 Under-Sampling Software Defect Prediction Process for Unbalanced Datasets

The main purpose of this section is to use the multi-objective bat algorithm's optimization ability to select the non-defective modules of the unbalanced datasets, and simultaneously optimize the SVM parameters. Therefore, the under-sampling software defect prediction process for unbalanced datasets can be expressed as follows:

Step 1: Input the software defect prediction datasets.
Step 2: Remove the duplicate modules in the datasets.
Step 3: Initialize the population, front solution set and related parameters, and the position of each bat is set to randomly selected non-defective modules and SVM parameters.
Step 4: Bring the selected non-defective module and the defective module into the SVM, train the support vector machine with the training samples, and then use the test sample to calculate the fitness value corresponding to the population position, and save the population position corresponding to the non-dominated solution to front set, and randomly select a location as the global optimal location.
Step 5: Update the position and speed of the population bats.
Step 6: Calculate the fitness value of the updated population, compare the fitness value and the crowding distance of the contemporary population with the previous generation population, and judge whether the new solution is accepted.
Step 7: Update the position and speed of the population bats using local disturbances.
Step 8: Calculate the fitness value and determine whether the new solution is accepted, and update the population.
Step 9: Update the Front solution set and the global optimal position.
Step 10: Determine whether the algorithm reaches the maximum number of iterations and meet the output non-dominated defect prediction model. Otherwise, perform step 5.

According to the under-sampling software defect prediction process of the above unbalanced datasets, The corresponding flow chart is shown in Fig. 1.

4 Simulation Experiment

This section will test the multi-objective bat algorithm on under-sampling software defect prediction by the promise datasets and analyze the experimental results.

(1) Test suite
In the experiment of this section, the selected defect datasets are all from Promise public database which is commonly used in defect prediction, and there are different degrees of class imbalance in the tested datasets. Details of each datasets are shown in Table 4.

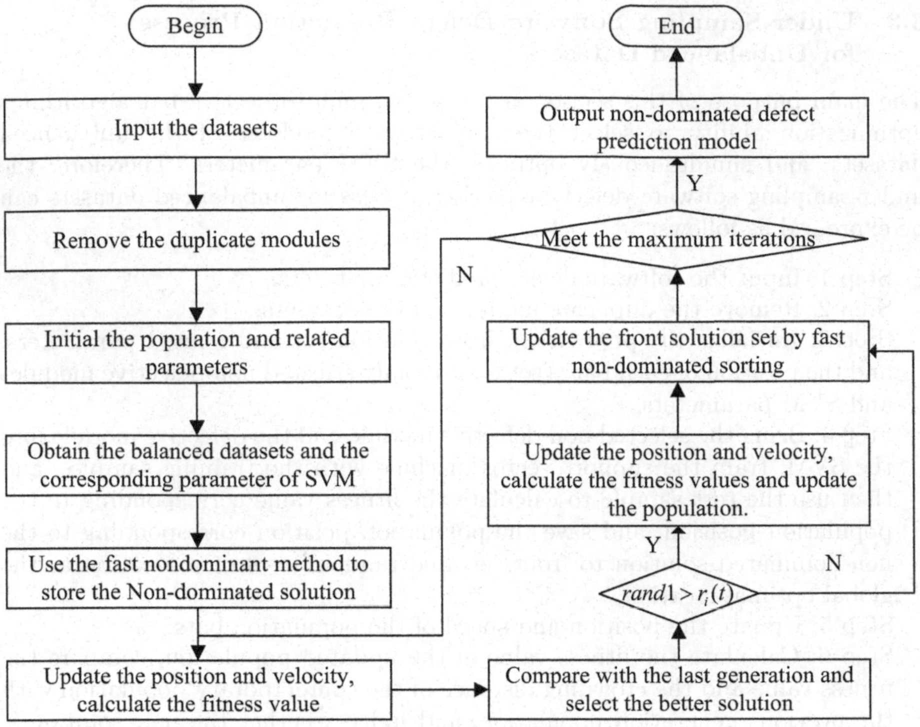

Fig. 1. The flowchart of MOBA for software defect prediction.

(2) Evaluation indicators
Evaluation indicators is critical to evaluate the model. In the software defect prediction, the comprehensive evaluation indicators (G-mean) is a kind of indicators that can comprehensively evaluate *pd* and *pf*. Its distribution range is between 0 and 1. It is often used to evaluate the datasets class imbalance model. Larger means that the classifier performance is better.

$$G - mean = \sqrt{pd \times (1 - pf)} \tag{14}$$

(3) Environment settings
In order to verify the performance of the proposed MOBA algorithm software defect prediction model, the software defect prediction model parameters are set to be: the maximum number of iterations is 200, and the population size is 50. The number of runs is 10 times. In addition, the expectation of the MOBA is to find suitable non-defective software modules and relatively good SVM parameters synchronously. Therefore, the initial population setting of the multi-objective bat algorithm is as follows: $x_i = (x_{i,1}, x_{i,2}, ..., x_{i,n}, x_{i,n+1}, x_{i,n+2})$, where $n + 2$ is the dimension (n is the number of selected non-defective software modules, its size is same as the number of defective modules in the datasets. The first dimensions are

respectively placed the selected non-defective modules, and the last two dimensions are respectively placed the parameters C and σ of the SVM [17]).The non-defective module is indexed so that when MOBA chooses the non-defective module, it only chooses its index number to determine which module to choose. The first $n - dimensional$ value [25] of the algorithm is an integer, and its lower bound is 1, and the upper limit is the number of non-defective modules in the datasets. The latter two-dimensional values are set to $[-1000, 1000]$. For the other algorithms involved in the section, detailed settings can be referenced and verified by cross-checking methods.

- Software defect prediction based on multi-objective bat algorithm (MOBA)
- Naive Bayes (NB) [26]
- Random Forest (RF) [4]
- Based on class overlap and imbalance prediction model (COIM) [11]
- Combine SMOTE's NB model (SMT+NB) [15]
- Combine RUS's NB model (RUS+NB) [12]
- SMOTE Boosting algorithm (SMTBST) [27]
- RUS Boosting algorithm (RUSBST) [28]

(4) Analysis of results
In order to visualize the experimental results, Fig. 2 shows the general evaluation index G-mean of the non-dominated solution of the MOBA algorithm on 8 different datasets. Overall, MOBA has a large fluctuation range on 8 datasets, and the resource waste rate pf fluctuates in (0, 0.7), (0, 0.8), (0, 0.75), (0, 0.8), (0, 0.9), (0, 0.55), (0, 0.7) and (0, 1), undetected defect rate 1-pd fluctuates in (0, 0.45), (0, 0.8), (0, 0.25), (0, 0.3), (0, 0.3), (0, 0.5), (0, 0.85) and (0, 0.75). In the datasets KC3 and mw1, although we did not get a lot of solutions, the solutions of both objectives are within a good prediction range. Specifically, in the datasets CM1 and KC2, the false alarm rate is basically within an acceptable range, and the performance of defect detection is also good. In the datasets KC3 and mw1, although we did not get a lot of solutions, the solutions of both objectives are within a good prediction range. Other remaining datasets have gotten more solutions, and their pareto front are closer to (0, 0) points, and both targets can get relatively better solutions.

For Table 5, the G-mean value of MOBA achieved the best performance. Specifically, although it is only slightly worse than COIM 0.008 on KC1, the best performance is obtained for cm1, KC2, KC3, MW1, PC1, PC3 and PC4. And higher than the second-best value by 0.068, 0.131, 0.174, 0.131, 0.103, 0.122, 0.073. The effectiveness of the algorithm is proved.

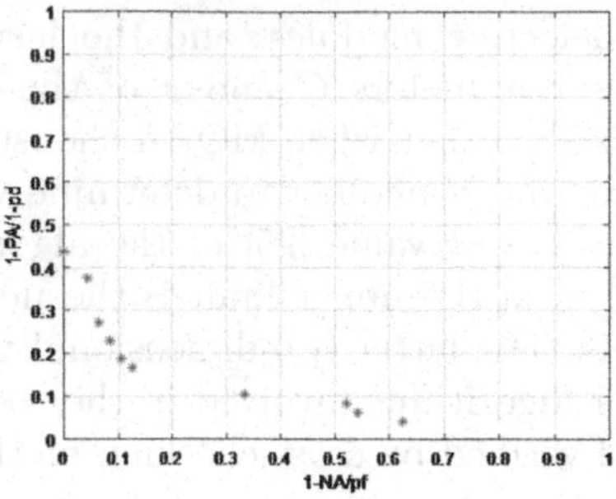

(a) CM1.

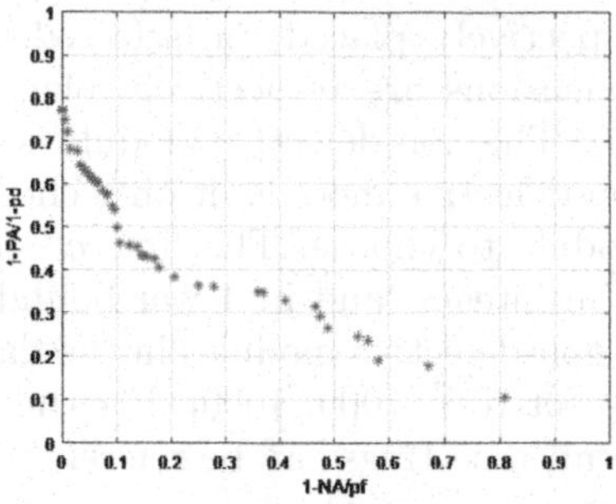

(b) KC1.

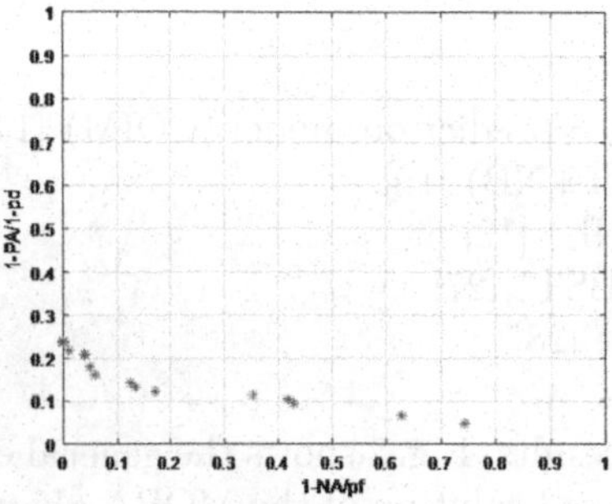

(c) KC2.

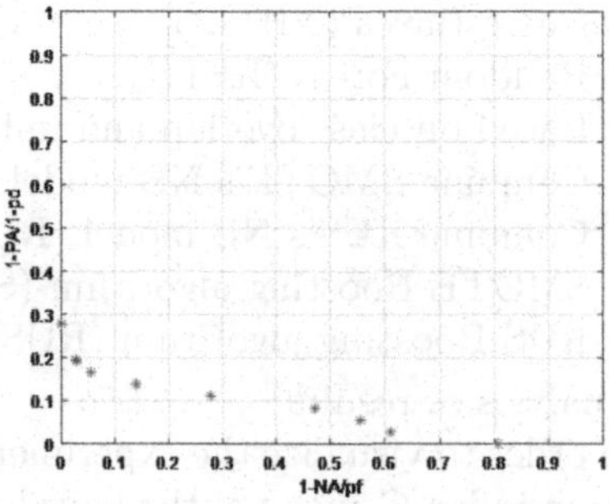

(d) KC3.

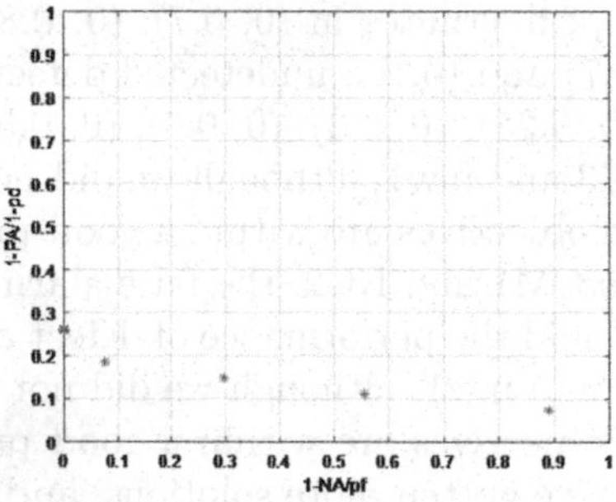

(e) mw1.

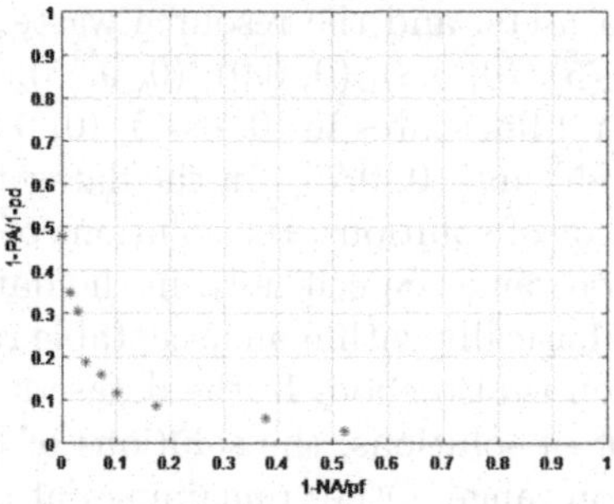

(f) PC1.

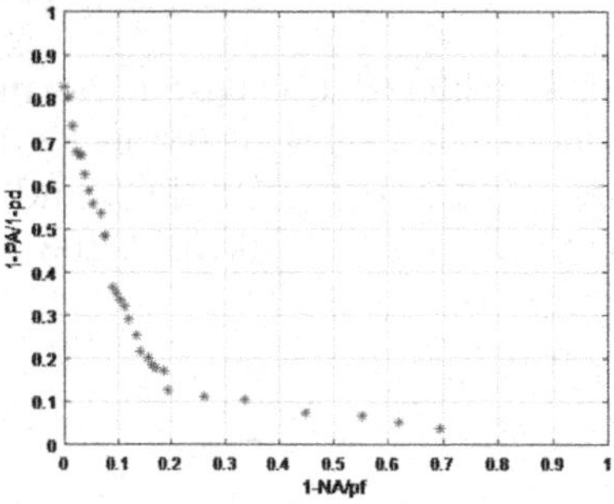

(g) PC3.

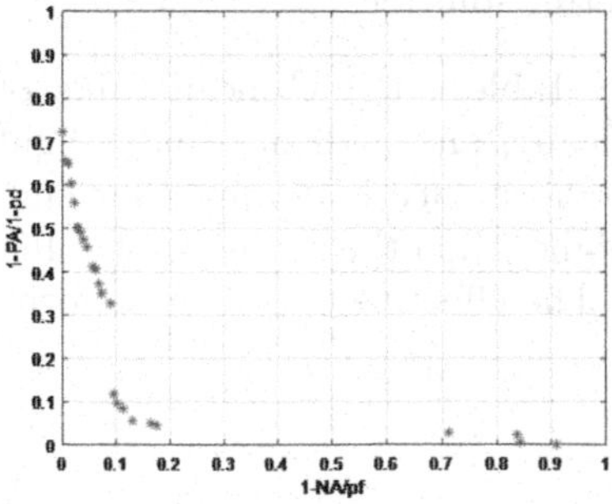

(h) PC4.

Fig. 2. MOBA in different test Non-dominated solution map on different data sets

Table 4. Eight datasets in promise.

Algorithm	Metric number	Before removing duplicate sample			After removing duplicate samples		
		Model number	Defects number	Unbalance rate	Model number	Defects number	Unbalance rate
CM_1	22	498	49	9.8%	442	48	10.86%
KC_1	22	2109	326	15.46%	1116	219	19.62%
KC_2	22	522	107	20.5%	375	105	28%
KC_3	40	194	36	18.56%	194	36	18.56%
mw_1	38	253	27	10.67%	253	27	10.67%
PC_1	22	1109	77	6.94%	952	69	7.24%
PC_3	38	1077	134	12.44%	1077	134	12.44%
PC_4	38	1458	178	12.2%	1344	177	13.17%

Table 5. Comparision results of G-mean on software defect prediction.

	CM1	KC1	KC2	KC3	mw1	PC1	PC3	PC4
MOBA	**0.853**	0.700	**0.889**	**0.887**	**0.869**	**0.891**	**0.839**	**0.906**
NB	0.314	0.581	0.624	0.651	0.652	0.511	0.375	0.585
RF	0.062	0.518	0.622	0.270	0.316	0.497	0.506	0.689
COIM	0.785	**0.708**	0.758	0.713	0.738	0.788	0.716	0.762
SMT+NB	0.438	0.602	0.659	0.604	0.661	0.527	0.406	0.716
RUS+NB	0.410	0.597	0.651	0.616	0.658	0.531	0.578	0.628
SMTBST	0.308	0.621	0.666	0.562	0.514	0.611	0.580	0.738
RUSBST	0.342	0.675	0.725	0.677	0.685	0.710	0.717	0.833

5 Conclusion

A multi-objective under-sampling bat algorithm is proposed to solve the under-sampling software defect prediction problem of the software defect prediction datasets imbalance and classifier SVM parameter selection. In the MOBA, fast triangle flip BA algorithm is extended to solve multi-objective optimization problem. For the multi-objective software defect prediction problem, the non-defective module of the defect datasets and the parameters of the SVM are searched as the optimized parameters of the MOBA. Meanwhile, the defect detection rate and the defect false alarm rate are selected as the optimization objectives. In the future work, other practical problems are solved by employing the designed MOBA.

Acknowledgments. This work is supported by the National Natural Science Foundation of China under Grant No. 61806138, Natural Science Foundation of Shanxi Province under Grant No. 201801D121127, Taiyuan University of Science and Technology Scientific Research Initial Funding under Grant No. 20182002. Postgraduate education Innovation project of Shanxi Province under Grant No. 2019SY495.

References

1. Sommerville, I.: Softw. Eng. **8**(12), 1226–1241 (2008)
2. Elish, K.O., Elish, M.O.: Predicting defect-prone software modules using support vector machines. J. Syst. Softw. **81**(5), 649–660 (2008)
3. Canfora, A.D.L.G., Penta, M.D., Oliveto, R., Panichella, A.: Multi-objective cross-project defect prediction. In: IEEE Sixth International Conference on Software Testing, Verification and Validation, pp. 252–261. IEEE Computer Society, Luxembourg (2013)
4. Malhotra, R.: A systematic review of machine learning techniques for software fault prediction. Appl. Soft Comput. J. **27**, 504–518 (2015)
5. Vapnik, V., Golowich, S.E., Smola, A.: Support vector method for function approximation, regression estimation, and signal processing. Adv. Neural Inf. Process. Syst. **9**, 281–287 (1996)
6. Cui, Z.B., Tang, G.M.: Software reliability prediction model based on support vector machine optimized by genetic algorithm. Comput. Eng. Appl. **45**(36), 71–74 (2009)
7. Yan, J.H.: Research of software defect prediction model based on ACO-SVM. Chin. J. Comput. **34**(6), 1148–1154 (2011)
8. Dong, H., Jian, G.: Parameter selection of a support vector machine, based on a chaotic particle swarm optimization algorithm. Cybern. Inf. Technol. **15**(3), 739–743 (2015)
9. Rong, X.T., Cui, Z.H.: Hybrid algorithm for two-objective software defect prediction problem. Int. J. Innovative Comput. Appl. **8**(4), 207–212 (2017)
10. Shan, C.: Software Defect Distribution Prediction Techniques and Application. Beijing Institute of Technology (2015)
11. Chen, L., Fang, B., Shang, Z., Tang, Y.: Tackling class overlap and imbalance problems in software defect prediction. Softw. Qual. J. **26**(1), 97–125 (2016). https://doi.org/10.1007/s11219-016-9342-6
12. Khoshgoftaar, T.M., Gao, K., Seliya, N.: Attribute selection and imbalanced data: Problems in software defect prediction. In: 22nd IEEE International Conference on Tools with Artificial Intelligence, vol. 1, pp. 137–144 (2010)
13. Wang, Q.: Distance metric learning based software defect prediction. Nanjing University of Posts and Telecommunications (2016)
14. Drown, D.J., Khoshgoftaar, T.M., Seliya, N.: Evolutionary sampling and software quality modeling of high-assurance systems. IEEE Trans. Syst. Man Cybern.-Part A: Syst. Hum. **39**(5), 1097–1107 (2009)
15. Pelayo, L., Dick, S.: Applying novel resampling strategies to software defect prediction. In: NAFIPS 2007-2007 Annual Meeting of the North American Fuzzy Information Processing Society, pp. 69–72 (2007)
16. Yang, L.: Research of the software defect prediction method for imbalanced data. China University of Petroleum (East China) (2014)
17. Cai, X.J., et al.: An under-sampled software defect prediction method based on hybrid multi-objective cuckoo search. Concurr. Comput. Pract. Exp. (2019). https://doi.org/10.1002/cpe.5478
18. Cortes, C., Vapnik, V.: Support-vector networks. Mach. Learn. **20**(3), 273–297 (1995)

19. Pan, L., Li, L., He, C., Tan, K.C.: A subregion division-based evolutionary algorithm with effective mating selection for many-objective optimization. IEEE Trans. Cybern. (2019). https://doi.org/10.1109/TCYB.2019.2906679
20. Zhang, M., Wang, H., Cui, Z., Chen, J.: Hybrid multi-objective cuckoo search with dynamical local search. Memetic Comput. **10**(2), 199–208 (2017). https://doi.org/10.1007/s12293-017-0237-2
21. Pan, L., He, C., Tian, Y., Su, Y., Zhang, X.: A region division based diversity maintaining approach for many-objective optimization. Integr. Comput.-Aided Eng. **24**(3), 279–296 (2017)
22. Wang, G.G., Cai, X.J., Cui, Z.H., Min, G.Y., Chen, J.J.: High performance computing for cyber physical social systems by using evolutionary multi-objective optimization algorithm. IEEE Trans. Emerg. Top. Comput. (2017)
23. He, C., Tian, Y., Jin, Y., Zhang, X., Pan, L.: A radial space division based evolutionary algorithm for many-objective optimization. Appl. Soft Comput. **61**, 603–621 (2017)
24. Cui, Z.H., Du, L., Wang, P.H., Cai, X.J., Zhang, W.S.: Malicious code detection based on CNNs and multi-objective algorithm. J. Parallel Distrib. Comput. **129**, 50–58 (2019)
25. Pan, L., He, C., Tian, Y., Wang, H., Zhang, X., Jin, Y.: A classification-based surrogate-assisted evolutionary algorithm for expensive many-objective optimization. IEEE Trans. Evol. Comput. **23**(1), 74–88 (2018)
26. Wu, D., Zhang, J.J., Geng, S.J., Ren, Y.Q., Cai, X.J., Zhang, G.Y.: A parallel computing optimization algorithm based on many-objective. In: IEEE 21st International Conference on High Performance Computing and Communications (2019). https://doi.org/10.1109/HPCC/SmartCity/DSS.2019.00254
27. Hall, T., Beecham, S., Bowes, D., Gray, D., Counsell, S.: A systematic literature review on fault prediction performance in software engineering. IEEE Trans. Softw. Eng. **38**(6), 1276–1304 (2012)
28. Chawla, N.V., Bowyer, K.W., Hall, L.O., Kegelmeyer, W.P.: SMOTE: synthetic minority over-sampling technique. J. Artif. Intell. Res. **16**(1), 321–357 (2002)
29. Seiffert, C., Khoshgoftaar, T.M., Hulse, J.V.: Improving software-quality predictions with data sampling and boosting. IEEE Trans. Syst. Man Cybern.-Part A: Syst. Hum. **39**(6), 1283–1294 (2009)

Mutation Strategy Selection Based on Fitness Landscape Analysis: A Preliminary Study

Jing Liang[1(✉)], Yaxin Li[1(✉)], Boyang Qu[2], Kunjie Yu[1], and Yi Hu[1]

[1] School of Electrical Engineering, Zhengzhou University, Zhengzhou 450001, China
{liangjing,yukunjie}@zzu.edu.cn, zzuliyaxin@163.com, eehuyi@163.com
[2] School of Electronic and Information Engineering, Zhongyuan University of Technology, Zhengzhou 450007, China
qby1984@hotmail.com

Abstract. Different algorithms and strategies behave disparately for different types of problems. In practical problems, we cannot grasp the nature of the problem in advance, so it is difficult for the engineers to choose a proper method to solve the problem effectively. In this case, the strategy selection task based on fitness landscape analysis comes into being. This paper gives a preliminary study on mutation strategy selection on the basis of fitness landscape analysis for continuous real-parameter optimization based on differential evolution. Some fundamental features of the fitness landscape and the components of standard differential evolution algorithm are described in detail. A mutation strategy selection framework based on fitness landscape analysis is designed. Some different types of classifiers which are applied to the proposed framework are tested and compared.

Keywords: Mutation strategy selection · Fitness landscape analysis · Classifier · Differential evolution algorithm

1 Introduction

In the context of Evolutionary Algorithms (EA), the option of previous search strategies often relies on the experience of the algorithm designer, the prior knowledge or parameters of the problem to be optimized. However, based on the No Free Lunch theory, different algorithms may have advantages over different problems and researches cannot ask an algorithm to perform well for all types. It is difficult for the engineers to find a proper optimization algorithm to solve a certain problem. Therefore, we consider the fitness landscape analysis (FLA) and search for the suitable strategy based on the nature and characteristics of optimization problems. According to the definition of fitness landscape [1], it is a triple $L = (S, V, F)$, where S is a collection of all solutions, V: S $\rightarrow 2^s$ is a specified neighborhood function, for each $s \in S$, the set of its neighbors $V(s)$, and f: S $\rightarrow \mathbb{R}$ is the fitness function. The role of the fitness landscape is to contrast

L. Pan et al. (Eds.): BIC-TA 2019, CCIS 1159, pp. 284–298, 2020.
https://doi.org/10.1007/978-981-15-3425-6_23

with the real landscape, so that we can understand the working principle of the algorithm and solve the practical problems better [2]. FLA refers to a collection of data technologies used to extract descriptive or numerical metrics related to fitness landscape attributes [3].

Several descriptive features have been proposed in the earlier researches of FLA. For instance, the concept closely related to multimodality is smoothness, which refers to the magnitude of change in fitness within the neighborhood. Besides, the rugged landscape has great fluctuations between the neighborhoods, showing steep ascents and descents. Furthermore, the neutral landscape has large flat areas or stepped surfaces whose input changes do not produce significant output changes. Of course, there are some other accessorial measures. For example, information landscape hardness (ILH) by Borenstein and Poli [4,5] with extensions [6] was focused on deception in terms of difference from a landscape with perfect information for search. The result of the evaluation is a value in the range of [0,1], where 0 indicates no misleading information and 1 indicates maximum misleading information. In addition, fitness cloud (FC) by Verel et al. [7] with extensions [8] was concentrated on evolvability. It uses a scatter plot to represent the relationship between parents and the child. What is more, the negative slope coefficient (NSC) has been defined to capture some of the characteristics of FC with a single number. It is known from the classification hypothesis [8]: If NSC $= 0$, the problem is easy to solve; if NSC < 0, its value quantifies this difficulty: the smaller the value, the harder the problem. Vanneschi et al. [8,9] discussed the pros and cons of this measure.

These measures can help us find characteristics of the fitness landscape. It is worth noting that the ultimate goal of FLA in this work is to find the correlation between the properties of the fitness landscape and the performance of algorithms. Some studies have made constructive progress in this regard.

(1) Discrete fitness landscape analysis and its application on practical industrial problems

Information on perfect landscapes through the means of discrete time fourier transform (DTFT) and dynamic time warping (DTW) distances was obtained by Lu et al. [10]. In order to analyze the fitness landscape deeply, the authors proposed five methods, including the stability of amplitude variation, keenness, periodicity, similarity and the degree of change in average fitness. In addition, the author applied these criteria to task scheduling issues to illustrate fairness and adaptability.

(2) Improvement and statistical study on the characteristics of fitness landscape

In the evolutionary computation community, the properties of dispersion metric were studied by Morgan et al. [11]. In order to improve the defects of dispersion metric, the author proposed three independent modifications to the basic methodology, namely, the standardization with the dispersive boundary, the LP norm of fraction p, and the random walk with fixed steps. Finally, the results demonstrated that these improvements can promote convergence and increase the separability of problems. A theorem was proved in this paper by deriving the formula: Given $t = 1$ and $S = [0, 1]^D$, the dispersion of solutions sampled uniform randomly from S will converge to $1/\sqrt{6}$ as $D \to \infty$. In the meantime, t is the number of subsamples, S is the value in the interval [0,1] and D is the dimension of space.

(3) Improving adaptive algorithms by using the characteristics of fitness landscape

Li et al. [12] proposed a new self-feedback DE algorithm (SFDE), which selected the optimal mutation strategy by extracting the features of local fitness landscape. The probability distributions of single mode and multimodality are calculated in each local fitness landscape. The advantage of the self-feedback control mechanism is that when the group falls into the local optimal solution, the inferior solution is more conducive to the population which can help to jump out of the local optima.

(4) Analysis and research on dynamic fitness landscape

Static FLA focuses on extracting the attributes of a problem without considering any information about the optimization algorithm. In contrast, dynamic FLA combines the behavior of the algorithm with the attributes of optimization problems to determine the effectiveness of a given algorithm to solve the problem. Wang et al. [3] used the concept of population evolvability as an important basis for dynamic FLA. The authors utilized the evolutionary of the population (evp) to represent the evolution of the entire population. Finally, the effectiveness of the proposed algorithm selection framework was proved by experiments.

The main difference between this work and the above researches is that we find that the spatial topography of the optimization problem can directly reflect characteristics of the problem. Since Differential Evolution (DE) algorithm is a very popular EA, we take it as an example and design mutation strategy selection based on DE to solve optimization problems. Therefore, the purpose of this work is to find the proper strategy of DE algorithm for each problem. Then, we establish a strategy selection model by learning the relationship between excellent strategies and features of the landscape to improve the intelligent solving ability of optimization problems.

The rest of this paper is organized as follows. Section 2 is devoted to explaining the measures of FLA and DE algorithm summarized in this paper. Section 3 discusses the mutation strategy selection task based on the black box optimization problem in detail. Further experimental analyses are presented in Sect. 4. Finally, Sect. 5 provides the conclusion.

2 Measures of Fitness Landscape Analysis

2.1 Global Sampling and Local Sampling

Sampling refers to the process of converting continuous variables in the spatial domain into discrete variables. Global sampling is to taking values over the entire spatial region while local sampling is to collecting data in a partial region [13]. We reduced the problem space to 2 dimensions and used function $F = x^2$ to better demonstrate both sampling methods. The schematic diagram is shown in Fig. 1. It should be mentioned that the two sampling methods are applied to the characteristics of different fitness landscapes according to the dissimilar situation.

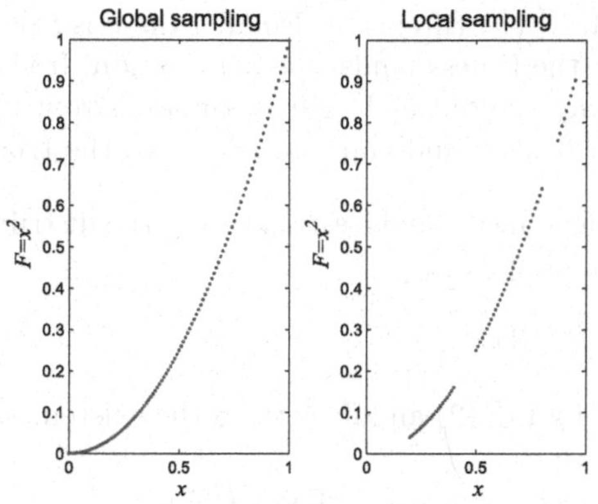

Fig. 1. The schematic diagram of global sampling and local sampling

2.2 Fitness Landscape Characteristics

The extraction of landform information helps to analyze characteristics of the problem to be optimized. As a preliminary study, this paper selects four characteristics to measure continuous problems with unknown optima based on sampling points of fitness landscape.

(1) The number of optimal values (NUM)

We have improved a simple method to measure NUM in FLA which is based on the achievements of Sheng et al. [14]. Its implementation is described as follows:

For a random sample $x_1, \ldots, x_u$.

1) Finding the best point of the sample and set it to x^*. Then, calculating the distance (d_i) between each $x_i (i = 1, \ldots, u)$ and x^* as Eq. (1):

$$d_i = \sum_{j=1}^{n} \left| x_{i,j} - x_j^* \right|. \tag{1}$$

The difference between the above distance and the Euclidean distance is that the former performs fast calculation by removing the square root.

2) Sorting the individuals based on the distance value from low to high, and denoting the order by $k_1, k_2, \ldots, k_u$.

3) Setting $c = 0$ initially. Then, the value of c will be increased by 1, if $x(k_m)(m = 1, \ldots, u)$ is better than $x(k_{m-1})(m = 1, \ldots, u)$ (if exists) and $x(k_{m+1})(m = 1, \ldots, u)$ (if exists). Finally, the c is taken as the number of optimal value in the fitness landscape observation. It should be emphasized that $x(k_m)$ is only the optimal value estimated from the sample to reflect attributes of the fitness landscape, which is not the true optimum.

Intuitively, the ruggedness can be estimated by the distribution of the optimal values in the current sample.

(2) Basin size ratio (BSR)

BSR is caculated by Eq. (2) and it pictures the existence of a dominant basin [15].

$$BSR = \frac{\max_x |B(x)|}{\min_x |B(x)|}. \tag{2}$$

where $\max_x |B(x)|$ is the maximum fitness value in the local sampling points, $\min_x |B(x)|$ is the minimum value. Due to the wide range in fitness values of various problems, normalized BSR is employed in our work.

(3) Keenness (KEE)

Lu et al. [16] proposed a method to describe keenness of the topography. It is computed by Eq. (3):

$$\begin{aligned} KEE = a_{\text{sum}} \times (-1) + b_{\text{sum}} \times (-0.6) + c_{\text{sum}} \\ \times (-0.2) + d_{\text{sum}} \times (-0.2) + e_{\text{sum}} \times (+1). \end{aligned} \tag{3}$$

where the coefficients for a_{sum}, b_{sum}, c_{sum}, d_{sum} and e_{sum} are allocated according to the contribution to the keenness degree. The larger the value of KEE, the sharper the solution space.

(4) Fitness distance correlation (FDC)

FDC was proposed by Jones et al. [17] to measure the relationship between the parameter space and the fitness values. It is designed to evaluate whether the landscape is unimodal or multimodal and whether it has a strong global structure. It is shown by Eq. (4):

$$FDC = \frac{1}{n-1} \sum_{i=1}^{n} \left(\frac{y_i - \overline{y}}{\hat{\varepsilon}_y} \right) \left(\frac{d_i - \overline{d}}{\hat{\varepsilon}_d} \right). \tag{4}$$

where $\overline{y}$ and $\overline{d}$ are the mean fitness and the mean distance between x_0 and x_i, $\widehat{\varepsilon_y}$ and $\widehat{\varepsilon_d}$ are the sample standard deviation of the fitness and the distance, respectively. It is worth mentioning that FDC is invariant to shifts and rotations on the parameter space and the fitness values, because they are global isometries of the Euclidean space.

2.3 Differential Evolution Algorithm

In EA, there always are many different offspring generating operators, such as different crossover operators, different mutation operators and so on. Usually, different operators have the different performance on different types of optimization problems. Since DE has a variety of mutation strategies [18,19], we take it as an example in this paper. Let D is the dimension of problems, x_j^{U} and x_j^{L} are the upper and lower bounds of the constraint range of individual x_j, respectively. Then the minimization problem can be described as: $f_{\min}\left(x_1, x_2, \ldots, x_D\right)$, where $x_j^{\mathrm{L}} \leq x_j \leq x_j^{\mathrm{U}}, j = 1, 2, \ldots, D$. DE is committed to continuously improving the ability of populations to adapt to the external environment through personal communication, competition and iteration to achieve the goal of getting the best solution [20]. The flow chart of the standard DE is shown in Fig. 2.

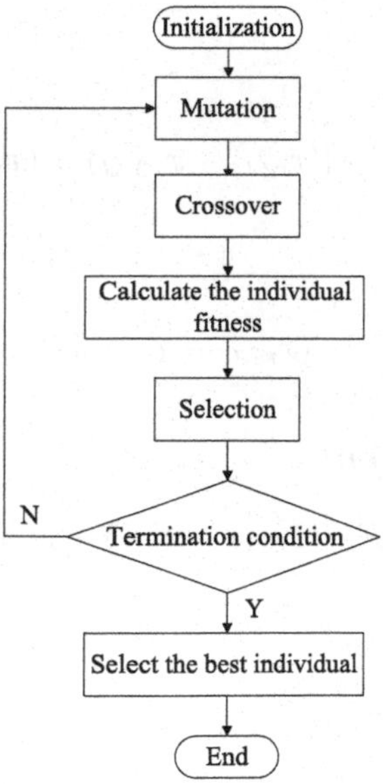

Fig. 2. The flow chart of the standard DE

(1) Initialization

We use Eq. (5) to generate initial individuals which is satisfying constraints in the D-dimensional space as the $0th$ generation population.

$$x_{i,j,0} = x_{i,j}^{\mathrm{L}} + rand \cdot \left(x_{i,j}^{\mathrm{U}} - x_{i,j}^{\mathrm{L}}\right). \tag{5}$$

where $i = 1, \ldots, NP; j = 1, \ldots, D$; $rand$ is a uniformly generated random number in [0,1]; NP is the size of the population.

(2) Mutation

The six most common used mutation strategies are listed from Eqs. (6) to (11).

DE/rand/1/bin:

$$v_{i,G} = x_{r1,G} + F \cdot (x_{r2,G} - x_{r3,G}). \tag{6}$$

DE/best/1/bin:

$$v_{i,G} = x_{best,G} + F \cdot (x_{r1,G} - x_{r2,G}). \tag{7}$$

DE/rand/2/bin:

$$v_{i,G} = x_{r1,G} + F \cdot [(x_{r2,G} - x_{r3,G}) + (x_{r4,G} - x_{r5,G})]. \tag{8}$$

DE/best/2/bin:

$$v_{i,G} = x_{best,G} + F \cdot [(x_{r1,G} - x_{r2,G}) + (x_{r3,G} - x_{r4,G})]. \tag{9}$$

DE/current-to-rand/1/bin:

$$v_{i,G} = x_{i,G} + F \cdot [(x_{r1,G} - x_{i,G}) + (x_{r2,G} - x_{r3,G})]. \tag{10}$$

DE/current-to-best/1/bin:

$$v_{i,G} = x_{i,G} + F \cdot [(x_{best,G} - x_{i,G}) + (x_{r1,G} - x_{r2,G})]. \tag{11}$$

where $r1, r2, r3, r4, r5$ represent the random numbers between 1 and NP, which are different from each other and not equal to the number of target individual. $x_{best,G}$ symbolizes the optimal individual in the Gth generation population. $v_{i,G}$ stands for the individual after the mutation operation. F is the scaling factor that controls the amplification of the bias variable. The value of F is generally set to 0.5.

(3) Crossover

There are two common crossover means in the DE algorithm. The binomial crossover is usually preferred than exponential crossover [21,22], which is expressed as Eq. (12):

$$u_{i,G} = \begin{cases} v_{i,G} & \text{if } rand[0,1] \leq CR \text{ or } j = j_{rand} \\ x_{i,G} & \text{otherwise} \end{cases}. \tag{12}$$

where j_{rand} is a random number in $[1, 2, ..., D]$ to ensure that at least one dimension component of the intersecting individual will be different from the target individual. CR is called crossover probability which is generally recommended to 0.9.

(4) Selection

The selection is to determine if there are individuals in the parents who can become members of the next generation. The rules for selecting operation are as Eq. (13), where $f(\cdot)$ is the value of the objective function.

$$x_{i,G+1} = \begin{cases} u_{i,G} & \text{if } f(u_{i,G}) \leq f(x_{i,G}) \\ x_{i,G} & \text{otherwise} \end{cases}. \tag{13}$$

(5) Repeat steps (2)–(4), until the stopping criterion is satisfied.

3 Mutation Strategy Selection for Black-Box Optimization Problems Based on FLA

3.1 Overall Framework

Figure 3 depicts an overall framework of the mutation strategy selection task based on FLA, where the features of candidate function A are used as input portion, and the best strategy B is recommended as its output. This framework is composed of three related components.

Sampling: For the sake of generality, uniform sampling is used in this work to generate samples that can represent various states (ie, $P_1, \ldots, P_n$, which are presented in Fig. 3). For some features, global sampling (ie, BSR, KEE, etc.) is not applicable. Therefore, this paper performs local sampling and normalization to make these features more effective and persuasive.

Calculating the Features: It should be reminded that this paper is a preliminary study, so the benchmark functions are used for convenience. Since their structures are known, it is easy to get various characteristics of the problem to improve the performance of algorithms. What is more, the study of single objective optimization algorithms is the basis of more complex optimization algorithms [23–27].

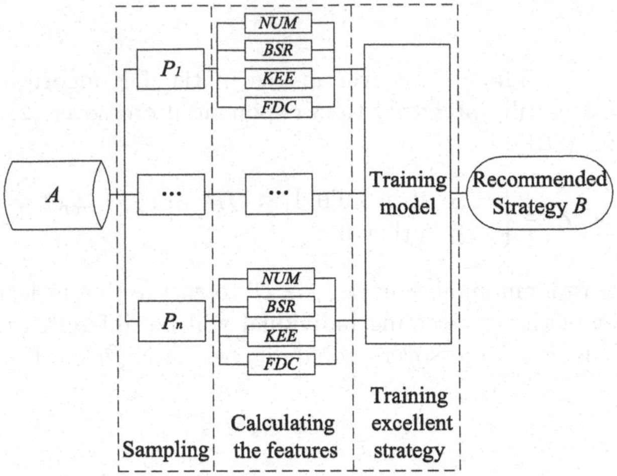

Fig. 3. Overall framework of the mutation strategy selection

In the CEC2005, the focus is on low and medium dimensional problems [23]. The benchmark set consists of 25 test functions which is divided in four groups: unimodal, basic multimodal, expanded multimodal, and hybrid composition. In the CEC2013 test set [24], the previously proposed composition functions are improved and additional test functions are included. In the CEC2014 [25], CEC2015 [26] and CEC2017 [27], Liang and Suganthan et al. developped benchmark problems with some novel features such as novel basic problems, composing test problems by extracting features dimension-wise from several problems, graded level of linkages, rotated trap problems, and so on.

It is worth noting that CEC2017, CEC2015 and CEC2014 with a total of 75 functions are treated as the black box optimization problems in this paper. The individual features computed for each function are used as the input of our model.

Training the Excellent Strategy: The benchmark functions are tentatived under the standard operating conditions in combination with six strategies of the standard DE algorithm, respectively. Furthermore, K-Nearest Neighbor (KNN) and Random Forest (RF) are used as learners for the training model of all functions with 4 features, respectively. The core idea of KNN [28] is that if most of the k nearest neighbors in the characteristical space belong to a certain category, the sample also belongs to this category and has the characteristics of the neighboring samples on this type. RF is a classifier with multiple decision trees whose output categories are determined by the mode of the multi-tree output class [29]. For each test function, we recommend the best mutation strategy based on the proposed strategy selection framework.

4 Experimental Analysis

4.1 Feature Values and Benchmark Functions

The population size of all benchmark functions is set to 100 while the sampling points of the feature is tune into 3000. Moreover, the search range is $[-100, 100]^D$ and the FEs budget is preset to $10^4 \times D$, where D is the problem dimension.

The mean and variance for each feature of 75 functions were obtained by running 20 times. The values are shown in Fig. 4. From the figure, we can see that each feature presents diversity among different functions, especially in complex functions. In the meanwhile, the value of variance for BSR and FDC are far less than others in the range of 0 to 0.4, so they are more stable than NUM and KEE.

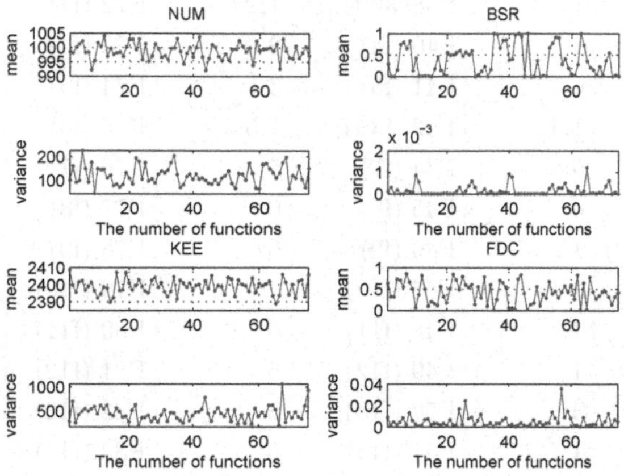

Fig. 4. Feature values for each function

In addition, in order to compare the significance of difference between various strategies, the assessment is performed using the Friedman non-parametric test method [30]. In that case, the parameters of standard DE are: $F = 0.5, CR = 0.9$. The standard DE with six strategies are run independently 51 times to find the standard deviation, respectively. When the average ranks obtained by each strategy in the Friedman test are the same small, we hold them to be equally good strategies. The judgments are concluded in Table 1, where the number in the even columns refers to the used mutation strategy corresponding to Sect. 2 and the number in parentheses refers to the location of the function in this benchmark problem.

As we can see from the table, for a given optimization problem, each function is not limited to an excellent strategy, such as f1, f2 in CEC2017. We put the same excellent strategies for the function as the different data in this situation (ie. F1–2, F3–4, etc.). We can also find that DE/rand/1/bin which is recorded as the first strategy has been proven to perform well on most functions.

4.2 Results and Discussion

The results of Table 1 which are obtained by performance evaluation can be considered as the gold standard for mutation strategy selection. We can evaluate the outcomes gained from the proposed framework at the same time. The evaluation criteria are as follows: For the test functions, if the excellent strategy selected based on the performance evaluation contains the strategy recommended by the

Table 1. The excellent strategy for benchmark functions

CEC2017	Excellent	CEC2014	Excellent	CEC2015	Excellent
F1–2 (f1)	1,4	F35–36 (f1)	1,4	F69–70(f1)	1,4
F3–4 (f2)	1,2	F37 (f2)	1	F71 (f2)	1
F5 (f3)	1	F38–39 (f3)	1,2	F72 (f3)	3
F6 (f4)	1	F40 (f4)	2	F73 (f4)	6
F7 (f5)	6	F41 (f5)	3	F74 (f5)	4
F8–9 (f6)	1,4	F42–43 (f6)	1,5	F75 (f6)	1
F10 (f7)	6	F44 (f7)	5	F76 (f7)	6
F11 (f8)	6	F45 (f8)	6	F77 (f8)	1
F12–13 (f9)	1,2	F46 (f9)	6	F78 (f9)	6
F14 (f10)	6	F47 (f10)	4	F79 (f10)	4
F15 (f11)	1	F48 (f11)	6	F80 (f11)	6
F16 (f12)	1	F49 (f12)	6	F81 (f12)	6
F17 (f13)	4	F50 (f13)	6	F82 (f13)	2
F18 (f14)	1	F51 (f14)	6	F83 (f14)	1
F19 (f15)	1	F52 (f15)	5	F84 (f15)	1
F20 (f16)	1	F53 (f16)	4		
F21 (f17)	1	F54 (f17)	1		
F22 (f18)	1	F55 (f18)	1		
F23 (f19)	1	F56 (f19)	6		
F24 (f20)	1	F57 (f20)	1		
F25 (f21)	6	F58 (f21)	1		
F26 (f22)	5	F59 (f22)	1		
F27 (f23)	5	F60–61 (f23)	1,2		
F28 (f24)	4	F62 (f24)	4		
F29 (f25)	1	F63 (f25)	5		
F30 (f26)	6	F64 (f26)	6		
F31 (f27)	2	F65 (f27)	6		
F32 (f28)	1	F66 (f28)	2		
F33 (f29)	1	F67 (f29)	1		
F34 (f30)	1	F68 (f30)	1		

proposed framework, the recommended strategy can be deemed as a correct answer; otherwise, it is a mistake.

On the basis of this evaluation method, the output accuracy is a statistical indicator in the proposed framework. It refers to the number of correct classifications divided by the total number of test functions. It is employed to KNN and RF to manifest which learner is more suitable for the suggested framework.

The Selection of the Number of Classifier Nodes: The proposed strategy selection framework is based on the appropriate number of classifier nodes and data set folds. Under the circumstance of CEC2015 as a testing set, we put up the nodes number from 1 to 10 for each classifier and then run the suggested framework in this research. The results are demonstrated in Fig. 5.

We show the impact of nodes number on the output accuracy of the classifier as a line graph. Experimental results prove that the value of RF is higher than KNN for each node from the figure. Moreover, $K = 4$ behaves better in the case of less consumption.

The Selection of the Number of Data Set Folds: After that, we set up M from 1 to 10 to verify the effect of folds number on the output accuracy on the basis of $K = 4$. The specific details are shown in Fig. 6.

It can be seen that, for RF, the classification accuracy is as high as 0.525 if $M = 3$. In the same time, the best value is up to 0.571 in KNN, which manifests that the KNN algorithm successfully satisfies the intrinsic mapping relationship of the proposed framework. However, it can also be seen that the values of KNN are more volatile than RF. Because the output accuracy of KNN is as low as 0 when M = 1, which is closely related to the working principle of KNN. At the same time, although the lowest value of RF at M = 9 is close to 0.1, it is still

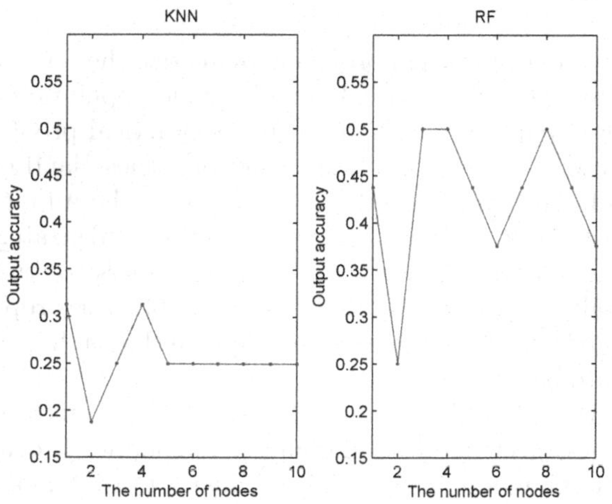

Fig. 5. The impact of nodes number on the output accuracy of each classifier

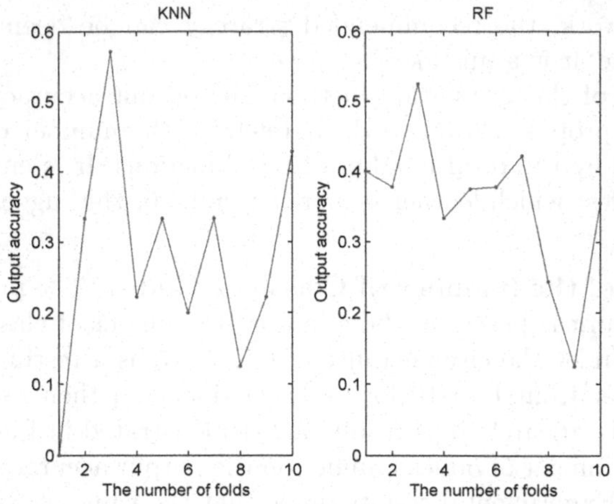

Fig. 6. The impact of folds number on the output accuracy of each classifier

Table 2. The effect of the classifier on the output accuracy

Classifier	Output accuracy
KNN	**0.571**
RF	0.525

stable overall. The final results are shown in Table 2. And the best value is bold. It must be admitted that the results are not as good as we expected, but this framework has proven to be effective.

5 Conclusion

As the complexity of optimization problems increases, there is an urgent need to develop learning-based methods to adaptively guide population evolution based on the multifaceted requirements and nature of practical problems. This paper proposes a mutation strategy selection based on fitness landscape analysis to meet the demand. The sample model are designed, where four features of the fitness landscape are used as inputs and the recommended strategy after training are applied as the output. Then, we use some classifiers to study the mapping relationship. Finally, experimental analyses show that the proposed framework can efficiently match excellent strategies to improve the intelligent solving ability of optimization problems.

In the future, we will conduct extended research based on the proposed framework. From the prospect of frame designers, further work can be focused on expanding the sampling size and adding verified sets to demonstrate effectiveness of the proposed work. From the perspective of these measures, the next step is to increase features of the fitness landscape to improve the classified accuracy.

Acknowledgement. This work is supported by the National Natural Science Foundation of China (61922072, 61876169, 61673404, 61976237).

References

1. Stadler, P.F.: Fitness landscapes. In: Lässig, M., Valleriani, A. (eds.) Biological Evolution and Statistical Physics, pp. 183–204. Springer, Heidelberg (2002). https://doi.org/10.1007/3-540-45692-9_10
2. Pitzer, E., Affenzeller, M.: A comprehensive survey on fitness landscape analysis. In: Fodor, J., Klempous, R., Suárez Araujo, C.P. (eds.) Recent Advances in Intelligent Engineering Systems, pp. 161–191. Springer, Heidelberg (2012). https://doi.org/10.1007/978-3-642-23229-9_8
3. Wang, M., Li, B., Zhang, G., et al.: Population evolvability: dynamic fitness landscape analysis for population-based metaheuristic algorithms. IEEE Trans. Evol. Comput. **22**, 550–563 (2018)
4. Borenstein, Y., Poli, R.: Information landscapes. In: GECCO 2005: Proceedings of the 2005 Conference on Genetic and Evolutionary Computation, pp. 1515–1522. ACM Press, New York (2005)
5. Borenstein, Y., Poli, R.: Information landscapes and problem hardness. In: GECCO 2005: Proceedings of the 2005 Conference on Genetic and Evolutionary Computation, pp. 1425–1431. ACM Press, New York (2005)
6. Borenstein, Y., Poli, R.: Decomposition of fitness functions in random heuristic search. In: Stephens, C.R., Toussaint, M., Whitley, D., Stadler, P.F. (eds.) Foundations of Genetic Algorithms, pp. 123–137. Springer, Heidelberg (2007). https://doi.org/10.1007/978-3-540-73482-6_8
7. Verel, S., Collard, P., Clergue, M.: Where are bottlenecks in NK fitness landscapes? In: The 2003 Congress on Evolutionary Computation, pp. 273–280 (2003). https://doi.org/10.1109/CEC.2003.1299585
8. Vanneschi, L.: Theory and practice for efficient genetic programming. Ph.D. thesis, Faculty of Sciences, University of Lausanne, Switzerland (2004)
9. Vanneschi, L., Tomassini, M., Collard, P., Vérel, S.: Negative slope coefficient: a measure to characterize genetic programming fitness landscapes. In: Collet, P., Tomassini, M., Ebner, M., Gustafson, S., Ekárt, A. (eds.) EuroGP 2006. LNCS, vol. 3905, pp. 178–189. Springer, Heidelberg (2006). https://doi.org/10.1007/11729976_16
10. Lu, H., Shi, J., Fei, Z., et al.: Measures in the time and frequency domain for fitness landscape analysis of dynamic optimization problems. Soft Comput. **51**, 192–208 (2017)
11. Morgan, R., Gallagher, M.: Sampling techniques and distance metrics in high dimensional continuous landscape analysis: limitations and improvements. IEEE Trans. Evol. Comput. **18**(3), 456–461 (2013)
12. Li, W., Li, S., Chen, Z., et al.: Self-feedback differential evolution adapting to fitness landscape characteristics. Soft Comput. **23**(4), 1151–1163 (2019)
13. Morris, M.D., Mitchell, T.J.: Exploratory designs for computational experiments. J. Stat. Plann. Infer. **43**(3), 381–402 (1992)
14. Shen, L., He, J.: A mixed strategy for evolutionary programming based on local fitness landscape. In: 2010 IEEE Congress on Evolutionary Computation, Barcelona, Spain, pp. 1–8. IEEE (2010)

15. Munoz, M.A., Kirley, M., Halgamuge, S.K.: Landscape characterization of numerical optimization problems using biased scattered data. In: 2012 IEEE Congress on Evolutionary Computation, Brisbane, QLD, Australia, pp. 1–8. IEEE (2012)
16. Lu, H., Shi, J., Fei, Z., et al.: Analysis of the similarities and differences of job-based scheduling problems. Eur. J. Oper. Res. **270**(3), 809–825 (2018)
17. Munoz, M.A., Kirley, M., Smith-Miles, K.: Reliability of exploratory landscape analysis (2018). https://doi.org/10.13140/RG.2.2.23838.64327
18. Storn, R., Price, K.: Differential evolution - a simple and efficient adaptive scheme for global optimization over continuous spaces. Technical report, International Computer Sciences Institute, Berkeley, California, USA (1995)
19. Storn, R., Price, K.: Differential evolution: a simple and efficient heuristic for global optimization over continuous spaces. J. Global Optim. **11**(4), 341–359 (1997)
20. Otieno, F.A.O., Adeyemo, J.A., Abbass, H.A., et al.: Differential evolution: a simple and efficient adaptive scheme for global optimization over continuous spaces. Trends Appl. Sci. Res. **5**(1), 531–552 (2002)
21. Das, S., Suganthan, P.N.: Differential evolution: a survey of the state-of-the-art. IEEE Trans. Evol. Comput. **15**(1), 4–31 (2010)
22. Das, S., Mullick, S.S., Suganthan, P.N.: Recent advances in differential evolution - an updated survey. IEEE Trans. Evol. Comput. **27**, 1–30 (2016)
23. Suganthan, P.N., Hansen, N., Liang, J.J., et al.: Problem definitions and evaluation criteria for the CEC 2005 special session on real-parameter optimization. Technical report, Nanyang Technological University, Singapore, KanGAL Report, IIT Kanpur, India (2005)
24. Liang, J.J., Qu, B.Y., Suganthan, P.N., et al.: Problem definitions and evaluation criteria for the CEC 2013 special session and competition on real-parameter optimization. Technical report, Zhengzhou University, Zhengzhou, China and Nanyang Technological University, Singapore (2013)
25. Liang, J.J., Qu, B.Y., Suganthan, P.N.: Problem definitions and evaluation criteria for the CEC 2014 special session and competition on single objective real-parameter numerical optimization. Technical report, Zhengzhou University, Zhengzhou, China and Nanyang Technological University, Singapore (2013)
26. Liang, J.J., Qu, B.Y., Suganthan, P.N., et al.: Problem definitions and evaluation criteria for the CEC 2015 competition on learning-based real-parameter single objective optimization. Technical report, Zhengzhou University, Zhengzhou, China and Nanyang Technological University, Singapore (2014)
27. Wu, G., Mallipeddi, R., Suganthan, P.N.: Problem definitions and evaluation criteria for the CEC 2017 competition on constrained real-parameter optimization. Technical report, National University of Defense Technology, Changsha, China and Kyungpook National University, Daegu, South Korea and Nanyang Technological University, Singapore (2017)
28. Friedman, M.: The use of ranks to avoid the assumption of normality implicit in the analysis of variance. J. Am. Stat. Assoc. **32**(200), 675–701 (1937)
29. Altman, N.: An introduction to Kernel and nearest-neighbor nonparametric regression. Am. Stat. **46**(3), 175–185 (1992)
30. Breiman, L.: Random forests. Mach. Learn. **45**(1), 5–23 (2001)

Ensemble Learning Based on Multimodal Multiobjective Optimization

Jing Liang[1], Panpan Wei[1(✉)], Boyang Qu[2], Kunjie Yu[1], Caitong Yue[1], Yi Hu[1], and Shilei Ge[1]

[1] School of Electrical Engineering, Zhengzhou University, Zhengzhou, China
{liangjing,yukunjie}@zzu.edu.cn, ppwei0218@163.com, zzuyuecaitong@163.com, eehuyi@163.com, shileige1001@163.com

[2] School of Electronic and Information Engineering, Zhongyuan University of Technology, Zhengzhou, China
qby1984@hotmail.com

Abstract. In ensemble learning, the accuracy and diversity are two conflicting objectives. As the number of base learners increases, the prediction speed of ensemble learning machines drops significantly and the required storage space also increases rapidly. How to balance these two goals for selective ensemble learning is an extremely essential problem. In this paper, ensemble learning based on multimodal multiobjective optimization is studied in detail. The great significance and importance of multimodal multiobjective optimization algorithm is to find these different classifiers ensemble by considering the balance between accuracy and diversity, and different classifiers ensemble correspond to the same accuracy and diversity. Experimental results show that multimodal multiobjective optimization algorithm can find more ensemble combinations than unimodal optimization algorithms.

Keywords: Selective ensemble learning · Ensemble learning · Multimodal multiobjective optimization · Learner diversity

1 Introduction

Ensemble learning accomplishes the task of learning by building multiple individual-learners and is a general term for a method (it combines inducers to make decisions and the inducers are the individual-learners). The algorithms commonly used by individual-learners can be any type of machine learning algorithm (such as decision tree, neural network, linear regression model, etc.). Ensemble learning generate a set of individual learners by training data and then combine individual learners by combining strategies (simple averaging, majority voting, weighted averaging, majority voting, plurality voting, etc.). The main premise of ensemble learning is that by combining multiple individual-learners, the error of a single individual-learner may be compensated by other individual-learners. Therefore, the overall predictive performance will be better than the

L. Pan et al. (Eds.): BIC-TA 2019, CCIS 1159, pp. 299–313, 2020.
https://doi.org/10.1007/978-981-15-3425-6_24

overall predictive performance of a single individual-learner. So ensemble learning has become increasingly popular in the field of machine learning, and has been widely used in prediction of peer-to-peer lending [1], spam filtering [2], search engine [3], natural language processing [4], pattern classification, data mining [5], and other fields of learning tasks.

Ensemble learning can use multiple learners to achieve better results than a single learner, so a more intuitive idea is to achieve better integration through a large number of base learners. Currently, most the ensemble learning algorithm adopts this kind of thinking, the most representative of which is Boosting, Bagging, AdaBoost, and Random Forest. A complete description of these algorithms can be found in [6]. With the increase of the number of learners participating in integration, the following problems are inevitable in the process of ensemble learning: the speed of ensemble is due to a large number of base learners. The learner generation and aggregation process become slow; the scale of the ensemble is too large to take up a large amount of storage space and to consume the time of the prediction process. Therefore, people began to consider whether using a smaller number of base learners can achieve better performance. In 2002, Zhou et al. proposed the concept of "selective ensemble" and reached a positive conclusion on the above problems [7]. Relevant theoretical analysis and experimental research show that it will not be effective or performance from existing base learners. Bad individual deletion, only select some individual learners for building ensemble can really get better prediction results. However, it is important that how to design the classifier ensemble selection and obtain an ensemble system by trade-off between accuracy and diversity of ensemble classifiers. Evolutionary algorithms are way to design for finding the optimal trade-off between these objectives.

In the current research, evolutionary algorithms have long been applied to individual-learner optimization tasks [8–12], and it is often used to optimize learner algorithm parameters, individual learner internal connection weights, network structure or feature selection. The optimal solution output of the population after iteration will be stopped to construct a learner. Some scholars have introduced cluster intelligence and other evolutionary algorithms into ensemble learning, which is called evolutionary ensemble learning. In evolutionary ensemble learning, most of the work uses cluster intelligence algorithms or other evolutionary algorithms to optimize individual learner internal connection weights [13,14], network structure [15,16], and learning parameters [17]. Some work also uses intelligent optimization algorithms to optimize the choice of individual learners [18] and the weights of multiple body learners [19]. Liu et al. proposed a negative correlation learning (NCL) for ensemble learning to measure the differences between individual learners, and better balance the accuracy and difference of individual learners [20]. Wu et al. proposed an attribute-oriented combined classifier based on niche gene expression programming (AO-ECNG), which can improve the accuracy of sub-classifiers while maintaining their diversity. Sheng et al. proposed a niche-based negative correlation neural network emsemble algorithm that combines negative correlation learning and evolutionary algorithms [21].

There are many studies on multiobjective evolutionary algorithms in ensemble learning. The idea of designing neural networks using a multiobjective technique was initially proposed by Kottathra and Attikiouzel [22]. Kupinski and Anastasio proposed the Niched Pareto GA in which the sensitivity and specificity are the two objectives to optimize. Chandra et al. suggested the neural network ensembles based on evolutionary multiobjective algorithms [23], which optimize the two objectives of accuracy and diversity.

Multiple ensemble subsets with a similar classification accuracy and the same diversity, but there are different ensemble combinations. So classifier ensemble selective problems have the multimodal multiobjective attribute. Multimodal multiobjective optimization is the way to solve the problem of generating ensemble learning models. Solving multimodal multiobjective optimization (MMO) problems is still a challenging task in the field of evolutionary algorithms, MMO not only provides the number of ensemble combinations, but also shows the different picking for decision-makers in the paper. So decision-makers can make a decision according to the preferences.

In this paper, decision variables in MMO are a combination of individual-learners (i.e., a population represents a combination of individual learners). Ensemble strategy of learners combination uses majority voting method. Through the combination of individual-learners input by the population, the accuracy and diversity of each combination are obtained, and then MMO evolves according to these two conflicting objectives for a specified maximum iteration. Experimental results show that the MMO for addressing the classifier ensemble learning is extremely effective.

The rest of this paper is organized as follows. In Sect. 2, the related work are selective ensemble learning, multimodal multiobjective algorithms, and the method of the diversity measurement. Section 3 describes the proposed ensemble learning based on MMO in detail. In Sect. 4, the experimental results and discuss are shown. Eventually, Sect. 5 concludes summary of this paper and plans.

2 Related Work

This section presents related work on selective ensemble learning, multimodal multiobjective optimization, and diversity measurement.

2.1 Selective Ensemble Learning

In recent years, many methods in selective ensemble learning have been proposed. They are divided into selective ensemble learning algorithms based on classification and regression.

Selective ensemble learning algorithms based on regression have less research, it is probably caused by the following factors: the purpose of selective ensemble learning reduce the number of base classifier participating in ensemble learning as much as possible without reducing or even further improve the prediction accuracy of ensemble learning. To greatly reduce the demand for storage space and speed up the prediction, it adopt a certain strategy to select base classifier in the regression problems, however, the deleted base classifier is often less, and the effectivess on the improvement of prediction accuracy is not obvious.

In the selective ensemble learning based on classification, the main algorithms include clustering, sorting, selection, and optimization and so on. The optimization algorithms include GA [24,25], DE [26], and swarm intelligence algorithms for multi-classifier selection [27], regularization theory [28] are used to address the problems. And there are some multiobjective algorithms [24,27] to address the selective ensemble learning. These approachs get the ensemble learning model with regulation. However, only the multiobjective algorithms were used, which can obtain the optimal ensemble learning model by the trade-off between these objectives. The drawbacks is that it cannot provide decision-makers with more choices. Multiple ensemble subsets with a similar classification accuracy and the same diversity, but there are different ensemble combinations. So multimodal multiobjective algorithms can overcome the drawbacks, so it's used to address the problems.

2.2 Multimodal Multiobjective Optimization

The multimodal multiobjective [29–31] problems that have more than one Pareto set, generally speaking, there are at least two Pareto sets in the decision space correspond to the same Pareto front in the objective space. As shown in Fig. 1. where there are two Pareto sets corresponding to the same Pareto front.

It is extremely essential to find all the Pareto sets. Firstly, decision-makers may prefer different PSs. Generally speaking, it can provide the more selections. Secondly, equivalent PSs may have better performance than other PSs. Plenty of studies on MMO have been proposed. Deb [29] proposed the Omni-optimizer algorithms to solve both unimodal and multimodal optimization problems, and then H Ishibuchi et al. Liang et al. carry out many test problems [30–32], consequently many new algorithms have been proposed to solve the MMO problems [32,33].

There are many MMO problems in the real world, for example, the construction of the ensemble model is a multimodal multiobjective optimization problem in this paper. Different combinations of base learners may achieve the same target value.

2.3 Diversity Measurement

In the study of ensemble learning, how to access the diversity of the learner is a pivotal question. The typical approach is to consider the diversity between individual classifiers. Method of diversity measurement includes pairwise approach and non-pairwise. The pairwise diversity measure firstly calculates the diversity value between each pair of classifiers and then uses the average to measure the diversity of the ensemble learning system.

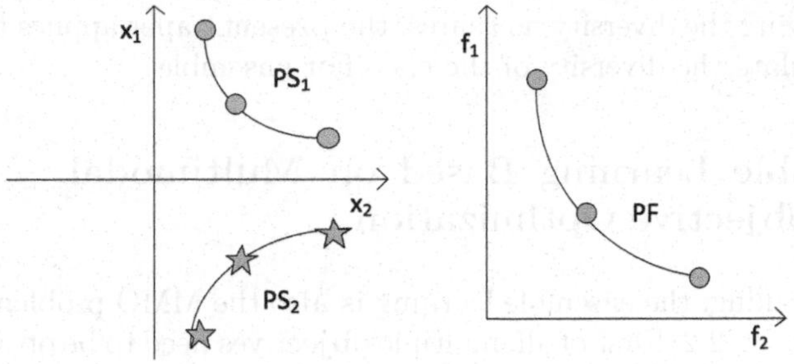

Fig. 1. Multimodal multiobjective optimization problem.

The first diversity measure based on pairwise for the two classifiers is calculated by the formula (1).

$$Q_{ij} = \frac{N^{11}N^{00} - N^{10}N^{01}}{N^{11}N^{00} + N^{10}N^{01}} \quad -1 \leq Q \leq 1 \tag{1}$$

The meaning of N is shown in the Table 1 below. Where N^{11} is the number samples which are correct classified by both C_i and C_j classifier, N^{00} is the inversion of N^{11}. N^{10} is the number classifier C_i correctly classified and classifier C_j incorrectly classified. N^{01} is the number classifier C_i incorrectly classified and C_j incorrectly classified. Where, C_i and C_j are two different classifier in Fig. 2. The table show the detail of the formula.

Table 1. Classification results of two classifiers.

	C_i correct	C_i incorrect
C_j correct	N^{11}	N^{01}
C_j incorrect	N^{10}	N^{00}

The second and third diversity measures between the two classifiers C_i and C_j is defined as *dis* (disagreement, as calculated by formula (2)) and *df* (double fault, as to calculated by formula (2)). It can be seen from the formula (2) that *dis* pays attention to the samples with different classification results of the two classifiers. The double fault represents the number of the sample which is incorrectly classified by both classifiers, respectively.

$$dis_{ij} = \frac{N^{10} + N^{01}}{N} \quad df_{ij} = \frac{N^{00}}{N} \quad (N = N^{00} + N^{01} + N^{10} + N^{11}) \tag{2}$$

The last diversity measures based on the pairwise is defined for two classifiers is correlation coefficient (ρ) that is calculated by formula (3).

$$\rho_{ij} = \frac{N^{11}N^{00} - N^{10}N^{01}}{\sqrt{(N^{11} + N^{10})(N^{01} + N^{00})(N^{11} + N^{01})(N^{10} + N^{00})}} \quad -1 \leq \rho \leq 1 \tag{3}$$

For observing the diversity measures, the present paper applies the first measures to calculate the diversity of the classifier ensemble.

3 Ensemble Learning Based on Multimodal Multiobjective Optimization

Generally speaking, the ensemble learning is also the MMO problem, the details are shown in Sect. 2.2. First of all, multiple objectives need to be optimized simultaneously, for instance, the accuracy, diversity, weight summation, and negative correlation phase. Consequently, there are multiobjective optimization problems. In addition to this, many ensemble learning models have the same objectives, so, it can be regarded as multimodal optimization problems. And the details are presented in the next section.

3.1 Algorithm Introduction

In this section, we will first introduce the chromosome encoding. As shown in Fig. 2. Assume we have nine classifiers, the 9 bits are used as the selecting classifiers. If one of the genes in the 1–9 norms is set to '1', this shows that all nine classifiers are selected for combining. It is shown in Fig. 3. The '1' represent the classifier is selected, and the '0' represents the classifier is unselected. For our instance, classifiers #2, #3, #5, #6, and #8 is selected for combination.

1	2	3	4	5	6	7	8	9
0	1	1	0	1	1	0	1	0

Fig. 2. Chromosome encoding (9 classifiers).

Two objectives, error and diversity, are considered in the ensemble learning based on multimodal multiobjective optimization. The first objective (f_1) is the error of selected classifiers for the ensemble in formula (4), and the second objective (f_2) is the mean of Q - statistic (classifiers are selected for ensemble) in formula (5).

$$f_1 = \frac{N_{incorrect}}{N} \tag{4}$$

$$f_2 = \exp\left(\frac{2}{n(n-1)} \sum_{i=1}^{n-1} \sum_{j=i+1}^{n} \ln\left(Q_{i,j}\right)\right) \tag{5}$$

where $N_{incorrect}$ represents the number of errors that are classified for ensemble with selected classifiers in the training set. In the second objective, n represent the number of selected classifiers, and the details has been introduced in Sect. 2.

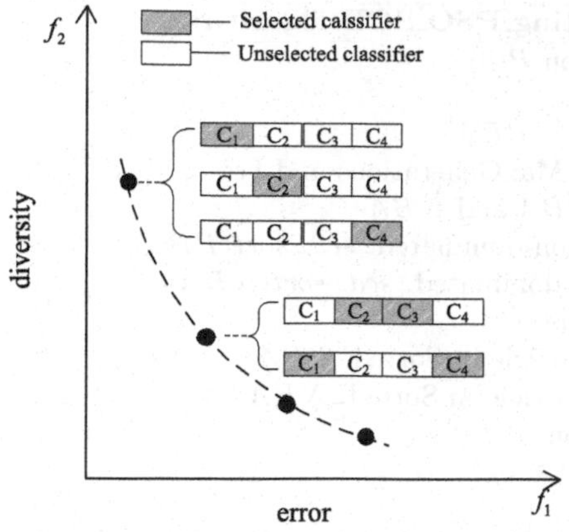

Fig. 3. Ensemble learning based on multimodal multiobjective optimization.

Figure 3 demonstrates an MMO situation in selective ensemble learning. The f_1 and f_2 represent the two objective: error and diversity of ensemble learning, the dotted line represents the PF of this bio-objective optimization problem. $C_1 - C_4$ represent the four personal classifiers, and the white rectangles indicate the classifier is not select, and the green rectangles are selected classifier. As shown in the figures, their subsets have the same target values, when the different classifier is selected for ensemble learning, such as C_1, or C_2, or C_3 is selected as ensemble learning model, there are the same error and diversity. They have other situation, when the $\{C_2, C_3\}$, or $\{C_1, C_4\}$ are selected as the ensemble learning model, they have also the same the objective. That's why multimodal multiobjective is used to solve selective ensemble learning problems. The MMO algorithms can provide more than situations. It provides more possibilities for the ensemble learning model with high prediction accuracy and good generalization performance.

3.2 Algorithm Framework

As described in Sect. 3. The MMO can provide more choices of ensemble learning model. First, the ring topology helps fully search the feasible space. Then, the SCD can help keep the equivalent ensemble learning model. Figure 4 shows the execution flow of the MMO for selective classifier ensemble, using the flow chart display method, the termination condition is set to 100 generations, so the steps 2, 3, and 4 are run 100 times. The population $P(0)$ is randomly initialized in the range $[0, 1]$. Algorithm 1 shows the fitness evaluation of MMO algorithms. Extreme Learning Machine (ELM) is used as the base learner. The details of the 'non-dominated_scd_sort' and 'Update NBA using ring topology' is the same with the original MMO [31].

Algorithm 1: MO_Ring_PSO_SCD algorithm.

1.//Initialize Population $P(0)$
2. Evaluation $(P(0))$
3. $PBA = P(0), NBA = PBA$
4. while Generation < MaxGenerations and Fes < MaxFEs do
5.//Sort particles in PBA and NBA
6. Sorted_$PBA\{i\}$ =non-dominated_scd_$sort(PBA)$
7. Sorted_NBA =non-dominated_scd_$sort(NBA)$
8.//Select pbest and nbest
9. pbest = The first particle in Sorted_PBA
10. nbest= The first particle in Sorted_NBA
11.//Update population
12. Update $P(t)$ to $P(t+1)$
13.// Evaluation $(P(t+1))$
14.// Update PBA Put $p(t+1)$ into PBA and remove particles dominated by $P(t+1)$
15.// Update NBA Update NBA using ring topology
16.end while
17.Output the nondominated particles in NBA

This algorithm is composed of seven steps as follows:

Step 1. Normalize the input data
The preprocessing procedure is normalization to zero mean and unit standard deviation.
Step 2. Divide the dataset into the training set and test set.
The dataset is divided into the training set and testing set, 80% of which are used for model training and 20% for model testing.
Step 3. Train the base classifiers.
The classifier is trained using the bootstrap sampling of the training set.
Step 4. Perform the MMO algorithms.
The details of the MMO algorithms as shown in Fig. 5. The chromosome encodes a subset of the classifier. The encodes is shown in the Sect. 3.1.
Step 5. Select the best Pareto optimal.
According to MMO algorithms, the best Pareto situation are selected from the two objectives of error and diversity.
Step 6. Evaluate solutions.
The Pareto optimal solution of the best choice on the training set is evaluated on the corresponding test set. Each solution represents a subset of classifiers.
Step 7. Get the best solution.
The solution is the best pareto optimal.

The benefits of solving problems of ensemble learning based on MMO are listed as follows. First, it can provide more selection for decision-makers. Second, it can help search redundant ensemble learning model. Third, it can reduce the number of classifiers for ensemble learning model and the amount of required storage space.

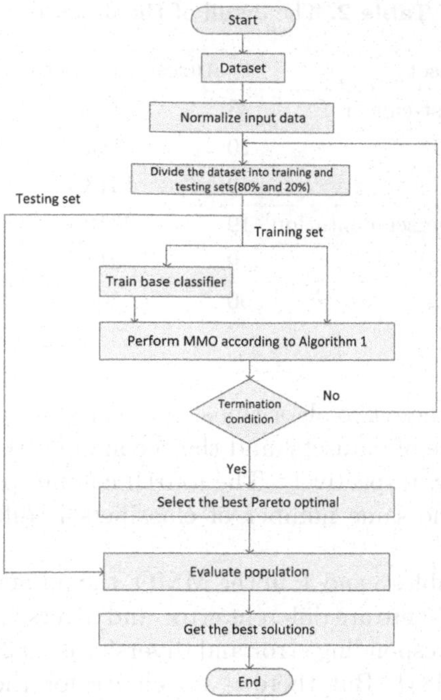

Fig. 4. Flow chart of ensemble learning based on MMO.

4 Experiment

In this study, we use the six datasets with different characteristics. The datasets are selected from the Machine Learning Repository (UCI) [34]. The details of the datasets are given in Table 2 Each dataset is divided into training (80%) and testing (20%). The classifier used the ELM to classify the data. The error and diversity of selective classifier ensemble as the two objective values, in the MMO, the population and the number of maximum evaluation is set to 100 and 10000, respectively. Each data set normalized to zero mean and unit standard deviation. In order to design the ensemble by MMO, 100 base classifiers are used in this study, as shown in Sect. 3, each chromosome includes 100 bits (it present 100 base classifiers).

In the MMO algorithms, the ring topology and special crowding distance can provide more selection for users. In addition to this, it can find redundant classifier ensemble selection. The different classifier can get the same accuracy if you do not have a particular preference for the different classifiers ensemble. So the different classifiers ensemble is redundant. Furthermore, it means that the same number of classifiers in the decision space, but there are different error and diversity.

Two algorithms is used in this study, NSGAII [35] and MO_Ring_PSO_SCD for address problems of clarifies ensemble learning are compared. Tables 3 and 4

Table 2. The detail of the datasets.

Datasets	Features	Instances	Class
Breast-cancer	33	194	2
ILPD	10	583	2
CMC	9	1473	3
Image_segmentation	19	2310	7
Glass	9	214	6
Libras	90	360	15

present the results of these two algorithms, there are four columns. The first column presents the name of datasets, and the second and third column represents the error and diversity, respectively. The fourth column presents the number of selective classifiers (the same number of classifiers is different in the decision space).

As shown in the Tables 3 and 4, in the MMO, the number of classifiers is 31 in the CMC dataset, but there are different error and diversity. When the number of classifier is 31, the corresponding error and diversity is {0.3913, 0.0405}, {0.3820, 0.0508}, {0.3829, 0.0484}. But there is no choice for the different number of classifiers in the NAGAII. In the CMC dataset of MMO, when the error and diversity is {0.3786, 0.0524}, the number of classifiers is 26, 29, 30, 25, 27. It shows that different base clasifiers in the decision space correspond to one point in the objective space.

In the libras dataset, there are no different errors and diversity corresponding to the same number of classifiers in the NSGAII, so solution of NSGAII is less than MMO in the libras dataset. it can demonstrate that MMO can provide more selection for decision-makers, such as CMC, libras, glass, and breast_cancer datasets.

Figure 5 shows the PFs obtained by the two algorithms on four datasets. The horizontal and the vertical axis correspond to the error and diversity respectively. In Fig. 5(a), (b) the MMO algorithms can provide more solutions, and get better accuracy than the NSGAII. In Fig. 5(c), (d), performance of NSGAII and MMO is similar, accuracy of NSGAII is less than accuracy of MMO, but the MMO can provide the more spread PF. In general, the classification accuracy do not deteriorate significantly throughout the process.

This is just the preliminary study on classifier ensemble selection, the accuracy is not an obvious improvement. But the paper aims at drawing attention to find the different classifier combines (in the decision space) corresponding to the same error and diversity (in the objective space). Ordinary algorithms may discard this combination. Therefore, when building classifier combines, it can provide more choices for the decision-makers.

Table 3. Results of the MMO (Partial data results).

Dataset	Error	Diversity	The number of classifiers
CMC	0.3913	0.0415	31
	0.3786	0.0524	26
	0.3786	0.0524	29
	0.3786	0.0524	30
	0.3786	0.0524	25
	0.3786	0.0524	27
	0.3820	0.0508	24
	0.3820	0.0508	28
	0.3820	0.0508	31
	0.3820	0.0508	32
	0.3829	0.0484	25
	0.3829	0.0484	32
	0.3829	0.0484	31
	0.3829	0.0484	37
	0.3829	0.0484	32
	0.3761	0.0543	35
	0.3727	0.0619	33
	0.3727	0.0619	34
	0.3727	0.0619	42
	0.3727	0.0619	42
	0.3871	0.0459	15
	0.3871	0.0459	21
	0.3871	0.0459	30
	0.3913	0.0415	33
	0.3913	0.0415	32
Libras	0.0278	0.7925	25
	0.0278	0.7925	28
	0.0278	0.7925	31
	0.0278	0.7925	32
	0.0417	0.7586	17
	0.0139	0.8697	25
	0.0139	0.8697	28
	0.0417	0.7586	21
	0.0417	0.7586	34
	0.0417	0.7586	34
	0.0417	0.7586	35
	0.0486	0.7510	19
	0.0208	0.8042	28
	0.0313	0.7614	26
	0.0313	0.7614	28
	0.0313	0.7614	15
	0.0313	0.7614	16
	0.0313	0.7614	21
	0.0139	0.8697	33
	0.0139	0.8697	36
	0.0139	0.8697	37
	0.0174	0.8556	33
	0.0174	0.8556	27
	0.0174	0.8556	32
	0.0174	0.8556	39
	0.0174	0.8556	32
	0.0486	0.7510	35

Table 4. Results of the NSGAII (Partial data results).

Dataset	Error	Diversity	The number of classifiers
CMC	0.3345	0.0730	13
	0.3413	0.0593	17
	0.3404	0.0647	12
	0.3438	0.0531	11
	0.3387	0.0645	16
	0.3302	0.0708	17
	0.3396	0.0594	19
	0.3404	0.0591	16
	0.3370	0.0648	14
	0.3362	0.0733	20
Libras	0.0258	0.7772	10
	0.0194	0.9105	15
	0.0194	0.9017	14
	0.0194	0.9002	13

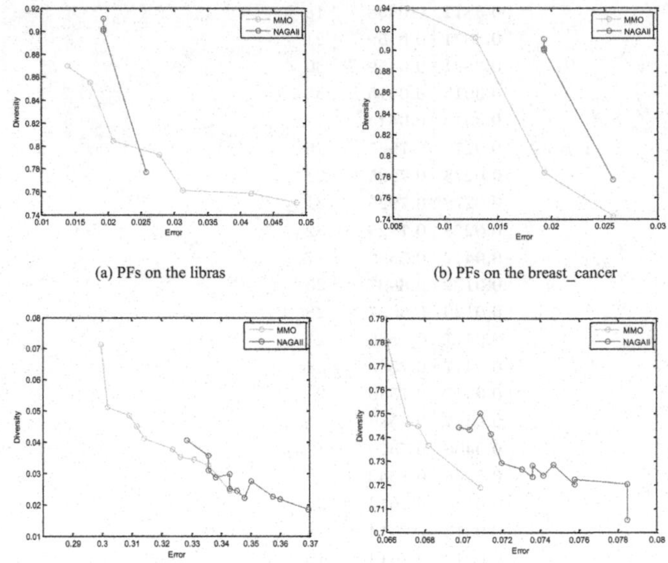

(a) PFs on the libras

(b) PFs on the breast_cancer

(c) PFs on the glass

(d) PFs on the image_segmentation

Fig. 5. The comparisons of PFs obtained by different algorithms.

5 Conclusion

The paper proposed a new challenge, the MMO algorithm is used to address problem of ensemble learning, and also analyzed the effectiveness and essence of the MMO algorithm to address the selective the classifiers for the ensemble. The experimental results shown that the MMO can obtain more solutions in the ensemble learning, and thus more options are available for future work.

However, the paper only focuses on obtaining the solutions, but how to use them is not considered, and the generation of classifiers remains to be studied. Consequently, in future work, we will make use of these solutions and generate the base classifiers to construct a better ensemble model.

Acknowledgments. This work is supported by the National Natural Science Foundation of China (61976237,61922072, 61876169, 61673404).

References

1. Li, W., Ding, S., Wang, H., Chen, Y., Yang, S.: Heterogeneous ensemble learning with feature engineering for default prediction in peer-to-peer lending in China. World Wide Web **23**, 23–45 (2020). https://doi.org/10.1007/s11280-019-00676-y
2. Barushka, A., Hajek, P.: Spam filtering in social networks using regularized deep neural networks with ensemble learning. In: Iliadis, L., Maglogiannis, I., Plagianakos, V. (eds.) AIAI 2018. IAICT, vol. 519, pp. 38–49. Springer, Cham (2018). https://doi.org/10.1007/978-3-319-92007-8_4
3. Bekiroglu, K., Duru, O., Gulay, E., Su, R., Lagoa, C.: Predictive analytics of crude oil prices by utilizing the intelligent model search engine. Appl. Energy **228**, 2387–2397 (2018)
4. Lemaître, G., Nogueira, F., Aridas, C.K.: Imbalanced-learn: a python toolbox to tackle the curse of imbalanced datasets in machine learning. J. Mach. Learn. Res. **18**(1), 559–563 (2017)
5. Oh, S., Lee, M.S., Zhang, B.T.: Ensemble learning with active example selection for imbalanced biomedical data classification. IEEE/ACM Trans. Comput. Biol. Bioinf. **8**(2), 316–325 (2010)
6. Zhang, C., Ma, Y.: Ensemble Machine Learning: Methods and Applications. Springer, Boston (2012). https://doi.org/10.1007/978-1-4419-9326-7
7. Zhou, Z.H., Wu, J., Wei, T.: Ensembling neural networks: many could be better than all. Artif. Intell. **137**(1), 239–263 (2002)
8. Islam, M.M., Xin, Y.: Evolving artificial neural network ensembles. IEEE Comput. Intell. Mag. **3**(1), 31–42 (2008)
9. Pan, L., He, C., Tian, Y., Wang, H., Zhang, X., Jin, Y.: A classification-based surrogate-assisted evolutionary algorithm for expensive many-objective optimization. IEEE Trans. Evol. Comput. **23**(1), 74–88 (2018)
10. Pan, L., He, C., Tian, Y., Su, Y., Zhang, X.: A region division based diversity maintaining approach for many-objective optimization. Integr. Comput. Aided Eng. **24**(3), 279–296 (2017)
11. He, C., Tian, Y., Jin, Y., Zhang, X., Pan, L.: A radial space division based evolutionary algorithm for many-objective optimization. Appl. Soft Comput. **61**, 603–621 (2017)

12. Pan, L., Li, L., He, C., Tan, K.C.: A subregion division-based evolutionary algorithm with effective mating selection for many-objective optimization. IEEE Trans. **2019**, 1–14 (2019)
13. Bui, L.T., Dinh, T.T.H., et al.: A novel evolutionary multi-objective ensemble learning approach for forecasting currency exchange rates. Data Knowl. Eng. **114**, 40–66 (2018)
14. Ojha, V.K., Abraham, A., Snášel, V.: Ensemble of heterogeneous flexible neural trees using multiobjective genetic programming. Appl. Soft Comput. **52**, 909–924 (2017)
15. Mocanu, D.C., Mocanu, E., Stone, P., Nguyen, P.H., Gibescu, M., Liotta, A.: Scalable training of artificial neural networks with adaptive sparse connectivity inspired by network science. Nat. Commun. **9**(1), 2383 (2018). https://doi.org/10.1038/s41467-018-04316-3
16. Hu, J., Li, T., Luo, C., Fujita, H., Yang, Y.: Incremental fuzzy cluster ensemble learning based on rough set theory. Knowl.-Based Syst. **132**, 144–155 (2017)
17. Yang, D., Liu, Y., Li, S., Li, X., Ma, L.: Gear fault diagnosis based on support vector machine optimized by artificial bee colony algorithm. Mech. Mach. Theory **90**, 219–229 (2015)
18. Ni, Z., Zhang, C., Ni, L.: Haze forecast method of selective ensemble based on glowworm swarm optimization algorithm. Int. J. Pattern Recognit. Artif. Intell. **29**(2), 143–153 (2016)
19. Yong, Z., Bo, L., Fan, Y.: Differential evolution based selective ensemble of extreme learning machine. In: Trustcom/BigDataSE/ISPA, pp.1327–1333 (2017)
20. Liu, Y., Yao, X., Higuchi, T.: Evolutionary ensembles with negative correlation learning. IEEE Trans. Evol. Comput. **4**(4), 380–387 (2000)
21. Sheng, W., Shan, P., Chen, S., Liu, Y., Alsaadi, F.E.: A niching evolutionary algorithm with adaptive negative correlation learning for neural network ensemble. Neurocomputing **247**, 173–182 (2017)
22. Kottathra, K., Attikiouzel, Y.: A novel multicriteria optimization algorithm for the structure determination of multilayer feedforward neural networks. J. Netw. Comput. Appl. **19**(2), 135–147 (1996)
23. Kupinski, M.A., Anastasio, M.A.: Multiobjective genetic optimization of diagnostic classifiers with implications for generating receiver operating characteristic curves. IEEE Trans. Med. Imaging **18**(8), 675–685 (1999)
24. Chandra, A., Yao, X.: Ensemble learning using multi-objective evolutionary algorithms. J. Math. Model. Algorithms **5**(4), 417–445 (2006). https://doi.org/10.1007/s10852-005-9020-3
25. Thompson, S.: Pruning boosted classifiers with a real valued genetic algorithm. In: Miles, R., Moulton, M., Bramer, M. (eds.) Research and Development in Expert Systems XV, pp. 133–146. Springer, London (1999). https://doi.org/10.1007/978-1-4471-0835-1_9
26. Zhou, Z.-H., Tang, W.: Selective ensemble of decision trees. In: Wang, G., Liu, Q., Yao, Y., Skowron, A. (eds.) RSFDGrC 2003. LNCS (LNAI), vol. 2639, pp. 476–483. Springer, Heidelberg (2003). https://doi.org/10.1007/3-540-39205-X_81
27. Mao, W., Tian, M., Cao, X., Xu, J.: Model selection of extreme learning machine based on multi-objective optimization. Neural Comput. Appl. **22**(3–4), 521–529 (2013). https://doi.org/10.1007/s00521-011-0804-2
28. Pavelski, L.M., Delgado, M.R., Almeida, C.P., Gonçalves, R.A., Venske, S.M.: Extreme learning surrogate models in multi-objective optimization based on decomposition. Neurocomputing **180**, 55–67 (2016)

29. Deb, K.: Multi-objective genetic algorithms: problem difficulties and construction of test problems. Evol. Comput. **7**(3), 205–230 (1999)
30. Liang, J., Yue, C., Qu, B.: Multimodal multi-objective optimization: a preliminary study. In: 2016 IEEE Congress on Evolutionary Computation (CEC), pp. 2454–2461. IEEE (2016)
31. Yue, C., Qu, B., Liang, J.: A multiobjective particle swarm optimizer using ring topology for solving multimodal multiobjective problems. IEEE Trans. Evol. Comput. **22**(5), 805–817 (2017)
32. Liang, J., Guo, Q., Yue, C., Qu, B., Yu, K.: A self-organizing multi-objective particle swarm optimization algorithm for multimodal multi-objective problems. In: Tan, Y., Shi, Y., Tang, Q. (eds.) ICSI 2018. LNCS, vol. 10941, pp. 550–560. Springer, Cham (2018). https://doi.org/10.1007/978-3-319-93815-8_52
33. Yue, C., Qu, B., Yu, K., Liang, J., Li, X.: A novel scalable test problem suite for multimodal multiobjective optimization. Swarm Evol. Comput. **48**, 62–71 (2019)
34. Dua, D., Graff, C.: UCI machine learning repository (2017). http://archive.ics.uci.edu/ml
35. Deb, K., Pratap, A., Agarwal, S., Meyarivan, T.: A fast and elitist multiobjective genetic algorithm: NSGA-II. IEEE Trans. Evol. Comput. **6**(2), 182–197 (2002)

Aircraft Scheduling Problems Based on Genetic Algorithms

Jingzhi Ding(✉)

National University of Defence Technology, Changsha 410073,
Hunan, People's Republic of China
7488112@qq.com

Abstract. The problem of aircraft scheduling is a typical scheduling problem, by abstracting the problem into a typical personnel on duty, the support aircraft is regarded as a person who need to be laid off, requiring that at any time, there are sufficient number of support aircraft in the patrol area to ensure the safe flight of special aircraft, but also to meet the constraints. In order to solve the problem of shift discharge efficiently, an adaptive genetic algorithm is proposed. Especially in the algorithm, the two-layer coding method of individual coding is introduced innovatively, in which the first layer of coding represents the take-off airport serial number and the second layer of coding represents the take-off aircraft serial number. The traditional individual coding method directly composes the take-off time into chromosomes, which makes it difficult to know the number of schedules, increases the complexity of calculation, and the direct use of take-off time coding cannot guarantee that the initialized individual meets the constraints, and too much non-feasible generation will seriously reduce the efficiency of iterative evolution. The problem can be effectively avoided by using the two-layer coding method of individual coding.

Keywords: Scheduling problem · Genetic algorithm · Self-adaptive method

1 Introduction

As an important social public resource, the airport needs to be secure enough to keep the airport's internal affairs running [1]. In our case, there are three airports in an area, and in order to ensure the normal operation of the airport, a special type of aircraft is now required to patrol the area of the identified patrol area without interruption, and two aircraft are required to carry out the protection of the patrol process at all times [2]. Our goal is to ensure that special aircraft complete patrol tasks and meet the conditions of personnel, aircraft, oil and other constraints, require the oil surplus of each airport, the workload of the two types of pilots as far as possible balanced, and get the support solutions for each airport [3].

L. Pan et al. (Eds.): BIC-TA 2019, CCIS 1159, pp. 314–324, 2020.
https://doi.org/10.1007/978-981-15-3425-6_25

2 Problem Description

The aircrafts which carry out the protection of the patrol process, also called support aircrafts, fly at 100 km/h and the coordinates of the three airports were (0, 0), (400, 0) and (300, 300). The patrol area of the special aircraft is two separate circular areas with the center coordinates of Y_1 (596.62, 553.28) and Y_2 (614.84, 409.71) separately. The number of support aircraft to choose from at each airport is 24, 20 and 18 respectively, the number of pilots is 32, 26 and 24 respectively, of which 20, 17 and 13 are pilots, and the other is the second type of pilots, and the initial residual fuel of the three airports is 12500 tons, 9800 tons, and 9000 tons respectively. It is stipulated that the maximum number of flights per aircraft per day shall be 2, that there shall be at least one type of pilot per flight and one flight per pilot per day. The length of the class A, Class B, Class C and Class D life control units of the aircraft is 30, 60, 120 and 360 h, respectively, and periodic inspection is required, during which the aircraft needs to be grounded for 24 h, 24 h, 48 h, and 5 days, respectively. The maximum number of aircraft capable of being in simultaneous maintenance inspections at the three airports is 6, 5 and 4.

3 Problem Analysis and Modeling

Under the premise of ensuring that special aircraft can meet the patrol mission, it is necessary to arrange the escort dispatch plan of the aircraft reasonably, so that the fuel consumption, support personnel and pilot's workload at the airports need to be as balanced as possible. This problem can be seen as a shift scheduling problem. Taking the support aircraft as personnel need to be lined up, requiring that at any time, the patrol area should have a sufficient number of aircraft to ensure the flight of special aircraft, and can meet the non-linear constraints. Due to the number of pilots and the ability of the airports, when arranging for the aircraft to fly the same flight, oil, life control units consumption can keep in balance, but at this time is bound to cause pilots, security personnel per capita workload differences. Therefore, for the airport material, fuel consumption, security personnel, pilots can only be as balanced as possible. Based on the above analysis, the nonlinear optimization model can be described as follows:

Optimization goal:

$$min\sum_{n=1}^{n^3}(n_{max}^i+n_{min}^i)+|\frac{n_{total}^1}{n_f^1}-\frac{n_{total}^2}{n_f^2}|+|\frac{n_{total}^2}{n_f^2}-\frac{n_{total}^3}{n_f^3}|+|\frac{n_{total}^1}{n_f^1}-\frac{n_{total}^3}{n_f^3}| \quad (1)$$

In which n_{max}^1 is the maximum number of flights per unit for support aircraft in airport i, and n_{total}^1 denotes the number of departures for support aircraft in a week at A airport. n_f^1 is the total number of aircrafts at A airport. The first term indicates that the number of missions performed by aircraft at the same airport is as equal as possible, the last three terms indicate that the number of missions performed at the three airports is proportional to the number of aircraft and is as balanced as possible.

And the constraints:

First, The remaining oil storage at each airport is the difference between the original storage capacity and the fuel consumption of the fighter aircraft, and the remaining oil after the N escort mission has been completed at the k airport within 7 days:

$$O_{k-rest} = O_k - O_{one} - O_{one} * \sum_{i=1}^{N} N_k^m \tag{2}$$

Second, there are two types of restrictions on the use of support materials at each airport, namely, the time already in flight and the total number of times that have been mounted. It has been calculated that if the total number of mounted limits have been reached, have exceeded the air flight time limit, so the use of support materials at each airport can be considered in the air flight time, so here the actual use of two types of support materials at each airport is calculated, the calculation formula is as follows:

$$D_{kz}^{frequency} = \sum_{i=1}^{6} D_{kz}^{i} * f(Maxz - Time_{D_{kz}^{i}}) \tag{3}$$

$$D_{kj}^{frequency} = \sum_{i=1}^{j} D_{kz}^{i} * f(Maxz - Time_{D_{kj}^{i}}) \tag{4}$$

Third, every time after the N_m^k flight takes off, the remaining support materials at each airport is as follows:

$$D_{kz}^{frequency} = D_{kz}^{frequency} - 2 \tag{5}$$

$$D_{kj}^{frequency} = D_{kj}^{frequency} - 2 \tag{6}$$

There are limits on the maximum mount flight time or the maximum number of mounts per day at each airport, but in actual combat missions, most missiles cannot reach the scrap conditions after using near the maximum mount flight time (the remaining time is less than 3 h, not enough to complete a escort), calculating that the scrap limit will not be violated.

Four, While the Nth support aircraft took off and returned;

If this aircraft has flown twice this day:

$$Time_N = Time_N + 24 \tag{7}$$

Check if the service hours or cumulative working hours of the life controls of this particular aircraft have reached the service time limit; End if.

If this support aircraft life control units have reached the service time limit after inspection:

$$Time_{N_{pi}}^{check} = 0 \tag{8}$$

$$Time_{N_{pi}}^{Total} = Time_{N_{pi}}^{Total} + 3 \tag{9}$$

$$Time_N = Time_N + Time_p \tag{10}$$

Else if the cumulative operating time of this support aircraft life control units reach the service time limit:

$$Time_{N_{pi}}^{check} = Time_{N_{pi}}^{check} + 3 \tag{11}$$

$$Time_{N_{pi}}^{Check} = 0 \tag{12}$$

$$Time_N = Time_N + Time_p \tag{13}$$

Else if this support aircraft life control units do not meet the service time limit:

$$Time_{N_{pi}}^{Check} = Time_{N_{pi}}^{Check} + 3 \tag{14}$$

$$Time_{N_{pi}}^{Total} = Time_{N_{pi}}^{Total} + 3 \tag{15}$$

END IF

END WHILE

Fifth, the remaining number of airport life control units after maintenance also determines whether support aircraft can resume escort capability.

While the Nth aircraft took off and returned, when the life control unit pi of the support aircraft at airport k has reached the service time limit, the remaining number of this life control unit is:

$$k_{pi} = k_{pi} - 1 \tag{16}$$

if $k_{pi} < 0$:

$$Time_N = Time_N + \propto \tag{17}$$

In this case, we consider the number of support aircrafts at airport k which can carry out the escort missions has been reduced by one.

4 Problem Solving

This paper uses adaptive genetic algorithm to solve the problem. Genetic algorithm [4] is a kind of randomized search method which has evolved from the evolutionary laws of the biological world. It was first proposed by Professor J. Holland of the United States in 1975, and its main features are direct operation of structural objects, without the limitation of seeking and functional continuity, with inherent hidden parallelism and better global search capability, and the use of a probabilistic approach to the search space that automatically acquires and guides optimization. Adjust the search direction adaptively without the need for a definite rule. In the chromosome coding, taking binary coding as an example, after the emergence of the first generation of population, according to the principle of survival and survival of the fittest, generation by generation evolution produces a better and better approximation, in each generation, according to the size of individual adaptation in the problem domain to select individuals, and with the help of natural genetic ally combination cross and variation, Produces a population that represents a new set of solutions. This process will lead to the population like natural evolution of the post-generation population more adapted to the environment than the previous generation, the optimal individual in the last generation of population decoded, can be used as the problem approximation of optimal solution.

For the problem of premature convergence of genetic algorithm, it is difficult to jump out of the local optimal solution, and we use adaptive genetic algorithm (AGA) adaptive adjustment of cross over and variation probability strategy to solve this problem. During the algorithm search process, changes in the probability of cross over and variation can be calculated by the following formula:

$$\begin{aligned} pc &= k_1 * \frac{f_{max} - f}{f_{max} - f_{avg}}, k_1 \leq 1.0 \\ pm &= k_2 * \frac{f_{max} - f}{f_{max} - f_{avg}}, k_2 \leq 1.0 \end{aligned} \tag{18}$$

Where f_{max} is the optimal adaptation value of the contemporary population, f_{avg} is the average adaptation value, and is the individual's adaptation value.

For both cases, the number of special aircraft is 1 or 2, with 2 or 4 support aircraft required accordingly. The following is to optimize the ranking of the two cases, in the two cases, the common algorithm framework and coding ideas are as follows:

4.1 Coding Scheme

For the scheduling problems, coding methods have a great impact on the solution space of the problem. In this article, we use two-layer coding to [5] solve the problem and optimize it. Described as follows, assuming a total of n jobs are required over a week, the schedule for the week can be described as follows:

Each post has two variables, representing the airport number of the aircraft sent, and the number of the aircraft sent. The first n codes represent the airport number, and the last n codes represent the airport aircraft number [6]. Given an individual, then the order of aircraft layout can be given for a week.

4.2 Constraints Processing

In this paper, when the initial solution and cross-variation produce new solutions, the solution may not satisfy the constraints, which are manifested in two kinds: the constraints of pilots and security personnel, and the constraints of the take-off time of the aircraft [7]. The following are described separately as the constraint processing method that produces the initial solution and the processing method that produces the constraint solution in the evolutionary process.

First, when the initial solution is generated, each time the island and aircraft number slot is randomly selected, and when the aircraft is unable to take off, it is re-selected at random until the aircraft can take off. Repeat the process until all the posts are complete. Second, after the new solution is produced, traverse each post at a time. When the post cannot be dispatched, randomly select the airports and aircraft that can send support aircraft. If none of them can be sent out, discard the solution.

4.3 Crossover

This paper uses two-layer coding to describe the problem, and the second layer of coding is based on the first layer of coding [8]. Therefore, in the crossover process, we will only cross the first layer of coding in order to produce a new solution. The diagram is as follows (Fig. 1):

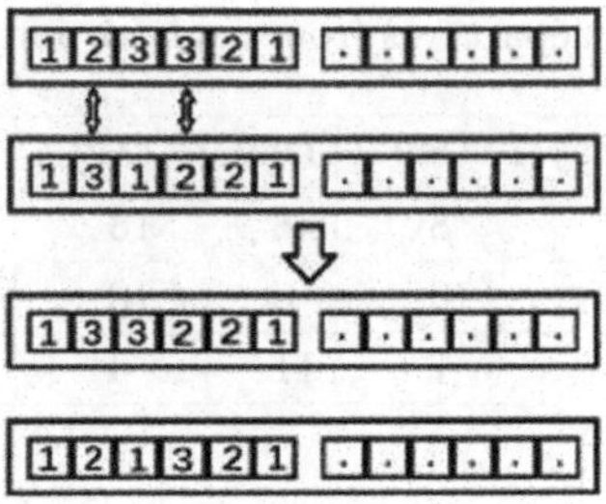

Fig. 1. The crossover process diagram

4.4 Mutation

When the mutation occurs, we randomly select n posts and then randomly re-designate the airport and aircraft numbers for that post, as shown below: t is important to note that after cross-variation, the newly generated solution needs to be processed so that the solution can meet the constraint [9]. When the number of early warning aircraft n is 1, the time spent supporting the aircraft arriving in the patrol area is initialized according to the identified patrol area, and the time taken to reach the mission area by the support aircraft departing from the three airports of A, B and C can be calculated as: (0.9714, 0.5460, 0.5000) (Fig. 2).

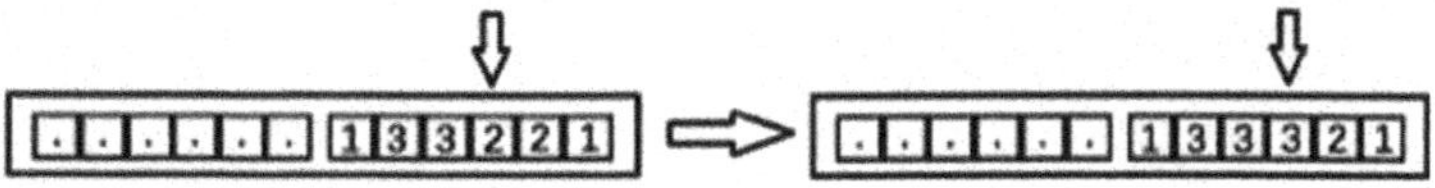

Fig. 2. The process of the mutation

5 Results

5.1 When $n = 1$

When n = 1, the optimization results are as follows: In the figure above, the upper subplot shows the change in the value of the target function during the iteration, showing that the algorithm has converged around 68 generations. The final scheduling situation at n – 1 is shown in the following side map. The horizontal coordinates of the triangle are the shift order number, and the ordinates are the shift departure time. The different colors of the triangle represent the aircraft that took off from different airports, as shown in the Fig. 3.

The Table 1 below provides the total number of flights and life control units replacement information for each airport within a week when n = 1.

Table 1. Total departures from each airport during a week

Departures	Day 1	Day 2	Day 3	Day 4	Day 5	Day 6	Day 7
Departures from A	4	8	8	13	11	5	8
Departures from B	9	10	6	10	7	8	8
Departures from C	9	7	11	4	8	11	9

5.2 When $n = 2$

When n = 2, according to the patrol area determined before, the time spent by the aircraft arriving in the patrol area is initialized, and the time it takes for the aircraft departing from the three airports of A, B and C to arrive at the mission area is calculated:

(0.8381, 0.4127, 0.3667)

The optimization results are as follows (Table 2 and Fig. 4):

Similarly, the upper subplot shows the change in the value of the target function during the iteration, showing that the algorithm has converged around 62 generations. The final schedule at n = 2 is shown in the lower subplot (Table 3).

The table below provides the total number of flights and life control units replacement information for each airport on each day of the week when n = 2 (Table 4).

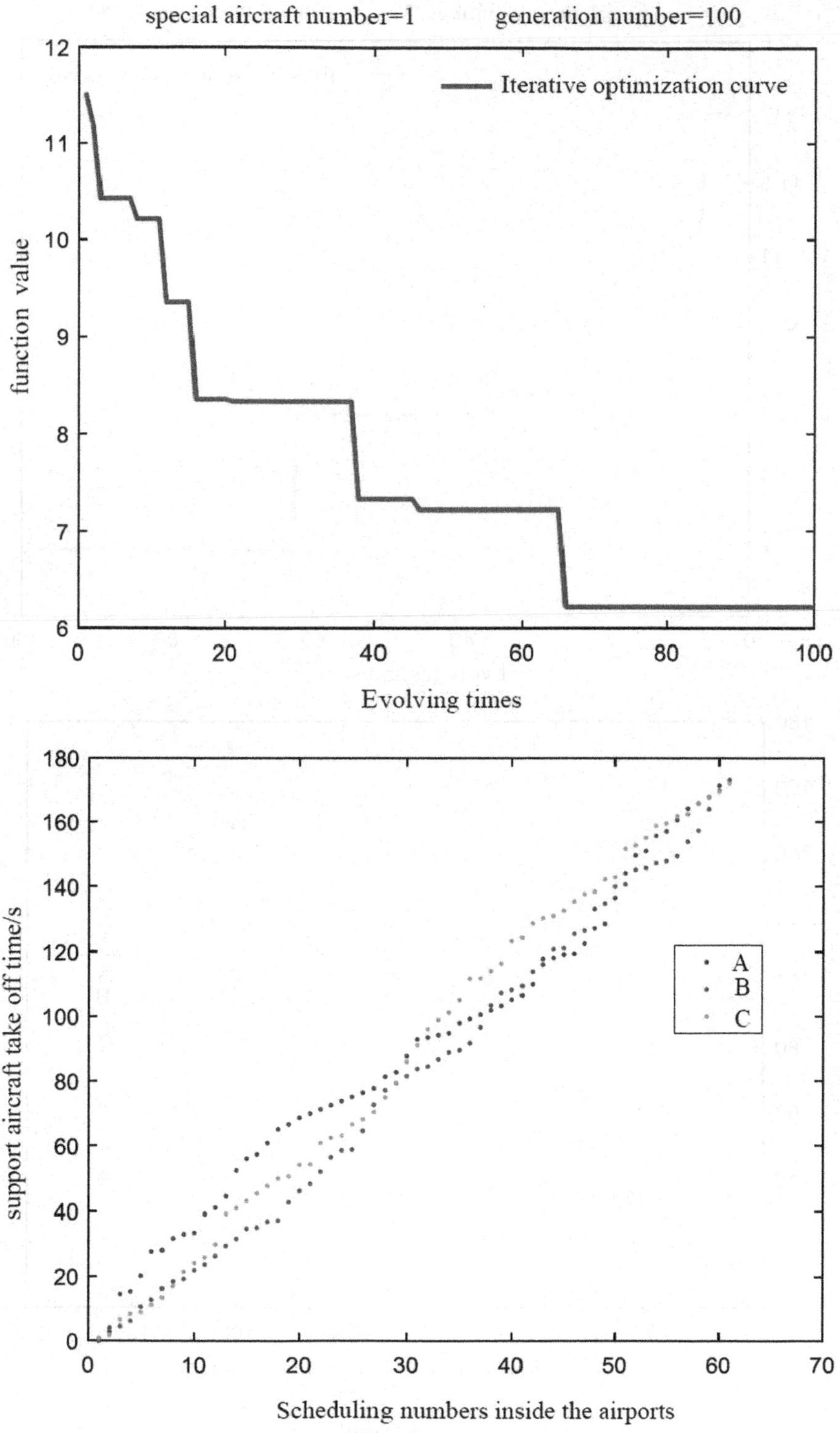

Fig. 3. The optimization results when n = 1

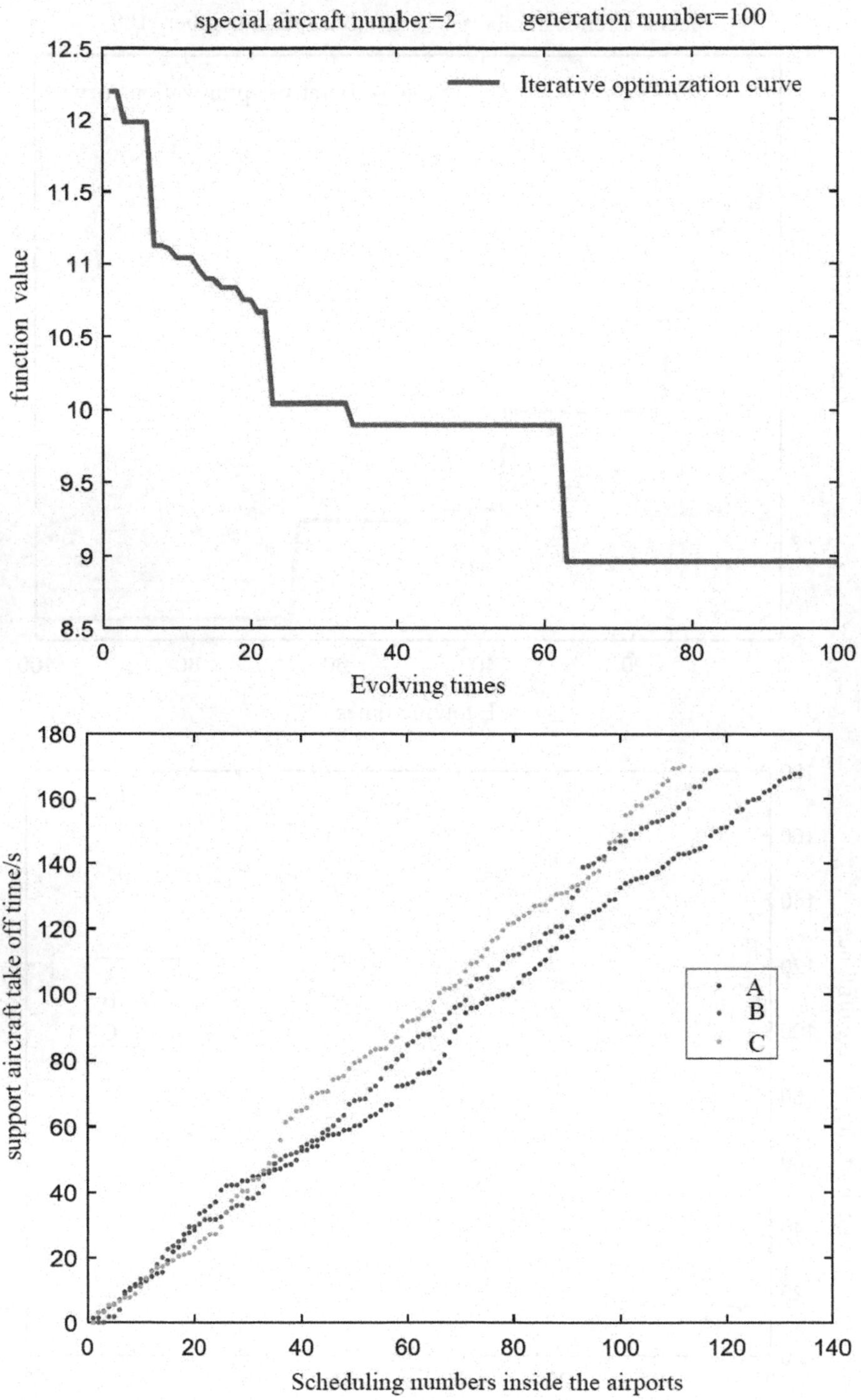

Fig. 4. The optimization results when n = 2

Table 2. The replacement information of life control units

Airport code	1	1	3	2	2
Aircraft code	19	4	18	5	4
Take-off time	14.57	20.24	24.21	77.37	140.70
Replacement type	4	4	2	1	3
Replaced type	3	3	4	1	2

Table 3. Total departures from each airport during a week

Departures	Day 1	Day 2	Day 3	Day 4	Day 5	Day 6	Day 7	Total	Avg
Departures from A	4	8	8	13	11	5	8	132	5.5
Departures from B	9	10	6	10	7	8	8	113	5.65
Departures from C	9	7	11	4	8	11	9	108	6

Table 4. The replacement information of life control units

Airport code	1	3	1	2	2	2	3	2	3	2	1	1
Aircraft code	4	18	19	5	4	20	11	20	5	12	15	3
Take-off time	10.6	13.7	14.2	48.6	55.5	84.8	104	115.7	127.4	129.6	145.3	165.4
Replacement type	4	2	4	1	3	2	3	2	3	4	2	2
Replaced type	3	4	3	1	2	4	2	2	2	2	4	4

6 Conclusions

In this paper, the problem of aircraft scheduling is abstracted into a typical personnel shift, in order to efficiently solve the problem of shift, the paper puts forward an adaptive genetic algorithm. In the algorithm, the two-layer coding of individual coding methods is introduced innovatively, avoiding the complexity and other problems caused by traditional coding, and finally a satisfactory aircraft scheduling scheme can be better solved. The method proposed in this paper also has certain universality.

References

1. Ren, Z., San, Y.: A hybrid optimized algorithm based on simplex method and genetic algorithm. In: 2006 6th World Congress on Intelligent Control and Automation, Dalian, pp. 3547–3551 (2006)
2. Haupt, R., Haupt, S.E.: The creative use of genetic algorithms. Computers evolve into the artistic realm. IEEE Potentials **19**(2), 26–29 (2000)
3. Frenzel, J.F.: Genetic algorithms. IEEE Potentials **12**(3), 21–24 (1993)
4. Haupt, R.L., Werner, D.H.: Anatomy of a genetic algorithm. In: Genetic Algorithms in Electromagnetics. IEEE (2007)

5. Abd-Alhameed, R.A., Zhou, D., See, C.H., Excell, P.S.: A wire-grid adaptive-meshing program for microstrip-patch antenna designs using a genetic algorithm [EM Programmer's Notebook]. IEEE Antennas Propag. Mag. **51**(1), 147–151 (2009)
6. Asafuddoula, M., Ray, T., Sarker, R.: A decomposition-based evolutionary algorithm for many objective optimization. IEEE Trans. Evol. Comput. **19**(3), 445–460 (2015)
7. Li, S., Zhou, H., Hu, J., Ai, Q., Cai, C.: A fast path planning approach for unmanned aerial vehicles. Concurrency Comput. Pract. Experience **27** (2014). https://doi.org/10.1002/cpe.3291
8. Farouki, R.T.: The elastic bending energy of pythagorean-hodograph curves. Comput. Aided Geom. Des. **13**(3), 227–241 (1996)
9. Carnahan, J., Sinha, R.: Nature's algorithms [genetic algorithms]. IEEE Potentials **20**(2), 21–24 (2001)

Estimating Approximation Errors of Elitist Evolutionary Algorithms

Cong Wang[1], Yu Chen[1(✉)], Jun He[2], and Chengwang Xie[3(✉)]

[1] School of Science, Wuhan University of Technology, Wuhan 430070, China
ychen@whut.edu.cn

[2] School of Science and Technology, Nottingham Trent University, Nottingham NG11 8NS, UK

[3] School of Computer and Information Engineering, Nanning Normal University, Nanning 530299, China
chengwangxie@nnnu.edu.cn

Abstract. When evolutionary algorithms (EAs) are unlikely to locate precise global optimal solutions with satisfactory performances, it is important to substitute alternative theoretical routine for the analysis of hitting time/running time. In order to narrow the gap between theories and applications, this paper is dedicated to perform an analysis on approximation error of EAs. First, we proposed a general result on upper bound and lower bound of approximation errors. Then, several case studies are performed to present the routine of error analysis, and theoretical results show the close connections between approximation errors and eigenvalues of transition matrices. The analysis validates applicability of error analysis, demonstrates significance of estimation results, and then, exhibits its potential to be applied for theoretical analysis of elitist EAs.

Keywords: Evolutionary algorithm · Approximation error · Markov chain · Budget analysis

1 Introduction

For theoretical analysis, convergence performance of evolutionary algorithms (EAs) is widely evaluated by the expected first hitting time (FHT) and the expected running time (RT) [1], which quantify the respective numbers of iteration and function evaluations (FEs) to hit the global optimal solutions. General methods for estimation of FHT/RT have been proposed via theories of Markov chains [2,3], drift analysis [4,5], switch analysis [6] and application of them with partition of fitness levels [7], etc.

Although popularly employed in theoretical analysis, simple application of FHT/RT is not practical when the optimal solutions are difficult to hit. One of these "difficult" cases is optimization of continuous problems. Optimal sets of continuous optimization problems are usually zero-measure set, which could not be hit by generally designed EAs in finite time, and so, FHT/RT could be infinity

L. Pan et al. (Eds.): BIC-TA 2019, CCIS 1159, pp. 325–340, 2020.
https://doi.org/10.1007/978-981-15-3425-6_26

for most cases. A remedy to this difficulty is to take a positive-measure set as the destination of population iteration. So, it is natural to take an approximation set for a given precision as the hitting set of FHT/RT estimation [8–11]. Another "difficult" case is the optimization of NP-complete (NPC) problems that cannot be solved by EAs in polynomial FHT/RT. For this case, it is much more interesting to investigate the quality of approximate solutions obtained in polynomial FHT/RT. In this way, researchers have estimated approximation ratios of approximate solutions that EAs can obtain for various NPC combinatorial optimization problems in polynomial expected FHT/RT [12–17].

However, the aforementioned methods could be impractical once we have little information about global optima of the investigated problems, and then, it is difficult to "guess" what threshold can result in polynomial FHT/RT. Since the approximation error after a given iteration number is usually employed to numerically compared performance of EAs, some researchers tried to analyze EAs by theoretically estimating the expected approximation error. Rudolph [18] proved that under the condition $e^{[t]}/e^{[t-1]} \leq \lambda < 1$, the sequence $\{e^{[t]}; t = 0, 1, \cdots\}$ converges in mean geometrically to 0, that is, $\lambda^t e^{[t]} = o(1)$. He and Lin [19] studied the geometric average convergence rate of the error sequence $\{e_t; t = 0, 1, \cdots\}$, defined by $R^{[t]} = 1 - \left(e^{[t]}/e^{[0]}\right)^{1/t}$. Starting from $R^{[t]}$, it is straightforward to claim that $e^{[t]} = (1 - R^{[t]})^t e^{[0]}$.

A close work to analysis of approximation error is the fixed budget analysis proposed by Jansen and Zarges [20,21], who aimed to bound the fitness value $f(X^{[t]})$ within a fixed time budget t. However, Jansen and Zarges did not present general results for any time budget t. In fixed budget analysis, a bound of approximation error holds for some small t but might be invalid for a large one. He [22] made a first attempt to obtain an analytic expression of the approximation error for a class of elitist EAs. He proved if the transition matrix associated with an EA is an upper triangular matrix with unique diagonal entries, then for any $t \geq 1$, the approximation error $e^{[t]}$ is expressed by $e^{[t]} = \sum_{k=1}^{L} c_k \lambda_k^{t-1}$, where λ_k are eigenvalues of the transition matrix. He *et al.* [23] also demonstrated the possibility of approximation estimation by estimating one-step convergence rate e_t/e_{t-1}, however, it was not sufficient to validate its applicability to other problems because only two studied cases with trivial convergence rates were investigated.

This paper is dedicated to present an analysis on estimation of approximation error depending on any iteration number t. We make the first attempt to perform a general error analysis of EAs, and demonstrate its feasibility by case studies. Rest of this paper is presented as follows. Section 2 presents some preliminaries. In Sect. 3, a general result on the upper and lower bounds of approximation error is proposed, and some case studies are performed in Sect. 4. Finally, Sect. 5 concludes this paper.

2 Preliminaries

In this paper, we consider a combinatorial optimization problem

$$\max \quad f(\mathbf{x}), \tag{1}$$

where $\mathbf{x}$ has only finite available values. Denote its optimal solution as $\mathbf{x}^*$, and the corresponding objective value as f^*. Quality of a feasible solution $\mathbf{x}$ is quantified by its approximation error $e(\mathbf{x}) = |f(\mathbf{x}) - f^*|$. Since there are only finite solutions of problem (1), there exist finite feasible values of $e(\mathbf{x})$, denoted as $e_0 \le e_1, \dots, \le e_n$. Obviously, the minimum value e_0 is the approximation error of the optimal solution $\mathbf{x}^*$, and so, takes the value 0. We call that $\mathbf{x}$ is *located at the status* i if $e(\mathbf{x}) = e_i$. Then, there are totally $n+1$ statuses for all feasible solutions. Status 0 consists of all optimal solutions, called the *optimal status*; other statuses are the *non-optimal statuses.*

Suppose that an feasible solution of problem (1) is coded as a bit-string, and an elitist EA described in Algorithm 1 is employed to solve it. When the one-bit mutation is employed, it is called a *random local search (RLS)*; if the bitwise mutation is used, it is named as a *(1+1) evolutionary algorithm ((1+1)EA).* Then, the error sequence $\{e(\mathbf{x}_t), t = 0, 1, \dots\}$ is a *Markov Chain.* Assisted by the initial probability distribution of individual status $(q_0, q_1, \dots, q_n)^T$, the evolution process of (1+1) elitist EA can be depicted by the transition probability matrix

$$\mathbf{P} = \begin{pmatrix} p_{0,0} & p_{0,1} & \cdots & p_{0,n} \\ \vdots & \vdots & \vdots & \vdots \\ p_{n,0} & p_{n,1} & \cdots & p_{n,n} \end{pmatrix}, \tag{2}$$

where $p_{i,j}$ is the probability to transfer from status j to status i.

Algorithm 1. A Framework of the Elitist EA

counter $t = 0$;
randomly initialize a solution $\mathbf{x}_0$;
while the stopping criterion is not satisfied **do**
  generate a new candidate solution $\mathbf{y}_t$ from $\mathbf{x}_t$ by mutation;
  set individual $\mathbf{x}_{t+1} = \mathbf{y}_t$ if $f(\mathbf{y}_t) > f(\mathbf{x}_t)$; otherwise, let $\mathbf{x}_{t+1} = \mathbf{x}_t$;
  $t = t + 1$;
end while

Since the elitist selection is employed, the probability to transfer from status j to status i is zero when $i > j$. Then, the transition probability matrix is upper triangular, and we can partition it as

$$\mathbf{P} = \begin{pmatrix} p_{0,0} & \mathbf{p}_0 \\ \mathbf{0} & \mathbf{R} \end{pmatrix}, \tag{3}$$

where $\boldsymbol{p}_0 = (p_{0,1}, p_{0,2}, \ldots, p_{0,n})$, $\mathbf{0} = (0, \ldots, 0)^T$,

$$\mathbf{R} = \begin{pmatrix} p_{1,1} & p_{1,2} & \cdots & p_{1,n} \\ & p_{2,2} & \cdots & p_{2,n} \\ & & \ddots & \vdots \\ & & & p_{n,n} \end{pmatrix}. \tag{4}$$

Thus, the expected approximation error at the tth iteration is

$$e^{[t]} = \mathbf{e}\mathbf{R}^t\mathbf{q}, \tag{5}$$

where $\mathbf{e} = (e_1, \ldots, e_n)$, $\mathbf{q} = (q_1, \ldots, q_n)^T$, $\mathbf{R}$ is the sub-matrix representing transition probabilities between non-optimal statuses [24]. Because sum of each column in $\mathbf{P}$ is equal to 1, the first row $\boldsymbol{p}_0$ can be confirmed by $\mathbf{R}$, and in the following, we only consider the transition submatrix $\mathbf{R}$ for estimation of approximation error. According to the shape of $\mathbf{R}$, we can further divide searching process of elitist EAs into two different categories.

1. **Step-by-step Search**: If the transition probability satisfies

$$\begin{cases} p_{i,j} = 0, & \text{if } i \neq j-1, j, \\ p_{j-1,j} + p_{j,j} = 1, & \end{cases} \quad j = 1, \ldots, n. \tag{6}$$

   it is called a **step-by-step search**. Then, the transition submatrix is

$$\mathbf{R} = \begin{pmatrix} p_{1,1} & p_{1,2} & & \\ & \ddots & \ddots & \\ & & p_{n-1,n-1} & p_{n-1,n} \\ & & & p_{n,n} \end{pmatrix}, \tag{7}$$

   which means the elitist EA cannot transfer between non-optimal statues that are not adjacent to each other;
2. **Multi-step Search**: If there exists some $i, j > i+1$ such that $p_{i,j} \neq 0$, we called it a **multi-step search**. A multi-step search can transfer between inconsecutive statuses, which endows it with better global exploration ability, and probably, better convergence speed.

Note that this classification is problem-dependent because the statuses depend on the problem to be optimized. So, the RLS could be either a step-by-step search or a multi-step search. However, the (1+1)EA is necessarily a multi-step search, because the bitwise mutation can jump between any two statuses. When $\boldsymbol{p}_0$ in (3) is non-zero, column sums of $\mathbf{R}$ is less than 1, which means it could jump from at least one non-optimal status directly to the optimal status. So, a step-by-step search represented by (7) must satisfies

$$p_{j-1,j} + p_{j,j} = 1, \ \forall j \in \{1, \ldots, n\}.$$

3 Estimation of General Approximation Bounds

3.1 General Bounds of the Step-by-step Search

Let $\mathbf{R}$ be the submatrix of a step-by-step search. Its eigenvalues are

$$\lambda_i = p_{i,i}, \quad i = 1, \dots, n, \tag{8}$$

which represents the probability of remaining at the present status after one iteration. Then, it is very natural to declare that greater the eigenvalues are, slower the step-by-step search converges. Inspired by this idea, we can estimate general bounds of a step-by-step search by enlarging or reducing the eigenvalues. Achievement of the general bounds is based on the following lemma.

Lemma 1. *Denote*

$$\mathbf{f}_t(\mathbf{e}, \lambda_1, \dots, \lambda_n) = (f_{t,1}(\mathbf{e}, \lambda_1, \dots, \lambda_n), \dots, f_{t,n}(\mathbf{e}, \lambda_1, \dots, \lambda_n)) = \mathbf{e}\mathbf{R}^t. \tag{9}$$

Then, $f_{t,i}(\mathbf{e}, \lambda_1, \dots, \lambda_n)$ is monotonously increasing with λ_j, $\forall t > 0, i, j \in \{1, \dots, n\}$.

Proof. This lemma could be proved by mathematical induction.

1. When $t = 1$, we have

$$f_{1,i}(\mathbf{e}, \lambda_1, \dots, \lambda_n) = \begin{cases} e_1\lambda_1, & i = 1, \\ e_{i-1}(1 - \lambda_i) + e_i\lambda_i, & i = 2, \dots, n. \end{cases} \tag{10}$$

   Note that λ_j is not greater than 1 because it is an element of the probability transition matrix $\mathbf{P}$. Then, from the truth that $0 = e_0 \le e_1 \le \dots \le e_n$, we conclude that $f_{1,i}(\mathbf{e}, \lambda_1, \dots, \lambda_n)$ is monotonously increasing with λ_j, $\forall i, j \in \{1, \dots, n\}$. Meanwhile, (10) also implies that

$$0 \le f_{1,1}(\mathbf{e}, \lambda_1, \dots, \lambda_n) \le e_1 \le \dots \le f_{1,n}(\mathbf{e}, \lambda_1, \dots, \lambda_n) \le e_n. \tag{11}$$

2. Suppose that when $t = k \ge 1$, $f_{k,i}(\mathbf{e}, \lambda_1, \dots, \lambda_n)$ is monotonously increasing with λ_j for all $i, j \in \{1, \dots, n\}$, and it holds that

$$0 \le f_{k,i}(\mathbf{e}, \lambda_1, \dots, \lambda_n) \le f_{k,i+1}(\mathbf{e}, \lambda_1, \dots, \lambda_n), \quad \forall i \in \{1, \dots, n-1\}. \tag{12}$$

   First, the monotonicity indicated by (12) implies that

$$\frac{\partial}{\partial \lambda_j} f_{k,i}(\mathbf{e}, \lambda_1, \dots, \lambda_n) \ge 0, \quad \forall i, j \in \{1, \dots, n\}. \tag{13}$$

   Meanwhile,
   according to Eq. (9) we know $\mathbf{f}_{t+1}(\mathbf{e}, \lambda_1, \dots, \lambda_n) = \mathbf{f}_t(\mathbf{e}, \lambda_1, \dots, \lambda_n)\mathbf{R}$, that is,

$$f_{k+1,i}(\mathbf{e}, \lambda_1, \dots, \lambda_n) = \begin{cases} f_{k,1}(\mathbf{e}, \lambda_1, \dots, \lambda_n)\lambda_1, & i = 1, \\ f_{k,i-1}(\mathbf{e}, \lambda_1, \dots, \lambda_n)(1 - \lambda_i) + f_{k,i}(\mathbf{e}, \lambda_1, \dots, \lambda_n)\lambda_i, & i = 2, \dots, n. \end{cases}$$

So, $\forall\, j \in \{1, \ldots, n\}$,

$$\frac{\partial}{\partial \lambda_j} f_{k+1,i}(\mathbf{e}, \lambda_1, \ldots, \lambda_n)$$
$$= \begin{cases} \dfrac{\partial}{\partial \lambda_j} f_{k,1}(\mathbf{e}, \lambda_1, \ldots, \lambda_n)\lambda_1 + f_{k,1}(\mathbf{e}, \lambda_1, \ldots, \lambda_n)\dfrac{\partial \lambda_1}{\partial \lambda_j}, & i = 1, \\ \dfrac{\partial}{\partial \lambda_j} f_{k,i-1}(\mathbf{e}, \lambda_1, \ldots, \lambda_n)(1-\lambda_i) + \dfrac{\partial}{\partial \lambda_j} f_{k,i}(\mathbf{e}, \lambda_1, \ldots, \lambda_n)\lambda_i & \\ \quad + (f_{k,i}(\mathbf{e}, \lambda_1, \ldots, \lambda_n) - f_{k,i-1}(\mathbf{e}, \lambda_1, \ldots, \lambda_n))\dfrac{\partial \lambda_i}{\partial \lambda_j}, & i = 2, \ldots, n. \end{cases} \tag{14}$$

Combining (12), (13) and 14, we know that

$$\frac{\partial}{\partial \lambda_j} f_{k+1,i}(\mathbf{e}, \lambda_1, \ldots, \lambda_n) \geq 0, \quad \forall\, i, j \in \{1, \ldots, n\},$$

which means $f_{k+1,i}(\mathbf{e}, \lambda_1, \ldots, \lambda_n)$ is monotonously increasing with λ_j for all $i, j \in \{1, \ldots, n\}$.

In conclusion, $f_{t,i}(\mathbf{e}, \lambda_1, \ldots, \lambda_n)$ is monotonously increasing with λ_j, $\forall t > 0$, $i, j \in \{1, \ldots, n\}$. □

Denote

$$\mathbf{R}(\lambda) = \begin{pmatrix} \lambda \; 1-\lambda & & \\ & \ddots \; \ddots & \\ & & \lambda \; 1-\lambda \\ & & \lambda \end{pmatrix}, \tag{15}$$

If we enlarge or shrink all eigenvalues of $\mathbf{R}$ to the maximum value and the minimum value, respectively, we can get two transition submatrices $\mathbf{R}(\lambda_{max})$ and $\mathbf{R}(\lambda_{min})$, where $\lambda_{max} = \max_i \lambda_i$, $\lambda_{min} = \min_i \lambda_i$. Then, $\mathbf{R}(\lambda_{max})$ depicts a searching process converging slower than the one $\mathbf{R}$ represents, and $\mathbf{R}(\lambda_{min})$ is the transition submatrix of a process converging faster than what $\mathbf{R}$ represents.

Theorem 1. *The expected approximation error $e^{[t]}$ of a step-by-step search represented by $\mathbf{R}$ and $\mathbf{q}$ is bounded by*

$$\mathbf{e}\mathbf{R}^t(\lambda_{min})\mathbf{q} \leq e^{[t]} \leq \mathbf{e}\mathbf{R}^t(\lambda_{max})\mathbf{q}. \tag{16}$$

Proof. Note that

$$e^{[t]} = \mathbf{e}\mathbf{R}^t\mathbf{q} = \mathbf{f}_t(\mathbf{e}, \lambda_1, \ldots, \lambda_n)\mathbf{q},$$

where $\mathbf{q}$ is a non-zero vector composed of non-negative components. Then, by Lemma 1 we can conclude that $e^{[t]}$ is also monotonously increasing with λ_j, $\forall\, j \in \{1, \ldots, n\}$. So, we can get the result that

$$\mathbf{e}\mathbf{R}^t(\lambda_{min})\mathbf{q} \leq e^{[t]} \leq \mathbf{e}\mathbf{R}^t(\lambda_{max})\mathbf{q}.$$

□

Theorem 1 provides a general result about the upper and the lower bounds of approximation error. From the above arguments we can figure out that the lower bounds and the upper bounds can be achieved once the transition submatrix $\mathbf{R}$ degenerates to $\mathbf{R}(\lambda_{max})$ and $\mathbf{R}(\lambda_{min})$, respectively. That is to say, they are indeed the "best" results about the general bounds. Recall that $\lambda_i = p_{i,i}$. Starting from the ith status, $p_{i,i}$ is the probability that the (1+1) elitist EA stays at the ith status after one iteration. Then, greater λ_i is, harder the step-by-step search transfers to the sub-level status $i-1$. So, performance of a step-by-step search depicted by $\mathbf{R}$, for the worst case, would not be worse than that of $\mathbf{R}(\lambda_{max})$; meanwhile, it would not be better than that of $\mathbf{R}(\lambda_{min})$, which contributes to a bottleneck for improving performance of the step-by-step search.

3.2 General Bounds of the Multi-step Search

Denoting the transition submatrix of a multi-step search as

$$\mathbf{R_M} = \begin{pmatrix} p_{1,1} & p_{1,2} & \cdots & p_{1,n-1} & p_{1,n} \\ & & \ddots & \ddots & \vdots \\ & & & p_{n-1,n-1} & p_{n-1,n} \\ & & & & p_{n,n} \end{pmatrix}, \tag{17}$$

we can bound its approximation error by defining two transition matrices

$$\mathbf{R_{S_u}} = \begin{pmatrix} p_{1,1} & \sum_{k=0}^{1} p_{k,1} & & \\ & \ddots & \ddots & \\ & & p_{n-1,n-1} & \sum_{k=0}^{n-1} p_{k,n} \\ & & & p_{n,n} \end{pmatrix} \tag{18}$$

and

$$\mathbf{R_{S_l}} = diag(p_{1,1}, \dots, p_{n,n}). \tag{19}$$

Lemma 2. *Let $\mathbf{R_M}$, $\mathbf{R_{S_u}}$ and $\mathbf{R_{S_l}}$ be the transition matrix defined by (17), (18) and (19), respectively. Given any nonnegative vector $\mathbf{e} = (e_1, \dots, e_n)$ satisfying $e_1 \le \cdots \le e_n$ and the corresponding initial distribution $\mathbf{q} = (q_1, \dots, q_n)$, it holds that*

$$\mathbf{e}\mathbf{R_{S_l}}^t\mathbf{q} \le \mathbf{e}\mathbf{R_M}^t\mathbf{q} \le \mathbf{e}\mathbf{R_{S_u}}^t\mathbf{q}, \ \forall t > 0. \tag{20}$$

Proof. It is trivial to prove that $\mathbf{e}\mathbf{R_{S_l}}^t\mathbf{q} \le \mathbf{e}\mathbf{R_M}^t\mathbf{q}$. Because $\mathbf{R_{S_l}}$ has part of non-zero elements of $\mathbf{R_M}$, $\mathbf{e}\mathbf{R_{S_l}}^t\mathbf{q}$ is a partial sum of $\mathbf{e}\mathbf{R_M}^t\mathbf{q}$. Since all elements included in $\mathbf{e}\mathbf{R_M}^t\mathbf{q}$ are nonnegative, it holds that $\mathbf{e}\mathbf{R_{S_l}}^t\mathbf{q} \le \mathbf{e}\mathbf{R_M}^t\mathbf{q}$.

Moreover, the second inequality can be proved by mathematical induction. Denote

$$\mathbf{a} = (a_1, \dots, a_n) = \mathbf{e}\mathbf{R_M}, \tag{21}$$
$$\mathbf{b} = (b_1, \dots, b_n) = \mathbf{e}\mathbf{R_{S_u}}, \tag{22}$$

where $a_i = \sum_{j=1}^{i} e_j p_{j,i}, b_i = \sum_{j=0}^{i-1} e_{i-1} p_{j,i} + e_i p_{i,i},\ i = 1, \ldots, n$. Combining with the fact that $e_1 \le e_2 \le \cdots \le e_n$, we know that

$$0 \le a_i \le b_i,\ i = 1, \ldots, n. \tag{23}$$

1. When $t = 1$, (21), (22) and (23) imply that

$$\mathbf{eR_M q} = \sum_{i=1}^{n} a_i p_i \le \sum_{i=1}^{n} b_i p_i = \mathbf{eR_{S_u} q}.$$

2. Assume that (20) holds when $t = k \ge 1$. Then, (23) implies that

$$\mathbf{eR_M}^{k+1}\mathbf{q} = \mathbf{eR_M R_M}^{k}\mathbf{q} = \mathbf{aR_M}^{k}\mathbf{q} \le \mathbf{bR_M}^{k}\mathbf{q}. \tag{24}$$

   Meanwhile, because $e_1 \le e_2 \le \cdots \le e_n$, we know $b_1 \le b_2 \le \cdots \le b_n$. Then, the assumption implies that

$$\mathbf{bR_M}^{k}\mathbf{q} \le \mathbf{bR_{S_u}}^{k}\mathbf{q}.$$

   Combining it with (24), we can conclude that

$$\mathbf{eR_M}^{k+1}\mathbf{q} \le \mathbf{bR_M}^{k}\mathbf{q} \le \mathbf{bR_{S_u}}^{k}\mathbf{q} = \mathbf{eR_{S_u}}^{k+1}\mathbf{q}.$$

   So, the result also holds for $t = k + 1$.

In conclusion, it holds that $\mathbf{eR_M}^{t}\mathbf{q} \le \mathbf{eR_{S_u}}^{t}\mathbf{q},\ \forall t > 0$. □

Theorem 2. *The approximation error of the multi-step search defined by (17) is bounded by*

$$\mathbf{eR_{S_l}}^{t}\mathbf{q} \le e^{[t]} \le \mathbf{eR}^{t}(\lambda_{max})\mathbf{q}, \tag{25}$$

where $\lambda_{max} = \max_i \lambda_i = \max_i p_{i,i}$.

Proof. From Lemma 2 we know that

$$\mathbf{eR_{S_l}}^{t}\mathbf{q} \le \mathbf{eR_M}^{t}\mathbf{q} \le \mathbf{eR_S}^{t}\mathbf{q},\ \forall t > 0. \tag{26}$$

Moreover, by Theorem 1 we know that

$$e^{[t]} = \mathbf{eR_S}^{t}\mathbf{q} \le \mathbf{eR}(\lambda_{max})\mathbf{q}. \tag{27}$$

Combing (26) and (27) we get the theorem proved. □

3.3 Analytic Expressions of General Bounds

Theorems 1 and 2 show that computation of general bounds for approximation errors is based on the computability of $\mathbf{eR}^{t}(\lambda)\mathbf{q}$ and $\mathbf{eR_{S_l}}^{t}\mathbf{q}$, where $\mathbf{R}^{t}(\lambda)$ and $\mathbf{R_{S_l}}$ are defined by (15) and (19), respectively.

1. **Analytic Expression of $\mathbf{eR}^t(\lambda)\mathbf{q}$:** The submatrix $\mathbf{R}(\lambda)$ can be split as $\mathbf{R}(\lambda) = \boldsymbol{\Lambda} + \mathbf{B}$, where

$$\boldsymbol{\lambda} = \begin{pmatrix} \lambda & & & \\ & \ddots & & \\ & & \lambda & \\ & & & \lambda \end{pmatrix}, \quad \mathbf{B} = \begin{pmatrix} 0 & 1-\lambda & & \\ & \ddots & \ddots & \\ & & 0 & 1-\lambda \\ & & & 0 \end{pmatrix}.$$

Because multiplication of $\boldsymbol{\Lambda}$ and $\mathbf{B}$ is commutative, the binomial theorem [25] holds and we have

$$\mathbf{R}^t(\lambda) = (\boldsymbol{\Lambda} + \mathbf{B})^t = \sum_{i=0}^{t} C_t^i \boldsymbol{\Lambda}^{t-i} \mathbf{B}^i, \tag{28}$$

where

$$\boldsymbol{\Lambda}^{t-i} = diag\{\lambda^{t-i}, \ldots, \lambda^{t-i}\}, \tag{29}$$

Note that $\mathbf{B}$ is a nilpotent matrix of index n [1], and

$$\mathbf{B}^i = \begin{pmatrix} 0 \ldots (1-\lambda)^i & & & \\ \ddots & \ddots \ddots & & \\ & \ddots \ddots & (1-\lambda)^i & \\ & \ddots & \vdots & \\ & & 0 & \end{pmatrix}, \quad i < n. \tag{30}$$

Then, from (29), (30) and (28) we know
(a) if $t < n$,

$$\begin{aligned}\mathbf{eR}^t(\lambda)\mathbf{q} = & \sum_{j=1}^{t} \sum_{i=1}^{j} e_i C_t^{j-i} \lambda^{t-(j-i)} (1-\lambda)^{j-i} q_j \\ & + \sum_{j=t+1}^{n} \sum_{i=j-t}^{j} e_i C_t^{j-i} \lambda^{t-(j-i)} (1-\lambda)^{j-i} q_j. \end{aligned} \tag{31}$$

(b) if $t \geq n$,

$$\mathbf{eR}^t(\lambda)\mathbf{q} = \sum_{j=1}^{n} \sum_{i=1}^{j} e_i C_t^{j-i} \lambda^{t-(j-i)} (1-\lambda)^{j-i} q_j. \tag{32}$$

2. **Analytic Expression of $\mathbf{eR_{S_1}}^t\mathbf{q}$:** For the diagonal matrix $\mathbf{R_{S_1}}$, it holds that

$$\mathbf{eR_{S_1}}^t\mathbf{q} = \sum_{i=1}^{n} e_i p_{i,i}^t q_i = \sum_{i=1}^{n} e_i \lambda_i^t q_i. \tag{33}$$

---

[1] In linear algebra, a nilpotent matrix is a square matrix M such that $N^k = 0$ for some positive integer k. The smallest such k is called the index of M [26].

4 Case-by-Case Estimation of Approximation Error

In Sect. 3 general bounds of approximation error are obtained by ignoring most of elements in the sub-matrix **R**. Thus, these bounds could be very general but not tight. In this section, we would like to perform several case-by-case studies to demonstrate a feasible routine of error analysis, where the RLS and the (1+1)EA are employed solving the popular OneMax problem and the Needle-in-Haystack problem.

Problem 1. (**OneMax**)

$$\max f(\mathbf{x}) = \sum_{i=1}^{d} x_i, \quad \mathbf{x} = (x_1, \ldots, x_n) \in \{0,1\}^n.$$

Problem 2. (**Needle-in-Haystack**)

$$\max f(\mathbf{x}) = \begin{cases} 1, \text{ if } \sum_{i=1}^{d} x_i = 0, \\ 0, \quad \text{otherwise.} \end{cases} \qquad \mathbf{x} = (x_1, \ldots, x_n) \in \{0,1\}^n.$$

4.1 Error Estimation for the OneMax Problem

Application of RLS on the unimodal OneMax problem generates a step-by-step search, the transition submatrix of which is

$$\mathbf{R_S} = \begin{pmatrix} 1-1/n & 2/n & & & \\ & 1-2/n & 3/n & & \\ & & \ddots & \ddots & \\ & & & 1/n & 1 \\ & & & & 0 \end{pmatrix}. \tag{34}$$

Eigenvalues and corresponding eigenvectors of $\mathbf{R_S}$ are

$$\begin{aligned} &\lambda_1 = 1-1/n, \boldsymbol{\eta}_1 = (C_1^1, 0, \ldots, 0)^T, \\ &\lambda_2 = 1-2/n, \boldsymbol{\eta}_2 = (-C_2^1, C_2^2, 0, \ldots, 0)^T, \\ &\ldots, \\ &\lambda_n = 0, \qquad \boldsymbol{\eta}_n = ((-1)^{n+1}C_n^1, (-1)^{n+2}C_n^2, \ldots, (-1)^{2n)}C_n^n)^T. \end{aligned} \tag{35}$$

Theorem 3. *The expected approximation error of RLS for the OneMax problem is*

$$e^{[t]} = \frac{n}{2}\left(1-\frac{1}{n}\right)^t. \tag{36}$$

Proof. Denote $\mathbf{Q} = (q_{i,j})_{n\times n} = (\boldsymbol{\eta}_1, \boldsymbol{\eta}_2, \ldots, \boldsymbol{\eta}_n)$. Then we know that

$$\mathbf{Q}^{-1} = (q'_{i,j})_{n\times n} = \begin{pmatrix} C_1^1 & C_2^1 & C_3^1 & \cdots & C_n^1 \\ & C_2^2 & C_2^3 & \cdots & C_2^n \\ & & & \ddots & \vdots \\ & & & & C_n^n \end{pmatrix}. \tag{37}$$

$\mathbf{R_S}$ has n distinct eigenvalues, and so, can be diagonalized as $\boldsymbol{\Lambda} = diag(\lambda_1, \ldots, \lambda_n) = \mathbf{Q}^{-1}\mathbf{R_S}\mathbf{Q}$ [27]. Then, we have

$$e^{[t]} = \mathbf{e}{\mathbf{R_S}}^t\mathbf{q} = \mathbf{e}\mathbf{Q}\boldsymbol{\Lambda}^t\mathbf{Q}^{-1}\mathbf{q} = \mathbf{a}\boldsymbol{\Lambda}^t\mathbf{b}, \tag{38}$$

where $\mathbf{a} = \mathbf{eQ} = (a_1, \ldots, a_n)$, $\mathbf{b} = \mathbf{Q}^{-1}\mathbf{q} = (b_1, \ldots, b_n)$,

$$\begin{aligned} a_k &= \sum_{i=1}^{k} e_i q_{i,k} = \sum_{i=1}^{k} i(-1)^{i+k} C_k^i = \begin{cases} 1, & k = 1, \\ 0, & k = 2, \ldots, n, \end{cases} \\ b_k &= \sum_{j=k}^{n} q_{i,k} p_j = \sum_{j=k}^{n} C_j^k C_n^j \frac{1}{2^n} = \frac{C_n^k}{2^k}, \quad k = 1, \ldots, n. \end{aligned} \tag{39}$$

Substituting (39) into (38) we get the result

$$e^{[t]} = a_1\lambda_1^t b_1 = \frac{n}{2}\left(1 - \frac{1}{n}\right)^t.$$

□

Theorem 4. *The expected approximation error of (1+1)EA for the OneMax problem is bounded from above by*

$$e^{[t]} \leq \frac{n}{2}\left[1 - \frac{1}{ne}\right]^t. \tag{40}$$

Proof. According to the definition of population status, we know that the status index i is the number of 0-bits in $\mathbf{x}$. Once one of i 0-bits is flip to 1-bit and all 1-bits keep unchanged, the generated solution will be accepted, and the status transfers from i to $i-1$. Recalling that the probability this case happen is $\frac{i}{n}\left(1 - \frac{1}{n}\right)^{n-i}$, we know that

$$p_{i-1,i} \geq \frac{i}{n}\left(1 - \frac{1}{n}\right)^{n-i} \geq \frac{i}{ne}, \quad i = 1, \ldots, n.$$

Denote

$$\mathbf{R_S} = \begin{pmatrix} 1 - \frac{1}{ne} & \frac{2}{ne} & & \\ & \ddots & \ddots & \\ & & 1 - \frac{n-1}{ne} & 1 \\ & & & 0 \end{pmatrix},$$

and we know that

$$e^{[t]} \leq \mathbf{e}\mathbf{R_S}^t\mathbf{q}. \tag{41}$$

With n distinct eigenvalues, $\mathbf{R_S}$ can be diagonalized:

$$\mathbf{P}^{-1}\mathbf{R_S}\mathbf{P} = \boldsymbol{\Lambda}, \tag{42}$$

where $\boldsymbol{\Lambda} = diag(\lambda_1, \ldots, \lambda_n)$, $\mathbf{P} = (\boldsymbol{\eta}_1, \cdots, \boldsymbol{\eta}_n)$. λ_i and $\boldsymbol{\eta}_i$ are the eigenvalues and the corresponding eigenvectors:

$$\begin{array}{ll} \lambda_1 = 1 - 1/(ne),\ \boldsymbol{\eta}_1 = (C_1^1, 0, \ldots, 0)^T, & \\ \lambda_2 = 1 - 2/(ne),\ \boldsymbol{\eta}_2 = (-C_2^1, C_2^2, 0, \ldots, 0)^T, & \\ \ldots, & \\ \lambda_n = 0, & \boldsymbol{\eta}_n = ((-1)^{n+1}C_n^1, (-1)^{n+2}C_n^2, \ldots, (-1)^{2n)}C_n^n)^T. \end{array} \tag{43}$$

It is obvious that $\mathbf{P}$ is invertible, and its inverse is

$$\mathbf{P}^{-1} = \begin{pmatrix} C_1^1 & C_2^1 & C_3^1 & \cdots & C_n^1 \\ & C_2^2 & C_2^3 & \cdots & C_2^n \\ & & & \ddots & \vdots \\ & & & & C_n^n \end{pmatrix}. \tag{44}$$

Similar to the result illustrated in (39), we know that

$$\begin{array}{l} \mathbf{eP} = (1, 0, \ldots, 0)^T, \\ \mathbf{P}^{-1}\mathbf{q} = (\frac{C_n^2}{2}, \frac{C_n^2}{2^2}, \ldots, \frac{C_n^{n-1}}{2^{n-1}}, \frac{1}{2^n})^T. \end{array} \tag{45}$$

Combing (41), (42), (43), (44) and (45) we know that

$$e^{[t]} \leq \mathbf{eP}\boldsymbol{\Lambda}^t\mathbf{P}^{-1}\mathbf{q} = \frac{n}{2}\left[1 - \frac{1}{ne}\right]^t.$$

□

4.2 Error Estimation for the Needle-in-Haystack Problem

Landscape of the Needle-in-Haystack problem has a platform where all solutions have the same function value 0, and only the global optimum $\mathbf{x}^* = (0, \ldots, 0)$ has a non-zero function value 1. For this problem, the status i is defined as total number of 1-bits in a solutions $\mathbf{x}$.

Theorem 5. *The expected approximation error of RLS for the Needle-in-Haystack problem is bounded by*

$$\left(1 - \frac{1}{n}\right)^t + 1 - \frac{n+1}{2^n} \leq e^{[t]} \leq \left(1 - \frac{1}{en}\right)^t + 1 - \frac{n+1}{2^n}. \tag{46}$$

Proof. When the RLS is employed to solve the Needle-in-Haystack problem, the transition submatrix is

$$\mathbf{R_S} = diag\left(1 - \frac{1}{n}\left(1 - \frac{1}{n}\right)^{n-1}, 1, \ldots, 1\right). \tag{47}$$

Then,

$$e^{[t]} = \mathbf{eR_S q} = \sum_{i=1}^{n} e_i p_{i,i}^t p_i = \left[1 - \frac{1}{n}\left(1 - \frac{1}{n}\right)^{n-1}\right]^t + \sum_{i=2}^{n} \frac{C_n^i}{2^n}. \tag{48}$$

Since

$$\left(1 - \tfrac{1}{n}\right)^t \leq \left[1 - \tfrac{1}{n}\left(1 - \tfrac{1}{n}\right)^{n-1}\right]^t \leq \left(1 - \tfrac{1}{en}\right)^t,$$
$$\textstyle\sum_{i=2}^{n} \frac{C_n^i}{2^n} = 1 - \frac{C_n^0}{2^n} - \frac{C_n^1}{2^n} = 1 - \frac{n+1}{2^n},$$

we can conclude that

$$\left(1 - \frac{1}{n}\right)^t + 1 - \frac{n+1}{2^n} \leq e^{[t]} \leq \left(1 - \frac{1}{en}\right)^t + 1 - \frac{n+1}{2^n}.$$

□

Theorem 5 indicates that both the upper bound and the lower bound converge to the positive $1 - \frac{n+1}{2^n}$ when $t \to \infty$, which implies the fact that RLS cannot converge in mean to global optimal solution of the Needle-in-Haystack problem. Because the RLS searches adjacent statuses and only better solutions can be accepted, it cannot converge to the optimal status once the initial solution is not located at the status 1.

Theorem 6. *The expected approximation error of (1+1)EA for the Needle-in-Haystack problem is bounded by*

$$\frac{n}{2}\left(1 - \frac{1}{n}\right)^t \leq e^{[t]} \leq \frac{n}{2}\left(1 - \frac{1}{n^n}\right)^t. \tag{49}$$

Proof. When the (1+1)EA is employed to solve the Needle-in-Haystack problem, the transition probability submatrix is

$$\mathbf{R_S} = diag\left(1 - \frac{1}{n}\left(1 - \frac{1}{n}\right)^{n-1}, \ldots, 1 - \left(\frac{1}{n}\right)^n\right). \tag{50}$$

Then,

$$e^{[t]} = \mathbf{eR_S q} = \sum_{i=1}^{n} e_i p_{i,i}^t p_i = \sum_{i=1}^{n} i \left[1 - \left(\frac{1}{n}\right)^i \left(1 - \frac{1}{n}\right)^{n-i}\right]^t \frac{C_n^i}{2^n}. \tag{51}$$

Since

$$\sum_{i=1}^{n} i\left[1-\left(\tfrac{1}{n}\right)^{i}\left(1-\tfrac{1}{n}\right)^{n-i}\right]^{t}\tfrac{C_n^i}{2^n} \geq \left(1-\tfrac{1}{n}\right)^{t}\sum_{i=1}^{n} i\tfrac{C_n^i}{2^n} = \tfrac{n}{2}\left(1-\tfrac{1}{n}\right)^{t},$$
$$\sum_{i=1}^{n}\left[1-\left(\tfrac{1}{n}\right)^{i}\left(1-\tfrac{1}{n}\right)^{n-i}\right]^{t}\tfrac{C_n^i}{2^n} \leq \left(1-\tfrac{1}{n^n}\right)^{t}\sum_{i=1}^{n} i\tfrac{C_n^i}{2^n} = \tfrac{n}{2}\left(1-\tfrac{1}{n^n}\right)^{t},$$

we can conclude that

$$\frac{n}{2}\left(1-\frac{1}{n}\right)^{t} \leq e^{[t]} \leq \frac{n}{2}\left(1-\frac{1}{n^n}\right)^{t}.$$

□

5 Conclusion

To make theoretical results more instructional to algorithm developments and applications, this paper proposes to investigate performance of EAs by estimating approximation error for any iteration budget t. General bounds included in Theorems 1 and 2 demonstrate that bottlenecks of EAs' performance are decided by the maximum and the minimum eigenvalues of transition submatrix $\mathbf{R}$. Meanwhile, Theorems 3, 4, 5, and 6 present estimations of approximation error for RLS and (1+1)EA for two benchmark problems, which shows that our analysis scheme is applicable for elitist EAs, regardless the shapes of transition matrices. Moreover, the estimation results demonstrate that approximation errors are closely related to eigenvalues of the transition matrices, which provide useful information for performance improvements of EAs. Our future work is to further perform error analysis on real combinatorial problems to show its applicability in theoretical analysis of EAs.

Acknowledgements. This work was supported in part by the National Nature Science Foundation of China under Grants 61303028 and 61763010, in part by the Guangxi "BAGUI Scholar" Program, and in part by the Science and Technology Major Project of Guangxi under Grant AA18118047.

References

1. Oliveto, P., He, J., Yao, X.: Time complexity of evolutionary algorithms for combinatorial optimization: a decade of results. Int. J. Autom. Comput. **4**(3), 281–293 (2007). https://doi.org/10.1007/s11633-007-0281-3
2. He, J., Yao, X.: Towards an analytic framework for analysing the computation time of evolutionary algorithms. Artif. Intell. **145**(1–2), 59–97 (2003)
3. Ding, L., Yu, J.: Some techniques for analyzing time complexity of evolutionary algorithms. Trans. Inst. Meas. Control. **34**(6), 755–766 (2012)
4. He, J., Yao, X.: Drift analysis and average time complexity of evolutionary algorithms. Artif. Intell. **127**(1), 57–85 (2001)
5. Doerr, B., Johannsen, D., Winzen, C.: Multiplicative drift analysis. Algorithmica **64**(4), 673–697 (2012). https://doi.org/10.1007/s00453-012-9622-x

6. Yu, Y., Qian, C., Zhou, Z.H.: Switch analysis for running time analysis of evolutionary algorithms. IEEE Trans. Evol. Comput. **19**(6), 777–792 (2014)
7. Droste, S., Jansen, T., Wegener, I.: On the analysis of the (1+1) evolutionary algorithm. Theor. Comput. Sci. **276**(1–2), 51–81 (2002)
8. Chen, Y., Zou, X., He, J.: Drift conditions for estimating the first hitting times of evolutionary algorithms. Int. J. Comput. Math. **88**(1), 37–50 (2011)
9. Huang, H., Xu, W., Zhang, Y., Lin, Z., Hao, Z.: Runtime analysis for continuous (1+1) evolutionary algorithm based on average gain model. Scientia Sinica Informationis **44**(6), 811–824 (2014)
10. Yushan, Z., Han, H., Zhifeng, H., Guiwu, H.: First hitting time analysis of continuous evolutionary algorithms based on average gain. Clust. Comput. **19**(3), 1323–1332 (2016). https://doi.org/10.1007/s10586-016-0587-4
11. Akimoto, Y., Auger, A., Glasmachers, T.: Drift theory in continuous search spaces: expected hitting time of the (1+1)-ES with 1/5 success rule. In: Proceedings of the Genetic and Evolutionary Computation Conference, pp. 801–808. ACM (2018)
12. Yu, Y., Yao, X., Zhou, Z.H.: On the approximation ability of evolutionary optimization with application to minimum set cover. Artif. Intell. **180–181**, 20–33 (2012)
13. Lai, X., Zhou, Y., He, J., Zhang, J.: Performance analysis of evolutionary algorithms for the minimum label spanning tree problem. IEEE Trans. Evol. Comput. **18**(6), 860–872 (2014)
14. Zhou, Y., Lai, X., Li, K.: Approximation and parameterized runtime analysis of evolutionary algorithms for the maximum cut problem. IEEE Trans. Cybern. **45**(8), 1491–1498 (2015)
15. Zhou, Y., Zhang, J., Wang, Y.: Performance analysis of the (1+1) evolutionary algorithm for the multiprocessor scheduling problem. Algorithmica **73**(1), 21–41 (2015). https://doi.org/10.1007/s00453-014-9898-0
16. Xia, X., Zhou, Y., Lai, X.: On the analysis of the (1+1) evolutionary algorithm for the maximum leaf spanning tree problem. Int. J. Comput. Math. **92**(10), 2023–2035 (2015)
17. Peng, X., Zhou, Y., Xu, G.: Approximation performance of ant colony optimization for the TSP (1, 2) problem. Int. J. Comput. Math. **93**(10), 1683–1694 (2016)
18. Rudolph, G.: Convergence rates of evolutionary algorithms for a class of convex objective functions. Control. Cybern. **26**, 375–390 (1997)
19. He, J., Lin, G.: Average convergence rate of evolutionary algorithms. IEEE Trans. Evol. Comput. **20**(2), 316–321 (2016)
20. Jansen, T., Zarges, C.: Fixed budget computations: a different perspective on run time analysis. In: Proceedings of the 14th Annual Conference on Genetic and Evolutionary Computation, pp. 1325–1332. ACM (2012)
21. Jansen, T., Zarges, C.: Performance analysis of randomised search heuristics operating with a fixed budget. Theor. Comput. Sci. **545**, 39–58 (2014)
22. He, J.: An analytic expression of relative approximation error for a class of evolutionary algorithms. In: Proceedings of 2016 IEEE Congress on Evolutionary Computation (CEC 2016), pp. 4366–4373, July 2016
23. He, J., Jansen, T., Zarges, C.: Unlimited budget analysis. In: Proceedings of the Genetic and Evolutionary Computation Conference Companion, pp. 427–428. ACM (2019)

24. He, J., Chen, Y., Zhou, Y.: A theoretical framework of approximation error analysis of evolutionary algorithms. arXiv preprint arXiv:1810.11532 (2018)
25. Aigner, M.: Combinatorial Theory. Springer, New York (2012)
26. Herstein, I.N.: Topics in Algebra. Wiley, Hoboken (2006)
27. Lay, D.C.: Linear Algebra and Its Applications. Pearson Education, London (2003)

Research on Two-Level Inventory Optimization Algorithm for Repairable Spare Parts Based on Improved Differential Evolution

Tao Gu[1](✉), Sujian Li[1], Jie Li[2], and Jie Jiao[2]

[1] Institute of Mechanical Engineering, Beijing University of Science and Technology, Beijing 100083, China
babygo1003@163.com

[2] Provincial Key Laboratory of Food Logistics Equipment Technology Research, Zhejiang University of Science and Technology, Hangzhou 310023, China

Abstract. For solving the two-level inventory optimization model of precious repairable spare parts, an improved differential evolution algorithm is proposed to solve the model in view of the shortcomings of the traditional marginal analysis method. The difference between the worst solution and the best solution of the algorithm is 1.93% after 30 runs, the standard deviation is 0.6% of the mean, and the result is reduced by 4.85% and 6.24%, respectively, compared with the traditional marginal analysis method and the VMETRIC software calculation result. It shows that the algorithm is stable and has certain superiority.

Keywords: Repairable spare parts · Multilevel inventory · Marginal analysis method · Improved differential evolution algorithms

1 Foreword

Maintainability of repairable spare parts determines the link of a maintenance in the equipment guarantee system. In the traditional spare parts guarantee, it is not necessary for most parts to be repaired by original parts. Thus there is basically no repair links and no special repair organization. Most of the replaced parts are treated as wastes. The traditional spare parts system only covers "raising, storage, supply and management". When the repairable spare parts are damaged, the basic performance can be recovered by replacing the components to which they belong, and the repairable spare parts are expensive. Therefore, the replaced spare parts should be recovered by a specific repair organization, and then they should be transferred into the corresponding warehouse as new spare parts in order to reduce the amount of repairable spare parts raised from manufacturers and effectively decrease equipment guarantee costs under the new maintenance mode.

The high price of repairable spare parts determines that inventory optimization control must be carried out in spare parts guarantee. Statistics show that the proportion of vulnerable and wearable repairable spare parts in all kinds of spare parts is about 10% and the proportion of cost is about 50%. Since the ratio of direct replacement of

L. Pan et al. (Eds.): BIC-TA 2019, CCIS 1159, pp. 341–354, 2020.
https://doi.org/10.1007/978-981-15-3425-6_27

repairable spare parts to the cost of replacement is more than 5:1, the traditional "the more the better", the "just in case" inventory control mode will be difficult to achieve in the financial guarantee [1]. Therefore, on the premise of certain guarantee cost, to establish a multi-level inventory model of repairable spare parts and to study the optimal inventory technology of warehouses at all levels in the guarantee system is an important brace to improve the efficiency-cost ratio of equipment guarantee and to achieve "in the right time, in the right place and with the right quantity", according to the requirements of maintenance guarantee, raising cycle, repair cycle and other conditions. The existing equipment repairable spare parts support system is mainly based on two-level inventory model including rear warehouse and base-level sites. The traditional method to solve this model is the marginal analysis method for the first time [2] proposed by O. Gross, which is applied by Sherbrooke to the solution of METRIC model [3], whose principle is to add one piece to each spare parts inventory of warehouses at all levels, and to divide the marginal increment of the expected shortage by the unit price of spare parts, then to find out a spare parts with the greatest unit cost effect and the corresponding inventory location, where one piece of inventory is added at this location, and cycled repeatedly until the constraints are reached. Finally it generates the optimal relationship curve between the equipment spare parts shortage and cost. This method is a greedy local search algorithm, which only traverses the marginal benefits of increasing inventory by one in each location. It belongs to the incomplete exhaustive method, whose advantage is the high optimization speed, and whose corresponding disadvantage is the low-level of quality of the optimal solution of the algorithm.

Differential Evolution (DE) is first proposed by Storn and Price in 1995 [4], which is mainly used to solve real-valued optimization problems. This algorithm is a group-based adaptive global optimization algorithm and belongs to evolutionary algorithm. It is widely used because of the features of simple structure, easy implementation, fast convergence and strong robustness [5]. In this paper, differential evolution algorithm is used to solve the two-level inventory model of repairable spare parts for the first time. Various mutation methods are considered on the basis of traditional differential evolution algorithm. Aiming at the problem of algorithm prematurity and unavailability of better solution, a premature mechanism of self-processing algorithm is designed, which improves the optimization ability of the algorithm. At the same time, a local search algorithm is designed in each iteration, which improves the search speed of the algorithm.

2 Two-Level Inventory Model for Repairable Spare Parts

2.1 Model Description

Based on a 2X2 two-level inventory model, the support site is divided into two levels. A rear warehouse supplies multiple base-level sites. Both levels can store repairable spare parts and have certain maintenance capability for some repairable spare parts. Repairable spare parts are divided into two levels, and an external field replaceable part (Line-Replacement Unit, LRU) consists of multiple internal field replaceable parts (Shop-Replacement Unit, SRU) [6, 7].

Its mode of operation is shown in Fig. 1, when 1 LRU requirement occurs at the base-level site, it's replaced if the base-level has inventory and it's applied to the rear warehouse for 1 LRU (1 → 3) if there is no inventory. When repairing the LRU, it is found that its SRU is causing the fault. The SRU needs to be replaced if the base-level has SRU inventory (1 → 2). If there is no inventory, it's applied for 1 SRU (2 → 4) from the rear warehouse. When the rear warehouse repairs the LRU, it is found that the SRU causes the fault, and then there is a requirement for 1 SRU which can replace it (3 → 4).

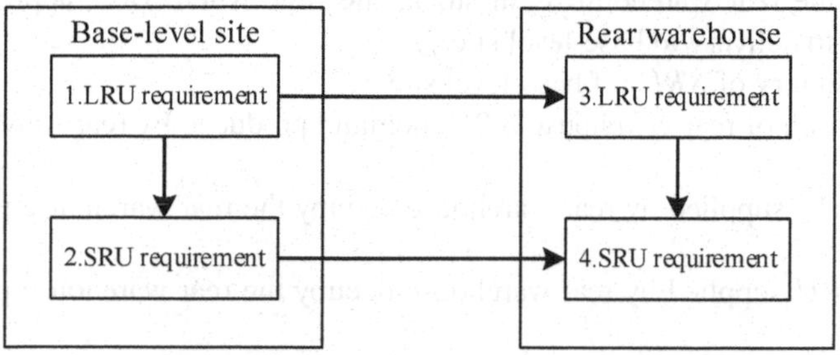

Fig. 1. Operation mode of safeguard system.

2.2 Assumed Condition

There are eight important assumptions for the multi-level inventory problem of repairable spare parts [8, 9]:

- (s – 1, s) the ordering strategy, the ordering strategy between the rear warehouse and the base-level adopts a one-by-one ordering strategy, no batch ordering;
- Infinite channel queuing hypothesis, there is no queue or interaction in repairing spare parts, and the average maintenance time does not consider its distribution;
- Repair of faulty parts in the base-level depends on maintenance capability of the base-level site, regardless of its inventory level and work load;
- Each base-level site only replenishes goods from the rear warehouse, and there is no horizontal supply between base-level sites;
- All spare parts can be repaired regardless of the case of scrap;
- Regardless of the spare parts storage, transportation, management and other additional costs, only to consider the purchase price of spare parts themselves;
- All LRU are of the same importance and will be unavailable in case of shortage;
- Multiple faults do not occur at the same time.

2.3 Symbol Definition

The symbols used in the multi-level inventory model of repairable spare parts are defined as follows:

- i: project number of SRU, $i = 1, 2, \ldots, I$, $i = 0$ represents LRU;
- k: project number of LRU, $k = 1, 2, \ldots, K$;

- j: project number of base-level site, $j = 1, 2, \ldots, J$, $j = 0$ represents the rear warehouse;
- m_{ij}: average annual demand of the SRU_i of base-level site j;
- T_{ij}: average annual repair time of the SRU_i of base-level site j;
- r_{ij}: the probability of repairing faulty parts at base-level site of the SRU_i of base-level site j;
- q_{ij}: conditional probability of LRU generating fault isolation SRU_i in base-level site j;
- O_j: when the rear warehouse is in stock, the time when SRU applies from rear warehouse to delivery to base-level site j;
- s_{ij}: the inventory of SRU_i of base-level site j;
- f_{i0}: proportion of rear warehouse SRU_i demand produced by rear warehouse repair LRU;
- f_{ij}: the SRU_i supplied by rear warehouse occupy the rear warehouse proportion of demand;
- f_{0j}: the LRU supplied by rear warehouse occupy the rear warehouse proportion of demand;
- A: system availability;
- A_m: system availability constraint value;
- A_j: the availability of base-level site j;
- N_j: the number of equipment of base-level site j;
- Z_k: the number of installations of a certain LRU_k on an equipment, representing single machine installation number
- c: unit price of spare parts;
- X_{ij}: the quantity of SRU_i that the base-level site is repairing or replenishing at any time, and the quantity of supply channels is a random variable.

2.4 Mathematical Derivation

- Mean and variance of the quantity of LRU under repair in the rear warehouse

The quantity of LRU being repaired in the rear warehouse consists of two parts: (1) the quantity of which is supplied by the rear warehouse when the rear warehouse is in stock; (2) when the rear warehouse is out of stock, the quantity of LRU being repaired in the rear warehouse which is caused by the delay of SRU_i. Its mean $E[X_{00}]$ and variance $Var[X_{00}]$ are:

$$E[X_{00}] = m_{00}T_{00} + \sum_{i=1}^{I} f_{i0}EBO(s_{i0}|m_{i0}T_{i0}) \tag{1}$$

$$Var[X_{00}] = m_{00}T_{00} + \sum_{i=1}^{I} f_{i0}(1 - f_{i0})EBO(s_{i0}|m_{i0}T_{i0}) + \sum_{i=1}^{I} f_{i0}^2 VBO(s_{i0}|m_{i0}T_{i0}) \tag{2}$$

in which,

$$f_{i0} = m_{00}q_{i0}/m_{i0} \tag{3}$$

- Mean and variance of the quantity of SRU being repaired or replenished at the base

The quantity of SRU_i being repaired or replenished at the base consists of two parts: (1) The quantity of SRU_i being repaired at the base and replenishing from the rear warehouse when the SRU_i in the rear warehouse is in stock; (2) The quantity of SRU_i causing replenishment delays when there is no SRU_i inventory in the rear warehouse. Its mean $E[X_{0j}]$ and variance $Var[X_{0j}]$ are:

$$E[X_{ij}] = m_{ij}[(1-r_{ij})O_j + r_{ij}T_{ij}] + f_{ij}EBO(s_{i0}|m_{i0}T_{i0}) \tag{4}$$

$$\begin{aligned} Var[\mathrm{X}_{00}] = m_{ij}[(1-r_{ij})O_j + r_{ij}T_{ij}] + f_{ij}(1-f_{ij})EBO(s_{i0}|m_{i0}T_{i0}) \\ + f_{ij}^2 VBO(s_{i0}|m_{i0}T_{i0}) \end{aligned} \tag{5}$$

in which,

$$f_{ij} = m_{ij}(1-r_{ij})/m_{i0} \tag{6}$$

- Mean and variance of the quantity of LRU being repaired or replenished at the base

The quantity of LRU being repaired or replenished at the base consists of three parts: (1) The quantity of LRU being repaired at the base and replenishing from the rear warehouse when the LRU in the rear warehouse is in stock; (2) The quantity of LRU that caused delays in replenishment when there is no LRU inventory in the rear warehouse; (3) The quantity of LRU that caused delays in repairing when there is no SRU_i inventory in the rear warehouse. Its mean $E[X_{0j}]$ and variance $Var[X_{0j}]$ are:

$$\begin{aligned} E[X_{0j}] = m_{0j}[(1-r_{0j})O_j + r_{0j}T_{0j}] + f_{0j}EBO(s_{00}|E[X_{00}], Var[X_{00}]) \\ + \sum_{i=1}^{I} EBO(s_{ij}|E[X_{ij}], Var[X_{ij}]) \end{aligned} \tag{7}$$

$$\begin{aligned} Var[\mathrm{X}_{0j}] = m_{0j}[(1-r_{0j})O_j + r_{0j}T_{0j}] + f_{0j}(1-f_{0j})EBO(s_{00}|E[X_{00}], Var[X_{00}]) \\ + f_{0j}^2 VBO(s_{00}|E[X_{00}], Var[X_{00}]) + \sum_{i=1}^{I} VBO(s_{ij}|E[X_{ij}], Var[X_{ij}]) \end{aligned} \tag{8}$$

in which,

$$f_{0j} = m_{0j}(1-r_{0j})/m_{00} \tag{9}$$

- Availability

The formula for the equipment availability of base j resulting from expectation shortage of LRU and SRU is,

$$A_j = 100\prod_{k=1}^{K}\{1 - EBO(s_{kj}|E[X_{kj}], Var[X_{kj}])/(N_j Z_k)\}^{Z_k} \tag{10}$$

The system availability is equal to available quantity of equipment at bases, and is divided by the sum of all equipment in the system, in which the available amount of equipment at each base is equal to the quantity of equipment at corresponding base multiplied by A_j:

$$A = (\sum_{j=1}^{J} N_j A_j) / \sum_{j=1}^{J} N_j \tag{11}$$

- Mathematical model

In this paper, the system availability as a constraint, and the total cost of repairable spare parts inventory as an optimization objective, the mathematical model is as follows:

$$\begin{cases} \min \sum_{i=1}^{I} \sum_{j=0}^{J} c_i \cdot s_{ij} \\ \text{s.t.A} \geq \text{A}_\text{m} \end{cases} \tag{12}$$

3 Design of Model Solving Algorithms

3.1 Algorithmic Flow

In this paper, local search and precocious function processing are added based on differential evolution algorithm. The algorithm flow is shown in Fig. 2, and the steps are as follows [10]:

Step 1: Configuration parameters, including maximum iteration times *maxIteration*, current iteration algebra *Generation*, search upper bound *Xmax*, search lower bound *Xmin*, individual dimension *Dim*, population size *NP*, inventory matrix *Inventory*, scaling factor *F*, crossover probability *CR*, mutation strategy *Mutation_Strategy*, cross strategy *Cross_Strategy*, maximum availability *Max_A*, population individual similarity *Similarity_Degree*, reset ratio *Reset_Proportion*.
Step 2: Initialize the population and randomly generate the initial population according to the *Xmax* and *Xmin* ranges, but the system availability of each individual in the population is required to be greater than *Max_A*, and optimize from a high-cost inventory location to a low-cost one.
Step 3: Calculate the fitness value of each individual in the population, find out the optimal value of the population, and assign it to the temporary variable *Best_Inventory*.
Step 4: Randomly select the mutation algorithm, and perform mutation calculation for each individual of the population. After the calculation is completed, the individual values of the mutations are sorted out. The value less than 0 is assigned to 0, and the decimal is rounded.
Step 5: Perform cross-operation on each individual of the population. After each crossover, judge whether the system availability of the individual is greater than *Max_A*. If it is greater, continue and if it is smaller, re-cross-operate.

Step 6: Calculate the fitness value of each individual in the population after cross-mutation. If it is better than the original population, replace it. If it is better than the optimal value of the original population, assign the value to *Best_Inventory*.
Step 7: Local search for the existing optimal solution to obtain a better value.
Step 8: Determine whether the population is premature. If the similarity of the population is greater than *Similarity_Degree*, the individual whose proportion in the population is *Reset_Proportion* is randomly selected and re-initialized.
Step 9: Determine whether the end condition is satisfied, and if it is satisfied, calculate the inventory cost of *Best_Inventory*. If not, proceed to step 3 and continue iterating.

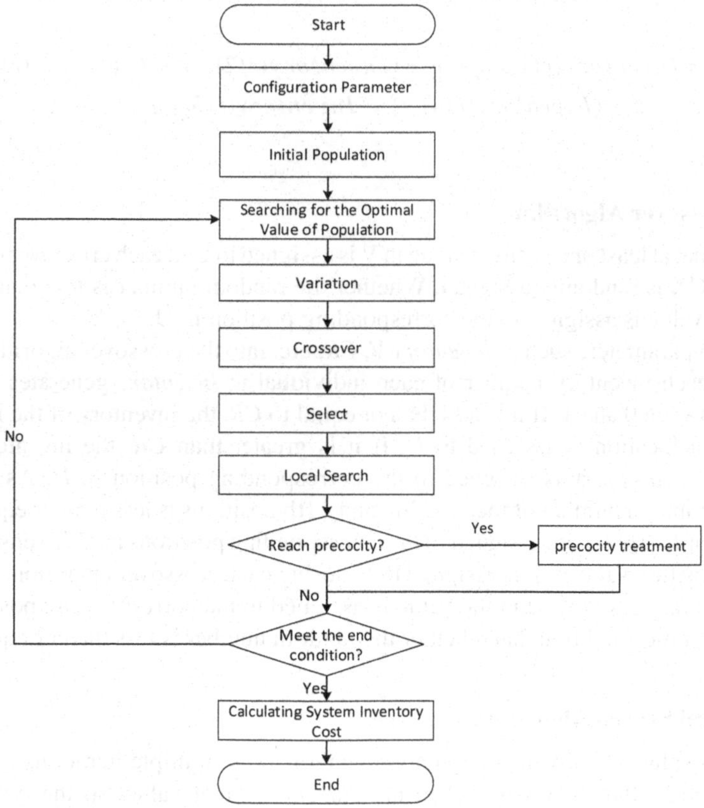

Fig. 2. Flow chart of improved differential evolution algorithm

3.2 Detailed Design of the Algorithm

3.2.1 Mutation Algorithm

Parameters such as *Inventory*, *Best_Inventory*, and F are passed to the mutation algorithm, and one variant algorithm formula in Eqs. 13–17 is randomly selected to operate on the mutation result population V.

$$V(i,:) = Inventory(r(1),:) + F * (Inventory(r(2),:) - Inventory(r(3),:)) \quad (13)$$

$$V(i,:) = Best\_Inventory + F * (Inventory(r(1),:) - Inventory(r(2),:)) \quad (14)$$

$$V(i,:) = Inventory(i,:) + F * (Best\_Inventory - Inventory(i,:)) + F * (Inventory(r(1),:) - Inventory(r(2),:)) \quad (15)$$

$$V(i,:) = Best\_Inventory + F * (Inventory(r(1),:) - Inventory(r(2),:)) + F * (Inventory(r(3),:) - Inventory(r(4),:)) \quad (16)$$

$$V(i,:) = Inventory(r(1),:) + F * (Inventory(r(2),:) - Inventory(r(3),:)) + F * (Inventory(r(4),:) - Inventory(r(5),:)) \quad (17)$$

3.2.2 Crossover Algorithm

To ensure that at least one position value in V is assigned to U in each crossover operation, a position CX is randomly assigned. Whether the random number is less than or equal to CR, the value is assigned to the corresponding position in U.

Passing parameters such as *Inventory V*, *CR*, etc. into the crossover algorithm, cross-operating each inventory number of each individual in *Inventory* generates a random number between 0 and 1. If it is less than or equal to *CR*, the inventory of the individual in *V* at that location is assigned to *U*. If it is greater than *CR*, the inventory of the individual in *Inventory* is assigned to the corresponding position in *U*. As shown in Fig. 3, the random number of the 3rd, 8th, and 11th positions is less than or equal to *CR*, the values of I_3, I_8, I_{11} are assigned to the corresponding positions in *U*. To ensure that at least one position value in *V* is assigned to *U* during each crossover operation, a position *CX* is randomly assigned, and the value is assigned to the corresponding position in *U* regardless of the condition that whether the random number is less than or equal to *CR*.

3.2.3 Local Search Algorithm

The fitness value of individual population generated by multiple iterations of ordinary difference algorithm may not exceed the current optimal value, so the optimization speed is slow. A local search algorithm is added at the end after each iteration selection operation. The algorithm finds that the unit price of spare parts with the highest inventory position in the current optimal solution decreases the inventory in this location by one, and to judge whether the constraints are met. If it is satisfied, the local search exists. If not, the number of stocks in the position is restored, and to find out the second-high price inventory position to decrease the inventory in this location by one, and to judge whether the constraint is satisfied. The local search is terminated by iterating until a better value satisfying the constraints is found or traversing each inventory location.

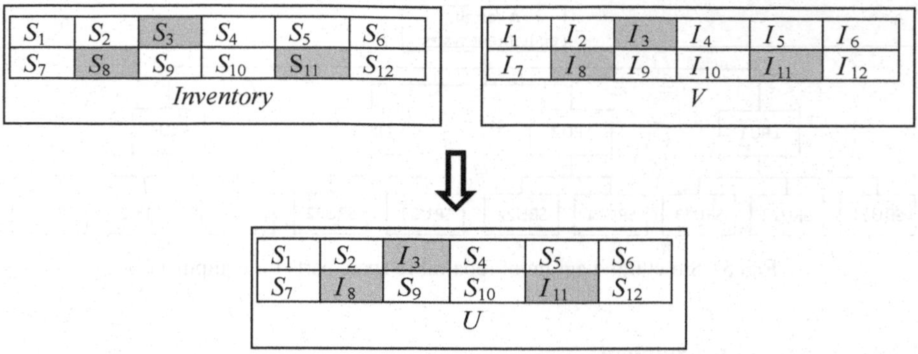

Fig. 3. Schematic diagram of crossover algorithm

3.2.4 Self-processing Premature Algorithm

Because the variation principle of the difference algorithm leads to the convergence of individuals in the population at the later stage of the optimization process, which makes it more difficult to produce more optimal values. Therefore, in each iteration, if the similarity ratio of individuals in the population is larger than *Similarity_Degree*, the algorithm is considered premature, the individual in the population with the ratio of *Reset_Proportion* is reinitialized, and then the optimization is continued.

4 Calculation Example

The objective of the experiment is to use the examples in reference [11] to compare the results calculated by the improved differential evolution algorithm with those calculated by the marginal optimization method in the reference.

4.1 Case Design

The support sites are shown in Fig. 4. There are three base-level sites in a rear warehouse. The number of equipments deployed at the base-level sites are 18, 12 and 15 respectively. The average service time of each site is 1. The structure of repairable spare parts is shown in Fig. 5, and the parameters of repairable spare parts are shown in Table 1. The system availability is required to be no less than 95%. Based on experience, the parameter configuration of the improved differential evolution algorithm is shown in Table 2.

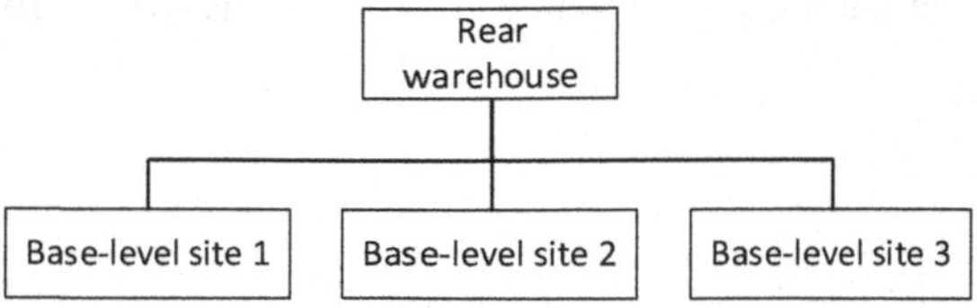

Fig. 4. Structural diagram of safeguard sites

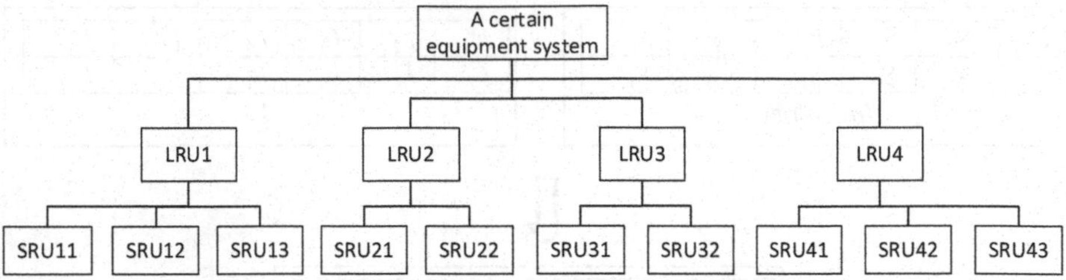

Fig. 5. Structural diagram of repairable spare parts for equipment

4.2 Operating Environment

Main configuration of operating computer: Intel Core i5 2.4 GHz CPU, 8GBRAM, Operating system: Windows 7, Data analysis and processing software: MATLAB R2018a.

4.3 Experimental Results

The experimental data are obtained by running the 30-time algorithm as shown in Table 3. The calculation results are the optimized iteration times and the total inventory cost, among which the iteration times are the iteration times of the first stable calculation results. The results show that the optimal solution is obtained between the 104th iterations and 253th iterations. The average of total inventory cost is 5,178,437 yuan. The optimal solution is 5,154,700 yuan. The difference between the worst solution and the optimal solution is 1.93%. The standard deviation is 33,602 yuan, and the standard deviation is 0.6% of the mean, which indicates the high stability of the algorithm.

As shown in Fig. 6, the minimum total inventory amount of the initial population randomly generated by the improved differential evolution algorithm is 7,471,700 yuan, which is optimized to 5,154,700 yuan at the 132nd iteration, and no better value is found at the 368 iterations after that. The algorithm converges to 5,154,700 yuan, and the inventory allocation of repairable spare parts is shown in Table 4.

As shown in Table 5, with 95% availability as a constraint, the total inventory amount corresponding to the optimal inventory allocation scheme of the improved differential evolution algorithm after 500 iterations is 5,154,700, which is reduced by 4.85% compared with 5,417,700 yuan in reference [11], and which is 6.24% lower than 5,497,700 yuan optimized by the VMERTIC software optimization mentioned in reference [11]. It can be seen that the optimization ability of the algorithm in this paper is better than the marginal analysis method used in reference [11] and VMETRIC software.

Table 1. Repairable spare parts parameter table

Spare parts items	Rear warehouse demand rate (pieces/year)	Site 1 demand rate (pieces/year)	Site 2 demand rate (pieces/year)	Site 3 demand rate (pieces/year)	Maintenance time (days)	Repairable rate	Isolation rate	Unit price (yuan)	Installation Number
LRU1	108.336	77.383	51.589	64.486	10	0.44	–	113400	1
LRU2	146.48	94.503	63.002	78.753	11	0.38	–	78800	2
LRU3	91.687	53.934	35.956	44.945	13	0.32	–	53400	1
LRU4	313.401	232.149	154.766	193.458	9	0.46	–	92300	3
SRU11	42.120	8.672	5.782	7.227	5	0.33	0.2547	16700	2
SRU12	29.237	5.782	3.854	4.818	7	0.25	0.1698	38800	1
SRU13	22.954	4.956	3.304	4.130	7	0.42	0.1455	19300	1
SRU21	52.839	8.988	5.992	7.490	8	0.28	0.2503	11200	1
SRU22	83.364	13.827	9.218	11.523	4	0.22	0.3850	9700	2
SRU31	59.183	8.410	5.606	7.008	5	0.31	0.4873	6700	2
SRU32	24.899	3.440	2.294	2.867	3	0.23	0.1993	8800	1
SRU41	226.459	50.215	33.477	41.846	7	0.37	0.4702	12300	2
SRU42	45.378	10.529	7.019	8.774	6	0.45	0.0986	21800	1
SRU43	49.365	11.657	7.771	9.714	7	0.48	0.1092	18800	1

Table 2. Algorithm parameter configuration table

maxIteration	*Xmax*	*Xmin*	*Dim*	*NP*	*F*	*CR*	*Similarity_Degree*	*Reset_Proportion*
500	7	0	56	50	0.5	0.3	0.8	0.4

Table 3. Experimental result data table

Serial number	Calculation results (yuan)	Iteration times	Serial number	Calculation results (yuan)	Iteration times	Serial number	Calculation results (yuan)	Iteration times
1	5,154,700	166	11	5,254,200	117	21	5,163,500	154
2	5,181,100	153	12	5,198,700	189	22	5,154,700	159
3	5,154,700	113	13	5,154,700	200	23	5,154,700	128
4	5,225,100	253	14	5,154,700	104	24	5,154,700	155
5	5,154,700	193	15	5,154,700	156	25	5,154,700	151
6	5,254,200	139	16	5,207,500	129	26	5,216,300	230
7	5,154,700	130	17	5,154,700	182	27	5,154,700	176
8	5,154,700	166	18	5,189,900	109	28	5,234,800	185
9	5,154,700	148	19	5,181,100	135	29	5,154,700	289
10	5,244,500	183	20	5,172,300	177	30	5,154,700	144

Table 4. Repairable spare parts inventory allocation table

Repairable spare parts number	Rear warehouse inventory	Site 1 Inventory	Site 2 Inventory	Site 3 Inventory
LRU1	2	2	2	2
LRU2	5	3	3	2
LRU3	3	3	2	4
LRU4	7	6	4	5
SRU11	2	1	1	0
SRU12	1	0	0	0
SRU13	2	0	0	0
SRU21	4	0	0	1
SRU22	3	1	0	0
SRU31	1	0	0	1
SRU32	2	1	0	0
SRU41	4	3	2	2
SRU42	1	0	1	0
SRU43	3	0	1	1

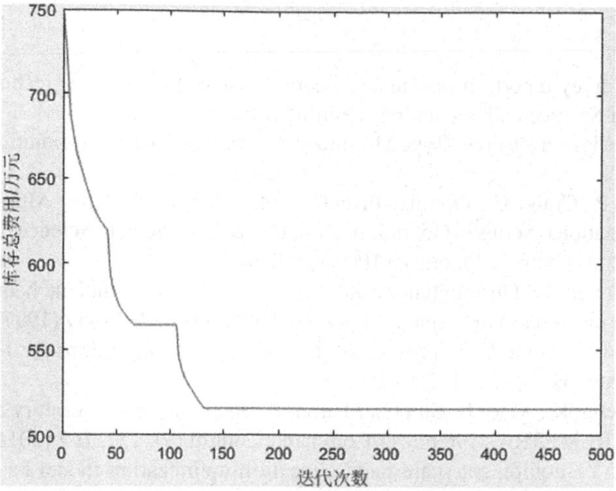

Fig. 6. Convergence curve of improved differential evolution algorithm

Table 5. Comparison table of experimental results

Content	Reference [11]	The result of VMETRIC	The result of this paper
Availability	0.952	0.956	0.950
Cost (ten thousand yuan)	541.77	549.77	515.47
Cost reduction margin	4.85%	6.24%	–

5 Conclusion

- Aiming at solving the two-level inventory model of valuable repairable spare parts for equipment, this paper proposes and designs an improved differential evolution algorithm to solve the problem of poor quality of the traditional marginal analysis method.
- The experimental results show that the proposed algorithm not only has high stability, but also reduces 4.85% and 6.24% compared with the results of reference [11] and VMETRIC software, respectively, which demonstrates the superiority of the proposed algorithm.
- Although the calculation time of this algorithm is longer than that of marginal analysis method, in reality, it is usually calculated once when making annual procurement plan, and the requirements for computing time are not high. The shortcoming has no effects in practical application.
- In the future, the two-level model of support structure is to be extended to three-level model to study the effectiveness of the proposed algorithm under complex model conditions.

References

1. Zhang, L.: Survey report on maintenance equipment support and key technologies. Beijing University of Science and Technology, Beijing (2018)
2. Gross, O.: A Class of Discrete-Type Minimization Problems. Rand Corporation, Santa Monica (1956)
3. Sherbrooke, P., Craig, C.: Optimal Inventory Modeling of Systems: Multi-Echelon Techniques. International Series in Operations Research & Management Science, vol. 72. Springer, New York (2004). https://doi.org/10.1007/b109856
4. Storn, R., Price, K.: Differential evolution – a simple and efficient heuristic for global optimization over continuous spaces. J. Glob. Optim. **11**(4), 341–359 (1997)
5. Ning, G., Cao, D., Zhou, Y.: Improved differential evolution algorithms for 0-1 programming. Syst. Sci. Math. **39**(1), 120–132 (2019)
6. Zhang, S., Teng, K., Xiao, F.: Shipboard aircraft repairable parts inventory allocation model based on VARI-METRIC. Firepower Command Control (9), 157–162 (2015)
7. Yu, B., Wei, Y.: Equipment spare parts allocation optimization model based on two-level inventory. J. Armored Forces Eng. Coll. **28**(6), 23–27 (2014)
8. Zhou, W., Liu, Y., Guo, B., et al.: Initial configuration model of weapon equipment valuables based on two-level supply relationship. Syst. Eng. Theor. Pract. **31**(6), 1056–1061 (2011)
9. Ruan, M., Zhai, Z., Peng, Y., Li, Q., et al.: Optimization of three-level inventory scheme for equipment spare parts based on system guarantee degree. Syst. Eng. Theor. Pract. **32**(7), 1623–1630 (2012)
10. Qin, A., Huang, V., Suganthan, P.N.: Differential evolution algorithm with strategy adaptation for global numerical optimization. IEEE Trans. Evol. Comput. **13**(2), 398–417 (2009)
11. Yu, X.: Research and Application of Multi-level Guarantee Inventory Optimization for Maintainable Spare Parts

A Clustering-Based Multiobjective Evolutionary Algorithm for Balancing Exploration and Exploitation

Wei Zheng[1], Jianyu Wu[1], Chenghu Zhang[2], and Jianyong Sun[1(✉)]

[1] School of Mathematics and Statistics, Xi'an Jiaotong University, Xi'an 710049, China
jy.sun@xjtu.edu.cn

[2] Xi'an Satellite Control Center, Xi'an 710043, China

Abstract. This paper proposes a simple but promising clustering-based multi-objective evolutionary algorithm, termed as CMOEA. At each generation, CMOEA first divides the current population into several subpopulations by Gaussian mixture clustering. To generate offsprings, the search stage, either exploration or exploitation, is determined by the relative difference between the subpopulations' hypervolumes of two adjacent generations. CMOEA selects the parents from different subpopulations in case of exploration stage, and from the same subpopulation in case of exploitation stage. In the environmental selection phase, the hypervolume indicator is used to update the population. Simulation experiments on nine multi-objective problems show that CMOEA is competitive with five popular multi-objective evolutionary algorithms.

Keywords: Multi-objective evolutionary algorithm · Gaussian mixture clustering method · Exploration · Exploitation

1 Introduction

A box-constrained continuous multi-objective optimization problem (MOP) can be defined as follows:

$$\begin{aligned} &\text{minimize} \;\; F(x) = (f_1(x), f_2(x), ..., f_m(x))^\intercal \\ &\text{subject to} \;\; x \in \Omega \subset \mathbb{R}^n \end{aligned} \tag{1}$$

where $\Omega = \prod_i^n [a_i, b_i]$ is the search space, $x = (x_1, \ldots, x_n) \in \Omega$ is the decision vector, $F : \Omega \rightarrow \mathbb{R}^m$ consists of m real-valued objective functions. For an MOP, it usually involves at least two conflicting objectives. No single solution can simultaneously achieve optimum of all objectives, the optimal solution of such problems is a solution set. Therefore, the goal of MOP is to find a set of optimal solutions called the Pareto Set (PS) in Ω while the image of the PS in objective space $\mathbb{R}^m$ is called the Pareto Front (PF) [8].

L. Pan et al. (Eds.): BIC-TA 2019, CCIS 1159, pp. 355–369, 2020.
https://doi.org/10.1007/978-981-15-3425-6_28

Nowadays, multi-objective evolutionary algorithm (MOEA) has becoming a promising tool for solving MOP. Compared with traditional optimization methods, MOEAs could obtain an approximation set of PS and PF through a single run. The goal of MOEAs is to find a set of solutions to approximate the true PF with well-balanced convergence and diversity. Numerous MOEAs have been proposed, please see [30] for a comprehensive survey. MOEAs can be roughly divided into four categories: (1) Pareto domination-based MOEAs [6,7]; (2) performance indicator-based MOEAs [1,2]; (3) decomposition-based MOEAs [14,15,28]; (4) learning model-based MOEAs [13,16,18,27].

Over the last several years, learning model-based MOEAs have attracted great attentions. For example, regularity model-based multiobjective estimation of distribution algorithm (RM-MEDA) was proposed in [27], in which local principal component analysis (LPCA) is used to model the PS in the decision space and an improved version of RM-MEDA was proposed to solve many objective optimization problems (i.e. multiobjective optimization problems with more than 4 objectives) [19]. Multiobjective evolutionary algorithm based on decomposition (MOEA/D) with support vector machine (SVM) model, termed as MOEA/D-SVM, was proposed in [12], which is to filter and pre-select all new generated solutions in decision space and only evaluate those promising ones for reducing real function evaluation costs during the search process. In [26], a classification based preselection (CPS) coupled with MOEA/D (MOEA/D-CPS) was proposed. In MOEA/D-CPS, the k-nearest neighbor (KNN) model was used. On the other hand, the classification and regression tree (CART) and KNN were used in [24] to train a classifier with Pareto domination based algorithm framework and an extended version of [24] was proposed in [25]. Very recently, Gaussian process regression was used to model the PF to realize adjusting the decomposition method in MOEA/D dynamically [21].

Clustering is one of the basic tasks in machine learning. A great variety of clustering methods have been proposed, such as K-means clustering method [9], Gaussian mixture clustering (GMC) method [4], fuzzy c-means clustering method [3], and so on. In the context of solving MOPs by MOEA, the use of clustering can help us deal with the balance of exploitation and exploration, which is a key issue in evolutionary algorithm. K-means clustering method has been investigated in many papers for both MOPs and many objective optimization problems (MaOPs) [11,17,18,23,29] well. However, we cannot find the successful use of GMC in MOEAs. Therefore, we propose to apply this method to MOEAs for MOPs in this paper.

The proposed algorithm, called clustering-based MOEA (dubbed as CMOEA), is developed in which GMC is used to partition population into some subpopulations. In CMOEA, to generate offsprings, parents are selected either from different subpopulations or the same subpopulation depending on the search stage. Judging the search stage is based on the hypervolume indicator [2] of the centers of subpopulations between two adjacent generations. While generating a new individual, it will be merged into an archive. Once the size of archive is equal to the size of population, CMOEA combines these two sets as a merged population. In selection phase, CMOEA updates the population by the hypervolume indicator of the merged population. CMOEA is experimentally tested on F1–F9 [27]. Five well-known MOEAs: MOPSO [5], GrEA [22], SMS-EMOA [2], NSGA-II [7] and PESA-II [6] are compared with CMOEA. The final results reflect the effectiveness of CMOEA for solving MOPs.

The rest of this paper is organized as follows. Section 2 introduces the Gaussian mixture clustering method briefly. Section 3 presents the developed algorithm CMOEA in detail. Experimental results are presented in Sect. 4. Section 5 concludes the paper.

2 Gaussian Mixture Clustering

There are a series of clustering methods in machine learning and GMC is widely used in data mining, pattern recognition, machine learning, and statistical analysis [4]. In this paper, at each generation, we utilize the Gaussian mixture method (GMM) to partition the population into a certain number of subpopulations. In GMM, the decision vector x is assumed to follow a finiite Gaussian mixture distribution, which can be stated as follows:

$$p(x) = \sum_{k=1}^{K} \pi_k \mathcal{N}(x|\mu_k, \Sigma_k) \tag{2}$$

where K is the number of clusters. π_k is the probability of the data point x which belongs to the $k-$th cluster. μ_k and Σ_k are the mean and the covariance matrix for the $k-$th cluster respectively. The above parameters π_k, μ_k and Σ_k, $k = 1, \cdots, K$ can be obtained by maximizing the log of the likelihood function with the following form:

$$\ln p(\mathbf{X}|\pi, \mu, \Sigma) = \sum_{n=1}^{N} \ln \left\{ \sum_{k=1}^{K} \pi_k \mathcal{N}(x^n|\mu_k, \Sigma_k) \right\} \tag{3}$$

where $\mathbf{X} = \{x^1, \cdots, x^n, \cdots, x^N\}$. To achieve this purpose, we use the traditional expectation-maximization (EM) method. There are four main steps:

1. Initialize the parameters: π_k, μ_k, Σ_k, $k = 1, \cdots, K$, and evaluate the initial value of the log likelihood.
2. E step: Evaluate the responsibilities using the current parameter values:

$$\gamma_{nk} = \frac{\pi_k \mathcal{N}(x_n|\mu_k, \Sigma_k)}{\sum_{k=1}^{K} \pi_k \mathcal{N}(\mu_k, \Sigma_k)} \tag{4}$$

3. M step: Re-estimate the parameters using the current responsibilities:

$$N_k = \sum_{n=1}^{N} \gamma_{nk} \tag{5}$$

$$\mu_k^{new} = \frac{1}{N_k} \sum_{n=1}^{N} \gamma_{nk} x^n \tag{6}$$

$$\pi_k^{new} = \frac{N_k}{N} \tag{7}$$

$$\Sigma_k^{new} = \frac{1}{N_k} \sum_{n=1}^{N} \gamma_{nk} (x^n - \mu_k^{new})(x^n - \mu_k^{new})^\top \tag{8}$$

4. Evaluate the log likelihood (3) and check for convergence of either the parameters or the log likelihood. If the convergence criterion is not satisfied return to Step 2.

3 The Clustering-Based Multiobjective Evolutionary Algorithm: CMOEA

In this section, CMOEA is presented in detail. The general framework of CMOEA is shown in Algorithm 1. Firstly, a population $\mathcal{P}= \{x^1, ..., x^N\}$ is initialized randomly where N is the population size. An empty external archive $\mathcal{A}$ is set and set the control parameter $\delta = 0$. At each generation, the population $\mathcal{P}$ is partitioned into the K subpopulations: $\{Subpop_1, Subpop_2, ..., Subpop_K\}$ by GMM. Assume the centers of these subpopulations are $\{center_1, center_2, ..., center_K\}$. Until the stop condition is satisfied, the algorithm either perform exploration or exploitation. $\delta \geq 0$ implies that the current search is in exploration stage. Algorithm 2 is applied to update the subpopulations. On the other hand, Algorithm 3 is used to update the subpopulations for the purpose of exploitation.

To perform exploration, in Algorithm 2, for every individual in each subpopulation, two parent individuals will be selected in exclusively different two subpopulations randomly (lines 5–6). In line 7, new solutions are generated by Algorithm 5. When a new solution is generated, this solution will be merged into archive $\mathcal{A}$. Once all individuals of population are generated, the original population $\mathcal{P}$ and archive $\mathcal{A}$ will be merged into $\mathcal{A}$. Furthermore, Algorithm 6 performs on $\mathcal{A}$ to choose the next population $\mathcal{P}$ with size N. In the end, new population $\mathcal{P}$ is partitioned into K subpopulations by GMM again.

To perform exploitation, in Algorithm 3, three parent individuals are to be selected from a single subpopulation. Before selection, we should ensure how many individuals are in a single subpopulation. If the size of a subpopulation is one, we choose to select another individual from a subpopulation which has the closest clusters to the current subpopulation in terms of the Euclidean distance between the subpopulations' centers. If the size of one subpopulation is two, both two individuals are selected, while the last individual is selected in closest subpopulation randomly. If the size of one subpopulation is three or more, all of the three individuals are selected in this subpopulation randomly.

In Algorithm 5, the differential evolutional (DE) [10] and polynomial mutation (PM) [10] operators are applied. In addition, the repair operator is carried out if it is necessary. In Algorithm 6, it performs the environmental selection on the combined population $\mathcal{A}$. The hypervolume indicators of all the individuals are computed. The N individuals with the highest hypervolume indicators are reformed new population $\mathcal{P}$.

The update of the control parameter δ is performed in Algorithm 4. Firstly, the hypervolume indicators of centers of the last oldest and newest clusters $centerOldHV = \{centeroldHV_1, \cdots, centeroldHV_K\}$ and $centerNewHV = \{centernewHV_1, \cdots, centernewHV_K\}$ are obtained. Then, δ is computed as $\delta = \sum_{k=1}^{K} \frac{centeroldHV_k - centernewHV_k}{centeroldHV_k}$. If $\delta \geq 0$, CMOEA performs the exploration stage, otherwise, exploitation stage is carried out.

Meanwhile, the flow chart of CMOEA has been given in Fig. 1, which gives a clear understanding of CMOEA.

Algorithm 1: CMOEA

Input: MOP (1); Population size N; Number of clusters: K; Stop criterion: Maxevaluations (MaxFES); An external archive $\mathcal{A}$.
Output: Approximation to the PF.

Initialize the population $\mathcal{P}$: $x^1, ..., x^N$;
Partition $\mathcal{P}$ into $\{Subpop_1, Subpop_2, ..., Subpop_K\}$ with their center are $\{center_1, center_2, ..., center_K\}$ by using GMC method;
Set $\mathcal{A} = \emptyset$ and $\delta = 0$;
while $FES \leq MaxFES$ **do**
  Update δ by using algorithm 4;
  if $\delta \geq 0$ **then**
    Update newly clusters $\{Subpop_1, Subpop_2, ..., Subpop_K\}$ by using algorithm 2;
  else
    Update newly clusters $\{Subpop_1, Subpop_2, ..., Subpop_K\}$ by using algorithm 3;
  end
end

---

4 Experimental Studies

In this paper, test instances F1–F9 from [27] are used as benchmark. Through [27], we can find that the test instances F1–F4 are four continuous MOPs with linear variable linkages and F5–F8 are the test instances with non-linear variable linkages. The test instances F3 and F8 have three objectives, and the others have two objectives. All of them are challenging to MOEAs. For more detailed information of these nine test instances, please refer to [27]. Our experiments are realized in the PlatEMO [20] platform of Matlab 2017b on a personal computer. CMOEA is compared with five popular MOEAs: MOPSO [5],

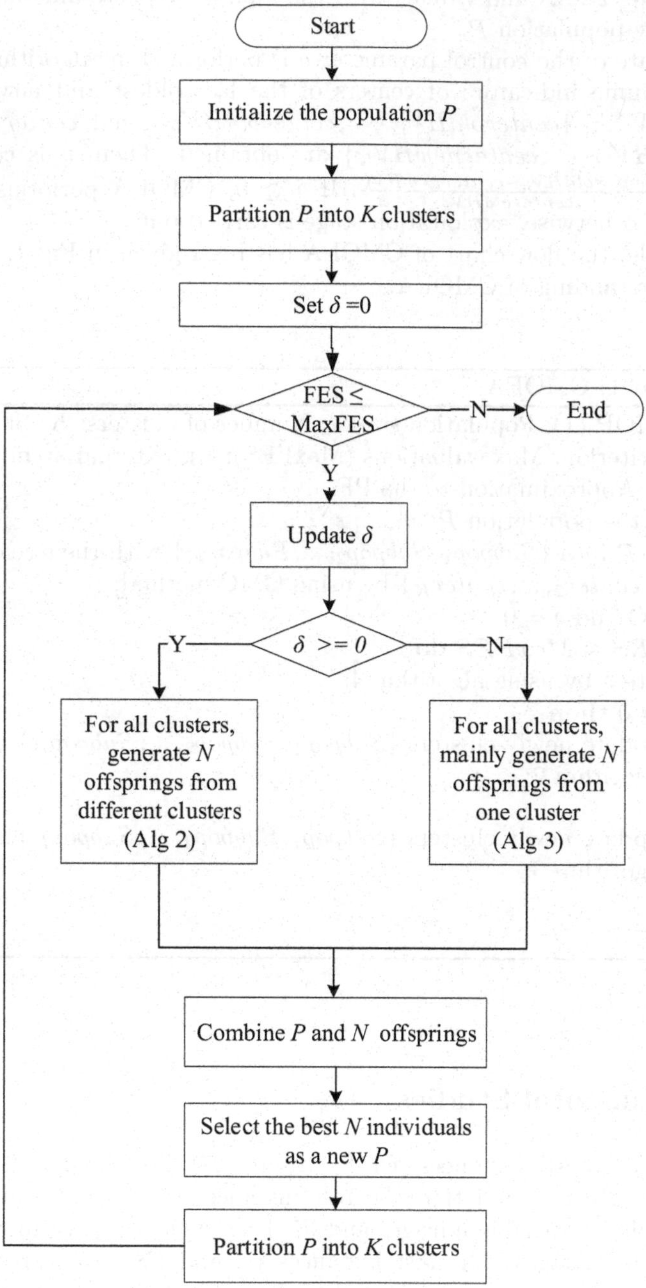

Fig. 1. The flow chart of CMOEA

Algorithm 2: Exploration Stage

Input: $\{Subpop_1, Subpop_2, ..., Subpop_K\}$; External archive $\mathcal{A}$.
Output: Newly clusters $\{Subpop_1, Subpop_2, ..., Subpop_K\}$.

for $1 \le i \le K$ **do**
  Calculate the size of cluster $|Subpop_i|$;
  for $1 \le j \le |Subpop_i|$ **do**
    $x^1 = individual_j$;
    Randomly select two indices h and l from $\{1, \ldots, K\} \setminus i$;
    Randomly select two individuals x^2 and x^3 from $Subpop_h$ and $Subpop_l$, respectively;
    $y \leftarrow GenerateOffspring(x^1, x^2, x^3)$ (Algorithm 5);
    $\mathcal{A} = \mathcal{A} \cup y$;
  end
end
Combine $\mathcal{A}$ and $\mathcal{P}$ to reform $\mathcal{A}$, i.e. $\mathcal{A} = \mathcal{P} \cup \mathcal{A}$;
$\mathcal{P}$ = Environmental selection $\mathcal{A}$ (Algorithm 6);
Partition new population $\mathcal{P}$ into $\{Subpop_1, Subpop_2, ..., Subpop_K\}$ with their centers are $\{center_1, center_2, ..., center_K\}$ by using GMC method.

GrEA [22], SMS-EMOA [2], NSGA-II [7] and PESA-II [6]. MOPSO is an multi-objective particle swarm optimization. GrEA is a grid-based MOEA for many-objective problems. SMS-EMOA belongs to the category of indicator-based MOEAs which is based on the hypervolume indicator. Both NSGA-II and PESA-II belong to the Pareto domination-based MOEAs. All of the compared MOEAs use the same parameters as found in the respective original papers. Specifically, for all MOEAs, the population size of nine test instances is set as $N = 100$, the max number of evaluations is set as $MaxFES = 800,000$. The DE and PM operators used in the MOEAs are set as $F = 0.5$, $CR = 1.0$, $p_m = 0.9$ and $\eta_m = 20$. In addition, we set the cluster number $K = 10$.

To validate the performance of MOEAs, two commonly-used performance metrics *IGD* metric [32] and *HV* metric [31] are used. Both can measure the convergence and diversity of the obtained approximation set well. The larger *HV* metric value, as well as the smaller *IGD* metric value, the better performance of an MOEA. All MOEAs are run 20 times on all test instances to reflect their performances.

4.1 Experimental Results and Analysis

Tables 1 and 2 summarize the *IGD* metric and *HV* metric obtained by all compared MOEAs. In each table, the mean values and standard deviation values in the brackets are shown and the best mean metric values of each row are shown in bold. To have statistically comprehensive conclusions, the Wilcoxons Rank test [33] at the 0.05 significance level is adopted to test the significant difference between the data obtained by compared algorithms. '+/−/=' means the compared MOEA's results are better than/worse than/equal to the CMOEA's results. In the last row, the sum of each symbol of each column is shown.

Algorithm 3: Exploitation Stage

Input: $\{Subpop_1, Subpop_2, ..., Subpop_K\}$; External archive $\mathcal{A}$.
Output: Newly clusters $\{Subpop_1, Subpop_2, ..., Subpop_K\}$.

for $1 \leq i \leq K$ **do**
  Calculate the size of cluster $|Subpop_i|$;
  Find two closest centers of clusters $Subpop_{i1}$ and $Subpop_{i2}$ based on the Euclidean distance between any two centers $center_i$;
  if $|Subpop_i| < 3$ **then**
    if $|Subpop_i| = 1$ **then**
      $x^1 = center_i$;
      Randomly select an individual x^2 from $Subpop_{i1}$;
      Randomly select an individual x^3 from $Subpop_{i2}$;
      $y \leftarrow GenerateOffspring(x^1, x^2, x^3)$;
      $\mathcal{A} = \mathcal{A} \cup y$;
    else
      for $1 \leq j \leq |Subpop_i|$ **do**
        $x^1 = individual_j$;
        $x^2 = Subpop_i \setminus individual_j$;
        Randomly select an individual x^3 from $Subpop_{i1}$;
        $y \leftarrow GenerateOffspring(x^1, x^2, x^3)$ (Algorithm 5);
        $\mathcal{A} = \mathcal{A} \cup y$;
      end
    end
  else
    for $1 \leq j \leq |Subpop_i|$ **do**
      $x^1 = individual_j$;
      Randomly select two individuals x^2 and x^3 from $Subpop_i$;
      $y \leftarrow GenerateOffspring(x^1, x^2, x^3)$ (Algorithm 5);
      $\mathcal{A} = \mathcal{A} \cup y$;
    end
  end
end
Combine $\mathcal{A}$ and $\mathcal{P}$ to reform $\mathcal{A}$, i.e. $\mathcal{A} = \mathcal{P} \cup \mathcal{A}$;
$\mathcal{P}$ = Environmental selection $\mathcal{A}$ (Algorithm 6);
Partition new population $\mathcal{P}$ into $\{Subpop_1, Subpop_2, ..., Subpop_K\}$ with their centers are $\{center_1, center_2, ..., center_K\}$ by using GMC method.

Algorithm 4: Update δ

Input: The centers obtained by the last oldest and newest clusters: $CenterOld$, $CenterNew$; Number of clusters: K.
Output: Parameter δ.

Calculate the Hypervolume indicators of $CenterOld$ and $CenterNew$ as: $centerOldHV = \{centeroldHV_1, \cdots, centeroldHV_K\}$ and $centerNewHV = \{centernewHV_1, \cdots, centernewHV_K\}$;

$\delta = \sum_{k=1}^{K} \frac{centeroldHV_k - centernewHV_k}{centeroldHV_k}$.

Algorithm 5: Generate Offspring

Input: Three parents: x^1, x^2 and x^3; Parameters: F, CR, p_m and η_m.
Output: An offspring y.

Generate a trial solution $y = (y_1, ..., y_n)^\mathsf{T}$ by

$$y_i = \begin{cases} x_i^1 + F \times (x_i^2 - x_i^3), & \text{if } rand() \leq CR, \\ x_i^1, & \text{otherwise.} \end{cases}$$

Repair y : as for $i \in \{1, ..., n\}$,$y_i < a_i$ then $y_i = a_i$, otherwise as for i, $y_i > b_i$, then $y_i' = b_i$;

Mutate y:

$$y_i = \begin{cases} y_i + \delta_i \times (b_i - a_i), & \text{if } rand() < p_m, \\ y_i, & \text{otherwise} \end{cases}$$

where

$$\delta_i = \begin{cases} [2r + (1 - 2r)(\frac{b_i - y_i}{b_i - a_i})^{\eta_m}]^{\frac{1}{\eta_m}} - 1, & \text{if } r < 0.5, \\ 1 - [2 - 2r + (2r - 1)(\frac{y_i - a_i}{b_i - a_i})^{\eta_m}]^{\frac{1}{\eta_m}}, & \text{otherwise} \end{cases}$$

Repair y if necessary.

return y

Algorithm 6: Environmental Selection

Input: Combined population $\mathcal{A}$.
Output: New population $\mathcal{P}$.

Calculate the Hypervolume indicators of all the individuals of $\mathcal{A}$, sort them descending to a set $tempHV$;

Arrange N individuals with the highest Hypervolume indicators in $tempHV$ to new population $\mathcal{P}$.

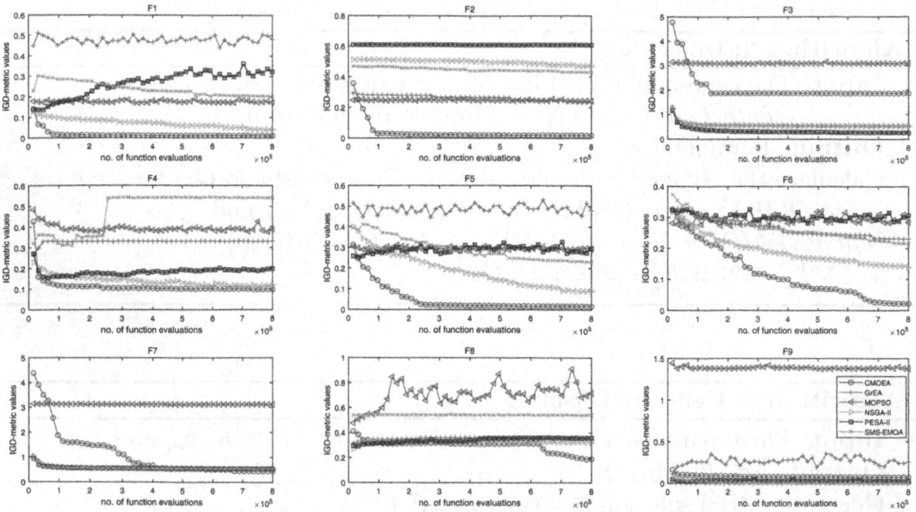

Fig. 2. 50 results during the evolution of the best IGD-metric values for F1–F9

Table 1. The IGD metric results of CMOEA and the other compared MOEAs.

Problem	MOPSO	GrEA	SMS-EMOA	NSGA-II	PESA-II	CMOEA
F1	1.1655e-1 (4.59e-2) –	4.7164e-1 (1.98e-2) –	1.6502e-1 (3.02e-2) –	4.7734e-2 (1.96e-2) –	3.0865e-1 (1.29e-2) –	**1.0435e-2** (2.26e-4)
F2	2.7673e-1 (2.70e-2) –	4.3920e-1 (1.34e-1) –	4.0436e-1 (1.03e-1) –	2.4893e-1 (2.04e-1) –	5.2985e-1 (6.65e-2) –	**1.8681e-2** (7.72e-4)
F3	3.0894e+0 (2.27e-1) –	**4.9103e-1** (1.10e-1) +	5.1473e-1 (8.25e-2) +	5.4443e-1 (9.84e-3) +	5.1576e-1 (8.77e-2) +	1.8708e+0 (1.07e-1)
F4	4.3825e-1 (8.38e-2) –	3.4077e-1 (1.94e-2) –	5.4866e-1 (2.01e-2) –	2.7540e-1 (7.11e-2) –	2.9914e-1 (7.01e-2) –	**9.8986e-2** (2.89e-3))
F5	3.0625e-1 (1.97e-2) –	4.8512e-1 (1.34e-2) –	2.1126e-1 (2.07e-2) –	8.2529e-2 (1.13e-2) –	3.1471e-1 (1.68e-2) –	**1.6598e-2** (4.56e-4)
F6	2.9224e-1 (1.38e-2) –	2.1783e-1 (8.54e-3) –	2.1357e-1 (7.11e-3) –	1.3426e-1 (6.51e-3) –	3.0266e-1 (9.76e-3) –	**7.5671e-2** (1.37e-2)
F7	3.2350e+0 (1.91e-1) –	5.3501e-1 (1.86e-3) –	5.3753e-1 (2.23e-3) –	5.3777e-1 (1.30e-3) –	5.3781e-1 (2.57e-3) –	**4.8906e-1** (2.76e-2)
F8	8.0567e-1 (3.05e-1) –	3.4291e-1 (1.64e-2) ≈	5.4264e-1 (1.33e-3) –	3.1168e-1 (2.93e-2) ≈	3.1258e-1 (7.06e-2) ≈	**2.4276e-1** (8.14e-2)
F9	1.2812e+0 (2.17e-1) –	2.8046e-1 (2.29e-2) –	**5.1362e-3** (2.31e-3) +	1.2419e-2 (9.11e-3) +	3.1159e-2 (6.10e-3) +	6.8009e-2 (2.51e-3)
+/-/≈	0/9/0	1/7/1	2/7/0	2/6/1	2/6/1	

From Table 1, we see that MOPSO performs the worst, all of the *IGD* results obtained by CMOEA are better than them. GrEA performs also poor. The rest three MOEAs perform slightly worse than CMOEA in general. However, CMOEA performs worse on F3: it ranks only the fifth. The same situation happens in Table 2. The *HV* metric of F3 of CMOEA is 0, which indicates that CMOEA does not solve this instance well. Nevertheless, CMOEA performs better among most of the test instances. Overall, we may conclude that CMOEA performs well in the light of both convergence and diversity.

To show the convergence speed of the six MOEAs, we have recorded 50 results during the evolutionary process of each MOEA, and plotted the evolutionary curve with the best *IGD* metric value in Fig. 2. In these subfigures of Fig. 2, the horizontal axis represents the number of function evaluations, and the vertical axis represents the value of the *IGD* metric. As can be seen from the figures, the convergence speed of CMOEA is always the fastest in most cases and can achieve the minimum value of six MOEAs, which indicates that CMOEA can balance the exploration and exploitation well during the evolutionary process. However,

Table 2. The HV metric results of CMOEA and the other compared MOEAs.

Problem	MOPSO	GrEA	SMS-EMOA	NSGA-II	PESA-II	CMOEA
F1	5.8987e-1 (3.98e-2) −	4.1426e-1 (8.47e-3) −	6.1991e-1 (1.82e-2) −	6.8511e-1 (1.09e-2) −	5.2749e-1 (7.44e-3) −	**7.0992e-1** (3.17e-4
F2	1.8662e-1 (1.77e-2) −	1.3672e-1 (5.46e-2) −	1.4266e-1 (5.23e-2) −	2.4783e-1 (1.30e-1) −	1.0198e-1 (1.29e-2) −	**4.2607e-1** (4.54e-4
F3	0.0000e+0 (0.00e+0) ≈	**6.5911e-2** (4.50e-2) +	5.6364e-2 (2.97e-2) +	4.3444e-2 (9.67e-3) +	5.5063e-2 (3.40e-2) +	0.0000e+0 (0.00e+0)
F4	1.5627e-1 (5.70e-2) −	3.9304e-1 (3.57e-3) −	3.1620e-1 (7.92e-2) −	4.2763e-1 (3.17e-2) −	4.1149e-1 (2.64e-2) −	**5.1012e-1** (1.16e-3)
F5	5.0387e-1 (1.18e-2) −	4.1031e-1 (7.54e-3) −	5.9152e-1 (1.27e-2) −	6.6452e-1 (5.47e-3) −	5.2604e-1 (1.05e-2) −	**7.0303e-1** (5.43e-4)
F6	1.7679e-1 (1.20e-2) −	2.3918e-1 (5.95e-3) −	2.4330e-1 (4.71e-3) −	3.0307e-1 (5.30e-3) −	1.8399e-1 (3.97e-3) −	**3.4698e-1** (2.59e-2)
F7	0.0000e+0 (0.00e+0) −	**5.2863e-2** (1.90e-3) +	5.0299e-2 (2.26e-3) +	5.0054e-2 (1.31e-3) +	5.0022e-2 (2.62e-3) +	3.1827e-2 (9.09e-3)
F8	4.2421e-2 (1.06e-1) −	3.9269e-1 (3.04e-3) ≈	3.3860e-1 (4.08e-3) −	**4.1328e-1** (7.60e-3) ≈	4.1056e-1 (3.02e-2) ≈	4.0979e-1 (3.74e-2)
F9	2.0724e-4 (6.22e-4) −	5.4372e-1 (1.52e-2) −	**7.1751e-1** (3.98e-3) +	7.0711e-1 (5.52e-3) +	6.8415e-1 (3.64e-3) +	6.4337e-1 (2.66e-3)
+/ − / ≈	0/8/1	2/6/1	3/6/0	3/5/1	3/5/1	

the performance of other MOEAs are unsatisfactory. For example, in subfigure F1, the curve of PESA-II has not decreased, but increased in the twists and turns. It has the same tendency of GrEA in this subfigure. It should be pointed out that CMOEA can get more prefer results while the number of evaluations increasing, such as test instance F6. Therefore, we may conclude that CMOEA has its own advantage in solving MOPs.

4.2 Parameter K Sensitivity Analysis

To investigate the sensitivity of the number of cluster K, we have tested CMOEA with $K = 3, 6, 8, 12$ and original $K = 10$ on all of the test instances. We mark CMOEA with $K = 3, 6, 8, 12$ as "CMOEA ($K = 3$)", "CMOEA ($K = 6$)", "CMOEA ($K = 8$)" and "CMOEA ($K = 12$)", respectively. All of them have run 20 times too. The final results of the IGD results have been shown in Table 3. The mean, standard deviation and the p-values obtained from the Wilcoxon's Rank test are also calculated. The best mean metric values of each row are shown in bold too. As we can see from the table, the results of CMOEA ($K = 3$), CMOEA ($K = 6$), CMOEA ($K = 8$) and CMOEA ($K = 12$) are slightly worse than CMOEA. Particularly, the Wilcoxon's Rank test performance of CMOEA ($K = 6$) are equal to CMOEA. However, most of the mean values of CMOEA are best in the corresponding rows. Hence, we may conclude that CMOEA has relatively strong robustness for K.

4.3 Parameter Population Size N Sensitivity Analysis

To study the sensitivity of population size N, we have tested CMOEA with $N = 50$, $N = 150$ and $N = 200$ on test instance F1 to compare with CMOEA. Each algorithm has been run 20 times. In order to make the results more clearly, we have plotted the line charts of IGD metric and HV metric in Fig. 3, respectively. Note that we use the mean value of 20 runs with different population size N. From Fig. 3, we can see that the population size of CMOEA is not very sensitive and in this test instance, CMOEA with $N = 100$ has achieved the best performance.

Table 3. The IGD metric results of CMOEA with different values of K.

Problem	CMOEA($K = 3$)	CMOEA($K = 6$)	CMOEA($K = 8$)	CMOEA($K = 12$)	CMOEA
F1	1.2465e-2 (3.17e-4) –	1.0443e-2 (2.27e-4) ≈	1.2382e-2 (2.61e-4) –	1.2438e-2 (2.93e-4) –	**1.0435e-2** (2.26e-4)
F2	2.3056e-2 (1.00e-3) –	**1.8475e-2** (6.37e-4) ≈	2.2795e-2 (7.74e-4) –	2.3038e-2 (1.10e-3) –	1.8681e-2 (7.72e-4)
F3	1.9133e+0 (1.21e-1) ≈	**1.8601e+0** (1.20e-1) ≈	1.9115e+0 (1.15e-1) ≈	1.8960e+0 (9.33e-2) ≈	1.8708e+0 (1.07e-1)
F4	1.1956e-1 (1.06e-2) –	**9.8425e-2** (1.99e-3) ≈	1.2500e-1 (3.52e-2) –	1.1829e-1 (3.88e-3) –	9.8986e-2 (2.89e-3)
F5	1.8902e-2 (4.36e-4) –	**1.6587e-2** (3.47e-4) ≈	1.8840e-2 (4.34e-4) –	1.8633e-2 (3.36e-4) –	1.6598e-2 (4.56e-4)
F6	8.5390e-2 (2.13e-2) –	**7.2272e-2** (2.21e-2) ≈	9.2338e-2 (1.84e-2) –	8.4518e-2 (2.43e-2) ≈	7.5671e-2 (1.37e-2)
F7	4.8927e-1 (2.80e-2) ≈	4.7720e-1 (3.42e-2) ≈	4.8580e-1 (1.74e-2) ≈	**4.7571e-1** (2.68e-2) ≈	4.8906e-1 (2.76e-2)
F8	2.8366e-1 (8.48e-2) ≈	2.7676e-1 (1.31e-1) ≈	2.3800e-1 (1.11e-1) ≈	**2.3077e-1** (9.16e-2) ≈	2.4276e-1 (8.14e-2)
F9	9.4218e-2 (5.98e-3) ≈	9.8535e-2 (1.47e-2) ≈	9.4953e-2 (4.72e-3) ≈	9.5176e-2 (9.17e-3) ≈	**6.8009e-2** (2.51e-3)
+/-/≈	0/5/4	0/0/9	0/5/4	0/4/5	

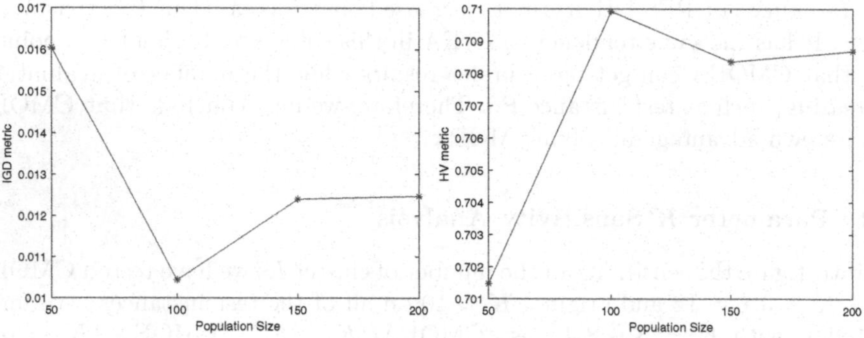

Fig. 3. Line charts of CMOEA with different population size

5 Conclusion

In this paper, we proposed a clustering based MOEA, named CMOEA, in which strategies based on Gaussian mixture clustering were proposed to balance exploration and exploitation. The search stage was decided by the hypervolume indicator comparison between two adjacent generations, while the indicators are computed based on the centers obtained through GMM clustering. The generation of offspring was designed based on the GMM clustering results, which are different according to the search stage. Our experiments on nine test instances with test capability verified the effectiveness of the proposed CMOEA.

Acknowledgment. This work was supported by the Fundamental Research Funds for the Central Universities (No. xzy022019074).

References

1. Bader, J., Zitzler, E.: HypE: an algorithm for fast hypervolume-based many-objective optimization. Evol. Comput. **19**(1), 45–76 (2011). https://doi.org/10.1162/evco_a_00009
2. Beume, N., Naujoks, B., Emmerich, M.: SMS-EMOA: multiobjective selection based on dominated hypervolume. Eur. J. Oper. Res. **181**(3), 1653–1669 (2007). https://doi.org/10.1016/j.ejor.2006.08.008
3. Bezdek, J.C., Ehrlich, R., Full, W.: FCM: the Fuzzy C-Means clustering algorithm. Comput. Geosci. **10**(2), 191–203 (1984). https://doi.org/10.1016/0098-3004(84)90020-7
4. Bishop, C.: Pattern Recognition and Machine Learning. Springer, New York (2006). https://doi.org/10.1007/978-1-4615-7566-5
5. Coello Coello, C.A., Lechuga, M.S.: MOPSO: a proposal for multiple objective particle swarm optimization. In: Proceedings of the 2002 Congress on Evolutionary Computation, CEC 2002 (Cat. No. 02TH8600), vol. 2, pp. 1051–1056, May 2002. https://doi.org/10.1109/CEC.2002.1004388
6. Corne, D.W., Jerram, N.R., Knowles, J.D., Oates, M.J.: PESA-II: region-based selection in evolutionary multiobjective optimization. In: Proceedings of the 3rd Annual Conference on Genetic and Evolutionary Computation, GECCO 2001, pp. 283–290. Morgan Kaufmann Publishers Inc., San Francisco (2001). http://dl.acm.org/citation.cfm?id=2955239.2955289
7. Deb, K., Pratap, A., Agarwal, S., Meyarivan, T.: A fast and elitist multiobjective genetic algorithm: NSGA-II. IEEE Trans. Evol. Comput. **6**(2), 182–197 (2002). https://doi.org/10.1109/4235.996017
8. Deb, K.: Multi-Objective Optimization Using Evolutionary Algorithms. Wiley, New York (2001)
9. Jain, A.K., Murty, M.N., Flynn, P.J.: Data clustering: a review. ACM Comput. Surv. **31**(3), 264–323 (1999). https://doi.org/10.1145/331499.331504
10. Li, H., Zhang, Q.: Multiobjective optimization problems with complicated Pareto sets, MOEA/D and NSGA-II. IEEE Trans. Evol. Comput. **13**(2), 284–302 (2009). https://doi.org/10.1109/tevc.2008.925798
11. Li, X., Zhang, H., Song, S.: A self-adaptive mating restriction strategy based on survival length for evolutionary multiobjective optimization. Swarm Evol. Comput. **43**, 31–49 (2018). https://doi.org/10.1016/j.swevo.2018.02.009
12. Lin, X., Zhang, Q., Kwong, S.: A decomposition based multiobjective evolutionary algorithm with classification. In: 2016 IEEE Congress on Evolutionary Computation (CEC), pp. 3292–3299, July 2016. https://doi.org/10.1109/CEC.2016.7744206
13. Pan, L., He, C., Tian, Y., Wang, H., Zhang, X., Jin, Y.: A classification-based surrogate-assisted evolutionary algorithm for expensive many-objective optimization. IEEE Trans. Evol. Comput. **23**(1), 74–88 (2019). https://doi.org/10.1109/TEVC.2018.2802784
14. Pan, L., Li, L., He, C., Tan, K.C.: A subregion division-based evolutionary algorithm with effective mating selection for many-objective optimization. IEEE Trans. Cybern. 1–14 (2019). https://doi.org/10.1109/TCYB.2019.2906679
15. Shi, J., Zhang, Q., Sun, J.: PPLS/D: Parallel Pareto local search based on decomposition. IEEE Trans. Cybern. 1–12 (2018). https://doi.org/10.1109/TCYB.2018.2880256
16. Sun, J., et al.: Learning from a stream of nonstationary and dependent data in multiobjective evolutionary optimization. IEEE Trans. Evol. Comput. **23**(4), 541–555 (2019). https://doi.org/10.1109/TEVC.2018.2865495

17. Sun, J., Zhang, H., Zhang, Q., Chen, H.: Balancing exploration and exploitation in multiobjective evolutionary optimization. In: Proceedings of the Genetic and Evolutionary Computation Conference Companion, GECCO 2018, pp. 199–200. ACM, New York (2018). https://doi.org/10.1145/3205651.3205708
18. Sun, J., Zhang, H., Zhou, A., Zhang, Q., Zhang, K.: A new learning-based adaptive multi-objective evolutionary algorithm. Swarm Evol. Comput. **44**, 304–319 (2019). https://doi.org/10.1016/j.swevo.2018.04.009
19. Sun, Y., Yen, G.G., Yi, Z.: Improved regularity model-based EDA for many-objective optimization. IEEE Trans. Evol. Comput. **22**(5), 662–678 (2018). https://doi.org/10.1109/TEVC.2018.2794319
20. Tian, Y., Cheng, R., Zhang, X., Jin, Y.: PlatEMO: a MATLAB platform for evolutionary multi-objective optimization (educational forum). IEEE Comput. Intell. Mag. **12**(4), 73–87 (2017). https://doi.org/10.1109/MCI.2017.2742868
21. Wu, M., Li, K., Kwong, S., Zhang, Q., Zhang, J.: Learning to decompose: a paradigm for decomposition-based multiobjective optimization. IEEE Trans. Evol. Comput. **23**(3), 376–390 (2019). https://doi.org/10.1109/TEVC.2018.2865931
22. Yang, S., Li, M., Liu, X., Zheng, J.: A grid-based evolutionary algorithm for many-objective optimization. IEEE Trans. Evol. Comput. **17**(5), 721–736 (2013). https://doi.org/10.1109/TEVC.2012.2227145
23. Zhang, H., Song, S., Zhou, A., Gao, X.: A clustering based multiobjective evolutionary algorithm. In: 2014 IEEE Congress on Evolutionary Computation (CEC), pp. 723–730, July 2014. https://doi.org/10.1109/CEC.2014.6900519
24. Zhang, J., Zhou, A., Zhang, G.: A classification and Pareto domination based multiobjective evolutionary algorithm. In: 2015 IEEE Congress on Evolutionary Computation (CEC), pp. 2883–2890, May 2015. https://doi.org/10.1109/CEC.2015.7257247
25. Zhang, J., Zhou, A., Tang, K., Zhang, G.: Preselection via classification: a case study on evolutionary multiobjective optimization. Inf. Sci. **465**, 388–403 (2018). https://doi.org/10.1016/j.ins.2018.06.073
26. Zhang, J., Zhou, A., Zhang, G.: A multiobjective evolutionary algorithm based on decomposition and preselection. In: Gong, M., Pan, L., Song, T., Tang, K., Zhang, X. (eds.) BIC-TA 2015. CCIS, vol. 562, pp. 631–642. Springer, Heidelberg (2015). https://doi.org/10.1007/978-3-662-49014-3_56
27. Zhang, Q., Zhou, A., Jin, Y.: RM-MEDA: a regularity model-based multiobjective estimation of distribution algorithm. IEEE Trans. Evol. Comput. **12**(1), 41–63 (2008). https://doi.org/10.1109/TEVC.2007.894202
28. Zhang, Q., Li, H.: MOEA/D: a multiobjective evolutionary algorithm based on decomposition. IEEE Trans. Evol. Comput. **11**, 712–731 (2008). https://doi.org/10.1109/TEVC.2007.892759
29. Zhang, X., Tian, Y., Cheng, R., Jin, Y.: A decision variable clustering-based evolutionary algorithm for large-scale many-objective optimization. IEEE Trans. Evol. Comput. **22**(1), 97–112 (2018). https://doi.org/10.1109/TEVC.2016.2600642
30. Zhou, A., Qu, B.Y., Li, H., Zhao, S.Z., Suganthan, P.N., Zhang, Q.: Multiobjective evolutionary algorithms: a survey of the state of the art. Swarm Evol. Comput. **1**(1), 32–49 (2011). https://doi.org/10.1016/j.swevo.2011.03.001
31. Zitzler, E., Thiele, L.: Multiobjective evolutionary algorithms: a comparative case study and the strength Pareto approach. IEEE Trans. Evol. Comput. **3**(4), 257–271 (1999). https://doi.org/10.1109/4235.797969

32. Zitzler, E., Thiele, L., Laumanns, M., Fonseca, C.M., Fonseca, V.G.D.: Performance assessment of multiobjective optimizers: an analysis and review. IEEE Trans. Evol. Comput. **7**(2), 117–132 (2003). https://doi.org/10.1109/TEVC.2003.810758
33. Zitzler, E., Knowles, J., Thiele, L.: Quality assessment of Pareto set approximations. In: Branke, J., Deb, K., Miettinen, K., Słowiński, R. (eds.) Multiobjective Optimization. LNCS, vol. 5252, pp. 373–404. Springer, Heidelberg (2008). https://doi.org/10.1007/978-3-540-88908-3_14

An Improved Squirrel Search Algorithm with Reproduction and Competition Mechanisms

Xuncai Zhang(✉) and Kai Zhao

School of Electrical and Information Engineering,
Zhengzhou University of Light Industry, Zhengzhou 450002, China
zhangxuncai@163.com

Abstract. The performance of the recently-proposed Squirrel Search Algorithm (SSA) is improved in this paper. SSA is a swarm intelligence algorithm that simulates the dynamic foraging behavior of squirrels. The traditional SSA is prone to premature convergence when solving optimization problems. This work proposed a propagation and diffusion search mechanism to alleviate these drawbacks by expand the search space using the Invasive Weed Algorithm (IWO). The proposed algorithm, which called SSIWO, has high ability to improve the exploration and local optimal avoidance of SSA. In order to investigate the performance proposed SSIWO algorithm, several experiments are conducted on eight benchmark functions and using three algorithms. The experimental results show the superior performance of the proposed SSIWO algorithm to determine the optimal solutions of the benchmark function problems.

Keywords: Squirrel Search Algorithm · Hybrid algorithm · Reproduction mechanism · Global searching capability

1 Introduction

In 1975, Professor John Holland published an article entitled "Adaptation in Natural and Artificial System", which described the adaptive mechanisms of intelligent systems and nature. Genetic algorithm (GA) was proposed for the first time in this paper. It is the pioneering work of Meta-heuristic Algorithm (MHA) [1]. The MHA is a random search Algorithm that simulates natural phenomena or biological population behavior. Compared with the traditional optimization method, MHA has the advantages of simple parameters, easy programming, no gradient information and strong global search ability. Therefore, MHA is widely used in parameter estimation [2], classification [3], scheduling [4], DNA sequence coding [5], feature selection [6], medical image [7] and multi-objective optimization [8], to name a few. In recent years, a number of new MHA with replication and competition mechanisms have emerged including fireworks algorithm (FWA) [9], invasive weed algorithm (IWO) [10], etc. Artificial

L. Pan et al. (Eds.): BIC-TA 2019, CCIS 1159, pp. 370–383, 2020.
https://doi.org/10.1007/978-981-15-3425-6_29

fish swarm algorithm (AFSA) [11] imitates the foraging, gathering, and rear-tailing behavior of fish swarm by constructing artificial fish to achieve optimal results; ant colony optimization (ACO) [12] is inspired by the behavior of ants in finding the optimal path; cuckoo search (CS) [13] effectively solves the optimization problem by simulating the cuckoo species' brood parasitism; krill herd algorithm (KH) [14] simulates the response behavior of krill for the biochemical processes and environments evolve; fruit fly optimization algorithm (FOA) [15] is a method for seeking global optimization based on fruit fly foraging behavior; chicken optimization algorithm (COA) [16] simulates the chicken hierarchy and behavior. Recently, some new swarm intelligence algorithms have emerged, such as grey wolf optimization algorithm (GWO) [17], grasshopper optimization algorithm (GOA) [18], future search algorithm (FSA) [19], butterfly optimization algorithm (BOA) [20], artificial raindrop algorithm (ARA) [21] etc.

More and more optimization models appear to help solve optimization problems. In the era of big data, however, optimization problems are more and more complex. MHA need to be constantly updated and improved to adapt to the new complex optimization problems. MHA can be improved on two aspects. One is to combine the advantages of other algorithms to compensate for the shortcoming of the original algorithms, and the other is to modify the operator of original iterative formulas. In general, Intelligent computing there are two core elements: "Exploration" and "Exploitation". The key to improving the performance of the algorithm is to balance these two elements. "Exploration" is to find the unknown domain. "Exploitation" is to mine the information more deeply when it is known that there may be an optimal solution in an area. Engelbrecht et al. analyzed various definitions of population diversity and quantified the ratio of "Exploration" and "Exploitation" [22]. Singh et al. illustrate the limitations of the unbalanced artificial bee colony algorithm for exploration and development in dealing with multidimensional problems [23]. Similarly, this mechanism also exists completely in the hybrid algorithm. Most of the hybrid algorithms combine the advantages of the two algorithm and introduce the sharing mechanism between the two algorithms. Abbattista et al. first proposed a hybrid algorithm of ant colony algorithm and genetic algorithm [24]. Shieh et al. proposed the algorithm combining particle swarm optimization algorithm and simulated annealing algorithm has excellent mechanism of two source algorithms. Trivedi et al. adopted a hybrid algorithm combining genetic algorithms with differential evolution algorithm to solve the unit combination scheduling problem [25]. The continuous variables are optimized by differential evolution algorithms. They have achieved good results in the optimization of unit combination scheduling. Compared with a single intelligent algorithm, the hybrid algorithm achieved better efficiency in most cases.

Recently a novel optimization algorithm, Squirrel Search Algorithm (SSA) [26], is proposed. It's inspired by the dynamic foraging behavior of squirrels. The Squirrel's daily activities consume different energy in different seasons. Therefore, squirrels store high-energy food in autumn to replenish their energy in cold winters. Based on this strategy, the SSA expand the search space by defining multiple suboptimal solution in each iteration. This search method can accelerate the speed of convergence. However the SSA tend to stagnate when squirrel found the hickory tree. In this case, the convergence rate gradually slows down and the convergence accuracy is low in the later stage of the search. For the problems mentioned above, the invasive weed optimization and squirrel search mixed algorithm (SSIWO) is proposed in this paper. SSIWO algorithm contains two major improvements. Firstly, weed reproduction and dispersal mechanisms were introduced to coordinate "Exploration" and "Exploitation" capabilities. Then, squirrel competition mechanism were introduced to improve convergence accuracy. And an elitist strategy optimization algorithm is designed to ensure that every generation of squirrels is closer to the optimal solution. The Invasive Weed Optimization (IWO) was first proposed by Mehrabian et al. to solve numerical optimization problems in 2006 [27]. The algorithm simulates the colonization process of weeds. In general, weeds are reproduced according to certain rules, and then sorted according to fitness. When the population reaches a certain number after reproduction, the solution with low fitness will be deleted according to the competition mechanism, and the optimal solution will be calculated. In IWO, weeds with lower fitness still have the opportunity to reproduce, which can improve the global convergence performance of the algorithm. As a new optimization method, IWO algorithm has been applied in many fields, such as robust controller optimization [28], image clustering problem [29], constrained engineering design problems [30], a design problem of antenna array for multi-input and multi-output system [31], etc. In essence, the evolutionary algorithm is probabilistic. The SSA algorithm ignores the position of the squirrel with low fitness and may have more useful information than the higher fitness, thus giving up the development of the area near the inferior solution. As a result, the convergence rate on complex function is slower, and even the optimal solution is not obtained at all. IWO retains the information of inferior solution, and gets a new solution with a certain probability, so that the search space near the inferior solution can be developed. Based on this, the SSIWO is proposed in this paper. It introduces the reproduction and competition mechanisms of IWO into the SSA. We tested the three algorithms with eight benchmark functions. The experiments show that the hybrid algorithm effectively improves the global search ability and convergence speed.

The remainder of the paper is organized as follows. Basic algorithm are given in Sect. 2 to introduce the SSA and IWO algorithm. In the Sect. 3, an improved squirrel algorithm based on reproduction and dispersal search mechanisms is proposed. In the Sect. 3.2 the experimental results and analysis are presented. Finally, this paper is concluded in Sect. 4.

2 Algorithm Introduction

2.1 Squirrel Search Algorithm

SSA simulated the foraging behavior of squirrels. In the coming winter, squirrels forage frequently. To meet their energy needs and store food for the winter, each squirrel leaves its place and glides to find a source of food [32–34]. In the process of searching for food, squirrels have their unique foraging behavior. In the autumn, when the temperature is not too low, squirrels only eat some resource-rich acorns nuts to meet their daily nutritional needs, while hickory nuts with higher energy are stored. During winters when nutritional demands are higher due to low temperature, hickory nuts are eaten promptly at the site of discovery during foraging and are also taken out from reserve food stores. Therefore, according to nutritional needs, selectively eat some nuts and store other nuts to make the best use of the two available nuts. The dynamic foraging behavior of squirrels is the main idea of SSA algorithm.

The framework of SSA algorithm is as follows:

(1) Define input parameters.
(2) The fitness F of each squirrel was calculated by randomly generating position $X = \begin{pmatrix} X_{11} & \dots & X_{1d} \\ \vdots & \ddots & \vdots \\ X_{n1} & \dots & X_{nd} \end{pmatrix}$ of N number of squirrels, and the position of each squirrel was arranged in descending order according to fitness value. The squirrel with highest fitness is declared on the hickory nut tree. The next three best squirrels are considered to be on the acorn nuts trees and they are assumed to move towards hickory nut tree. The remaining squirrels are supposed to be on normal trees.
(3) Update new locations. Randomly select a part of the squirrel from the normal tree to move to the hickory nut tree and the rest to the acorn trees. The squirrel on the acorn tree moves towards the pecan tree.
(4) Calculate seasonal constant S_C and the minimum value of seasonal constant S_{min}.

$$S_{min} = \frac{10E^{-6}}{(365)^{t/(t_m/2.5)}} \tag{1}$$

$$S_c^t = \sqrt{\sum_{k=1}^{d}(FS_{at,k}^t - FS_{ht,k})^2} \tag{2}$$

Where t is the current generation and t_m is the maximum number of iterations.
(5) Whether the seasonal detection conditions are satisfied or not, if the conditions are satisfied, formula (3) is used to randomly reset the position of the squirrel on the normal tree.

$$FS_{nt}^{new} = FS_L + levy(n) \times (FS_U - FS_L) \tag{3}$$

Where Levy (n) is Levy distribution. FS_U and FS_L are the upper and lower bounds of squirrel position.
(6) Preserve the individuals with the highest fitness.

(7) Repeat 2–6 steps until the maximum number of iterations. According to the above steps, the flow chart of SSA algorithm can be drawn as shown in Fig. 1.

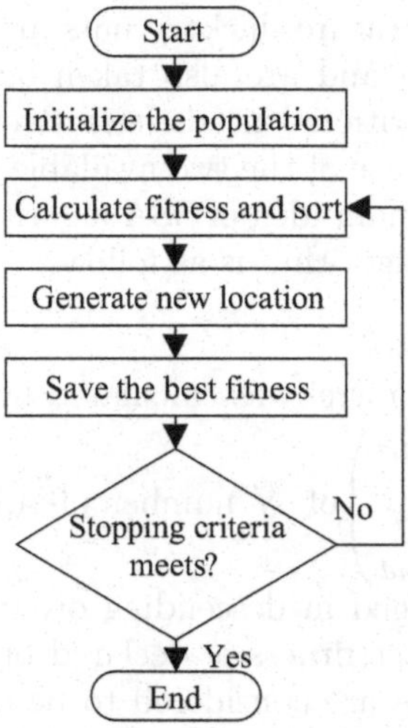

Fig. 1. SSA algorithm flow chart.

2.2 Invasive Weed Optimization

Weeds produce seeds by reproduction in the process of invasion. The seeds are diffused in search space. When weeds grow to a certain number, they compete for space to survive. Adaptable weeds will gain more chances of survival. Therefore, plant birth, growth, and reproduction activities are largely affected by adaptation. The framework of IWO algorithm is as follow:

(1) Initialize the weeds population and the positions of N number of D-dimensional weeds are randomly generated. Calculate the fitness of each weed.
(2) Growth and reproduction. Each weed seed grows and then produces seeds according to its fitness (reproductive ability). The seed breeding formula is shown in formula (4). There was a linear relationship between the number of seeds produced by parent weeds and the adaptability of parent weeds. It is shown in Fig. 2.

$$N_s = \frac{f - f_{min}}{f_{max} - f_{min}}(S_{max} - S_{min}) + S_{min} \tag{4}$$

where f is the current population fitness; f_{max} and f_{min} are the maximum and minimum fitness of the population; S_{max} and S_{min} respectively represent the maximum and minimum size of the population.

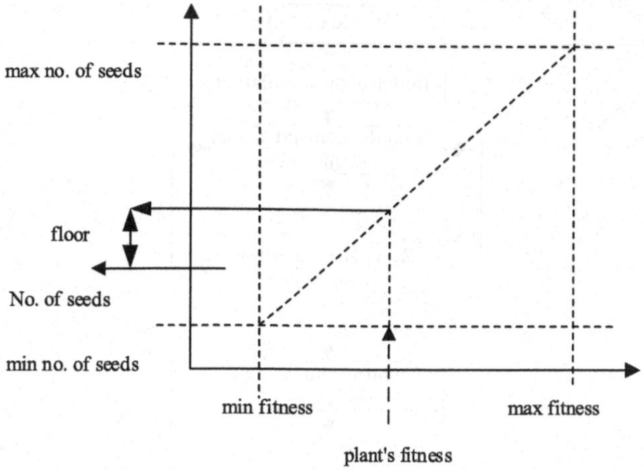

Fig. 2. Seed production method.

(3) Spatial diffusion. Parent weeds spread in space according to Gaussian distribution. It is shown in formula (5).

$$\sigma_{iter} = \frac{(iter_{max} - iter)^n}{(iter_{max})^n}(\sigma_{initial} - \sigma_{final}) + \sigma_{final} \tag{5}$$

where σ_{iter} is the standard deviation at the current iteration; $\sigma_{initial}$ is the initial standard deviation and σ_{final} is the final standard deviation; $iter_{max}$ is the maximum iteration number and n is the nonlinear harmonic index.

(4) According to the competitive survival rules, the best adaptability of weeds was evaluated.

(5) If the maximum number of generations has not been reached, go to step 2, otherwise exit the algorithm execution process and output the optimal solution.

According to the above steps, the flow chart of the IWO algorithm can be drawn as shown in Fig. 3.

3 Implementation of SSIWO

The fitness of the algorithm describes the effective use of food by squirrels. The higher the fitness, the higher the effective utilization of food by squirrels. The SSA algorithm ignores that squirrels with lower fitness may have more useful information than squirrels with higher fitness, resulting in slower convergence. In order to make up for this defect, the SSIWO algorithm introduces the reproduction mechanism of weeds in IWO. The number of squirrels is not fixed again, but over time, the number of squirrels will get more and more. And introduce a competitive mechanism to limit the number of squirrels.

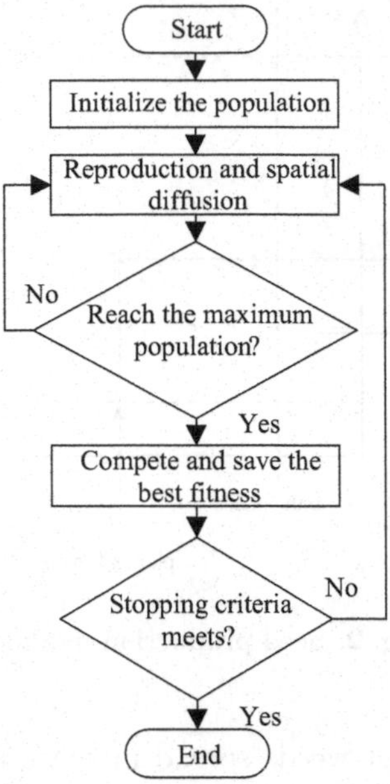

Fig. 3. The flow chart of the IWO algorithm.

3.1 The Principle of SSIWO

The higher the fitness of the squirrel, the stronger the reproductive ability. The update of population in SSIWO algorithm is divided into three cases according to reference [26]. Inspired by the IWO reproduction mechanism, and taking into account that every individual can reproduce. The propagation mechanism of prize IWO algorithm is introduced into SSA algorithm. According to the way of updating the population in reference [26], the new population is determined by the reproduction mechanism of IWO algorithm. Based on this, the design steps of the SSIWO algorithm are as follows:

(1) Initialize the squirrel population location, population size, maximum number of iterations, etc.
(2) Calculate the fitness of the population, and arrange them in descending order, stating that the solution with the greatest fitness is the hickory nut tree, followed by the acorn nuts trees and the normal trees.
(3) Growing and reproduction to generate new locations. Because of the different types of food sources and the current nutritional needs of squirrels, there are three types of squirrel location updates:

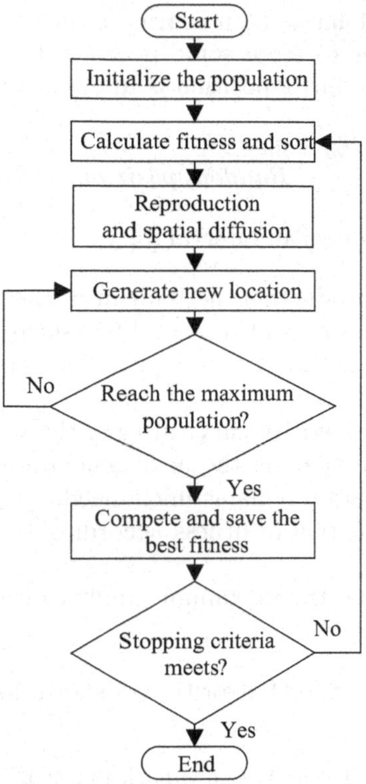

Fig. 4. SSIWO algorithm flow chart.

Case 1: After the squirrels on the acorn nut trees gets the daily life needed, it will glide on the hickory nut tree. In this case, the updated formula of the squirrel position is as follows:

$$FS_{at}^{t+1} = \begin{cases} FS_{at}^{t} + d_g \times G_c \times (FS_{ht}^{t} - FS_{at}^{t}),\ R1 \geq p_{dp} \\ \quad Random\ position, \qquad\qquad other \end{cases} \tag{6}$$

where d_g is the random gliding distance and R is the random number in interval $[0, 1]$; FS_{at} is the position of the squirrel on the hickory nut tree and t is the current iteration number; G_C is the squirrel gliding constant and P_{dp} is the probability of a predator appearing.

Case 2: The squirrels on the normal trees may move to the oak because it needs to find the daily source of daily food. In this case, the update formula for the pine position is as follows:

$$FS_{nt}^{t+1} = \begin{cases} FS_{nt}^{t} + d_g \times G_c \times (FS_{at}^{t} - FS_{nt}^{t}),\ R2 \geq p_{dp} \\ \quad Random\ position, \qquad\qquad other \end{cases} \tag{7}$$

where FS_{nt} is the position of the squirrels on the acorn nuts trees and R_2 is the random number in interval $[0, 1]$.

Case 3: Squirrels that have eaten acorns on normal trees may move to the hickory nut tree in order to store some pecan nuts during the food-deficient season. In this case, the squirrel position is updated as follows:

$$FS_{nt}^{t+1} = \begin{cases} FS_{nt}^{t} + d_g \times G_c \times (FS_{ht}^{t} - FS_{nt}^{t}), & R3 \geq p_{dp} \\ Random\ position, & other \end{cases} \tag{8}$$

where R_3 is a random number in interval [0, 1].

(4) According to the reproduction mechanism in the formula (4), the number of squirrels was generated and spread. The squirrel's position is affected by predators, in the case of predators, squirrels through random walks to find hidden places.
(5) According to the competitive survival rule, the value with better fitness is selected as the initial value of the next generation iteration to ensure that the size of the population remains unchanged.
(6) Evaluate the best solution of fitness according to the competitive survival rule.
(7) Repeat steps 2–6 until the maximum number of iterations and output the optimal solution.

The flow chart of the SSIWO algorithm is shown in Fig. 4.

Table 1. Benchmark function

Benchmark function	Formula	Search interval	d	f_{min}
Sphere	$f_1 = \sum_{i=1}^{30} x_i^2$	$[-100, 100]$	30	0
Rosenbrock	$f_2 = \sum_{i=1}^{30} (100(x_{i+1} - x_i^2)^2) + (x_i - 1)^2$	$[-30, 30]$	30	0
Griewank	$f_3 = 1 + \sum_{i=1}^{30} \frac{x_i^2}{4000} + \prod_{i=1}^{30} cos(\frac{x_i}{\sqrt{i}})$	$[-600, 600]$	30	0
Rastrigin	$f_4 = \sum_{i=1}^{30} (x_i^2 - 10cos(2\pi x_i) + 10)$	$[-5.12, 5.12]$	30	0
Schwefel	$f_5 = \sum_{i=1}^{30} (\sum_{j}^{i} x_j)^2$	$[-100, 100]$	30	0
Ackley	$f_6 = -20exp(-0.2\sqrt{\frac{1}{30} \sum_{i=1}^{30} x_i^2}) + 20 + e$	$[-32, 32]$	30	0
Axis parallel	$f_7 = \sum_{i=1}^{30} ix_i^2$	$[-5.12, 5.12]$	30	0
Rotated hyper	$f_8 = \sum_{i=1}^{30} (\sum_{j}^{i} x_j^2)^2$	$[-65.536, 65.536]$	30	0

3.2 Experimental Design and Results

The benchmark functions used in this paper are all high-dimensional functions. The purpose is to test the performance of the algorithm in a high-dimensional environment. The eight benchmark functions selected in this paper are easy to construct high-dimensional benchmark functions. So we chose the test functions listed in Table 1. The simulation test platform is Matlab R2014a. The parameters are set as follows: (1) SSA algorithm: the maximum number of iterations is 1000, the population size is 50, and the sliding constant is 0.8; (2) IWO algorithm: the maximum number of iterations is 1000, the minimum population is 20, and the maximum population is 50. The "minimum population" and the "maximum population" refer to the population before reproduction and the population after reproduction, respectively. In this paper, We mainly test the convergence performance of the algorithm in high-dimensional environment, so when we select the parameters, we deliberately keep the parameters of the three SSA, IWO, SSIWO algorithms consistent; (3) SSIWO algorithm: the maximum number of iterations is 1000, the minimum population is 20, the maximum population is 50, and the sliding constant is 0.8. The three algorithms run independently 100 times and take their average. The results are shown in Table 2. In the comparative experiment of this paper, we consider the comparison of the convergence performance of different algorithms under the same number of iterations. Therefore, in this paper, we choose a unified standard, that is, 1000 iterations to complete the experiment.

Table 2. Convergence characteristics test of three algorithms (average of 50 runs).

Benchmark function	SSIWO	SSA	IWO
Sphere	2.3102E−20	5.6412E−06	3.8435E−03
Rosenbrock	4.9829E−43	2.0438E−04	3.6245E−03
Griewank	2.2689E−03	3.2579E−02	1.5693E−02
Rastrigin	6.8956E−20	0.0125	0.3215
Schwefel	3.2549E−80	2.0594E−50	1.0259E−19
Ackley	1.3597E−46	5.2136E−12	3.2179E−03
Axis parallel	5.7915E−16	3.1587E−05	0.4359
Rotated hyper	3.2589E−26	2.1896E−12	4.0159E−10

It can be seen from Table 2 that the convergence accuracy of SSIWO algorithm is higher than that of the other two algorithms. In Fig. 5(a), the convergence rate of the SSIWO algorithm is significantly higher than the other two algorithms. The Griewank function is often used to test the balance performance of the algorithm for global and local search capabilities. As seen in Fig. 5(c), both the IWO algorithm and the SSA algorithm are trapped in local optimal, while the SSIWO algorithm quickly jumps out of the local minimum and finds the

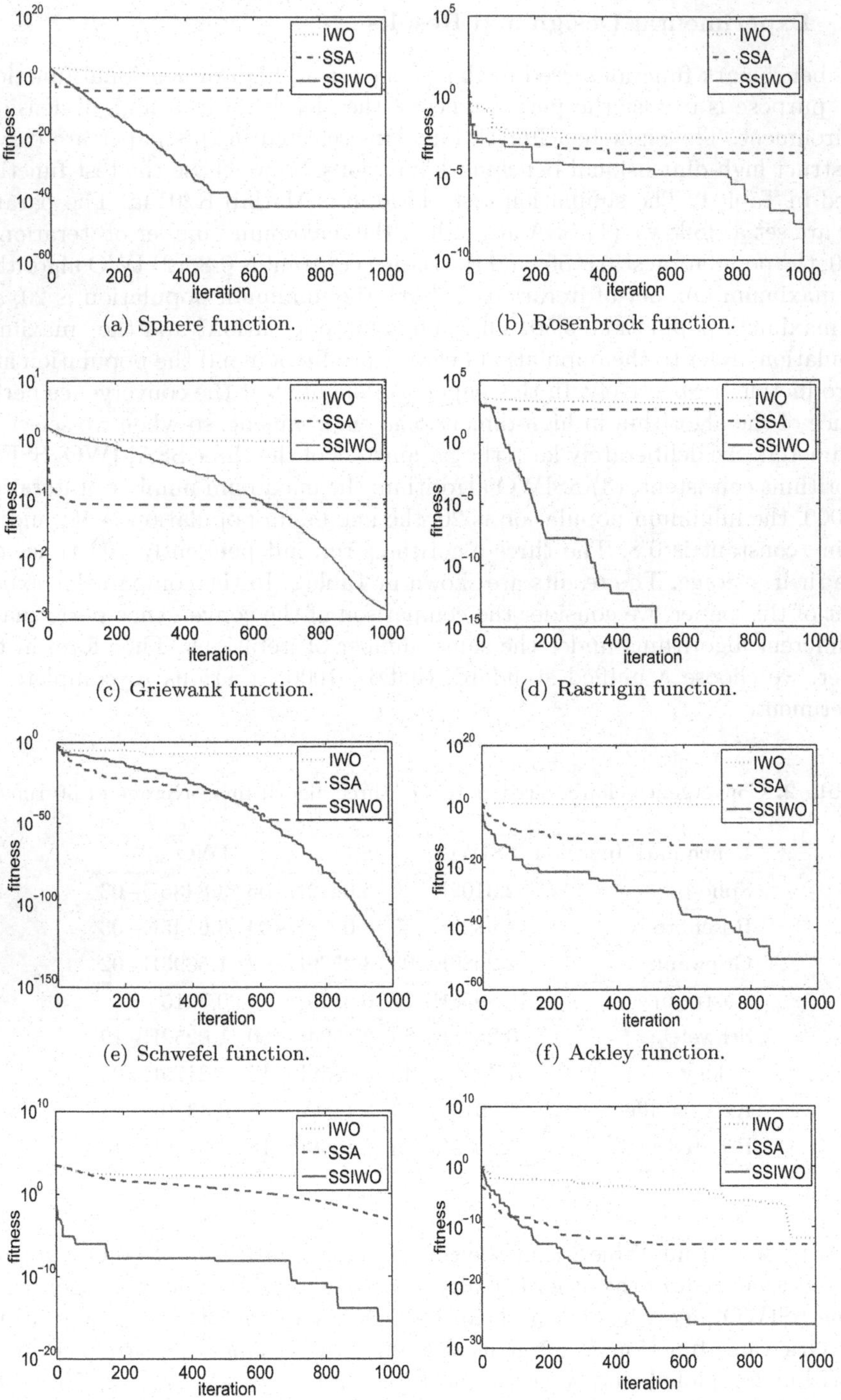

(a) Sphere function.

(b) Rosenbrock function.

(c) Griewank function.

(d) Rastrigin function.

(e) Schwefel function.

(f) Ackley function.

(g) Axis Parallel function.

(h) Rotated Hyper function.

Fig. 5. Test results of three algorithms.

most excellent solution. The Rastrigin function is often used to test the global search ability of the algorithm. In the early iteration, as shown in Fig. 5(d), the convergence rate of the three algorithms is equivalent. As the number of iterations increases, the "premature" feature of the IWO algorithm begins to appear. It is trapped in a local minimum, and the SSIWO algorithm can quickly jump out of the local minimum. Ackley function is a continuous, rotating and indivisible multimodal function, which is often used to test the ability of the algorithm to jump out of the local optimal value. By testing the Ackley function, it is proved that the SSIWO algorithm has good convergence. As seen in Fig. 5(f), the IWO algorithm can never find a global optimal value. The SSA algorithm falls into a local optimal value early, and after a few iterations, it barely jumps out of the local area. The SSIWO algorithm greatly enhances the ability to resist "premature" while maintaining convergence rate.

4 Conclusion

In this paper, a hybrid algorithm of squirrel search algorithm and invasive weeds optimization is proposed. The search strategy of growth and reproduction and competition exclusion in IWO algorithm is applied to the SSA optimization process. We adopt eight benchmark functions to show the performance of SSIWO by comparing our algorithm with other approaches. The experimental results show that the proposed algorithm is superior to the basic SSA algorithm and IWO algorithm in both convergence speed and global search ability, and can effectively solve the multivariate and multi-mode function optimization problems. The experimental data also shows that the SSIWO algorithm is also superior to SSA and IWO in terms of search accuracy. The algorithm can be further applied to practical problems.

Acknowledgments. The work for this paper was supported by the National Natural Science Foundation of China (Grant nos. 61572446, 61602424, and U1804262), Key Scientific and Technological Project of Henan Province (Grant nos. 174100510009, 192102210134), and Key Scientific Research Projects of Henan High Educational Institution (18A510020).

References

1. Holland John, H.: Adaptation in Natural and Artificial Systems. University of Michigan Press, Ann Arbor (1975)
2. Oliva, D., El Aziz, M.A., Hassanien, A.E.: Parameter estimation of photovoltaic cells using an improved chaotic whale optimization algorithm. Appl. Energy **200**, 141–154 (2017)
3. Lin, K.-C., Zhang, K.-Y., Huang, Y.-H., Hung, J.C., Yen, N.: Feature selection based on an improved cat swarm optimization algorithm for big data classification. J. Supercomput. **72**(8), 3210–3221 (2016). https://doi.org/10.1007/s11227-016-1631-0

4. Tang, J., Yang, Y., Qi, Y.: A hybrid algorithm for urban transit schedule optimization. Phys. A **512**, 745–755 (2018)
5. Zhang, X., Wang, Y., Cui, G., Niu, Y., Xu, J.: Application of a novel IWO to the design of encoding sequences for DNA computing. Comput. Math. Appl. **57**(11–12), 2001–2008 (2009)
6. Kabir, M.M., Shahjahan, M., Murase, K.: A new hybrid ant colony optimization algorithm for feature selection. Expert Syst. Appl. **39**(3), 3747–3763 (2012)
7. Wang, S.-H., et al.: Single slice based detection for Alzheimer's disease via wavelet entropy and multilayer perceptron trained by biogeography-based optimization. Multimed. Tools Appl. **77**(9), 10393–10417 (2016). https://doi.org/10.1007/s11042-016-4222-4
8. Zhang, X., Tian, Y., Cheng, R., Jin, Y.: An efficient approach to nondominated sorting for evolutionary multiobjective optimization. IEEE Trans. Evol. Comput. **19**(2), 201–213 (2014)
9. Tan, Y., Zhu, Y.: Fireworks algorithm for optimization. In: Tan, Y., Shi, Y., Tan, K.C. (eds.) ICSI 2010. LNCS, vol. 6145, pp. 355–364. Springer, Heidelberg (2010). https://doi.org/10.1007/978-3-642-13495-1_44
10. Zhang, X., Niu, Y., Cui, G., Wang, Y.: A modified invasive weed optimization with crossover operation. In: 2010 8th World Congress on Intelligent Control and Automation, pp. 11–14. IEEE (2010)
11. Shen, W., Guo, X., Wu, C., Wu, D.: Forecasting stock indices using radial basis function neural networks optimized by artificial fish swarm algorithm. Knowl.-Based Syst. **24**(3), 378–385 (2011)
12. Marzband, M., Yousefnejad, E., Sumper, A., Domínguez-García, J.L.: Real time experimental implementation of optimum energy management system in standalone microgrid by using multi-layer ant colony optimization. Int. J. Electr. Power Energy Syst. **75**, 265–274 (2016)
13. Zhang, M., Wang, H., Cui, Z., Chen, J.: Hybrid multi-objective cuckoo search with dynamical local search. Memetic Comput. **10**(2), 199–208 (2017). https://doi.org/10.1007/s12293-017-0237-2
14. Wang, G.-G., Gandomi, A.H., Alavi, A.H., Gong, D.: A comprehensive review of krill herd algorithm: variants, hybrids and applications. Artif. Intell. Rev. **51**(1), 119–148 (2017). https://doi.org/10.1007/s10462-017-9559-1
15. Mitić, M., Vuković, N., Petrović, M., Miljković, Z.: Chaotic fruit fly optimization algorithm. Knowl.-Based Syst. **89**, 446–458 (2015)
16. Wu, D., Xu, S., Kong, F.: Convergence analysis and improvement of the chicken swarm optimization algorithm. IEEE Access **4**, 9400–9412 (2016)
17. Mirjalili, S., Saremi, S., Mirjalili, S.M., Coelho, L.d.S.: Multi-objective grey wolf optimizer: a novel algorithm for multi-criterion optimization. Expert Syst. Appl. **47**, 106–119 (2016)
18. Mirjalili, S.Z., Mirjalili, S., Saremi, S., Faris, H., Aljarah, I.: Grasshopper optimization algorithm for multi-objective optimization problems. Appl. Intell. **48**(4), 805–820 (2017). https://doi.org/10.1007/s10489-017-1019-8
19. Elsisi, M.: Future search algorithm for optimization. Evol. Intel. **12**(1), 21–31 (2018). https://doi.org/10.1007/s12065-018-0172-2
20. Arora, S., Singh, S.: Butterfly optimization algorithm: a novel approach for global optimization. Soft. Comput. **23**(3), 715–734 (2018). https://doi.org/10.1007/s00500-018-3102-4
21. Jiang, Q., Wang, L., Hei, X.: Parameter identification of chaotic systems using artificial raindrop algorithm. J. Comput. Sci. **8**, 20–31 (2015)

22. Mirjalili, S., Lewis, A.: The whale optimization algorithm. Adv. Eng. Softw. **95**, 51–67 (2016)
23. Singh, A., Deep, K.: Exploration–exploitation balance in Artificial Bee Colony algorithm: a critical analysis. Soft. Comput. **23**(19), 9525–9536 (2018). https://doi.org/10.1007/s00500-018-3515-0
24. Abbattista, F., Abbattista, N., Caponetti, L.: An evolutionary and cooperative agents model for optimization. In: Proceedings of 1995 IEEE International Conference on Evolutionary Computation, vol. 2, pp. 668–671. IEEE (1995)
25. Trivedi, A., Srinivasan, D., Biswas, S., Reindl, T.: A genetic algorithm–differential evolution based hybrid framework: case study on unit commitment scheduling problem. Inf. Sci. **354**, 275–300 (2016)
26. Jain, M., Singh, V., Rani, A.: A novel nature-inspired algorithm for optimization: squirrel search algorithm. Swarm Evol. Comput. **44**, 148–175 (2019)
27. Mehrabian, A.R., Lucas, C.: A novel numerical optimization algorithm inspired from weed colonization. Ecol. Inform. **1**(4), 355–366 (2006)
28. Chen, Z., Wang, S., Deng, Z., Zhang, X.: Tuning of auto-disturbance rejection controller based on the invasive weed optimization. In: 2011 Sixth International Conference on Bio-Inspired Computing: Theories and Applications, pp. 314–318. IEEE (2011)
29. Pan, G., Li, K., Ouyang, A., Zhou, X., Xu, Y.: A hybrid clustering algorithm combining cloud model IWO and K-means. Int. J. Pattern Recognit Artif Intell. **28**(06), 1450015 (2014)
30. Zhou, Y., Luo, Q., Chen, H., He, A., Wu, J.: A discrete invasive weed optimization algorithm for solving traveling salesman problem. Neurocomputing **151**, 1227–1236 (2015)
31. Karimkashi, S., Kishk, A.A., Kajfez, D.: Antenna array optimization using dipole models for mimo applications. IEEE Trans. Antennas Propag. **59**(8), 3112–3116 (2011)
32. Bishop, K.L.: The relationship between 3-D kinematics and gliding performance in the southern flying squirrel, Glaucomys volans. J. Exp. Biol. **209**(4), 689–701 (2006)
33. Vernes, K.: Gliding performance of the northern flying squirrel (*Glaucomys Sabrinus*) in mature mixed forest of eastern Canada. J. Mammal. **82**(4), 1026–1033 (2001)
34. Thomas, R.B., Weigl, P.D.: Dynamic foraging behavior in the southern flying squirrel (Glaucomys volans): test of a model. Am. Midl. Nat. **140**(2), 264–271 (1998)

Modified Self-adaptive Brain Storm Optimization Algorithm for Multimodal Optimization

Ze-yu Dai[1(✉)], Wei Fang[2], Qing Li[1], and Wei-neng Chen[3]

[1] Hong Kong Polytechnic University, Hung Hom, Kowloon, Hong Kong
ze-yu.dai@connect.polyu.hk
[2] Jiangnan University, Wuxi, China
[3] South China University of Technology, Guangzhou, China

Abstract. Multimodal Optimization is one of the most challenging tasks for optimization, since many real-world problems may have multiple acceptable solutions. Different from single objective optimization problem, multimodal optimization needs to both find multiple optima/peaks at the same time, and maintain these found optima until the end of a run. A novel swarm intelligent method, Modified Self-adaptive Brain Storm Optimization (MSBSO) algorithm is proposed to solve multimodal optimization problems in this paper. In order to find potential multiple optima, a modified disruption strategy is used for BSO algorithms to maintain the identified optima until the end of the search. Besides, the self-adaptive cluster number control is applied to improve Max-fitness Clustering Method with no need for a predefined subpopulation size M. Eight multimodal benchmark functions are used to validate the performance and effectiveness. Compared with the other swarm intelligent algorithms reported in the literature, the new algorithm can outperform others on most of the test functions.

Keywords: Multimodal Optimization · Brain Storm Optimization Algorithm · Modified Disruption Strategy · Self-adaptive Cluster Number Control

1 Introduction

Optimization refers to finding the optimal feasible solution(s) for a given optimized problem. An optimization problem is mapping from decision space to objective space. And multimodal optimization is a most challenging task in the area of optimization. Since many real-world problems tend to have multiple optima, which are called multimodal optimization problems (MMOPs) [31], multimodal optimization is needed to locate multiple optimal solutions simultaneously. The advantages can be summarized as follows [1]:

- In the real world, there will be some factors that are hard to modal mathematically, such as manufacturing degree of difficulty, maintenance degree of

L. Pan et al. (Eds.): BIC-TA 2019, CCIS 1159, pp. 384–397, 2020.
https://doi.org/10.1007/978-981-15-3425-6_30

difficulty, reliability, etc. Finding multiple solutions with similar quality provides the decision maker with alternative options to be further determined, alleviating the impact of uncertainty.

- If the search ability of the algorithm cannot guarantee that the global optimal solution be found, then locating multiple optima can improve the chance to jump out of the local optima and increase the possibility of finding the global optimal solution.
- Multiple solutions with similar quality are important for finding a robust solution and useful for the sensitivity analysis of a problem.

Unlike classical optimization techniques that use different starting points to obtain different solutions, evolutionary algorithms (EAs), due to their population-based approach, provide a population of potential solutions processed at every iteration. However, the original EAs are usually designed for single objective optimization problems (SOPs). These algorithms often converge to single final solution because of the greedy selection scheme and global recombination operators [23]. Therefore, the difficulty lies in how to modify EAs so that they can locate multiple global and local optima and maintain these identified solutions until the end of the search.

To alleviate this issue, a variety of methods commonly known as niching have been developed to address MMOPs. In multimodal EAs, niche represents the partial solution landscape where only one peak resides [1]. Therefore, the niching methods take effect by forming multiple niches within a single population, and each subpopulation is used to locate one or multiple optima, enhancing the diversity of the population [28]. Some of the proposed techniques include crowding [14,27], fitness sharing [9,12], clearing [22], speciation [16] and parallelization [2], etc. These methods have been incorporated into various EA paradigms to deal with the multiple optima finding task in a multimodal landscape. Nevertheless, these methods generally involve certain sensitive niching parameters, which are problem-dependent and need to be set properly in order to have a good performance [17].

In addition, clustering, considered to be one of the most effective techniques, is popular among researchers to solve multimodal optimization [19]. Clustering-based approaches [3,11,13,15,20,29] divide the population into a set of different groups in every generation, which is highly likely to maintain multiple optimal solutions. Alessandro and Antonina [20] presented a new method in particle swarm optimization to identify niches with clustering particles and the resulting algorithm keeps the same structure as the standard PSO. However, the clustering procedure costs a high computational overhead. Bošković and Brest [3] also use a new clustering mechanism based on small subpopulations with the best strategy to improve the differential evolution algorithm's efficiency, but it needs two priori parameters (M and ϵ) within the algorithm. In [11,13,15], an effective clustering method is involved and it selects nearest individuals measured by Euclidean distance with a fixed small subpopulation size rather than niching radius. However, the main difficulty of all these clustering methods is how to define the area for each subdomain in the search space and how to determine the number of sub-populations [11].

In this paper, we focus on Brain Storm Optimization (BSO) algorithm, proposed by Shi [24,25], which is inspired by human idea generation process to solve SOPs originally. But two features make BSO show a very promising searching ability for MMOPs. One is the clustering operator that divides the whole population into numbers of clusters, so that these clusters can locate distinct optima in corresponding subpopulations. Another is the creating operator that creates new individual learning from the self-cluster or other clusters, which can maintain the diversity of the population [13].

Recently, adaptive or self-adaptive parameter control, has attracted a lot of attention. If designed appropriately, adaptive strategy can improve the robustness and accuracy of an algorithm by dynamically adjusting the parameters over the evolution process. Some kinds of adaptive or self-adaptive strategies have been applied in [21,30], and the experimental results show good performance of BSO. However, these approaches mainly concentrate on SOPs. Thus, can we try to design a self-adaptive strategy for predefined parameters needed in multimodal optimization?

Motivated by the above issues, this paper proposes a modified cluster-based BSO with a self-adaptive strategy for multimodal optimization. The proposed algorithm has the following two features:

1. In order to find potential multiple optima, a modified disruption strategy is used for BSO algorithms. As long as the optima are found in each cluster during the search, the identified solutions can definitely be maintained until the end of a run.
2. A self-adaptive cluster number control is applied for Max-fitness Clustering Method [13] with no need for a predefined subpopulation size M. And the experimental results show that it can not only reduce the requirement for parameter, but also improve the convergence accuracy.

The rest of the paper is organized as follows. Section 2 briefly reviews the related works about BSO. In Sect. 3, the proposed algorithm is described in details. The experiment setting and results are presented in Sect. 4. Finally, the conclusion and further research are given in Sect. 5.

2 Brain Storm Optimization and Multimodal Optimization

2.1 Brain Storm Optimization

The Brain Storm Optimization (BSO) algorithm, based on the collective behavior of human being, is a young and promising algorithm in swarm intelligence [24,25]. Swarm intelligence algorithms should have two kind of ability: capability learning and capacity developing [26]. The capacity developing focuses on moving the algorithm's search to the area(s) where higher search potential may exist, which is also called the exploration ability. While the capability learning focuses on its concrete search to find better solution(s) from the current solution(s), which is also called the exploitation ability. If the algorithm is overly dependent on the exploration ability, the convergence rate is likely to be affected. In contrast, if the algorithm overly relies on the exploitation ability, the search tends to be trapped into local optima [7]. Therefore, how to balance the exploration ability and the exploitation ability simultaneously is the key to solve multimodal optimization problems.

The algorithm is given in Algorithm 1. In the initialization, N potential individuals are randomly generated. Then the original BSO uses k-means clustering method to diverge the whole population into several clusters. And a randomly selected cluster center will be replaced by a randomly generated individual, in order to avoid the premature convergence and help individuals "jump out" of the local optima. In creating operator, the new individual can be generated on the mutation of one or two individuals in clusters. The best solution in the population will be kept if the newly generated solution at the same index is not better.

Algorithm 1. Procedure of the brain storm optimization algorithm

Initialization: Randomly generate N potential individuals, and evaluate the N individuals;
while *have not found "good enough" solution or not reached the pre-determined maximum number of iterations* **do**
  Clustering: Cluster N individuals into M clusters by a cluster algorithm;
  Disruption: Randomly select a cluster center and replace the it with a randomly generated individual;
  New individual's generation: Randomly select one or two clusters(s) to generate new individual;
  Seletion: The newly generated individual is compared with the existing individual with the same individual index; the better one is kept and recorded as the new individual;
  Evaluate the N individuals;

The new individuals are generated according to the Eqs. 1 and 2.

$$x_{new}^{i} = x_{old}^{i} + \xi(t) \times N(\mu, \sigma^2) \tag{1}$$

$$\xi(t) = logsig(\frac{0.5 \times T - t}{k}) \times rand() \tag{2}$$

2.2 BSO for Multimodal Optimization Problems

Some researchers have attempted to apply BSO to cope with MMOPs and some multimodal optimization BSO algorithms have been proposed in the literatures [5,6,8,13]. In [13], a self-adaptive BSO (SBSO) is proposed to solve

MMOPs with a max-fitness clustering method and self-adaptive parameter control. To reduce the computational cost in clustering, the brain storm optimization in objective space (BSO-OS) is utilized to solve MMOPs [8]. Then more variants of BSO algorithms, including BSO-OS with Gaussian random variable and BSO-OS with Cauchy random variable, are applied for MMOPs. They are conducted on more benchmark problems and nonlinear equation system problems to validate their performance and effectiveness [5]. In [6], BSO algorithm is also introduced to solve dynamic multimodal optimization (DMO) problem, which is a combination of dynamic optimization and multimodal optimization.

3 Proposed Algorithm

For multimodal optimization, the objective is to locate multiple local and global optima during the searching process, and to keep these found optima until the end of a run. The experimental results in [3,11,13,15,20,29] show that clustering the operator is an ideal technique. Plenty of EAs have introduced a variety of clustering strategies into the evolutionary process to solve MMOPs. And BSO algorithm just adopts the clustering operator as its converging process, which makes BSO very suitable for MMOPs. It is known that different clustering methods show different advantages. The k-means method in the original BSO is a typical clustering method based on distance. But it faces a biggest drawback that the choice of initial cluster center has a considerable influence on its clustering result. Once the initial cluster center is not chosen well, it could have a bad impact on its clustering result.

Hence in [13], a Max-fitness Clustering Method (MCM) is designed to replace k-means clustering method in BSO. MCM firstly selects the individual with the best fitness value in the population as the first cluster center. Then its nearest individuals measured by Euclidean distance are classified into this cluster. Finally, the process is repeated until all individuals are classified. The complete MCM step is given in Algorithm 2. The improvement is that each cluster center in MCM is the best individual of all remaining individuals, so each cluster center has a lager probability to become an extreme point, which enhances the ability to find multiple optima. And the experimental results show good performance and robustness. However, MCM still needs a predefined parameter M to denote the subpopulation size, which is problem-dependent and hard to choose.

In this paper, the proposed algorithm has two key components. The first one is a modified disruption strategy, which can maintain the identified optima until the end of a run. Another is a self-adaptive cluster number control, eliminating the need for a predefined subpopulation size M in MCM. Detailed explanation will be given as follow.

Algorithm 2. Max-fitness clustering method

Initialize a population P of N individuals randomly;
do
  Find the best individual in P as the cluster center X;
  Combine $M-1$ individuals, which are nearest to X, with X to form a cluster;
  Eliminate these M individuals from the current population;
while *P is successfully divided into N/M clusters*;

3.1 Modified Disruption Strategy

The original disruption process in BSO is to randomly select a cluster center and replace it with a randomly generated individual. But for multimodal optimization, multiple global optima exist, so the original disruption may cause the identified optima being replaced. Meanwhile, different individuals possibly point to the same optimum, which is superfluous and wasteful.

Therefore, to address the above issues, the disruption strategy is modified. The replacement is also selected from the cluster centers, but for those cluster centers having the same optimal fitness value and different positions in the solution space, they won't be selected to replace, because they tend to be distinct optima. While for the remaining cluster centers, one of them will be selected to replace with a randomly generated individual. The modified disruption strategy is presented in Algorithm 3. In this way, as long as certain global optimal solutions are found during the search process, they can be kept to the end of the run, avoiding the problems that original disruption strategy causes.

Algorithm 3. Modified disruption stategy

Find the individuals(s) with the best fitness value in cluster centers and assign them into set A, and assign others into set B;
while *A is not empty* **do**
  Take an individual from A, record its position as X, and delete it from A;
  if *there is an individual whose position equals X* **then**
    assign it into set C, and delete it from A;
  else
    keep it in A unchanged;
Combine B and C into set D, randomly select one individual from D, and replace it with a randomly generated individual;

3.2 Self-adaptive Cluster Number Control

The max-fitness clustering method in [13] has a drawback that the subpopulation size M needs to be predefined. But for real-world MMOPs, we cannot know in advance how many global optimal solutions the problem has, so it is hard to preset the subpopulation size and the number of subpopulations. In this paper, a self-adaptive cluster number control is used to avoid the predefined parameter M by updating the cluster number iteration over iteration. Detailed procedure is given in Algorithm 4.

The above steps 2–10 are added after the selection step of each iteration, and the cluster number n_c is not updated in the first iteration. Since a multimodal optimization problem has at least 2 global optimal solutions, the n_c should not be less than 2. As for the subpopulation size M needed in MCM, it can be calculated by dividing population size N by n_c.

Because the cluster center in MCM is highly likely to be an extreme point, the average distance between each cluster center and the other, can be used to measure the distance change among the extreme points. If the distance change increases, the extreme points could be further, and more cluster centers are needed to represent them, and vice versa. When almost all global optimal solutions have been found, namely, all cluster centers represent nearly all global optimal solutions, then the average distance will no longer change, and the number of cluster centers will also stabilize. Through the self-adaptive cluster number control, we can continuously update the cluster number as the search progresses without knowing the number of global optimal solutions in advance. The BSO algorithm with the modified disruption strategy and the self-adaptive cluster number control is called the modified self-adaptive BSO (MSBSO) algorithm.

Algorithm 4. Self-adaptive cluster number control

---

1 Initialize a cluster number n_c at the start of the run;
2 Traverse n_c cluster centers and calculate the distances from the remaining $n\_c - 1$ cluster centers;
3 Sum up the above n_c^2 distances and divide it by n_c^2 to obtain the average distance d_i (i denotes the current number of iterations) between each cluster center and the other one;
4 **if** $d_i > d_{i-1}$ **then**
5   The cluster number $n\_c + 1$;
6 **else**
7   **if** $n\_c > 2$ **then**
8   The cluster number $n\_c - 1$;
9   **else**
10   Keep the cluster number n_c unchanged;

---

4 Experimental Study

4.1 Benchmark Functions and Parameters Setting

Eight benchmark functions [18] are applied to evaluate the performance of the proposed algorithm for MMOPs. $f_1 - f_3$ are simple 1D multimodal functions; f_4

and f_5 are 2D multimodal functions; $f_6 - f_8$ are scalable multimodal functions, whose number of global optima are determined by the dimension D. The benchmark functions and their settings are given in Tables 1, 2 and 3. The accuracy level ϵ is 1.0E−04 and all the algorithms will be tested for 50 runs. The parameter settings of the compared algorithms, based on BSO [5] and PSO [4,10], are listed as follow:

- Original BSO algorithm: $p_{clustering} = 0.2$, $p_{generation} = 0.6$, $p_{oneCluster} = 0.4$, $p_{twoCluster} = 0.5$. The slope for $logsig()$ is 10, and the parameter k in k-means algorithm is 25.
- BSO-OS algorithm: $p_{elitist} = 0.2$, $p_{one} = 0.8$, and the slop is 10.
- PSO algorithms: $\omega = 0.72984$, $c_1 = c_2 = 1.496172$.

Table 1. The benchmark functions, where D is the dimension of each function

Function	Function name	D	Optima (global/local)
f_1	Five-Uneven-Peak Trap	1	2/3
f_2	Equal Maxima	1	5/0
f_3	Uneven Decreasing Maxima	1	1/4
f_4	Himmelblau	2	4/0
f_5	Six-Hump Camel Back	2	2/4
f_6	Shubert	2/3	$D \cdot 3^D$/many
f_7	Vincent	2/3	$6^D/0$
f_8	Modified Rastrigin - All Global Optima	2	$\prod_{i=1}^{D} k_i/0$

Table 2. The settings of benchmark functions

Function	r	Maximum	No. of global optima
$f_1(1D)$	0.01	200.0	2
$f_2(1D)$	0.01	1.0	5
$f_3(1D)$	0.01	1.0	1
$f_4(2D)$	0.01	200.0	4
$f_5(2D)$	0.5	4.126513	2
$f_6(2D)$	0.5	186.73090	18
$f_6(3D)$	0.5	2709.09350	81
$f_7(2D)$	0.2	1.0	36
$f_7(3D)$	0.2	1.0	216
$f_8(2D)$	0.01	−2.0	12
$f_8(3D)$	0.01	−16.0	48

Table 3. The population size, number of iterations and maximum number of fitness evaluation for BSO and PSO algorithm

Function	BSO Algo.		PSO Algo.		*MaxFEs*
	Popu.	Iter.	Popu.	Iter.	
$f_1(1D)$	500	100	100	500	5.0E+04
$f_2(1D)$	500	100	100	500	5.0E+04
$f_3(1D)$	500	100	100	500	5.0E+04
$f_4(2D)$	500	100	100	500	5.0E+04
$f_5(2D)$	500	100	100	500	5.0E+04
$f_6(2D)$	500	400	400	500	2.0E+05
$f_6(3D)$	500	800	800	500	4.0E+05
$f_7(2D)$	500	400	400	500	2.0E+05
$f_7(3D)$	500	800	800	500	4.0E+05
$f_8(2D)$	500	400	400	500	2.0E+05
$f_8(3D)$	500	800	400	1000	4.0E+05

4.2 Performance Criteria

Two criteria are used to assess the performance of the multimodal algorithms: peak ratio (PR) and success rate (SR) [18]. PR measures the average percentage of global optima found over multiple runs:

$$PR = \frac{\sum_{run=1}^{NR} NPF_i}{NKP \times NR} \tag{3}$$

where NPF_i is the number of global optima found in the end of the i-th run. NKP denotes the number of global optima and NR is the number of runs.

SR measures the percentage of successful runs in which all the global optima are found:

$$SR = \frac{NSR}{NR} \tag{4}$$

where NSR denotes the number of successful runs.

4.3 Experimental Results and Analysis

Comparison Between the Original SBSO and the SBSO with Modified Disruption Strategy. Since the original BSO disruption is likely to replace the identified optima, the modified disruption strategy is applied to BSO with max-fitness clustering method (SBSO) [13]. Table 4 shows the experimental results of the original SBSO and the SBSO with modified disruption strategy. The best results are highlighted in bold. Obviously the SBSO with modified disruption strategy performs better than the original SBSO, so it can be proved that the modified disruption strategy takes effect by keeping the found global optima until the end of the run. In addition, the modified disruption strategy can be applied to any BSO algorithm that uses the disruption process to solve MMOPs, and is not limited to the SBSO algorithm here.

Table 4. *PR* and *SR* of original SBSO and SBSO with modified disruption strategy (with accuracy level $\epsilon = 1.0E{-}04$)

Function	SBSO		SBSO-1	
	PR	SR	PR	SR
$f_1(1D)$	0.65	0.3	**1**	1
$f_2(1D)$	0.956	0.78	**1**	1
$f_3(1D)$	**1**	1	**1**	1
$f_4(2D)$	**0.905**	0.64	0.815	0.26
$f_5(2D)$	0.99	0.98	**1**	1
$f_6(2D)$	0.0589	0	**0.1789**	0
$f_6(3D)$	0.0126	0	**0.0390**	0
$f_7(2D)$	0.1922	0	**0.4517**	0
$f_7(3D)$	0.0178	0	**0.0345**	0
$f_8(2D)$	0.5517	0	**0.9533**	0.44
$f_8(3D)$	0.03	0	**0.1583**	0

Comparison of MSBSO with Different Initial Cluster Numbers. In order to solve the problem that the MCM in BSO needs a predefined subpopulation size M, a self-adaptive cluster number control is added to the proposed MSBSO algorithm, so that the cluster number can update automatically as the iteration proceeds. Of course, in the initialization of the run, we should set an initial value for the cluster number n_c. Then in each iteration, n_c will plus 1 or minus 1 according to the change of the average distance among the cluster centers. Different initial values of the cluster number are compared in Table 5.

In general, the experimental results of different initial cluster numbers n_c have little difference. And the trends of PR values for distinct benchmark functions vary slightly, which could result from the various distribution of global optima in different benchmark functions.

Comparison of Algorithms. The SBSO with modified disruption strategy (SBSO-1) and the MSBSO with both modified disruption strategy and self-adaptive cluster number control ($n\_c = 15$) are compared with the other seven algorithms for eight benchmark functions. The experimental results are shown in Table 6.

It can be observed that the MSBSO algorithm performs better than the SBSO-1 algorithm without self-adaptive cluster number control. The MSBSO algorithm and PSO-ring algorithm outperform other algorithms among all the

Table 5. *PR* and *SR* of MSBSO with different n_c (with accuracy level $\epsilon = 1.0E{-}04$)

Function	MSBSO ($n_c = 5$)		MSBSO ($n_c = 10$)		MSBSO ($n_c = 15$)		MSBSO ($n_c = 20$)		MSBSO ($n_c = 30$)	
	PR	SR	PR	SR	PR	SR	PR	SR	PR	SR
$f_1(1D)$	1	1	1	1	1	1	1	1	1	1
$f_2(1D)$	1	1	1	1	1	1	1	1	1	1
$f_3(1D)$	1	1	1	1	1	1	1	1	1	1
$f_4(2D)$	0.995	0.98	0.96	0.84	**1**	1	**1**	1	0.99	0.96
$f_5(2D)$	1	1	1	1	1	1	1	1	1	1
$f_6(2D)$	0.142	0	0.148	0	0.156	0	0.176	0	**0.232**	0
$f_6(3D)$	0.025	0	0.020	0	0.025	0	0.027	0	**0.028**	0
$f_7(2D)$	**0.558**	0	0.458	0	0.474	0	0.434	0	0.427	0
$f_7(3D)$	**0.173**	0	0.139	0	0.124	0	0.118	0	0.108	0
$f_8(2D)$	0.967	0.72	0.97	0.6	**0.977**	0.78	0.975	0.76	0.977	0.72
$f_8(3D)$	0.075	0	0.044	0	0.071	0	0.08	0	**0.135**	0

Table 6. *PR* and *SR* of nine algorithms (with accuracy level $\epsilon = 1.0E{-}04$)

Function	Original BSO PR	Original BSO SR	SBSO-1 PR	SBSO-1 SR	MSBSO PR	MSBSO SR	BSOOS-Gaussian PR	BSOOS-Gaussian SR	BSOOS-Cauchy PR	BSOOS-Cauchy SR
$f_1(1D)$	1	1	**1**	1	**1**	1	1	1	1	1
$f_2(1D)$	1	1	**1**	1	**1**	1	1	1	1	1
$f_3(1D)$	1	1	**1**	1	**1**	1	1	1	1	1
$f_4(2D)$	0.955	0.82	0.815	0.26	**1**	1	0.66	0.06	0.655	0.08
$f_5(2D)$	1	1	**1**	1	**1**	1	1	1	1	1
$f_6(2D)$	0.812	0.02	0.179	0	0.156	0	0.22	0	0.256	0
$f_6(3D)$	0.124	0	0.039	0	0.025	0	0.013	0	0.013	0
$f_7(2D)$	0.227	0	0.452	0	**0.474**	0	0.028	0	0.028	0
$f_7(3D)$	0.026	0	0.035	0	0.124	0	0.005	0	0.005	0
$f_8(2D)$	0.992	0.9	0.953	0.44	0.977	0.78	0.44	0	0.402	0
$f_8(3D)$	0.021	0	0.158	0	0.071	0	0.002	0	0.001	0

Function	PSO-Star PR	PSO-Star SR	PSO-FourCluster PR	PSO-FourCluster SR	PSO-vonNeumann PR	PSO-vonNeumann SR	PSO-Ring PR	PSO-Ring SR
$f_1(1D)$	0.55	0.1	0.87	0.74	0.99	0.98	**1**	1
$f_2(1D)$	0.2	0	0.604	0	0.836	0.34	0.98	0.92
$f_3(1D)$	1	1	1	1	1	1	**1**	1
$f_4(2D)$	0.25	0	0.65	0.1	0.715	0.16	0.96	0.82
$f_5(2D)$	0.5	0	0.94	0.88	0.98	0.96	**1**	1
$f_6(2D)$	0.056	0	0.199	0	0.813	0	**0.98**	0.66
$f_6(3D)$	0.012	0	0.049	0	0.470	0	**0.750**	0
$f_7(2D)$	0.029	0	0.093	0	0.203	0	0.427	0
$f_7(3D)$	0.005	0	0.017	0	0.082	0	**0.192**	0
$f_8(2D)$	0.083	0	0.292	0	0.928	0.36	**0.998**	0.98
$f_8(3D)$	0.021	0	0.075	0	0.181	0	**0.498**	0

test variants. And MSBSO can completely find all global optima of the five functions $f_1 - f_5$, which is the best across all the test algorithms. Nevertheless, PSO-ring shows better performance for the problems with high dimensions because of the large number of neighborhoods. While PSO-star is a global topology, so all the solutions will converge to one optimum at the end of each run. Thus, from PSO-star to PSO-FourCluster to PSO-vonNeumann to PSO-ring, the size of the neighborhood is getting smaller and the number of neighborhoods is increasing, resulting in more effectiveness for MMOPs.

Based on the experimental results, we can summarize that MSBSO can perform well for problems with low dimension, but for high dimension problems, the global search ability of MSBSO still needs to be enhanced.

5 Conclusion

In this paper, we develop a BSO algorithm with the modified disruption strategy and a self-adaptive cluster number control for solving multimodal optimization problems. The modified disruption strategy is used to maintain the found global optima during the search process to the end of the run. And the self-adaptive cluster number control improves the max-fitness clustering method with no need for a predefined parameter M. The performance of the new algorithm is validated on eight benchmark functions, compared with other algorithms. The experimental results show that the proposed MSBSO algorithm can outperform other algorithms for low dimension problems, but it performs worse than PSO-ring algorithm for high dimension problems, so its global search ability needs to be enhanced.

For future work, MSBSO may need to be combined with some global search strategies to solve more multimodal problems with high dimensionality and constraints. Moreover, actually I also did some experiments to track the change of the cluster number n_c as the iteration proceeds. I find that in some runs, the cluster number n_c eventually changes to the actual number of global optima for the corresponding benchmark function, while in other runs, n_c changes to a number which is far from the actual value. But it still provides a direction: via some methods, we can not only find the optimal solutions, but also know the approximate number of global optima to the multimodal optimization problem.

Acknowledgments. This work is supported by the Guangdong-Hong Kong Joint Innovation Platform of Big Data and Computational Intelligence under Grant 2018B050502006.

References

1. Yu, X., Gen, M.: Multimodal optimization. In: Introduction to Evolutionary Algorithms. DECENGIN, vol. 0, pp. 165–191. Springer, London (2010). https://doi.org/10.1007/978-1-84996-129-5_5
2. Bessaou, M., Pétrowski, A., Siarry, P.: Island model cooperating with speciation for multimodal optimization. In: Schoenauer, M., et al. (eds.) PPSN 2000. LNCS, vol. 1917, pp. 437–446. Springer, Heidelberg (2000). https://doi.org/10.1007/3-540-45356-3_43

3. Bošković, B., Brest, J.: Clustering and differential evolution for multimodal optimization. In: 2017 IEEE Congress on Evolutionary Computation (CEC), pp. 698–705, June 2017. https://doi.org/10.1109/CEC.2017.7969378
4. Bratton, D., Kennedy, J.: Defining a standard for particle swarm optimization. In: 2007 IEEE Swarm Intelligence Symposium, pp. 120–127, April 2007. https://doi.org/10.1109/SIS.2007.368035
5. Cheng, S., Chen, J., Lei, X., Shi, Y.: Locating multiple optima via brain storm optimization algorithms. IEEE Access **6**, 17039–17049 (2018). https://doi.org/10.1109/ACCESS.2018.2811542
6. Cheng, S., Lu, H., Song, W., Chen, J., Shi, Y.: Dynamic multimodal optimization using brain storm optimization algorithms. In: Qiao, J., et al. (eds.) BIC-TA 2018. CCIS, vol. 951, pp. 236–245. Springer, Singapore (2018). https://doi.org/10.1007/978-981-13-2826-8_21
7. Cheng, S., Qin, Q., Chen, J., Shi, Y.: Brain storm optimization algorithm: a review. Artif. Intell. Rev. **46**(4), 445–458 (2016). https://doi.org/10.1007/s10462-016-9471-0
8. Cheng, S., Qin, Q., Chen, J., Wang, G.G., Shi, Y.: Brain storm optimization in objective space algorithm for multimodal optimization problems. In: Tan, Y., Shi, Y., Niu, B. (eds.) ICSI 2016. LNCS, vol. 9712, pp. 469–478. Springer, Cham (2016). https://doi.org/10.1007/978-3-319-41000-5_47
9. Della Cioppa, A., De Stefano, C., Marcelli, A.: Where are the niches? Dynamic fitness sharing. IEEE Trans. Evol. Comput. **11**(4), 453–465 (2007)
10. Eberhart, R.C., Shi, Y.: Comparing inertia weights and constriction factors in particle swarm optimization. In: Proceedings of the 2000 Congress on Evolutionary Computation. CEC00 (Cat. No.00TH8512), vol. 1, pp. 84–88, July 2000. https://doi.org/10.1109/CEC.2000.870279
11. Gao, W., Yen, G.G., Liu, S.: A cluster-based differential evolution with self-adaptive strategy for multimodal optimization. IEEE Trans. Cybern. **44**(8), 1314–1327 (2014). https://doi.org/10.1109/TCYB.2013.2282491
12. Goldberg, D.E., Richardson, J.: Genetic algorithms with sharing for multimodal function optimization. In: Proceedings of the Second International Conference on Genetic Algorithms on Genetic Algorithms and Their Application, pp. 41–49. L. Erlbaum Associates Inc., Hillsdale (1987). http://dl.acm.org/citation.cfm?id=42512.42519
13. Guo, X., Wu, Y., Xie, L.: Modified brain storm optimization algorithm for multimodal optimization. In: Tan, Y., Shi, Y., Coello, C.A.C. (eds.) ICSI 2014. LNCS, vol. 8795, pp. 340–351. Springer, Cham (2014). https://doi.org/10.1007/978-3-319-11897-0_40
14. Harik, G.: Finding multimodal solutions using restricted tournament selection. In: Proceedings of the Sixth International Conference on Genetic Algorithms, pp. 24–31. Morgan Kaufmann (1995)
15. Huang, T., Zhan, Z., Jia, X., Yuan, H., Jiang, J., Zhang, J.: Niching community based differential evolution for multimodal optimization problems. In: 2017 IEEE Symposium Series on Computational Intelligence (SSCI), pp. 1–8, November 2017. https://doi.org/10.1109/SSCI.2017.8280801
16. Li, J.P., Balazs, M., Parks, G., Clarkson, P.: A species conserving genetic algorithm for multimodal function optimization. Evol. Comput. **10**, 207–234 (2002). https://doi.org/10.1162/106365602760234081
17. Li, X., Epitropakis, M.G., Deb, K., Engelbrecht, A.: Seeking multiple solutions: an updated survey on niching methods and their applications. IEEE Trans. Evol. Comput. **21**(4), 518–538 (2017). https://doi.org/10.1109/TEVC.2016.2638437

18. Li, X., Engelbrecht, A., Epitropakis, M.G.: Benchmark functions for CEC'2013 special session and competition on niching methods for multimodal function optimization (2013). http://goanna.cs.rmit.edu.au/~xiaodong/cec13-niching/competition/
19. Mehmood, Y., Aziz, N., Riaz, F., Iqbal, H., Shahzad, W.: PSO-based clustering techniques to solve multimodal optimization problems: a survey. In: 2018 1st International Conference on Power, Energy and Smart Grid (ICPESG), pp. 1–6, April 2018. https://doi.org/10.1109/ICPESG.2018.8417315
20. Passaro, A., Starita, A.: Particle swarm optimization for multimodal functions: a clustering approach. J. Artif. Evol. Appl. **2008**, 8 (2008). https://doi.org/10.1155/2008/482032
21. Peng, H., Deng, C., Wu, Z.: SPBSO: self-adaptive brain storm optimization algorithm with pbest guided step-size. J. Intell. Fuzzy Syst. **36**, 5423–5434 (2019). https://doi.org/10.3233/JIFS-181310
22. Petrowski, A.: A clearing procedure as a niching method for genetic algorithms. In: Proceedings of IEEE International Conference on Evolutionary Computation, pp. 798–803, May 1996. https://doi.org/10.1109/ICEC.1996.542703
23. Qu, B.Y., Suganthan, P.N., Liang, J.J.: Differential evolution with neighborhood mutation for multimodal optimization. IEEE Trans. Evol. Comput. **16**(5), 601–614 (2012). https://doi.org/10.1109/TEVC.2011.2161873
24. Shi, Y.: Brain storm optimization algorithm. In: Tan, Y., Shi, Y., Chai, Y., Wang, G. (eds.) ICSI 2011. LNCS, vol. 6728, pp. 303–309. Springer, Heidelberg (2011). https://doi.org/10.1007/978-3-642-21515-5_36
25. Shi, Y.: An optimization algorithm based on brainstorming process. Int. J. Swarm Intell. Res. (IJSIR) **2**(4), 35–62 (2011). https://doi.org/10.4018/ijsir.2011100103
26. Shi, Y.: Developmental swarm intelligence: developmental learning perspective of swarm intelligence algorithms. Int. J. Swarm Intell. Res. (IJSIR) **5**(1), 36–54 (2014). https://doi.org/10.4018/ijsir.2014010102
27. Thomsen, R.: Multimodal optimization using crowding-based differential evolution. In: Proceedings of the 2004 Congress on Evolutionary Computation (IEEE Cat. No.04TH8753), vol. 2, pp. 1382–1389, June 2004. https://doi.org/10.1109/CEC.2004.1331058
28. Wang, X., et al.: A multilevel sampling strategy based memetic differential evolution for multimodal optimization. Neurocomputing **334**, 79–88 (2019). https://doi.org/10.1016/j.neucom.2019.01.006
29. Yin, X., Germay, N.: A fast genetic algorithm with sharing scheme using cluster analysis methods in multimodal function optimization. In: Albrecht, R.F., Reeves, C.R., Steele, N.C. (eds.) Artificial Neural Nets and Genetic Algorithms, pp. 450–457. Springer, Vienna (1993). https://doi.org/10.1007/978-3-7091-7533-0_65
30. Yu, Y., Wu, L., Yu, H., Li, S., Wang, S., Gao, S.: Brain storm optimization with adaptive search radius for optimization. In: 2017 International Conference on Progress in Informatics and Computing (PIC), pp. 394–398, December 2017. https://doi.org/10.1109/PIC.2017.8359579
31. Zhan, Z., Wang, Z., Lin, Y., Zhang, J.: Adaptive radius species based particle swarm optimization for multimodal optimization problems. In: 2016 IEEE Congress on Evolutionary Computation (CEC), pp. 2043–2048, July 2016. https://doi.org/10.1109/CEC.2016.7744039

Recent Bio-inspired Algorithms for Solving Flexible Job Shop Scheduling Problem: A Comparative Study

Dongsheng Yang[1], Xianyu Zhou[1], Zhile Yang[2(✉)], and Yanhui Zhang[2(✉)]

[1] Intelligent Electrical Science and Technology Research Institute, Northeastern University, Shenyang 110819, China

[2] Shenzhen Institute of Advanced Technology, Chinese Academy of Sciences, Shenzhen 518000, Guangdong, China

{zl.yang,yh.zhang}@siat.ac.cn

Abstract. Flexible job shop scheduling problem (FJSP) is an extended formulation of the classical job shop scheduling problem, endowing great significance in the modern manufacturing system. The FJSP defines an operation that can be processed by any machine from a given set, which is a strong constrained NP-hard problem and intractable to be solved. In this paper, three recent proposed meta-heuristic optimization algorithms have been employed in solving the FJSP aiming to minimize the makespan, including moth-flame optimization (MFO), teaching-learning-based optimization (TLBO) and Rao-2 algorithm. Two featured FJSP cases are carried out and compared to evaluate the effectiveness and efficiency of the three algorithms, also associated with other classical algorithm counterparts. Numerical studies results demonstrate that the three algorithms can achieve significant improvement for solving FJSP, and MFO method appears to be the most competitive solver for the given cases.

Keywords: Flexible job shop scheduling · Makespan · MFO · TLBO · Rao-2 algorithm

1 Introduction

The key component of production management for modern enterprises is effective production planning and scheduling. It is of great significance for reducing production costs, shortening production cycle and improving production efficiency. The job shop scheduling problem has the major characteristics including modeling complexity, computer complexity, dynamic randomness, multiple constraints, and multi-objectiveness.

The production planning and scheduling problem is to arrange the production tasks delivered on the equipment according to the sequence. It discusses how to arrange the processing resources and sequence of the operations under the premise of satisfying the processing constraints, aiming to minimize the product

L. Pan et al. (Eds.): BIC-TA 2019, CCIS 1159, pp. 398–407, 2020.
https://doi.org/10.1007/978-981-15-3425-6_31

manufacturing time and the consumption cost. Due to the complexity of the production operation management system and various real-world constraints, the production scheduling problem becomes a NP-hard problem [6]. The classical job shop scheduling problem (JSP) assumes that there is no flexibility of the resources (including machines and tools) for each operation of the corresponding job. In another word, the problem requires that one machine only processes one type of the operation. However, in the real world application, many flexible manufacturing systems are used to improve the production efficiency in modern manufacturing enterprises [14].

In the light of this, the FJSP attracts increasing attentions from both research and industrial areas [2]. The FJSP can be divided into two sub-problems: the machine selection (MS) and the operations sequencing (OS), adding a more complicated scenario MS to the conventional JSP problem. Many methods have been proposed to solve this problem so far, including exact algorithm [3], dispatching rules [1], evolutionary algorithm (EA) [18], local search algorithms [7] and so on. For exact algorithm, Torabi et al. proposed a mixed integer nonlinear program for deterministic FJSP [16]. Roshanaei et al. presented two MILP models [13]. However, exact algorithm cannot obtain good results for large scale FJSP. So Tay and Ho used dispatching rules for multi-objective FJSP [15]. Ziaee proposed a construction procedure based heuristic for FJSP [19]. As for EA, the most used method is genetic algorithm, Pezzella et al. integrated GA with different strategies to solve FJSP [10]. Driss et al. proposed a new chromosome representation method and some novel crossover and mutation strategies for FJSP [4]. However, EA is lack of local search ability. So local search method is used in FJSP and tabu search (TS) is the most effective method for FJSP. Vilcot and Billaut used TS for the objective of minimum makespan and maximum lateness [17]. Jia and Hu proposed a TS based pathrelinking algorithm for multi-objective FJSP [7].

The remainder of this paper is structured as follows: the problem formulation of FJSP is discussed in Sect. 2, followed by the three compared algorithms that are briefly introduced in Sect. 3, where the encoding and decoding method are also given. Experimental studies and the corresponding discussion are reported in Sect. 4. Finally, Sect. 5 concludes the paper.

2 Problem Formulation

The formulation of FJSP is described as follows: Assuming that there are M machines in the workshop, which need to process N jobs within the overall time period. Each job consists of a series of operations that allows them to be processed in a set of available machines. In this paper, the objective function is to minimize the maximal completion time, i.e. makespan (C_{max}), which is denoted as follows:
Notations for the formulation:

i number of jobs
j operations of the jobs
k number of machines

B_{ijk} starting time of operation j of job i on machine k
P_{ijk} processing time of operation j of job i on machine k
F_{ijk} completion time of operation j of job i on machine k
F_i completion time of job i

Objective function: Minimize C_{max}
Constraints:

- Jobs are independent and preemption or cancellation of jobs is not permitted.

$$B_{ijk} + P_{ijk} = F_{ijk} \tag{1}$$

$$\sum_k B_{ijk} \geq \sum_k F_{i(j-1)k} \tag{2}$$

- Every machine can only process one job at a time.

$$B_{ik} + P_{ik} \leq B_{(i+1)k} \quad \forall j \tag{3}$$

- One operation of each job can be processed by only one machine at a time.

$$B_{ij} + P_{ij} \leq B_{i(j+1)} \quad \forall k \tag{4}$$

- All jobs and machines are available at time zero and the transmission time between machines is ignored.

$$B_{ijk} \geq 0 \tag{5}$$

$$F_{ijk} \geq 0 \tag{6}$$

- Processing time is deterministic and includes other elements of set-up, transportation and inspection. the makespan is the maximal completion time of all jobs.

$$F_i \geq F_{ij} \quad \forall j \tag{7}$$

$$C_{max} \geq F_i \quad \forall i \tag{8}$$

3 Three Algorithms for FJSP

3.1 Algorithm Introduction

Moth-Flame Optimization Algorithm. Moth-flame optimization algorithm is a modern intelligent bio-inspired optimization algorithm proposed in 2015 [8]. It mimics the navigation mechanism of the moth in the space during the flight. Several advantages could be found in the algorithm design. First of all, due to that several flames is formulated and moths are considered flying around the individual flame and do not interfere with each other, the parallel optimization ability of the algorithm is strong. In addition, due to the spiral wrap path of the moth is assumed in the algorithm, as the number of iterations increases, it gradually approaches the contemporary flame center with a certain random amount. Such scheme avoids the whole population to be easily falling into the local optimal solution, and therefore guarantees the global optimal solution of the algorithm with excellent search performance and robustness.

The position update mechanism of each moth relative to the flame can be expressed by an equation:

$$M_i = S(M_i, F_j) \tag{9}$$

where M_i represents the ith moth, F_j represents the jth flame, and S represents the spiral function. The spiral function of the moth flight path is defined as follows:

$$S(M_i, F_j) = D_i e^{bt} \cos(2\pi t) + F_j \tag{10}$$

where D_i represents the linear distance between the ith moth and the jth flame, b is the defined logarithmic spiral shape constant, and the path coefficient t is a random number in $[-1, 1]$. The expression of D_i is as follows:

$$D_i = |F_j - M_i| \tag{11}$$

In order to reduce the number of flames in the iterative process and balance the global search ability and local development ability of the algorithm in the search space, an adaptive mechanism for the number of flames is proposed. The formula is as follows:

$$flame.number = round(N - G * \frac{N-1}{Gm}) \tag{12}$$

where G is the current number of iterations, N is the initial maximum number of flames, and Gm is the maximum number of iterations. Due to the reduction in flames, the moths corresponding to the reduced flames in each generation update their position based on the flame with the worst fitness value.

Teaching-Learning-Based Optimization Algorithm. The optimization algorithm based on 'teaching and learning' simulates the interaction of student-teacher in a class, which is a group intelligent optimization algorithm proposed in 2011 [12]. The improvement of the grades of students in the class requires the teacher's 'teaching'. In addition, the students need to 'learn' to promote the absorption of knowledge. Among them, teachers and students are both individuals in the evolutionary algorithms, and the teacher is the best individual in each iteration. The following formula shows the process of 'teaching':

$$X_i^{new} = X_i^{old} + difference \tag{13}$$

$$difference = r_i * (X_{best} - F_i * Popmean) \tag{14}$$

where X_i^{old} and X_i^{new} represent the values of the ith student before and after learning; X_{best} is the student who gets the best grades, e.g. the teacher; $Popmean = \frac{1}{N}\sum_{i=1}^{N}(X_i)$ is the average value of all students; N is the number of students; the teaching factor $F_i = round[1 + rand(0,1)]$ and the learning step

$r_i = rand(0, 1)$. After the 'teaching' phase is completed, the students are updated according to their grades e.g. fitness value. In the process of 'learning', for each student X_i, a learning object $X_j (j \neq i)$ in the class is randomly selected, and X_i adjusts himself by analyzing the difference between himself and the student X_j, with the formula is as follows:

$$X_i^{new} = \begin{cases} X_i^{old} + r_i * (X_i - X_j) & f(X_j) < f(X_i) \\ X_i^{old} + r_i * (X_j - X_i) & f(X_i) < f(X_j) \end{cases} \tag{15}$$

Also the students are updated according to their fitness values.

Rao-2 Algorithm. Rao-2 algorithm is a simple metaphor-less optimization algorithm proposed in 2019 for solving the unconstrained and constrained optimization problems [11]. The algorithm is based on the best and worst solutions obtained during the optimization process and the random interactions between the candidate solutions. The algorithm requires only common control parameters like population size and number of iterations and does not require any algorithm-specific control parameters. The individuals are updated according to the following formula:

$$\begin{aligned} X_{j,k,i}^{new} &= X_{j,k,i} + r_{1,j,i} * (X_{j,best,i} - X_{j,worst,i}) \\ &+ r_{2,j,i} * (|X_{j,k,i} or X_{j,h,i}| - |X_{j,h,i} or X_{j,k,i}|) \end{aligned} \tag{16}$$

where $X_{j,best,i}$ and $X_{j,worst,i}$ are the value of the variable j for the best and worst candidate during the ith iteration. $X_{j,k,i}^{new}$ is the updated value of $X_{j,k,i}$. $r_{1,j,i}$ and $r_{2,j,i}$ are the two random numbers for the jth variable in the range [0, 1]. $X_{j,k,i}$ and $X_{j,h,i}$ are the candidate solution k and any randomly picked candidate solution h. If the fitness value of the kth solution is better than that of the hth solution, the term '$X_{j,k,i} or X_{j,h,i}$' becomes $X_{j,k,i}$. And if the fitness value of the hth solution is better than that of the kth solution, the term '$X_{j,h,i} or X_{j,k,i}$' becomes $X_{j,k,i}$.

Then update the individuals according to their fitness values.

3.2 Encoding and Decoding

The method of encoding is shown as follows:

- The individuals are corresponding to the solutions of the FJSP, where each individual is a matrix of m rows and n columns, m is the number of the jobs, n is the number of operations for each job.
- Each element in the individual represents the machine used in the corresponding process.
- Initialize the individuals by selecting from the alternative machines randomly.
- During the iterations of the algorithm, each row of the individual is treated as a variable.

Table 1. The parameters settings of optimization algorithms and benchmark

Parameters	Value
Size of the population	400
Total number of generations	200
b of MFO	1
t of MFO	$(-1 - \frac{G}{Gm} - 1) * rand + 1$
Scale factor of DE	0.7
Crossover constant of DE	0.9
Inerita weight of PSO	$0.9 - 0.4 * \frac{G}{Gm}$

The procedure of decoding is as follows:

Step 1: Obtain the best individual which has the information on which machine to use for each operation of each job.
Step 2: Determine the allowable starting time for each operation which satisfies all the constrains mentioned above. Specifically, the starting time is the completion time of previous job processed on the same machine, but if the completion time of the last operation for the same job is longer than the completion time of previous job processed on the same machine, the starting time should be the completion time of the last operation.
Step 3: Determine the completion time for each operation, which should be the sum of starting time and processing time.
Step 4: Use the set of starting and completion time to paint the Gantt chart. The ith occurrences of the job number in the square represents the ith operation of the job.

4 Experimental Results and Discussions

The objective of this paper is to minimize the maximal completion time. The comparisons among the three algorithms and other algorithms are provided to compare the optimization performance. These algorithms are compared on one medium and one large size FJSP (MFJS01 and MFJS10). MFSJ01 represents that this problem has 5 jobs with 3 operations and 6 machines, MFSJ10 represents that this problem has 12 jobs with 4 operations and 8 machines. The algorithm terminates when the number of iteration reaches to the maximum generation value. The parameters for the two experiments are shown in Table 1.

The data of the experiment is adopted from literature [5]. Table 2 shows the experimental results and comparisons of these algorithms. To eliminate the randomness, 10 independent run is implemented for each problem. 'Best' represents the minimum value of makespan, and 'Mean' represents the average value of makespan. The results of HSA/SA, HSA/TS, HTS/TS, HTS/SA, ISA and ITS are adopted from literature [5] and [9]. The results with * are the best result for the given problem among these algorithms.

Table 2. The statistical results obtained by algorithms

Algorithm	MFSJ01 best	MFSJ01 mean	MFSJ10 best	MFSJ10 mean
MFO	469*	469*	1294*	1340.8*
TLBO	469*	469*	1368	1384.9
Rao-2	469*	469*	1445	1491.1
GWO	469*	469.8	1507	1553.6
DE	469*	469*	1430	1466.8
PSO	469*	469.8	1500	1549.7
HSA/SA [5]	479	503.2	1538	1621.8
HSA/TS [5]	491	504	1615	1693.4
HTS/TS [5]	469*	502.8	1404	1511.8
HTS/SA [5]	469*	499	1384	1428.2
ISA [9]	488	517.8	1546	1733.3
ITS [9]	548	584.2	1737	1737

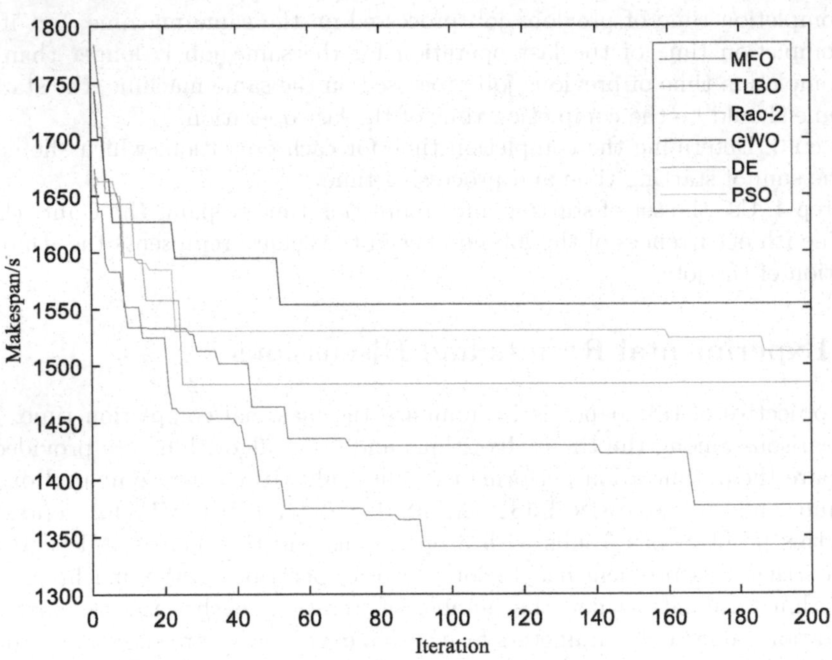

Fig. 1. Convergence results of the makespan for all the compared algorithms

It could be found in the Table 2 that MFO obtains all the best results for the two problems. Although other 7 algorithms can obtain the same results for problem MFSJ01 by 469, the average value may not as good as MFO. MFO also

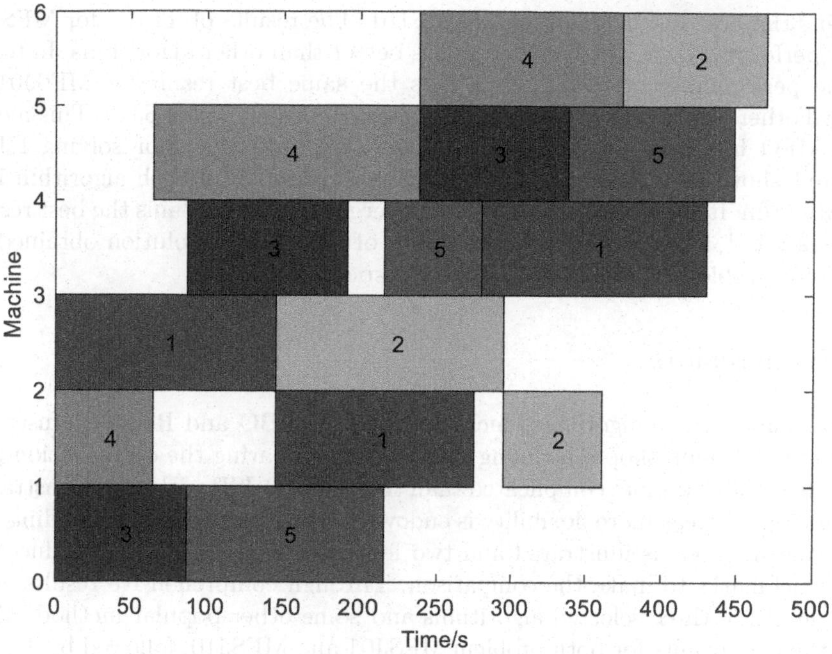

Fig. 2. Gantt chart of problem MFSJ01 of MFO

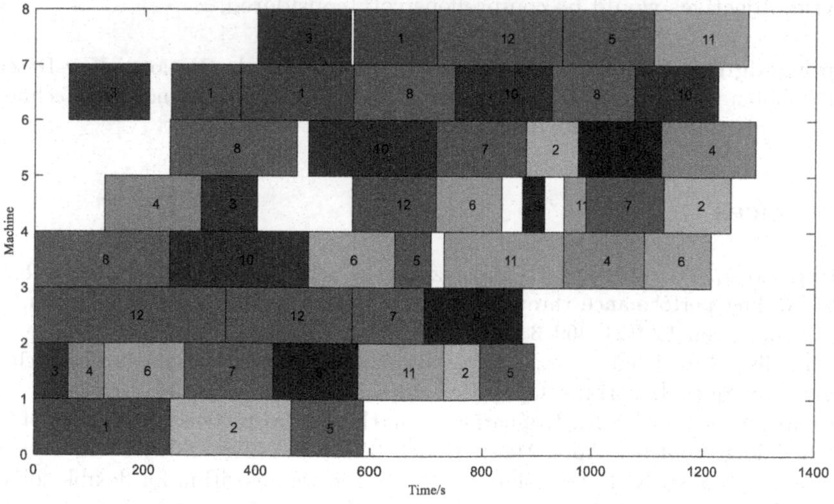

Fig. 3. Gantt chart of problem MFSJ10 of MFO

obtains the best result for problem MFSJ10. The results of TLBO for MFSJ10 is outperformed by MFO, but its result is better than other algorithms. In terms of the performance of Rao-2, it obtains the same best result for MFSJ01 as several other algorithms, it does not perform well in solving MFSJ10. This means that MFO has both good effectiveness and high efficiency for solving FJSP. Figure 1 shows the convergence results of the makespan of each algorithm in a featured run. It again shows that MFO converges fast and obtains the best result. Figures 2 and 3 illustrate the Gantt charts of the optimal solution obtained by MFO for problem MFSJ01 and MFSJ10 respectively.

5 Conclusion

In this paper, three algorithms including MFO, TLBO and Rao-2 are used for solving flexible job shop scheduling problem in comparing the optimization performance. FJSP is more complicated than the classical JSP with more constraints considered whereas more flexibility is endowed. The corresponding encoding and decoding methods is illustrated and two featured scales FJSP is introduced as the benchmarks to make the comparison. Through comprehensive results comparison of the three selected algorithms and some other popular method, MFO gets the best results for both problem MFSJ01 and MFSJ10, followed by TLBO. Rao-2 get the best result for problem MFSJ01 while performs pool in the larger scale problem MFSJ10. Future research will be addressing the further improvement for the algorithms variants and constraint handling methods, and more realistic objectives would be comprehensively considered.

Acknowledgment. This research work is supported by the National Key Research and Development Project under Grant 2018YFB1700500, and Science and Technology Project of Shenzhen (JSGG20170823140127645).

References

1. Baykasoğlu, A., Özbakır, L.: Analyzing the effect of dispatching rules on the scheduling performance through grammar based flexible scheduling system. Int. J. Prod. Econ **124**(2), 369–381 (2010)
2. Chaudhry, I.A., Khan, A.A.: A research survey: review of flexible job shop scheduling techniques. Int. Trans. Oper. Res. **23**(3), 551–591 (2015)
3. Demir, Y., İşleyen, S.K.: Evaluation of mathematical models for flexible job-shop scheduling problems. Appl. Math. Model. **37**(3), 977–988 (2013)
4. Driss, I., Mouss, K.N., Laggoun, A.: A new genetic algorithm for flexible job-shop scheduling problems. J. Mech. Sci. Technol. **29**(3), 1273–1281 (2015). https://doi.org/10.1007/s12206-015-0242-7
5. Fattahi, P., Mehrabad, M.S., Jolai, F.: Mathematical modeling and heuristic approaches to flexible job shop scheduling problems. J. Intell. Manuf. **18**(3), 331–342 (2007). https://doi.org/10.1007/s10845-007-0026-8
6. Garey, M.R., Johnson, D.S., Sethi, R.: The complexity of flowshop and jobshop scheduling. Math. Oper. Res. **1**(2), 117–129 (1976)

7. Jia, S., Hu, Z.H.: Path-relinking tabu search for the multi-objective flexible job shop scheduling problem. Comput. Oper. Res. **47**, 11–26 (2014)
8. Mirjalili, S.: Moth-flame optimization algorithm: a novel nature-inspired heuristic paradigm. Knowl.-Based Syst. **89**, 228–249 (2015)
9. Özgüven, C., Özbakır, L., Yavuz, Y.: Mathematical models for job-shop scheduling problems with routing and process plan flexibility. Appl. Math. Model. **34**(6), 1539–1548 (2010)
10. Pezzella, F., Morganti, G., Ciaschetti, G.: A genetic algorithm for the flexible job-shop scheduling problem. Comput. Oper. Res. **35**(10), 3202–3212 (2008)
11. Rao, R.: Rao algorithms: three metaphor-less simple algorithms for solving optimization problems. Int. J. Ind. Eng. Comput. **11**(1), 107–130 (2020)
12. Rao, R.V., Savsani, V.J., Vakharia, D.P.: Teaching–learning-based optimization: a novel method for constrained mechanical design optimization problems. Comput.-Aided Des. **43**(3), 303–315 (2011)
13. Roshanaei, V., Azab, A., Elmaraghy, H.: Mathematical modelling and a meta-heuristic for flexible job shop scheduling. Int. J. Prod. Res. **51**(20), 6247–6274 (2013)
14. Seebacher, G., Winkler, H.: Evaluating flexibility in discrete manufacturing based on performance and efficiency. Int. J. Prod. Econ. **153**(4), 340–351 (2014)
15. Tay, J.C., Ho, N.B.: Evolving dispatching rules using genetic programming for solving multi-objective flexible job-shop problems. Comput. Ind. Eng. **54**(3), 453–473 (2008)
16. Torabi, S.A., Karimi, B., Ghomi, S.M.T.F.: The common cycle economic lot scheduling in flexible job shops: The finite horizon case. Int. J. Prod. Econ. **97**(1), 52–65 (2005)
17. Vilcot, G., Billaut, J.C.: A tabu search algorithm for solving a multicriteria flexible job shop scheduling problem. Int. J. Prod. Res. **49**(23), 6963–6980 (2011)
18. Yuan, Y., Xu, H.: Multiobjective flexible job shop scheduling using memetic algorithms. IEEE Trans. Autom. Sci. Eng. **12**(1), 336–353 (2013)
19. Ziaee, M.: A heuristic algorithm for solving flexible job shop scheduling problem. The International Journal of Advanced Manufacturing Technology **71**(1–4), 519–528 (2013). https://doi.org/10.1007/s00170-013-5510-z

Unidirectional Cyclic Network Architecture for Distributed Evolution

Jingsong He(✉)

School of Microelectronics,
University of Science and Technology of China, Hefei, China
hjss@ustc.edu.cn

Abstract. How to distribute individuals in a population spatially is the most important issue in distributed evolutionary computation. In this paper, a novel distributed evolutionary model named UC model is proposed, in which the network architecture is circular, and the information propagation is unidirectional other than bidirectional. The proposed model is essentially a neighborhood model, but the communication mode between individuals is quite different, and the number of neighborhoods available is more free. As a specific algorithm for model implementation, differential evolution (DE) algorithm is incorporated into the UC model, and a distributed evolutionary algorithm called *uc*DE is formed. In the experiment, the benchmark functions of CEC'05 are used to test the performance. Experimental results show that the proposed model is a competitive and promising one.

Keywords: Evolutionary algorithms · Distributed evolutionary computation · Differential evolution

1 Introduction

Evolutionary algorithms (EAs) are optimal search method suitable for distributed parallel computing. In the past decades, many distributed evolutionary computation methods have been studied and proposed [1,6,8–12,14,17]. The typical distributed models include master-slave model, fine-grained neighborhood model (or cell model), coarse-grained island model and pool structure model. Distributed evolutionary computation can not only share the computing load, but also improve the convergence performance of evolutionary computation.

Distributed evolutionary algorithm differs from evolutionary algorithm in that it divides the population spatially and isolates direct communication between some individuals. From the architectures of the typical distributed evolutionary models at present, it can be known that the determination of the architecture of a distributed evolutionary model depends on the way in which the model divides the spatial location of individuals. The division of spatial position of individuals in a population determines whether individuals can communicate directly with each other.

J. He—This work was supported by the National Natural Science Foundation of China through Grant No. 61273315.

L. Pan et al. (Eds.): BIC-TA 2019, CCIS 1159, pp. 408–421, 2020.
https://doi.org/10.1007/978-981-15-3425-6_32

Looking at the architectures of the distributed evolutionary models, it is not difficult to find the fact that in the existing models, information exchange between two individuals who can communicate directly is bidirectional. That is, if individual A can get information from individual B, then individual B can also get information from individual A.

This paper discusses a novel distributed evolutionary model in which information propagates in a single direction. In the proposed model, the information transfer between two individuals who can communicate directly is unidirectional. That is, if individual A can get information from individual B, then individual B can not get information directly from individual A. The author proposes this model in the hope that this work will be helpful to understand and enrich the distributed evolutionary model from different perspectives.

2 The Architecture of Evolutionary Models from the Perspective of Communication

Distributed evolutionary computation has two typical models that fine-grained neighborhood model (also known as cell model) and coarse-grained island model. In order to clearly compare the communication modes between individuals in these two distributed evolutionary models, the architecture of communication between individuals in evolutionary algorithm is also given here, as shown in Fig. 1(1). The architecture of the cell model is shown in Fig. 1(2), and the architecture of the island model is shown in Fig. 2.

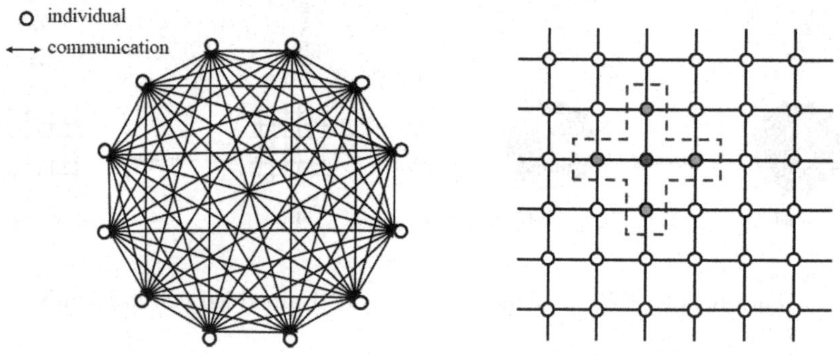

(1) Communication network of EAs (2) Communication network of cell model

Fig. 1. Communication network of EAs and cell model. (Color figure online)

Simply put, if there is a possibility of direct communication between two individuals, then a connecting line is used to connect the two individuals. Thus, the communication topology of the evolutionary algorithm can be obtained, as shown in Fig. 1(1).

In terms of the cell model, it is fine-grained and spatially structured. It has only one population, but individuals are arranged on the grid. The cell model implements interaction between individuals through communication defined by network topology. Everyone can only compete and mate in their own community. Good individuals can spread to the whole population through their neighbors. The cell model in Fig. 1(2). shows four neighbors (individuals labeled green) of an individual labeled red.

In terms of the island model, it is a kind of model whose spatial distribution is circular, and its topological structure is shown in Fig. 2. The island model is coarse-grained, in which the global population is divided into several sub-populations, each of which is processed by a processor. When some individuals on one island migrate to another at regular intervals, their communication takes place. Information exchange occurs between individuals migrating from the source island and individuals on the target island. The migration of individuals on the source islands leads to the migration of information. The migration mechanism includes the frequency and scope of migration, the selection strategy of the source island and the replacement strategy of the target island, which also determines the exchange rate of information between the source island and the target island.

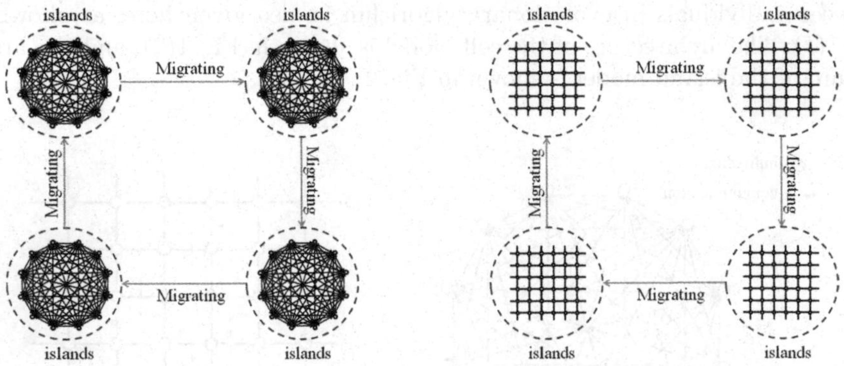

(1)Embedded EAs in islands (2)Embedded cell in islands

Fig. 2. Communication network of the island model.

From the above-mentioned typical distributed evolutionary models and the topological structure of communication between individuals, it can be clearly seen that the information transmission of any existing distributed evolutionary model is bidirectional.

3 Unidirectional Cyclic Network Architecture

The topological structure of the network of the distributed evolutionary model proposed in this paper is illustrated in Fig. 3, where the direction of arrow indicates the direction of information dissemination. In Fig. 3, the ith array (or subpopulation) is represented by P_i, $i = 1, 2, \cdots, N$. Each array contains M individuals. That is to say, the number of individuals in the global population is $N \times M$. All arrays are circled in ring shape. The transmission of information in the network moves in a single direction, from one of the arrays to the next, but the information is not transmitted back. There is no communication between individuals within the same array. For convenience, the distributed evolutionary model of unidirectional cyclic network shown in Fig. 3 is called UC model.

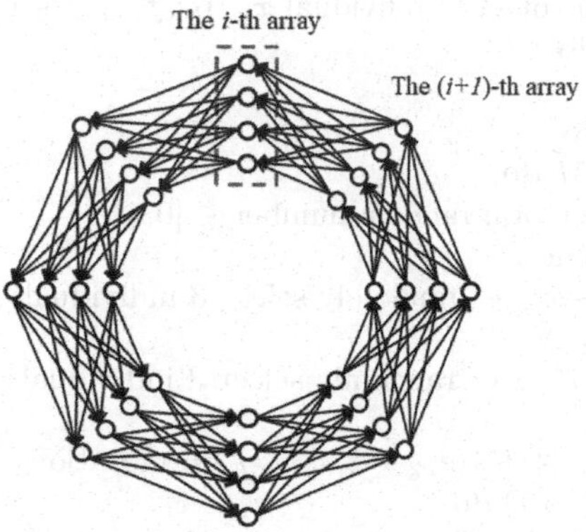

Fig. 3. The unidirectional cyclic network architecture.

The UC model is essentially a neighborhood model. In the cell model, information exchange between neighbor individuals is bidirectional, while information transmission between neighbor individuals in the UC model is unidirectional.

Another difference between the UC model and the cell model shown in Fig. 1 is that the individual in the cell model is grid, while the individual in the UC model is arranged in ring shape. In the UC model, individuals in the ith array (or sub-population) P_i can only get information from P_{i+1} array, that is to say, individuals in P_{i+1} array are neighbors of individuals in P_i. This means that individuals in P_i have common neighbors, and the number of individuals in P_{i+1} is the number of neighbors of individuals in P_i.

Compared with the cell model with grid distribution, the number of neighbors an individual can have in the UC model is arbitrary positive integer, while in the cell model, the number of neighbors an individual can have is limited. The UC model has more choices in the number of neighbors.

4 Differential Evolution Based on the UC Model

Differential evolution (DE) [19] is a well-known simple and effective numerical optimization algorithm. The crossover operation of DE algorithm is characterized by only one offspring. Because of this feature, DE can easily be incorporated into the UC model. For convenience, the algorithm that combines DE with the UC model is called *uc*DE. The pseudocode of *uc*DE algorithm is described as follows.

Algorithm 1. The *uc*DE Algorithm

Initialization: M: the number of individuals in an array; N: the number of total arrays; x_{ij}: the jth individual in the ith array; P_i: the ith array; $f(x)$: fitness function; D: dimensions of goal; G_{max}: the number of maximum generations.

Iteration:

```
calculate fitness value of each individual x_{ij} by f(x_{ij}); set g = 0;
while (g ≤ G_{max}) do
    g ← g + 1;
    for i = 1 to N do
        for j = 1 to M do
            F ← an uniform random number ∈ [0, 1].
            if i < N then
                x_{r1}, x_{r2}, x_{r3} ← randomly select 3 individuals from P_{i+1};
            else// i = N
                x_{r1}, x_{r2}, x_{r3} ← randomly select 3 individuals from P_1;
            end if
            V_{ij,G} = x_{r1} + F · (x_{r2} − x_{r3})   // do mutation
            for k = 1 to D do
                CR ← an uniform random number ∈ [0, 1].
                CRx ← an uniform random number ∈ [0, 1].
                if CRx < CR then // do crossover
                    U_{ijk} ← V_{ijk}; // U_{ij} is the new individual
                else
                    U_{ijk} ← X_{ijk};
                end if
            end for
            if f(U_{ij}) < f(x_{ij}) then // do selection
                x_{ij} ← U_{ij}; // do replacement to update x_{ij}
            end if
        end for
    end for
    x_{best} ← update the best individual;
end while
return x_{best}
```

It should be pointed out that F and CR factors will affect the performance of DE. In each crossover and mutation of *uc*DE algorithm above, F and CR are taken as uniform random numbers between 0 and 1. This setting of F and CR factors is a simple and general description.

5 Experiments

The purpose of the experiment is to observe and test the performance of the proposed UC model. It is well known, for a distributed evolutionary model, it can have many versions of changes when combined with specific evolutionary algorithms, and the performances of the distributed evolutionary algorithms are also different. In order to obtain the general test results, the DE algorithm used in the UC model here takes F and CR factors as randomization settings, i.e., the *uc*DE described in Sect. 4.

Some statements related to the experiment are as follows.

- *About test functions:* The CEC'05 benchmark is proposed in the technical report of Suganthan et al. in [20]. The test suite includes 25 functions, some of which are shifted and/or rotated versions of classical functions, plus others that are a hybridization of some of those functions. The first five functions are unimodal, while the rest are multimodal functions.
  The benchmark function of CEC'05 is challenging for performance testing of evolutionary algorithms, especially for multimodal functions. Considering that if a benchmark function is too difficult to test, it can hardly get discriminatory results, so the multimodal functions $f_6 - f_{17}$ in CEC'05 are selected for observation and comparison.
- *About distributed evolutionary models and algorithms for comparison:* There are two typical distributed evolutionary models, one is the coarse-grained island model, the other is the fine-grained neighborhood model (the cell model).

  (a) For the island model, literature [2] presents an island-based DE algorithm named dDE_2, which showed competitive performance compared with other excellent algorithms such as G-CMA-ES [4], L-CMA-ES [3], BLXMA [13], BLX-GL50 [7], CoEVO [15], DE [16], K-PCX [18], and SPC-PNX [5]. Based on this fact, it is very suitable to compare the experimental results of dDE_2 [2] which combines the island model with DE.
  (b) For the cell model, it can be incorporated into DE algorithm like the UC model. For convenience, the DE combined with the cell model is called *grid*-DE. The comparison of DE algorithm in the UC model and the cell model is due to the simplicity and effectiveness of DE, so it is appropriate.

The experiments are divided into three parts. Firstly, the *uc*DE and *grid*-DE are compared because they both belong to the neighborhood model. Therefore this comparison is necessary. Secondly, the *uc*DE is compared with the island-based DE, namely dDE_2 algorithm. The third part is about the analysis and discussion of the *uc*DE.

In experiments, the dimension of functions $f_6 - f_{17}$ in CEC'05 are taken as 30. In experiments for comparing with *grid*-DE and dDE_2, the global population size of the *uc*DE is set to 100, and the total number of iterations is 3,000, so the maximum number of fitness evaluation is 300,000. Such a setting is exactly the same as that of other algorithms in the number of fitness evaluations, so it is a fair setting.

5.1 Comparison with Cell-Based DE

For the *grid*-DE, the matrix of grid points is set to 10×10, and the calculation is performed in four neighborhoods as shown in Fig. 1(2). For the *uc*DE, the matrix form of $N \times M$ of spatial distribution of individuals in population, where N is the number of arrays and M is the number of individuals in arrays, is 4×25. M is also the number of neighbors. For both the *uc*DE and *grid*-DE, the F and CR factors in the DE algorithm are taken as random numbers between 0 and 1, in every evolutionary operation of each individual. For each benchmark function, the *uc*DE and *grid*-DE run 25 times independently. The experimental results are summarized as shown in Table 1.

Table 1. Results of 25 runs of the *uc*DE and *grid*-DE on CEC'05 benchmark functions from $f_6 - f_{17}$ (30 dimensions). The maximum number of fitness evaluation is 3e + 5.

Functions	*grid*-DE			*uc*-DE		
	Mean	Best	Worst	Mean	Best	Worst
f_6	*28.7793*	11.9482	75.2597	39.3410	6.8877e − 05	78.1133
f_7	4.696e + 03	4.696e + 03	4.696e + 03	*0.0295*	0.0099	0.0688
f_8	20.9508	20.8146	21.0183	*20.9468*	20.8529	21.0026
f_9	*0.0398*	0	0.9950	0.2786	0	0.9950
f_{10}	46.9222	32.8336	76.6117	*41.1913*	24.8740	70.6420
f_{11}	18.0292	14.4792	22.7432	*10.6237*	2.7660	19.2071
f_{12}	1.514e + 05	1.031e + 05	1.877e + 05	*5.433e + 03*	1.758e + 2	1.886e + 04
f_{13}	2.0685	1.3623	3.1863	*1.9716*	1.2037	2.8045
f_{14}	12.7548	12.2098	12.9782	*12.6928*	12.1019	13.2112
f_{15}	350	300	400	*308.1645*	82.2014	402.5233
f_{16}	*66.2949*	41.8320	96.9378	72.4742	53.0613	101.6634
f_{17}	75.2655	56.1718	109.5296	*72.0754*	40.5034	92.4787

From the results listed in Table 1, it can seen clearly that from the mean error of 12 benchmark functions, the *uc*DE performs better than the *grid*-DE on 9 functions which are f_7, f_8, $f_{10} - f_{15}$, and f_{17}. The mean errors of the *uc*DE on f_6, f_9 and f_{16} are worse than those of the *grid*-DE.

The mean error reflects the approximate performance of a model or algorithm, while the best result reflects the possibility of further development of a model or algorithm. In particular, by observing the best results on f_6, it is easy to find that the *uc*DE is more likely to achieve the global optimum than the *grid*-DE in 25 tests. However, the *grid*-DE does not see much hope of achieving the global optimum. On the other hand, from the mean results and the best results on f_{12}, it is easy to find that the results of the *grid*-DE are much worse than the *uc*DE, although the results of the *uc*DE are unsatisfactory.

Because the *uc*DE and *grid*-DE use the same DE setting, there can be such a simple and direct comparison conclusion. That is to say, the UC model with single direction (or unidirectional) information transmission is more suitable for isolating individuals in the population than the cell model with mutual direction information transmission, and this isolation is conducive to multimodal function optimization.

5.2 Comparison with Island-Based DE

As mentioned earlier in this section, compared with other excellent algorithms, the island-based DE algorithm dDE_2 [2] has competitive performance. Comparing the proposed *uc*DE with dDE_2 is helpful to measure the general performance of the *uc*DE.

Similarly, because of that test on the most difficult benchmark functions do not provide valuable information, the functions $f_6 - f_{17}$ are still taken as observation and comparison. Since only the average test results of these functions are shown in the literature [2], no more detailed comparison can be made. According to the data provided by the literature [2], the results of the *uc*DE and dDE_2 are summarized in Table 2.

Table 2. Averaging results of 25 runs of the *uc*DE and island-based dDE_2 [2] on CEC'05 benchmark functions $f_6 - f_{17}$ (30 dimensions). The maximum number of fitness evaluation is 3e + 5.

Functions	f_6	f_7	f_8	f_9	f_{10}	f_{11}
dDE_2 [2]	2.16e + 02	3.76e − 02	2.10e + 01	1.19e − 01	9.04e + 01	3.90e + 01
*uc*DE	**3.93e + 01**	**2.95e − 02**	**2.09e + 01**	0.2786	**4.12e + 01**	**1.06e + 01**
Functions	f_{12}	f_{13}	f_{14}	f_{15}	f_{16}	f_{17}
dDE_2 [2]	7.27e + 03	9.12e − 01	1.28e + 01	1.43e + 02	2.26e + 02	1.50e + 02
*uc*DE	**5.43e + 03**	1.9716	**1.27e + 01**	3.08e + 02	**7.25e + 01**	**7.21e +01**

From the results of Table 2, it can be seen directly that 9 of the 12 benchmark functions of the *uc*DE have better average results than that of dDE_2. Although this result can not be sure that the *uc*DE has better performance than dDE_2 in all aspects, it can at least be explained that *uc*DE's performance will not be inferior to dDE_2's overall performance. This fact can indirectly prove that the *uc*DE also has competitive performance compared with other excellent algorithms.

5.3 Observation and Discussion on *uc*DE

Firstly, the effect of spatial distribution of individuals in the UC model on the performance of the *uc*DE is observed. For the UC model, its individual distribution is determined by two parameters, M and N, where N represents the number of arrays divided by global individuals and M represents the number of individuals contained in a array.

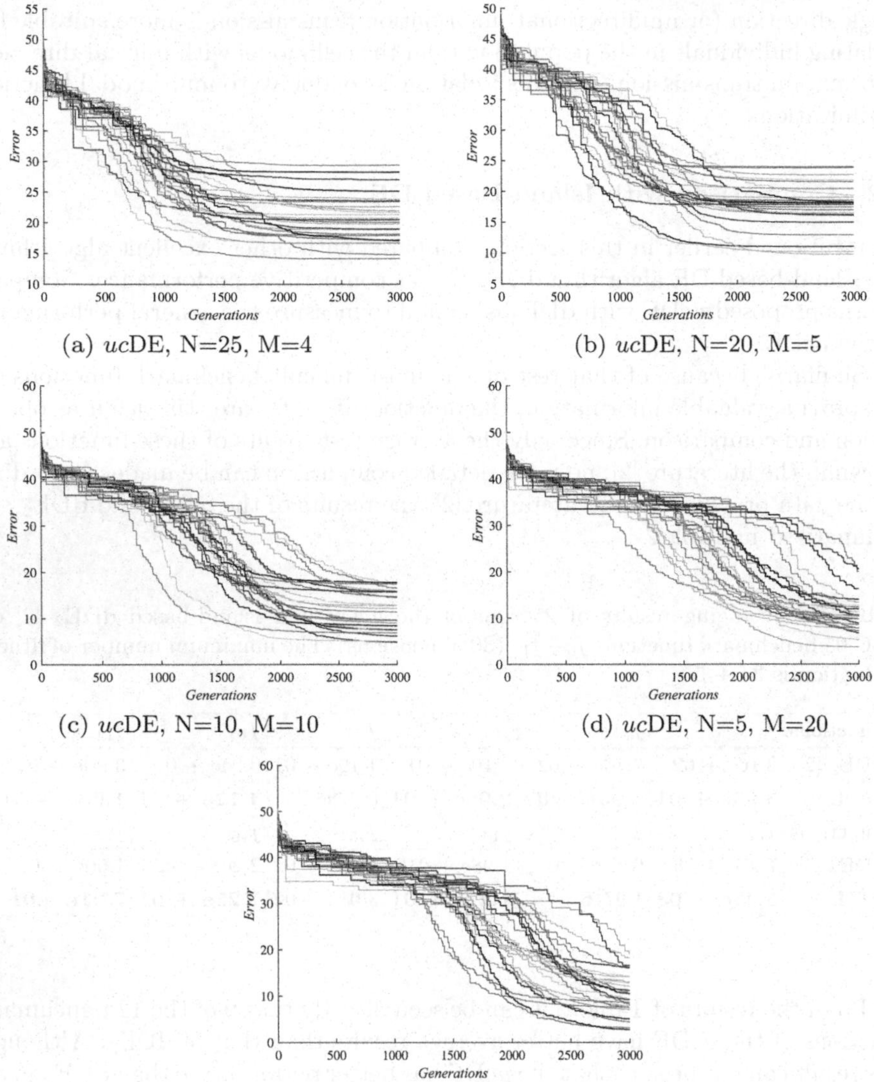

(a) *uc*DE, N=25, M=4

(b) *uc*DE, N=20, M=5

(c) *uc*DE, N=10, M=10

(d) *uc*DE, N=5, M=20

(e) *uc*DE, N=4, M=25

Fig. 4. Fitness curves of 25 runs recorded in the experiment on f_{11}.

Here, an example observation on f_{11} is presented for illustration. Given the global population size of 100, taking different combinations of M and N, the detailed experimental results are shown in Table 3, and the corresponding fitness curves are recorded as shown in Fig. 4.

From the experimental results in Table 3 and the fitness curves of 25 runs recorded in Fig. 4, it can be seen that the number of neighbors significantly affects the performance of the *uc*DE. When the number of neighbors increases

Table 3. Results of 25 runs of *uc*DE on f_{11} (30 dimensions). The maximum number of fitness evaluation is 3e + 5.

*uc*DE settings	N = 25, M = 4	N = 20, M = 5	N = 10, M = 10	N = 5, M = 20	N = 4, M = 25
Mean	20.7413	17.7029	11.5374	10.6223	10.6237
Best	14.1261	9.7541	4.8548	3.6925	2.7660
Worst	29.3268	24.0319	17.7528	20.4807	19.2071

gradually, the performance of *uc*DE with more neighbors is significantly better than that of *uc*DE with less neighbors. When the *uc*DE takes 25 neighbors (i.e., $M = 25$), its performance is slightly better than that of *uc*DE takes 4 neighbors (i.e., $M = 4$). The experimental results show that the spatial distribution of individuals in the UC model can be arranged flexibly, and the flexible arrangement of spatial distribution of individuals also makes the corresponding performance changeable. Appropriate number of neighbors is an important factor affecting the performance of the UC model.

Secondly, the reason why the maximum fitness evaluation times of 30 dimensional function optimization problem are limited to 3e + 5 comes from the rules of CEC'05 algorithm competition. The results of such regulations lead researchers to find that the performance of evolutionary algorithms on some difficult function optimization problems is not optimistic, and promote the research of evolutionary algorithms.

Table 4. Results of 25 runs of the *uc*DE on $f_6 - f_{13}$ (30 dimensions), where $N = 8$ and $M = 50$ in the UC model. The maximum number of fitness evaluation is 1.2e + 7.

Functions	f_6	f_7	f_8	f_9	f_{10}	f_{11}	f_{12}	f_{13}
Mean	4.3690e − 26	0.0060	20.7079	0	36.4951	7.2166	2.2042e + 03	0.9445
Best	0	0	20.0008	0	16.9143	0.0781	3.2890	0.6192
Worst	1.3168e − 25	0.0246	20.8271	0	56.7126	15.5706	8.9012e + 03	1.2709

It is obvious, when how to obtain global optimum becomes the main problem, there is no need to consider the number of fitness evaluation, if enough computing resources are available. From this understanding, it can be said that as long as an algorithm can significantly improve the optimization performance of some functions, even if there is only one good result in many runs, regardless of the upper limit of the number of fitness evaluation, it will be encouraging.

Relax the limitation of maximum fitness evaluation times of 3e + 5, and reset the M and N parameters of *uc*DE. Let $M = 50$ and $N = 8$, i.e., the global population size is 400. The maximum generation is set to 3e + 4, that is, the maximum fitness evaluation times are 1.2e + 7 (i.e., 400 × 3e + 4). A relatively large population is conducive to maintaining the diversity of the population, while allowing more fitness evaluations is conducive to continuously observing the performance of an evolutionary algorithm.

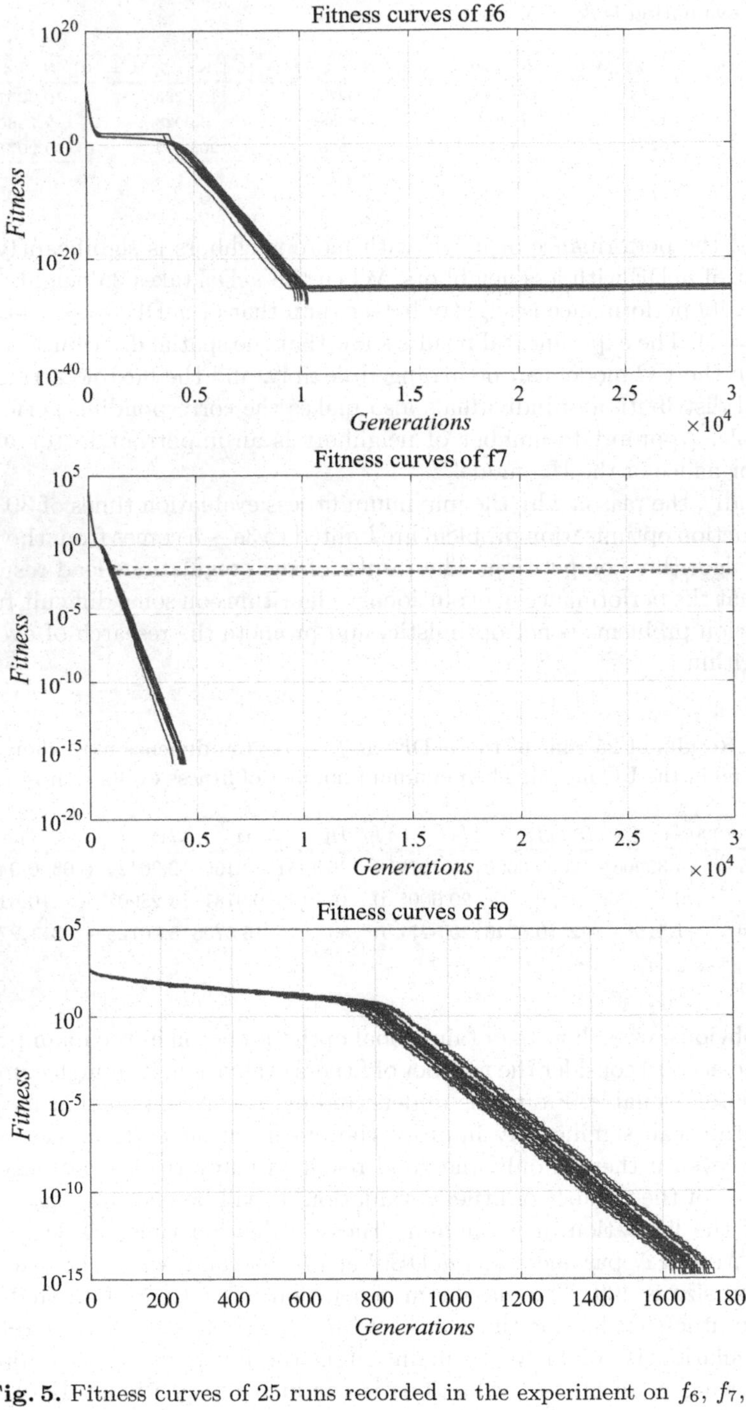

Fig. 5. Fitness curves of 25 runs recorded in the experiment on f_6, f_7, f_9.

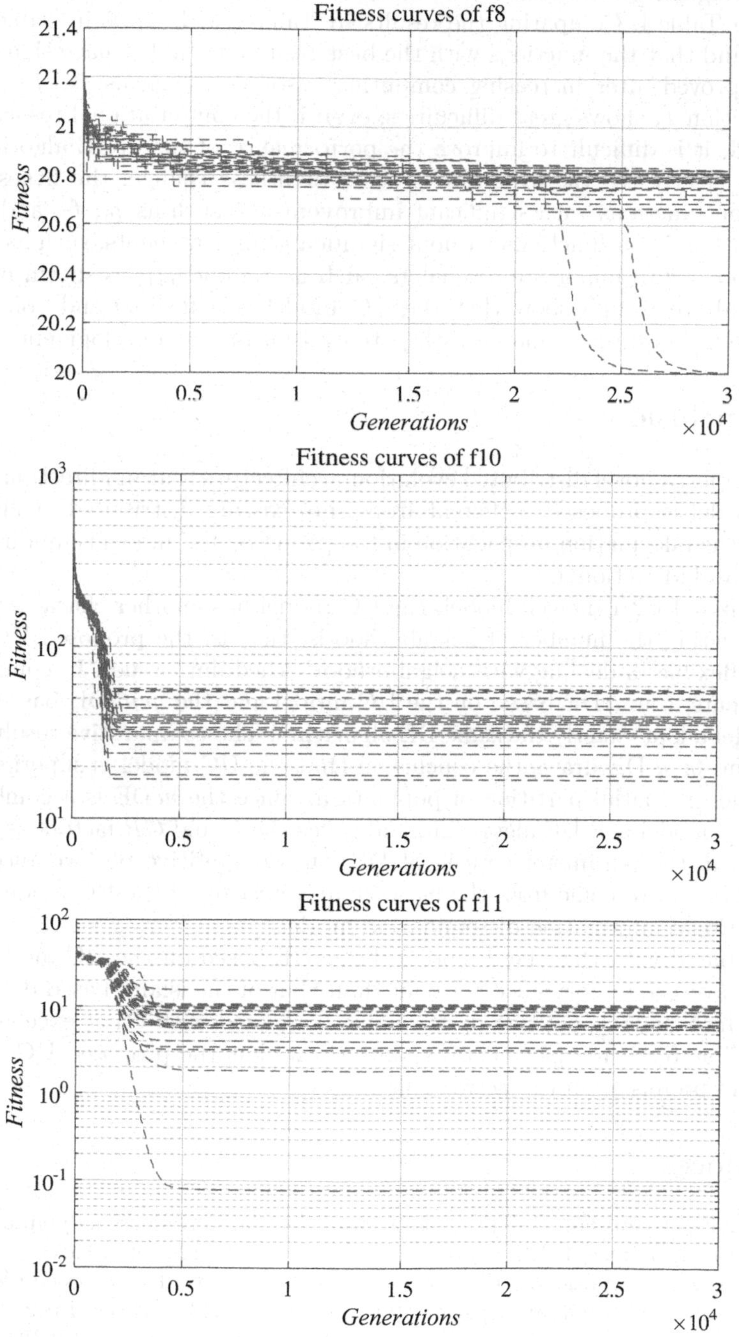

Fig. 6. Fitness curves of 25 runs recorded in the experiment on f_8, f_{10}, f_{11}.

Running the *uc*DE again on the benchmark functions $f_6 - f_{13}$, the results are shown in Table 4. Comparing the results in Table 4 with those in Table 1, it is easy to find that the functions with the best results in Table 1 have significantly been improved after increasing computing resources, such as f_6, f_7, and f_9. The function f_8 shows great difficulties, even if the computational resources are increased, it is difficult to improve the performance of the *uc*DE algorithm. In order to clearly see the evolution process of fitness change, the fitness curve records on functions with significant improvements, such as f_6, f_7, and f_9, are shown in Fig. 5, on functions without significant improvements, such as f_8, and on functions with minor improvements, such as f_{10} and f_{11}, as shown in Fig. 6.

The above results show that the UC model is a flexible and competitive distributed evolutionary model with potential for further development.

6 Conclusions

In this paper, a novel distributed evolution architecture that unidirectional cyclic (UC) model is proposed. Different from the existing distributed evolutionary models, the information propagation in the proposed UC model is unidirectional other than bidirectional.

Compared with the cell model, the UC model has another characteristics of not restricting the number of neighborhoods, that is, the proposed UC model is more flexible in dealing with neighborhood scheduling issues. In comparative experiments, the *uc*DE based on the UC model and the *grid*-DE based on the cell model adopt exactly the same DE algorithm. The comparative results of the *uc*DE and *grid*-DE draws the conclusion that the UC model is superior to the cell model in spatial partition of population. Since the *uc*DE is a combination of the UC model and DE algorithm, and takes the F and CR factors as random numbers in the parameter setting of DE, the competitive performance of the *uc*DE is bound to come from the network architecture of the UC model and its unidirectional information dissemination mode.

Compared with the island model, the *uc*DE based on the UC model shows that the average performance is better than that of the island-based dDE_2 algorithm which has competitive performance compared with other excellent algorithms. The further experimental results show that the proposed UC model is novel and promising in the network architecture.

References

1. Alba, E., Dorronsoro, B.: The exploration/exploitation tradeoff in dynamic cellular genetic algorithms. IEEE Trans. Evol. Comput. **9**(2), 126–142 (2005)
2. Apolloni, J., Leguizamón, G., García-Nieto, J., Alba, E.: Island based distributed differential evolution: an experimental study on hybrid testbeds. In: 2008 Eighth International Conference on Hybrid Intelligent Systems, pp. 696–701. IEEE (2008)
3. Auger, A., Hansen, N.: Performance evaluation of an advanced local search evolutionary algorithm. In: 2005 IEEE Congress on Evolutionary Computation, vol. 2, pp. 1777–1784. IEEE (2005)

4. Auger, A., Hansen, N.: A restart cma evolution strategy with increasing population size. In: 2005 IEEE Congress on Evolutionary Computation, vol. 2, pp. 1769–1776. IEEE (2005)
5. Ballester, P.J., Stephenson, J., Carter, J.N., Gallagher, K.: Real-parameter optimization performance study on the CEC-2005 benchmark with SPC-PNX. In: 2005 IEEE Congress on Evolutionary Computation, vol. 1, pp. 498–505. IEEE (2005)
6. Folino, G., Pizzuti, C., Spezzano, G.: Training distributed GP ensemble with a selective algorithm based on clustering and pruning for pattern classification. IEEE Trans. Evol. Comput. **12**(4), 458–468 (2008)
7. García-Martínez, C., Lozano, M.: Hybrid real-coded genetic algorithms with female and male differentiation. In: 2005 IEEE Congress on Evolutionary Computation, vol. 1, pp. 896–903. IEEE (2005)
8. Ge, Y.F., Yu, W.J., Lin, Y., Gong, Y.J., Zhan, Z.H., Chen, W.N., Zhang, J.: Distributed differential evolution based on adaptive mergence and split for large-scale optimization. IEEE Trans. Cybern. **48**(7), 2166–2180 (2017)
9. Giacobini, M., Tomassini, M., Tettamanzi, A.G.B., Alba, E.: Selection intensity in cellular evolutionary algorithms for regular lattices. IEEE Trans. Evol. Comput. **9**(5), 489–505 (2005)
10. Gong, Y.J., Chen, W.N., Zhan, Z.H., Zhang, J., Li, Y., Zhang, Q., Li, J.J.: Distributed evolutionary algorithms and their models: a survey of the state-of-the-art. Appl. Soft Comput. **34**, 286–300 (2015)
11. Herrera, F., Lozano, M.: Two-loop real-coded genetic algorithms with adaptive control of mutation step sizes. Appl. Intell. **13**(3), 187–204 (2000). https://doi.org/10.1023/A:1026531008287
12. Marc, D., Christian, G., Marc, P.: Analysis of a master-slave architecture for distributed evolutionary computations. IEEE Trans. Syst. Man Cybern. Part B Cybern. **36**(1), 229–235 (2006). A Publication of the IEEE Systems Man and Cybernetics Society
13. Molina, D., Herrera, F., Lozano, M.: Adaptive local search parameters for real-coded memetic algorithms. In: 2005 IEEE Congress on Evolutionary Computation, vol. 1, pp. 888–895. IEEE (2005)
14. Pierreval, H., Paris, J.L.: Distributed evolutionary algorithms for simulation optimization. IEEE Trans. Syst. Man Cybern. Part A: Syst. Hum. **30**(1), 15–24 (2002)
15. Posik, P.: Real-parameter optimization using the mutation step co-evolution. In: 2005 IEEE Congress on Evolutionary Computation, vol. 1, pp. 872–879. IEEE (2005)
16. Ronkkonen, J., Kukkonen, S., Price, K.V.: Real-parameter optimization with differential evolution. In: 2005 IEEE Congress on Evolutionary Computation, vol. 1, pp. 506–513. IEEE (2005)
17. Roy, G., Lee, H., Welch, J.L., Zhao, Y., Pandey, V., Thurston, D.: A distributed pool architecture for genetic algorithms. In: 2009 IEEE Congress on Evolutionary Computation, pp. 1177–1184. IEEE (2009)
18. Sinha, A., Tiwari, S., Deb, K.: A population-based, steady-state procedure for real-parameter optimization. In: 2005 IEEE Congress on Evolutionary Computation, vol. 1, pp. 514–521. IEEE (2005)
19. Storn, R., Price, K.: Differential evolution-a simple and efficient heuristic for global optimization over continuous spaces. J. Global Optim **11**(4), 341–359 (1997). https://doi.org/10.1023/A:1008202821328
20. Suganthan, P.N., Hansen, N., Liang, J.J., Deb, K., Chen, Y.P., Auger, A., Tiwari, S.: Problem definitions and evaluation criteria for the CEC 2005 special session on real-parameter optimization. KanGAL report 2005005 (2005)

A Re-initialization Clustering-Based Adaptive Differential Evolution for Nonlinear Equations Systems

Zuowen Liao, Wenyin Gong(✉), and Zhihua Cai

School of Computer Science, China University of Geosciences, Wuhan 430074, China
wygong@cug.edu.cn

Abstract. Solving nonlinear equations systems is one of the most challenges for evolutionary algorithms, especially to locate multiple roots in a single run. In this paper, a new approach which combines speciation clustering with dynamic cluster sizing, adaptive parameter control and re-initiation mechanism is proposed to deal with this optimization problem. The major advantages are as follows: (1) the speciation clustering with dynamic cluster sizing can alleviate the trivial task to set proper cluster size; (2) to improve the search ability in each species and avoid the trivial work of parameter setting, adaptive parameter control is employed; and (3) re-initialization mechanism motivates the search algorithm to find new roots by increasing population diversity. To verify the performance of our approach, 30 nonlinear equations systems are selected as the test suite. Experiment results indicate that the speciation clustering with dynamic cluster size, adaptive parameter control, and re-initialization mechanism can work effectively in a synergistic manner and locate multiple roots in a single run. Moreover, comparison of other state-of-the-art methods, the proposed method is capable of obtaining better results in terms of peak ratio and success rate.

Keywords: Nonlinear equations system · Dynamic cluster size · Adaptive parameter setting · Re-initialization mechanism

1 Introduction

Many real-world optimization problems can be transformed into nonlinear equations systems (NESs), such as physics [1], engineerings [2,3], economics [4], and so on. Generally, an NES contains multiple equally important roots. The researchers can select a suitable root according to the environmental or conditional constraints to make a final decision [5]. Due to the crucial importance of solving

W. Gong—This work was partly supported by the National Natural Science Foundation of China under Grant Nos. 61573324, and 61673354, in part by the National Natural Science Fund for Distinguished Young Scholars of China under Grant 61525304, and the Fundamental Research Funds for the Central Universities, China University of Geosciences (Wuhan) under Grant No. CUG160603.

L. Pan et al. (Eds.): BIC-TA 2019, CCIS 1159, pp. 422–438, 2020.
https://doi.org/10.1007/978-981-15-3425-6_33

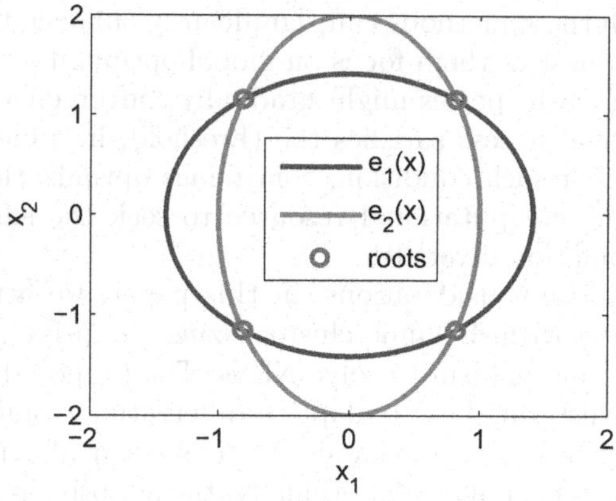

Fig. 1. An example of a NES problem with four roots

NESs, it receives increasing attention in the optimization community during the last few decades. Solving NESs is to locate multiple roots in a single run, which is a challenging work in the mathematical field.

Differential evolution (DE) [6] that is extensively used to resolve different kinds of optimization problems owning to its simplicity and easy implementation. However, the direct use of the differential evolution algorithm will suffer from the following problems: (i) due to the lack of diversity preserving mechanism, DE is difficult to be used to obtain multiple roots of NESs in a single run; (ii) several DE variants can find multiple roots, but they encounter the problems of setting parameters, such as cluster size [8], repulsion radius [7]; (iii) how to obtain a better balance between population diversity and convergence. Population diversity helps to find multiple roots, while convergence ensures that the roots with higher accuracy can be obtained with less computational resources.

Clustering techniques are considered to be effective methods to maintain population diversity, which can partition the whole population into different species. Inspired by this, several clustering-based algorithms have been designed to find the roots of NESs, such as [8–10]. However, they need to be given the number of clusters before the run. Too small the number of clusters may lose some roots. For example, suppose an NES has 4 roots, as shown in Fig. 1, if the clustering number is set to 2, it may lead to losing two roots due to unreasonable the number of clusters. In contrast, too large cluster number may result in the roots with low accuracy since the search algorithm cannot make use of enough computational source to exploit in each cluster. Therefore, the optimal setting of the number of cluster is difficult and problem-dependent.

Parameter control in evolutionary computation plays a vital role to the robustness of the search algorithm [11]. In recent years, several adaptive or self-adaptive parameter control methods [12–16] were proposed to dynamically adjust the parameters according to different fitness landscapes of optimization problems. Although these methods can significantly enhance the performance of original algorithm, most of them focus on global optimum.

After clustering, each species might gradually converge to a narrow range of space. If an individual fitness satisfies the threshold, it is considered as a candidate solution. Under such condition, continuous optimization in such species may waste more the computational resource to seek the same root, which is harmful to the population diversity.

Based on the above considerations, in this paper, we attempt to combine the species clustering with dynamic cluster sizing, adaptive parameter control, and re-initialization mechanism to solve NESs. The proposed method names as a re-initialization clustering-based adaptive differential evolution (RCADE). In RCADE, species clustering with dynamic cluster size can alleviate the trivial task to give a proper cluster number. Additionally, the adaptive parameter control is dynamically adapt the control parameters of the algorithm, thereby improving the search efficiency in each species. Furthermore, re-initialization mechanism is employed to enhance the performance of locating multiple roots of NESs.

2 Background

2.1 Problem Statement

Generally, a NES can be denoted as follows:

$$\mathbf{e}(\mathbf{x}) = \begin{cases} e_1(x_1, x_2, \ldots, x_D) = 0 \\ e_2(x_1, x_2, \ldots, x_D) = 0 \\ \vdots \\ e_n(x_1, x_2, \ldots, x_D) = 0 \end{cases} \tag{1}$$

where n is the number of equations, $\mathbf{x} = (x_1, x_2, \ldots, x_D)$ represents an m-dimensional decision vector, $\mathbf{x} \in \mathbb{S}$, and $\mathbb{S} \subseteq \mathbb{R}^D$ is the search space. Generally,

$$\mathbb{S} = [\underline{x}_j, \overline{x}_j]^m$$

where $j = 1, \cdots, D$, $\underline{x}_j$ and $\overline{x}_j$ are the lower bound and upper bound of x_j, respectively.

Before solving a NES via optimization algorithm, it is commonly transformed into a multimodal single-objective optimization problem as follows:

$$\text{minimize} \quad f(\mathbf{x}) = \sum_{i=1}^{n} e_i^2(\mathbf{x}) \tag{2}$$

Subsequently, solving NES is equivalent to locate the global minimizers of the transformed optimization problem in (2).

2.2 Differential Evolution

The differential evolution (DE) mainly takes advantage of mutation, crossover, and selection technique to dispose of the population during the search process. In general, the population is composed of NP real-valued vectors: $X = \{\mathbf{x}_1, \mathbf{x}_2, ..., \mathbf{x}_{NP}\}$. NP is the population size. In what follows, we briefly introduce three operations in the species.

Mutation. Mutation operator is to generate a mutant vector $\mathbf{v}_i$ according to the current parent population. The following are two common mutation strategies.

- "DE/rand/1"

$$\mathbf{v}_i = \mathbf{x}_{r1} + F \cdot (\mathbf{x}_{r2} - \mathbf{x}_{r3}) \tag{3}$$

- "DE/best/1"

$$\mathbf{v}_i = \mathbf{x}_{best} + F \cdot (\mathbf{x}_{r2} - \mathbf{x}_{r3}) \tag{4}$$

r_1, r_2 and r_3 respectively denote different random indices selected from the population, and they are distinct from the base index i. F is the scale factor that controls the difference vectors. $\mathbf{x}_{best}$ is the vector with best fitness value in the population.

Crossover. The crossover operator is to generate trial vectors by recomposing the current vector and the mutant vector. The trial vector $\mathbf{u}'_{i,j}$ is depicted as follow:

$$\mathbf{x}'_i(j) = \begin{cases} \mathbf{v}_i(j), \text{if } \text{rand}_j(0,1) \leq CR_i(j) \quad \text{or} \quad j = j_{rand} \\ \mathbf{x}_i(j), \text{otherwise} \end{cases} \tag{5}$$

where $CR_{i,j} \in (0,1)$ is referred to as the crossover rate. $\text{rand}_j(0,1)$ is a random value within $[0,1]$; $j = 1, ..., D$; $j_{rand} \in \{1, 2, ..., D\}$ indicates a random index.

Selection. In original DE algorithm, greedy selection operator is employed to chose the individual with better fitness values $f(\cdot)$ between parent vector and the trial vector. Thus, if the trial vector $\mathbf{x}'_i$ is better than $\mathbf{x}_i$, $\mathbf{x}_i$ is set to $\mathbf{x}'_i$; otherwise, $\mathbf{x}_i$ keeps unchanged.

$$\mathbf{x}_i = \begin{cases} \mathbf{x}'_i, \text{if } f(\mathbf{x}'_i) \leq f(\mathbf{x}_i) \\ \mathbf{x}_i, \text{otherwise} \end{cases} \tag{6}$$

3 Related Work

Besides the classical methods, some stochastic methods were presented to locate multiple roots of NESs, which can be roughly divided into three categories, *i.e.*, clustering-based methods, repulsion-based methods, and multiobjective optimization-based methods.

3.1 Clustering-Based Methods

Clustering [17] is used to divide the whole population into different species to promote the search algorithm to find different solutions. Owing to the advantage of clustering, some researchers proposed clustering-based methods that can be exploited to locate multiple roots of NESs. In [8], the clustering technique was employed to separate the estimated locations of solutions and the exact solution can be detected in each cluster via invasive weed optimization. In [10], the author presented a hybrid approach, where the Luus-Jaakola random search, Fuzzy Clustering Means and the Nelder-Mead method were combined to locate multiple roots of NESs. In [9], a Multistart and Minfinder method based on clustering technique was proposed for NESs.

3.2 Repulsion-Based Methods

The repulsion technique can generate the regions of repulsion around the obtained roots. It can drive the search algorithm to detect new promising ares to find new roots of NESs. Based on the repulsion techniques, there are some methods have been designed to solving NESs. In [18], repulsion technique combines with simulated annealing (SA) to compute critical points in binary systems. Henderson *et al.* [19] presented a methodology, which employs the continuous SA to locate different roots of double retrograde vaporization. In [20], C-GRASP method cooperates with the repulsion technique to solve NESs. In [8], the repulsion technique was employed to drive the algorithm to seek new roots. In [21], a biased random-key genetic algorithm (BRKGA) is executed many times to find different roots. In [7], the improved harmony search algorithm combines with the repulsion methods to solve NESs. In [22], a new approach consists of the repulsion technique, adaptive parameter control, and diversity preserving mechanism, named RADE, was designed to locate multiple roots. Further, Liao *et al.* [23] collaborated the dynamic repulsion technique with evolutionary algorithm to deal with NESs.

3.3 Multiobjective Optimization Based-Methods

The task of multiobjective optimization is to gain a group of Pareto optimal solutions, which is analogous to find different roots of NESs. Therefore, solving NESs by using multiobjective optimization-based methods has been gained growing attention. In [24], the NES is transformed into a multiobjective optimization problem, and then solve the problem via the evolutionary algorithm. A bi-objective transformation technique was presented in [25], where the two objective functions respectively include the location function and the system function. Gong *et al.* [26] designed a weighted bi-objective transformation technique (A-WeB) for NESs. In [27], Naidu and Ojha proposed a hybrid multiobjective optimization algorithm to solve NESs.

4 Our Approach

In this section, we firstly expound the motivations of the proposed approach. Afterward, a species clustering with dynamical cluster sizing, adaptive parameter control, and re-initialization mechanism are respectively presented in Sects. 4.2, 4.3, and 4.4. Finally, the framework, RCADE, will be given in detail.

4.1 Motivations

Generally, before the clustering operation, the clustering size should be given in advance. However, it is difficult to set the appropriate cluster size and problem-dependent. Besides, after clustering, there are different species in the search space. Different species represent different promising regions. Thus, how to enhance the search ability of the algorithm to locate the root in each species is also crucial. Moreover, as the search proceeds, each species gradually converges to excessively narrow range. Continuing to optimize in such species will bring about the loss of diversity.

Based on the above considerations, we present a re-initialization clustering-based adaptive DE, named RCADE, to solve NESs. In RCADE, the problem of sensitivity to cluster size is handled by the dynamical cluster sizing. Meanwhile, adaptive parameter setting is applied to improve the search ability of DE in each species and avoid the trivial task of parameter setting. Moreover, the re-initialization mechanism will be triggered if a root has found in a species. For one thing, the population diversity can be preserved due to re-initialization mechanism; for another, it is a benefit for exploration ability of the search algorithm and increases the probability to detect new roots in other promising regions. The dynamic clustering size, adaptive parameter control, and the re-initialization mechanism when combined in a synergistic manner can enhance the performance of solving NESs.

4.2 Dynamic Clustering Size

In [28], a speciation clustering was proposed to partition the whole population into different species. To reduce the sensitivity of the cluster size, dynamic cluster sizing technique [29] was introduced into speciation clustering. The effectiveness of this simple scheme was verified by experiments.

The process of speciation clustering with dynamic cluster sizing (DCS) is outlined in Algorithm 1. First, the population is sorted according to fitness value in ascending order. Second, a random integer M is selected from the cluster size set C. Based on such integer, $M-1$ individuals that is close to species seed are combined with the species seed to form a species. Finally, the entire population is divided into several species.

4.3 Adaptive Parameter Control

After clustering, different species distribute in the search space. Since different species include different features of the landscape, thus, it is not suitable to set

Algorithm 1. Speciation Clustering with Dynamic Cluster Size

Input: population P, cluster size set C
Output: a set of species
Sort P in ascending order according to fitness value;
while *P is not empty* **do**
    Random select a cluster size M from C;
    Select the best individual in P as a new seed;
    Find $M-1$ individuals closest to the species seed and combine them as a species;
    Remove these M individuals from P

the fixed parameter for DE during the run. From the literature, several parameter control methods [12–14] have been proposed to improve the search ability and avoid the trivial task of parameter setting. In this subsection, we make a minor modifications based on JADE [14] to accommodate the evolution of each species.

At each generation, the crossover rate $CR_{i,j}$ of each individual $\mathbf{x}_i$ in j-cluster is independently generated according to a normal distribution of mean CR_j and standard deviation 0.1.

$$CR_{i,j} = randn(\mu_{CR}, 0.1) \tag{7}$$

and truncated to [0,1], μ_{CR} is a value which is used to generate $CR_{i,j}$. It is updated as follows:

$$\mu_{CR} = (1-c)\cdot\mu_{CR} + c\cdot \text{mean}_L(S_{CR_j}) \tag{8}$$

where c is constant number between 0 and 1 and S_{CR_j} is the set of the crossover rates $CR_{i,j}$ in each species. $\text{mean}_L(.)$ is the Lehmer mean:

$$\text{mean}_L(S_{CR_j}) = \frac{\sum S_{CR_j}^2}{\sum S_{CR_j}} \tag{9}$$

Similarly, The mutation factor $F_{i,j}$ of each individual $\mathbf{x}_i$ in species j is independently generated according to a Cauchy distribution.

$$F_{i,j} = randc(\mu_F, 0.1) \tag{10}$$

It is regenerated if $F_{i,j} \le 0$ or truncated to be 1 if $F_{i,j} \ge 1$. The parameter μ_F of Cauchy distribution is given to be 0.5 in advance and updated as

$$\mu_F = (1-c)\cdot\mu_F + c\cdot \text{mean}_A(S_{F_j}) \tag{11}$$

where $\text{mean}_A(.)$ is the arithmetic mean; and S_{F_j} is the set of mutation factor $F_{i,j}$ at each species.

Comparison with JADE, we make minor modifications. S_{F_j} is updated via arithmetic mean whereas S_{CR_j} is revised using Lehmer mean. The reasons are two-fold: (i) the adaptation S_{F_j} put a greater emphasis on normal mutation factor by using the arithmetic mean instead of a Lehmer mean. The arithmetic mean is helpful to propagate average mutation factors, which improve the exploitation ability in each species; (ii) to improve population diversity in the species and avoid trapping in local optima, Lehmer mean is used to update S_{CR_j}.

4.4 Re-initialization Mechanism

As the search proceeds, the candidate solution can be found in the species. When the evolution process continues to execute, such species pays more attention to exploitation rather than exploration. It may result in the loss of population diversity and the waste of computational resource. Therefore, the re-initialization mechanism is triggered to deal with these problems.

If a species detects the candidate solution, it is considered as a convergence situation during the run and the candidate solution is stored into a archive $\mathcal{A}$. Subsequently, all of the individuals in the species will re-initialize in the search place for maintaining the diversity of the population. In addition, after reinitialization, F and CR of each individual are respectively set to 0.5 and 0.9.

It should be pointed out that our approach may locate the same solution during the run. Thus, a way to update an archive is employed to avoid encountering this dilemma. Algorithm 2 describes the process of updating the archive. In Algorithm 2, $\mathbf{x}^*$ is one of the root in $\mathcal{A}$, and ϵ is a small fixed value to avoid storing the same root in $\mathcal{A}$. In this paper, ϵ is set to 0.01.

In this work, we initialize the entire species instead of the found candidate solution. On the one hand, it uses computational resource efficiently and benefits population diversity. On the other hand, to some extent, this mechanism prevents the remaining individuals in the species from continuing to find the same root.

4.5 The Proposed Framework: RCADE

By integrating the improved clustering technique, adaptive parameter control and re-initialization mechanism, the framework of RCADE is outlined in Algorithm 3, where NP is the population size; NFE is the number of function evaluations; NFE_{max} is the maximal number of function evaluations; M is the size of j-th species. In what follows, we mainly describe the proposed approach in detail.

In line 2–5, the works respectively include generating a random population, initialization of F and CR, calculating the fitness value, and updating NFE. Subsequently, the whole population is partitioned into different species. In line 12–14, a mutation vector produces by using Eqs. (3) or (4). In line 15, the crossover operation is used to generate the trial vector according to the crossover rate. In line 16, the fitness value of trial vector is evaluated via Eq. (2). Next, if the trial vector has better fitness value, the parent vector is replaced and its values of $F_{i,j}$ and $CR_{i,j}$ are also updated. In line 20, μ_F and μ_{CR} in j-th species are modified through Eqs. (11) and (8). Following this, $F_{i,j}$ and $CR_{i,j}$ of each

Algorithm 2. Archive updating

```
Input: Solution x and ϵ > 0
Output: The updated archive A
if s_A = 0 then                                   // The archive is empty
    if f(x) < τ then
        A = A ∪ x;
        s_A = s_A + 1;
else
    if f(x) < τ then
        Find the closest root x* to x in A
        if || x − x* ||< ϵ and f(x*) < f(x) then  // Update the found root
            x = x*;
        else if || x − x* ||> ϵ then              // A new root is found
            A = A ∪ x;
            s_A = s_A + 1;
```

individual in j-th species are revised by using Eqs. (10) and (7). In line 24–26, suppose a candidate root is detected, which saves in the archive; meanwhile, all the individuals in such species are re-initialized; $F_{i,j}$ and $CR_{i,j}$ of each individual are also respectively set to 0.5 and 0.9. Finally, the value of NFE is updated. If the termination criterion is met, $\mathcal{A}$ will be output.

One should take care to note that we adopt a hybrid mutation strategy to generate a mutation vector. There are two main reasons: (i) to accelerate the convergence in each species, the strategy of "DE/best/1" is employed; (ii) however, there is still the possibility of falling into local optima during the run. Thus, "DE/rand/1" is used to escape the local optima.

5 Experimental Results and Analysis

In this section, we mainly focuses on experimental results and discussions, including the impact of different parts in RCADE, comparing RCADE with state-of-the-art methods.

5.1 Performance Criteria

To verify the performance of different methods, we select 30 NESs with different features from the literature [22] as a test suite. Moreover, to validate the performance of different methods effectively, two performance criteria in [22] are employed in this paper.

Algorithm 3. The framework of RCADE

```
Input: Control parameters: NP, NFE, NFEs_max
Output: The final archive A
Set NFE = 0 and the archive A = ∅;
Randomly generate the population P;
F and CR of each individual are set to 0.5 and 0.9;
Calculate the fitness value of x via Eq.(2);
NFE = NFE + NP;
while NFE < NFEs_max do
    Partition the whole population into different species via Algorithm 1
    for j-th species do
        for each individual in j-th species do
            Select r1, r2, r3 from the j-th species
            if rand < 0.5 then
                Generate mutation vector using Eq.(3)
            else
                Generate mutation vector using Eq.(4)
            Produce the trial vector x'_i via Eq.(5).
            Evaluate offspring x'_i using Eq.(2).
            if f(x'_i) < f(x_i) then
                x_i = x'_i
                Respectively update F_{i,j} and CR_{i,j} in S_{F_j} and S_{CR_j}
            Update μ_F and μ_CR via Eq.(11) and Eq.(8);
            Implement (10) and (7) to produce F_{i,j} and CR_{i,j}.
        Find the minimal fitness value f(min) in j-th species
        if f(min) < τ then
            Record the corresponding individual in the archive via Algorithm 2
            Re-initialize the individuals in j-th species
            Reset the F_{i,j} and CR_{i,j} to 0.5 and 0.9 in j-th species
            NFE = NFE + M;
    NFE = NFE + NP;
```

- Root ratio (RR): It computes the average ratio of the found roots over multiple runs

$$RR = \frac{\sum_{i=1}^{N_r} N_{f,i}}{NoR \cdot N_r} \tag{12}$$

where N_r is the number of runs; $N_{f,i}$ is the number of the found roots in the i-th run; NoR is the number of the known optima of a NES. To obtain a fair comparison, each algorithm conducts over 30 independent runs. In addition, for a individual $\mathbf{u}$, if its fitness value $f(\mathbf{u}) < \tau$, it can be considered a root. τ is the accuracy level, if $D \leq 5$, $\tau = 1e-06$; otherwise, $\tau = 1e-04$.

- Success rate (SR): It calculates the percentage of successful runs[1] out of all run:

$$SR = \frac{N_{r,s}}{N_r} \tag{13}$$

where $N_{r,s}$ is the number of successful runs.

5.2 Influence of Different Parts in RCADE

As mentioned in Sect. 4.5, a framework of RCADE is proposed to solve NESs. It contains three parts, *i.e.*, an improving clustering technique, adaptive parameter control, and the re-initialization mechanism. This subsection mainly dedicates to discuss the impact of different parts of RCADE on performance of solving NESs. Due to the poor results obtained by adaptive parameter control and re-initialization mechanism alone, the RR and SR values obtained by these two methods will not show in Table 1.

(1) DCS: In [29], a species clustering method with dynamic cluster size is used to partition the whole population into different species.
(2) RCADE/DA, in which the re-initialization mechanism was removed from RCADE;
(3) RCADE/DR, where we eliminated the adaptive parameter control of RCADE and F and CR were respectively fixed to 0.5 and 0.9;
(4) RCADE/AR, in which the dynamic clustering size was not used;
(5) RCADE, where the dynamic cluster size, adaptive parameter control, and the re-initialization mechanism are combined.

The detailed experiment result in terms of RR and SR is shown in Table 1. It can be seen clearly that RCADE obtained the best average RR value, *i.e.*, 0.9951 and the best average SR value, *i.e.*, 0.9556. Additionally, RCADE successfully solves 26 out of 30 NESs over 30 independent runs. In contrast, RCADE/AR, RCADE/DR, RCADE/DA and DCS successfully solve 23, 20, 15, 13 NESs over 30 independent runs, respectively.

The statistical test results acquired by the multiple-problem Wilcoxon's test are showed in Table 3. In addition, the ranking results came from the Friedman's test are illustrated in Table 2. From Table 3, RCADE consistently offers observably better results than RCADE/DR, RCADE/DA, and DCS due to the fact that p-values are less than 0.05 in all the cases. Additionally, RCADE also obtains the best ranking as revealed in Table 2. In what follows, we try to analyze the influence of different parts of RCADE on the performance of solving NESs. In what follows, we try to analyze the influence of different parts of RCADE on the performance of solving NESs:

- DCS: from Table 1, DCS can successfully solve 13 out of 30 NESs, *i.e.*, F01, F05, F06, F09-F11, F18,F20, F21, F26-F30. However, one feature of them is that they contain a small number of the known roots. For example, the known

[1] A successful run is considered as a run where all known optima of a NES are found.

roots of this kind of NESs are no more than 8. Thus, DCS is more suitable for solving these NESs.
- RCADE/DA: The combination of dynamic clustering size and adaptive parameter control can improve the performance of the algorithm to some extent. As shown in Table 1, *RR* and *SR* values obtained by RCADE/DA were higher than DCS.
- RCADE/DR: From Table 1, the re-initialization method can improve the performance of the algorithm. For example, *RR* and *SR* values obtained by RCADE/DR were 0.9365 and 0.8156, respectively, significantly higher than those obtained by DCS.

Table 1. Influence of different parts in RCADE with respect to the peak ratio and success rate.

Prob.	RR					SR				
	RCADE	RCADE/AR	RCADE/DR	RCADE/DA	DCS	RCADE	RCADE/AR	RCADE/DR	RCADE/DA	DCS
F01	**1.0000**	0.8667	**1.0000**	0.9667	**1.0000**	**1.0000**	0.7333	**1.0000**	0.9333	**1.0000**
F02	**1.0000**	**1.0000**	**1.0000**	0.9515	0.9273	**1.0000**	**1.0000**	**1.0000**	0.6667	0.4667
F03	**1.0000**	**1.0000**	**1.0000**	0.9911	0.9733	**1.0000**	**1.0000**	**1.0000**	0.8667	0.6000
F04	**1.0000**	**1.0000**	0.7538	0.8103	0.7949	**1.0000**	**1.0000**	0.0000	0.0000	0.0667
F05	1.0000	1.0000	1.0000	1.0000	1.0000	1.0000	1.0000	1.0000	1.0000	1.0000
F06	**1.0000**	**1.0000**	0.9667	0.9667	0.8833	**1.0000**	**1.0000**	0.7333	0.7333	0.3333
F07	1.0000	1.0000	1.0000	1.0000	1.0000	1.0000	1.0000	1.0000	1.0000	1.0000
F08	**1.0000**	**1.0000**	**1.0000**	0.9905	0.9619	**1.0000**	**1.0000**	1.0000	0.9333	0.7333
F09	1.0000	1.0000	1.0000	1.0000	1.0000	1.0000	1.0000	1.0000	1.0000	1.0000
F10	1.0000	1.0000	1.0000	1.0000	1.0000	1.0000	1.0000	1.0000	1.0000	1.0000
F11	1.0000	1.0000	1.0000	1.0000	1.0000	1.0000	1.0000	1.0000	1.0000	1.0000
F12	0.9867	**0.9933**	0.9467	0.8667	0.8667	0.9333	**0.9667**	0.7333	0.2000	0.2000
F13	0.9167	**0.9778**	0.9333	0.9500	0.9556	0.4667	**0.7667**	0.4667	0.6667	0.6000
F14	**1.0000**	**1.0000**	0.9926	0.9481	0.9259	**1.0000**	**1.0000**	0.9333	0.5333	0.4000
F15	**1.0000**	**1.0000**	1.0000	0.0000	0.0000	**1.0000**	**1.0000**	**1.0000**	0.0000	0.0000
F16	0.9744	**1.0000**	0.7026	0.9128	0.9641	0.6667	**1.0000**	0.0000	0.0667	0.6000
F17	**1.0000**	0.6125	0.9542	0.8750	0.7417	**1.0000**	0.0000	0.4667	0.0667	0.0000
F18	1.0000	1.0000	1.0000	1.0000	1.0000	1.0000	1.0000	1.0000	1.0000	1.0000
F19	**1.0000**	0.8833	**1.0000**	**1.0000**	0.4333	**1.0000**	0.7667	**1.0000**	**1.0000**	0.3333
F20	1.0000	1.0000	1.0000	1.0000	1.0000	1.0000	1.0000	1.0000	1.0000	1.0000
F21	1.0000	1.0000	1.0000	1.0000	1.0000	1.0000	1.0000	1.0000	1.0000	1.0000
F22	**1.0000**	**1.0000**	**1.0000**	**1.0000**	0.9889	**1.0000**	**1.0000**	**1.0000**	**1.0000**	0.9333
F23	0.9750	0.8979	0.2583	0.4000	0.3875	0.6000	0.0333	0.0000	0.0000	0.0000
F24	**1.0000**	**1.0000**	**1.0000**	0.8917	0.9167	**1.0000**	**1.0000**	**1.0000**	0.1333	0.3333
F25	**1.0000**	0.9667	0.6000	**1.0000**	0.9333	**1.0000**	0.9333	0.2000	1.0000	0.8667
F26	1.0000	1.0000	1.0000	1.0000	1.0000	1.0000	1.0000	1.0000	1.0000	1.0000
F27	1.0000	1.0000	1.0000	1.0000	1.0000	1.0000	1.0000	1.0000	1.0000	1.0000
F28	1.0000	1.0000	1.0000	1.0000	1.0000	1.0000	1.0000	1.0000	1.0000	1.0000
F29	**1.0000**	**1.0000**	0.9867	0.9867	0.9867	**1.0000**	**1.0000**	0.9333	0.9333	0.9333
F30	1.0000	**1.0000**	1.0000	1.0000	1.0000	1.0000	1.0000	1.0000	1.0000	1.0000
Avg	**0.9951**	0.9733	0.9365	0.9169	0.8880	**0.9556**	0.9067	0.8156	0.7244	0.6800

- RCADE/AR: Similarly, the combination of adaptive parameter control and re-initialization mechanism can also improve the performance of the algorithm.
- RCADE: These three methods were combined to get the best *RR* and *SR* values. The reasons are as follows: (1) dynamic clustering size can balance the capacity of exploitation and exploration; (2) adaptive parameter control reduces the tedious task of parameter settings and improves the search ability of the algorithm; (3) the re-initialization mechanism can improves the population diversity and detect the new roots.

Table 2. Average rankings of RCADE, RCADE/CA, RCADE/C, and DCS-DE obtained by the Friedman test for both *RR* and *SR* criteria.

Algorithm	Ranking (*RR*)	Ranking (*SR*)
RCADE	**2.3667**	**2.3500**
RCADE/AR	2.5500	2.5333
RCADE/DR	3.0667	3.0333
RCADE/DA	3.3167	3.3667
DCS	3.7000	3.7167

Table 3. Results obtained by the Wilcoxon test for algorithm RCADE in terms of *RR* and *SR* compared with RCADE/AR, RCADE/DR, RCADE/DA and DCS.

VS	*RR*			*SR*		
	R^+	R^-	*p*-value	R^+	R^-	*p*-value
RCADE/AR	265.5	199.5	4.90E−01	260.5	204.5	5.57E−01
RCADE/DR	337.0	128.0	**3.07E−02**	345.0	120.0	**1.42E−02**
RCADE/DA	378.5	86.5	**1.36E−03**	378.5	86.5	**1.28E−03**
DCS	393.0	72.0	**4.99E−04**	394.5	70.5	**2.93E−04**

5.3 Comparison Among Different Methods

This subsection mainly focuses on comparison between RCADE and different methods for NESs. These comparison algorithms are briefly introduced as follows.

- NCDE [30]: the neighborhood mutation-based crowding DE;
- NSDE [30]: the neighborhood mutation-based speciation DE;
- I-HS [7]: the repulsion-based harmony search approach;
- MONES [25]: the multiobjective optimization for NESs;
- A-WeB [26]: the weighted bi-objective transformation technique for NESs;
- RADE [22]: the repulsion-based adaptive DE;
- DR-JADE [23]: the dynamic repulsion technique based adaptive DE with optional external archive.

For the seven compared methods, the parameter setting are listed in Table 4. For a fair comparison, each test problem was conducted over 30 independent runs.

The statistical result obtained by Wilcoxon's test are showed in Table 6. Furthermore, the ranking results obtained from the Friedman's test are reported in Table 5. From Table 6, RCADE consistently obtains remarkable better results than NCDE, NSDE, I-HS, MONES, A-WeB, RADE, and DR-JADE due to the fact that *p*-values are less than 0.05. Besides, RCADE also acquires the highest ranking as revealed in Table 5.

Table 4. Parameter settings for different methods.

Method	Parameter settings
RCADE	$NP = 100, u_{CR} = 0.5, u_F = 0.5, c = 0.1, C = \{5, 6, 7, 8, 9, 10\}$
NCDE	$NP = 100, F = 0.5, CR = 0.9$
NSDE	$NP = 100, F = 0.5, CR = 0.9$
I-HS	$NP = 10, \text{HMCR} = 0.95, \text{PAR}_{\text{min}} = 0.35,$ $\text{PAR}_{\text{max}} = 0.99, \text{BW}_{\text{min}} = 10^{-6}, \text{BW}_{\text{max}} = 5$
MONES	$NP = 100, H_m = NP$
A-WeB	$NP = 100, H_m = NP$
RADE	$NP = 100, H_m = 200$
DR-JADE	$u_{CR} = 0.5, u_F = 0.5, c = 0.1$

Table 5. Average rankings of RCADE, NCDE, NSDE, I-HS, MONES, A-WeB, RADE, and DR-JADE obtained by the friedman test for both RR and SR.

Algorithm	Ranking (*RR*)	Ranking (*SR*)
RCADE	**2.3833**	**2.3833**
NCDE	6.2167	6.1167
NSDE	4.3500	4.3833
I-HS	5.0500	5.2667
MONES	6.0833	5.9500
A-WeB	4.6833	4.4833
RADE	3.4833	3.5833
DR-JADE	3.7500	3.8333

From the above comparison, it is clear that RCADE exhibit the significant performance for locating multiple roots. As a consequence, RCADE can be seen as an effective approach for solving NESs.

Table 6. Results obtained by the Wilcoxon test for algorithm RCADE in terms of RR and SR compared with NCDE, NSDE, I-HS, MONES, A-WEB, RADE, and DR-JADE.

VS	*RR*			*SR*		
	R^+	R^-	*p*-value	R^+	R^-	*p*-value
NCDE	430.0	5.0	**3.01E−06**	430.0	5.0	**0.00E+00**
NSDE	426.0	39.0	**6.60E−05**	426.0	39.0	**4.80E−05**
I-HS	405.0	30.0	**4.60E−05**	406.0	29.0	**1.01E−06**
MONES	430.0	5.0	**4.01E−06**	430.0	5.0	**0.00E+00**
A-WeB	395.5	39.5	**1.04E−04**	391.5	43.5	**1.55E−04**
RADE	375.0	60.0	**6.13E−04**	374.0	61.0	**6.64E−03**
DR-JADE	333.5	131.5	**3.63E−02**	334.0	131.0	**1.97E−02**

6 Conclusions and Future Work

Solving NESs is a challenging task since it needs to locate multiple roots of NESs in a single run. To address this issue, we propose a re-initialization clustering based adaptive differential evolution, named RCADE, in which the improved clustering technique, the adaptive parameter control, and re-initialization mechanism were combined together to solve NESs effectively. The performance of RCADE was verified by 30 NESs selected from the literature. Experimental results demonstrated that RCADE is able to locate multiple roots in a single run. In addition, comparison with other state-of-the-art methods, our approach also obtains significant performance. MTo address this issue, a re-initialization clustering-based adaptive differential evolution, named RCADE, is proposed. It consists of three parts: the dynamic clustering size, adaptive parameter control, and re-initialization mechanism. Experimental results demonstrated that RCADE is able to locate multiple roots in a single run. In addition, comparison with other state-of-the-art methods, our approach also obtains significant performance.

The re-initialization mechanism in RCADE plays an important role. However, it also loses some useful information of individual. In the near future, we will dedicate to develop other more suitable mechanism for avoiding the loss of information. In addition, how to detect the promising regions is also an urgent issue to be solved. Moreover, we will employ RCADE to solve other complex re-world NESs.

References

1. Kastner, M.: Phase transitions and configuration space topology. Rev. Mod. Phys. **80**, 167–187 (2008)
2. Guo, D., Nie, Z., Yan, L.: The application of noise-tolerant ZD design formula to robots' kinematic control via time-varying nonlinear equations solving. IEEE Trans. Syst. Man Cybern. Syst. **48**, 2188–2197 (2017)

3. Chiang, H.D., Wang, T.: Novel homotopy theory for nonlinear networks and systems and its applications to electrical grids. IEEE Trans. Control Netw. Syst. **5**, 1051–1060 (2017)
4. Facchinei, F., Kanzow, C.: Generalized Nash equilibrium problems. 4OR **5**(3), 173–210 (2007)
5. Sun, Z., Wu, J., Pei, J., Li, Z., Huang, Y., Yang, J.: Inclined geosynchronous spacebornec-airborne bistatic SAR: performance analysis and mission design. IEEE Trans. Geosci. Remote Sens. **54**(1), 343–357 (2016)
6. Das, S., Suganthan, P.N.: Differential evolution: a survey of the state-of-the-art. IEEE Trans. on Evol. Comput. **15**(1), 4–31 (2011)
7. Ramadas, G.C.V., Fernandes, E.M.G.P., Rocha, A.M.A.C.: Multiple roots of systems of equations by repulsion merit functions. In: Murgante, B., et al. (eds.) ICCSA 2014. LNCS, vol. 8580, pp. 126–139. Springer, Cham (2014). https://doi.org/10.1007/978-3-319-09129-7_10
8. Pourjafari, E., Mojallali, H.: Solving nonlinear equations systems with a new approach based on invasive weed optimization algorithm and clustering. Swarm Evol. Comput. **4**, 33–43 (2012)
9. Tsoulos, I.G., Stavrakoudis, A.: On locating all roots of systems of nonlinear equations inside bounded domain using global optimization methods. Nonlinear Anal. Real World Appl. **11**(4), 2465–2471 (2010)
10. Sacco, W.F., Henderson, N.: Finding all solutions of nonlinear systems using a hybrid metaheuristic with fuzzy clustering means. Appl. Soft Comput. **11**(8), 5424–5432 (2011)
11. Karafotias, G., Hoogendoorn, M., Eiben, A.E.: Parameter control in evolutionary algorithms: trends and challenges. IEEE Trans. Evol. Comput. **19**(2), 167–187 (2015)
12. Brest, J., Greiner, S., Boskovic, B., Mernik, M., Zumer, V.: Self-adapting control parameters in differential evolution: a comparative study on numerical benchmark problems. IEEE Trans. Evol. Comput. **10**(6), 646–657 (2006)
13. Qin, A.K., Huang, V.L., Suganthan, P.N.: Differential evolution algorithm with strategy adaptation for global numerical optimization. IEEE Trans. Evol. Comput. **13**(2), 398–417 (2009)
14. Zhang, J., Sanderson, A.C.: JADE: adaptive differential evolution with optional external archive. IEEE Trans. Evol. Comput. **13**(5), 945–958 (2009)
15. Tanabe, R., Fukunaga, A.: Evaluating the performance of SHADE on CEC 2013 benchmark problems. In: IEEE Congress on Evolutionary Computation (CEC), pp. 1952–1959 (2013)
16. Mallipeddi, R., Suganthan, P.N., Pan, Q.K., Tasgetiren, M.F.: Differential evolution algorithm with ensemble of parameters and mutation strategies. Appl. Soft Comput. **11**(2), 1679–1696 (2011)
17. Jain, A.K., Murty, M.N., Flynn, P.J.: Data clustering: a review. ACM Comput. Surv. **31**(3), 264–323 (1999)
18. Freitas, L., Platt, G., Henderson, N.: Novel approach for the calculation of critical points in binary mixtures using global optimization. Fluid Phase Equilib. **225**, 29–37 (2004)
19. Henderson, N., Sacco, W.F., Platt, G.M.: Finding more than one root of nonlinear equations via a polarization technique: an application to double retrograde vaporization. Chem. Eng. Res. Des. **88**(5–6), 551–561 (2010)
20. Hirsch, M.J., Pardalos, P.M., Resende, M.G.C.: Solving systems of nonlinear equations with continuous grasp. Nonlinear Anal. Real World Appl. **10**(4), 2000–2006 (2009)

21. Silva, R.M.A., Resende, M.G.C., Pardalos, P.M.: Finding multiple roots of a box-constrained system of nonlinear equations with a biased random-key genetic algorithm. J. Global Optim. **60**(2), 289–306 (2013). https://doi.org/10.1007/s10898-013-0105-7
22. Gong, W., Wang, Y., Cai, Z., Wang, L.: Finding multiple roots of nonlinear equation systems via a repulsion-based adaptive differential evolution. IEEE Trans. Syst. Man Cybern. Syst. PP(99), 1–15 (2018)
23. Liao, Z., Gong, W., Yan, X., Wang, L., Hu, C.: Solving nonlinear equations system with dynamic repulsion-based evolutionary algorithms. IEEE Trans. Syst. Man Cybern. Syst. 1–12 (2018)
24. Grosan, C., Abraham, A.: A new approach for solving nonlinear equations systems. IEEE Trans. Syst. Man Cybern. Part A Syst. Hum. **38**(3), 698–714 (2008)
25. Song, W., Wang, Y., Li, H.-X., Cai, Z.: Locating multiple optimal solutions of nonlinear equation systems based on multiobjective optimization. IEEE Trans. Evol. Comput. **19**(3), 414–431 (2015)
26. Gong, W., Wang, Y., Cai, Z., Yang, S.: A weighted biobjective transformation technique for locating multiple optimal solutions of nonlinear equation systems. IEEE Trans. Evol. Comput. **21**(5), 697–713 (2017)
27. Naidu, Y.R., Ojha, A.K.: Solving multiobjective optimization problems using hybrid cooperative invasive weed optimization with multiple populations. IEEE Trans. Syst. Man Cybern. Syst. **48**, 821–832 (2016)
28. Gao, W., Yen, G.G., Liu, S.: A cluster-based differential evolution with self-adaptive strategy for multimodal optimization. IEEE Trans. Cybern. **44**(8), 1314–1327 (2014)
29. Yang, Q., Chen, W.N., Li, Y., Chen, C.L., Xu, X.M., Zhang, J.: Multimodal estimation of distribution algorithms. IEEE Trans. Cybern. **47**(3), 636–650 (2017)
30. Qu, B.Y., Suganthan, P.N., Liang, J.J.: Differential evolution with neighborhood mutation for multimodal optimization. IEEE Trans. Evol. Comput. **16**(5), 601–614 (2012)

Ensemble Learning via Multimodal Multiobjective Differential Evolution and Feature Selection

Jie Wang, Bo Wang, Jing Liang(✉), Kunjie Yu, Caitong Yue, and Xiangyang Ren

School of Electrical Engineering, Zhengzhou University, Zhengzhou, China
liangjing@zzu.edu.cn

Abstract. Ensemble learning is an important element in machine learning. However, two essential tasks, including training base classifiers and finding a suitable ensemble balance for the diversity and accuracy of these base classifiers, are need to be achieved. In this paper, a novel ensemble method, which utilizes a multimodal multiobjective differential evolution (MMODE) algorithm to select feature subsets and optimize base classifiers parameters, is proposed. Moreover, three methods including minimum error ensemble, all Pareto sets ensemble, and error reduction ensemble are employed to construct ensemble classifiers for executing classification tasks. Experimental results on several benchmark classification databases evidence that the proposed algorithm is valid.

Keywords: Multimodal multiobjective optimization · Feature selection · Ensemble learning · Classifier parameter

1 Introduction

The advent of the information age prompts us to mine valuable knowledge from big data and complete diverse classification tasks, forming an extensive research field, i.e., machine learning that includes a variety of methods. Thereinto, ensemble learning receives widespread attention from researchers owing to its more dependable accuracy and generalization performance than an individual classifier. Hence, numerous ensemble learning algorithms have been employed in a variety of areas, such as texture image classification [1], medical information analysis [2] and synthetic aperture radar image classification [3].

Ordinarily, ensemble learning consists of two steps, training a set of base learners and integrating predictions of these learners. As for training base classifiers, the most prevailing strategies are Bagging [4], Adaboost [5], random forest [6], rotation forest [7]. Recently, many studies focus on employing feature selection to train different classifiers. For example, in [8], an optimal feature and instance subsets were obtained by embedding both parameters searching in a multiobjective evolutionary algorithm with a wrapper approach. And in [9],

L. Pan et al. (Eds.): BIC-TA 2019, CCIS 1159, pp. 439–453, 2020.
https://doi.org/10.1007/978-981-15-3425-6_34

the Pareto sets of image features obtained from a multiobjective evolutionary trajectory transformation algorithm was utilized for generating base classifiers. Meanwhile, an increasing number of researches have coped with feature selection (FS), which can be generally classified into three sorts: filter [10,11], wrapper [12–14] and embedded methods [15]. As for filter, feature selection is independent of the generalization performance of the learning algorithms by scoring and ranking features. Thus, the selected feature subsets may not enhance the performance of the classification algorithms. In contrast, wrapper methods utilize search strategies to quest the optimal feature subsets and evaluate these subsets by learning algorithms. Obviously, wrapper methods have a larger amount of computation than filter approaches but more credible. As for embedded approaches, feature selection and learner are incorporated in a single model, such as decision tree learner. In feature selection, studies always focus on two aspects: classification accuracies of learners and the size of selected feature subsets. Actually, multiple feature subsets of the same number of features can be able to achieve the same accuracy. If unimodal multiobjective evolutionary methods are utilized to deal with these problems, only one of them may be retained, which may cause some excellent feature subsets to be lost. These studies do not consider the multimodal [16] of Pareto sets in multiobjective optimization problems. Specifically, different solutions could have the same objective results.

Motivated to solve this problem, we utilize the evolutionary algorithms (EAs), which are highly popular on multimodal multiobjective issues. Usually, they are called multimodal multiobjective EAs (MMOEAs). Recently, there are numerous multi-objective evolutionary algorithms [17,18]. Meanwhile, several MMOEAs [19–21] are proposed to solve multimodal multiobjective optimization (MMO) issues that may exist multiple Pareto sets which corresponds to the same Pareto front (PF) point. In [19], a multiobjective PSO by means of ring topology was proposed, which could produce stable niches and employ a special crowding distance. Here, we concentrate on utilizing an MMOEA for generating the base classifiers ELMs by optimizing feature subsets and the number of ELMs hidden nodes simultaneously then constructing an ensemble model in different ways.

In this paper, we present an ensemble method via multimodal multiobjective differential evolution (EMMODE), a novel approach that performs MMOEA to optimize the size of feature sets and the performance of ELM, by way of feature selection and selecting the number of hidden nodes. Due to the characteristics of MMODE, we are able to get a series of non-dominated solutions from it. As for the strategies of combining the base classifiers, EMMODE fulfills an operation on the Pareto sets for constructing them into an ensemble. This intent is accomplished by three strategies that are: (1) minimum error ensemble; (2) all Pareto sets ensemble, and; (3) error reduction ensemble approach. The experiments are conducted for classification problems. The experiments results of benchmark datasets from the UCI Machine Repository [22] show the effectiveness of our proposal, being capable to obtain multiple solutions with the same number of features and similar classification accuracy. Meanwhile, the EMMODE is also able to achieve solutions with an excellent tradeoff between the reduction rate in the number of features and accuracy.

The rest of this paper is organized as follows. Section 2 describes the related works. Section 3 details the EMMODE methods. The experimental settings and results are introduced in Sect. 4. Finally, Sect. 5 is the conclusion.

2 Related Works

This section describes related works on multimodal multiobjective optimization and feature selection for ensemble learning.

2.1 Multimodal Multiobjective Optimization

MMO problems are those which have multiple Pareto sets corresponding to the same PF [23]. Evidently, it is significant to find all Pareto solutions which are equivalent to PF. Give an example, decision-makers can use more Pareto sets to solve the real-world tasks. The MMO problem is vividly demonstrated in Fig. 1, where three Pareto solutions with similar objective values.

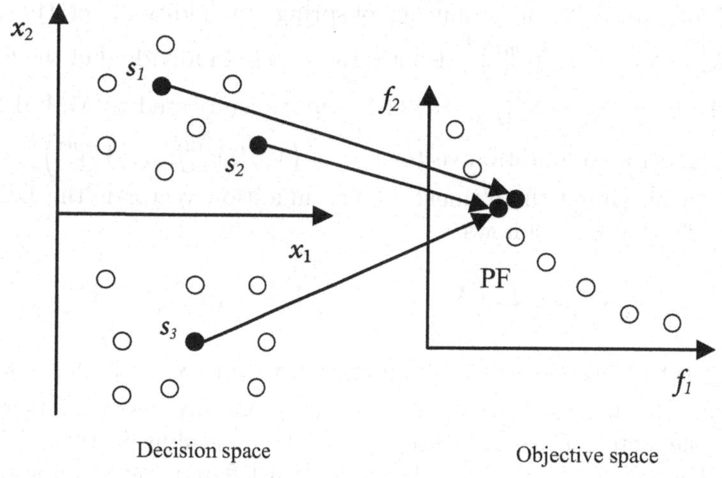

Fig. 1. Illustration of the multimodal multiobjective optimization problem.

In the real world, many applications belong to MMO problems [24], which conclude optimization of truss-structures, metabolic network modeling, femtosecond laser pulse shaping problem, automatic determination of point, and so on. To solve these problems, many MMOEAs have been proposed. In [25], a multimodal multiobjective differential evolution (MMODE) algorithm which formulated a decision variable preselection strategy was proposed. The niching mechanism was employed in Niching-CMA [26] and MO-Ring-PSO-SCD [19]. In this paper, we use the MMODE, due to its good performance on MMO problems [27].

MMODE is an enhanced version of differential evolution (DE). The process of MMODE is as follows. Firstly, users define a possible solution search space $\Omega_d \subseteq \Omega$. The boundary can be just defined as the endpoint of the actual range of values for the decision variables. In particular, the bounds of each element $X^{(e)}$ in the decision variables $x = \left(X^{(1)}, X^{(2)}, \ldots, X^{(m)}\right)^{\mathrm{T}} \in R^m$ are indicated as $X^{(e),\mathrm{L}} \leq X^{(e)} \leq X^{(e),\mathrm{U}}$, in which the $X^{(e),\mathrm{L}}$ and $X^{(e),\mathrm{U}}$ are applied to express the lower and upper values of the solution space for the element in the decision vectors respectively, and the variable m is defined as problem dimension [25]. Secondly, initialize the population P which consists of N individuals. There is a common method for initialization.

$$X_{i,1}^{(e)} = X_i^{(e),\mathrm{L}} + rand(0,1)\left(X_i^{(e),\mathrm{U}} - X_i^{(e),\mathrm{L}}\right) \tag{1}$$

where the sub index 1 in $X_{i,1}^{(e)}$ is utilized to express the element of an initial decision solution, $i = 1, 2, \ldots, N$, and $e = 1, 2, \ldots, m$. Meanwhile, the $rand(0,1)$ produces random real numbers between 0 and 1.

The next step is the preselection scheme, which applies both an objective space crowding distance (SCD) and a decision SCD indicators, to select population Q of size $N/2$ for producing offspring. In addition, let the notation $x_{i,G} = \left(X_{i,G}^{(1)}, X_{i,G}^{(2)}, \ldots, X_{i,G}^{(m)}\right)^{\mathrm{T}}$ denotes the selected individual of the G-th generation whose elements $X_{i,G}^{(e)}$, $e = 1, 2, \ldots, m$ are subjected to MMODE mutation. Then generate a mutation vector $v_{i,G} = \left(V_{i,G}^{(1)}, V_{i,G}^{(2)}, \ldots, V_{i,G}^{(m)}\right)^{\mathrm{T}}$. One possible way for obtaining the elements of the mutation vector is the DE/rand/2 technique [25] that is as follows.

$$V_{i,G}^{(e)} = X_{r_1^i,G}^{(e)} + F_1\left(X_{r_2^i,G}^{(e)} - X_{r_3^i,G}^{(e)}\right) + F_2\left(X_{r_4^i,G}^{(e)} - X_{r_5^i,G}^{(e)}\right) \tag{2}$$

where $V_{i,G}^{(e)}$ is the e-th element of the mutation vector, $X_{r_s^i,G}^{(e)}$ is the e-th element of $x_{r_r^i,G}$, and the indices r_k^i, $k = 1, 2, \ldots, 5$, are randomly selected integers in the [1, $N/2$]. The factors $F_1 \in (0,1)$ and $F_2 \in (0,1)$ are scaling factors of difference terms, and set $F_1 = F_2 = F$ in MMODE. If a left-hand side value of Eq. (2) was outside the decision space boundary, the MMODE would implement an alternative mutation bound scheme

$$V_{i,G}^{(e)} = X_{r_1^i,G}^{(e)} - F\left(X_{r_2^i,G}^{(e)} - X_{r_3^i,G}^{(e)}\right) - F\left(X_{r_4^i,G}^{(e)} - X_{r_5^i,G}^{(e)}\right) \tag{3}$$

Then, use a common method to implement crossover process:

$$U_i^{(e)} = \begin{cases} V_{i.G}^{(e)} & \text{if } rand(0,1) < Cr \\ X_{i,G}^{(e)} & \text{otherwise} \end{cases} \tag{4}$$

where the cross probability $Cr \in (0,1)$ is set by users. And a vector $u_i = \left(U_i^{(1)}, U_i^{(2)}, \ldots, U_i^{(m)}\right)^{\mathrm{T}} \in R^m$ stores the crossover results. For the case of MMO problems, MMODE applies the following selection offspring generation method:

$$c_{i,G} = \begin{cases} u_i & \text{if } u_i \text{ dominates } x_{i,G} \\ x_{i,G} & \text{if } x_{i,G} \text{ dominates } u_i \end{cases} \tag{5}$$

In addition, when neither branch is true in i-th individual, vector u_i would be added to the $c_{i,G}$.

Finally, splice $c_{i,G}$ and $x_{i,G}$, and use a nondominated sorting scheme on the spliced vector to generate the $(G+1)$ – st generation. A complete algorithm for conducting MMODE is shown in [25].

2.2 Feature Selection for Ensemble Learning

In this subsection, we classify feature selection methods in two kinds: ordinary feature selection algorithms and feature selection by evolutionary algorithms.

On the one hand, ordinary feature selection algorithms exist several defects, such as difficult to set the value of important parameters, nesting effect, falling into local optima. For instance, common feature selection methods, Sequential Forward Selection (SFS) [28] and Sequential Backward Selection (SBS) [29], affect by the nesting effect [30].

On the other hand, evolutionary algorithms [31,32] supply a valid strategy for coping with feature selection owing to the three reasons: (1) We can acquire quite acceptable feature subsets without searching the entire decision space. (2) They are capable to search the decision space comprehensively. (3) They get over falling into local optima and nesting effect for they set no restriction on selecting features. Recently, an increasing number of studies apply evolutionary approaches for feature selection. For example, classic EAs such as PSO [33], DE [34], GA [35], and ACO [36] were used. The above-mentioned methods utilized a single objective or multiobjective EAs to select feature subsets. In addition, there also exist some studies optimizing both the learners parameters and feature selection. For example, [35] encoded the parameters of support vector machine (SVM) and the feature subsets into GA chromosomes.

Moreover, to evaluate feature subsets, we employ Extreme Learning Machine (ELM) as base classifier of an ensemble. ELM [37] is an efficient method for single-hidden layer feedforward neural networks (SLFNs). ELM randomly generates the parameters of hidden nodes and input weights, while the output weights are determined analytically. In [38], the universal approximation and effective generalization performance of ELM are proved. Compared with conventional learning methods, such as the back-propagation algorithm (BP) and SVM, ELM can learn extremely fast because it need not adjust network parameters iteratively. For ELM, it is vital that hidden layer parameters, especially the number of hidden nodes for the generalization performance. Thus, Huang et al. proposed an Incremental Extreme Learning Machine (I-ELM) method by adding the hidden nodes one by one [38]. Another method called Error Minimized ELM (EM-ELM) is proposed in [39]. The difference from I-ELM is that EM-ELM adjusts all output weights iteratively when it adds one or more new hidden nodes. In [40], an improved method of EM-ELM called Incremental Regularized Extreme Learning Machine (IR-ELM) is proposed by utilizing the RELM. However, the expected termination accuracy may be difficult to set in the real world, which can cause the overfitting or fail to achieve the desired testing accuracy.

Based on the above, the purposes of feature selection and the classifiers model selection are obtaining the representation of datasets and appropriate model parameters that are adequate for classification tasks. In addition, there exists more than one such solution in the same objectives. Thus, MMODE is utilized to overcome these drawbacks in this paper.

3 The EMMODE Method

In this section, we present the EMMODE approach formulating the feature and model parameters selection as a multimodal multiobjective one. The flow chart of the EMMODE method is depicted in Fig. 2. Like DE, this process begins with the initial population generation whose each individual encodes a possible solution. For each individual, calculate its error rate and feature selection rate on the datasets. After that, new individuals are generated by means of differential evolution operations over the existing ones. Then, repeated this process iteratively until a termination condition is satisfied. The detail of the EMMODE approach is introduced as follows.

3.1 Encoding

MMODE works by chromosomes whose each chromosome encodes a potentially feasible solution for the optimization task, i.e., the number of hidden nodes for an ELM and the selected feature subsets. The first process is to encode a potentially feasible solution for the task. In this paper, the feature is encoded in a binary variable demonstrating if the corresponding feature is selected. As for the parameter of ELM, the number of selected hidden nodes is encoded with an integer variable. The encoding of the chromosome is shown in Fig. 3.

3.2 Evolutionary Operators and Fitness Functions

This subsection presents the evolutionary operators which are different from MMODE, namely the mutation process and mutation-bound process. Meanwhile, the fitness functions for determining the quality of a solution are explained. For the model and feature selection task, the range of decision variables is relatively small, thus this paper uses fewer perturbation vectors and a mutation vector is generated by the DE/rand/1 method, which is shown in the mutation equation.

$$V_{i,G}^{(e)} = X_{r_1^i,G}^{(e)} + F\left(X_{r_2^i,G}^{(e)} - X_{r_3^i,G}^{(e)}\right) \tag{6}$$

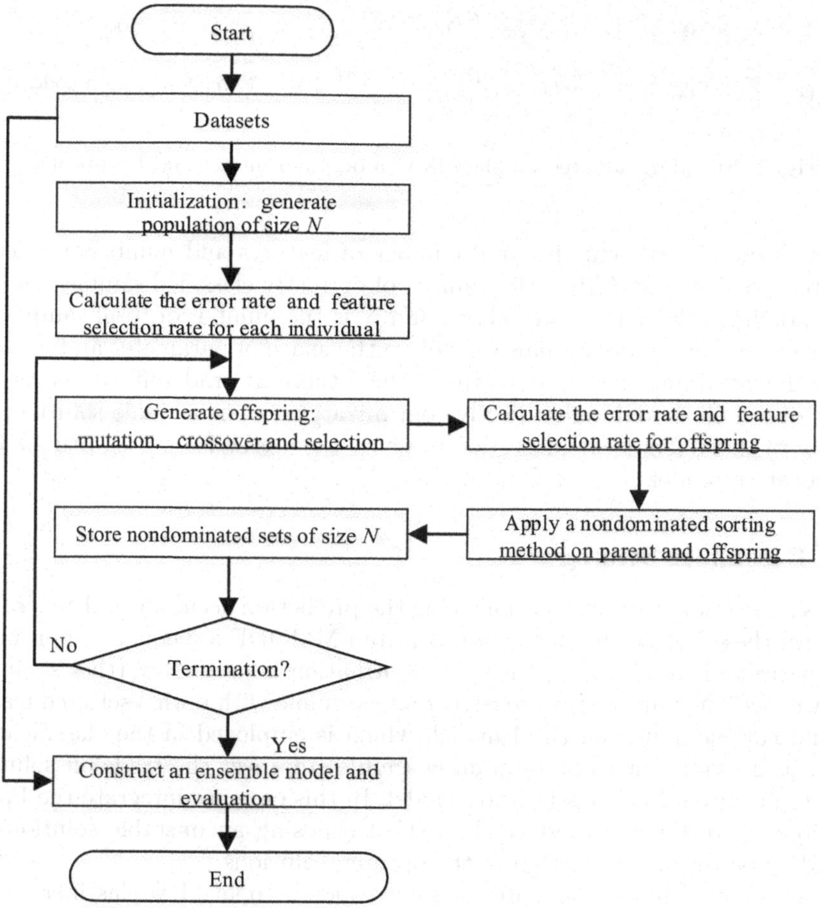

Fig. 2. The process of EMMODE.

where $V_{i,G}^{(e)}$, $X_{r_s^i,G}^{(e)}$, r_ℓ^i, $\ell = 1, 2, 3$, and $F \in (0, 1)$ represent the same meaning as mentioned in the Eq. (2) formula.

As for the mutation-bound process, if a left-hand side value of Eq. (6) was outside the decision space boundary, the EMMODE would implement an alternative mutation bound scheme

$$V_{i,G}^{(e)} = X_{r_1^i,G}^{(e)} - F\left(X_{r_2^i,G}^{(e)} - X_{r_3^i,G}^{(e)}\right) \tag{7}$$

Our aim is to attain optimal feature subsets and a corresponding number of the hidden nodes. We optimize the following functions: the feature selection rate (f_1) and the error rate (f_2).

$$\begin{cases} f_1 = \frac{|S|}{|\mathcal{F}|} \\ f_2 = 1 - \frac{T(\boldsymbol{\theta})}{N} \end{cases} \tag{8}$$

1	0	...	0	1	50
$F_{(1)}$	$F_{(2)}$	...	$F_{(m-2)}$	$F_{(m-1)}$	Number of hidden nodes $F_{(m)}$

Fig. 3. Encoding adopted for the MMO model parameter and FS problem.

where $\mathcal{F}$ and $\mathcal{S}$ represent the total number of features and number of selected features, respectively. $T(\boldsymbol{\theta})$ is the number of correctly classified samples for each corresponding selected feature subset, and N is the number of total samples.

The proposed approach aims to explore the space of parameter and FS techniques for attaining the solutions that suffice the best trade-off. Moreover, the result of a multimodal multiobjective optimization is not a single solution, but a series of them. The next subsection presents the methods of integrating a final classification model.

3.3 Ensembles Strategy

In this subsection, we focus on enhancing the prediction accuracy and generalization. All these Pareto solutions obtained from MMODE are equally appropriate for the task when no other preference information is used. Nevertheless, in the task we face, the purpose is to construct an ensemble with both a selected feature set and corresponding an ELM model, which is employed in the classification. Thus, it is significant to perform an ensemble step over the trade-off solutions so as to acquire a final classification model. In this case, we integrate the Pareto solutions of MMODE to reduce the risk of choosing an unstable solution and provide a better approximation to the optimal solutions.

Each solution in nondominated sets corresponds to an ELM-classifier trained with different parameter and different subsets of the original feature set. An ensemble of classifiers can combine the individual information acquired from each model and provide more information on the predicted label than a single classifier. In this regard, we study three different strategies of combining the ELMs which are described in the following.

(1) All Pareto Sets Ensemble (APSE): The basic method here is to construct an ensemble applying all Pareto solutions of MMODE.
(2) Minimum Error Ensemble (MEE): The opinion of this method does not use all Pareto solutions, but a subset of them. The ensemble consists of n solutions which have low error rate. As recommended in [41], it satisfies to combine 5 to 35 ELMs for most practical applications. In view of this, we set the n as 11. The ensemble $\mathcal{E}$ is defined by the equation

$$\mathcal{E} = \underset{n}{\operatorname{argmin}}\, \mathcal{PS}_2 \tag{9}$$

where the index n is the number of the selected ELMs, and $\mathcal{PS}_2$ indicates the second target value of the Pareto solutions, namely error rate.

(3) Error Reduction Ensemble (ERE): This approach is also not to employ all solutions. First, the solutions in the Pareto sets are sequenced in ascending order according to the corresponding error rate, and the solutions with the error greater than 0.5 are eliminated. Second, the misclassification samples numberings of each classifier constructed by the Pareto solutions in the dataset classification are stored in the matrix $Misnum$. Third, set the matrices $Misnum_1, Misnum_2$ of the first two solutions in the sorted Pareto solution set as the reference matrices and decipher the two as part of the integration. And then, operate on each of the remaining solutions. For example, for the i-th ($i > 2$) solution, calculate the number of identical elements of its matrix $Misnum$ and two reference matrices.

$$\begin{cases} a_{i1} = Misnum_i \cap Misnum_1 \\ a_{i2} = Misnum_i \cap Misnum_2 \\ q_i = numel(a_{i1}) + numel(a_{i2}) \end{cases} \tag{10}$$

where the a_{i1} is the intersection matrix of matrix $Misnum_1$ and matrix $Misnum_i$, a_{i2} denotes the intersection matrix of matrix $Misnum_2$ and matrix $Misnum_i$. And q_i represents the sum of the number of elements of the matrix a_{i1} and the number of elements of the matrix a_{i2}.

Next, sort the quantity values of identical elements in ascending order and select the solutions corresponding to the first $n-2$ values as part of the integration. At last, these n base classifiers form the final ensemble model. The main framework is demonstrated in Algorithm 1.

Algorithm 1. Error Reduction Ensemble

1.Sort the solutions according to error rate in ascending order, and eliminate the solutions with $error > 0.5$

2.**For** each solution in Pareto sets

Calculate its the misclassification samples numberings

Store them in $Misnum_i$

End for

3.Set the matrices $Misnum_1, Misnum_2$ of the first two solutions in the sorted Pareto solution set as the reference matrices and decipher the two as part of the integration

4.**For** each solution in PS except for the first two

Calculate the number of identical elements by Eq.(10)

$$\begin{cases} a_{i1} = Misnum_i \cap Misnum_1 \\ a_{i2} = Misnum_i \cap Misnum_2 \\ q_i = numel(a_{i1}) + numel(a_{i2}) \end{cases}$$

End for

5.Sort the quantity values of identical elements in ascending order and select the solutions corresponding to the first $n-2$ values

6.Form the final ensemble model with the n solutions

These schemes introduce different approaches to select the ELM classification model from Pareto sets. The next procedure is that integrating the results obtained by base learners to acquire a final prediction. We deal with this problem in the following. When we combine the models into an ensemble model, we take a majority voting to acquire the final prediction of the model.

4 Experiments and Results

In this section, we present the experiments implemented and the results acquired by the proposed approach by means of different classification datasets.

4.1 Experimental Settings

For our study, we used 6 datasets available in the UCI repository. Table 1 shows the characteristics of these datasets, such as the number of samples, the number of classes and the number of features. In our experiments, the results are the mean values by ten executes of ten-fold cross-validation. The process of EMMODE is a nested loop: as for inner loop, one-third of the training dataset is set as a validation set randomly to estimate each solution, while the rest is applied to train learners. In outer loop, these datasets are divided into ten subsets previously using the k-fold cross-validation method. In ten-fold cross-validation, a dataset is partitioned into ten subsets [42], and other processes are similar to the above.

Table 1. The attributes of 6 datasets.

Datasets	Features	Classes	Samples
Vehicle	18	4	752
Wine	13	3	178
Ionosphere	34	2	351
Image segmentation	18	7	2310
Sonar	60	2	208
SPECT	22	2	267

We apply two standards to evaluate the performance of the EMMODE. One is the testing accuracy, and the other is the selection rate attained in the FS. In our experiments, the population size and maximum fitness evaluation are set to 100 and 5000. While the mutation rate and crossover rate are set to 0.9 and 0.6. For different datasets, the upper and lower bounds of the number of hidden layer nodes of the ELM are as shown in Table 2.

4.2 Experimental Results

This subsection introduces the experimental assessment of the EMMODE. First, we compare the performance of the three ensembles strategies, which aims at comparing among the different ensemble strategies to find one of them that performs best. Second, we compare the performance of EMMODE with traditional feature selection method and standard learning algorithms.

Table 2. The settings of EMMODE for 6 datasets.

Datasets	Upper bound of nodes	Lower bound of nodes
Vehicle	20	100
Wine	5	40
Ionosphere	10	60
Image segmentation	10	200
Sonar	5	40
SPECT	5	30

Tables 3 and 4 illustrate the obtained results by each of the ensembles. The displayed results are the average testing accuracy and the selection rate in feature set. These results are the mean and standard deviation values obtained by the algorithm running 10 times in the dataset. For each case, the best result is highlighted in boldface.

Table 3. Average accuracy by the different ensemble strategies.

Datasets	EMMODE-APSE	EMMODE-MEE	EMMODE-ERE
Vehicle	76.86 ± 3.36	77.25 ± 4.20	$\mathbf{78.59 \pm 3.74}$
Wine	99.44 ± 1.76	98.89 ± 2.68	$\mathbf{99.44 \pm 1.76}$
Ionosphere	92.87 ± 2.79	93.44 ± 3.32	$\mathbf{93.72 \pm 2.64}$
Image segmentation	97.32 ± 1.24	97.40 ± 1.34	$\mathbf{97.93 \pm 1.24}$
Sonar	81.76 ± 9.52	81.29 ± 8.58	$\mathbf{83.69 \pm 9.84}$
SPECT	86.41 ± 7.75	85.41 ± 7.76	$\mathbf{87.26 \pm 7.75}$

From the results in Tables 3 and 4, we can see that ERE is the excellent ensemble strategy among the three methods. It achieves the best performance when classifying test sets while reducing the feature set size. Hence, the ERE is used to compare with other methods, namely single ELM [37], wrapper feature selection method: PSO-SVM [43], whose SVM is the implementation of LibSVM [44]. and standard ensemble learning algorithms: random forest (RF) [6] and

Table 4. The feature selection rate by the different ensemble strategies.

Datasets	EMMODE-APSE	EMMODE-MEE	EMMODE-ERE
Vehicle	77.78 ± 0	77.78 ± 0	$\mathbf{77.78 \pm 0}$
Wine	38.46 ± 0	38.46 ± 0	$\mathbf{38.46 \pm 0}$
Ionosphere	35.29 ± 0	35.29 ± 0	$\mathbf{35.29 \pm 0}$
Image segmentation	52.63 ± 0	52.63 ± 0	$\mathbf{52.63 \pm 0}$
Sonar	$\mathbf{68.33 \pm 0}$	73.33 ± 0	73.33 ± 1.31
SPECT	59.09 ± 0	59.09 ± 0	$\mathbf{59.09 \pm 0}$

Adaboost [5]. For the random forest and Adaboost, their base classifiers are decision trees and the number of trees is set to 100.

In Table 5, we compare EMMODE-ERE with ELM, PSO-SVM, random forest (RF), Adaboost. The ELM is utilized as a baseline for comparing the performance of other methods. From the results indicated in the table, we can see the following. (1) EMMODE-ERE are capable of enhancing the performance of classification. (2) Traditional FS and ensemble approaches outperform the standard ELM.

Therefore, EMMODE is a competitive method for performing feature reduction and parameter selection for an ELM and can be adopted to far-going supervised learning problems. Meanwhile, EMMODE is an intensely efficient classification algorithm when compare it with traditional learning algorithms.

Table 5. Comparisons the performance of EMMODE-ERE against traditional learning algorithms.

Datasets	ELM	PSO-SVM	RF	Adaboost	EMMODE-ERE
Vehicle	74.88 ± 6.27	76.44 ± 3.44	70.74 ± 2.54	49.48 ± 5.72	$\mathbf{78.59 \pm 3.74}$
Wine	97.22 ± 3.93	98.87 ± 1.54	98.31 ± 3.75	97.15 ± 4.08	$\mathbf{99.44 \pm 1.76}$
Ionosphere	87.74 ± 4.08	91.22 ± 2.77	93.16 ± 2.59	92.61 ± 3.78	$\mathbf{93.72 \pm 2.64}$
Image segmentation	89.74 ± 2.23	96.32 ± 1.53	97.86 ± 1.35	81.43 ± 1.86	$\mathbf{97.93 \pm 1.24}$
Sonar	77.43 ± 8.03	81.28 ± 7.25	79.37 ± 6.58	82.19 ± 7.31	$\mathbf{83.69 \pm 9.84}$
SPECT	82.75 ± 9.35	85.18 ± 6.41	82.92 ± 1.93	82.41 ± 6.98	$\mathbf{87.26 \pm 7.75}$

5 Conclusion

In this paper, we have presented EMMODE. The significance and importance of solving MMO problems of selecting features and the parameter are analyzed. EMMODE deals with the MMO problem by selecting features and the parameter of an ELM simultaneously. Moreover, it also presents three different strategies, including the APSE, MEE, and ERE, for combining the Pareto solutions into an ensemble. Experimental results prove the effectiveness of the proposed EMMODE approach.

The datasets used in this paper relatively small-scale. When the dimension of the dataset is higher, the result of ELM may be unstable. In the future, utilizing our method on unbalanced classification datasets and improving the performance of our method on large-scale datasets will be studied.

Acknowledgments. This work is supported by the National Natural Science Foundation of China (61976237, 61922072, 61876169, 61673404).

References

1. Song, Y., et al.: Gaussian derivative models and ensemble extreme learning machine for texture image classification. Neurocomputing **277**, 53–64 (2018)
2. Piri, S., Delen, D., Liu, T., Zolbanin, H.M.: A data analytics approach to building a clinical decision support system for diabetic retinopathy: developing and deploying a model ensemble. Decis. Support Syst. **101**, 12–27 (2017)
3. Zhao, Z., Jiao, L., Liu, F., Zhao, J., Chen, P.: Semisupervised discriminant feature learning for SAR image category via sparse ensemble. IEEE Trans. Geosci. Remote Sens. **54**(6), 3532–3547 (2016)
4. Breiman, L.: Bagging predictors. Mach. Learn **24**(2), 123–140 (1996)
5. Freund, Y., Schapire, R.E.: A decision-theoretic generalization of on-line learning and an application to boosting. J. Comput. Syst. Sci. **55**(1), 119–139 (1997)
6. Breiman, L.: Random forests. Mach. Learn. **45**(1), 5–32 (2001)
7. Rodriguez, J.J., Kuncheva, L.I., Alonso, C.J.: Rotation forest: a new classifier ensemble method. IEEE Trans. Pattern Anal. Mach. Intell. **28**(10), 1619–1630 (2006)
8. Fernández, A., Carmona, C.J., Jose del Jesus, M., Herrera, F.: A Pareto-based ensemble with feature and instance selection for learning from multi-class imbalanced datasets. Int. J. Neural Syst. **27**(06), 1750028 (2017)
9. Albukhanajer, W.A., Jin, Y., Briffa, J.A.: Classifier ensembles for image identification using multi-objective Pareto features. Neurocomputing **238**, 316–327 (2017)
10. Lyu, H., Wan, M., Han, J., Liu, R., Wang, C.: A filter feature selection method based on the maximal information coefficient and Gram-Schmidt orthogonalization for biomedical data mining. Comput. Biol. Med. **89**, 264–274 (2017)
11. Guyon, I., Elisseeff, A.: An introduction to variable and feature selection. J. Mach. Learn. Res. **3**, 1157–1182 (2003)
12. Kohavi, R., John, G.H.: Wrappers for feature subset selection. Artif. Intell. **97**(1–2), 273–324 (1997)
13. Xue, X., Yao, M., Wu, Z.: A novel ensemble-based wrapper method for feature selection using extreme learning machine and genetic algorithm. Knowl. Inf. Syst. **57**(2), 389–412 (2017). https://doi.org/10.1007/s10115-017-1131-4
14. Zhang, Y., Gong, D., Cheng, J.: Multi-objective particle swarm optimization approach for cost-based feature selection in classification. IEEE/ACM Trans. Comput. Biol. Bioinf. (TCBB) **14**(1), 64–75 (2017)
15. Quinlan, J.R.: Improved use of continuous attributes in C4.5. J. Artif. Intell. Res. **4**, 77–90 (1996)
16. Kamyab, S., Eftekhari, M.: Feature selection using multimodal optimization techniques. Neurocomputing **171**, 586–597 (2016)
17. Pan, L., Li, L., He, C., Tan, K.C.: A subregion division-based evolutionary algorithm with effective mating selection for many-objective optimization. IEEE Trans. Cybern. (2019). https://doi.org/10.1109/TCYB.2019.2906679

18. He, C., Tian, Y., Jin, Y., Zhang, X., Pan, L.: A radial space division based evolutionary algorithm for many-objective optimization. Appl. Soft Comput. **61**, 603–621 (2017)
19. Yue, C., Qu, B., Liang, J.: A multiobjective particle swarm optimizer using ring topology for solving multimodal multiobjective problems. IEEE Trans. Evol. Comput. **22**(5), 805–817 (2017)
20. Deb, K., Tiwari, S.: Omni-optimizer: a procedure for single and multi-objective optimization. In: Coello Coello, C.A., Hernández Aguirre, A., Zitzler, E. (eds.) EMO 2005. LNCS, vol. 3410, pp. 47–61. Springer, Heidelberg (2005). https://doi.org/10.1007/978-3-540-31880-4_4
21. Liang, J., Guo, Q., Yue, C., Qu, B., Yu, K.: A self-organizing multi-objective particle swarm optimization algorithm for multimodal multi-objective problems. In: Tan, Y., Shi, Y., Tang, Q. (eds.) ICSI 2018. LNCS, vol. 10941, pp. 550–560. Springer, Cham (2018). https://doi.org/10.1007/978-3-319-93815-8_52
22. Dua, D., Graff, C.: UCI machine learning repository (2017). http://archive.ics.uci.edu/ml
23. Tanabe, R., Ishibuchi, H.: A review of evolutionary multimodal multiobjective optimization. IEEE Trans. Evol. Comput. **24**(1), 193–200 (2020). ISSN 1941-0026
24. Li, X., Epitropakis, M.G., Deb, K., Engelbrecht, A.: Seeking multiple solutions: an updated survey on niching methods and their applications. IEEE Trans. Evol. Comput. **21**(4), 518–538 (2017)
25. Liang, J., et al.: Multimodal multiobjective optimization with differential evolution. Swarm Evol. Comput. **44**, 1028–1059 (2019)
26. Shir, O.M., Preuss, M., Naujoks, B., Emmerich, M.: Enhancing decision space diversity in evolutionary multiobjective algorithms. In: Ehrgott, M., Fonseca, C.M., Gandibleux, X., Hao, J.-K., Sevaux, M. (eds.) EMO 2009. LNCS, vol. 5467, pp. 95–109. Springer, Heidelberg (2009). https://doi.org/10.1007/978-3-642-01020-0_12
27. Sikdar, U.K., Ekbal, A., Saha, S.: MODE: multiobjective differential evolution for feature selection and classifier ensemble. Soft Comput. **19**(12), 3529–3549 (2015). https://doi.org/10.1007/s00500-014-1565-5
28. Whitney, A.W.: A direct method of nonparametric measurement selection. IEEE Trans. Comput. **100**(9), 1100–1103 (1971)
29. Marill, T., Green, D.: On the effectiveness of receptors in recognition systems. IEEE Trans. Inf. Theory **9**(1), 11–17 (1963)
30. Yusta, S.C.: Different metaheuristic strategies to solve the feature selection problem. Pattern Recogn. Lett. **30**(5), 525–534 (2009)
31. Pan, L., He, C., Tian, Y., Wang, H., Zhang, X., Jin, Y.: A classification-based surrogate-assisted evolutionary algorithm for expensive many-objective optimization. IEEE Trans. Evol. Comput. **23**(1), 74–88 (2018)
32. Pan, L., He, C., Tian, Y., Su, Y., Zhang, X.: A region division based diversity maintaining approach for many-objective optimization. Integr. Comput. Aided Eng. **24**(3), 279–296 (2017)
33. Wang, X., Yang, J., Teng, X., Xia, W., Jensen, R.: Feature selection based on rough sets and particle swarm optimization. Pattern Recogn. Lett. **28**(4), 459–471 (2007)
34. Yu, K., Qu, B., Yue, C., Ge, S., Chen, X., Liang, J.: A performance-guided jaya algorithm for parameters identification of photovoltaic cell and module. Appl. Energy **237**, 241–257 (2019)
35. Huang, C.L., Wang, C.J.: A GA-based feature selection and parameters optimizationfor support vector machines. Expert Syst. Appl. **31**(2), 231–240 (2006)

36. Wan, Y., Wang, M., Ye, Z., Lai, X.: A feature selection method based on modified binary coded ant colony optimization algorithm. Appl. Soft Comput. **49**, 248–258 (2016)
37. Huang, G.B., Zhu, Q.Y., Siew, C.K.: Extreme learning machine: theory and applications. Neurocomputing **70**(1–3), 489–501 (2006)
38. Huang, G.B., Chen, L., Siew, C.K., et al.: Universal approximation using incremental constructive feedforward networks with random hidden nodes. IEEE Trans. Neural Networks **17**(4), 879–892 (2006)
39. Feng, G., Huang, G.B., Lin, Q., Gay, R.: Error minimized extreme learning machine with growth of hidden nodes and incremental learning. IEEE Trans. Neural Networks **20**(8), 1352–1357 (2009)
40. Xu, Z., Yao, M., Wu, Z., Dai, W.: Incremental regularized extreme learning machine and it's enhancement. Neurocomputing **174**, 134–142 (2016)
41. Cao, J., Lin, Z., Huang, G.B., Liu, N.: Voting based extreme learning machine. Inf. Sci. **185**(1), 66–77 (2012)
42. Rosales-Perez, A., Garcia, S., Gonzalez, J.A., Coello, C.A.C., Herrera, F.: An evolutionary multi-objective model and instance selection for support vector machines with Pareto-based ensembles. IEEE Trans. Evol. Comput. **21**(6), 863–877 (2017)
43. García-Nieto, J., Alba, E., Jourdan, L., Talbi, E.: Sensitivity and specificity based multiobjective approach for feature selection: application to cancer diagnosis. Inf. Process. Lett. **109**(16), 887–896 (2009)
44. Chang, C.C., Lin, C.J.: LIBSVM: a library for support vector machines. ACM Trans. Intell. Syst. Technol. (TIST) **2**(3), 27 (2011)

A Knee Point Based NSGA-II Multi-objective Evolutionary Algorithm

Jing Liang[1], Zhimeng Li[1(✉)], Boyang Qu[2], Kunjie Yu[1], Kangjia Qiao[1], and Shilei Ge[1]

[1] School of Electrical Engineering, Zhengzhou University, Zhengzhou 450001, China
liangjing@zzu.edu.cn, mengzhili@gs.zzu.edu.cn
[2] School of Electronic and Information Engineering, Zhongyuan University of Technology, Zhengzhou 450007, China
qby1984@hotmail.com

Abstract. Many evolutionary algorithms (EAs) can't select the solution which can accelerate the convergence to the Pareto front and maintain the diversity from a group of non-dominant solutions in the late stage of searching. In this article, the method of finding knee point is embedded in the process of searching, which not only increases selection pressure solutions in later searches but also accelerates diversity and convergence. Besides, niche strategy and special crowding distances are used to solve multimodal features in test problems, so as to provide decision-makers with multiple alternative solutions as much as possible. Finally, the performance indicators of knee point are compared with the existing algorithms on 14 test functions. The results show that the final solution set of the proposed algorithm has advantages in coverage area of the reference knee regions and convergence speed.

Keywords: Knee point · Niche · Special crowding distances · Performance indicators of knee point

1 Introduction

Over the last few decades, a group of uniform distributed solutions approximating the optimal Pareto-optimal front (POF) can be found by many evolutionary computation based on domination relations, include genetic algorithm (NSGA-II) [1], strength Pareto evolutionary algorithm (SPEA2) [2], niched Pareto Genetic Algorithm (NPGA) [3], Pareto envelope-based selection algorithm (PESA) [4], knee point Driven evolutionary algorithm (KnEA) [5]. While the study of MOPs has made great progress, there are still some challenges remain to be solved. For example, in objective space a small amount of solutions can't quickly represent the total Pareto-optimal solution. Moreover, it is difficult to determine the preferred solution from the final obtained solution set.

People can choose the final Pareto solution set based on prior knowledge or posterior knowledge. But when the decision-maker (DM) has no special preference, the geometric attribute of the Pareto-optimal solution obtained by multi-objective optimization algorithm is the first choice of the DM. For example, the

L. Pan et al. (Eds.): BIC-TA 2019, CCIS 1159, pp. 454–467, 2020.
https://doi.org/10.1007/978-981-15-3425-6_35

knee point of the Pareto-optimal solution set is preferred for DM. In [6] and [7], there are different interpretations about knee points.

So far, many ways of find the knee point on the Pareto solution was proposed. Such as, the reflex angle [8]; extended angle dominance [9,10]; expected marginal utility [10,11]; distance-based strategy [7,12]; the ratio between the improvement and deterioration is changed when exchanging the objectives of two solutions [6]; and the niching-based method [13,14]. However, a few methods of finding knee point are embedded into the search process. And most of them only consider the diversity and convergence of solution sets, rather than the multi-modal properties of the problem. Multi-modal optimization can determine a group of local optimal solutions [14]. The niching technique [15] is inspired by the way organisms evolve in nature and is the most popular has the call means to address multimodal problems.

In this paper, the parent selection and environment selection mechanism of KnEA algorithm are modified. Niche technology is embedded in parent selection to solve the multi-modal feature of the test problem. In environmental selection, the method of finding the knee point and the special crowded distance are used as the selection condition to update the next generation and increase the selection pressure. The rest of this paper will be arranged as follows. The related works is introduced in Sect. 2. The proposed algorithm is detailed introduction in Sect. 3. Experimental results are displayed in Sect. 4. Finally, the conclusion is given in Sect. 5.

2 Related Works

2.1 Knee Point

In the multi-objective optimization algorithm, knee points are part of the solutions of the Pareto optimal solution set, as shown in Fig. 1. This type of solution is visually represented as the most "concave" part of the Pareto front [5]. In the vicinity of the knee points, any one-dimensional object value change will led to a substantial increase for the other dimensions of the object value. The solutions on the Pareto front are non-dominated solutions, but where the knee points to be better than the other solution to a certain extent. From the perspective of DM, knee points are most likely to be preferred if no specific preference points are specified. From the perspective of theory, convergence performance can be increased by knee points, the verification process can be referred to [5].

2.2 Niche Strategy

In the multi-objective evolutionary algorithm NSGA-II, mating is completely random, especially in the late search period, where a large number of individuals are concentrated at extreme points. This makes the algorithm very inefficient when solving multi-modal problems. However, in local space, the evolution of solutions can be achieved through niching methods. And niching methods can

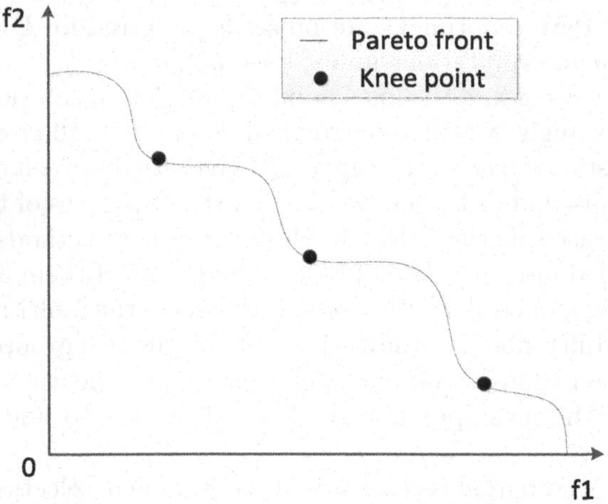

Fig. 1. Illustrative example of the knee points

promote the formation of subpopulations and maintain them. There are many different niche methods, such as crowding [17], fitness sharing [18], clearing [19] and speciation [20]. In the traditional algorithm, maintenance of different solutions is very difficult in decision space. Therefore, the niche method based on crowding mechanism [21] is embedded into the multi-objective optimal algorithm in this paper. It's better to preserve as many optimal solution sets when considering the characteristics of multimodality.

2.3 Special Crowding Distance

It's a common method to improve the distribution of the resulting Pareto solution set by considering the diversity of decision spaces. However, it's difficult to obtain an evenly distributed Pareto-optimal solution set by considering both the decision space and the objective space. This paper uses SCD [22] instead of CD in environment selection. The SCD is mainly calculated with two steps. Firstly, formula (1), (2) and (3) represent the CD value of each particle in the decision space and the objective space. Then, a SCD value is assigned to each particle in the population using formula (4) and the previously calculated CD value. In decision space and objective space:

$$CD_{i,x} = \frac{|x_{1,i+1} - x_{1,i-1}|}{|x_{1,max} - x_{1,min}|} + \frac{|x_{j,i+1} - x_{j,i-1}|}{|x_{j,max} - x_{j,min}|} + \cdots + \frac{|x_{m,i+1} - x_{m,i-1}|}{|x_{m,max} - x_{m,min}|}. \quad (1)$$

Where $CD_{i,x}$ is the CD value of the particle i $(2 \le i \le n-1)$ in the decision space, $x_{j,i}$ is the value of particle i in the jth $(1 \le j \le m)$ dimension of decision space, $x_{j,max}$ is the maximum in the jth dimension of decision space, $x_{j,min}$ is the minimum in the jth dimension of decision space, n is the number of population, m is the number of dimensions in the decision space.

The method of calculating $CD_{i,f}$ in the objective space is implemented in an analogous fashion for the decision space. Where range of j is $(1 \leq j \leq M)$in the objective space, M is the number of objective in the objective space. However, the calculation methods are different when considering the boundary points in the decision space and the objective space.

In decision space (Minimization problem):

$$CD_{i,x} = 2 * (\frac{|x_{1,2} - x_{1,1}|}{|x_{1,max} - x_{1,min}|} + \frac{|x_{j,2} - x_{j,1}|}{|x_{j,max} - x_{j,min}|} + \cdots + \frac{|x_{m,2} - x_{m,1}|}{|x_{m,max} - x_{m,min}|}). \tag{2}$$

The method of calculating another boundary points is implemented in an analogous fashion for formula (2).

In objective space (Minimization problem):

$$CD_{i,f} = \begin{cases} 1, i = f_{M,\min} \\ 0, i = f_{M,\max} \end{cases} \tag{3}$$

The boundary point of the maximization problem is treated in the opposite way to formula (3).

Distribution of SCD:

$$SCD_i = \begin{cases} \max\left(CD_{i,x}, CD_{i,f}\right), CD_{i,x} > CD_{\text{arg},x} \quad \text{or} \quad \mathrm{CD_{i,f} > CD_{arg,f}} \\ \min\left(CD_{i,x}, CD_{i,f}\right), CD_{i,x} \leq CD_{\text{arg},x} \quad \text{or} \quad \mathrm{CD_{i,f} \leq CD_{avg,f}} \end{cases} \tag{4}$$

According to the CD value calculated above, make the distribution as shown in formula (4). Where, $CD_{\text{arg},x}$ is the average CDs value in the decision space, $CD_{\text{arg},f}$ express analogous meaning in objective space.

3 The Proposed Algorithm Kn_NSGA-II_SCD

In this section, Kn_NSGA-II_SCD is proposed for the problem of multi-modal characteristics in multi-objective optimization. In Algorithm 1, we mainly describe the framework of the proposed algorithm Kn_NSGA-II_SCD. In the following work, we introduce the tournament selection and environment selection in detail, which is an important part of the algorithm that proposed in this paper.

3.1 A General Framework of Kn_NSGA-II_SCD

In Algorithm 1, the initial population P is randomly generated with N different individuals. After that, the niche strategy is embedded into the binary tournament mechanism to select the mating pool, referring to lines 6 in Algorithm 1. Next, the offspring population is generated by the method of simulating binary intersection and the operation of polynomial mutation. The initial population and offspring population are merged to perform non-dominated sorting, find the knee point and calculate the SCD distance, referring to lines 7–9 in Algorithm 1. Finally, environment selection is carried out to obtain N individuals to enter the next generation. This process is repeated until the termination condition is satisfied, referring to lines 10–12 in Algorithm 1.

Algorithm 1. Framework of Kn_NSGA2_SCD.

Require: N: population size; Maxiter: the maximum number of iterations; T: rate of knee points in population;
Ensure: The final population P;
1. $k \leftarrow \varnothing$;
2. $r \leftarrow 1, t \leftarrow 0$ /* adaptive parameters for finding knee points */
3. $P \leftarrow$ Initialize (N);
4. *iter*=1;
5. **while** *iter* $\leq$ Maxiter **do**
6. Pool←Tournament_selection(P, N); /* Using the niche strategy to select the Pool */
7. P← P $\bigcup$ Genetic(Pool,N);
8. F←Nondominated_sort(P); /* Calculate the SCD value of each individual */
9. $[k, r, t]$ ←Finding_knee point(F, T, r);
10. P←Environmental_selection(F,k, N);
11. *iter*=*iter*+1;
12. **end while**

3.2 Tournament Selection

In Algorithm 2, the entire tournament selection mechanism is described in detail. Firstly, the size of the niche and the mating pool are set to c. Then the collection of candidates is filled with different individuals of randomly select from the population P, referring to lines 3–5 in Algorithm 2. The distance between Candidate_dis (1), an individual in the binary tournament, and all candidates after removing the first Candidate is calculated. The smallest Candidate_dis individual is selected as another individual in the binary tournament, referring to lines 6–8 in Algorithm 2. If one of the two individuals in the binary tournament mechanism dominates the other, then the former is selected into the mating pool. Moreover, if the two individuals are non-dominate each other, the individuals with large CD value in the decision space is selected into the mating pool. If the two individuals have the same CD value, put Candidate (1) into the mating pool, referring to lines 9–21 in Algorithm 2. The operation is terminated until the mating pool is filled.

3.3 Environmental Selection

The purpose of environment selection is to select a more suitable solution for the next generation as the parent to update the particle. The detailed description is shown in Algorithm 3. Firstly, in this generation, a combination of the parent and offspring populations is carried out performed non-dominated sorting, forming N_f non-dominated fronts. Then, Look for the first non-dominated fronts F_j that can fulfill $F_1 \bigcup \cdots \bigcup F_j \geq N$, and put the individuals of the former $j-1$ non-dominated fronts into the collection Y , referring to lines 1–2 in Algorithm 3. If the sum of Y and $k \bigcap F_j$ greater than N, we will select $|Y| - N$ individuals with a large SCD distance from $k \bigcap F_j$ into the Y, referring to lines 3–5 in Algorithm 3. If the sum less than N, we will select $N - |Y|$ individuals with a large SCD distance from $F_j \backslash (F_j \cap K)$ into the Y, referring to lines 6–8 in Algorithm 3.

Algorithm 2. Tournament_selection(P, N, $CD_{i,x}$).

Require: N: population size; P: population; $CD_{i,x}$;
Ensure: Pool;
1. Pool_size $\leftarrow$ $(N/2)$, Niche_size $\leftarrow$ $(N/2)$, $Pool \leftarrow \varnothing$, $Candidate \leftarrow \varnothing$;
2. **while** i <Pool_size **do**
3. **if** j <Niche_size **then**
4. Candidate(j) $\leftarrow$ Randomly select an individual from the population P /* No identical individuals in the Candidate */
5. **end if**
6. a $\leftarrow$ Candidate(1);
7. Candidate_dis $\leftarrow$ /* Calculate the distance between remove the first individual in Candidate and Candidate (1) */
8. b $\leftarrow$ Candidate(find(Candidate_dis == min(Candidate_dis) +1);
9. **if** a $\prec$ b **then**
10. Pool $\leftarrow$ Pool $\bigcup$ {a}
11. **else if** b $\prec$ a **then**
12. Pool $\leftarrow$ Pool $\bigcup$ {b}
13. **else**
14. **if** $CD_{a,x}$ >$CD_{b,x}$ **then**
15. Pool $\leftarrow$ Pool $\bigcup$ {a}
16. **else if** $CD_{b,x} > CD_{a,x}$ **then**
17. Pool $\leftarrow$ Pool $\bigcup$ {b}
18. **else**
19. Pool $\leftarrow$ Pool $\bigcup$ {a}
20. **end if**
21. **end if**
22. **end while**
23. **return** Pool

Algorithm 3. Environmental selection(F, N, k, SCD_i).

Require: N: population size; F: sorted population; k: knee points; SCD_i: distance of the particle i ($1 \le i \le F_{size}$);
Ensure: Y /* next population */
1. $Y \leftarrow \varnothing$;
2. $Y \leftarrow F_1 \bigcup \cdots \bigcup F_{j-1}$
3. $Y \leftarrow Y \bigcup (k \bigcap F_j)$;
4. **if** $|Y|>N$ **then**
5. select $|Y|-N$ solutions from $k \bigcap F_j$ to Y which have the maximum distances of SCD
6. **else if** $|Y|<N$ **then**
7. add $N-|Y|$ solutions from $F_j \backslash (F_j \cap K)$ to Y which have the maximum distances of SCD
8. **end if**
9. **return** Y

Table 1. The properties of test function

Problems	Shifting	Separability	Multi-modal	Degeneration	Differentiation	Symmetry
PMOP1	L	NS	Uni	F	T	T/F
PMOP2	NL	S	Uni	F	T	T/F
PMOP3	L	S	Multi	F	T	T/F
PMOP4	L	S	Multi	F	F	F
PMOP5	L	NS	Multi	F	F	F
PMOP6	NL	S	Multi	F	T	T/F
PMOP7	L	NS	Multi	F	T	T/F
PMOP8	NL	S	Multi	F	T	T/F
PMOP9	L	NS	Uni	F	T	T/F
PMOP10	NL	NS	Multi	F	T	F
PMOP11	NL	NS	Uni	F	T	T/F
PMOP12	NL	S	Multi	F	T	T/F
PMOP13	L	NS	Uni	T	T	T/F
PMOP14	NL	S	Multi	T	T	T/F

4 Experimental Results and Discussion

4.1 Test Functions

The Properties of 14 test functions are detailed in Table 1. Way of shifting represents the way of link in decision space and objective space, where L is a linear link and NL is non-linear link. The separable or non- separable of problem is displayed by S/NS. Uni/ Multi shows that the test problem is unimodal or multimodal. T/F of the Table 1 mainly shows whether the related feature is true of false.

4.2 Experimental Settings

The Kn_NSGA2_SCD algorithm proposed in this paper will conduct a comparative analysis on the four performance indexes of IGD, KD, KGD, KIGD with the six methods mentioned in Guo [23]. The parameter settings of the comparison algorithm are consistent with the original text, and in this paper the specific parameter settings are assigned in Table 2. In addition, the population size is set as 105, and 132 for 3-objective and 5-objective PMOP test problems, respectively. In the experiments, the distribution index is set to 20 in both the simulated binary crossover operator and polynomial mutation. The parameter settings of looking for knee points can refer to Zhang [5].

4.3 Experimental Results and Discussion

According to the above parameter settings, Tables 3, 4, 5, and 6 express the experimental results on PMOP test suite. The IGD mainly evaluates the degree

Table 2. Parameter settings

Problems	Generations	Number of runs	A	B	s	p	l
PMOP1	3000	30	4	1	−1	1	–
PMOP2	3000	30	4	1	2	1	–
PMOP3	3000	30	4	1	2	1	–
PMOP4	10000	30	6	1	−1	1	–
PMOP5	10000	30	1	1	2	1	12
PMOP6	3000	30	2	1	2	1	–
PMOP7	3000	30	4	1	2	1	–
PMOP8	3000	30	4	1	2	1	–
PMOP9	3000	30	2	1	2	1	–
PMOP10	5000	30	1	1	2	1	12
PMOP11	5000	30	4	1	2	1	–
PMOP12	5000	30	4	1	2	1	–
PMOP13	3000	30	2	1	−2	1	–
PMOP14	5000	30	2	1	−1	1	–

of approximation between the obtained solution set and the known Pareto-optimal solution set. The purpose of KGD is to evaluate the distance between the solution and the knee regions. It also assesses their ability to identify solutions in the knee regions. The KIGD value represents the coverage degree of the solution to the knee regions, and mainly evaluates the distribution of the solution in the knee regions. When the DM favors the solutions near the knee points, KD can evaluate the capacity of finding all knee points and whether there is at least one solution in the solution set near the knee point. Therefore, KD is an evaluation of the algorithm's ability to recognize solutions close to the knee points. IGD primarily evaluates the degree of approximation between the obtained solution set and the known Pareto-optimal solution set.

In this paper, we mainly adopt niche technology to maintain of different solutions in decision space and objective space. And the method of finding the knee point and the special crowded distance are used as the selection condition to update the next generation and increase the selection pressure. Finally, in the final experimental results, smaller values of IGD, KGD, KIGD, and KD indicate better performance. As can be seen from Tables 3, 5 and 6, the algorithm proposed in this paper is very competitive in finding knee point on the Pareto fronts, and compared with the previous six methods, the final solution set has advantages in covering the reference knee regions and converging to the optimal Pareto front. However, the ability to recognize solution sets within the knee regions is not so ideal.

Table 3. The IGD results obtained by seven algorithms on PMOP test suit

Problem(m)	RVEA+WD	RVEA+Dis	RVEA+EMU	NSGA_II+WD	NSGA_II+Dis	NSGA_II+EMU	Kn_NSGA-II_SCD
PMOP1(3)	5.52E-01 (4.08E-03)	4.69E-01 (6.46E-03)	7.00E-01 (4.09E-02)	7.25E-01 (1.79E-02)	1.09E+00 (2.23E-01)	1.07E+00 (1.23E-01)	**3.16E-01(5.07E-02)**
PMOP1(5)	2.03E+00 (1.80E-02)	2.34E+00 (4.28E-01)	2.05E+00 (1.32E-01)	6.59E+00 (1.76E+01)	9.12E+00 (1.74E+01)	8.53E+00 (1.50E+01)	**6.87E-01(4.37E-02)**
PMOP2(3)	1.60E-01 (5.18E-04)	**1.49E-01 (8.13E-04)**	3.08E-01(1.50E-04)	3.48E-01 (1.71E-05)	3.41E-01(1.47E-03)	4.04E-01 (3.49E-04)	1.44E-01(1.47E-02)
PMOP2(5)	2.48E-01 (1.90E-03)	**2.27E-01 (2.69E-03)**	2.71E-01 (8.79E-04)	1.90E+00 (1.90E+01)	6.76E+00 (1.49E+02)	3.99E+00 (2.49E+01)	2.35E-01(1.42E-02)
PMOP3(3)	7.12E-01 (1.77E-02)	6.41E-01 (1.78E-02)	9.99E-01 (3.96E-03)	1.15E+00 (3.23E-02)	1.75E+00 (1.79E-01)	1.39E+00 (4.79E-05)	**1.48E-01(1.15E-02)**
PMOP3(5)	8.37E-01 (4.07E-03)	9.01E-01 (1.36E-03)	9.27E-01 (1.29E-03)	8.83E+00 (5.30E+00)	1.02E+01 (1.48E+01)	9.07E+00 (1.08E+01)	**1.85E-01(1.72E-02)**
PMOP4(3)	4.98E+00 (1.72E+01)	1.09E+00 (2.63E-01)	1.19E+00 (2.83E-01)	8.89E-01 (1.45E-02)	9.01E-01 (9.23E-03)	8.69E-01 (1.43E-02)	**5.02E-01(3.03E-01)**
PMOP4(5)	1.04E+00 (2.31E-02)	1.41E+00 (4.71E-01)	**1.03E+00 (2.72E-02)**	4.65E+02 (5.62E+04)	9.53E+02 (2.13E+05)	8.68E+02 (1.18E+05)	1.05E+00(2.19E-01)
PMOP5(3)	9.93E+00 (3.18E+01)	6.13E+00 (1.60E+01)	5.45E+00 (6.20E+00)	1.54E+01 (5.50E+01)	1.34E+01 (1.54E+01)	1.28E+01 (5.10E+01)	**1.04E-01(3.09E-02)**
PMOP5(5)	3.16E+01 (1.90E+02)	3.89E+01 (3.85E+02)	5.46E+01 (9.26E+02)	3.99E+01 (1.06E+03)	6.69E+01 (2.24E+03)	5.80E+01 (1.73E+03)	**2.37E-01(2.01E-02)**
PMOP6(3)	6.15E-01 (5.91E-03)	7.47E-01 (1.58E-01)	6.67E-01 (1.63E-02)	8.74E-01 (6.67E-03)	1.28E+00 (1.99E-01)	8.99E-01 (9.21E-05)	**1.71E-01(4.61E-02)**
PMOP6(5)	1.06E+00 (3.54E-03)	1.04E+00 (3.97E-03)	1.11E+00 (2.18E-03)	3.34E+01 (2.59E+02)	4.39E+01 (4.04E+02)	5.63E+01 (7.11E+02)	**2.95E-01(1.46E-02)**
PMOP7(3)	2.57E-01(5.14E-04)	4.60E-01 (9.53E-03)	3.98E-01 (1.09E-03)	3.33E-01 (2.15E-03)	6.98E-01 (7.45E-02)	5.52E-01 (1.42E-02)	**1.14E-01(9.11E-03)**
PMOP7(5)	3.69E-01 (2.52E-03)	4.13E-01 (2.00E-02)	3.26E-01 (1.93E-03)	8.89E-01 (7.95E-03)	9.51E-01 (5.83E-02)	1.18E+00 (3.13E-02)	**1.70E-01(6.41E-03)**
PMOP8(3)	1.34E-01 (2.35E-04)	2.45E-01 (5.80E-03)	2.17E-01 (1.65E-03)	2.45E-01 (4.75E-04)	2.39E-01 (9.77E-04)	3.28E-01 (1.97E-03)	**7.88E-02(1.63E-02)**
PMOP8(5)	1.17E-01 (4.69E-05)	1.43E-01 (2.60E-03)	1.26E-01 (6.45E-05)	6.91E-01 (1.46E-01)	1.54E+00 (1.15E+00)	1.70E+00 (3.32E-01)	**1.15E-01(1.96E-02)**
PMOP9(3)	2.32E-01 (9.90E-04)	1.89E-01 (1.32E-04)	3.25E-01 (4.33E-04)	2.14E-01 (1.07E-03)	1.54E-01 (2.28E-04)	3.00E-01 (5.70E-03)	**6.48E-02(5.36E-03)**
PMOP9(5)	4.95E-01 (1.21E-03)	4.74E-01 (2.02E-03)	4.70E-01 (7.42E-04)	2.71E+00 (2.83E-01)	3.71E+00 (6.41E-01)	3.36E+00 (2.64E-01)	**1.68E-01(1.25E-02)**
PMOP10(3)	1.57E+00 (2.04E-03)	1.52E+00 (4.48E-03)	1.64E+00 (2.41E-04)	1.51E+00 (7.94E-02)	7.92E-01 (2.37E-02)	1.42E+00 (7.56E-03)	**2.63E-01(9.82E-03)**
PMOP10(5)	1.44E+00 (4.16E-03)	1.45E+00 (9.52E-03)	1.50E+00 (5.00E-03)	2.36E+00 (7.74E-02)	4.18E+00 (4.11E+00)	2.47E+00 (2.07E-01)	**4.16E-01(1.87E-02)**
PMOP11(3)	7.94E-01 (8.61E-03)	3.79E-01 (1.93E-04)	8.50E-01 (3.89E-03)	5.64E+00 (6.97E+01)	5.54E-01 (4.16E-03)	3.47E+01 (1.24E+02)	**2.86E-01(4.37E-02)**
PMOP11(5)	1.16E+00 (2.03E-02)	1.08E+00 (2.61E-02)	1.06E+00 (2.43E-02)	1.45E+00 (6.52E-03)	2.32E+00 (4.33E+00)	1.95E+00 (6.72E-01)	**5.62E-01(3.19E-02)**
PMOP12(3)	2.03E-01 (7.27E-04)	1.26E-01 (3.02E-05)	1.53E-01 (2.05E-04)	8.96E-02 (4.52E-04)	8.80E-02 (2.95E-04)	1.26E-01 (1.10E-03)	**2.36E-02(2.02E-03)**
PMOP12(5)	3.60E-02 (1.22E-06)	3.58E-02 (1.15E-06)	3.48E-02 (4.37E-07)	5.99E-01 (5.28E-02)	7.15E-01 (1.22E-01)	6.93E-01 (8.45E-02)	**1.08E-02(1.43E-03)**
PMOP13(3)	5.66E-01 (6.62E-03)	8.16E-01 (3.76E-02)	7.15E-01 (1.71E-03)	6.77E-01 (2.63E-02)	9.76E-01(1.11E-01)	1.45E+00 (3.92E-01)	**1.54E-01(1.16E-02)**
PMOP13(5)	2.17E+00 (1.96E-02)	3.14E+00 (5.66E-01)	1.99E+00 (2.07E-02)	1.11E+01 (4.66E+01)	1.47E+01 (4.18E+01)	1.61E+01 (5.76E+01)	**1.18E+00(2.57E-02)**
PMOP14(3)	1.78E+00 (1.07E-03)	1.16E+00 (1.22E-04)	1.81E+00 (1.82E-05)	7.80E-01 (2.33E-02)	1.02E+00 (4.24E-02)	1.21E+00 (8.32E-02)	**1.25E-01(2.61E-02)**
PMOP14(5)	1.03E+00 (2.07E-02)	9.54E-01 (5.40E-03)	9.08E-01 (7.87E-05)	1.74E+01 (5.43E+01)	4.23E+01 (5.27E+02)	2.51E+01 (1.27E+02)	**2.74E-01(5.43E-03)**

Table 4. The KGD results obtained by seven algorithms on PMOP test suit

Problem(m)	RVEA+WD	RVEA+Dis	RVEA+EMU	NSGA_II+WD	NSGA_II+Dis	NSGA_II+EMU	Kn_NSGA-II_SCD
PMOP1(3)	2.51E-01 (1.99E-02)	2.92E-01 (1.78E-02)	**1.36E-01 (5.08E-03)**	1.50E-01 (1.46E-03)	5.91E-01 (7.89E-01)	1.61E-01 (1.32E-04)	4.98E-02 (3.98E-03)
PMOP1(5)	4.03E-01 (5.10E-02)	7.63E-01 (8.00E-02)	**2.93E-01 (1.68E-02)**	9.21E+00 (7.73E-01)	2.28E+01 (9.78E+00)	1.09E+01 (2.44E+00)	7.15E-01 (1.34E-01)
PMOP2(3)	5.07E-02 (3.33E-04)	3.16E-02 (3.76E-05)	4.56E-02 (9.35E-06)	9.80E-01 (1.73E+01)	9.93E-01 (6.48E+00)	3.79E-02 (6.67E-05)	**1.18E-02 (2.09E-03)**
PMOP2(5)	**4.37E-02 (2.06E-04)**	4.44E-02 (2.59E-03)	4.87E-02 (5.30E-04)	3.49E+01 (2.70E+02)	2.01E+02 (9.88E+02)	2.32E+01 (3.53E+01)	1.41E-01 (9.91E-03)
PMOP3(3)	1.52E+00 (1.10E+00)	7.41E-02 (3.03E-03)	**7.55E-03 (4.59E-05)**	8.43E-02 (4.87E-03)	1.24E+00 (1.03E+00)	8.43E-03 (2.25E-06)	1.47E-01 (1.22E-01)
PMOP3(5)	2.50E-01 (3.13E-02)	3.03E-02 (2.98E-04)	**1.79E-02 (2.64E-06)**	6.09E+01 (1.28E+02)	3.42E+02 (4.01E+03)	6.43E+01 (1.08E+02)	1.82E+00 (1.50E+00)
PMOP4(3)	1.36E+02 (1.66E+04)	8.48E-01 (1.88E+00)	3.66E-01 (7.10E-02)	2.25E+01 (1.52E+01)	6.14E+01 (3.38E+02)	**5.97E-03 (1.16E-06)**	3.62E-02 (4.67E-03)
PMOP4(5)	1.58E+02 (2.20E+04)	1.79E+00 (1.74E+01)	**2.36E-01 (9.01E-02)**	1.28E+04 (3.99E+07)	6.88E+04 (2.05E+08)	7.17E+03 (2.08E+07)	3.75E+00 (7.08E+00)
PMOP5(3)	2.77E+01 (5.73E+02)	2.80E+01 (2.96E+01)	2.40E+01 (1.37E+02)	2.40E+01 (5.52E+00)	5.51E+01 (2.75E+02)	2.53E+01 (5.82E+00)	**2.06E-02(3.38E-03)**
PMOP5(5)	2.63E+01 (4.50E+01)	1.96E+01 (7.03E+01)	2.87E+01 (1.19E+01)	1.39E+03 (6.39E+06)	4.45E+05 (1.01E+12)	1.27E+03 (2.91E+07)	**2.23E-02(3.22E-03)**
PMOP6(3)	3.25E+00 (1.35E+00)	4.65E-01 (1.14E-01)	2.34E-01 (9.95E-03)	4.32E-01 (1.04E-01)	2.36E+00 (4.96E+00)	1.20E-01 (3.00E-06)	**4.69E-02(6.42E-03)**
PMOP6(5)	2.56E-01 (7.46E-02)	2.30E-01 (7.60E-03)	**9.67E-02 (2.19E-03)**	4.45E+01 (7.84E+02)	8.07E+02 (2.78E+04)	1.31E+02 (4.12E+03)	5.09E-01(3.46E-02)
PMOP7(3)	6.48E-02 (4.59E-04)	1.81E-01 (1.03E-03)	1.02E-01 (1.20E-04)	1.16E-01 (1.96E-04)	2.39E-01 (1.76E-02)	1.42E-01 (2.41E-04)	**4.12E-02(1.91E-03)**
PMOP7(5)	**5.97E-02 (2.34E-04)**	9.09E-02 (3.27E-03)	9.04E-02 (4.11E-04)	3.15E-01 (1.03E-03)	3.50E-01 (1.54E-02)	4.55E-01 (2.51E-03)	3.68E-01(2.15E-02)
PMOP8(3)	6.22E-02 (1.16E-03)	1.13E-01 (2.66E-03)	**3.14E-02 (1.11E-05)**	3.25E-02 (1.37E-06)	6.94E-02 (5.36E-02)	4.23E-02 (1.25E-06)	3.60E-02(6.05E-02)
PMOP8(5)	4.45E-02 (1.06E-03)	5.63E-02 (2.85E-03)	**2.51E-02 (1.20E-04)**	1.29E+00 (1.15E-02)	2.60E+00 (1.36E-01)	1.10E+00 (1.11E-02)	6.08E-01(1.10E+00)
PMOP9(3)	9.25E-02 (3.44E-04)	5.51E-02 (1.06E-04)	**4.28E-02 (1.86E-05)**	1.06E-01 (4.16E-04)	7.58E-02 (4.68E-05)	9.22E-02 (5.03E-05)	6.24E-02(8.36E-03)
PMOP9(5)	**3.05E-02 (1.06E-04)**	1.14E-01 (1.95E-03)	4.05E-02 (2.93E-04)	1.19E+00 (2.38E-02)	5.03E+00 (1.77E+00)	1.59E+00 (5.07E-02)	7.90E-01(1.10E-01)
PMOP10(3)	2.27E-01 (1.43E-05)	2.91E-01 (8.74E-04)	2.26E-01 (1.05E-05)	1.71E+01 (3.74E+01)	9.29E+00 (5.30E+02)	1.13E+01 (8.83E+00)	**6.16E-02(1.86E-03)**
PMOP10(5)	1.49E-01 (1.11E-03)	1.43E-01 (1.56E-03)	1.56E-01 (2.69E-03)	1.63E+01 (1.12E+02)	2.22E+02 (4.15E+03)	1.25E+01 (6.70E+01)	**7.05E-02(2.34E-03)**
PMOP11(3)	3.86E-02 (2.32E-04)	1.74E-01 (7.47E-03)	8.32E-02 (4.05E-03)	2.44E+02 (6.73E+03)	7.17E+00 (9.51E+01)	8.22E+01 (2.43E+02)	**2.90E-02(7.67E-04)**
PMOP11(5)	1.04E-01 (2.72E-02)	1.69E-01 (2.60E-02)	1.00E-01 (9.66E-03)	9.77E+01 (9.46E+02)	4.22E+02 (7.84E+03)	3.85E+00 (6.52E+00)	**6.04E-02(2.88E-03)**
PMOP12(3)	6.61E-02 (7.91E-05)	4.06E-02 (3.15E-05)	1.91E-02 (1.36E-05)	1.70E-02 (1.94E-04)	4.40E-02 (1.32E-03)	**1.15E-02 (4.76E-07)**	3.26E-02(2.89E-02)
PMOP12(5)	6.58E-03 (4.60E-06)	9.06E-03 (7.70E-06)	**3.18E-03 (7.74E-07)**	9.24E-01 (2.20E-01)	6.64E+00 (1.49E+01)	5.06E-01 (2.83E-03)	3.46E-02(2.30E-02)
PMOP13(3)	2.16E-01 (2.56E-02)	5.38E-01 (6.87E-02)	6.79E-02 (4.34E-04)	4.33E-01 (8.20E-04)	6.80E-01 (5.16E-01)	4.63E-01 (1.31E-03)	**1.39E-01(3.42E-03)**
PMOP13(5)	1.44E+00 (8.14E-02)	2.93E+00 (4.79E-01)	1.12E+00 (2.65E-02)	1.64E+01 (1.59E+00)	6.23E+01 (1.44E+01)	2.41E+01 (1.54E+01)	**7.82E+00(7.23E-01)**
PMOP14(3)	5.38E-01 (9.90E-04)	3.58E-01 (2.07E-03)	4.25E-01 (2.48E-04)	4.13E-01 (1.53E-01)	1.55E+00 (2.10E+00)	3.02E-01 (6.07E-06)	**8.61E-02(1.42E-02)**
PMOP14(5)	5.14E-01 (1.88E-02)	6.99E-01 (7.00E-03)	4.54E-01 (7.25E-04)	2.41E+01 (2.02E+02)	5.79E+02 (5.80E+04)	2.04E+01 (3.54E+02)	**1.90E-01(4.47E-02)**

Table 5. The KIGD results obtained by seven algorithms on PMOP test suit

Problem(m)	RVEA+WD	RVEA+Dis	RVEA+EMU	NSGA_II+WD	NSGA_II+Dis	NSGA_II+EMU	Kn_NSGA-II_SCD
PMOP1(3)	4.72E-01 (8.17E-03)	9.12E-01 (2.57E-03)	9.37E-01 (2.17E-03)	1.01E+00 (4.19E-02)	1.19E+00 (2.94E-01)	1.45E+00 (7.71E-02)	**1.81E-01(3.04E-02)**
PMOP1(5)	2.03E+00 (1.80E-02)	**3.58E-01 (1.43E-02)**	5.39E-01 (2.90E-04)	5.92E-01 (1.84E-04)	8.80E+00 (1.76E+01)	5.78E-01 (2.64E-03)	6.45E-01(4.66E-02)
PMOP2(3)	1.54E-01 (6.98E-04)	1.73E-01 (9.64E-04)	2.83E-01 (2.59E-04)	3.27E-01 (5.21E-05)	3.31E-01 (2.38E-03)	3.79E-01 (3.88E-04)	**9.97E-02(1.99E-02)**
PMOP2(5)	2.17E-01 (1.05E-03)	2.25E-01 (3.69E-03)	2.41E-01 (6.98E-04)	1.83E+00 (1.91E+01)	6.69E+00 (1.49E+02)	3.93E+00 (2.50E+01)	**2.16E-01(1.50E-02)**
PMOP3(3)	6.69E-01 (2.39E-02)	6.23E-01 (2.68E-02)	9.27E-01 (3.47E-03)	1.15E+00 (2.91E-02)	1.80E+00 (1.91E-01)	1.34E+00 (2.42E-05)	**1.48E-01(1.47E-02)**
PMOP3(5)	7.32E-01 (3.63E-03)	7.95E-01 (1.21E-03)	8.14E-01 (1.22E-03)	8.89E+00 (5.30E+00)	1.02E+01 (1.48E+01)	9.13E+00 (1.08E+01)	**1.97E-01(1.89E-02)**
PMOP4(3)	5.08E+00 (1.67E+01)	1.26E+00 (2.56E-01)	1.34E+00 (2.98E-01)	9.00E-01 (7.83E-02)	8.27E-01 (5.36E-02)	8.62E-01 (7.59E-02)	**4.23E-01(2.51E-01)**
PMOP4(5)	1.07E+00 (5.03E-02)	1.47E+00 (5.45E-01)	1.05E+00 (6.61E-02)	4.65E+02 (5.61E+04)	9.53E+02 (2.13E+05)	8.68E+02 (1.18E+05)	**9.79E-01(3.35E-01)**
PMOP5(3)	1.00E+01 (3.17E+01)	6.18E+00 (1.58E+01)	5.51E+00 (6.20E+00)	1.55E+01 (5.50E+01)	1.35E+01 (1.54E+01)	1.29E+01 (5.10E+01)	**4.73E-02(6.06E-03)**
PMOP5(5)	3.16E+01 (1.90E+02)	3.89E+01 (3.85E+02)	5.46E+01 (9.25E+02)	4.00E+01 (1.06E+03)	6.70E+01 (2.23E+03)	5.81E+01 (1.73E+03)	**2.18E-01(1.66E-02)**
PMOP6(3)	4.32E-01 (2.67E-02)	8.53E-01 (3.07E-01)	8.68E-01 (9.76E-02)	1.03E+00 (2.48E-03)	1.72E+00 (3.45E-01)	1.02E+00 (7.27E-09)	**6.39E-02(5.00E-03)**
PMOP6(5)	7.41E-01 (3.22E-03)	8.41E-01 (1.44E-02)	8.25E-01 (1.05E-02)	3.35E+01 (2.59E+02)	4.38E+01 (4.04E+02)	5.63E+01 (7.11E+02)	**2.25E-01(7.84E-03)**
PMOP7(3)	1.49E-01 (5.66E-04)	4.98E-01 (5.63E-03)	3.25E-01 (2.70E-03)	3.20E-01 (4.04E-03)	6.83E-01 (7.89E-02)	5.41E-01 (1.63E-02)	**7.78E-02(7.94E-03)**
PMOP7(5)	3.81E-01 (4.28E-03)	4.45E-01 (2.67E-02)	3.34E-01 (4.24E-03)	8.44E-01 (9.74E-03)	8.54E-01 (7.64E-02)	1.01E+00 (3.66E-02)	**1.46E-01(5.73E-03)**
PMOP8(3)	1.22E-01 (3.61E-04)	2.31E-01 (6.94E-03)	2.00E-01 (5.09E-03)	2.10E-01 (8.13E-04)	2.25E-01 (1.32E-03)	3.12E-01 (5.35E-03)	**6.86E-02(1.88E-02)**
PMOP8(5)	1.17E-01 (4.69E-05)	1.35E-01 (3.22E-03)	1.01E-01 (1.33E-04)	6.49E-01 (1.51E-01)	1.51E+00 (1.15E+00)	1.67E+00 (3.35E-01)	**1.06E-01(1.96E-02)**
PMOP9(3)	1.85E-01 (4.05E-03)	1.07E-01 (5.98E-04)	1.13E-01 (5.15E-04)	1.74E-01 (2.28E-03)	1.94E-01 (2.41E-03)	2.53E-01 (2.15E-02)	**7.37E-02(1.41E-02)**
PMOP9(5)	3.40E-01 (3.82E-04)	4.21E-01 (5.19E-03)	3.60E-01 (1.50E-03)	2.72E+00 (2.95E-01)	3.73E+00 (7.36E-01)	3.36E+00 (2.84E-01)	**1.72E-01(1.77E-02)**
PMOP10(3)	1.47E+00 (4.39E-03)	1.40E+00 (1.07E-02)	1.54E+00 (2.89E-04)	1.39E+00 (9.02E-02)	1.07E+00 (3.76E-02)	1.29E+00 (8.88E-03)	**2.90E-01(2.14E-02)**
PMOP10(5)	1.22E+00 (4.88E-03)	1.23E+00 (1.33E-02)	1.28E+00 (6.42E-03)	2.26E+00 (8.90E-02)	4.16E+00 (4.38E+00)	2.41E+00 (2.54E-01)	**5.16E-01(2.41E-02)**
PMOP11(3)	6.54E-01 (6.14E-03)	3.12E-01 (5.29E-04)	7.21E-01 (5.23E-04)	5.29E+00 (7.11E+01)	5.50E-01 (4.73E-03)	3.47E+01 (1.25E+02)	**2.50E-01(5.41E-02)**
PMOP11(5)	9.05E-01 (8.43E-03)	8.31E-01 (1.32E-02)	8.00E-01 (1.88E-02)	1.16E+00 (7.32E-03)	2.04E+00 (4.42E+00)	1.67E+00 (7.07E-01)	**4.63E-01(4.65E-02)**
PMOP12(3)	1.99E-01 (9.31E-04)	1.16E-01 (5.73E-05)	1.42E-01 (2.10E-04)	9.45E-02 (9.07E-04)	1.04E-01 (5.80E-04)	1.27E-01 (1.39E-03)	**2.00E-02(2.43E-03)**
PMOP12(5)	2.84E-02 (1.90E-06)	2.83E-02 (2.04E-06)	2.70E-02 (3.95E-07)	6.02E-01 (5.30E-02)	7.18E-01 (1.22E-01)	6.96E-01 (8.47E-02)	**9.90E-03(1.47E-03)**
PMOP13(3)	3.16E-01 (1.80E-02)	5.88E-01 (1.12E-01)	1.45E-01 (1.91E-03)	6.02E-01 (1.46E-01)	1.16E+00 (4.57E-01)	1.48E+00 (6.20E-01)	**9.06E-02(8.75E-03)**
PMOP13(5)	1.24E+00 (1.02E-01)	3.46E+00 (1.60E+00)	1.19E+00 (7.98E-02)	1.17E+01 (5.02E+01)	1.54E+01 (4.42E+01)	1.69E+01 (6.14E+01)	**7.33E-01(2.37E-02)**
PMOP14(3)	1.71E+00 (1.14E-03)	1.10E+00 (4.23E-05)	1.74E+00 (2.07E-05)	1.10E+00 (1.68E-02)	1.36E+00 (4.33E-02)	1.36E+00 (2.85E-02)	**8.00E-02(1.71E-02)**
PMOP14(5)	9.47E-01 (3.03E-02)	8.40E-01 (1.11E-02)	8.02E-01 (9.33E-05)	1.76E+01 (5.45E+01)	4.23E+01 (5.30E+02)	2.52E+01 (1.27E+02)	**1.28E-01(1.59E-02)**

Table 6. The KD results obtained by seven algorithms on PMOP test suit

Problem(m)	RVEA+WD	RVEA+Dis	RVEA+EMU	NSGA_II+WD	NSGA_II+Dis	NSGA_II+EMU	Kn_NSGA-II_SCD
PMOP1(3)	5.01E-01 (1.96E-02)	3.51E-01 (4.94E-03)	6.31E-01 (8.87E-02)	6.76E-01 (3.63E-02)	1.22E+00 (3.17E-01)	1.19E+00 (1.80E-01)	**1.60E-01(3.86E-02)**
PMOP1(5)	1.85E+00 (1.72E-02)	2.41E+00 (7.46E-01)	1.97E+00 (1.94E-01)	6.35E+00 (1.95E+01)	8.97E+00 (1.88E+01)	8.37E+00 (1.64E+01)	**3.22E-01(4.34E-02)**
PMOP2(3)	1.47E-01 (1.24E-03)	1.67E-01 (1.05E-03)	2.81E-01 (2.80E-04)	3.04E-01 (1.21E-04)	3.09E-01 (2.54E-03)	3.70E-01 (3.55E-04)	**8.83E-02(2.21E-02)**
PMOP2(5)	2.22E-01 (2.07E-03)	2.08E-01 (4.31E-03)	2.74E-01 (1.31E-03)	1.88E+00 (2.01E+01)	6.76E+00 (1.57E+02)	3.99E+00 (2.63E+01)	**2.04E-01(1.26E-02)**
PMOP3(3)	6.62E-01 (3.18E-02)	6.03E-01 (4.55E-02)	8.92E-01 (1.64E-03)	1.16E+00 (3.89E-02)	1.80E+00 (1.75E-01)	1.35E+00 (1.08E-05)	**1.43E-01(4.06E-02)**
PMOP3(5)	7.33E-01 (7.02E-03)	8.03E-01 (1.72E-03)	8.23E-01 (2.00E-03)	8.85E+00 (5.59E+00)	1.02E+01 (1.56E+01)	9.08E+00 (1.14E+01)	**1.83E-01(3.84E-02)**
PMOP4(3)	5.08E+00 (1.78E+01)	1.22E+00 (2.77E-01)	1.31E+00 (3.27E-01)	8.80E-01 (7.29E-02)	8.19E-01 (4.98E-02)	8.38E-01 (6.79E-02)	**4.22E-01(2.56E-01)**
PMOP4(5)	9.83E-01 (3.87E-02)	1.43E+00 (6.39E-01)	1.01E+00 (5.80E-02)	4.65E+02 (5.91E+04)	9.53E+02 (2.25E+05)	8.68E+02 (1.24E+05)	**9.57E-01(2.98E-01)**
PMOP5(3)	1.00E+01 (3.34E+01)	6.19E+00 (1.66E+01)	5.52E+00 (6.52E+00)	1.56E+01 (5.79E+01)	1.36E+01 (1.62E+01)	1.29E+01 (5.37E+01)	**3.44E-02(7.29E-03)**
PMOP5(5)	3.17E+01 (2.00E+02)	3.89E+01 (4.05E+02)	5.46E+01 (9.72E+02)	4.00E+01 (1.11E+03)	6.71E+01 (2.35E+03)	5.81E+01 (1.82E+03)	**1.13E-01(1.28E-02)**
PMOP6(3)	4.29E-01 (4.37E-02)	9.40E-01 (3.07E-01)	8.82E-01 (1.53E-01)	1.03E+00 (6.25E-03)	1.82E+00 (4.91E-01)	1.01E+00 (8.07E-08)	**1.88E-02(7.48E-03)**
PMOP6(5)	6.73E-01 (5.74E-03)	7.92E-01 (3.04E-02)	7.88E-01 (1.64E-02)	3.37E+01 (2.72E+02)	4.40E+01 (4.26E+02)	5.65E+01 (7.48E+02)	**8.13E-02(5.02E-02)**
PMOP7(3)	1.22E-01 (9.46E-04)	4.94E-01 (4.14E-03)	3.00E-01 (2.76E-03)	3.12E-01 (7.30E-03)	6.48E-01 (7.66E-02)	5.11E-01 (1.49E-02)	**2.99E-02(1.31E-02)**
PMOP7(5)	4.19E-01 (7.60E-03)	4.40E-01 (2.77E-02)	3.43E-01 (9.66E-03)	8.03E-01 (1.30E-02)	8.00E-01 (9.21E-02)	9.35E-01 (4.08E-02)	**6.55E-02(1.69E-02)**
PMOP8(3)	1.11E-01 (5.20E-04)	2.29E-01 (7.74E-03)	1.98E-01 (5.86E-03)	2.04E-01 (1.21E-03)	2.28E-01 (1.61E-03)	3.12E-01 (5.44E-03)	**6.36E-02(1.95E-02)**
PMOP8(5)	**7.58E-02 (1.90E-04)**	1.32E-01 (4.31E-03)	8.20E-02 (2.89E-04)	6.36E-01 (1.63E-01)	1.50E+00 (1.22E+00)	1.67E+00 (3.54E-01)	1.03E-01(2.07E-02)
PMOP9(3)	2.20E-01 (5.52E-03)	1.29E-01 (3.00E-04)	1.32E-01 (1.89E-04)	2.10E-01 (4.40E-03)	2.33E-01 (4.59E-03)	2.83E-01 (1.86E-02)	**1.66E-02(3.47E-03)**
PMOP9(5)	3.15E-01 (2.03E-04)	4.53E-01 (1.39E-02)	3.38E-01 (1.71E-03)	2.84E+00 (3.23E-01)	3.84E+00 (8.09E-01)	3.47E+00 (3.03E-01)	**1.02E-01(8.46E-02)**
PMOP10(3)	1.46E+00 (5.14E-03)	1.38E+00 (1.24E-02)	1.53E+00 (3.07E-04)	9.28E-01 (3.02E-02)	1.06E+00 (4.46E-02)	1.11E+00 (6.33E-03)	**3.01E-01(2.52E-02)**
PMOP10(5)	1.15E+00 (8.45E-03)	1.14E+00 (2.96E-02)	1.23E+00 (1.18E-02)	2.30E+00 (1.00E-01)	4.22E+00 (4.65E+00)	2.46E+00 (2.90E-01)	**5.57E-01(3.85E-02)**
PMOP11(3)	4.79E-01 (2.88E-03)	3.16E-01 (4.61E-04)	5.43E-01 (1.02E-03)	3.60E-01 (1.29E-03)	4.54E-01 (1.11E-02)	4.31E-01 (6.04E-05)	**1.72E-01(3.29E-02)**
PMOP11(5)	6.17E-01 (5.21E-03)	5.56E-01 3.66E-03)	5.29E-01 (8.06E-03)	8.71E-01 (1.08E-02)	1.81E+00 (4.84E+00)	1.43E+00 (8.27E-01)	**3.10E-01(4.84E-02)**
PMOP12(3)	2.06E-01 (1.00E-03)	1.23E-01 (7.00E-05)	1.48E-01 (2.26E-04)	9.40E-02 (1.42E-03)	1.06E-01 (8.88E-04)	1.33E-01 (2.00E-03)	**1.52E-02(4.17E-03)**
PMOP12(5)	3.02E-02 (3.93E-06)	3.05E-02 (3.55E-06)	2.87E-02 (7.47E-07)	6.02E-01 (5.60E-02)	7.15E-01 (1.29E-01)	6.93E-01 (8.99E-02)	**7.78E-03(2.16E-03)**
PMOP13(3)	3.16E-01 (1.89E-02)	5.88E-01 (1.18E-01)	1.45E-01 (2.01E-03)	6.02E-01 (1.53E-01)	1.16E+00 (4.81E-01)	1.48E+00 (6.52E-01)	**9.06E-02(8.75E-03)**
PMOP13(5)	1.24E+00 (1.07E-01)	3.46E+00 (1.68E+00)	1.19E+00 (8.40E-02)	1.17E+01 (5.28E+01)	1.54E+01 (4.66E+01)	1.69E+01 (6.47E+01)	**7.33E-01(2.37E-02)**
PMOP14(3)	1.71E+00 (1.20E-03)	1.10E+00 (4.45E-05)	1.74E+00 (2.18E-05)	1.10E+00 (1.77E-02)	1.36E+00 (4.56E-02)	1.36E+00 (3.00E-02)	**8.00E-02(1.71E-02)**
PMOP14(5)	9.47E-01 (3.19E-02)	8.40E-01 (1.17E-02)	8.02E-01 (9.82E-05)	1.76E+01 (5.73E+01)	4.23E+01 (5.58E+02)	2.52E+01 (1.34E+02)	**1.28E-01(1.59E-02)**

5 Conclusion

The paper mainly uses the niche strategy and SCD mechanism to solve the multimodal characteristics in the PMOP test suite. Moreover, the method of finding the knee point is embedded into the search process to increase the select pressure of the later search process and improve the convergence and diversity. The experimental results shown that obtain the solution of Kn_NSGA2_SCD can close to the Pareto-optimal solution. And the reference knee areas provided by the test suite is better covered. However, it is still worth studying about knee point questions in the future. For example, the search ability of the reference knee regions is not high, and the running time grows exponentially with the increase of dimension.

Acknowledgment. This work is supported by the National Natural Science Foundation of China (61922072, 61876169, 61673404, 61976237).

References

1. Deb, K., Member, A., Pratap, A., et al.: A fast and elitist multi-objective genetic algorithm NSGAII. IEEE Trans. Evol. Comput. **6**(2), 182–197 (2002)
2. Zitzler, E., Laumanns, M., Thiele, L.: SPEA2_ improving the strength Pareto evolutionary algorithm for multiobjective optimization. In: Evolutionary Methods for Design, Optimization and Control with Applications to Industrial Problems, pp. 95–100 (2001)
3. Horn, J., Nafpliotis, N., Goldberg, D.E.: A niched Pareto genetic algorithm for multi-objective optimization. In: IEEE Conference on Evolutionary Computation IEEE World Congress on Computational Intelligence (1994)
4. Corne, D.W., Knowles, J.D., Oates, M.J.: The Pareto envelope-based selection algorithm for multiobjective optimization. In: Schoenauer, M., et al. (eds.) PPSN 2000. LNCS, vol. 1917, pp. 839–848. Springer, Heidelberg (2000). https://doi.org/10.1007/3-540-45356-3_82
5. Zhang, X., Tian, Y., Jin, Y.: A knee point driven evolutionary algorithm for many-objective optimization. IEEE Trans. Evol. Comput. **19**(6), 761–776 (2015)
6. Rachmawati, L., Srinivasan, D.: Multiobjective evolutionary algorithm with controllable focus on the knees of the Pareto front. IEEE Trans. Evol. Comput. **13**(4), 810–824 (2009)
7. Schütze, O., Laumanns, M., Coello, C.A.C.: Approximating the knee of an MOP with stochastic search algorithms. In: Rudolph, G., Jansen, T., Beume, N., Lucas, S., Poloni, C. (eds.) PPSN 2008. LNCS, vol. 5199, pp. 795–804. Springer, Heidelberg (2008). https://doi.org/10.1007/978-3-540-87700-4_79
8. Deb, K., Gupta, S.: Understanding knee points in bicriteria problems and their implications as preferred solution principles. Eng. Optim. **43**(11), 1175–1204 (2011)
9. Sudeng, S., Wattanapongsakorn, N.: Adaptive geometric angle-based algorithm with independent objective biasing for pruning Pareto-optimal solutions. In: Science & Information Conference (2013)
10. Branke, J., Deb, K., Dierolf, H., Osswald, M.: Finding knees in multi-objective optimization. In: Yao, X., et al. (eds.) PPSN 2004. LNCS, vol. 3242, pp. 722–731. Springer, Heidelberg (2004). https://doi.org/10.1007/978-3-540-30217-9_73

11. Bhattacharjee, K., Singh, H., Ryan, M., et al.: Bridging the gap: many-objective optimization and informed decision-making. IEEE Trans. Evol. Comput. **21**(5), 813–820 (2017)
12. Das, I.: On characterizing the "knee" of the Pareto curve based on normal-boundary intersection. Struct. Optim. **18**(2–3), 107–115 (1999)
13. Yu, G., Jin, Y., Olhofer, M.: A Method for a posteriori identification of knee points based on solution density. In: Presented at the Congress on Evolutionary Computation (2018)
14. Qu, B.Y., Suganthan, P.N.: Novel multimodal problems and differential evolution with ensemble of restricted tournament selection. In: Evolutionary Computation, pp. 3480–3486 (2010)
15. Preuss, M.: Niching methods and multimodal optimization performance. Multimodal Optimization by Means of Evolutionary Algorithms. NCS, pp. 115–137. Springer, Cham (2015). https://doi.org/10.1007/978-3-319-07407-8_5
16. While, L., Hingston, P., Barone, L., et al.: A faster algorithm for calculating hypervolume. IEEE Trans. Evol. Comput. **10**(1), 29–38 (2006)
17. Jong, D., Alan, K.: Analysis of the behavior of a class of genetic adaptive systems. Ph.D. thesis University of Michigan (1975)
18. Holland, J.H.: Adaptation in Natural and Artificial Systems, vol. 6, 2nd edn, pp. 126–137. MIT Press, Cambridge (1992)
19. Petrowski, A.: A clearing procedure as a niching method for genetic algorithms. In: Proceedings of the IEEE International Conference on Evolutionary Computation (1996)
20. Li, J.P., Balazs, M.E., Parks, G.T., et al.: A species conserving genetic algorithm for multimodal function optimization. Evol. Comput. **10**(3), 207–234 (2014)
21. Liang, J.J., Yue, C.T., Qu, B.Y.: Multimodal multi-objective optimization: a preliminary study. In: Evolutionary Computation (2016)
22. Yue, C.T., Qu, B., Jing, L.: A Multi-objective particle swarm optimizer using ring topology for solving multimodal multi-objective problems. IEEE Trans. Evol. Comput. **22**(5), 805–817 (2017)
23. Yu, G., Jin, Y., Olhofer, M.: Benchmark problems and performance indicators for search of knee points in multi-objective optimization. In: 2019 IEEE Congress on Evolutionary Computation (CEC), pp. 2410–2417 (2019)

A Cell Potential and Motion Pattern Driven Multi-robot Coverage Path Planning Algorithm

Meng Xu(✉), Bin Xin(✉), Lihua Dou, and Guanqiang Gao

School of Automation, State Key Laboratory of Intelligent Control and Decision of Complex Systems, Beijing Advanced Innovation Center for Intelligent Robots and Systems, Beijing Institute of Technology, Beijing 100081, China
xumengbit@outlook.com, brucebin@bit.edu.cn

Abstract. This paper proposes an intelligent "Cell Potential and Motion Pattern driven Coverage (CPMPC)" algorithm to solve a cooperative coverage path planning problem for multiple robots in two-dimensional target environment. The target environment is divided into cell areas according to the detection range of robot, and the cell matrix is given correspondingly. The values in the cell matrix are defined as cell potential, which represents the number of times each cell is detected by robots. The priority of the robot's neighbor cell is called the motion pattern. At different moments, robots can choose within different motion patterns. Genetic algorithm (GA) is used to optimize the combination of motion patterns. By taking account obstacle avoidance and collision avoidance into consideration, the CPMPC algorithm adopts a double-layer choice strategy driven by cell potential and motion pattern to generate the next waypoint. Furthermore, this algorithm contains two optimal strategies: avoiding collision and jumping out of the detected area. Compared with the pattern-based genetic algorithm, the results obtained by us show that the CPMPC algorithm could solve the multi-robot coverage path planning (MCPP) problem effectively with guarantee of complete coverage, and improved makespan.

Keywords: Cell potential and motion pattern driven coverage path planning algorithm · Multi-robot system · Genetic algorithm

1 Introduction

In recent decades, the coverage path planning (CPP) problem of robots has received increasing attention in both military and civil sectors. CPP is the task

This work was supported in part by the National Outstanding Youth Talents Support Program 61822304, the National Natural Science Foundation of China under Grant 61673058, the NSFC-Zhejiang Joint Fund for the Integration of Industrialization and Informatization under Grant U1609214, the National Key R&D Program of China (2018YFB1308000).

L. Pan et al. (Eds.): BIC-TA 2019, CCIS 1159, pp. 468–483, 2020.
https://doi.org/10.1007/978-981-15-3425-6_36

of determining a path that passes over all points of an area or volume of interest while avoiding obstacles [5]. A wide range of common use and specific applications of CPP include vacuum cleaning robots [11,15], painter robots [2], demining robots [1], harvesting and surveillance robots [3,7], and forest fire monitoring robots [12]. Therefore, some effective CPP algorithms to persistently search the target area and avoid obstacles are required. The CPP problem is related to the traveling salesman problem [5]. However, a very important constraint must be satisfied, that is, the city that robot visits each time must be its neighbor [5]. E.M. Arkin et al. have proved both the traveling salesman problem and CPP problem are NP-hard [17]. Therefore, there is no fixed optimal solution for CPP problem. However, in order to achieve better coverage effect, a robot should meet some requirements as follows [5]:

1. A robot should move through all the points in the target area covering it completely.
2. A robot should fill the region without overlapping paths.
3. Continuous and sequential operation without any repetition of paths is required.
4. A robot should avoid all obstacles.
5. Simple motion trajectories (e.g., straight lines or circles) should be used (for simplicity in control).
6. An "optima" path is desired under available conditions.

Satisfying all the above requirements is not always possible in most environment with obstacles. Therefore, the second and third requirements are not to be considered in this paper. This paper focuses on meeting the other four requirements while minimizing the overlap and repetition of paths. In the field of CPP, there have been many researches on single robot, so more and more scholars begin to focus on the research of multi-robot coverage path planning (MCPP) [9,10]. MCPP need to consider conflicts and collision avoidance compared to single robot coverage tasks, therefore, MCPP are relatively complex. However, MCPP have time advantages and are more robust. In general, MCPP algorithm can be classified as heuristic and complete [4].

Some heuristic MCPP algorithms which may work well, while do not have any provable guarantees ensuring complete coverage of the target environment. Several recent works about MCPP is to guarantee that the robots generate paths that completely cover the target environment. The coverage algorithm proposed in this paper is complete MCPP algorithm. All robots continue to explore until all cells in the target area have been detected. Therefore, in order to ensure high efficiency and low repetition, the target area is often divided into multiple subregions (called cells) according to the detection range of the robot [4]. Moravec et al. presented cell-based methods to decompose the target area into a collection of uniform cells [14]. Typically, each cell is a square, and can be viewed as obstacle or free space according to the complexity of the target area. Cell-based MCPP algorithm can be classified as off-line or on-line depending on whether or not the target environment is assumed to be known in advance [4]. On-line MCPP algorithms do not assume full prior knowledge of the environment. Therefore, to some extent, the online algorithm cannot get better coverage effect. In this paper, we adopt an off-line MCPP algorithm. Recently, many scholars have made great contributions to off-line MCPP algorithm.

Ahmet Yazici et al. first proposed a novel pattern-based genetic algorithm (PBGA) for a sensor-based multi-robot coverage path planning problem. The target area is modeled with cells. The size of every cell is equal to the range of sensing devices. Then the problem is defined as finding a sequence of cells for each robot to minimize the coverage completion time [9]. Then in 2014, Yazici et al. improved the above method, he used a generalized Voronoi diagram-based graph to model the environment with obstacles. Each robot in a team gets a sequence of cells by sensor-based coverage path planning. Every point in a given workspace is covered by at least one robot using its sensors [18]. The PBGA has many advantages. However, it does not guarantee complete coverage of the target area.

In this study, a novel off-line MCPP algorithm, based on two main choice strategies (cell potential and motion pattern), is proposed, named "Cell Potential and Motion Pattern driven Coverage (CPMPC)" algorithm. Hence, continuous and sequential of cells for each robot can be obtained. At the same time, two other optimal strategies are proposed to quickly jump out of the detected areas and avoid conflicts and collisions among robots. A* algorithm is used to plan path to get to the closest undetected cell when the robot's current neighbor cells have all been explored at least once. In order to avoid conflicts and collisions, the priority of robots is given in advance. Compared with previous works, there are three main contributions in this paper.

1. In this paper, the MCPP is solved from the global perspective and the obtained solutions can provide comparison for distributed algorithms.
2. The proposed CPMPC algorithm is complete and is suitable for many complex environments with or without obstacles.
3. The strategy of generating the next waypoint based on cell potential is proposed for the first time. This strategy can easily divide the obstacle area and free area, and quickly generate the next waypoint. Additionally, when two or more robots are adjacent to each other, the cell potential of cell where neighbor robot located at can be set to infinity to avoid collision.

The remainder of this article is organized as follows. Section 2 describes the target environment and formulates the objective function. Section 3 introduces the proposed CPMPC algorithm. Section 4 introduces the genetic algorithm used to solve the CPMPC algorithm. Section 5 focuses on simulation results, performance comparison and analysis to evaluate the performance of the proposed CPMPC algorithm. Finally, concluding remarks and directions for further works are given in Sect. 6.

2 Problem Description and Formulation

This section describes the target environment and defines the coverage problem. Firstly, the target environment is divided into cells according to the sensor detection range of the robot. And then consider the constraints and formulate the objective function.

2.1 Environment Description

In order to solve the MCPP, that is, making sure there are no undetected cells left, the robot should possess the following attributes [16]:

1. A way to represent the environment.
2. A waypoint selection method, i.e. to have an exploration strategy.
3. An efficient strategy to move from its current waypoint to the next waypoint.

In this paper, we use cell-based method to represent the environment, which is also a very important innovation proposed in this paper. The environment is decomposed into a collection of uniform cells [5], thus, the target area transforms into a graph ENV. Each cell is a square, and the size of each cell is equal to the size of a robot's detection range, the side length of each cell is equal to D. $Env = [env_1, env_2, env_3, ..., env_n]$ represents the reachable cells in the target area, $Obs = [obs_1, obs_2, obs_3, ..., obs_m]$ represents the obstacle cells, where m is the number of obstacles. Here we assume that a cell is either fully occupied or completely free. With this modeling, when a robot passes through the center of a cell, this cell is supposed to be fully covered. Additionally, each cell holds a value, corresponding to the number of times that this cell is detected, we define it as cell potential. The polygonal environment and obstacles can be treated in the same way, which is not discussed in detail here. In the initial situation, the cell potential of undetected cell is 0, the cell potential of obstacle cell is infinite. The cell matrix of the target map is an important basis for the proposed algorithm. Multiple robots share a cell matrix throughout the execution of the coverage task. That is, it is assumed that all robots keep communicating during the execution of the coverage task. Each value in the cell matrix corresponds to the cell potential of the corresponding cell.

2.2 Objective Function

Before building the optimization model, take some practical considerations into account, the following constraints are considered:

Constraint 1 (Collision). *For multi-robots, at time t, any two robots cannot choose the same cell as next position.*

$$Pos_t^i \cap Pos_t^j = \emptyset, \forall i, j \in \{1, ..., n\} \tag{1}$$

Constraint 2 (Search-stop). *Robots will stop searching only when all reachable areas are detected.*

$$POS^1 \cup POS^2 \cup \cdots \cup POS^n = Env \tag{2}$$

Constraint 3 (Obstacles). *Robots cannot cross obstacle areas.*

$$POS^i \cap Obs = \emptyset, \forall i \in \{1, ..., n\} \tag{3}$$

Constraint 4 (Path continuous). *For every robot, the final path sequence must be continuous. That is, there is only one unit step difference between any two adjacent path points.*

$$\left|x_{t+1}^i - x_t^i\right| + \left|y_{t+1}^i - y_t^i\right| = 1, \forall i \in \{1, ..., n\} \tag{4}$$

The objective is to minimize the maximal completion time of multiple robots. That is to minimize the completion time of the robot with the longest path sequence. According to above constraints, the CPP problem in this paper can be formulated as the following minimization problem:

$$\begin{aligned} &\Gamma = \arg \min_{POS} \max_{i \in \{1,2,...,n\}} T_i \\ &s.t.\ POS^1 \cup POS^2 \cup \cdots \cup POS^n = Env \\ &\forall i \in \{1, ..., n\}, POS^i \cap Obs = \emptyset \\ &\forall i \in \{1, ..., n\}, \left|x_{t+1}^i - x_t^i\right| + \left|y_{t+1}^i - y_t^i\right| = 1 \\ &\forall i, j \in \{1, ..., n\}, Pos_t^i \cap Pos_t^j = \emptyset \end{aligned} \tag{5}$$

where Γ represents the optimal solution. T_i represents the time to complete the coverage task of robot i. $\max_{i \in \{1,2,...,n\}} T_i$ represents the maximum time to complete the coverage task, that is the makespan of the coverage task.

3 The Proposed CPMPC Algorithm

In order to solve the above problem and optimize the objective function, this section introduces the proposed CPMPC algorithm. Two choice strategies and two optimize strategies are introduced in detail.

3.1 Two Choice Strategies

The two choice strategies are cell potential and motion pattern. The cell potential represents the numerical value corresponding to each cell in the cell map, and the value range is 0 to infinity. The motion pattern represents the different priorities of the neighbor cell of the robot. According to Sect. 2, after decomposition of the target environment, the cell matrix of the corresponding environment can be obtained. For the corresponding cell matrix, everytime the cell is detected by a robot, the cell potential of the cell add by 1. The initial position of each robot is randomly placed in the center of cell except for obstacles in the cell map, and they does not overlap. The initial velocity of each robot is v_{robot},

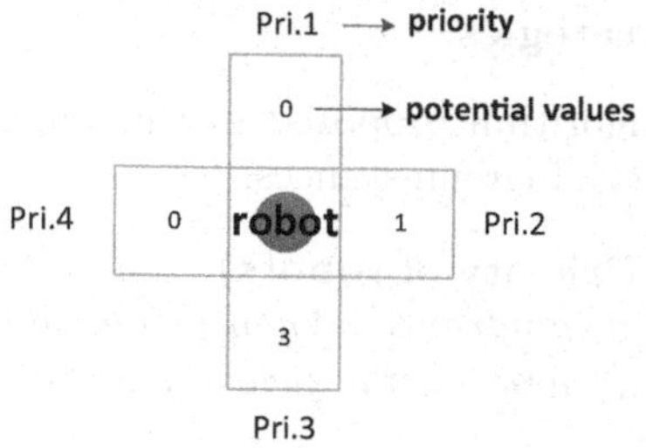

Fig. 1. The principle of cell potential and motion pattern

satisfying the equation $v_{robot} = D$, which means each time the robot moves, the step length of robot is equal to the side length of the cell in the decomposed environment. We assume that, the time for robots to take turn are small enough to be ignored. Therefore, turns do not have to be taken into consideration in this paper. $Pos_t^i = (x_t^i, y_t^i)$ denotes the position of the robot i at the moment t. The next position of robot is generated according to the cell potential of the four neighbors and the motion pattern of the robot at current moment. The motion trajectory L_i of robot i can be formulated as a list of consecutive path points $POS^i = [Pos_1^i, Pos_2^i, ..., Pos_t^i]$. The double-layer Choice Strategy driven by cell potential and motion pattern are described in detail below.

Choice Strategy 1 (Cell potential). *The robot compares the cell potential of all neighbor cells at its current position and selects one or more cells with the minimum cell potential as the alternative cell/cells of the waypoint at the next moment.*

When there is only one cell with the minimum cell potential, the next waypoint is generated. When there are multiple cells with the minimum cell potential, the only waypoint at the next moment can be generated by Choice Strategy 2 (Fig. 1).

Choice Strategy 2 (Motion pattern). *When the number of neighbors of the robot is nb, the different type of robot's alternative motion patterns is the different permutation and combination of nb. Different motion patterns represent different priorities of robot's neighbors.*

The number of neighbor cells of the robot can have different values according to different environments. In this paper, 4 neighbor cells are taken as examples. Therefore, the robot has $24(A_4^4 = 24)$ different choice, which we called it 24 motion patterns. Based on the choice strategies, the only next waypoint of robot is generated. The current pattern of the robot is optimized according to the genetic algorithm (GA) introduced in the following section, so as to achieve complete coverage while minimizing the makespan.

3.2 Two Optimal Strategies

Additionally, CPMPC algorithm proposed in this paper contains two kinds of optimal strategies for special circumstances.

Optimal Strategy 1 (Priority of robots). *Before the path planning, the priority of the robots is given in advance. When two or more robots encounter conflicts in the path planning process, the priority of the robot plays a important role in avoiding conflicts.*

Optimal Strategy 2 (A* Algorithm). *A* algorithm is an optimization strategy for the robot to quickly jump out of the detected area and find the nearest undetected cell.*

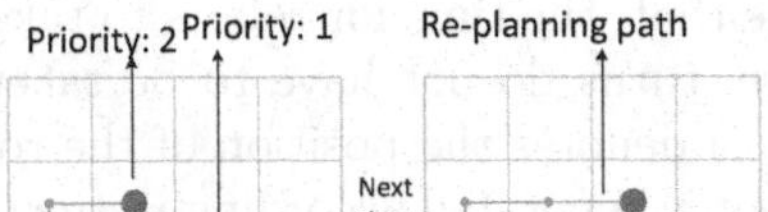

Fig. 2. Conflict avoidance path planning method 1.

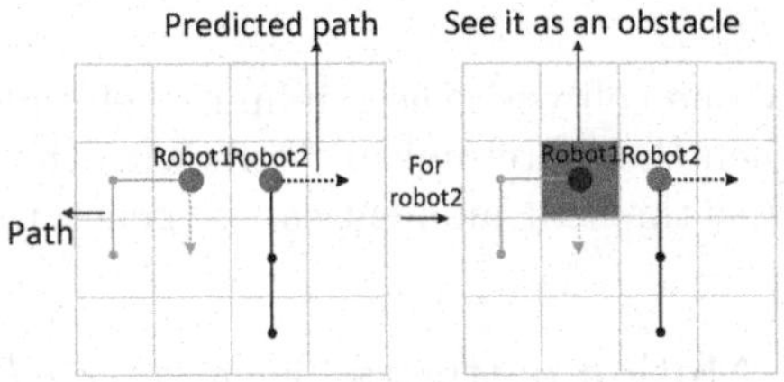

Fig. 3. Conflict avoidance path planning method 2.

The first kind of strategy for the priority of a given robot in advance, that is when two or more robots choose the same next waypoint at the same time, the robot with higher priority moves first, and the robot with lower priority will re-plan the path, see Fig. 2. When any two or more robots are neighbors, the surrounding robots are considered as obstacles in the planning process, see Fig. 3. In this paper, given the priority strategy of robots, collision between robots can be effectively avoided. The second strategy is using A* algorithm to jump out of the detected area. A* algorithm is the most effective direct search method to get the shortest path and an effective algorithm to solve many search problems.

A* algorithm is a kind of heuristic algorithm. Therefore, when all neighbor cell potential of the robot are greater than or equal to 1, in order to faster jump out of the detection area, continue to detect the undetected area, first using equation (6) and (7) to get the nearest undetected cell, and then adopting the strategy of A* algorithm to generate the next waypoint (Fig. 4).

The final goal is to cooperate with multiple robots to cover the whole target area as quickly as possible for a special mission, avoiding obstacles and collision between robots. The evaluation criteria to judge the implementation effect of the whole task is the makespan, which means the cost time of the robot when the task has been complete. The coverage algorithm described above is shown in Algorithm 1.

$$d^k = d(pos^i, R^k) = \left|x_t^i - x_k\right| + \left|y_t^i - y_k\right|, \quad \forall i \in \{1,2,...,n\}, k \in \{1,2,...,K\} \tag{6}$$

$$d^{nearest} = \arg\min[d^1, d^2, ..., d^k] \tag{7}$$

4 The Genetic Algorithm to Solve CPMPC

In order to solve the above optimal problem, Genetic Algorithm (GA) is used. In the work of Mitschke [13] and Kapanoglu [9] the GA is used to solve the CPP problem. GA is commonly used to generate high-quality solutions to optimization and search problems by relying on bio-inspired operators such as mutation, crossover and selection [8]. A sample chromosome for a solution is as Fig. 5.

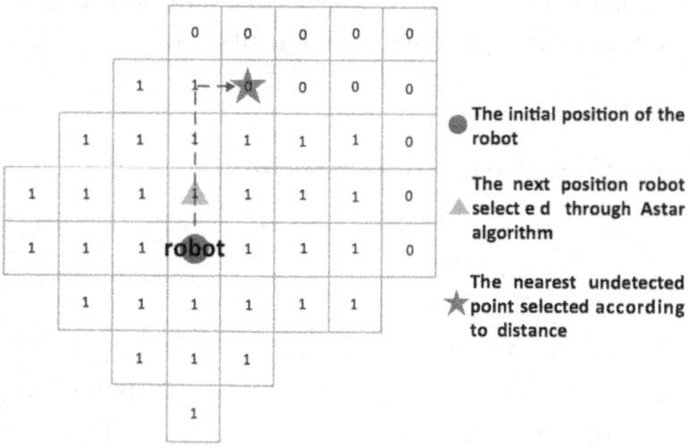

Fig. 4. The principle of A* algorithm

The chromosome is expressed as a matrix. The size of the chromosome not only depends on the number of robots, but also the number of motion patterns allowed. Take an individual in the initial population as an example. The individual adopts the coding method of four rows and four columns. The number of columns in the coding matrix is determined by the number of robots. The number of rows is determined by the number of times each robot can change

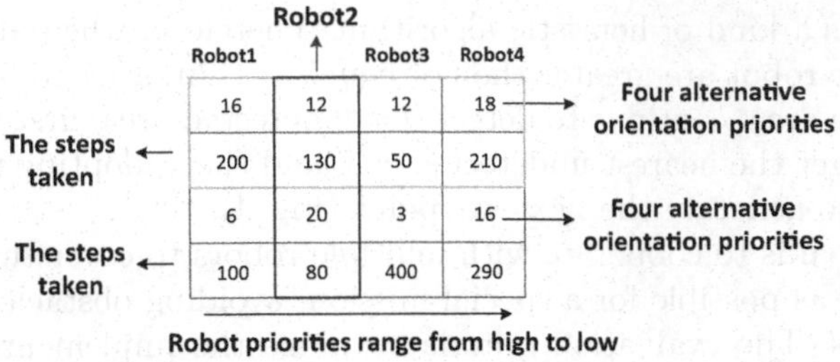

Fig. 5. A sample chromosome

Algorithm 1. CPMPC algorithm.

Input: Cell map, *Env*; The number of four alternative orientation priorities index, P_n; The number of robots, Rob_n; The position of robots, $Start_pos = [(x_rob1, y_rob1), (x_rob2, y_rob2), ..., (x_robn, y_robn)]$;
Output: Path sequences for each robot, $POS^i = [Pos_1^i, Pos_2^i, Pos_3^i, ..., Pos_t^i]$;

Loop to get the next waypoint, thus getting the whole path;
function GA-SEARCH($Env, Start_pos, Rob_n, P_n$)
    while 0 in *Env*==True **do**
        Strategy 1(Cell potential);
        if There are more than one minimum cell potential among robot's neighbor **then**
            Strategy 2(Motion pattern);
        end if
        if All neighbors have been detected at least once **then**
            Optimal-Strategy 1(A* algorithm jump out of detected area);
        end if
        if There are more than one robot choose the same next waypoint **then**
            Optimal-Strategy 2(Priority of robots);
        end if
    end while
    return $POS^i = [Pos_1^i, Pos_2^i, Pos_3^i, ..., Pos_t^i]$;
end function

---

its motion pattern during task execution. If a 4 × 4-gene-chromosome is divided among four robots, the first column-gene corresponds to the parameters of the first robot, second column to the second and so on. The gene of each column should be interpreted in pairs, in which the first element indicates the motion pattern index and the second is the number of steps taken by the corresponding robot with respect to the motion pattern index given in the first row, the third and the forth element is as similar. So the size of chromosomes are computed as {number of robots*[number of motion patterns allowed per robot*2]}. For this case, robots are allowed to build their route based on two motion patterns, that is, every robot can transform motion patterns only once.

The members of the initial population are generated by human control. The inputs are the population size, the number of motion pattern index, the number of robots, and the value of crossover operator and mutation operator. Motion pattern indexes are randomly initialized between 1 and $24(A_4^4)$, the steps taken by the corresponding robot are randomly initialized between 1 and the size of *Env*. As the iteration goes on, the process of crossover and mutation is shown in Figs. 6 and 7. Two robots in two chromosomes are randomly selected for crossover. The mutation occurs in a random robot on a chromosome. The occurrence of crossover and mutation is based on the magnitude of crossover operator (CR) and mutation operator (MR) respectively. And the crossover and mutation operations only apply to the two positions of the motion patterns and the number of motion steps corresponding to the motion patterns. The priority of the robot does not change during the operation of crossover and mutation, so as to avoid the priority of two or more robots getting same in the process of crossover and mutation. That is, the priority of the robot does not change with the progress of GA after the initial allocation. The crossover operation refers to randomly selecting two individuals from the initial population and randomly selecting the crossover location. Then to do content exchange of the cross location of two individuals. The mutation operation refers to the reinitialization of each individual randomly selected at a mutation location among the population after the crossover operation.

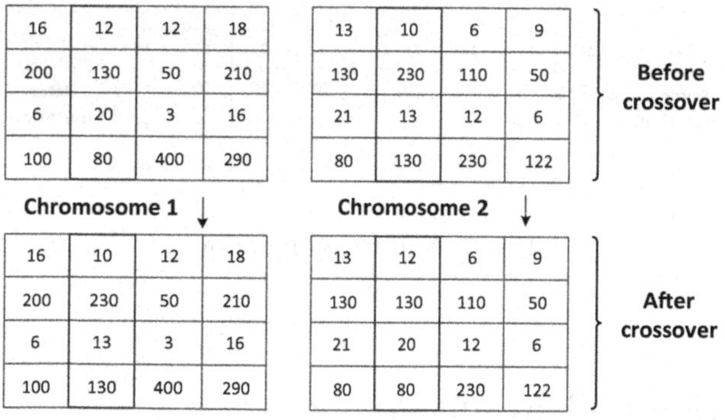

Fig. 6. Crossover operation

As shown in Fig. 6, when the random probability is less than the CR, two individuals in the initial population are randomly selected for crossover operation. The position of the crossover operation is determined by a random function. In this example, select 1 to 4 rows from the second column of two individuals to exchange. That is, swap the values of 1 to 4 rows in the second column of two individuals. Two new individuals are generated to replace the parent generation in the corresponding initial population, and the new population is generated for subsequent mutation operation.

As shown in Fig. 7, when the random probability is less than the MR, an individual in the new population after crossover is selected for mutation operation. The position of mutation operation is determined by random function. In this case, select 1 to 4 rows from the second column of the individual for mutation operations. The motion pattern and step in the mutation process are determined in the same way as they were initialized. The generation of new individuals replaces the parent individuals in the corresponding crossover operations and generates a new population for subsequent selection operations.

The next step after crossover and mutation is selection. The fitness of a chromosome indicates the makespan in seconds. Since an ideal solution would contain no repeated waypoint and every robot should cover an average number of cells. Through optimization, the optimization problem can balance the length of the trajectory sequence between robots. Therefore optimal fitness can be got by minimizing the number of repeated cells. Tournament selection with elitism is adopted for getting new population. The individuals in the initial population were competed with the individuals after crossover and mutation operations, and the elite of the two individuals were selected as the individuals in the new population. Based on the new population obtained, Continue to cross, mutate, and select, and then get the new population and the elite individuals. The cycle is terminated until the population evolution iteration reaches the given maximum iteration. Output the current optimal individual and get the paths for every robot.

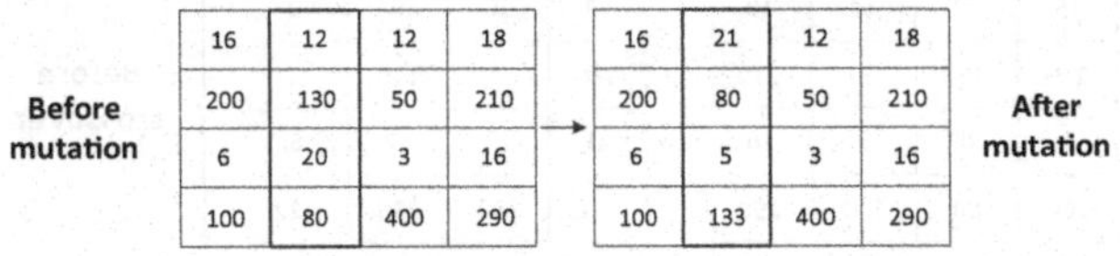

Fig. 7. Mutation operation

5 Simulation and Result Analysis

In this section, the simulation verification and comparison experiments of the proposed CPMPC algorithm are carried out. It is stipulated that all robots are isomorphic, and the robot can move one cell per movement. The MCPP problem, which is an NP-hard problem, cannot easily generate the optimal solution when there are obstacles in the target area. A low boundary (LB) of an MCMP problem's solution is defined as the minimum number of reachable cells that are covered by each robot [6]. LB is used in this section to evaluate a solution's proximity to the optimal solution. The LB is determined as "the number of cells to be covered ÷ the number of robot". First, the simulation verification of the coverage task of multiple robots was carried out in the target area with obstacles. Second, experiments on the impact of the robot's number were implemented and analyzed. Finally, The CPMPC algorithm is compared with PBGA [9] in terms of the performance of MCPP problems. All simulations are implemented with Python3.6 environment on a PC with 32 GB RAM and 2.10 GHz Intel(R) Xeon(R) E5-2620 CPU under Windows 10 operating system. The performance of the algorithm is tested on the environment given in Fig. 8(a).

Accordingly, the input parameters of the algorithm are as follows: The size of the environment (SE), number of obstacles (NO), number of robots (NR) and the initial position of robots ($IPRs$), population size (PS), maximum number of iterations (MNI), crossover operator (CR) and mutation operator (MR).

5.1 Implementation

The performance of the algorithm is tested on the environment given in Fig. 8(a). The relevant parameters are set as follows in this test: SE: 40 × 40, NO: 160, NR: 4, $IPRs$: (9, 36), (15, 0), (28, 25) and (29, 29), PS: 100, MNI: 400, CR: 65% and MR: 5%. The planned path for each robot is given in the Fig. 8(b).

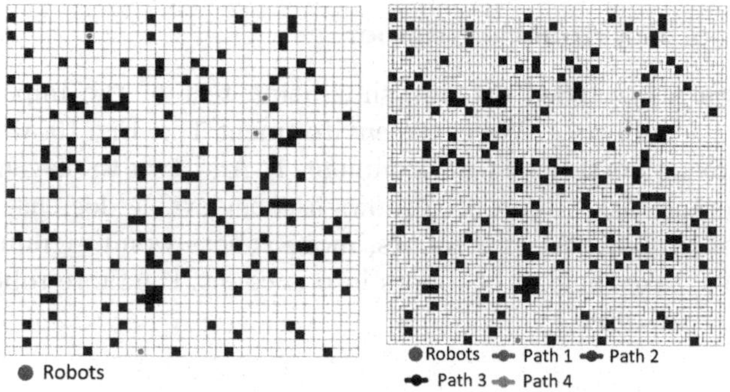

(a) Target area (b) Coverage result

Fig. 8. Target area and coverage result

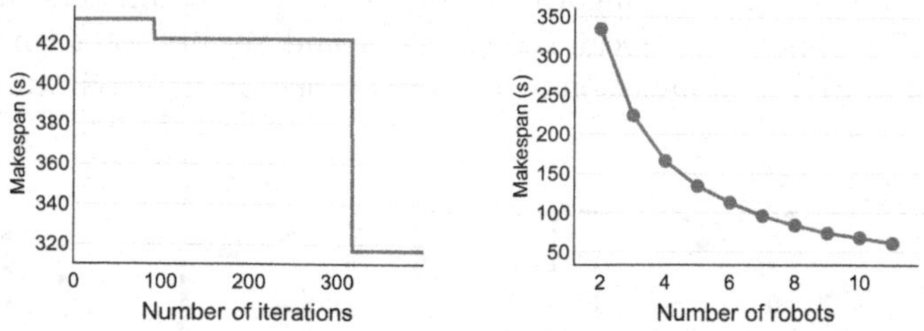

(a) Convergence curve (b) The impact of robot's number on the search time

Fig. 9. Convergence curve and the impact of robot's number on the search time

By iteration, the better solution is selected. The simulation results show that the GA finally optimizes four continuous paths. Each body successfully avoids obstacles and other robots. The cost time of the area to be covered by the four robots is 204 s. Since this problem is NP-hard, the theoretical optimal solution cannot be generated, but the LB can be determined to be 200.5 s. By analyzing the results, there are repeated paths in the solution, but the repetition rate is very small and the coverage rate is up to 100%, which means it is complete.

And Fig. 9(a) shows the convergence curve of the proposed algorithm. In the whole simulation process, a total of 400 iterations were carried out. There were two significant decreases about the fitness. From the 317th iteration, the convergence curve does not change anymore. However, due to the lack of more iterative verification, the convergence result at this time cannot be determined to be the optimal solution. However, it can be proved that GA is feasible to solve the coverage optimization strategy proposed in this paper. However, because the iteration will take a lot of time, greatly reduce the efficiency and the makespan will not be reduced too much, so in general consideration, more iterations is not a better solution.

5.2 Impact of the Robot's Number

In addition, we also carried out the simulation test of the impact of robot's number on the search time. We chose environment 1 in Fig. 10(a) as the target area. There are no obstacles in environment 1. And two to eleven robots were tested for coverage task. Figure 9(b) shows a considerable decrease of makespan when up to four robots. After 8 robots, more robots won't significantly reduce makespan. Also, take the cost of robots into account, the more robots there are, the higher the cost takes.

5.3 Performance Comparison

In order to analyze the performance of the proposed coverage algorithm and GA, we designed 5 different target areas, and test by 2 robots, 3 robots or 4 robots.

For the proposed CPMPC algorithm, GA's typical parameters are set as same as Sect. 5.1. The five typical target areas and results for each environment as shown in Fig. 10. Simulation tests were carried out on environment with

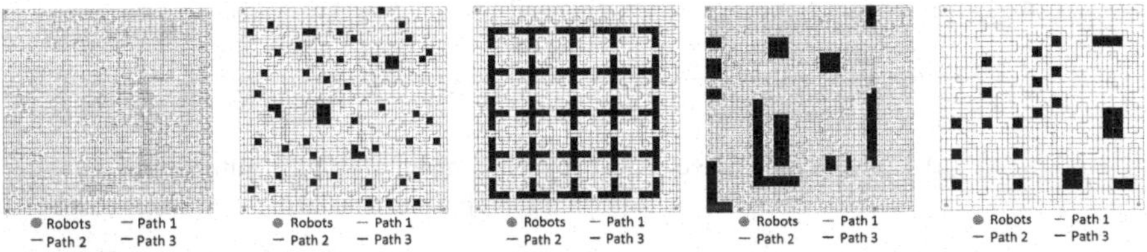

(a) Environment 1 (b) Environment 2 (c) Environment 3 (d) Environment 4 (e) Environment 5

Fig. 10. The five typical coverage results with multi-robot

Table 1. Performance comparison results of two methods.

Type	Number of robots	Space size	LB (s)	The proposed CPMPC algorithm		PBGA [9]	
				Makespan(s)	Coverage rate	Makespan(s)	Coverage rate
Env 1	2	40 * 40	800	800	1	800	1
	3	40 * 40	533.33	**533.5**	1	534	1
	4	40 * 40	400	**400**	1	401.3	1
Env 2	2	30 * 30	420.5	445.3	1	439.9	0.977
	3	30 * 30	280.33	**297.1**	1	297.5	0.985
	4	30 * 30	210.25	**227.1**	1	229.6	0.989
Env 3	2	30 * 30	357.5	405	1	387.9	0.928
	3	30 * 30	238.33	**271.5**	1	279.8	0.971
	4	30 * 30	178.75	**205.3**	1	220	0.987
Env 4	2	40 * 40	697.5	705.9	1	703.4	0.981
	3	40 * 40	465	**473.3**	1	476.9	0.992
	4	40 * 40	348.75	366.5	1	363.7	1
Env 5	2	20 * 20	185	191.2	1	185.3	0.974
	3	20 * 20	123.33	**126.4**	1	129.9	0.993
	4	20 * 20	92.5	**96.8**	1	100.2	1

and without obstacles respectively. Meanwhile, the placement of obstacles was divided into regular placement and irregular placement. The results obtained during simulations are presented in Table 1. The table provides two important comparison parameters which are coverage rate and makespan respectively. For the corresponding situation of each environment and the number of robots, the average value is obtained after 10 runs. The results of the comparison algorithm for the same target area are also listed in Table 1. Therefore, the effectiveness of the proposed algorithm is proved.

Through the experimental results, it can be concluded that the proposed CPMPC algorithm has a good ability to approximate the optimal solution. Compared with the PBGA in [9], the CPMPC algorithm proposed in this paper can ensure complete coverage. The makespan of PBGA is not a real makespan when the corresponding coverage rate is not equal to 1. The makespan is equal to the time when all robots are stuck. No matter how complex environments are, the makespan decreases as the number of robots becomes large. For the environment 1 (target area without obstacles), under the premise of ensuring complete coverage, the proposed CPMPC algorithm can achieve the optimal solution in the case of the three test situations. However, the PBGA does not reach the optimal solution under the same conditions. For the environment 2 to 5 (target environments with obstacles), the algorithm proposed in this paper can achieve better makespan in most cases, while ensuring complete coverage. In some cases, the algorithm proposed in this paper is relatively poor compared with PBGA. However, PBGA can not guarantee complete coverage, while the algorithm proposed in this paper can. In a few cases, under the premise that both algorithms can guarantee complete coverage, the algorithm proposed in this paper got worse makespan, but the difference of the makespan from the PBGA algorithm is particularly small.

6 Conclusion

In this paper, a novel and complete CPMPC algorithm for dealing with MCPP problem is presented. The CPMPC algorithm has two main choice strategies: cell potential and motion pattern. Through the two choice strategies, each robot can generate the only next waypoint and finally plan a path to complete the coverage task. In addition, two optimal strategies driven by A* algorithm and the priority of robots are used to quickly jump out of the detected areas and avoid conflicts and collisions between robots. The proposed CPMPC algorithm has two main contributions. Firstly, the method of generating the next waypoint based on cell potential is proposed for the first time to our knowledge. This method not only can easily distinguish the obstacle area and free area, but also can quickly generate the next waypoint. Secondly, the CPMPC algorithm is complete and can cover some typical environments with obstacles by multi-robot cooperation. The validity and the adapt ability of the CPMPC algorithm are verified by computational experiments. In addition, the proposed CPMPC algorithm is compared with the PBGA. And the results show that the proposed CPMPC algorithm has advantages in ensuring complete coverage. Moreover, the proposed CPMPC algorithm is faster in some cases.

The proposed CPMPC algorithm is an off-line and centralized algorithm. However, for practical applications, special situations should be considered when performing coverage tasks. Therefore, an effective on-line and distributed algorithm should be studied in future work.

References

1. Acar, E.U., Choset, H., Zhang, Y., Schervish, M.: Path planning for robotic demining: robust sensor-based coverage of unstructured environments and probabilistic methods. Int. J. Robot. Res. **22**(7–8), 441–466 (2003). https://doi.org/10.1177/02783649030227002
2. Atkar, P.N., Greenfield, A., Conner, D.C., Choset, H., Rizzi, A.A.: Uniform coverage of automotive surface patches. Int. J. Robot. Res. **24**(11), 883–898 (2005). https://doi.org/10.1177/0278364905059058
3. Boryga, M., Graboś, A., Kołodziej, P., Gołacki, K., Stropek, Z.: Trajectory planning with obstacles on the example of tomato harvest. Agric. Agric. Sci. Procedia **7**, 27–34 (2015). https://doi.org/10.1016/j.aaspro.2015.12.026
4. Choset, H.: Coverage for robotics - a survey of recent results. Ann. Math. Artif. Intell. **31**(1–4), 113–126 (2001). https://doi.org/10.1023/A:1016639210559
5. Galceran, E., Carreras, M.: A survey on coverage path planning for robotics. Robot. Auton. Syst. **61**(12), 1258–1276 (2013). https://doi.org/10.1016/j.robot.2013.09.004

6. Gao, G., Xin, B.: A-STC: auction-based spanning tree coverage algorithm formotion planning of cooperative robots. Front. Inf. Technol. Electron. Eng. **20**(1), 18–31 (2019). https://doi.org/10.1631/FITEE.1800551
7. Ghaffarkhah, A., Mostofi, Y.: Path planning for networked robotic surveillance. IEEE Trans. Signal Process. **60**(7), 3560–3575 (2012). https://doi.org/10.1109/tsp.2012.2194706
8. Goldberg, D.E.: Genetic algorithms and rule learning in dynamic system control. In: Proceedings of the 1st International Conference on Genetic Algorithms, pp. 8–15 (1985)
9. Kapanoglu, M., Alikalfa, M., Ozkan, M., Parlaktuna, O., et al.: A pattern-based genetic algorithm for multi-robot coverage path planning minimizing completion time. J. Intell. Manuf. **23**(4), 1035–1045 (2012). https://doi.org/10.1007/s10845-010-0404-5
10. Kapoutsis, A.C., Chatzichristofis, S.A., Kosmatopoulos, E.B.: DARP: divide areas algorithm for optimal multi-robot coverage path planning. J. Intell. Robot. Syst. **86**(3), 663–680 (2017). https://doi.org/10.1007/s10846-016-0461-x
11. Liu, Y., Lin, X., Zhu, S.: Combined coverage path planning for autonomous cleaning robots in unstructured environments. In: Proceedings of the 7th World Congress on Intelligent Control and Automation, pp. 8271–8276 (2008). https://doi.org/10.1109/WCICA.2008.4594223
12. Merino, L., Caballero, F., Martínez-De-Dios, J.R., Maza, I., Ollero, A.: An unmanned aircraft system for automatic forest fire monitoring and measurement. J. Intell. Robot. Syst. **65**(1–4), 533–548 (2012). https://doi.org/10.1007/s10846-011-9560-x
13. Mitschke, M., Uchiyama, N., Sawodny, O.: Online coverage path planning for a mobile robot considering energy consumption. In: Proceedings of 14th International Conference on Automation Science and Engineering (CASE), pp. 1473–1478 (2018). https://doi.org/10.1109/COASE.2018.8560376
14. Moravec, H., Elfes, A.: High resolution maps from wide angle sonar. In: Proceedings of the 1985 IEEE International Conference on Robotics and Automation, pp. 116–121 (1985). https://doi.org/10.1109/ROBOT.1985.1087316
15. Park, E., Kim, K.J., Del Pobil, A.P.: Energy efficient complete coverage path planning for vacuum cleaning robots. In: J. (Jong Hyuk) Park, J., Leung, V., Wang, C.L., Shon, T. (eds) Future Information Technology, Application, and Service. LNEE, vol. 164, no. 1, pp. 23–31. Springer, Dordrecht (2012). https://doi.org/10.1007/978-94-007-4516-2_3
16. Tsardoulias, E.G., Iliakopoulou, A., Kargakos, A., Petrou, L.: A review of global path planning methods for occupancy grid maps regardless of obstacle density. J. Intell. Robot. Syst. **84**(1–4), 829–858 (2016). https://doi.org/10.1007/s10846-016-0362-z
17. Wilfahrt, R., Kim, S.: Traveling salesman problem (TSP). Encyclopedia of GIS, pp. 1173–1176 (2008). https://doi.org/10.1007/978-0-387-35973-1_1406
18. Yazici, A., Kirlik, G., Parlaktuna, O., Sipahioglu, A.: A dynamic path planning approach for multirobot sensor-based coverage considering energy constraints. In: 2009 IEEE/RSJ International Conference on Intelligent Robots and Systems, pp. 5930–5935 (2009). https://doi.org/10.1109/IROS.2009.5354058

Task Set Scheduling of Airport Freight Station Based on Parallel Artificial Bee Colony Algorithm

Haiquan Wang[1(✉)], Jianhua Wei[2], Menghao Su[2], Zhe Dong[1,2], and Shanshan Zhang[2]

[1] Zhongyuan Petersburg Aviation College, Zhongyuan University of Technology, 41 Zhongyuan Road, Zhengzhou 450007, China
wanghq@zut.edu.cn

[2] School of Electric and Information Engineering, Zhongyuan University of Technology, 41 Zhongyuan Road, Zhengzhou 450007, China

Abstract. In order to improve the operation efficiency in airport freight station, the task scheduling problem of freight station is studied in this paper. Based on the mathematical model of the whole system, the integer encoding and continuous encoding methods are proposed to describe the sequence of tasks, then the parallel artificial bee colony algorithm is used to optimize the tasks set. The simulation results show that proposed improved artificial bee colony algorithm based on the two encoding methods are effective, and compared with the traditional bee colony algorithm, the parallel algorithm can reduce the optimization time and improve the optimization efficiency without affecting the optimization results.

Keywords: Artificial bee colony algorithm · Tasks set scheduling · Parallel processing · Task set encoding · Freight station

1 Introduction

The time for handling cargoes in airport freight station consumes 80% of air cargoes turnaround time. Thus, the efficiency of handling cargoes in airport freight station determines the efficiency of whole air transport system [1,2]. With the development of air logistics, the problem of cargoes scheduling in airport freight station has become more and more important, and such a complex optimization problem has become a hotspot in related fields, different algorithms [3–9] have been widely used. Ma et al. [7] employed ensemble multi-objective biogeography-based optimization algorithm to improve scheduling efficiency in

Supported by organization Program of Educational Committee of Henan Province (18A120005), Science & Technology Program of Henan Province (172102210588), and Science and Technology Key Project of Henan Province (162102410056).

L. Pan et al. (Eds.): BIC-TA 2019, CCIS 1159, pp. 484–492, 2020.
https://doi.org/10.1007/978-981-15-3425-6_37

automated warehouse scheduling. Ardjmand et al. [8] proposed an improved particle swarm optimization algorithm to study the allocation of warehouse orders and path planning of the picker. Nesello et al. [9] focused on the issue of scheduling single stacker, and an algorithm based on iterative arc-time-indexed models was applied. Different with the former references, artificial bee colony algorithm (ABC) which is easy to set multi-dimensional solution and possess few parameters is introduced and applied to multi-import and multi-task cargoes scheduling optimization problem in this paper. Based on the established model, different encoding methods are used for coding and ordering the task set, and in order to reduce the time complexity of optimization algorithm, parallelization strategy is introduced based on multi-core processor.

2 Analysis and Modeling of Scheduling Problem

2.1 Establish of Running Time Matrix of ETVS

The running time of elevating transfer vehicle (ETV) is determined by the larger one between the horizontal moving time T_x and vertical lifting time T_y. Two kinds of time from the starting cargo position to the target position can be solved by Eqs. 1 and 2:

$$\begin{cases} T_y = 2 * \sqrt{\frac{h*y}{a_j}}, h * y \leq D_j \\ T_y = T_j + (h * y - D_j)/V_j, h * y > D_j \end{cases} \tag{1}$$

$$\begin{cases} T_x = 2 * \sqrt{\frac{L*X}{a_j}}, L * x \leq D_i \\ T_x = T_i + (\mathrm{L} * x - D_i)/V_i, L*\mathrm{x} > D_i \end{cases} \tag{2}$$

where, x and y represent the number of columns and rows of shelves respectively, and the horizontal and vertical speeds of EVT are V_i and V_j respectively. a_i and a_j are ETV horizontal and vertical acceleration. The width and height of space for cargoes in shelves are L and h respectively. D_i and D_j in Eq. 3 indicate the distances that ETV finishes a complete process including acceleration and deceleration in horizontal and vertical directions, that is, it accelerates from static condition to the maximum speed and immediately decreases to 0.

$$D_i = 1/4a_i * T_i^2, D_j = 1/4a_j * T_j^2 \tag{3}$$

T_i, T_j in Eq. 4 represents the minimum time required to complete a complete acceleration and deceleration process in horizontal and vertical directions.

$$T_i = 2 * V_i/a_i, \quad T_j = 2 * V_j/a_j \tag{4}$$

A three-dimensional warehouse with 5 floors, 45 rows and 2 rows is assumed in this paper, the number of positions in the shelves is 45 * 5 * 2 = 450 [14], $V_i = 120$ m/min, Vertical speed $V_j = 20$ m/min. ETV horizontal acceleration $a_i = 0.5\,\mathrm{m/s^2}$, vertical acceleration $a_j = 0.3\,\mathrm{m/s^2}$, height of cargo space is h = 3750 mm, width is L = 3750 mm. The time for ETV to pick up the goods is

Table 1. Running time

	1	2	3	4	5	6
1	0	5.47	7.74	9.62	11.50	13.37
2	11.62	11.62	11.62	11.62	11.62	13.37
3	22.87	22.87	22.87	22.87	22.87	22.87
4	34.12	34.12	34.12	34.12	34.12	34.12
5	45.37	45.37	45.37	45.37	45.37	45.37

25 s. Thus the running time of ETV for different cargo spaces is shown in Table 1 (Only partial information is listed). The values in table show the horizontal and vertical arrival times from the initial position to the destination position.

2.2 Modeling of Task Set Scheduling Problem

Set H_i as the execution time of task I, i = 1, 2, 3,. . . , n.

$$H_i = H_{i_0} + H_{i_1} + 2\zeta \tag{5}$$

where H_{i0} represents the running time of ETV from current location to the entrance and exit; H_{i1} is the time for ETV running from the entrance and exit to the destination position; H_{i0}, H_{i1} can be obtained from the running time matrix shown in Table 1; ζ is the time for ETV to load or unload the cargoes.

For the scheduling problem with ETV, the objective is to obtain the shortest time for ETV to complete n in-out stock operations in scheduling sequence, the fitness function can be expressed as Eq. 6.

$$Fitness = \sum_{i=1}^{n} H_i = \sum_{i=1}^{n} (H_{i_0} + H_{i_1}) + 2n\zeta \tag{6}$$

3 Parallel ABC Scheduling Algorithm

For the above scheduling optimization problem, artificial bee colony algorithm is introduced and corresponding improvement is adopted.

3.1 Parallel ABC Algorithm

Artificial Bee Colony (ABC) algorithm has fewer parameters and stronger robustness than other algorithms, but during the optimization process, it needs repeated cycle to get the optimal solution which means it will take a lot of time. In order to improve the efficiency of ABC, parallelization is adopted with multi-core CPU for this typical swarm intelligence algorithm with natural and efficient parallel search capability.

The proposed parallel artificial bee colony (PABC) does not change the structure of standard algorithm, the master-slave parallel structure with one population is adopted. The parallelization is only performed in the initialization phase and onlooker bee phase which occupy most of time especially for the for-loop structure. They are divided into several equal parts, each part is executed by different cores in CPU [11]. The basic steps are set as follows

(1) Initialization: Initializes the number of food sources N, the maximum number of iterations, and the maxi-mum number of local optimization. The initial population is evenly distributed among six cores in CPU, parallel computation is performed to obtain the fitness values of all food sources.
(2) Employed bees: The employed bees are divided into six parts, and allocated to each core of CPU. The neighborhood search operation is adopted for each solution as Eq. 7, and the fitness of new solution is evaluated.

$$x'_{ik} = x_{ik} + \text{rank}(-1,1) * (x_{ik} - x_{jk}) \tag{7}$$

i, j are the index of solution in population size, $i, j \in 1, 2, \ldots, N$, $i \neq j$. k is the dimension of the population, $k \in 1, 2, \ldots, N$.
(3) Onlooker bees: With the information shared by employed bees, onlooker bees select employed bees to follow based on the probability defined as Eq. 8.

$$P_i = 0.9 * \frac{\text{fit}_i}{\max(\text{fit}_i)} + 0.1 \tag{8}$$

(4) Scout bees: If the iterations of search with employed bees, onlooker bees reach the threshold of limit, and no better solution is found, the solution will be abandoned, and corresponding bees will transform into scout bees, further search with a random new solution will be executed.
(5) The global optimal solution so far will be recorded and jumps to step 2 until reaching the maximum iteration number.

3.2 Encoding for the Scheduling Problem

In order to optimize the sequence of task set, it is necessary to establish the relationship between the solution of the scheduling problem and the food source of the algorithm, and two methods which are continuous coding and integer coding are proposed in this paper:

(1) Continuous encoding scheme (CES): This method assigns random numbers to each dimension of the generated solution, and then sort them in ascending order based on the size of the random number and the index values, the sequence generated by the index value is the corresponding scheduling scheme. The process of parallel algorithm with continuous task set encoding method is shown in Fig. 1. Table 2 shows the encoding result for a solution with 40 dimensions where the second row is the random number assigned to each dimension of solution, and the index value in the third row represents the result sequence of the task set.

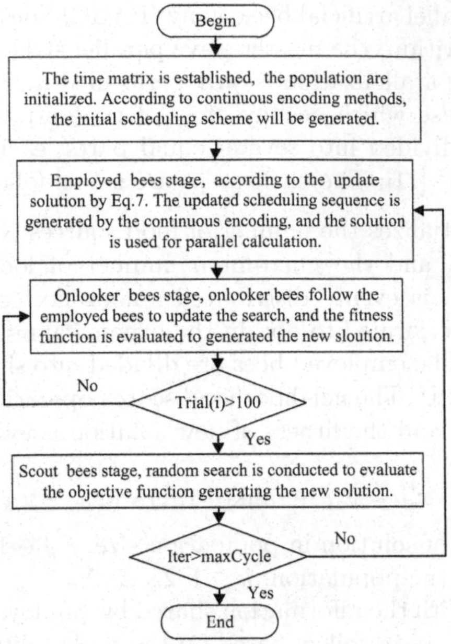

Fig. 1. Scheduling algorithm flow chart based on PABC and continuous coding

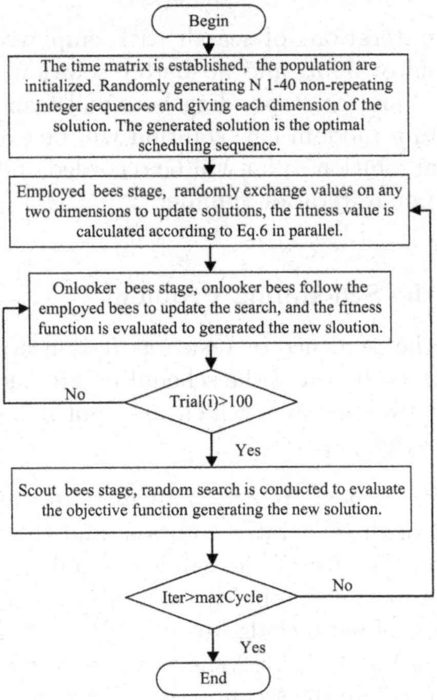

Fig. 2. Scheduling algorithm flow chart based on PABC and integer coding

(2) Integer encoding scheme (IES): Different integral number is generated and assigned to each dimension of the solution. Different from continuous encoding method, during the optimization process, dimensions less than 40 will be randomly selected and exchange the index value with each other, and the new solution will be obtained. The flowchart is shown as Fig. 2.

Table 2. Encoding results of a solution by continuous encoding method

1	2	3	4	5	6	7	8
6.294474	3.674311	3.445405	6.338224	−5.02095	−1.45875	−1.04221	5.002316
33	29	27	34	10	16	18	32
9	10	11	12	13	14	15	16
−0.09054	8.030761	3.582433	8.844962	6.892639	1.897304	−7.35739	1.624593
19	37	28	38	35	23	4	22
17	18	19	20	21	22	23	24
2.612078	2.803109	7.833893	0.961074	9.432063	−8.34092	−5.27861	−7.4108
24	25	36	20	39	2	9	3
25	26	27	28	29	30	31	32
−3.1582	−1.27551	3.915903	−6.66998	−9.84133	−1.96389	3.035783	−6.7554
13	17	30	7	1	14	26	6
33	34	35	36	37	38	39	40
−3.70332	1.198704	4.70441	9.491093	−6.9612	−5.78753	−4.31616	−1.61553
12	21	31	40	5	8	11	15

The encoding logic of the two methods is very different. Compared with CES, IES is simpler and easier to implement. When updating the solution, the new solution will cause the entire sequence to be rearranged by CES. However, IES only exchanges some dimensions, and the sequence does not change significantly. Therefore, in the process of updating the solution, the search space of CES is larger than that of IES.

4 Simulation and Analysis

4.1 Task Settings to Be Scheduled

In order to verify the effectiveness of the proposed algorithm, the scheduling task set was set according to Ref.8, and the simulation was carried out. There are thirteen exits and entrances in the whole system, and the coordinates of entrances (layer - column) are R1(1-5), R2(1-11), R3(1-19), R4(1-21), R5(1-32), R6(1-35), R7(1-45) respectively and the exits coordinates are: C1(1-8), C2(1-17), C3(1-27), C4(1-30), C5(1-40), C6(1-43). The number of task sets is forty, where the first twenty ones are inbound tasks and the last 20 ones are outbound tasks.

4.2 Scheduling Results and Analysis

PC used for simulation possesses Inter(R) Core (TM) i7-8750H CPU @ 2.20 GHz, clock speed is 2.21 GHz, memory is 16 GB. Different algorithm such as Nonlinear Learning Factors Adjusting-Particle Swarm Optimization (NLA-PSO), ABC and PABC are applied to test the effectiveness for task scheduling problem with continuous and integer encoding schemes. For ABC and PABC, the population size is set to be 800, dimension of each solution is 40, maximum iteration number is 2000. All of the experiments were repeated for 20 times. The results with different encoding methods and optimization algorithms are shown in Table 3 and part of results are listed as follows:

Table 3. Optimization results

	Min(s)	Max(s)	Avg(s)	CPU(s)
NLA-PSO with CES	3348.8	3490.4	3419.135	11.7719
ABC with CES	3318.4	3339	3328.255	455.1239
PABC with CES	3314.8	3344.5	3332.015	322.328
ABC with IES	3358.6	3361.5	3359.735	857.3289
PABC with IES	3358.6	3361.5	3359.79	466.3814

(1) PABC with continuous encoding scheme.
Optimal scheduling route:
28 → 1 → 14 → 15 → 13 → 3 → 33 → 24 → 26 → 40 → 34 → 17 → 16 → 20 → 11 → 4 → 7 → 9 → 18 → 27 → 37 → 38 → 8 → 25 → 30 → 10 → 21 → 6 → 39 → 35 → 29 → 31 → 32 → 36 → 22 → 2 → 5 → 23 → 19 → 12.
Shortest time required to finish the tasks: 3314.8 s.
Program running time: 319.982 s.

(2) ABC with continuous encoding scheme.
Optimal scheduling route:
28 → 10 → 23 → 38 → 7 → 26 → 12 → 29 → 15 → 37 → 3 → 25 → 27 → 20 → 40 → 36 → 16 → 11 → 39 → 8 → 17 → 21 → 30 → 31 → 4 → 33 → 24 → 13 → 14 → 6 → 5 → 19 → 9 → 32 → 18 → 22 → 34 → 2 → 1 → 35.
Shortest time required to finish the tasks: 3318.4 s.
Program running time: 451.687 s.

(3) PABC with integer encoding scheme.
Optimal scheduling route:
28 → 29 → 23 → 37 → 12 → 33 → 8 → 24 → 3 → 32 → 9 → 34 → 17 → 39 → 18 → 16 → 38 → 40 → 31 → 14 → 10 → 13 → 35 → 30 → 20 → 7 → 11 → 1 → 27 → 19 → 5 → 2 → 4 → 22 → 6 → 15 → 21 → 36 → 25 → 26.
Shortest time required to finish the tasks: 3358.6 s.
Program running time: 462.803954 s.

(4) ABC with integer encoding scheme
Optimal scheduling route:
15 → 23 → 8 → 32 → 18 → 1 → 38 → 14 → 10 → 13 → 35 → 30 → 31 → 40 → 16 → 20 → 7 → 11 → 27 → 26 → 34 → 12 → 37 → 6 → 36 → 4 → 22 → 21 → 29 → 17 → 39 → 19 → 5 → 3 → 24 → 2 → 9 → 33 → 28 → 25.
Shortest time required to finish the tasks: 3358.6 s.
Program running time: 861.162018 s.

According to the above results, all of the proposed algorithms could solve the problem of task scheduling in airport freight station. Compared with NLA-PSO, ABC-based scheduling method has better scheduling results. Regardless of the encoding scheme, compared with ABC algorithm, the average running time of PABC algorithm is reduced by 45.6%, and the average scheduling time of PABC increased by 0.1%. It means PABC can effectively improve the running efficiency without affecting the optimization results. On the other hand, the maximum, minimum, average and running time of the optimization results corresponding to CES are better than those of IES. The running time of ABC algorithm based on CES is reduced by 46.9%. Thus the efficiency of proposed algorithms could be proved in task set scheduling problem.

5 Conclusion

ABC algorithm is used to solve the problem of cargo scheduling in airport freight station in this paper. Two encoding schemes, including integer and continuous encoding, are adopted to assist in solving the fitness values. Meanwhile, in order to improve the operation efficiency of the algorithm, parallelization based on multi-core CPU is carried out. The results show the proposed PABC algorithm based on two different encoding schemes is able to finish the freight task set scheduling task effectively, and compared with ABC algorithm, the efficiency could be greatly improved.

Acknowledgement. The authors acknowledge the support of Program of Educational Committee of Henan Province (18A120005), Science & Technology Program of Henan Province (172102210588), and Science and Technology Key Project of Henan Province (162102410056).

References

1. Wang, H., Hu, Y., Liao, W.: Path planning algorithm based on improved artificial bee colony algorithm. Control Eng. China **23**(95), 1407–1411 (2016)
2. Henn, S.: Order batching and sequencing for the minimization of the total tardiness in picker-to-part warehouses. Flex. Serv. Manuf. J. **27**(1), 86–114 (2012). https://doi.org/10.1007/s10696-012-9164-1
3. Chen, R.-M., Shen, Y.-M., Wang, C.-T.: Ant colony optimization inspired swarm optimization for grid task scheduling. In: CONFERENCE 2016. LNCS, pp. 461–464 (2016)

4. Kundakci, N., Kulak, O.: Hybrid genetic algorithms for minimizing makespan in dynamic job shop scheduling problem. Comput. Ind. Eng. **96**(c), 31–51 (2016)
5. Cui, L., Li, G., Wang, X.: A ranking-based adaptive artificial bee colony algorithm for global numerical optimization. Inf. Sci. **417**(11), 169–185 (2017)
6. Ghambari, S., Rahati, A.: An improved artificial bee colony algorithm and its application to reliability optimization problems. Appl. Soft Comput. **62**(4), 736–767 (2018)
7. Ma, H., Su, S., Simon, D.: Ensemble multi-objective biogeography-based optimization with application to automated warehouse scheduling. Eng. Appl. Artif. Intell. **44**(9), 79–90 (2015)
8. Ardjmand, E., Shakeri, H., Singh, M.: Minimizing order picking makespan with multiple pickers in a wave picking warehouse. Int. J. Prod. Econ. **206**(C), 169–183 (2018)
9. Nesello, V., Subramanian, A., Battarra, M.: Exact solution of the single-machine scheduling problem with periodic maintenances and sequence-dependent setup times. Eur. J. Oper. Res. **266**(2), 498–507 (2018)
10. Cui, L., et al.: A ranking-based adaptive artificial bee colony algorithm for global numerical optimization. Inf. Sci. **417**(11), 169–185 (2017)
11. Wang, H., Wei, J., Wen, S.: Research on parallel optimization of artificial bee colony algorithm. In: CONFERENCE 2018. LNCS, pp. 125–129 (2018)
12. Asadzadeh, L.: A parallel artificial bee colony algorithm for the job shop scheduling problem with a dynamic migration strategy. Comput. Ind. Eng. **102**(12), 359–367 (2016)
13. Dell'Orco, M., Marinelli, M., Altieri, M.G.: Solving the gate assignment problem through the fuzzy bee colony optimization. Transp. Res. Part C Emerg. Technol. **80**(7), 424–438 (2017)
14. Qiu, J., Jiang, Z., Tang, M.: Research and application of NLAPSO algorithm to ETV scheduling optimization in airport cargo terminal. **34**(1), 65–70 (2015)

Water Wave Optimization with Self-adaptive Directed Propagation

Chenxin Wu, Yangyan Xu, and Yujun Zheng(✉)

Institute of Service Engineering, Hangzhou Normal University, Hangzhou 311121, China
wuchenxin@compintell.cn
http://www.compintell.cn

Abstract. Water wave optimization (WWO) is a recently proposed nature-inspired algorithm that mimics shallow water wave motions to solve optimization problems. In this paper, we propose an improved WWO algorithm with a new self-adaptive directed propagation operator, which dynamically adjusts the propagation direction of each solution according to the fitness change caused by the last propagation operation to improve local search ability. The new algorithm also adopts a nonlinear population reduction strategy to better balance the local search and global search abilities. Experimental results on 15 function optimization problems from the CEC2015 single-objective optimization test suite show that the improved algorithm exhibits significantly better performance than the original WWO and some other evolutionary algorithms including particle swarm optimization (PSO) and biogeography-based optimization (BBO), which validates the effectiveness and efficiency of the proposed strategies.

Keywords: Water wave optimization (WWO) · Global optimization · Self-adaptive directed propagation · Nonlinear population reduction

1 Introduction

With the increasing scale and complexity of optimization problems, the solution time of traditional optimization methods often exceed practically allocated computational time. In recent decades, there have been a variety of metaheuristics algorithms, such as genetic algorithm (GA) [2], particle swarm optimization (PSO) [3], biogeography-based optimization (BBO) [9], cuckoo search (CS) [15], etc, which are often capable of obtaining optimal or near-optimal solutions within an acceptable time.

Water wave optimization (WWO) algorithm is a recently proposed optimization algorithm [21]. It takes inspiration from the shallow water wave theory to

Supported by grants from National Natural Science Foundation of China under Grant No. 61872123 and 61473263.

L. Pan et al. (Eds.): BIC-TA 2019, CCIS 1159, pp. 493–505, 2020.
https://doi.org/10.1007/978-981-15-3425-6_38

solve global optimization problems. Owing to its characteristics of simple algorithm framework, fewer control parameters and smaller population size, WWO has aroused great research interest. Zheng and Zhang [24] proposed a simplified WWO (SimWWO) by removing the refraction operation and adding a linear population size reduction strategy, which can accelerate the convergence speed of the algorithm. Zhang et al. [16] enhanced WWO with a variable population size strategy and a comprehensive learning strategy, such that each solution can learn from more other solutions instead of just learning from the current best solution. Zhou and Wu [14] proposed an elite opposition-based WWO, which employs the elite opposition-based learning [12] to increase population diversity, integrates a local neighborhood search strategy to strengthen local search, and introduces the inertial weight of PSO to the propagation operator to balance exploration and exploitation. Zhang et al. [17] proposed an improved sine cosine WWO algorithm, which combines the position update operation of the sine cosine algorithm [6] with the propagation operator of WWO to improve the exploration ability, and also employs the elite opposition-based learning strategy to increase the diversity. Zhang et al. [18] proposed a wind-driven WWO, which mimics the motion of air parcels in the wind to explore the search space so as to improve the calculation accuracy of the basic WWO. WWO has also been successfully applied to many real-world optimization problems (e.g., [4,8,10,19,20,23,25]). Recently, Zheng et al. [22] proposed a systematic approach that consists of a set of basic steps and strategies for adapting WWO for different combinatorial optimization problems.

In this paper, we propose a new improved WWO algorithm, called SAWWO, which introduces a new self-adaptive propagation operator that dynamically adjusts the propagation direction of each solution according to the fitness change caused by the last propagation operation to improve local search ability. The algorithm also adopts a nonlinear population reduction strategy to accelerate the search. Numerical experiments on the CEC 2015 benchmark suite [5] demonstrate that the improved algorithm exhibits a better performance than the basic WWO, SimWWO, PSO, and BBO.

The rest of the paper is organized as follows. Section 2 briefly introduces the basic WWO algorithm, Sect. 3 describes the proposed improved WWO algorithm in details, Sect. 4 presents the numerical experiments, and Sect. 5 concludes with a discussion.

2 The Basic WWO Algorithm

Initially proposed by Zheng [21], WWO is a relatively new evolutionary algorithm inspired by the shallow water wave models. In WWO, the search space is analogous to the seabed area, each solution $\boldsymbol{x}$ is analogous to a "wave" with a wave height $h_{\boldsymbol{x}}$ (which is initialized to a constant integer $h_{\max}$) and a wavelength $\lambda_{\boldsymbol{x}}$ (which is initialized to 0.5), and the solution fitness is measured by its seabed depth: the shorter the distance to the still water level, the higher the fitness is, as illustrated by Fig. 1. When solving an optimization problem, WWO first initializes a population of waves, and then uses three operators including propagation, refraction, and breaking to evolve the waves to effectively search the solution space.

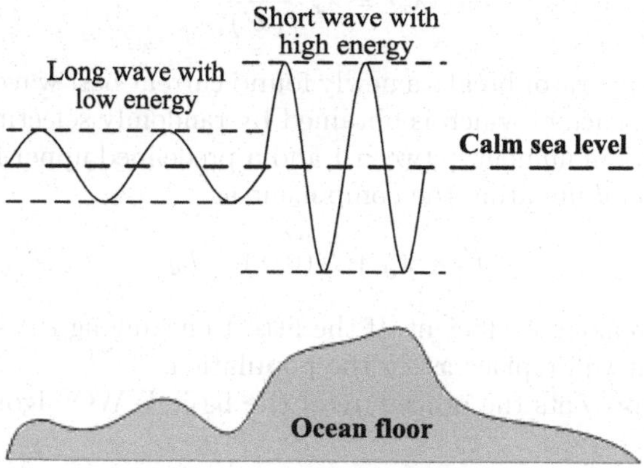

Fig. 1. Different wave shapes in deep and shallow water.

At each generation, each wave $\boldsymbol{x}$ propagates once to create a new wave $\boldsymbol{x}'$ by adding a different offset at each dimension d as:

$$x'_d = x_d + rand(-1, 1) \cdot \lambda_{\boldsymbol{x}} L_d \tag{1}$$

where *rand* is a function that generates a random number uniformly distributed within the specified range, and L_d is the length of the dth dimension of the problem. If $\boldsymbol{x}'$ is fitter than $\boldsymbol{x}$, the new wave will replace the old $\boldsymbol{x}$ in the population. Otherwise contrary, $\boldsymbol{x}$ is reserved and its wave height is reduced by one to simulate the loss of energy.

The wave wavelength of each solution is updated based on its fitness $f(\boldsymbol{x})$ at each generation as:

$$\lambda_{\boldsymbol{x}} = \lambda_{\boldsymbol{x}} \alpha^{-(f(\boldsymbol{x}) - f_{\min} + \varepsilon)/(f_{\max} - f_{\min} + \varepsilon)} \tag{2}$$

where $f_{\max}$ and $f_{\min}$ denote the maximum and minimum fitness values, respectively, α is the wavelength reduction coefficient, and ε is a very small value to avoid division by zero.

The refraction operator performs on any wave whose height (initialized as $h_{\max}$) decreases to zero to avoid search stagnation. It makes the stagnant wave learn from the current best wave $\boldsymbol{x}^*$ at each dimension d as:

$$x'_d = N\left(\frac{x^*_d + x_d}{2}, \frac{|x^*_d - x_d|}{2}\right) \tag{3}$$

where $N(\mu, \sigma)$ generates a Gaussian random number with mean μ and standard deviation σ. After refraction, its wavelength is updated according to the ratio of the new fitness to the original fitness as:

$$\lambda_{\boldsymbol{x}'} = \lambda_{\boldsymbol{x}} \frac{f(\boldsymbol{x})}{f(\boldsymbol{x}')} \tag{4}$$

The breaking operator breaks a newly found current best wave $\boldsymbol{x}^*$ into a series of solitary waves, each of which is obtained by randomly selecting k dimensions (where k is a random number between 1 and a predefined upper limit $k_{\max}$), and at each dimension d updating the component as:

$$x'_d = x^*_d + N(0,1) \cdot \beta L_d \tag{5}$$

where β is the breaking coefficient. If the fittest one among the solitary waves is better than $\boldsymbol{x}^*$, it will replace $\boldsymbol{x}^*$ in the population.

Algorithm 1 presents the procedure of the basic WWO algorithm.

Algorithm 1. The basic WWO algorithm.

Randomly initialize a population of n solutions (waves);
while the termination condition is not satisfied **do**
  for each wave $\boldsymbol{x} \in P$ **do**
    Propagate $\boldsymbol{x}$ to a new $\boldsymbol{x}'$ according to Eq. (1);
    if $f(\boldsymbol{x}') > f(\boldsymbol{x})$ **then**
      Replace $\boldsymbol{x}$ with $\boldsymbol{x}'$;
      if $f(\boldsymbol{x}) > f(\boldsymbol{x}^*)$ **then**
        Replace $\boldsymbol{x}^*$ with $\boldsymbol{x}$;
        $k = rand(1, k_{\max})$;
        for $i = 1$ to k **do**
          Break $\boldsymbol{x}^*$ into a new wave $\boldsymbol{x}_{new}$ according to Eq. (5);
          if $f(\boldsymbol{x}_{new}) > f(\boldsymbol{x}^*)$ **then**
            Replace $\boldsymbol{x}^*$ with $\boldsymbol{x}_{new}$;
          end if
        end for
      end if
    else
      $h_{\boldsymbol{x}} \leftarrow h_{\boldsymbol{x}} - 1$;
      if $h_{\boldsymbol{x}} = 0$ **then**
        Refract $\boldsymbol{x}$ to a new $\boldsymbol{x}'$ according to Eq. (3);
        Update $\lambda_{\boldsymbol{x}'}$ according to Eq. (4);
      end if
    end if
  end for
  Update the wavelengths of the waves according to Eq. (2);
end while
return the best wave found so far.

3 The Proposed SAWWO Algorithm

In this section, we propose a new improved WWO, called SAWWO, which utilizes the new self-adaptive directed propagation operator and the nonlinear population reduction strategy to enhance the performance of the algorithm.

3.1 A New Self-adaptive Directed Propagation Operator

The propagation operation of the basic WWO uses a random search direction, which does not utilize the historical information of evolution. To address this issue, we propose a new directed propagation operator, which equips each solution $\boldsymbol{x}$ with a direction vector $\boldsymbol{q}^{\boldsymbol{x}}$ that is iteratively updated according to the fitness change caused by the last propagation operation on $\boldsymbol{x}$. At the first generation, each dimension $q_d^{\boldsymbol{x}}$ is set to 1. At each iteration, when propagating $\boldsymbol{x}$, we generate the new wave $\boldsymbol{x}'$ by randomly selecting some dimensions and shifting each dimension d as:

$$x'_d = x_d + rand(0,1) \cdot \lambda_{\boldsymbol{x}} L_d q_d^{\boldsymbol{x}} \tag{6}$$

If the new wave $\boldsymbol{x}'$ is better than $\boldsymbol{x}$, the $\boldsymbol{x}$ will be replaced by the $\boldsymbol{x}'$; otherwise, we set $q_d^{\boldsymbol{x}} = -q_d^{\boldsymbol{x}}$ for all the selected dimensions. In this way, if a wave $\boldsymbol{x}$ propagates to a worse position, then it will change its propagation directions at the next generation; otherwise, it will keep the last direction to continue searching the potentially promising region, and thus improve the local search ability of the algorithm. The random selection of some dimensions can avoid the population converging to a local optimum too fast. The new directed propagation operator is also faster than the original WWO propagation operator because it only performs operations on probably half of the D dimensions.

3.2 Nonlinear Population Reduction

The mechanism of variable population size has been widely employed in many evolutionary algorithms and has demonstrated its effectiveness in balancing global search and local search [1,7,11,13] on many problems. In general, a large population size prefers to explore the whole solution space in early search stages, while a small population size can be used to facilitate local exploitation in later stages. Zheng and Zhang [24] introduced a linear population reduction mechanism in WWO. However, the population size controlled by the linear model is still insufficient to well fit the search requirements. Here, we employ a nonlinear model, which decreases the population size n from an upper limit $n_{\max}$ to a lower limit $n_{\min}$ as:

$$n = n_{\max} - (n_{\max} - n_{\min}) \cdot (\frac{t}{t_{\max}})^2 \tag{7}$$

where t is the number of fitness evaluations (NFEs) and $t_{\max}$ is the maximum allowable NFEs. Compared to the linear population reduction, this nonlinear strategy avoids the number of solutions decreasing too fast in early stages, and hence decreases the probability of being trapped into local optima.

When implementing the population reduction mechanism, we remove the worst individual(s) from the population.

In the original WWO, the refraction operator is used to update stagnant solutions. By using the nonlinear population reduction mechanism, bad and stagnant solutions will be automatically removed, and thus we remove the refraction from our improved WWO algorithm.

3.3 The Framework of SAWWO

Algorithm 2 presents the framework of the SAWWO algorithm.

Algorithm 2. The SAWWO algorithm.

Randomly initialize a population of n solutions (waves);
Initialize a direction vector $\boldsymbol{q}^{\boldsymbol{x}}$ for each solution $\boldsymbol{x}$ in the population;
while the termination condition is not satisfied **do**
  for each wave $\boldsymbol{x} \in P$ **do**
    Let $m = rand(1, D)$;
    for each selection dimension d **do**
      set $x'_d = x_d + rand(0, 1) \cdot \lambda_{\boldsymbol{x}} L_d q^{\boldsymbol{x}}_d$;
    end for
    if $f(\boldsymbol{x}') > f(\boldsymbol{x})$ **then**
      Replace $\boldsymbol{x}$ with $\boldsymbol{x}'$;
      if $f(\boldsymbol{x}) > f(\boldsymbol{x}^*)$ **then**
        Replace $\boldsymbol{x}^*$ with $\boldsymbol{x}$;
        $k = rand(1, k_{\max})$;
        for $i = 1$ to k **do**
          Break $\boldsymbol{x}^*$ into a new wave $\boldsymbol{x}_{new}$ according to Eq. (5);
          if $f(\boldsymbol{x}_{new}) > f(\boldsymbol{x}^*)$ **then**
            Replace $\boldsymbol{x}^*$ with $\boldsymbol{x}_{new}$;
          end if
        end for
      end if
    else
      for each selected dimension $d \in m$ dimensions **do**
        $q^{\boldsymbol{x}}_d = -q^{\boldsymbol{x}}_d$;
      end for
    end if
  end for
  Update the population size n according to Eq.(7);
  Update the wavelengths according to Eq. (2);
end while
return the best solution found so far.

4 Numerical Experiments

4.1 Experimental Parameter Setting

We conduct numerical experiments on the CEC15 learning-based benchmark suite [5], which consists of 15 benchmark functions, denoted by f_1–f_{15} (where f^* denotes the exact optimal function value). The functions are all scalable high-dimensional problems. f_1 and f_2 are unimodal functions, f_3–f_5 are simple multimodal functions, f_6–f_8 are hybrid functions, and f_9–f_{15} are composition functions. Interested readers can refer to [5] for more details of the test functions.

We compare the proposed SAWWO with WWO, SimWWO, PSO, BBO. To ensure a fair comparison, we set the maximum NFEs of all algorithms to 300,000 for each problem. The experimental environment is a computer of Intel Core i7-8700 processor and 16 GB memory. The control parameters of the algorithms are set as follows:

- PSO: Population size $n = 50$, inertia weight w ($w_{\max} = 0.9$, $w_{\min} = 0.4$), the acceleration coefficients $c_1 = 2$ and $c_2 = 2$.
- BBO: $n = 50$, maximum immigration rate $\lambda_{\max} = 1$, maximum emigration rate $\mu_{\max} = 1$, and mutation rate $m_p = 0.05$.
- WWO: $h_{\max} = 12$, $\alpha = 1.0026$, $k_{\max} = 12$, β linearly decreases from 0.25 to 0.001, and $n = 10$.
- SimWWO: $\alpha = 1.0026$, $k_{\max} = 6$, β linearly decreases from 0.25 to 0.001, and n linearly decreases from 50 to 3.
- SAWWO: $\alpha = 1.0026$, $k_{\max} = 12$, β linearly decreases from 0.25 to 0.001, n nonlinearly decreases from 50 to 3.

Note that we simply use a fixed parameter setting for each algorithm on all the test problems instead of fine-tuning the parameter values on each problem.

4.2 Comparative Experiment

We run each algorithm 50 times (with different random seeds) on each test function, and record the maximum, minimum, median, standard deviation of the results as presented in Tables 1 and 2. On each test function, the best median values among the five algorithms are marked in bold (some values appear to be the same as the bold values because the decimals are omitted). We perform t-test on the running results of SAWWO and other algorithms. An h value of 0 denotes there is no significant difference between SAWWO and the corresponding algorithm, 1^+ implies the performance of SAWWO is significantly better than the corresponding algorithm (with a confidence level of 95%), and 1^- vice versa.

According to the experimental results, on the two unimodal group functions, SAWWO and SimWWO obtain the best median values on f_1 and f_2, respectively. On the three simple multimodal functions, SAWWO obtains the best median values on f_3 and f_4, and BBO obtains the best median value on f_5. On the three hybrid functions, SAWWO obtains the best median values on f_7 and f_8, and WWO does so on f_6. On the seven composition functions, SAWWO obtains

Table 1. Experimental results on the unimodal and simple multimodal benchmark functions.

ID	Metric	PSO	BBO	WWO	SimWWO	SAWWO
f_1	max	3.72E+08	1.75E+07	3.25E+06	4.88E+06	2.78E+06
	min	7.53E+06	1.33E+06	3.89E+05	1.38E+06	3.48E+05
	median	5.72E+07	5.68E+06	1.62E+06	3.03E+06	**1.17E+06**
	std	6.06E+07	4.29E+06	6.88E+05	8.98E+05	5.60E+05
	h	1^+	1^+	1^+	1^+	
f_2	max	2.07E+10	3.84E+06	1.51E+04	4.98E+02	1.28E+04
	min	9.34E+07	3.45E+05	2.07E+02	2.00E+02	2.00E+02
	median	5.02E+09	9.48E+05	1.25E+03	**2.04E+02**	1.26E+03
	std	4.73E+09	5.80E+05	3.22E+03	5.28E+01	2.99E+03
	h	1^+	1^+	0	1^-	
f_3	max	3.21E+02	3.20E+02	3.20E+02	3.20E+02	3.20E+02
	min	3.20E+02	3.20E+02	3.20E+02	3.20E+02	3.20E+02
	median	3.21E+02	3.20E+02	**3.20E+02**	**3.20E+02**	**3.20E+02**
	std	1.34E-01	2.98E-02	3.13E-06	1.78E-05	4.27E-03
	h	1^+	1^+	0	0	
f_4	max	6.03E+02	5.02E+02	6.07E+02	4.99E+02	4.50E+02
	min	4.62E+02	4.27E+02	4.44E+02	4.49E+02	4.15E+02
	median	5.41E+02	4.50E+02	5.26E+02	4.74E+02	**4.32E+02**
	std	3.28E+01	1.38E+01	3.54E+01	1.18E+01	7.78E+00
	h	1^+	1^+	1^+	1^+	
f_5	max	5.19E+03	3.39E+03	6.39E+03	3.41E+03	3.21E+03
	min	2.76E+03	1.22E+03	2.39E+03	1.93E+03	1.45E+03
	median	4.06E+03	**2.40E+03**	3.96E+03	2.72E+03	2.67E+03
	std	5.96E+02	4.62E+02	7.08E+02	3.81E+02	4.32E+02
	h	1^+	0	1^+	1^+	
f_6	max	3.14E+07	7.74E+06	3.11E+05	1.80E+05	2.32E+05
	min	1.46E+05	1.66E+05	4.13E+03	3.13E+04	5.63E+03
	median	1.35E+06	3.24E+06	**3.53E+04**	8.47E+04	5.70E+04
	std	8.00E+06	1.67E+06	6.07E+04	3.82E+04	5.12E+04
	h	1^+	1^+	0	1^+	
f_7	max	9.41E+02	7.78E+02	7.89E+02	7.15E+02	7.11E+02
	min	7.16E+02	7.06E+02	7.13E+02	7.09E+02	7.07E+02
	median	7.23E+02	7.13E+02	7.18E+02	7.12E+02	**7.09E+02**
	std	3.86E+01	9.51E+00	1.03E+01	1.26E+00	1.02E+00
	h	1^+	1^+	1^+	1^+	
f_8	max	2.79E+06	2.28E+06	2.84E+05	7.56E+04	8.56E+04
	min	3.09E+04	6.18E+04	6.37E+03	1.13E+04	6.23E+03
	median	3.95E+05	1.01E+06	3.66E+04	3.93E+04	**2.98E+04**
	std	5.90E+05	5.94E+05	4.16E+04	1.64E+04	1.82E+04
	h	1^+	1^+	1^+	1^+	

Table 2. Experimental results on the hybrid and composition benchmark functions.

ID	Metric	PSO	BBO	WWO	SimWWO	SAWWO
f_9	max	1.28E+03	1.01E+03	1.01E+03	1.01E+03	1.01E+03
	min	1.01E+03	1.00E+03	1.01E+03	1.01E+03	1.01E+03
	median	1.04E+03	**1.00E+03**	1.01E+03	1.01E+03	1.01E+03
	std	5.29E+01	4.39E-01	1.13E+00	4.83E-01	2.76E-01
	h	1^+	1^-	1^+	1^+	
f_{10}	max	2.41E+08	5.62E+03	3.71E+05	2.13E+05	4.87E+04
	min	2.29E+03	1.86E+03	1.04E+04	3.80E+04	3.43E+03
	median	2.52E+07	**2.71E+03**	6.37E+04	9.86E+04	1.26E+04
	std	3.78E+07	5.80E+02	6.29E+04	4.52E+04	1.11E+04
	h	1^+	1^-	1^+	1^+	
f_{11}	max	2.31E+03	2.02E+03	2.31E+03	1.82E+03	1.61E+03
	min	1.47E+03	1.74E+03	1.41E+03	1.41E+03	1.40E+03
	median	2.13E+03	1.88E+03	2.02E+03	**1.43E+03**	1.50E+03
	std	1.73E+02	5.01E+01	1.59E+02	1.20E+02	6.44E+01
	h	1^+	1^+	1^+	0	
f_{12}	max	1.40E+03	1.31E+03	1.32E+03	1.31E+03	1.31E+03
	min	1.31E+03	1.31E+03	1.31E+03	1.31E+03	1.31E+03
	median	1.40E+03	1.31E+03	1.31E+03	1.31E+03	**1.31E+03**
	std	4.06E+01	1.26E+00	2.53E+00	7.39E-01	4.32E-01
	h	1^+	1^+	1^+	1^+	
f_{13}	max	1.30E+03	1.30E+03	1.30E+03	1.30E+03	1.30E+03
	min	1.30E+03	1.30E+03	1.30E+03	1.30E+03	1.30E+03
	median	1.30E+03	1.30E+03	1.30E+03	1.30E+03	**1.30E+03**
	std	1.01E-01	3.40E-03	1.31E-01	8.19E-03	4.03E-04
	h	1^+	1^+	1^+	1^+	
f_{14}	max	6.98E+04	3.64E+04	5.91E+04	4.88E+04	4.63E+04
	min	3.56E+04	3.26E+04	4.59E+04	4.31E+04	4.29E+04
	median	4.28E+04	**3.49E+04**	4.92E+04	4.77E+04	4.58E+04
	std	6.45E+03	8.70E+02	2.33E+03	1.11E+03	8.07E+02
	h	1^-	1^-	1^+	1^+	
f_{15}	max	3.95E+03	1.60E+03	1.60E+03	1.60E+03	1.60E+03
	min	1.61E+03	1.60E+03	1.60E+03	1.60E+03	1.60E+03
	median	1.63E+03	1.60E+03	**1.60E+03**	**1.60E+03**	**1.60E+03**
	std	4.18E+02	1.35E-01	0	0	0
	h	1^+	1^+	0	0	

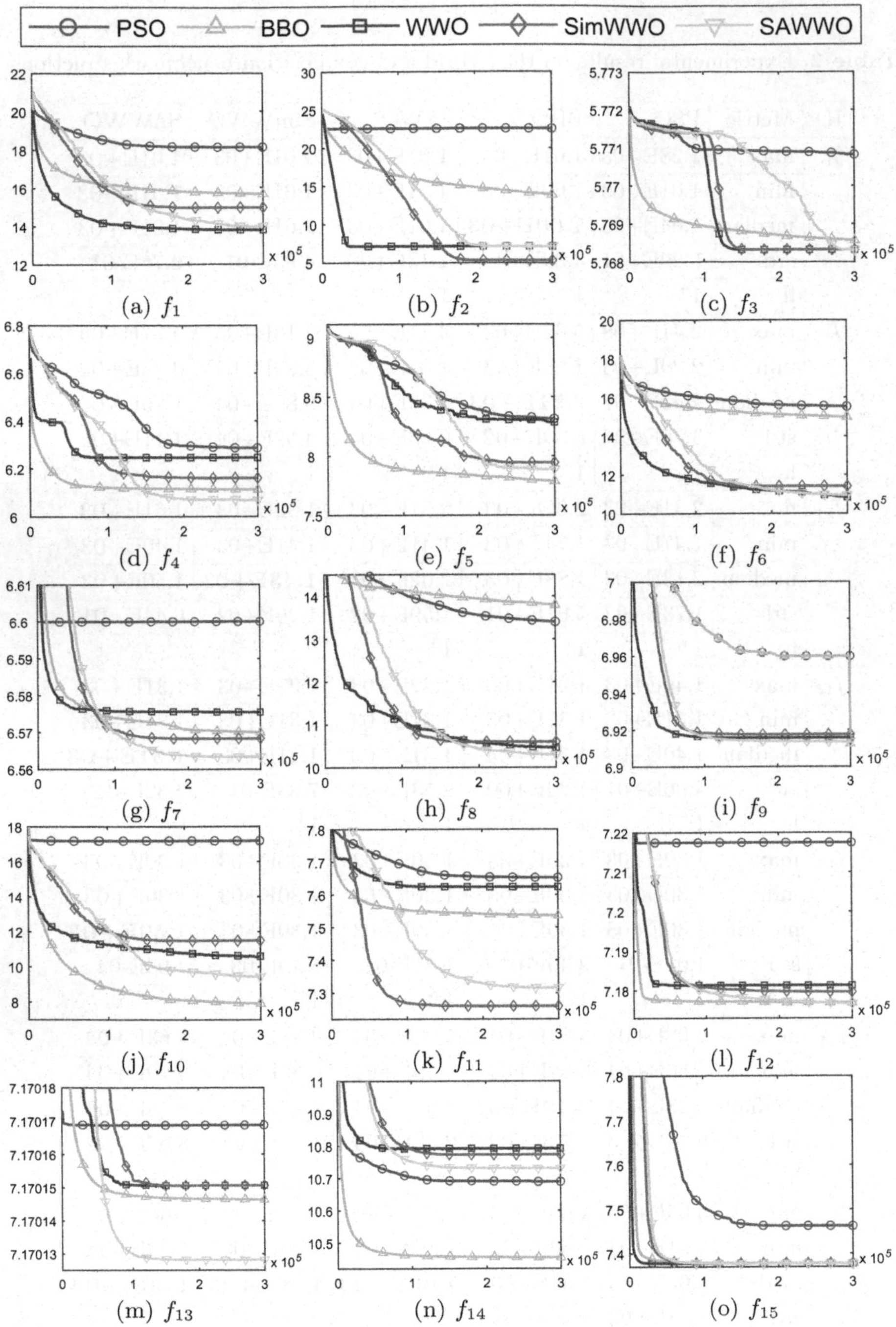

Fig. 2. Convergence curves of the algorithms on the CEC 2015 benchmark problems. The x-axis denotes NFEs, and the y-axis denotes the natural logarithm of the median values.

the best median values on f_{12}, f_{13} and f_{15}, BBO obtains the best median values f_9, f_{10} and f_{14}, and SimWWO obtains the best median value on f_{11}.

According to the statistical test results:

- SAWWO performs significantly better than PSO on 14 functions, while PSO performs significantly better than SAWWO only on one function (f_{14}).
- SAWWO performs significantly better than BBO on 11 functions, while BBO performs significantly better than SAWWO on three functions.
- SAWWO performs significantly better than the basic WWO on 11 functions, and there is no significant difference between SAWWO and the basic WWO on the remaining four functions.
- SAWWO performs significantly better than the SimWWO on 11 functions, while the SimWWO performs significantly better than SAWWO only on one function (f_2).

Figure 2(a)–(o) present the convergence curves of the algorithms on the 15 test functions, respectively. As we can observe, at early stages of the search, WWO often converges faster than SAWWO, and SAWWO converges to a better solution at the later stage. This is because SAWWO uses more individuals to explore wider area in the early stage to avoid being trapped into local optima, and uses less individuals to exploit narrow area to enhance local search in the later stage. Moreover, the self-adaptive directed propagation operator ability can further utilize historical search information to enhance solution accuracy.

In summary, the experimental results show that the overall performance of SAWWO is the best one among the five comparative algorithms on the benchmark suite, which demonstrates the effectiveness and efficiency of the proposed strategies.

5 Conclusion

In this paper, we present SAWWO, which uses the self-adaptive directed propagation operator to improve the local search ability and the nonlinear population reduction strategy to better balance local and global search abilities. According to the numerical results on the CEC15 test functions, the overall performance of SAWWO is best among the comparative algorithms.

There are still rooms for further improving the directed propagation operator and the nonlinear population reduction strategy. When setting the propagation directions, we can utilize the historical information of the propagation operations in more than one generations. We can also adapt the population size according to the search states of the algorithm.

References

1. Altinoz, O.T., Yilmaz, A.E.: A population size reduction approach for nondominated sorting-based optimization algorithms. Int. J. Comput. Intell. Appl. **16**(01), 1750005 (2017). https://doi.org/10.1142/S1469026817500055

2. Holland, J.H.: Genetic algorithms and classifier systems: foundations and future directions. Technical report, Michigan University, Ann Arbor, USA (1987)
3. Kennedy, J., Eberhart, R.: Particle swarm optimization. In: IEEE International Conference on Neural Networks, Perth, Australia, p. IV. IEEE Service Center, Piscataway (1942). https://doi.org/10.1109/ICNN.1995.488968
4. Lenin, K., Ravindhranath Reddy, B., Suryakalavathi, M.: Hybridization of firefly and water wave algorithm for solving reactive power problem. Int. J. Eng. Res. Afri. **21**, 165–171 (2016). https://doi.org/10.4028/www.scientific.net/JERA.21.165
5. Liang, J., Qu, B., Suganthan, P., Chen, Q.: Problem definitions and evaluation criteria for the CEC 2015 competition on learning-based real-parameter single objective optimization. Technical report 201411A, Computational Intelligence Laboratory, Zhengzhou University, Zhengzhou China and Technical report, Nanyang Technological University, Singapore (2014)
6. Mirjalili, S.: SCA: a sine cosine algorithm for solving optimization problems. Knowl.-Based Syst. **96**, 120–133 (2016). https://doi.org/10.1016/j.knosys.2015.12.022
7. Mohamed, A.K., Mohamed, A.W., Elfeky, E.Z., Saleh, M.: Enhancing AGDE algorithm using population size reduction for global numerical optimization. In: Hassanien, A.E., Tolba, M.F., Elhoseny, M., Mostafa, M. (eds.) AMLTA 2018. AISC, vol. 723, pp. 62–72. Springer, Cham (2018). https://doi.org/10.1007/978-3-319-74690-6_7
8. Shao, Z., Pi, D., Shao, W.: A novel discrete water wave optimization algorithm for blocking flow-shop scheduling problem with sequence-dependent setup times. Swarm Evol. Comput. **40**, 53–75 (2018). https://doi.org/10.1016/j.swevo.2017.12.005
9. Simon, D.: Biogeography-based optimization. IEEE Trans. Evol. Comput. **12**(6), 702–713 (2008). https://doi.org/10.1109/TEVC.2008.919004
10. Song, Q., Zheng, Y.J., Huang, Y.J., Xu, Z.G., Sheng, W.G., Yang, J.: Emergency drug procurement planning based on big-data driven morbidity prediction. IEEE Trans. Industr. Inform. (2018). https://doi.org/10.1109/TII.2018.2870879
11. Tanabe, R., Fukunaga, A.S.: Improving the search performance of shade using linear population size reduction. In: 2014 IEEE Congress on Evolutionary Computation (CEC), pp. 1658–1665. IEEE (2014). https://doi.org/10.1109/CEC.2014.6900380
12. Tizhoosh, H.: Opposition-based learning: a new scheme for machine intelligence. In: Computational Intelligence for Modelling, Control and Automation, vol. 1, pp. 695–701 (2005). https://doi.org/10.1109/CIMCA.2005.1631345
13. Viktorin, A., Pluhacek, M., Senkerik, R.: Network based linear population size reduction in shade. In: 2016 International Conference on Intelligent Networking and Collaborative Systems (INCoS), pp. 86–93. IEEE (2016). https://doi.org/10.1109/INCoS.2016.50
14. Wu, X., Zhou, Y., Lu, Y.: Elite opposition-based water wave optimization algorithm for global optimization. Math. Probl. Eng. **2017** (2017). https://doi.org/10.1155/2017/3498363
15. Yang, X.S., Deb, S.: Cuckoo search via Lévy flights. In: 2009 World Congress on Nature & Biologically Inspired Computing (NaBIC), pp. 210–214. IEEE (2009). https://doi.org/10.1109/NABIC.2009.5393690
16. Zhang, B., Zhang, M.-X., Zhang, J.-F., Zheng, Y.-J.: A water wave optimization algorithm with variable population size and comprehensive learning. In: Huang, D.-S., Bevilacqua, V., Prashan, P. (eds.) ICIC 2015. LNCS, vol. 9225, pp. 124–136. Springer, Cham (2015). https://doi.org/10.1007/978-3-319-22180-9_13

17. Zhang, J., Zhou, Y., Luo, Q.: An improved sine cosine water wave optimization algorithm for global optimization. J. Intell. Fuzzy Syst. **34**(4), 2129–2141 (2018). https://doi.org/10.3233/JIFS-171001
18. Zhang, J., Zhou, Y., Luo, Q.: Nature-inspired approach: a wind-driven water wave optimization algorithm. Appl. Intell. **49**(1), 233–252 (2018). https://doi.org/10.1007/s10489-018-1265-4
19. Zhao, F., Liu, H., Zhang, Y., Ma, W., Zhang, C.: A discrete water wave optimization algorithm for no-wait flow shop scheduling problem. Expert Syst. Appl. **91**, 347–363 (2018). https://doi.org/10.1016/j.eswa.2017.09.028
20. Zheng, Y., Zhang, B., Xue, J.: Selection of key software components for formal development using water wave optimization. J. Softw. **27**(4), 933–942 (2016)
21. Zheng, Y.J.: Water wave optimization: a new nature-inspired metaheuristic. Comput. Oper. Res. **55**, 1–11 (2015). https://doi.org/10.1016/j.cor.2014.10.008
22. Zheng, Y.J., Lu, X.Q., Du, Y.C., Xue, Y., Sheng, W.G.: Water wave optimization for combinatorial optimization: design strategies and applications. Appl. Soft Comput. **83**, 105611 (2019). https://doi.org/10.1016/j.asoc.2019.105611
23. Zheng, Y.J., Wang, Y., Ling, H.F., Xue, Y., Chen, S.Y.: Integrated civilian-military pre-positioning of emergency supplies: a multiobjective optimization approach. Appl. Soft Comput. **58**, 732–741 (2017). https://doi.org/10.1016/j.asoc.2017.05.016
24. Zheng, Y.J., Zhang, B.: A simplified water wave optimization algorithm. In: 2015 IEEE Congress on Evolutionary Computation (CEC), pp. 807–813. IEEE (2015). https://doi.org/10.1109/CEC.2015.7256974
25. Zhou, X.H., Zhang, M.X., Xu, Z.G., Cai, C.Y., Huang, Y.J., Zheng, Y.J.: Shallow and deep neural network training by water wave optimization. Swarm Evol. Comput. **50**, 100561 (2019)

An Unbiased Butterfly Optimization Algorithm

Gehan Ahmed Bahgat[1(✉)], Abd-Allah Fawzy[2], and Hassan M. Emara[2]

[1] Electronics Research Institute, Giza, Egypt
gehan@eri.sci.eg
[2] Cairo University, Giza, Egypt

Abstract. Several bio-inspired optimization techniques have been used to solve multidimensional multimodal optimization problems. Butterfly optimization algorithm (BOA) is a recent bio-inspired technique that emulates butterflies' foraging which is based on their sense of smell to determine the location of nectar. BOA shows fast convergence compared to other optimization techniques such as particle swarm optimization (PSO), genetic algorithm (GA) and biogeography based optimization (BBO) when compared over several standard optimization problems. In this paper, the basic algorithm of BOA is demonstrated to be biased to search for optimal points near origin. This feature is the main cause of the fast convergence of BOA in standard test problems where the optimal points is at (or near) the origin. Without this biasing effect, BOA suffers from slow convergence and may not converge to the global minimum even for unimodal low dimensionality problem. The evidence of biasing is demonstrated, and an unbiased version of the BOA (UBOA) is proposed to alleviate this problem. UBOA is proved to outperform BOA in case of shifted functions. Test results using several benchmark problems demonstrate that UBOA provides a competitive optimization method with respect to other bio-inspired methods.

Keywords: Butterfly optimization algorithm · Global optimization · Nature inspired · Metaheuristic · Unbiased algorithm

1 Introduction

Metaheuristic optimization techniques have become very popular over the last decades. These techniques have covered a wide range of applications due to their simplicity, flexibility, their ability to avoid local minima, and their derivation free nature [1].

Metaheuristic optimization methods involve single solution based methods like simulated annealing. However, most of the bio-inspired metaheuristic techniques are population based. In this case, the optimization process is initiated by a population randomly positioned in the search space, and the positions of the population are modified at each iteration in order to improve the solution.

Swarm intelligence is a concept that was first presented in [2]. As defined in [3], the swarm intelligence is a collective intelligence of groups of simple agents. Several swarm intelligence-based optimization techniques are proposed based on researchers mimicking natural creatures' behaviors. These techniques include: particle swarm optimization (PSO) [4], biogeographical-based optimization (BBO) [5], ant colony optimization (ACO) [6], gray wolf optimization [1], and chicken swarm optimization [7].

L. Pan et al. (Eds.): BIC-TA 2019, CCIS 1159, pp. 506–516, 2020.
https://doi.org/10.1007/978-981-15-3425-6_39

Many variants of these models are presented even for the same agent type. For example, honey bees swarm [8] is used as a motivator for the genetic algorithm to improve genetic algorithm by the queen bee egg laying strategy. In [9], the idea of information exchange between bees via dancing is used to develop an improved version of the genetic algorithm. In [10] and [11], the collective cooperative behavior of social honey bees and next site selection are used to develop optimization algorithms.

Several versions of butterflies based optimization algorithm have been proposed. In [12], a butterfly based PSO is proposed as an improvement of the PSO algorithm. The migration behavior of monarch butterflies inspires the development of monarch butterfly optimization algorithm [13]. The most recent butterfly inspired algorithm is the butterfly optimization algorithm (BOA) proposed by Aurora and Singh in [14]. This algorithm mimics the food searching behavior of the butterflies. The algorithm evaluation over standard test functions shows attractive performance that outperforms several other optimization algorithms especially in the speed of convergence. The BOA started by using Lèvy flight model in [15] and has been slightly improved in [16] that was combined with artificial bee colony (ABC) in [17].

The contribution of this paper is twofold. Firstly, this paper investigates the basic algorithm of BOA and demonstrates that it provides a biased estimate for the functions global minimum. The biasing effect attracts the butterflies to points that are closer to the origin than the butterfly swarm best point. This biasing effect results in a very fast conversion towards the minimum if it is close to the origin compared to other optimization methods, but causes very poor results otherwise. Secondly, an unbiased version of the BOA is proposed (UBOA). Without the biasing term, the BOA has a slow convergence. Hence in the developed UBOA, not only the biasing effect is removed, but also the algorithm is modified to provide faster conversion.

This paper is organized as follows: the basic algorithm of BOA is inspected in Sect. 2. The problems associated with the BOA basic algorithm is emphasized in Subsects. 2.2 and 2.3. Section 3 proposes the unbiased butterfly optimization algorithm (UBOA) as modified version of BOA that alleviated the problems mentioned in Sect. 2. Section 4 presents the experiments for UBOA performance over several test functions. Section 5 concludes this paper.

2 Butterfly Optimization Algorithm

2.1 BOA Basic Algorithm

The basic BOA algorithm was proposed by Arora and Singh in [14] as a novel approach for global optimization. Butterflies use their sense to find food and mating partner.

Butterflies sense includes smell, sight, touch, taste, and hearing. The most influencing sense in food locating is the smell. Smell helps butterflies to find nectar even from long distances. Special nerve cells called chemoreceptors are scattered on the butterfly's body parts (ex: legs, antennae) and are used as sense receptors for smell.

In BOA algorithm, it is assumed that each butterfly is able to emit fragrance (f) that depends on its fitness. As the butterfly moves, its fitness changes. This is based on the amount of food available in its current location. The presence of the food is sensed by the stimulus intensity (I), which is modeled by the objective function that is needed to be optimized. The relation between the fragrance and the stimulus intensity can be represented by:

$$f_i = cI^a \tag{1}$$

where f_i is the fragrance of butterfly i, c is called the sensory modality that presents the process of measuring the sensed energy and processing it, using the sensory organs. Its value is small, less than 1. And a is an exponent power that controls the degree of absorption of I. Its value is less than 1 to give reduction response.

The fragrance can propagate and can be sensed by other search agents (butterflies). The butterfly can accurately find the source of fragrance; they can separate different fragrances and sense their intensities. The BOA assumes that the motion of the butterfly is affected by either one of the two searches: First search is called the global search, when the butterfly senses fragrance from another one and moves towards it based on its position. This is given by the following equation:

$$x_i^{t+1} = x_i^t + \left(r^2 \times g^* - x_i^t\right) \times f_i \tag{2}$$

where x_i^{t+1} is the next position of butterfly i at time $t+1$, x_i^t is the current position at time t, r^2 is a random number ϵ [0, 1], g^* is the best solution (position) in the current iteration produced by the butterflies.

Second search is called the local search, the butterfly moves randomly using the following equation:

$$x_i^{t+1} = x_i^t + \left(r^2 \times x_j^t - x_k^t\right) \times f_i \tag{3}$$

where x_j^t and x_k^t are the positions of butterfly j and butterfly k at the current time t, respectively. These butterflies are randomly chosen.

The type of motion of the butterflies, at each iteration, follows a probability (p) that mimics the environmental conditions; such as the presence of rain and wind. It is calculated by comparing the generated r with p. Besides, this prevents the trap of local minima. The number of the butterflies is fixed through the iterations till the best solution is obtained. Their initial positions are randomly chosen.

2.2 BOA Biasing

The BOA algorithm that was proposed by Arora and Singh [14] has shown attractive results when compared to other bio-inspired techniques especially in its speed of convergence. For example, Table 1 summarizes the mean values of 30 Monte Carlo simulations testing BOA performance after 10,000 iterations for three different functions with dimension equal to 30, as reported in [11]. The convergence curves shown in the above mentioned paper demonstrates that the BOA can converge to values that are close to the global minimum after the first 1,000 iterations.

Table 1. The BOA results on 3 benchmark testing functions.

Function name	Sphere	Step	Schwefel 2.21
Best value for BOA	2.99E−24	0.0	2.87E−21

The performance of BOA is clearly superior for these cases. All these functions have a minimum at the origin. To investigate the performance of the BOA for functions having minima not at the origin, the above three functions are tested again but after shifting their minima location [18] to be as follows:

$$\text{Sphere\_s} : f_{1s}(x) = \sum\nolimits_{i=0}^{n} \left(x_i - x_i^*\right)^2, 1 \le i \le n \text{ domain} : [-100, 100] \quad (4)$$

$$\text{Step\_s} : f_{2s}(x) = \sum\nolimits_{i=0}^{n} \left\lfloor x_i - x_i^* \right\rfloor^2, 1 \le i \le n \text{ domain} : [-100, 100] \quad (5)$$

$$\text{Schwefel 2.21\_s} : f_{3s}(x) = max\left(\left|x_i - x_i^*\right|\right), 1 \le i \le n, \text{ Domain} : [-10, 10] \quad (6)$$

where the suffix s at the end of the function name denotes the shifting of the function, n is the dimension of the objective function, and x_i^* is the shift value for dimension i. The domain range of the functions is taken as given in [14]. Table 2 shows BOA results on the shifted functions with dimension equal to 30. The parameter values are: $p = 0.8$, $a = 0.1$ and $c = 0.01$. Each function is shifted by two different values.

Table 2. BOA Simulation results for shifted test functions.

Functions	Exact solution	Cost values f(x) at the optimal point				Cost f(x) at the origin ($x = 0$)
		Mean	Max	Min	Std	
Sphere_s1	$x_i^* = \pi,\ i = 1,\ n$	206.4	273.8	145.4	26.4	296.1
Sphere_s2	$x_i^* = 75,\ i = 1,\ n$	126.4E+3	142.6E+3	106.7E+3	9.1E+3	168.8E+3
Step_s1	$x_i^* = \pi,\ i = 1,\ n$	163.7	219.0	99.0	25.8	270.0
Step_s2	$x_i^* = 75,\ i = 1,\ n$	124.9E+3	143.1E+3	111.4E+3	8.3E+3	168.8E+3
Schwefel 2.21_s1	$x_i^* = \frac{\pi}{10},\ i = 1,\ n$	314.2E−3	314.2E−3	314.1E−3	4.5E−6	314.2E−3
Schwefel 2.21_s2	$x_i^* = 7.5,\ i = 1,\ n$	7.5	7.5	7.4999	25.6E−6	7.5

The Schwefel 2.21 function is the clearest demonstration for the biasing effect of the BOA. In the first case, the function was defined as: $f_{3s}(x) = max(|x_i - 0.314|), 1 \le i \le n$, domain: $[-10, 10]$, the optimal value of the function is zero at $x_i = 0.314$ for $i = 1, n$. However, the solution found by the BOA was very close to the origin (actually $max(|x_i|) = 1.76E{-4}$), such that x^* found by BOA is approximately equal to zero, while the optimal position is equal to 0.314. This indicates that the BOA attracts the butterflies towards the origin even if it is not the optimal solution. Similarly, the butterflies were also attracted by BOA to the origin for $f_{3s}(x) = max(|x_i - 7.5|), 1 \leq i \leq n$, giving a cost of 7.5, while the global best located at $x = 7.5$ for $i = 1, n$ has a cost of zero.

For the sphere and the step, the situation is similar. Investigating Table 2, it is clear that the butterflies best costs are closer to the cost of the origin with relatively low standard deviation, indicating that the butterflies are attracted nearly every run in the Monte Carlo simulation to the origin.

To analyze the cause of the attraction towards the origin in BOA, it is necessary to investigate the global search Eq. (2) of the BOA algorithm. The equation can be rewritten as

$$x_i^{t+1} = x_i^t \times (1 - f_i) + r^2 \times g^* \times f_i \tag{7}$$

If the system reached the global minimum point, any further motion of the butterflies cannot change g^*. In this case, the dynamics of this system can be analyzed dealing with g^* as a constant. The equilibrium point x^e of the system in this case is given by:

$$E(x_i^e) = E(x_i^e \times (1 - f_i)) + E\left(r^2 \times g^* \times f_i\right) \tag{8}$$

$$E(x_i^e) = E\left(r^2\right) \times g^* \tag{9}$$

Since r is a uniform random number $\in [0, 1]$, it is easy to show that the expectation of the square of uniform random number $E\left(r^2\right) = 1/3$. Hence, the butterflies are attracted (assuming the iterations are convergent) towards $1/3 \times g^*$. If g^* is the origin, then the butterflies are attracted to the global minimum points. If g^* is not located at the origin, then the butterflies are attracted towards a different point $(1/3 \times g^*)$, such that, the butterfly are searching for new solutions in an area that is far from the global minimum. It is important to note that this search cannot be considered an exploration phase since the butterflies is actually attracted to the same point.

In the more general case, where the butterflies didn't reach the global minimum point, g^* is just the best location found by the butterfly so far. During iterations, g^* is expected to change if a butterfly hits a location with better cost. However, at each iteration, each butterfly moves to a point that is a weighted average between its old position and x_i^t, and the point defined by $r^2 \times g^*$, where the weight of the later term is defined by the factor f_i. Hence, if $f_i < 1$ the butterflies moves finally towards a random point with expected value $(1/3 \times g^*)$ which is closer to the origin than the current best location g^*. If during this motion a better position is found, g^* is updated to $g^{*\prime}$ and the butterflies are attracted to $(1/3 \times g^{*\prime})$ which is closer to the origin than the previous attraction point. The process is repeated if the function has a global minimum at the origin. Hence the algorithm provides faster attraction towards the origin and hence has a superior performance in speed of convergence compared to other algorithms.

Compared to PSO, the standard equations of PSO provides attraction point of the swarm towards its best found position (g^*). Hence the swarm explores the area around g^* and can improve the solution if there is better solution in the neighborhood of g^*. The position of g^* is irrelevant.

For BOA, the attraction point is $g^*/3$, which improves the algorithm performance if the global optimal point is at the origin and deteriorate the performance otherwise.

2.3 Fragrance Function

Fragrance function maps the position of the butterfly to a real value that represents the fitness of this position. This function should be selected to be monotonic so that the fragrance emitted by the butterfly increases as the fitness increases.

The proposed fragrance function used in [14], and [15] is given by Eq. (1) where I represents the fitness of the butterfly's current position, and $c \in [0, 1]$, and $a \in [0, 1]$. This function is monotonic, but the proposed range of constants results in complex fragrance for negative values of I. This effect causes erroneous results when applying the BOA code to find the optimum value of shifted functions (4–6) having a global minimum at x*. This can be clearly found by inspecting the example given in Table 3 in reference [14], a function called Trid and labeled f28. This function has the best global minimum at −4940, while the results of BOA have a mean of -2.7×10^7 and standard deviation equal to 5×10^7.

3 Unbiased Butterfly Algorithm

The development of the Unbiased BOA (UBOA) algorithm requires two major modifications compared to the original BOA. These modifications are summarized as follows.

- Modifying the motion equation to avoid biasing and improve convergence speed.
- Modifying the fragrance definitions to avoid erroneous results for functions with negative values.

The following subsections present the proposed modifications and summarize the proposed UBOA algorithm.

3.1 Modified Fragrance Definition

The previous definition of the butterflies' fragrance, given in Eq. (1), could result in erroneous complex numbers. Therefore, the fragrance is redefined by:

$$f_i = \begin{cases} 0.01, & if |Ft_{max} - Ft_{min}| < |0.01 \times Ft_{min}| \\ \frac{Ft_i - Ft_{min}}{Ft_{max} - Ft_{min}}, & elsewhere \end{cases} \tag{10}$$

where Ft_i is the fitness of the butterfly i, Ft_{max} and Ft_{min} are the maximum and minimum fitness, respectively. The maximum fitness initial value is taken −1E7. The new definition ensures that $f_i \in [0, 1]$ and it approaches zero while the butterfly's fitness converges towards the global minimum (g).

3.2 Modified Exploitation and Exploration Movements

The propagation of each butterfly is classified into two types; exploitation and exploration. Switching between these two types is based on comparing a random number (r) with a factor (P) as follows:

$$Propagation\ Type = \begin{cases} Exploration, & r < P \\ Exploitation, & r \geq P \end{cases} \tag{11}$$

where,

$$P = max\left(\min\left(\log\left(\frac{|Ft_{max}|}{|Ft_{min}|}\right), 0.95\right), 0.1\right) \tag{12}$$

Equations (11) and (12) indicate that the butterfly mode selection, between either exploitation or exploration, is based on their fitness difference. Exploitation is encouraged if the fitness difference among swarm members is small.

In exploration phase, the butterfly position is updated by Eqs. (13) and (14). A new distance $\left(d_i^{t+1}\right)$ vector is formed based on the difference between the global minimum (g^t) and the current position (x_i^t). Random components of the distance vector are selected and the butterfly's position is updated by the new distance weighted with the predefined fragrance (10) in the randomly selected dimensions only.

$$d_i^{t+1} = \left(g^t - x_i^t\right) \tag{13}$$

$$\hat{x}_{ij}^{t+1} = x_{ij}^t + \varepsilon\, d_{ij}^{t+1} \times f_i \tag{14}$$

where, ε is a random number $\in [0, 1]$ and j is the selected component to be updated.

For the exploitation phase, the butterfly is attracted to a random position specified by the Kth component of the Jth butterfly (x_{JK}^t) as illustrated in Eq. (15). Moreover, Eqs. (15) and (16) declare that the position update term is weighted by the complementary of the fragrance instead of the fragrance itself.

$$d_i^{t+1} = \left(x_{JK}^t - x_{iK}^t\right) \tag{15}$$

$$\hat{x}_{iK}^{t+1} = x_{iK}^t + \varepsilon\, d_{iK}^{t+1}(1 - f_i) \tag{16}$$

3.3 Butterfly Position Update

The position of the butterfly is updated to the new proposed position $\left(\hat{x}_i^{t+1}\right)$ if it achieves a better fitness. However, to ensure better exploration and local best sticking avoidance the butterfly position is set to new proposed position if a random generated number is less than a small value (typically 0.1) as long as the affected butterfly is not located at the swarm best. The following equations illustrate these conditions:

$$x_i^{t+1} = \begin{cases} \hat{x}_i^{t+1}, & \widehat{Ft}_i < Ft_i\ or\ (r < 0.1\ and\ x_i^t \neq g^t) \\ x_i^t, & elsewhere \end{cases} \tag{17}$$

The flow chart in Fig. 1 illustrates the steps of the new algorithm and demonstrates the interaction between the previously mentioned governing equations.

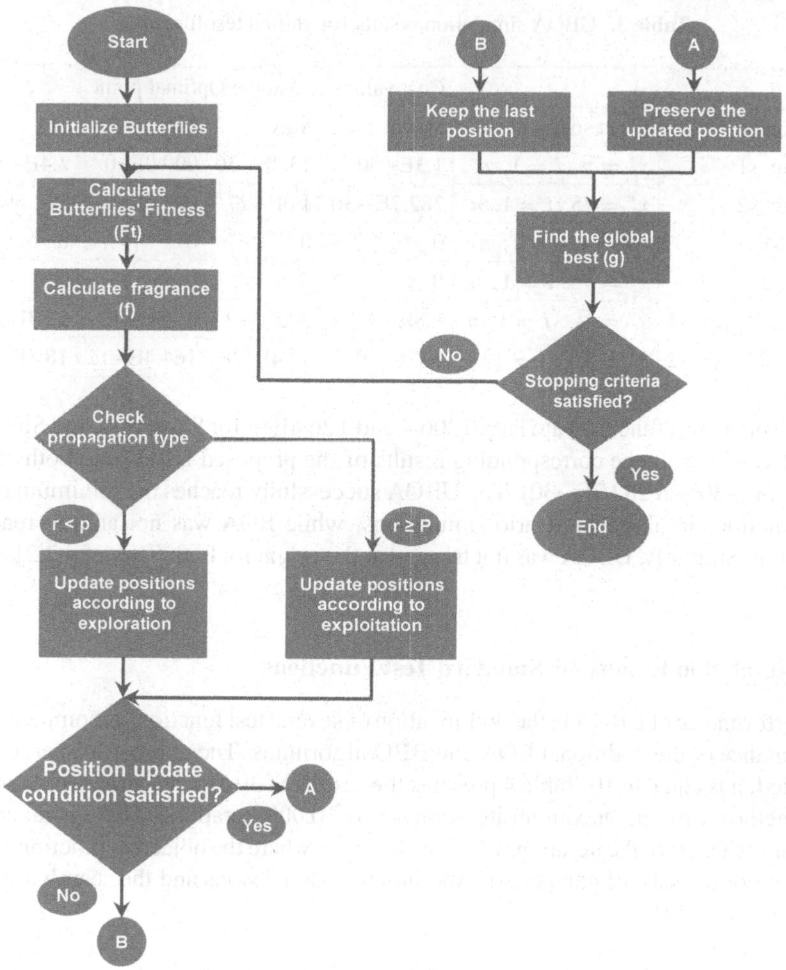

Fig. 1. The flowchart of the UBOA algorithm.

4 Results and Discussions

In order to evaluate the performance of the proposed UBOA algorithm, its performance for functions with shifted optimum is evaluated in Sect. 4.1 to demonstrate that it provides an unbiased optimization solution. In Sect. 4.2, the performance of UBOA is compared with BOA and BBO using standard test functions.

4.1 Performance Evaluation for Shifted Optimum Test Functions

Table 3 presents the simulation of the proposed UBOA for the same cases used in Sect. 2.2. Comparing the results presented in Table 3 with the corresponding values in Table 2, demonstrates that UBOA does not show any biased performance. For example, the mean of the cost function values of the BOA at the optimal point for the two

Table 3. UBOA simulation results for shifted test functions.

Functions	Exact solution	Cost values f(x) at the Optimal point			
		Mean	Max	Min	Std
Sphere_s1	$x_i^* = \pi,\ i = 1,\ n$	1.3E−30	13.2E−30	000.0E+0	2.4E−30
Sphere_s2	$x_i^* = 75,\ i = 1,\ n$	282.7E−30	1.0E−27	000.0E+0	288.5E−30
Step_s1	$x_i^* = \pi,\ i = 1,\ n$	0	0	0	0
Step_s2	$x_i^* = 75,\ i = 1,\ n$	0	0	0	0
Schwefel 2.21_s1	$x_i^* = \frac{\pi}{10},\ i = 1,\ n$	7.8E−9	59.8E−9	49.5E−12	12.4E−9
Schwefel 2.21_s2	$x_i^* = 7.5,\ i = 1,\ n$	13.0E−9	73.4E−9	164.4E−12	18.1E−9

shifted sphere test functions are large (206.4 and 126.4E+3 for Shpere_s1 and Sphere_s2 respectively), while the corresponding results of the proposed UBOA are both close to zero (1.3E−30 and 282.7E−30). The UBOA successfully reaches the minimum of both step functions in all Monte Carlo simulations, while BOA was not able to reach the minimum. Similarly, UBOA was not trapped at the origin for both Schwefel 2.21 shifted functions.

4.2 Simulation Results for Standard Test Functions

The performance of UBOA in the optimization of several test functions is compared to the performance of the traditional BOA and BBO algorithms. The number of search agents is unified, it is equal to 50. Table 4 presents the results of 30 Monte Carlo simulations for each method with the maximum iteration set to 10,000 iterations. There is an additive stopping criterion of the iteration, which is the value where the objective function reaches 1E−50. For the sake of comparison, the function dimensions and the search domain is the same as in [14].

Table 4. Simulation results of BOA, BBO, and the proposed UBOA.

Notation	Function name	BOA	BBO	UBOA
F1	Sphere	3.0E−24	1.7E−3	**38.2E−33**
F2	Beale	47.0E−3	76.2E−3	**9.5E−6**
F3	Cigar	**3.1E−24**	895.9	366.4E−3
F4	Step	**000.0**	**000.0**	**000.0**
F5	Quartic function with noise	**54.4E−6**	522.0E−6	65.9E−3
F6	Bohachevsky	**000.0**	**000.0**	**000.0**
F7	Ackley	**888.2E−18**	8.6E−3	19.6E−15
F8	Schwefel 2.21	**2.9E−21**	18.4E−3	7.6E−9

Compared to BOA, out of the 8 simulated functions, the BOA achieves better results than UBOA in 4 cases, namely: F3, F5, F7 and F8. This is due to the use of the standard form of the functions having the minimum at the origin. However, the results of UBOA is very close to that of BOA in F7, while UBOA outperforms BOA in the two cases of F1 and F2, and they reach the same value for F4 and F6. Tables 2, 3, and 4 indicate that UBOA provides unbiased estimate of the optimum without deteriorating the speed of performance of BOA in a sensible degree.

Compared to BBO, Table 4 shows that BBO outperforms UBOA in one function only (F5). UBOA outperforms BBO in 5 test functions (F1, F2, F3, F7 and F8) and they reach the same optimal value in F4 and F6.

5 Conclusions

This paper presents an unbiased version of the butterfly optimization algorithm. The original BOA version is investigated and shown to have a bias towards the origin. The proposed unbiased version (UBOA) modifies the search equations of the butterflies in order to avoid the biasing terms without greatly affecting the excellent fast convergence performance of the BOA. Simulation results using standard and shifted test functions demonstrate that UBOA provides a competitive optimization approach compared to other bio-inspired algorithms.

References

1. Mirjalili, S., Mirjalili, S.M., Lewis, A.: Grey wolf optimizer. Adv. Eng. Softw. **69**(3), 46–61 (2014)
2. Beni, G., Wang, J.: Swarm intelligence in cellular robotic systems. In: Dario, P., Sandini, G., Aebischer, P. (eds.) Robots and Biological Systems: Towards a New Bionics?. NATO ASI Series, vol. 102, pp. 703–712. Springer, Heidelberg (1993). https://doi.org/10.1007/978-3-642-58069-7_38
3. Bonabeau, E., Marco, D., Dorigo, M., Theraulaz, G.: Swarm Intelligence: From Natural to Artificial Systems, 1st edn. Oxford University Press, New York (1999)
4. Kennedy, J., Eberhart, R.: Particle swarm optimization (PSO). In: IEEE International Conference on Neural Networks, Perth, Australia, pp. 1942–1948 (1995)
5. Simon, D.: Biogeography-based optimization. IEEE Trans. Evol. Comput. **12**(6), 702–713 (2008)
6. Marco Dorigo, T.S.: Ant Colony Optimization. MIT Press, Cambridge (2004)
7. Meng, X., Liu, Yu., Gao, X., Zhang, H.: A new bio-inspired algorithm: chicken swarm optimization. In: Tan, Y., Shi, Y., Coello, C.A.C. (eds.) ICSI 2014. LNCS, vol. 8794, pp. 86–94. Springer, Cham (2014). https://doi.org/10.1007/978-3-319-11857-4_10
8. Jung, S.H.: Queen-bee evolution for genetic algorithms. Electron. Lett. **39**(6), 575–576 (2003)
9. Sato, T., Hagiwara, M.: Bee system: finding solution by a concentrated search. IEEJ Trans. Electron. Inf. Syst. **118**(5), 721–726 (1997)
10. Yonezawa, Y., Kikuchi, T.: Ecological algorithm for optimal ordering used by collective honey bee behavior. In: Proceedings of the 7th International Symposium on Micro Machine and Human Science, Nagoya, Japan. IEEE (2002)

11. Seeley, T.D., Visscher, P.K., Passino, K.M.: Group decision making in honey bee swarms: when 10,000 bees go house hunting, how do they cooperatively choose their new nesting site? Am. Sci. **94**(3), 220–229 (2006)
12. Bohre, A.K., Agnihotri, G., Dubey, M., Bhadoriya, J.S.: A novel method to find optimal solution based on modified butterfly particle swarm optimization. Int. J. Soft Comput. Math. Control **3**(4), 1–14 (2014)
13. Wang, G., Deb, S., Cui, Z.: Monarch butterfly optimization. Neural Comput. Appl. **31**(7), 1995–2014 (2019). https://doi.org/10.1007/s00521-015-1923-y
14. Arora, S., Singh, S.: Butterfly optimization algorithm: a novel approach for global optimization. Soft. Comput. **23**(3), 715–734 (2018). https://doi.org/10.1007/s00500-018-3102-4
15. Arora, S., Singh, S.: Butterfly algorithm with levy flights for global optimization. In: International Conference on Signal Processing, Computing and Control, India, pp. 220–224. IEEE (2015)
16. Arora, S., Singh, S.: An improved butterfly optimization algorithm for global optimization. Adv. Sci. Eng. Med. **8**(9), 711–717 (2016)
17. Arora, S., Singh, S.: An effective hybrid butterfly optimization algorithm with artificial bee colony for numerical optimization. Int. J. Interact. Multimedia Artif. Intell. **4**(4), 14–21 (2017)
18. Tang, K., et al.: Benchmark functions for the CEC'2008 special session and competition on large scale global optimization. Nature Inspired Computation and Applications Laboratory, USTC, China, vol. 24 (2007)

On-Chip Health Monitoring Based on DE-Cluster in 2.5D ICs

Libao Deng[1(✉)], Le Song[1], and Ning Sun[2]

[1] School of Information Science and Engineering, Harbin Institute of Technology at Weihai, Weihai, China
denglibao_paper@163.com

[2] Department of Automatic Test and Control, Harbin Institute of Technology, Harbin, China

Abstract. 2.5-dimensional integrated circuits (2.5D ICs) are considered today as a promising solution for overcoming the bottlenecks introduced by technology scaling. In 2.5D ICs, the circuit failures are also inevitable to arise, mainly because of timing variations induced by runtime and process variations as well as transistor aging which result in path delay. However, the increase of density and complexity of circuits in 2.5D ICs brings more challenges to the detection of in-field path delay. It is feasible to track the delay of every critical path due to the unaffordable overhead of sensors or test patterns. In this paper, we propose to adopt the DE-based clustering algorithm to select a set of representative critical paths. Therefore, by only testing the delay of the selected small number of paths, the concerning parameters of these circuits are derived and used to infer the delay of a large number of the rest paths which are probably to fail due to timing variations. We simulate the proposed approach on benchmark circuits, and the results demonstrate the effectiveness for the collaborative optimization of representative critical path number and delay prediction accuracy.

Keywords: 2.5D ICs · Health monitoring · DE-Cluster

1 Introduction

With the rapid development of semiconductor industry, the scale of integrated circuits is getting larger and larger while the feature size is becoming increasingly smaller, reaching nanometer order [1]. However, the interconnect length, especially in the worst case, cannot scale down as the chip size is expanding. As a consequence, interconnect delay will gradually dominate the delay of the entire chip. Moreover, the overall chip latency will increase with the improvement of

This work was supported by the Fundamental Research Funds for the Central Universities. (Grant No. HIT. NSRIF. 2019083) and Guangxi Key Laboratory of Automatic Detecting Technology and Instruments (Grant No. YQ19203).

L. Pan et al. (Eds.): BIC-TA 2019, CCIS 1159, pp. 517–526, 2020.
https://doi.org/10.1007/978-981-15-3425-6_40

manufacturing technique, which limits the further development of integrated circuits [2]. The novel circuit connection structure of 2.5D ICs that multiple dies are placed side-by-side on top of the passive silicon interposer and interconnected through it mitigates the problem effectively [3]. Due to the significant increase in the integration and complexity of 2.5D ICs, it is more difficult to implement on-chip health monitoring to guarantee the reliability of the ICs.

As in traditional 2D ICs, there are many factors that influence the performance and lifetime of 2.5D ICs. Especially in 2.5D ICs, the further reduction of feature size decreases the transistor gate oxide thickness, and better performance requirements lead to high operating temperature of high-frequency devices. These trends make the Bias Temperature Instability (BTI) degradation the major factor liming transistor lifetime. It gradually increases the threshold voltage and then degrades circuit speed, finally results in faulty operation once the delay in the critical path exceeds the pre-designed timing constraints [4]. The effect of BTI on path delay gets deeper with the increase of temperature, runtime, supply voltage and decrease of signal frequency. Moreover, process variations and transistor aging are also main sources of delay degradation [5]. Therefore, in order to predict timing delay information of circuits in 2.5D ICs accurately, the effects mentioned above that worsen reliability must be taken into account during timing analysis.

In this paper, since it is infeasible to monitor circuit delay of an endless number of long (critical) paths that are most likely to fail first over the chip lifetime, an on-chip health monitoring method based on DE-Cluster algorithm [6] for 2.5D ICs is proposed to appropriately select a small subset of critical paths to carry out delay detection, based on which the delay information of the rest large number of circuits can be predicted accurately (shown in Fig. 1). The proposed method considers all phenomena that may cause timing violation of circuits, including BTI degradation, process variations and transistor aging. The selected representative critical paths (RCPs) are extracted by analysing the topological, electrical, spatial, and functional similarities among different critical paths.

The rest of the paper is organized as follows. Section 2 shows an overview of the related prior works. Section 3 presents the process of selecting RCPs using DE-Cluster algorithm. Section 4 demonstrates the experimental results for benchmark. Finally, Sect. 5 draws a conclusion.

2 Related Prior Works

Up till now, there is few research on dealing with variation or aging induced detrimental effects in 2.5D ICs. Therefore, we have to learn from previous techniques proposed for traditional 2D ICs. The conventional methods that are used to avoid timing violations on circuits caused by several sources of variations usually assign additional timing margin, which is also called band-guarding, at the design stage of the chip, in order to provide extra maximum tolerance for signal propagation delay in the critical paths [7]. However, the conservative

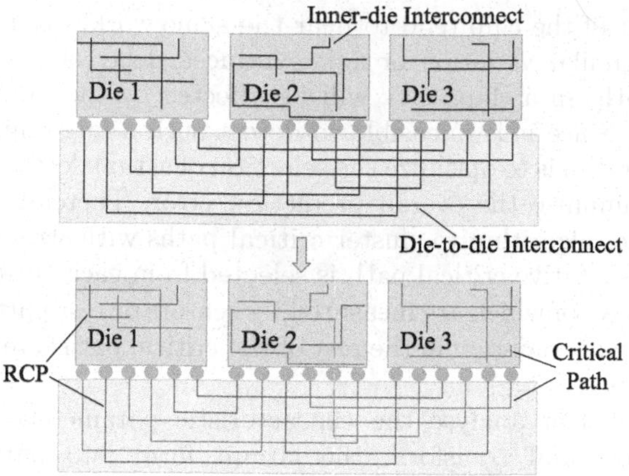

Fig. 1. Illustration of RCP.

band-guarding design is derived from the worst-case analysis without considering the runtime workload effects, which leads to the loss of chip performance considerably.

To improve the above shortcomings, several on-line detection and adaptation methods have been proposed to dynamically adjust chip operating parameters based on the variations and aging induced characteristics of the circuits which change over time. Such methods require on-chip health monitoring infrastructure. In order to implement periodic delay testing of the circuit at runtime, the normal execution of the chip has to be interrupted for test patterns application, which leads to significantly detrimental effects of chip performance [8]. Moreover, the storage of test patterns occupies on-chip memory resources. Another choice is utilizing the dedicated sensors to monitor the critical path delay during field-operation [9–11]. However, it requires too many sensors inserted in the circuits to detect the path delay accurately. Then, replica circuits are designed as representatives of aging-induced delay in the circuits to be actually measured [12]. However, the number of them is too small to cover the entire circuits.

Therefore, in order to implement the on-chip health monitoring efficiently, it is essential to select the appropriate critical paths to be detected, which are called representative critical paths [13]. By analysing various characteristics of the circuits, a minimum set of paths are selected as the most representative ones of the circuit delay. But the criteria for selection of the existing methods do not take into account all the influencing factors of the delay. The majority of them neglect the inevitable mechanisms that may lead to circuit parameter variations during fabrication process and runtime of the chips.

3 Proposed Methodology

Generally speaking, different critical paths in a chip often have a variety of similarities in topological, spatial and other aspects. As the circuits in the same functional layout of the chip tend to bear the same workload, they are likely to experience the similar variation or aging induced delay degradation. In terms of all critical paths in a chip, they will be affected by the same type but with varying degrees. Since it is infeasible to detect such a large number of circuits with sensors, our goal is to optimize the selection of actual detected critical paths, i.e. RCPs, to minimize the overall prediction error. Therefore, we propose to adopt DE-Cluster algorithm to cluster critical paths with stronger correlations. Then, one representative critical path is selected from each cluster to form a set of RCPs, the delays of which are measured by sensors or test patterns. According to that, the delay degradation in the rest of the critical paths are predicted based on the correlations.

In order to better analyse the characteristic parameters, we first model the critical paths and transform the circuit form into path matrix $P = \{p_1, p_2, ..., p_N\}^T$ and path information matrix $p_j = \{I_{j1}, I_{j2}, ..., I_{jM}\}$, where N and M denote the number of critical paths and features respectively. Each feature I_{ji} represents the sensitivity of the critical path to one source that is likely to cause path delay. The total delay of p_j, denoted as d_j, is derived from the following equation:

$$d_j = p_j F$$
$$F = \{F_{j1}, F_{j2}, ..., F_{jM}\}^T \tag{1}$$

where vector F is the value of features.

3.1 Concerning Features in Path Encoding

Note that the BTI effect has the dominant influence on circuit failure. We first study on the principle that BTI exacerbate the device deterioration by gradually increasing the threshold voltage. And then the specific factors that affect delay degradation can be analysed base on this.

BTI mechanism gradually increases threshold voltage of the transistors during lifetime, further causing transistor aging, and then circuit delay degradation. A set of typical equations are utilized here to model the BTI-induced threshold voltage shift at time t [14]:

$$\Delta V_{th}(t) = \left(\frac{\sqrt{{K_v}^2 \alpha T_{clk}}}{1 - {\beta_t}^{\frac{1}{2n}}}\right)^{2n} \tag{2}$$

$$K_v = f(V_{dd} - V_{th}) \tag{3}$$

$$\beta_t = f(T, T_{clk}, tox, t) \tag{4}$$

Where α represents duty cycle, T_{clk} is clock cycle, T is temperature, n and tox are constants of fabrication process, V_{dd} is supply voltage and V_{th} is threshold voltage. According to the threshold voltage degradation equation, the features in path information vector must contain the following aspects: temperature, supply voltage, circuit topology and process variations.

Temperature Features. Temperature, related closely to circuit power consumption, is strongly influenced by workload and operating conditions, which then results in runtime variations. To acquire temperature features, we divide the layout of the chip into multiple rectangular grids. Then, through inserting leakage and dynamic power in the gates of each grid, the corresponding power consumption can be calculated. The concerning information such as floorplan and power profile can be obtained from thermal-profiling tool [15], based on which the temperature features can be extracted. Finally, each critical path is encoded by the temperature features of the corresponding grids it passes through.

Supply Voltage Fluctuation Features. Since there exists current switching and resistance in power delivery network, the actual supply voltage on each gate during operation deviates from anticipated value. Voltage droop changes greatly over time, which has adverse impact on circuit delay degradation. Similar to the method of obtaining the temperature features, the power delivery network is modelled as a network with resistance distributed over the chip [16]. Therefore, the critical path is encoded as a vector that reflects the sensitivity of each critical path to the corresponding grid.

Topological Features. These features indicate the netlist and layout information about the types and locations of the gates, such as the affiliation of all gates with their corresponding critical paths, the location of each gate in the circuit.

Process-Variation Features. Since the process varies during fabrication of 2.5D ICs, the physical parameters of the components and interconnects deviate from designed values, which results in circuit delay [17,18]. In order to comprehensively analyse all possible factors, the process variation is modelled by parameters including die-to-die (Δ_{d2d}), that is the parameter of horizontal interconnects in the silicon interposer of a 2.5D IC, inner-die spatially-correlated (Δ_{cor}) and independent random (Δ_{rand}) variations. The total variation (Δ_{total}) is calculated as [15]:

$$\Delta_{total} = \Delta_{d2d} + \Delta_{cor} + \Delta_{rand} \tag{5}$$

To extract Δ_{cor}, the layout of the chip is partitioned into rectangular grids, the correlations among which are modelled by a diminishing function, e^{-d}. Parameter d is proportional to distance between grids. Then each critical path is encoded by the vector that reflects whether the path is sensitive to the grids.

3.2 DE-Cluster Method

We denote the delay of these N paths as the vector $D = \{d_1, d_2, ..., d_N\}^T$. In this paper, we propose to utilize DE-Cluster algorithm to select RCPs from set P due to its outstanding performance in unlabelled data clustering. The RCPs are expressed as the vector $P_R = \{p_{r1}, p_{r2}, ..., p_{rR}\}^T$, where $R << N$. The delay test results of RCPs are denoted as $D_R = \{d_{r_1}, d_{r2}, ..., d_{rR}\}^T$.

Most of the existing clustering methods consider the number of RCPs as an input, which decreases their effectiveness to a great extent. Too few selected RCPs cannot represent all features of the critical paths leading to high prediction error (Err) while too many selected RCPs requires unnecessary sensors. Thus, the number of RCPs is also an object that needs to be optimized in this situation. DE-Cluster algorithm achieves collaborative optimization of parameter R and Err on the run.

The classical DE (differential evolution), an effective and simple global optimization algorithm, has four main operations on the population of each generation: initialization, mutation, crossover, and selection [19]. DE-Cluster is improved on the basis of it for clustering application. The ith individual of the population in generation G is defined as:

$$\overrightarrow{X}_{i,G} = \{\overrightarrow{m}_{i,1}, \overrightarrow{m}_{i,2}, ..., \overrightarrow{m}_{i,K\max}, T_{i,1}, T_{i,2}, ..., T_{i,K\max}\} \tag{6}$$

K_{max} is the user-defined maximum number of clusters. $\overrightarrow{m}_{i,j}(j = 1, 2, ..., K_{max})$ is a randomly selected cluster center, which is M dimensional. $T_{i,j}(j = 1, 2, ..., K_{max})$ is a randomly selected parameters from [0, 1], which controls whether the corresponding cluster $\overrightarrow{m}_{i,j}$ is picked to be the real cluster center or not. The criterion of selecting the actual number of clusters is shown as following:

$$\begin{array}{l}\text{IF } T_{i,j} > 0.5, \text{ THEN } \overrightarrow{m}_{i,j} \text{ is ACTIVATED} \\ \text{ELSE } \overrightarrow{m}_{i,j} \text{ is INACTIVATED}\end{array} \tag{7}$$

We evaluate the similarity between the critical path p_i and different cluster centers by calculating its Euclidean distance from all active cluster centers in the ith individual $\overrightarrow{X}_{i,G}$. The critical path belongs to the particular cluster center that has the minimum Euclidean distance with it.

To create the new offspring of each individual, two other different individuals, i.e., $\overrightarrow{X}_{m,G}$ and $\overrightarrow{X}_{n,G}$, are selected from the same generation, based on the difference of which the mutation vector is calculated as:

$$\overrightarrow{V}_{i,G} = \overrightarrow{X}_{i,G} + F \bullet (\overrightarrow{X}_{m,G} - \overrightarrow{X}_{n,G}) \tag{8}$$

Constant F is the control parameter for scaling difference vector in order to maintain population diversity. It is derived from the following equation:

$$F = 0.5 \bullet (1 + rand(0, 1)) \tag{9}$$

Then the algorithm enters the crossover step to create trial offspring $\overrightarrow{U}_{i,G}$ of individual $\overrightarrow{X}_{i,G}$, the n_{th} element of which is calculated as:

$$u_{n,i,G} = \begin{cases} v_{n,i,G} & \text{if rand}(0,1) \leq Cr \\ x_{n,i,G} & \text{otherwise.} \end{cases} \tag{10}$$

$$Cr = (Cr_{\max} - Cr_{\min}) \bullet \frac{(G_{max} - G)}{G_{\max}} \tag{11}$$

Cr is another control parameter called *crossover rate*. Cr_{max} and Cr_{min} are maximum and minimum value of Cr. G_{max} and G represent the maximum generation and the current generation respectively.

Finally, the trial vector and the target vector compete based on the CS validity measure denoted as $f(\cdot)$ to decide who survives in the next generation.

$$\vec{X}_{i,G+1} = \begin{cases} \vec{U}_{i,G}, & \text{if } f\left(\vec{U}_{i,G}\right) > f\left(\vec{X}_{i,G}\right) \\ \vec{X}_{i,G}, & \text{otherwise.} \end{cases} \quad (12)$$

Note that the CS validity measure is a kind of clustering validity index to evaluate the quality of the clustering results. The pseudocode is shown in Table 1. After the algorithm finishes running, the optimal solution to clustering is obtained by the globally best individual that has the maximum value in CS validity index, and cluster centers in it are selected RCPs. The delay of the remaining critical paths are estimated using the following Eq. (13):

$$\overline{D} = PP_R{}^T(P_R P_R{}^T)^{-1} D_R \quad (13)$$

Table 1. Procedure for the DE-Clustering method.

Algorithm 1: DE-Cluster method
Input: P
Output: P_R, R
Initialize each individual of the first generation
Flag the active cluster centers in each individual following the rule of (7)
While $G < G_{max}$ do
Calculate the distance between critical path p_i and each active cluster center in the ith individual $\vec{X}_{i,G}$
Assign p_i to the cluster center $\vec{m}_{i,j}$ that has the minimum distance
Mutation step
Crossover step
Selection step
end while
return P_R and R

4 Simulation Results

Several simulations are performed on IWLS'05 and ITC'99 benchmark circuits [20,21] in order to evaluate the effectiveness of the proposed clustering method in the application of RCP selection in on-chip monitoring of 2.5D ICs. We synthesize the standard circuits using Nangate 45 nm library and synopsys design compiler [22]. BTI-induced transistor threshold voltage change is simulated by supposing a delay degradation of 10% in 3 years. The concerning parameters of circuit layout is extracted by Cadence SOC Encounter. Multiple random vectors are input to the circuits with a switching activity factor of 0.2 to obtain power profile. HotSpot analyse these data to extract the circuits' thermal profile [23]. The timing analysis of benchmark circuits is implemented by Prime-time under the influence of aging and variation induced path delay in order to obtain critical paths. The clustering optimization problem based on DE-Cluster algorithm is solved on the mathematical platform MATLAB R2016b.

The effectiveness of delay prediction is shown in Table 2. There are only top 5% critical paths are selected from millions of circuits based on the timing constrains. A smaller set of RCPs are further selected using DE-Cluster algorithm to be actually monitored for delay detection, based on which the delay of the rest critical paths are estimated using Eq. (13). The estimation error is evaluated by the following formula:

$$Err = \frac{\sqrt{\sum(\overline{D} - D)^2}}{N \bullet (\max(D) - \min(D))} \bullet 100\% \tag{14}$$

Table 2. Simulation results of DE-Cluster algorithm

Benchmark name	No. of gates	No. of critical paths	No. of RCPs	Err (1.5 years)	Err (3 years)
b17	27k	3340	35	3.59%	3.64%
b18	88k	2233	20	1.15%	1.25%
b19	165k	2612	22	0.53%	0.61%
b22	40k	2014	31	1.09%	1.53%
RISC	61k	3662	43	0.85%	0.85%
Vga	114k	1715	26	0.41%	0.25%
s5378	2390	831	36	1.06%	1.25%
s9234	5650	924	10	2.43%	2.72%

From the results we can see that the number of RCPs are significantly smaller compared to that of critical paths. The estimation error keeps quite low in all benchmark circuits, which indicates the high prediction accuracy by using the proposed method. Moreover, according to the variation trend between E_{rr} and R presented in [12], the prediction error drops fast with the increase of RCP number at first, and then they reach saturation. By comparing, the RCP number in each benchmark derived from the proposed algorithm is close to the inflection point of the corresponding trend. This means the DE-Cluster based on-chip health monitoring method achieves collaborative optimization of sensors and test accuracy.

Note that the proposed method only considers the four main factors that have influences on circuit delay degradation. But it is already the most comprehensive analysis with considerably low prediction deviation. Figure 2 demonstrates the

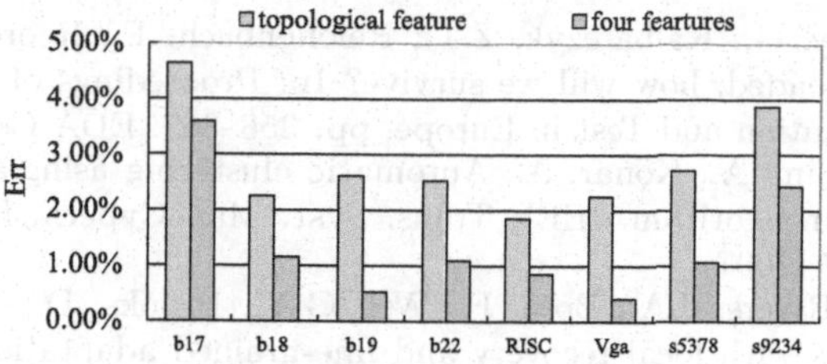

Fig. 2. Comparison between the DE-Cluster algorithm with only topological feature and with four features.

comparison of average E_{rr} between the same RCP selection method with only topological feature and with the four mentioned features. By taking process and runtime variations into consideration, the accuracy is improved significantly.

5 Conclusions

2.5D ICs are inevitably to be effected by the process and runtime variations as well as transistor aging, which further lead to the violation of timing constrains. Therefore, it is necessary to implement on-chip health monitoring efficiently and accurately to track the delay degradation in the circuits during chip lifetime. In this paper, we propose to adopt DE-Cluster algorithm to partition the critical paths into several clusters according to the similarities of them. And then the centers of each cluster are considered as a small set of representative critical path to be actually detected in order to reduce the required sensors. Based on that, the delay of the retaining large number of critical paths can be predicted. Simulation results for multiple benchmarks demonstrate the efficiency of the proposed method in RCP selection and delay prediction of the circuits with aging and variations presence.

References

1. Chaware, R., Nagarajan, K., Ramalingam, S.: Assembly and reliability challenges in 3D integration of 28 nm FPGA die on a large high density 65 nm passive interposer. In: IEEE 62nd Electronic Components and Technology Conference, pp. 279–283. IEEE, May 2012
2. Casale-Rossi, M., et al.: Panel: will 3D-IC remain a technology of the future... even in the future? In: Proceedings of the Conference on Design, Automation and Test in Europe, pp. 1526–1530. EDA Consortium (2013)
3. Wang, R., Chakrabarty, K., Eklow, B.: Scan-based testing of post-bond silicon interposer interconnects in 2.5-D ICs. IEEE Trans. Comput. Aided Des. Integr. Circuits Syst. **33**(9), 1410–1423 (2014)
4. Firouzi, F., Kiamehr, S., Tahoori, M., Nassif, S.: Incorporating the impacts of workload-dependent runtime variations into timing analysis. In: 2013 Design, Automation and Test in Europe Conference and Exhibition (DATE), pp. 1022–1025. IEEE (2013)

5. Aitken, R., Fey, G., Kalbarczyk, Z.T., Reichenbach, F., Reorda, M.S.: Reliability analysis reloaded: how will we survive? In: Proceedings of the Conference on Design, Automation and Test in Europe, pp. 358–367. EDA Consortium (2013)
6. Das, S., Abraham, A., Konar, A.: Automatic clustering using an improved differential evolution algorithm. IEEE Trans. Syst. Man Cybern. Part A Syst. Hum. **38**(1), 218–237 (2007)
7. Gupta, M.S., Rivers, J.A., Bose, P., Wei, G.Y., Brooks, D.: Tribeca: design for PVT variations with local recovery and fine-grained adaptation. In: Proceedings of the 42nd Annual IEEE/ACM International Symposium on Microarchitecture, pp. 435–446. ACM (2009)
8. Li, Y., Mutlu, O., Gardner, D.S., Mitra, S.: Concurrent autonomous self-test for uncore components in system-on-chips. In: 2010 28th VLSI Test Symposium (VTS), pp. 232–237. IEEE (2010)
9. Agarwal, M., et al.: Optimized circuit failure prediction for aging: practicality and promise. In: 2008 IEEE International Test Conference, pp. 1–10. IEEE (2008)
10. Wang, S., Tehranipoor, M., Winemberg, L.: In-field aging measurement and calibration for power-performance optimization. In: Proceedings of the 48th Design Automation Conference, pp. 706–711. ACM (2011)
11. Cabe, A.C., Qi, Z., Wooters, S.N., Blalock, T.N., Stan, M.R.: Small embeddable NBTI sensors (SENS) for tracking on-chip performance decay. In: 2009 10th International Symposium on Quality Electronic Design, pp. 1–6. IEEE (2009)
12. Firouzi, F., Ye, F., Chakrabarty, K., Tahoori, M.B.: Representative critical-path selection for aging-induced delay monitoring. In: 2013 IEEE International Test Conference (ITC), pp. 1–10. IEEE (2013)
13. Xie, L., Davoodi, A.: Representative path selection for post-silicon timing prediction under variability. In: Proceedings of the 47th Design Automation Conference, pp. 386–391. ACM (2010)
14. Bhardwaj, S., Wang, W., Vattikonda, R., Cao, Y., Vrudhula, S.: Predictive modeling of the NBTI effect for reliable design. In: IEEE Custom Integrated Circuits Conference 2006, pp. 189–192. IEEE (2006)
15. Xiong, J., Zolotov, V., He, L.: Robust extraction of spatial correlation. IEEE Trans. Comput. Aided Des. Integr. Circuits Syst. **26**(4), 619–631 (2007)
16. Haghdad, K., Anis, M.: Power yield analysis under process and temperature variations. IEEE Trans. Very Large Scale Integr. (VLSI) Syst. **20**(10), 1794–1803 (2011)
17. Rogachev, A., Wan, L., Chen, D.: Temperature aware statistical static timing analysis. In: 2011 IEEE/ACM International Conference on Computer-Aided Design (ICCAD), pp. 103–110. IEEE (2011)
18. Lu, Y., Shang, L., Zhou, H., Zhu, H., Yang, F., Zeng, X.: Statistical reliability analysis under process variation and aging effects. In: 2009 46th ACM/IEEE Design Automation Conference, pp. 514–519. IEEE (2009)
19. Storn, R., Price, K.: Differential evolution-a simple and efficient heuristic for global optimization over continuous spaces. J. Global Optim. **11**(4), 341–359 (1997). https://doi.org/10.1023/A:1008202821328
20. International Workshop on Logic and Synthesis Benchmark (IWLS 2005) (2005). http://iwls.org/
21. International Test Conference Benchmark (ITC 1999) (1999). http://www.cad.polito.it/downloads/tools/itc99.html
22. NANGATE. http://www.nangate.com
23. Huang, W., Ghosh, S., Velusamy, S., Sankaranarayanan, K., Skadron, K., Stan, M.R.: HotSpot: a compact thermal modeling methodology for early-stage VLSI design. IEEE Trans. Very Large Scale Integr. (VLSI) Syst. **14**(5), 501–513 (2006)

Multi-AGV Collision Avoidance Path Optimization for Unmanned Warehouse Based on Improved Ant Colony Algorithm

Yang Yang(✉), Jianmin Zhang(✉), Yilin Liu(✉), and Xin Song

School of Management, China University of Mining and Technology, Beijing 100083, China
bwu_yangyang@126.com, 17865613218@163.com, lylsansi@163.com,
18813062101@163.com

Abstract. In this paper, the problem of collision avoidance path optimization for multi-AGV systems in unmanned warehouses is studied. A multi-AGV collision avoidance path optimization strategy based on elastic time window and improved ant colony algorithm is proposed. In this paper, the traditional ant colony algorithm is improved by heuristic information and pheromone update strategy to improve the execution speed and optimization ability of the algorithm. The priority scheduling of AGV tasks and the improvement of conflict resolution strategies are proposed to solve the different path conflicts between multiple AGVs. Based on the environment of the e-commerce logistics unmanned warehouse, the MATLAB simulation software is used to model and analyze the multi-AGV collision avoidance path planning. The experimental results show that the multi-AGV collision avoidance path planning can be realized based on the elastic time window and the improved ant colony algorithm, and the optimal collision avoidance path can be found in a short time.

Keywords: Multi-AGV system · Unmanned warehouse · Ant colony algorithm · Elastic time window

1 Introduction

With the advent of the new retail era, the traditional logistics system is undergoing rapid changes. Big data technology, artificial intelligence, simulation technology and so on are widely used in the logistics industry, so that unmanned warehouse has become the innovation of these technologies in the logistics system. The goal of unmanned warehouse is to realize the unmanned operation of the warehouse operation process, including warehousing, storage, sorting, warehousing and so on. In China, unmanned warehouse technology started early and developed rapidly. At present, many e-commerce giants, both in China and abroad, have begun to build unmanned warehouse. Therefore, in the future development, by speeding up technology upgrades, reducing operating costs, and accurately serving consumers to improve their competitiveness, the unmanned warehouse technology application can be won [1]. At present, in the hardware facilities of unmanned cabin, the typical handling equipment includes conveying line, Automatic

L. Pan et al. (Eds.): BIC-TA 2019, CCIS 1159, pp. 527–537, 2020.
https://doi.org/10.1007/978-981-15-3425-6_41

Guided Vehicle (AGV), shuttle car, Kiva robot, unmanned forklift truck, etc. The application of intelligent handling equipment greatly improves the operation efficiency, but in operation, how to find the optimal path quickly and prevent collision becomes a problem worth studying in the current application. AGV is an intelligent handling equipment with automatic guidance device, electromagnetics or optics, and able to drive along the specified path. It also has security protection and transplanting function. As the key equipment of unmanned warehouse operation, its work efficiency directly affects the overall operation efficiency of unmanned warehouse. Reasonable AGV driving path can not only improve the product circulation efficiency and work efficiency in unmanned warehouse, but also reduce the number of AGV distribution and save energy consumption, and increase its effective working time [2]. Therefore, this paper studies the multi-AGV collision avoidance path optimization strategy based on the improved ant colony algorithm.

The increase of AGV quantity helps to improve the efficiency of the system transportation task. However, with the continuous expansion of the number of handling equipment, the problems such as collision and system deadlock will inevitably occur in the AGV path planning, which seriously affect the path optimization and reduce the flexibility and efficiency of the system. This is a hot spot and difficulty in the field of AGV research. The collision avoidance problem of AGV is mainly divided into two categories, one is to avoid static obstacles, the other is the other AGV in the system as dynamic obstacle, avoid collision [3]. Therefore, the task of the collision avoidance path optimization is that in an environment with static and dynamic obstacles, it is possible to find a collision-free route from the starting point to the destination according to the requirements of the least time and the shortest distance. The traditional optimal path planning methods are based on A* algorithm [4], Dijkstra. The algorithm [5], however, has the disadvantages of large amount of computation, slow convergence speed and no real-time performance. In recent years, neural network, ant colony algorithm and genetic algorithm have also been applied to path planning [6–10]. Among these methods, neural network has good adaptability and learning optimization ability, but its generalization ability is weak and not universal. Genetic algorithm has strong search ability, improves genetic operator and introduces applicable weight coefficient, which can find the global optimal solution well, but the efficiency is low.

Compared with the above algorithms, the ant colony algorithm exhibits good search ability, positive feedback characteristics, distributed computing, etc. in the path planning application process of mobile robots. It has strong enlightenment and stability, but the traditional ant colony in the application process, the algorithm still has the defects of slow convergence, easy to fall into local optimal solution and stagnation. In solving the AGV path planning, since the ant colony algorithm has the advantages of easy integration with other algorithms, fast speed, high precision, and quick finding of the optimal solution. Therefore, in the current research, a variety of improvement schemes for ant colony algorithm are proposed to adapt to the path planning problem of AGV. Such as: Li et al. combined ant colony algorithm with particle swarm optimization algorithm to solve the problem of multi-AGV path conflict based on driving time [11]. Based on the combination of ant colony algorithm and Bezier curve, the path planning of material transportation and maintenance in complex environment was studied by Bu Xinping

[12]. Xia et al. proposed an adaptive Tabu search algorithm, and studied the multi-AGV material allocation for order resolution. Sending path planning, considering distance, load, demand and other factors, makes the path planning more in line with the actual demand [13]. Based on the traditional ant colony algorithm, Hu changed its transfer probability through positive distribution, improves the convergence rate and avoids the problem of falling into the local optimal solution [14]. He and others used ant colony algorithm combined with Dijkstra algorithm to plan the initial path of AGV, introduce the node random selection mechanism to get the optimal path, or by changing the heuristic function, introduce the global information into the algorithm to get the global optimal solution [5]. In a word, the existing research has not solved the collision avoidance path optimization of multi-AGV system.

In this paper, based on ant colony algorithm and elastic time window principle [3, 15], the collision avoidance path problem of multi-AGV system in logistics picking is optimized, and the simulation experiment is carried out by using MATLAB software. It is proved that this method can optimize the time AGV path planning on the basis of ensuring that the system is deadlock-free and collision-free.

2 Description of the Problem

2.1 Description of Unmanned Warehouse Environment

At present, E-commerce logistics unmanned warehouse is a rapidly developing warehousing mode, in which logistics AGV undertakes the important "cargo arrival" handling task, which greatly improves the efficiency of order selection. Generally speaking, the unmanned warehouse of e-commerce includes three core areas: cargo storage area, sorting station and AGV parking/charging area. After receiving the order selection order, AGV proceeds from the parking place to the target shelf, sends its shelf from the cargo storage area to the picking station area, then sends the empty shelf back to its original position, and then the picking system can place an order. In order to improve the selection efficiency, the system needs to operate several AGVs at the same time. The AGV collision conflict rate soared. The path planning of this paper is that after receiving the order selection command, the multiple AGVs should look for the shortest path between the starting point, the target shelf and the picking station in the feasible path, and in this process, it is not only to avoid static obstacles such as shelves, picking stations, charging equipment, but also to prevent the collision between multiple AGV in operation.

According to the above description of unmanned warehouse, this paper uses the grid method to simplify the environment modeling. The actual working environment is represented by a two-dimensional plane graphic, which is called a grid map. The map contains several grids of unit size, and all the grids are divided into obstacle grids (non-passable) and feasible grids (freely Pass), shown in black and white, as shown in Fig. 1. In Fig. 1, the picking station is indicated by the black grids numbered 4, 5, 7, 8, 10, 11, 18, 19, 21, 22, 24, 25 and are not passable. The goods storage area is indicated by the black grids numbered 44, 45, 48, 59, 72, 73, 47, 48, 61, 62, 75, 76, 50, 51, 64, 65, 78, 79, 53, 54, 67, 68, 81, 82, 100, 101, 114, 115, 128, 129, 103, 104, 117, 118, 131, 132, 106, 107, 120, 121, 134, 135, 109, 110, 123, 124, 137, 138 and are not passable. The AGV parking/charging area is indicated by the black grids numbered 173, 174, 187,

188, 176, 177, 190, 191, 179, 180, 193, 194 and are not passable. The AGV workspace is represented by all the remaining white grids and are free to pass.

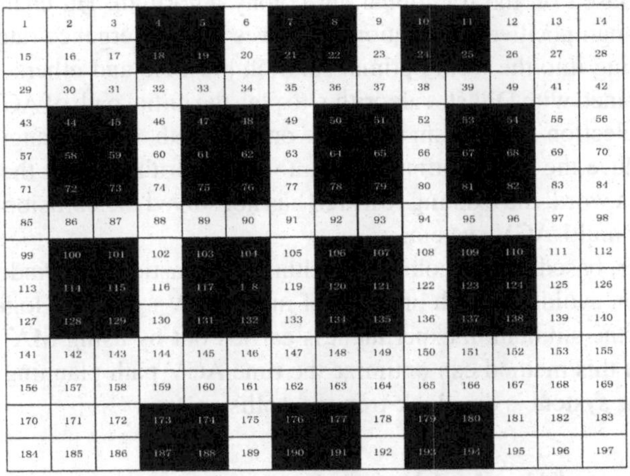

Fig. 1. Raster map

In this grid environment, assume that each AGV satisfies the following characteristics:

1. Each AGV occupies a grid, and the unit speed is equal to the edge length of the grid;
2. The speed of each AGV is the same, and the speed of no-load and load is the same, and the consumption is constant when turning;
3. The size of the unit feasible grid should ensure the smooth passage of AGV, and avoid the unnecessary influence of grid size on the path planning of AGV;
4. Each AGV can only accept one task at a time, and the next task can only be carried out after completion;
5. The minimum safe distance is defined between the two AGVs and is set as one body length.

2.2 Description of Multi-AGV Obstacle Avoidance Path Problem

In the actual unmanned cargo transportation process, if there are more goods orders, it will need multiple AGVs to run together in order to complete the order selection task more efficiently. For the multi-AGV dynamic system, the situation in the logistics transportation process is relatively complex, and with the increase of orders and the continuous addition of AGV, the path planning process becomes more complex. Therefore, due to system reasons, scheduling is unreasonable or unexpected occurs. It is likely to lead to AGV collision, path conflict, deadlock and so on. In order to solve the deadlock problem caused by collision between AGV, the minimum safe distance between AGV is specified. In the system, when one AGV path planning has been completed, there will be a corresponding time window through the node, and when the planning path of another AGV has the same node in a certain period of time, this node is called path conflict node. The path conflict is complex, including static conflict, cross conflict, directional conflict and pursuit conflict [16]. Because this paper assumes that the AGV speed is constant and the minimum safe distance between the two vehicles is maintained, then we mainly study how to avoid dynamic obstacles, so we mainly emphasize cross conflict and opposite direction conflict. The specific description of the two conflicts (see Fig. 2) is as follows:

1. Cross conflict.
2. As shown in Fig. 2(a), because each AGV runs at a consistent and stable speed and each grid is equal in size, AGV1 and AGV2 will reach the shadow grid in the figure at the same time, that is, AGV will have path conflicts in the shadow grid. This phenomenon is called the cross-conflict problem often encountered in multi-AGV system path planning.
3. Directional conflict.
4. As shown in Fig. 2(b), AGV1 runs in opposite directions with AGV2, and because a AGV occupies a grid, this path allows only one AGV to pass, that is, it cannot be passed by both AGV at the same time. Therefore, if the two AGV wants to continue to reach the shadow part, the path conflict will occur. This phenomenon is called the directional conflict problem which is often encountered in the path planning of multi-AGV systems.

3 Establishment of Model

3.1 Basic Model Based on Ant Colony Algorithm

Ant colony algorithm is an intelligent random search algorithm from nature, which has a strong ability to find a good solution. Its principle is that ants can release pheromones in the process of foraging for information transmission, the concentration of pheromones represents the distance length of the path, the concentration is inversely proportional

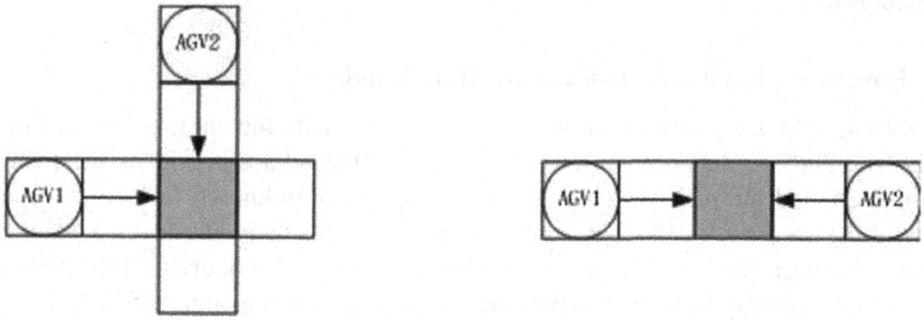

(a) Cross conflict (b) Directional conflict

Fig. 2. Path conflict schematic diagram

to the distance, and the distance with high concentration is short. Other ants determine the next transfer direction by judging the concentration of pheromones and heuristic information on each path, and gradually find food. Ant colony algorithm is robust, simple and easy to implement, so it has been widely used in solving path planning problems. Assuming that the number of ants is M, the optimal path is found in G iterations. At *t* time, ant *k* is located in grid *i*; $p_{ij}^k(t)$ is the transfer probability of ants from grid *i* to grid *j* at t time.

$$p_{ij}^k(t) = \begin{cases} \frac{[\tau_{ij}(t)]^\alpha \bullet [\eta_{ij}(t)]^\beta}{\sum\limits_{s \in J_k(i)} [\tau_{is}(t)]^\alpha \bullet [\eta_{is}(t)]^\beta}, & j \in J_k(i) \\ 0 & \end{cases} \tag{1}$$

k represents the k_{th} ant; $k = 1, 2, 3 \cdots$ M; *j* is the specific coordinates in the raster map; $\tau_{ij}(t)$ represents the pheromone from the *i* grid to the *j* grid at *t* time, and the initial value of the pheromone is generally set to be a small constant; η_{ij} is the heuristic information from grid *i* to grid j, $\eta_{ij} = 1/d_{ij}$, d_{ij} represents the distance from grid *i* to grid *j*; $J_k(i)$ represents the feasible domain of ant *k* in grid *i*; α, β indicates the importance of pheromone and heuristic information respectively; Taboo table $Tabu_k$ store the grid where ant *k* has passed. If an ant falls into a dead end and has no subsequent grid to choose from, the ant is dead by default, and the algorithm removes the ant and its path; In each iteration, when all ant routing is completed, the pheromones on all paths are updated.

$$\Delta\tau_{ij}(t+1) = (1-\rho)\tau_{ij}(t) + \Delta\tau_{ij}(t) \tag{2}$$

$$\Delta\tau_{ij}(t) = \sum_{k \in K_{ij}} \Delta\tau_{ij}^k(t) \tag{3}$$

$$\Delta\tau_{ij}^k(t) = Q/L_k \tag{4}$$

ρ is the pheromone volatilization coefficient; $\Delta\tau_{ij}(t)$ is the pheromone increment between the *ij* nodes of this loop; K_{ij} is all the ants passing through the node *ij*; $\Delta\tau_{ij}^k(t)$ is the amount of information left by the k ant; L_k is the length of the path sought by ant *k*, Q is a normal number. After the global pheromone update is completed, all the ants are put back to the starting point and re-searched, and after G iterations, the shortest path is calculated.

3.2 Improved Model Based on Elastic Time Window

In the basic ant colony algorithm, when searching the path, the ant takes the reciprocal of the straight-line distance between the grid and the adjacent grid as the heuristic information, and the path search is blind and has many unknown factors. In order to improve the accuracy of the algorithm search, this paper takes the inverse Manhattan distance between the ant's node and the target point as the heuristic information to improve the visibility of the ant to the end point, as shown in formula (5)[1].

$$\eta(\text{P}) = \frac{1}{|E(x) - P(x)| + |E(y) - P(y)|} \tag{5}$$

[1] E represents the end grid; P represents the grid where the ants are located.

Referring to the idea that Chen Guoliang and other experts set up ant colony pheromone updating mechanism to avoid the poor convergence and local optimal shortcomings of traditional ant colony algorithm [17], the formula of pheromone updating is improved in this paper. $\Delta\tau_{ij}^{k}(t)$ represents the amount of information left by the k ant. In order to reduce the interference of bad path pheromone and improve the search speed of ant, the shortest path is selected as the basis of pheromone updating. The revised formula is as follows:

$$\begin{aligned} \Delta\tau_{ij}^{k}(t) &= Q/L_k \\ L_k &= \min L \end{aligned} \tag{6}$$

In a multi-AGV system, each AGV in operation needs to be sorted by task priority. According to the time order of the command, that is, the time of the order task undertaken by AGV, the priority of the task before the time is high, and the priority of the task after the time is low. After ranking the priority level of tasks based on AGV, the problem of path conflict can be effectively solved. When the path cross conflict occurs at a certain node, the solution can be solved by waiting policy, the AGV with low task priority can stop and wait, and the AGV with high priority can pass first. When the path direction conflict occurs at a certain node, the waiting policy cannot solve the conflict, then the strategy adopted is to search the path of AGV with low task priority to find the optimal path under the premise of collision avoidance.

3.3 Implementation Steps of Improved Ant Colony Algorithm

1. **Environmental modeling.** The grid model of warehousing environment is established, the task is assigned to AGV according to the order of task priority, the target node of assigned task AGV is determined, the path search of AGV with the highest priority of task is carried out, and the time window table and other information parameters of each node are initialized.
2. **Preliminary path planning.** Set up the iterative counter G = 0, put M ants in the starting position, and add the initial points to the taboo table, calculate the probability of all transferable nodes of ant k according to the transfer probability formula, select the next node according to the roulette rule, update the taboo table, store the path length of all ants, save the path length, delete the dead ant, and the default path length is infinite. According to the pheromone update formula, the global pheromone is updated and the taboo table is emptied; whether G is equal to the maximum number of iterations, if the process is over, the optimal path of the AGV is searched out, if not, repeat. The above steps.
3. **Calculate the time window.** Calculate the time window of the AGV passing through the node and update the time window table of each node. Search the optimal path of the next AGV in order, calculate the time window of the node through which the AGV passes, update the time window table of all nodes, compare whether each node has time window conflict, if there is no conflict, carry on the path planning of the next AGV, if there is conflict, add the collision node to the taboo table, resolve the time window conflict, and compare it again.

4. **Research.** If in the process of path planning, there is a conflict with the adjacent nodes on the time window, which leads to the task not be executed, a new round of search will be repeated by expanding the search range of the time nodes.

4 Simulation Experiment and Result Analysis

In order to verify the effectiveness of the proposed algorithm, the simulation experiment is carried out by using MATLAB. Suppose there are now two different priority AGV1 and AGV2, starting grids of 33 and 37, respectively, and the target grid is 147. It is assumed that the task level of AGV1 takes precedence over AGV2.AGV1 and AGV2 moving to the target grid at the same time. Based on the parameter setting initiated in Table 1, the AGV collision avoidance experiment is carried out in this simulation environment.

Table 1. Description of simulation parameters.

Parameter	Short-cut process
Grid map size	14 * 14
Number of iterations G	100
Ant population K	80
Initial pheromone α	1
The importance of enlightening information β	7
Pheromone evaporation coefficient ρ	0.3
Constant Q	1
AGV quantity	2

After path search based on the basic model of ant colony algorithm, the path planning node of AGV1 is 33, 34, 49, 63, 77, 91, 105, 119, 133, 147; AGV2: 37, 36, 49, 63, 77, 91, 105, 119, 133, 147. During the execution of the mission, AGV1 and AGV2 collided at 49 grids, and the system reported an error and could not continue. As a result, AGV1 and AGV2 could not complete the task. Therefore, the task time was empty and recorded as "——". The experimental results are shown in Table 2. The corresponding path conflicts the two-dimensional plan and three-dimensional maps as shown in Figs. 3 and 4 respectively.

Based on the improved ant colony algorithm, the priority of AGV1 task is higher, its path does not need to be changed, and the task can continue to be executed. At this time, the priority of AGV2 task is low, and there are two strategies to choose when the perception is about to conflict with AGV1. Therefore, two kinds of coping strategies were compared.

Table 2. Task route of traditional ant colony algorithm (The completion of the mission time "—— " is indicated as "empty". Cause: AGV1 and AGV2 reach the grid 49 at the same time, a collision occurs, the handling task is no longer performed, and the task is not completed.)

Way	Time to complete the task
AGV1: 33,34,49,63,77,91,105,119,133,147	——
AGV2: 37,36,49,63,77,91,105,119,133,147	

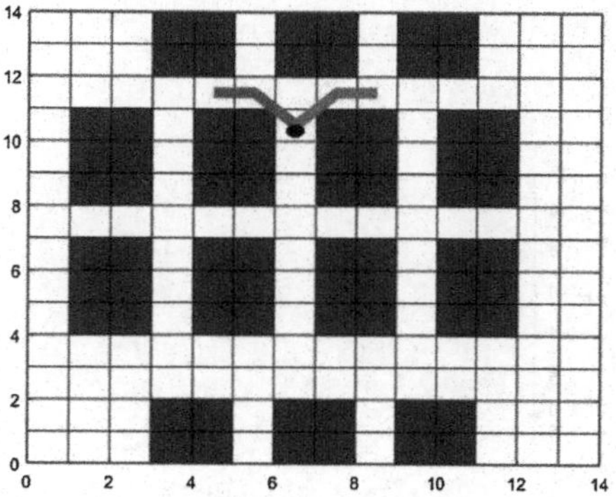

Fig. 3. Path conflict

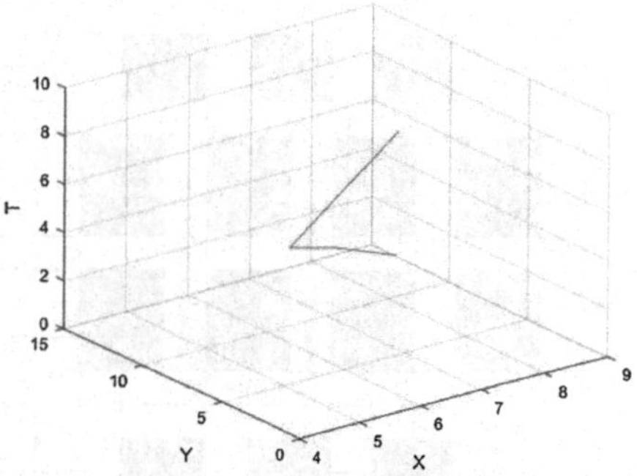

Fig. 4. Three-dimensional graph of path conflict

Strategy 1: AGV2 selects the re-routing strategy. At this time, AGV2 re-selects the other shortest path based on the improved ant colony algorithm. After the system search, the path is: 37, 52, 66, 80, 94, 108, 122, 136, 149 148,147, which takes 410 s. The results of MATLAB experiment are shown in Fig. 5 (two-dimensional) and Fig. 6 (three-dimensional);

Strategy 2: AGV2 selects the waiting strategy. At this time, AGV2 should wait at 36 grid, following AGV1 After passing through, the path of AGV2 is 37, 36, 36, 49, 77, 91, 105, 119, 133, 147, which takes 340 s. The results of MATLAB experiment are shown in Fig. 7 (two-dimensional) and Fig. 8 (three-dimensional).

See Table 3 for the path and task completion time of AGV1 and AGV2 under the two strategies.

Through the simulation results, it is found that the multi-AGV path conflict in the traditional ant colony algorithm will not be able to continue to execute the order task, which will cause the system to lock up and so on. After the improved ant colony algorithm, the search range can be reduced, the optimal path can be found in a shorter time, and the collision problem between AGV can be avoided completely in the event of multi-AGV path conflict, and the order task can be successfully completed, the locking phenomenon of the system can be avoided, and there are two different coping strategies to choose from, including waiting strategy and re-routing strategy. In this experiment, The experimental results show that the waiting strategy can avoid AGV collision. Complete the order task more efficiently under the premise of collision.

Table 3. Improved ant colony algorithm task route

Tactics	Way	Time to complete the task
Re-routing strategy	AGV1: 33,34,49,77,91,105,119,133,147	410 s
	AGV2: 37,52,66,80,94,108,122,136,149,148,147	
Waiting strategy	AGV1: 33,34,49,77,91,105,119,133,147	340 s
	AGV2: 37,36,36,49,77,91,105,119,133,147	

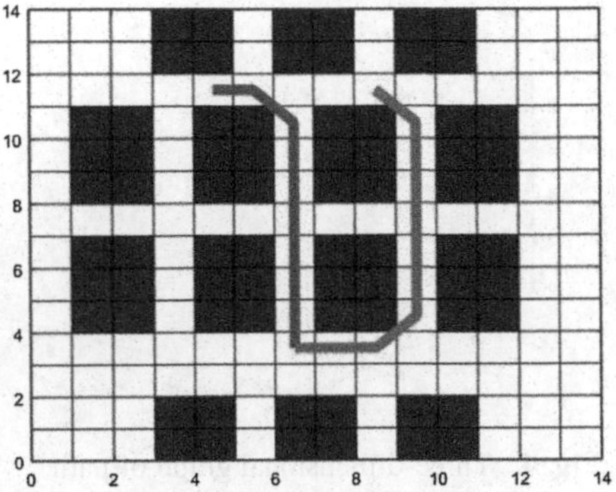

Fig. 5. Re-route strategy optimization path map

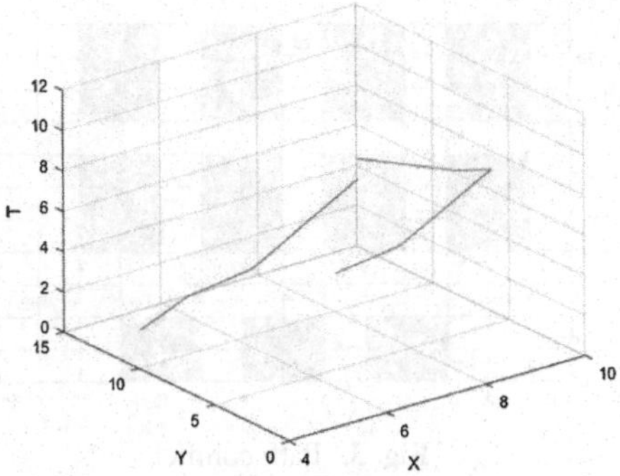

Fig. 6. Re-route strategy path 3D map

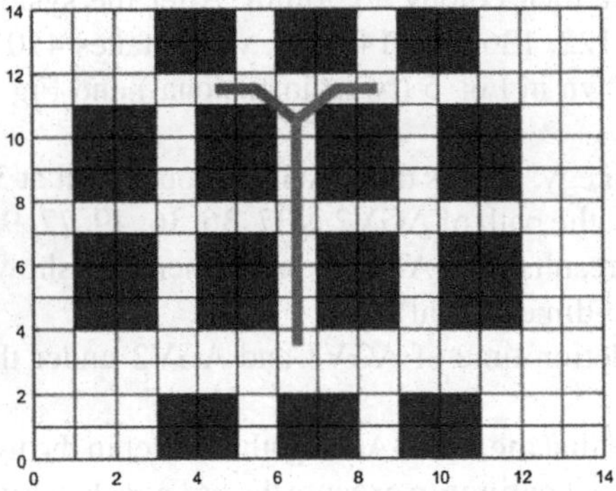

Fig. 7. Waiting strategy optimization path map

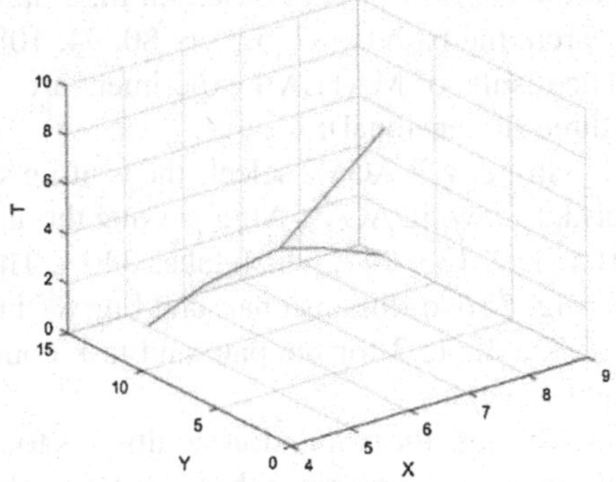

Fig. 8. Waiting strategy path 3D map

5 Conclusion

In this paper, a multi-AGV collision avoidance path optimization algorithm based on improved ant colony algorithm is proposed. The main contribution of this paper is to adjust the heuristic information to the inverse of the Manhattan distance between the node and the target point, and the basis of the pheromone update is adjusted to the shortest path. The task priority of the AGV is sorted, so that the path planning time can be shortened and the target of the path conflict can be effectively avoided. The reliability and effectiveness of the method are verified by the MATLAB simulation, which is better than the traditional ant colony algorithm in both the path planning time and the collision avoidance problem. With the continuous improvement of the algorithm, the development of transportation equipment can be considered in future research In the case of path conflict, the system can automatically identify the time used between different coping strategies and find the optimal path directly, which will make it more in line with the actual operation situation of unmanned warehouse.

References

1. http://it.people.com.cn/n1/2018/0530/c1009-30022080.html
2. Guo, X., Ji, M., Liu, S.: AGV path planning in unmanned warehouse combining multi-objective and energy consumption control. Comput. Integr. Manuf. Syst. 1–15 (2019). http://kns.cnki.net/kcms/detail/11.5946.tp.20190315.0932.012.html
3. Gu, B.: Research on path optimization of automatic guided vehicle system based on time window. Lanzhou Traffic Science (2015)
4. Tai, Y., Xing, K., Lin, Y., Zhang, W.: Research on multi-AGV path planning method. Comput. Sci. **44**(S2), 84–87 (2017)
5. He, C., Mao, J.: Research on the path of AGV based on improved ant colony algorithm. Logistics Sci. Technol. **42**(03), 60–65 (2019)
6. Meng, C., Ren, Y.: Multi-AGV scheduling based on multiple population genetic algorithm. Electron. Technol. **31**(11), 47–50+68 (2018)
7. Yuan, R., Wang, H., Sun, L., Li, J.: Research on Task scheduling of order picking system based on Logistics AGV. Oper. Res. Manage. **27**(10), 133–138 (2018)
8. Tang, S., Hong, Z.: Research on robot path planning based on ant colony algorithm. Electromech. Inf. (08), 46–47 (2019)
9. Zhao, Y., Zhang, S.: Intelligent traffic path planning based on improved ant colony algorithm. Industr. Instrum. Autom. Device (02), 30–32 (2019)
10. Wei, Y., Jin, W.: Intelligent vehicle path planning based on neural network Q-learning algorithm. Firepower Command Control **44**(02), 46–49 (2019)
11. Li, J., Xu, B., Yang, Y., Wu, H.: Induced ant swarm particle swarm optimization algorithm for multi-automatic guided vehicle path planning. Comput. Integr. Manuf. Syst. **23**(12), 2758–2767 (2017)
12. Pu, X., Su, H., Zou, W., Wang, P., Zhou, H.: Smooth path planning based on non-uniform environment modeling and third-order Bezier curve. J. Autom. **43**(05), 710–724 (2017)
13. Xia, Y., Fu, Z., Xie, J.: Multi-automatic guided vehicle material distribution path planning according to order resolution. Comput. Integr. Manuf. Syst. **23**(07), 1520–1528 (2017)
14. Hu, Q., Wang, T., Zhang, R.: Research on improved ant colony algorithm in AGV global path planning. Inf. Technol. Informatization (03), 116–118 (2019)
15. Fisher, M.L., Rnsten, J., et al.: Vehicle routing with time windows. Two Optim. Algorithms Oper. Res. **45**(3), 488–492 (1997)
16. Wang, S., Mao, Y., Yuan, X.: Multi-AGV path optimization strategy for finite state machines. J. Overseas Chin. Univ. (Natural Science Edition) **40**(02), 239–244 (2019)
17. Chen, G., Liu, J., Zhang, C.: Ant colony optimization with potential field based on grid map for mobile robot path planning. J. Donghua Univ. (English Edition) **33**(05), 764–767 (2016)

An Improved Competitive Swarm Optimizer for Large Scale Optimization

Zhenzu Liu, Lianghong Wu(✉), Hongqiang Zhang, and Panpan Mei

Hunan University of Science and Technology, Xiangtan 411201, China
lhwu@hnust.edu.cn

Abstract. In this paper, an improved competitive swarm optimizer (ICSO) for large scale optimization is proposed for the limited global search ability of paired competitive learning evolution strategies. The proposed algorithm no longer uses the competitive winner and the global average position of the current population to update the competitive loser position such a paired competitive learning evolution strategy. Three individuals are randomly selected without returning to compete, the compete failed individual update its speed and position by learning from the other two competing winners, thereby improving the global search ability of the algorithm. Theoretical analysis shows that the randomness of this improved competitive learning evolution strategy has been enhanced. In order to verify the effectiveness of the proposed strategy, 20 test functions from CEC'2010 large-scale optimization test set are selected to test the performance of the algorithm. Compared with the competitive swarm optimization (CSO) and the level-based learning swarm optimization (LLSO) two state-of-the-art algorithms, the experimental results show that ICSO has better performance than CSO and LLSO in solving large-scale optimization problems up to 1000 dimensions.

Keywords: Large-scale optimization · Competition · Learning evolution · Swarm optimization

1 Introduction

Since particle swarm optimization (PSO) was proposed by Kennedy and Eberhart in 1995 [1], it has been widely used to solve various practical problems [2,3] and has achieved very good results. In fact, PSO simulates the foraging behavior of animal groups such as flocks and ant colonies, and traverses the solution space to find the global optimal solution for the problem to be optimized. Specifically, each particle in the population is represented by position and velocity two attributes, and is updated individually in the manner expressed Eqs. (1) and (2):

$$v_i^d \leftarrow wv_i^d + c_1r_1(pbest_i^d - x_i^d) + c_2r_2(nbest_i^d - x_i^d) \tag{1}$$

$$x_i^d \leftarrow x_i^d + v_i^d \tag{2}$$

L. Pan et al. (Eds.): BIC-TA 2019, CCIS 1159, pp. 538–548, 2020.
https://doi.org/10.1007/978-981-15-3425-6_42

Where $X_i = [x_i^1, \ldots, x_i^d, \ldots, x_i^D]$ and $V_i = [v_i^1, \ldots, v_i^d, \ldots, v_i^D]$ represent the position vector and velocity vector of the ith particle, respectively. $pbest_i = [pbest_i^1, \ldots, pbest_i^d, \ldots, pbest_i^D]$ is the historical optimal position of the ith particle, and $nbest_i = [nbest_i^1, \ldots, nbest_i^d, \ldots, nbest_i^D]$ is the optimal position of the neighborhood under the specific topology adopted. Overall, each particle in the population continuously updates its speed and position by learning from the particle with the best fitness in the current population. This evolutionary learning strategy causes each particle in the population to evolve to a position with better fitness, so that the entire population may converge to a global optimal position.

Due to the simple concept of particle swarm optimization algorithm and its high search efficiency, it has attracted the interest of many researchers in the past few decades and has been successfully applied to many practical problems, such as distributed network design [4], Man-machine combat [5], resource allocation [6] and so on. However, the latest research results show that when there are a large number of local optimal solutions or high-dimensional variables in the optimization problem [7], these difficulties often lead to the premature convergence of the traditional particle swarm optimization algorithm and show very poor results [8].

In order to improve the optimization performance of the particle swarm optimization algorithm. Inspired by nature and human society, some researchers have proposed some new particle update strategies to deal with large-scale optimization problems [8–12]. Inspired by orthogonal experimental design, Zhan et al. proposed a particle swarm optimization algorithm based on orthogonal learning strategy by conducting orthogonal experiments on pbest and nbest to obtain better individuals [9]. Liang et al. [10] developed a comprehensive learning PSO (CLPSO) and further Lynn and Suganthan [11] devised heterogeneous CLPSO to enhance the exploration and exploitation of CLPSO. Qin et al. [12] proposed an interactive learning strategy that two particle in the population avoid algorithm's premature convergence by dynamically updating each particle when it detects an update stagnation. Chen et al. [8] introduced the aging mechanism into the evolution process of the population, providing opportunities for other individuals to challenge the status of the leader, thereby changing the diversity of the population and overcoming the premature convergence of the algorithm.

Although these improved particle swarm optimization algorithms show better performance than the basic particle swarm optimization algorithm, they are all aimed at low-dimensional optimization problems. As the dimensionality of the variables of the optimization problem continues to increase, the performance of such optimization algorithms drops dramatically. On the one hand, due to the increase of the optimization problem dimension, the size of the search space grows exponentially, which greatly limits the search ability of the current optimization algorithm. On the other hand, the increased dimension will bring a large number of local optimal solutions, and the optimization algorithm can easily fall into the local optimal solution in advance, which leads to the premature convergence of the algorithm. Therefore, when designing high-dimensional optimization algorithms, it is necessary to improve the randomness of the individual learning update strategy to overcome these difficulties.

Inspired by nature and human society, many scholars have achieved many important research results in the use of PSO to deal with large-scale problems. Liang and Suganthan [13] divides the population into small sub-populations, each of which is responsible for searching a local region, and proposes a dynamic multi-population particle swarm optimization algorithm to solve large-scale optimization problems. Cheng and Jin [14] developed a competitive swarm optimizer (CSO) for large scale optimization under the concept of not using the global optimal position and the optimal position of the individual history. Yang et al. [15] proposed a level-based learning swarm optimizer (LLSO) for large-scale optimization. In this paper, an improved competitive particle swarm optimization algorithm is proposed for the lack of randomness of individual competitive learning evolution strategy in CSO. This algorithm use three randomly selected particle to compete, the most disadvantaged individual learn from the other two individuals to determine the evolution direction, thereby improving the randomness of the evolution of the population during evolution, and maintaining the diversity of evolutionary of evolutionary directions to avoid premature convergence. 20 test functions from the CEC'2010 [16] large-scale test set are used to test the performance of the algorithm, and compared with 2 state-of-the-art algorithms CSO and LLSO. The experimental results show that ICSO is an effective algorithm for solving large-scale optimization problems.

2 Algorithm Principle

Without loss of generality, consider of the form of minimizing the problem as shown in Eq. (3).

$$\begin{aligned} &\min f(X), X = [x^1, x^2, \ldots, x^D] \\ &X \in \Omega \end{aligned} \tag{3}$$

where $f(X)$ denotes the optimization problem, $X = [x^1, x^2, \ldots, x^D]$ denotes the variable of the optimization problem, and D and Ω denote the dimension and feasible domain of the optimization variable, respectively. In addition, the function value $f(X)$ is taken as the fitness value of each particle.

2.1 Pairwise Competition Learning Evolution Strategy

In order to overcome the difficulties caused by large-scale optimization problems, many researchers are working on proposing new learning evolution strategies to improve the search ability of particles in the feasible region of the optimization problem. Among them, Cheng and Jin proposed a novel particle learning evolution strategy [14] by randomly selecting two individuals from the population without replaying back for pairwise competition, and the competing failed individual learn from the winner to update its position, while the winner enters the next generation directly. Specifically, the speed and position update methods of individual who fail to compete are shown in Eqs. (4) and (5), respectively:

$$v_l^d \leftarrow r_1 v_l^d + r_2(x_w^d - x_l^d) + \phi r_3(\bar{x}^d - x_l^d) \tag{4}$$

$$x_l^d \leftarrow x_l^d + v_l^d \tag{5}$$

where $X_l = [x_l^1, \ldots, x_l^d, \ldots x_{.l}^D]$ and $V_l = [v_l^1, \ldots, v_l^d, \ldots v_{.l}^D]$ represent the position and velocity of the failed individual in the pairwise competition, respectively. $X_w = [x_w^1, \ldots, x_w^d, \ldots x_{.w}^D]$ and $\bar{x} = [\bar{x}^1, \ldots, \bar{x}^d, \ldots, \bar{x}^D]$ represent the position of the winning individual in the pairwise competition and the average position of the entire population, respectively. In the pairwise competitive learning evolution strategy, compared with the traditional particle swarm optimization algorithm, the global optimal position of the population is no longer used to guide the individual's learning evolution, so the randomness of individual learning evolution can be improved, and the possibility of the algorithm searching for a better solution is increased.

2.2 Improved Competition Learning Evolution Strategy

In order to improve the randomness of the evolutionary strategies of inferior individual in the algorithm process and search for better solutions, this paper proposes a three-individual competitive learning evolution strategy base on the pairwise competitive learning strategy. Each iteration randomly selects three individuals from the population by repeating without replay, and competing failed individual learns from the other two wining individuals. The competitive learning idea of the algorithm is shown in Fig. 1. According to the fitness value, the individual with the lowest fitness value is recorded as the competitive failure individual X_l. The best fitness value of the two competing individuals is recorded as X_{w_1}, and the other is recorded as X_{w_2}. The speed and position learning evolution of the individual who fails the competition are as shown in Eqs. (6) and (7) respectively:

$$v_l^d \leftarrow r_1 v_l^d + r_2(x_{w_1}^d - x_l^d) + \phi r_3(x_{w_2}^d - x_l^d) \tag{6}$$

$$x_l^d \leftarrow x_l^d + v_l^d \tag{7}$$

Compared with the individual learning evolution strategy based on pairwise competition, the improved competitive learning evolution strategy no longer use the average position of the entire population. Because the average position of the population is shared by the entire population, it limits the randomness of individual learning evolution direction, which is not conductive to the exploration of the entire feasible domain. Therefore, the improved competitive learning evolution strategy proposed in this paper improves the randomness of individual learning evolution direction and enhances the ability of the population to explore the entire feasible domain.

2.3 Framework of ICSO

The algorithm framework of the improved competitive swarm optimizer for large-scale optimization in this paper is as follows:

Step1: Set the parameters of the ICSO, the range of the benchmark function variables, and the algorithm iteration termination condition (the maximum number of the fitness evolutions times).
Step2: Initialize the population P0 randomly as the initial population within the feasible region of the solution.
Step3: Calculate the fitness of each individual based on the benchmark function set.
Step4: Randomly select three individuals from the population to compete repeatedly, and the two dominant individuals who win will directly enter the next generation population, and the individual with the lowest fitness value will update to the next generation population according to formulas (6) and (7).
Step 5: Calculate the fitness of the new generation population.
Step 6: Determine whether the maximum fitness evaluation number is reached, and if it is reached, algorithm ends, otherwise, return to Step5 to continue execution.

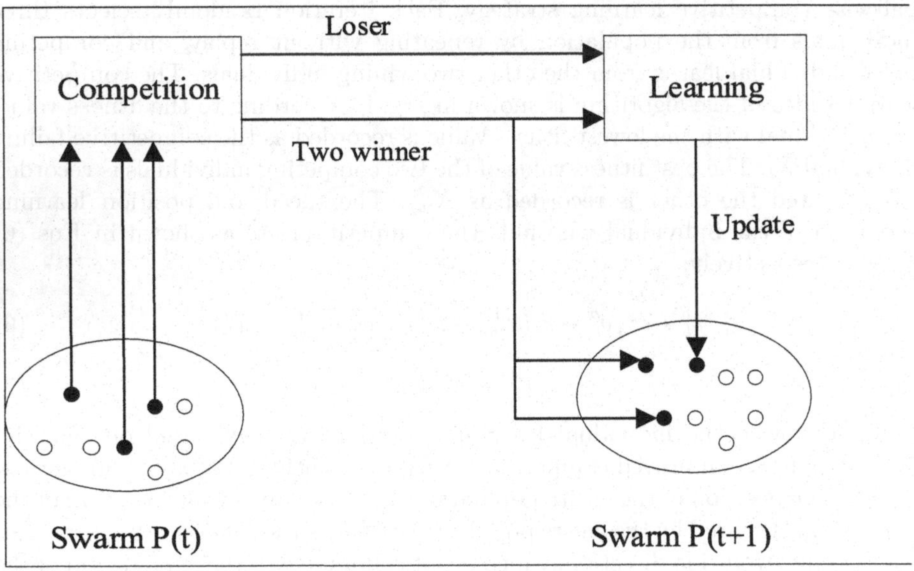

Fig. 1. General idea of the ICSO algorithm.

3 Theoretical Analysis

In this section, we will theoretically analyze the algorithm proposed n this paper in exploring feasible spaces and improving the randomness of individual learning evolution strategies.

In the large-scale optimization problem, the ability of the population to explore the feasible solution space has always played an important role, and the exploration ability is mainly determined by the individual diversity of the population. Therefore, the appropriate diversity is beneficial to the population search for better position. However, if the diversity of the initial evolution of the population is too large, the algorithm will not converge, and the algorithm takes too long. If the diversity is too small, the algorithm will converge too quickly to a local optimal position, it is therefore extremely important to maintaining the diversity of the population in a suitable range. In particular, for the multi-modal optimization problem, increasing the randomness of the individual learning evolution direction at the beginning of the population evolution is beneficial to the algorithm jumping out of the local optimal position and searching for a better solution. In short, the ability to explore the feasible solution space can be reflected by how the dominant individual can guide the learning evolution strategy of the disadvantaged individual.

In order to study the ability of ICSO to explore the feasible solution space and the randomness ability of individual learning update strategy, the speed update expression (6) of the failed individual in the ICSO algorithm competition strategy is transformed as shown in Eq. (8):

$$v_l^d \leftarrow r_1 v_l^d + \theta_1 (p_1 - x_l^d) \tag{8}$$

$$\theta_1 = r_2 + \phi r_3 \tag{9}$$

$$p_1 = \frac{r_2}{r_2 + \phi r_3} x_{w_1}^d + \frac{\phi r_3}{r_2 + \phi r_3} x_{w_2}^d \tag{10}$$

Similarly, the individual speed update expression in the CSO algorithm can be transformed as shown in (11):

$$v_l^d \leftarrow r_1 v_l^d + \theta_2 (p_2 - x_l^d) \tag{11}$$

$$\theta_2 = r_2 + \phi r_3 \tag{12}$$

$$p_2 = \frac{r_2}{r_2 + \phi r_3} x_w^d + \frac{\phi r_3}{r_2 + \phi r_3} \bar{x}^d \tag{13}$$

It can be seen from Eqs. (8) and (11) that p1 and p2 are the main difference between the two strategies, and they are also the main theoretical sources of evolutionary strategy randomness and population diversity, and it can be know from Eqs. (10) and (13) that ICSO has the potential to increase the randomness of individual learning evolution in the population. On the one hand, ICSO and CSO are the same for the first part of p_1 and p_2, both of which provide opportunities for disadvantaged individual to learn from dominant individuals. On the other hand, for the second part of p_1 and p_2, the CSO uses the global average position. Although the global average position is updated in each generation, it is shared by the entire population, and ICSO chooses to randomly select the other two. Obviously, ICSO algorithm has more randomness in particle evolution in the early stage of evolution, thus improving the randomness of learning of learning and evolution of competitive failure individual, which is beneficial to the ability of particle to explore feasible spaces. As such, this difference will also increase the rate of convergence of the population throughout the evolutionary process.

4 Experimental Verification and Analysis

In order to verify the feasibility and effectiveness of the ICSO algorithm proposed in this paper, the algorithm is tested by 20 benchmark functions $F_1 \sim F_{20}$ in CEC'2010 [16], which include single-modal and multi-modal functions of various properties.

In order to observe the diversity and overall distribution of the population during the evolution process, a diversity measure introduced in [14] is adopted here to indicate the change of diversity during the search process:

$$D(X) = \frac{1}{NP}\sum_{i=1}^{NP}\sqrt{\sum_{d=1}^{D}\left(x_i^d - \bar{x}^d\right)^2}$$
$$\text{with} \qquad (14)$$
$$\bar{x}^d = \frac{1}{NP}\sum_{i=1}^{NP} x_i^d$$

where $D(X)$ represent the distribution of the entire population X, and $\bar{x}$ represents the average position of the population.

4.1 Parameter Settings

In the algorithm ICSO proposed in this paper, NP = 504 and $\varphi = 0.4$ is adopted for ICSO on 1000-D problems, which also makes it fair to compare ICSO with LLSO and CSO that adopts the same setting of NP. In the experiment, the number of times of fitness evaluation reached $3 * 10^6$ times as the termination condition of the experimental iteration.

4.2 Comparisons with State-of-the-Art Methods

In order to analyze the dynamic process of the algorithm on the large-scale test problem, taking the single-modal test function test function F_1 and multi-modal test function F_2 as examples, compare the population dynamic evolution process of the ICSO algorithm proposed in this paper with the other two state-of-the-art large-scale optimization algorithms CSO [14] and LLSO [15]. The convergence of the population optimal individual fitness of the three algorithms on the F_1

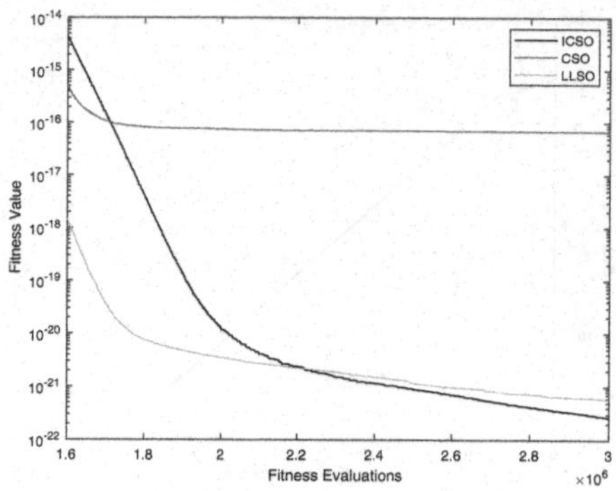

Fig. 2. Evolutionary process diagram of population optimal individual fitness on function F_1.

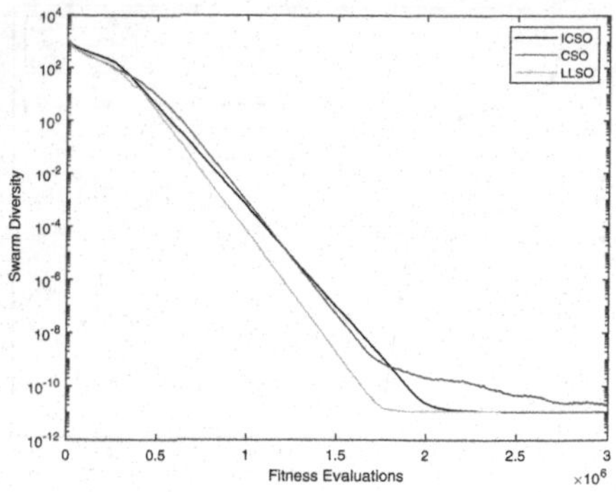

Fig. 3. Evolutionary process diagram of population diversity on function F_1.

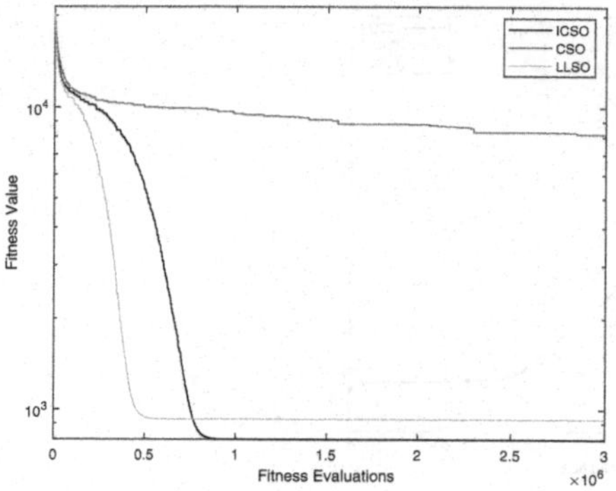

Fig. 4. Evolutionary process diagram of population optimal individual fitness on function F_2.

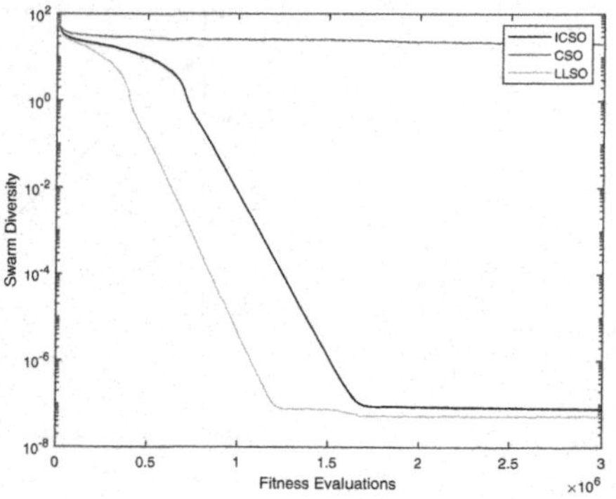

Fig. 5. Evolutionary process diagram of population diversity on function F_2.

and F_2 is shown in Figs. 2 and 4 respectively, the overall population distribution is shown in Figs. 3 and 5 respectively.

It can be seen from Figs. 2 and 4 that the ICSO algorithm proposed in this paper has better convergence accuracy than the other two state-of-the-art large-scale algorithms in the same number of times of fitness evaluation. In particular, for the multi-modal test function F_2, since the diversity of the CSO in the whole evolution process is basically maintained at a large value, the CSO cannot converge, and the finally only a local optimal solution can be searched. The ICSO algorithm maintains the appropriate diversity in the evolution process because it uses two random better individuals to guide the loser in the competition. As can be seen from Figs. 4 and 5, the diversity of ICSO is gradually reduced during the evolution process, but it maintains a better diversity CSO, so it finally finds a better solution than CSO and LLSO. To better reflect the performance of ICSO,

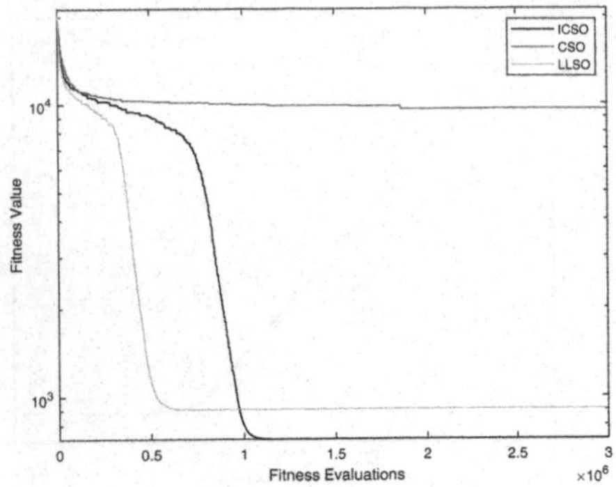

Fig. 6. Evolutionary process diagram of population optimal individual fitness on function F_{10}.

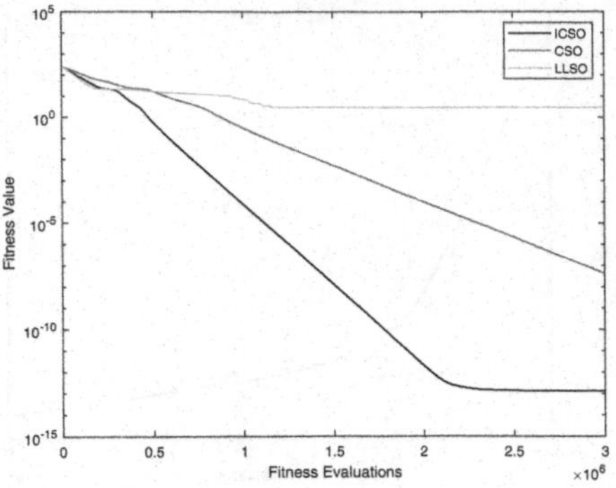

Fig. 7. Evolutionary process diagram of population optimal individual fitness on function F_{11}.

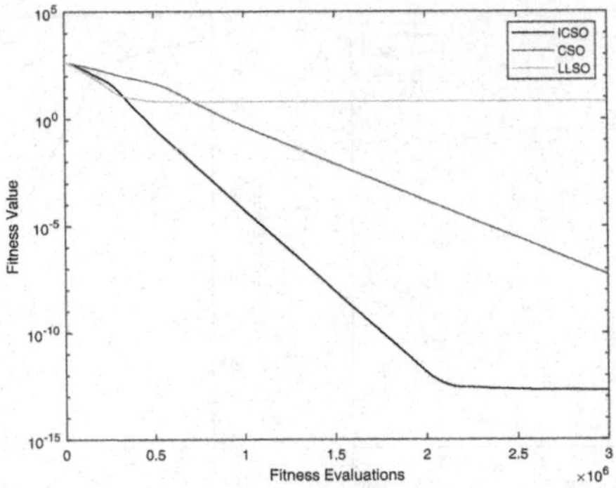

Fig. 8. Evolutionary process diagram of population optimal individual fitness on function F_{16}.

Figs. 6, 7, 8 show the evolution of the optimal individual of ICSO on the other three functions F_{10}, F_{11}, F_{16} respectively.

Finally, in order to verify the effectiveness of the ICSO algorithm proposed in this paper, the algorithm is tested using the CEC'2010 [16] benchmark set. Table 1 gives the test results for each algorithm, the data in the table given as mean(std). In Table 1, the last row w/l indicates the number of test problems that ICSO performed better and worse than the other two algorithms. As can be seen from Table 1, the ICSO algorithm can generally achieve better performance on large-scale problems of up to 1000-D. In addition, based on the improved competition learning evolution strategy, only three individuals need to be randomly selected for comparison, which is easier to implement than the hierarchical learning evolution strategy.

Table 1. Comparison of the results of the three algorithms.

Function	ICSO	CSO	LLSO
F_1	**2.47E−22(2.25e−23)**	3.13E−22(8.03E−23)	4.50E−16(5.94E−17)
F_2	**8.51E+02(3.30E+01)**	9.82E+02(4.39E+01)	7.42E+03(2.86E+02)
F_3	**2.32E−14(1.77E−15)**	2.76E−14(2.38E−15)	2.60E−09(2.62E−10)
F_4	5.43E+11(9.73E+10)	**4.40E+11(1.10E+11)**	7.25E+11(1.23E+11)
F_5	8.56E+06(3.12E+06)	1.22E+07(3.43E+06)	**2.86E+06(1.79E+06)**
F_6	**4.00E−09(8.41E−25)**	5.20E−01(7.46E−01)	8.21E−07(2.68E−08)
F_7	**4.26E+02(2.09E+02)**	7.19E+02(2.59E+03)	2.01E+04(3.86E+03)
F_8	2.92E+07(1.00E+05)	**2.34E+07(2.46E+05)**	3.87E+07(6.81E+04)
F_9	**3.11E+07(2.57E+06)**	4.36E+07(4.28E+06)	7.03E+07(5.73E+06)
F_{10}	**7.56E+02(3.76E+01)**	8.91E+02(3.66E+01)	9.60E+03(7.67E+01)
F_{11}	**1.35E−13(3.33E−15)**	5.80E+00(5.40E+00)	4.02E−08(5.12E−09)
F_{12}	2.62E+04(1.86E+03)	**1.25E+04(1.46E+03)**	4.37E+05(6.22E+04)
F_{13}	**5.85E+02(1.62E+02)**	7.35E+02(1.93E+02)	6.29E+02(2.32E+02)
F_{14}	**1.17E+08(6.14E+06)**	1.24E+08(7.38E+06)	2.49E+08(1.53E+07)
F_{15}	8.82E+03(2.18E+03)	**8.33E+02(4.31E+01)**	1.01E+04(5.23E+01)
F_{16}	**1.95E−13(5.80E−15)**	4.25E+00(2.41E+00)	5.89E−08(5.61E−09)
F_{17}	2.04E+05(9.00E+03)	**9.05E+04(3.53E+03)**	2.20E+06(1.55E+05)
F_{18}	1.95E+03(7.91E+02)	2.55E+03(8.32E+02)	**1.73E+03(5.22E+02)**
F_{19}	4.59E+06(2.89E+05)	**1.80E+06(9.96E+04)**	1.01E+07(5.64E+05)
F_{20}	1.30E+03(1.59E+02)	1.88E+03(1.90E+02)	**1.05E+03(1.49E+02)**
w/l	–	14/6	17/3

5 Conclusion

This paper proposes an improved large-scale competitive swarm optimization algorithm based on three individuals selected from the population. The most disadvantaged individual learns from the other two winning individuals to update its speed and position, which is beneficial to enhance the randomness of individual evolution to maintain proper population diversity, thereby improving the ability of the algorithm to converge to the better global optimal position of large-scale optimization problems. The final experimental results show that compared with the current two state-of-the-art algorithms CSO and LLSO, the algorithm proposed by this paper ICSO shows better performance overall and can effectively deal with large-scale problems with up to 1000-dimensional variables. In the future, we will also try to apply ICSO to multi-objective optimization problems and many-objective optimization problems, Combining other evolutionary algorithms (such as differential evolution or genetic algorithm) with three individuals competing learning evolution strategies will be a challenging task.

Acknowledgment. The work is supported by Hunan Graduate Research and Innovation Project (CX20190807), National Natural Science Foundation of China (Grant Nos. 61603132, 61672226), Hunan Provincial Natural Science Foundation of China (Grant

No. 2018JJ2137, 2018JJ3188), Science and Technology Plan of China (2017XK2302), and Doctoral Scientific Research Initiation Funds of Hunan University of Science and Technology (E56126).

References

1. Eberhart, R., Kennedy, J.: A new optimizer using particle swarm theory. In: Proceedings of the Sixth International Symposium on Micro Machine and Human Science, Nagoya, Japan, pp. 39–43. IEEE (1995)
2. Faria, P., Soares, J., Vale, Z.: Modified particle swarm optimization applied to integrated demand response and DG resources scheduling. IEEE Trans. Smart Grid **4**(1), 606–616 (2013)
3. Wen, X., Chen, W.N., Lin, Y.: A maximal clique based multiobjective evolutionary algorithm for overlapping community detection. IEEE Trans. Evol. Comput. **21**(3), 363–377 (2016)
4. Montalvo, I., Izquierdo, J., Pérez, R.: A diversity-enriched variant of discrete PSO applied to the design of water distribution networks. Eng. Optim. **40**(7), 655–668 (2008)
5. Fu, Y., Wang, Y.C., Chen, Z., Fan, W.L.: Target decision in collaborative air combats using multi-agent particle swarm optimization. J. Syst. Simul. **30**(11), 4151–4157 (2008)
6. Gong, Y.J., Zhang, J., Chung, S.H.: An efficient resource allocation scheme using particle swarm optimization. IEEE Trans. Evol. Comput. **16**(6), 801–816 (2012)
7. Yang, Y., Pedersen, J.O.: A comparative study on feature selection in text categorization. In: Proceedings of the International Conference on Machine Learning (1997)
8. Chen, W.N., Zhang, J., Lin, Y.: Particle swarm optimization with an aging leader and challengers. IEEE Trans. Evol. Comput. **17**(2), 241–258 (2013)
9. Zhan, Z.H., Zhang, J., Li, Y.: Orthogonal learning particle swarm optimization. IEEE Trans. Evol. Comput. **15**(6), 832–847 (2011)
10. Liang, J., Qin, A.K., Suganthan, P.N.: Comprehensive learning particle swarm optimizer for global optimization of multimodal functions. IEEE Trans. Evol. Comput. **10**(3), 281–295 (2006)
11. Lynn, N., Suganthan, P.N.: Heterogeneous comprehensive learning particle swarm optimization with enhanced exploration and exploitation. Swarm Evol. Comput. **24**, 11–24 (2015)
12. Qin, Q., Cheng, S., Zhang, Q.: Particle swarm optimization with interswarm interactive learning strategy. IEEE Trans. Cybern. **46**(10), 2238–2251 (2015)
13. Liang, J., Suganthan, P.N.: Dynamic multi-swarm particle swarm optimizer with local search. In: IEEE Congress on Evolutionary Computation, Edinburgh, Scotland, pp. 522–528 (2005)
14. Cheng, R., Jin, Y.: A competitive swarm optimizer for large scale optimization. IEEE Trans. Cybern. **45**(2), 191–204 (2015)
15. Yang, Q., Chen, W.N., Deng, J.D.: A level-based learning swarm optimizer for large scale optimization. IEEE Trans. Evol. Comput. **22**(4), 578–594 (2017)
16. Tang, K., Li, X., Suganthan, P.N., Yand, Z., Weise, T.: Benchmark functions for the CEC'2010 special session and competition on large-scale global optimization. Technical report, Nature Inspired Computation and Application Laboratory, USTC, China (2010)

MEAPCA: A Multi-population Evolutionary Algorithm Based on PCA for Multi-objective Optimization

Nan-jiang Dong and Rui Wang(✉)

College of System Engineering, National University of Defense Technology, Changsha, Hunan, People's Republic of China
ruiwangnudt@gmail.com

Abstract. The simulated binary crossover (SBX) and differential evolution operators (DE) are two most representative evolutionary operators. However, due to their different search pattens, they are found to face difficulty on multi-objective optimization problems (MOPs) with rotated Pareto optimal set (PS). The regularity model based multi-objective estimated distribution algorithm, namely, RM-MEDA that adopts a segmented PCA model to estimate the PS shows good performance on such problems. However, determining the offering number of segments (clusters) of the PCA model is difficult. This study therefore proposes a multi-population multi-objective evolutionary algorithm based on PCA (MEAPCA) in which the optimization process is divided into two phases. The first phase employs a multi-population method to quickly find a few well-converged solutions. In the second phase, new offspring are generated under the guidance of the PCA model. That is, the PCA model utilizes information of those well-converged solutions so as to ensure the generation of good offspring. The DTLZ1-5 with modified PS are used as test problems. The MEAPCA is then compared against RM-MEDA as well as two representative MOEAs, i.e., NSGA-II and MOEA/D, and is found to perform well on MOPs with complex Pareto optimal front.

Keywords: Multi-objective optimization · Multi-population · Decomposition · PCA · Evolutionary computation

1 Introduction

Multi-objective optimization problems (MOPs) [1,2] arise regularly in many engineering areas. Very often, the objectives in an MOP conflict with each other, and thus, no single solution can optimize all the objectives simultaneously. The Pareto set/front (PS/PF) refers to the set of all the optimal tradeoffs in the decision/objective space. Multi-objective evolutionary algorithms (MOEAs) [1,2] are

This work was supported by the National Natural Science Foundation of China (61773390), the Hunan Youth elite program (2018RS3081) and the key project of National University of Defense Technology (ZK18-02-09).

L. Pan et al. (Eds.): BIC-TA 2019, CCIS 1159, pp. 549–557, 2020.
https://doi.org/10.1007/978-981-15-3425-6_43

well-suited for addressing MOPs since their population based-nature enables to find multiple Pareto-optimal solutions in one single algorithm run. NSGA-II [3] and MOEA/D [4] are two of the most popular MOEAs.

When dealing with continuous MOPs, real coding is usually chosen to code the problem variables. The SBX [5,6] and DE [7] are two most representative evolutionary operators. Each operator has its own search patten, being suitable for different problems. Unlike SBX and DE, in RM-MEDA [8], the principal component analysis (PCA) [9] is applied to first segment the population and then the distribution model of segmented *PS* is built to produce new solutions.

The Karush-Kuhn-Tucker (KKT) [10,11] condition indicates that under mild conditions, the dimension of the *PS* for an MOP is $(m-1)$. This property is called regularity property of continuous MOPs. Upon this feature, we can produce the offspring more efficiently, thus speeding up the optimization of MOPs.

It should be mentioned that even if the shape of the *PS* is the same, the difficulty of the optimization problem varies with the rotation of the *PS* position.

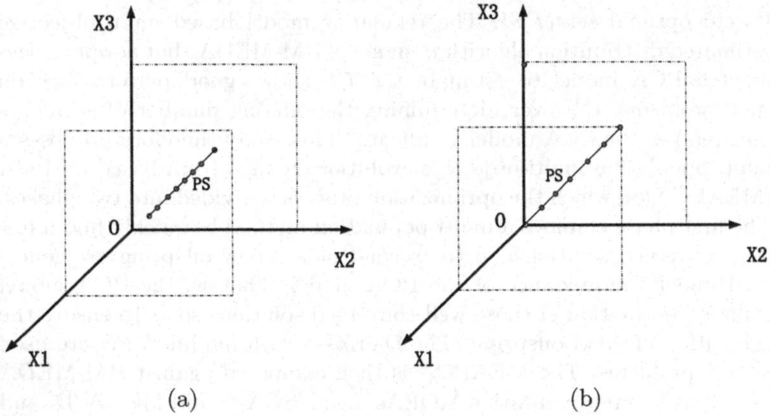

Fig. 1. Illustration of *PS* of a MOP.

Figure 1 illustrates simplified *PS* of MOPs. The *PS* in Fig. 1(a) is perpendicular or parallel to the coordinate axis. Experimental results show that SBX operator is more suitable for such problems. The *PS* in Fig. 1(b) is rotated compared to Fig. 1(a). In this case, the SBX becomes less effective while DE is found to be better. Besides, in general the DE operator is more robust on problems with complicated *PS*. However, it is usually difficult to know the problem property in advance, thus, the choice of good operator becomes challenging. Although RM-MEDA is less affected by the location of *PS*, determining the number of PCA clusters is difficult. Also, the accuracy of PCA model is not guaranteed in the early search process.

Based on the above analysis, we propose a multi-population multi-objective evolutionary algorithm based on PCA (MEAPCA), which is more robust to the rotation of PS. The rest of this study is organized as follows. Section 2 presents the basic idea and details of MEAPCA. Section 3 presents and discusses the experimental results. Finally, Sect. 4 concludes the paper.

2 Algorithm

2.1 Basic Idea

The idea of MEAPCA is based on the fact that under mild conditions the PS of the continuous MOPs is m-1 piece-wise continuous manifold. Specifically, the working principle of MEAPCA is as follows. First, find a few well-converged solutions. Then spread the diversity of the population through these well-converged solutions. As a result, the population can approach to and distribute along the PF appropriately. The MEAPCA is a two-phase MOEA.

- In the first phase, a multi-population method [12,13] is applied to quickly find a few well-converged solutions. That is, several sub-problems are separated by the decomposition strategy. Each sub-problem corresponds to a constructed sub-population, and each sub-population is independently optimized. Note that there is no information interaction and co-evolution between sub-populations. The information contained in each sub-population is more independent and less redundant. At the end of this phase, the sub-populations are combined and the diversity of the population is expanded to produce new solutions.
- In the second phase, the algorithm pays more attention to the distribution of the population. The PCA is used to analyze and extract features of individuals with good convergence, which then provides guidance for the generation of offspring, in particular, the location of offspring.

2.2 Detail of MEAPCA

The framework of MEAPCA is described in Algorithm 1.

- Initialmultipopulaiton(num, N_{subpop}): Decompose subproblems by weighting different objectives. Randomly generate $num * N_{subpop}$ individuals as initial population. The subpopulation S_i is formed by N_{subpop} individuals that are close to the subproblem. The Penalty Boundary Intersection (PBI) approach is used to evaluate individuals. The number of multi-population (num) is set as M, the number of objectives, and N_{subpop} is set as 30.
- Subpopoptimization($S_1, ..., S_{num}$): Each subpopulation S_i independently optimizes the corresponding subproblem. For each subpopulation, the optimization problem is a single objective optimization problem. There is no coordinated evolution between sub-populations.
- Mutation(Pop): By mutation, the diversity of the population is increased, which can enrich the sampled individuals for PCA.

Algorithm 1. The framework of MEAPCA

Require:
MOP, *num* (the number of multiple population), N_{subpop} (size of sub population), N_{pop} (size of population), *sumevs* (total number of evaluations), r (a ratio)
Ensure:
population : P
1: Initialmultipopulaiton(num, N_{subpop}) $->$ $(S_1, ..., S_{num})$
2: While $evs < sumevs$ (evs, the number of current evaluations)
3: If $evs < sumevs * r$
4: subpopoptimization$(S_1, , S_{num}) -> (S_1, , S_{num})$
5: End If
6: if $sumevs * r < evs < sumevs * r + N_{pop}$
7: Merge $(S1, ..., S_{num}) -> P_t$
8: Mutation$(P_t) -> Q_t$
9: Environmentselect$(P_t \bigcup Q_t) -> P_{t+1}$
10: End If
11: If $sumevs * r + N_{pop} < evs < sumevs$
12: PCAgenerate($Nondominateset$, N_{pop}, U, L) $->$ Q_t ; ($Nondominateset$ from P_t)
13: Environmentselect$(P_t, Q_t) -> P_{t+1}$
14: End If
15: End while
16: **return** P;

- Environmentselect(P, Q): In principle, any selection operators for MOEAs can be used in MEAPCA. The selection operator selects N_{pop} members from $P \cup Q$ to form P. Specifically, the selection operator of NSGA-II is adopted.

The detail of PCAgenerate($Nondominateset$, N_{pop}, U, L) is shown in Algorithm 2. The eigenvalues(D) and eigenvectors(V) are obtained by PCA analysis of non-dominant individuals. The eigenvectors(V) constitutes a new coordinate system, and the corresponding eigenvalue(D) reflects the extension degree of non-dominant individuals in the corresponding direction. V and D outline the shape of PS. Lines 7–9 show the process of generating a new solution in decision space. First map a solution in decision space to the coordinate system formed by V to get a new coordinate, called $Mapcoordinate$. Then disturb the $Mapcoordinate$ to get a new coordinate, called $Newdec$, and the $Newdec$ obeys the distribution of $N(Mapcoordinate, D)$. Finally map the $Newdec$ back to the decision space, and we get a new solution in the decision space.

The difference between MEAPCA and RM-MEDA on the use of PCA is outlined as follows. RM-MEDA divides PS into a few segments, and each segment is a space of $m-1$ dimensions. However MEAPCA considers PS as a whole, and tries to find a space with the smallest dimension that can contain $(m-1)$-dimensional manifold, reducing the search area and simultaneously improving the search efficiency. There is no doubt that the segmentation of RM-MEDA is more precise. However, it is difficult to determine a reasonable number of segments. Also, the model established based on poorly-converged individuals may

Algorithm 2. PCAgenerate($Nondominateset, N_{pop}, U, L$)

Require:
$Nondominateset$, N_{pop}, U (upper boundary of decision space), L (lower boundary of decision space)
Ensure:
Q

$PCA(Nondominateset) -> [V, D]$
$Q = []$
For $i = 1$ *to* N_{pop}
  $Dec = indv.dec$ ($indv$ is selected randomly from $Nondominateset$)
  $Offdec = []$
  While ($Offdec = NULL \;||\; U < Offdec \;||\; Offdec < L$)
    $Mapcoordinate = Dec * V$;
    $Newdec \sim N(Mapcoordinate, D)$
    $Offdec = Newdec * V^{-1}$
  End while
  $Offspr = Individual(Offdec)$
  $Q = Q \cup Offspr$
End for
return Q

mislead the evolution. On the contrary, in MEAPCA, first a few well-converged solutions are obtained based on which the PCA model is then built. As a result, the accuracy of the PCA model can be improved.

3 Experiment

3.1 Experiment Setup

The modified instances DTLZ1-5 [14,15] are used as test functions. Taking DTLZ1 as an example, the original distance function of DTLZ1 is as follows.

$$g(M) = 100\big[|X_M| + \sum_{x_i \in X_M} ((x_i - 0.5)^2 - cos(20\pi(x_i - 0.5)))\big], 0 \le x_i \le 1 (i = 1, 2, ..., n) \quad (1)$$

The modified distance function of DTLZ1 is as follows.

$$g(M) = 100\big[|X_M| + \sum_{x_i \in X_M} ((x_i - x_1)^2 - cos(20\pi(x_i - x_1)))\big], 0 \le x_i \le 1 (i = 1, 2, ..., n) \quad (2)$$

Specifically, the distance variable is linked with $x1$ instead of simply subtracting 0.5. Similar changes are made to DTLZ2-5 problems. The purpose of such change is to make the PS shift in the decision space [15].

In this study, the hypervolume (HV) metric [16] is chosen as performance metric, and is calculated in the normalized objective space where all results are normalized within the unit hypercube $[0, 1]^m$. The larger the HV the better the algorithm performs.

The MEAPCA is compared with NSGA-II, MOEA/D and RM-MEDA algorithms. The population size and the termination conditions are listed in Tables 1 and 2. The number of decision variables is set to $m + 7$, and m is the number of optimization objectives. Both NSGA-II and MOEA/D adopt the DE operator, and the control parameters F and CR of DE are set to F = 0.5 and CR = 1.0. The parameter r in MEAPCA is set as 0.6.

Table 1. The population size of the test algorithms.

m	3	5	8
MEAPCA, NSGA-II, RM-MEDA	100	100	100
MOEA/D	84	126	156

Table 2. The termination condition (number of function evaluations) for different problems.

m	3	5	8
DTLZ1, 3	40000	40000	60000
DTLZ2, 4, 5	20000	20000	30000

3.2 Experiment Result

In this subsection, the comparison results among MEAPCA, NSGA-II, MOEA/D and RM-MEDA are presented. The statistical results including the mean and standard deviation values of HV on all instances are shown in Table 3 where the best performance is shown in bold. The RankSum test is carried out to indicate significance between different results at the 0.05 significance level.

As can be seen from Table 3, the proposed method performs the best on most of problems. MEAPCA is not as good as NSGA-II in 3-dimensional DTLZ1 and DTLZ3, but the performance is very close. As the problem dimension increases, MEAPCA outperforms NSGA-II clearly. MEAPCA has poor performance on DTLZ4 problem. Part of the reason is that when using PCAgenerate() to generate offspring, the distribution of parents is not considered, which results in poor search ability for problems with deflection characteristics.

Figure 2 shows the *PS* obtained by the the algorithms for 3-objective DTLZ3. It is easy to know that the results in Fig. 2(a), obtained by the MEAPCA, is the

Table 3. Shape of a character in dependence on its position in a word

Problem	D	MEAPCA	NSGA-II	MOEA/D	RM-MEDA
DDTLZ1	3	7.3287e−1 (1.81e−1)	8.1981e−1≈ (4.71e−3)	4.3522e−1 − (6.71e−2)	1.4450e−1 − (2.71e−1)
	5	8.2960e−1 (3.00e−1)	4.5802e−1 − (4.59e−1)	4.6435e−1 − (1.43e−1)	6.5683e−1 − (2.83e−1)
	8	9.0611e−1 (2.49e−1)	8.4689e−2 − (2.21e−1)	5.2886e−1 − (8.23e−2)	7.6349e−2 − (2.43e−1)
DDTLZ2	3	5.4047e−1 (4.58e−3)	5.2237e−1 − (4.30e−3)	5.3218e−1 − (4.83e−3)	5.2785e−1 − (4.33e−3)
	5	7.0526e−1 (8.62e−3)	2.7058e−1 − (1.48e−1)	6.6918e−1 − (3.09e−2)	1.4687e−1 − (9.32e−2)
	8	6.5074e−1 (1.11e−1))	7.9288e−3 − (1.90e−2)	5.5837e−1 − (4.24e−2)	2.1404e−4 − (6.60e−4)
DDTLZ3	3	5.0716e−1 (9.85e−2)	5.3287e−1≈ (4.23e−3)	1.4312e−1 − (7.54e−2)	1.3086e−1 − (1.96e−1)
	5	5.6748e−1 (1.73e−1)	2.8557e−1 − (3.31e−1)	1.8402e−1 − (9.12e−2)	1.7934e−1 − (1.87e−1)
	8	7.2038e−1 (5.13e−2)	3.0303e−3 − (1.66e−2))	1.6660e−1 − (9.01e−2)	2.5303e−3 − (1.66e−2)
DDTLZ4	3	4.8423e−1 (3.60e−2)	4.7017e−1≈ (5.16e−2)	3.7818e−1≈ (1.70e−1)	4.7553e−1≈ (3.07e−2)
	5	5.6906e−1 (5.76e−2)	5.8128e−1≈ (9.78e−2)	5.0070e−1≈ (1.52e−1)	5.7431e−1 ≈ (7.76e−2)
	8	5.0204e−1 (7.87e−2)	6.2184e−2 − (8.36e−2)	5.6165e−1 + (1.86e−1)	2.1410e−2 − (4.89e−2)
DDTLZ5	3	1.9967e−1 (1.38e−4)	1.9779e−1 − (3.08e−4)	1.8138e−1 − (2.39e−3)	1.9898e−1 − (1.69e−4)
	5	1.2863e−1 (4.33e−4)	9.0873e−2 − (1.79e−2)	9.0212e−2 − (1.41e−2)	8.1628e−2 − (2.09e−2)
	8	1.0503e−1 (7.53e−4)	7.2565e−2 − (3.69e−2)	2.1892e−2 − (5.90e−3)	1.5828e−2 − (3.11e−2)

best. The results of NSGA-II is similar to Fig. 2(a). From Fig. 2(b), we can see that MOEA/D converges locally and it is difficult to cover the whole *PS* because of the loss of diversity in decision space. Figure 2(c) is the *PS* obtained by RM-MEDA from which we can see that the effect of PCA modeling in RM-MEDA is not very ideal.

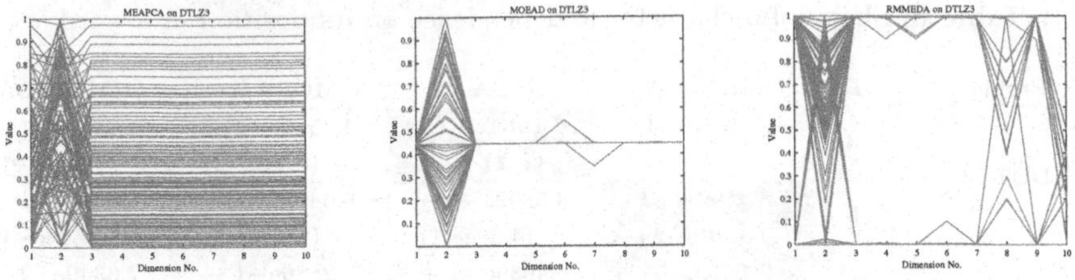

(a) (b) (c)

Fig. 2. The *PS* optimized by MOEAS

4 Conclusion

Given to the difficulty of SBX and/DE on problems with rotated *PS*, this study proposes a new method to produce high-quality offspring. The method is integrated into the general MOEA framework, resulting in a new algorithm, namely, MEAPCA. The algorithm contains two search phases. In the first phase, the MOP is decomposed into different subproblems. These subproblems are then solved by a multi-population method, producing a few well-converged solutions. Moreover, the use of decomposition improves the selection pressure, and enables the algorithm to perform well on high-dimensional problems [17–20]. In the second phase, new offspring are generated under the guidance of PCA model. The PCA model utilizes information of well-converged solutions so as to ensure the generation of good offspring. By doing so, the algorithm becomes less affected by the rotation of *PS* position. From the comparison results, MEAPCA shows the best performance compared to NSGA-II, MOEA/D and RM-MEDA on most instances.

With respect to the future studies, we would like to examine the performance of MEAPCA on more complex problems, e.g., many constraint optimization problems [21]. Also, the idea of MEAPCA would also be generalized to problems with mixed variables and some real problems [22].

References

1. Zheng, J.H., Juan, Z.: Multi-Objective Evolutionary Optimization. Science Press, Beijing (2017)
2. Cui, X.X.: Multi-Objective Evolutionary Algorithm and Its Application. National Defence Industry Press, Beijing (2006)
3. Deb, K., Pratap, A., Agarwal, S., et al.: A fast and elitist multiobjective genetic algorithm: NSGA-II. IEEE Trans. Evol. Comput. **6**(2), 182–197 (2002)
4. Zhang, Q., Li, H.: MOEA/D: a multiobjective evolutionary algorithm based on decomposition. IEEE Trans. Evol. Comput. **11**(6), 712–731 (2007)
5. Deb, K., Agrawal, R.B.: Simulated binary crossover for continuous search space. Complex Syst. **9**(3), 115–148 (1994)

6. Kumar, K.D.A., Deb, K.: Real-coded genetic algorithms with simulated binary crossover: studies on multimodal and multiobjective problems. Complex Syst. **9**, 431–454 (1995)
7. Storn, R., Price, K.: Differential evolution – a simple and efficient heuristic for global optimization over continuous spaces. J. Global Optim. **11**(4), 341–359 (1997)
8. Zhang, Q., Zhou, A., Jin, Y.: RM-MEDA: a regularity model-based multiobjective estimation of distribution algorithm. IEEE Trans. Evol. Comput. **12**(1), 41–63 (2008)
9. Kambhatla, N., Leen, T.K.: Dimension reduction by local principal component analysis. Neural Comput. **9**(7), 1493–1516 (1997)
10. Miettinen, K.: Nonlinear Multiobjective Optimization. Springer, Boston (2012). https://doi.org/10.1007/978-1-4615-5563-6
11. Hillermeier, C.: Nonlinear Multiobjective Optimization: A Generalized Homotopy Approach. Birkhauser, Boston (2001)
12. Helbig, M., Engelbrecht, A.P.: Heterogeneous dynamic vector evaluated particle swarm optimisation for dynamic multi-objective optimisation. In: IEEE Congress on Evolutionary Computation, pp. 3151–3159. IEEE (2014)
13. Goh, C.K., Tan, K.C.: A competitive-cooperative coevolutionary paradigm for dynamic multiobjective optimization. IEEE Trans. Evol. Comput. **13**(1), 103–127 (2009)
14. Deb, K., Thiele, L., Laumanns, M., et al.: Scalable test problems for evolutionary multiobjective optimization. In: Abraham, A., Jain, L., Goldberg, R. (eds.) Evolutionary Multiobjective Optimization, pp. 105–145. Springer, London (2005). https://doi.org/10.1007/1-84628-137-7_6
15. Deb, K., Sinha, A., Kukkonen, S.: Multi-objective test problems, linkages, and evolutionary methodologies. In: Proceedings of the 8th Annual Conference on Genetic and Evolutionary Computation, Seattle, Washington, USA, pp. 1141–1148 (2006)
16. While, L., Hingston, P., Barone, L., et al.: A faster algorithm for calculating hypervolume. IEEE Trans. Evol. Comput. **10**(1), 29–38 (2006)
17. Pan, L., He, C., Tian, Y., Su, Y., Zhang, X.: A region division based diversity maintaining approach for many-objective optimization. Integr. Comput. Aided Eng. **24**(3), 279–296 (2017)
18. He, C., Tian, Y., Jin, Y., Zhang, X., Pan, L.: A radial space division based evolutionary algorithm for many-objective optimization. Appl. Soft Comput. **61**, 603–621 (2017)
19. Pan, L., He, C., Tian, Y., Wang, H., Zhang, X., Jin, Y.: A classification-based surrogate-assisted evolutionary algorithm for expensive many-objective optimization. IEEE Trans. Evol. Comput. **23**(1), 74–88 (2018)
20. Pan, L., Li, L., He, C., Tan, K.C.: A subregion division-based evolutionary algorithm with effective mating selection for many-objective optimization. IEEE Trans. Cybern. (2019). https://doi.org/10.1109/TCYB.2019.2906679
21. Ming, M., Wang, R., Zhang, T.: Evolutionary many-constraint optimization: an exploratory analysis. In: Deb, K., et al. (eds.) EMO 2019. LNCS, vol. 11411, pp. 165–176. Springer, Cham (2019). https://doi.org/10.1007/978-3-030-12598-1_14
22. Wang, R., Li, G., Ming, M., et al.: An efficient multi-objective model and algorithm for sizing a stand-alone hybrid renewable energy system. Energy **141**, 2288–2299 (2017)

A Novel Genetic Algorithm with Population Perturbation and Elimination for Multi-satellite TT&C Scheduling Problem

Ming Chen(✉), Jun Wen(✉), Ben-Jie Pi, Hao Wang, Yan-Jie Song, and Li-Ning Xing

College of Systems Engineering, National University of Defense Technology, Changsha 410073, Hunan, China
chenming_nudt@163.com, jun_wen@aliyun.com

Abstract. Multi-Satellite Tracking Telemetry and Command Scheduling Problem is a multi-constrained, high-conflict complex combinatorial optimization problem. How to effectively utilize existing resources has always been an important topic in the satellite field. To solve this problem, this paper abstracts and simplifies the Multi-Satellite TT&C Scheduling problem and establishes the corresponding mathematical model. The hybrid goal of maximizing the profit and task completion rate is our objective function. Since the genetic algorithm has a significant effect in solving the problem of resource allocation, we have proposed an improved genetic algorithm with population perturbation and elimination (GA-PS) based on the characteristics of the Multi-Satellite TT&C Scheduling problem. A series of simulation experiments were carried out with the total profits and the task completion rate as the index of the algorithm. The experiment shows that compared with the other three comparison algorithms, our algorithm has better performance in both profit and task completion rate.

Keywords: Genetic algorithm · Multi-satellite TT&C scheduling · Intelligent optimization method · Bio-inspired computing

1 Introduction

An artificial satellite is an unmanned spacecraft orbiting the earth in space. According to the purpose, artificial satellites can be mainly divided into communication satellites, meteorological satellites, reconnaissance satellites, navigation satellites. With the continuous development of space technology and the increase in the number of satellites in orbit, satellites have gradually become an indispensable means of information acquisition in wartime information acquisition, disaster prevention, anti-terrorism security, and ship escort. It have been highly valued by countries all over the world. The increase in the number of satellites is often accompanied by the difficulty of satellite control. So it is significance to complete more tasks while ensuring the normal operation of the satellite.

This paper focuses on the ground tracking telemetry and command tasks in multi-satellite mission scheduling. Satellite tracking telemetry and command (TT&C) refers to the process of establishing a communication link with a ground station to complete

L. Pan et al. (Eds.): BIC-TA 2019, CCIS 1159, pp. 558–568, 2020.
https://doi.org/10.1007/978-981-15-3425-6_44

satellite telemetry data reception and command betting when the satellite passes over the ground station. At present, satellite command is dominated by ground station management, while it is often affected by various factors, such as weather and hardware conditions.

The multi-satellite TT&C scheduling problem is an inevitable problem with the development of aerospace technology. The essence is the conflict between the increasing demand and equipment of satellite TT&C. Therefore, how to use the existing and limited resources to maximize satellite utility is the core issue in multi-satellite mission scheduling. Establishing a suitable multi-satellite mission scheduling model is of great significance for making full use of resources and exerting satellite performance.

The multi-satellite TT&C scheduling problem is a kind of combinatorial optimization problem with multiple constraints and multiple objectives. [1] proposed the TT&C resource scheduling method for multi-satellite based on SDMA-CDMA (Space Division Multiple Access—Code Division Multiple Access) system. A kind of 'one ground station for multi-satellite' TT&C mode is realized. [2] put forward An algorithm of multi-satellite control resource scheduling problem based on ant colony optimization (MSCRSP-ACO). [3] described a tabu search heuristic for the problem of selecting and scheduling the requests to be satisfied, under operational constraints. An upper bounding procedure based on column generation is used to evaluate the quality of the solutions. [4] constructed an acyclic directed graph model for multi-satellite observation scheduling. Based on the graph model presented a novel hybrid ant colony optimization mixed with the iteration local search algorithm (ACO-ILS) to produce high quality schedules. [5] proposed a heuristic algorithm and a conflict-based back jumping algorithm to solve the model. [6] proposed an effective hybrid optimization method based on the combination of particle swarm optimization (PSO) and genetic algorithm (GA). [7] efficiently adapt the best ingredients of the graph colouring techniques to an NP-hard satellite range scheduling problem. [8] put a CSP model of the problem, and solves it with the help of ILOG Solver library. [9] pointed that single ant colony optimization (ACO) strategy has disadvantages of low efficiency and poor solution performance. For this reason, the genetic-ACO hybrid algorithm, (GA-ACO) which combines the ACO with genetic algorithm (GA) was proposed to solve this problem. [9] proposed an genetic-ant colony optimization hybrid algorithm for joint scheduling of space and ground tracking telemetry and command resources with multi-time window characteristics.

This paper constructs a mathematical model of multi-satellite mission scheduling, and proposes an improved genetic algorithm with population perturbation and elimination. Experiments show that compared with other algorithms, our algorithm has good performance in both profits and task completion rate.

The remainder of this paper is organized as follows. In the second part, we will put forward a mathematical description of the multi-satellite mission planning. After that, an improved genetic algorithm with population perturbation and elimination (GA-PS) will be introduced in detail. In the fourth part, there will use experiments to verify the effectiveness of the proposed algorithm. The conclusions of the study will be drawn in the last part.

2 Problem Description

This part is mainly a description of the multi-satellite TT&C scheduling problem. In practical engineering applications, it has multiple constraints and the complexity of problem is high. Therefore, we need to abstract and simplify the actual engineering problems before the research, and separate the objective function and constraints.

Satellite TT&C refers to the process of establishing a communication link with a ground station to complete satellite telemetry data reception and command betting when the satellite passes over the ground station. Traditional multi-satellite TT&C scheduling assumes that all satellite resources, ground station resources, and mission requests are known prior to the use of the scheduling algorithm.

The mission scheduling includes multiple satellites, each of them has a fixed operating state. This is also the case with ground stations. For a single satellite, the time to reach the ground station in a certain time range is limited, and the communication between satellite and the ground can only be completed within this period. Moreover, due to the limited resources of the ground station time window, it is necessary to coordinate all the requests and the ground station time window resources, use appropriate methods to ensure the smooth completion of the satellite tracking telemetry and command. The process of generating tracking telemetry and command plan according to resources is the process of satellite TT&C scheduling.

In order to abstract and simplify the actual problem, we make the following assumptions:

1. The ground station has abundant energy, storage and other resources, and does not consider the staff scheduling problem;
2. The time horizon of scheduling is limited, and the long-term scheduling scheme is a periodic repetition of a single one;
3. Each satellite is visible to at least one station;
4. Once the task begins to execute, it cannot be interrupted or preempted by other tasks.
5. Each task can only be executed once and cannot be executed repeatedly;
6. Regardless of the uncertainty of the on-board storage, battery and other load resource consumption;
7. Regardless of the failure of the satellite communication caused by equipment failure;

Given a set of tasks to be observed $O = \{o_i | i = 1, 2, \cdots, M\}$ contains M tasks and each task o_i contains four features, the earliest allowed execution time st_j, the latest end time et_i, the task duration d_i and the benefits p_i of the successful execution of task o_i. The ground station time window set $TW = \{tw_j | j = 1, 2, \cdots, N\}$ contains N available time windows, each of which available time window tw_j contains two attributes, the earliest satellite visible time stw_j and the latest satellite visible time etw_j. Since the latter task cannot be executed immediately after the previous task is completed, there is a transition time $tran$ between tasks. Due to the communication link between the satellite and the ground cannot be established immediately when the satellite and the ground station at the earliest visible time, there is also a given transition time, which is the attitude adjustment time adt.

At the same time, we define a 0–1 binary decision variable x_i to describe the execution of the task. When x_i is 1, it indicates that the task can be executed. And the task cannot be executed when x_i is 0.

As for the objective function, firstly, we consider to maximize the yield of the executable task sequence. Secondly, the total completion rate of the task is also an important evaluation index, so we weight the two to obtain the objective function of the model, such as the formulas (1) and (2) shown:

$$\max f(x) = w_1 \frac{\sum_{i=1}^{M} p_i x_i}{\sum_{i=1}^{M} p_i} + w_2 \frac{\sum_{i=1}^{M} x_i}{M} \tag{1}$$

$$w_1 + w_2 = 1 \tag{2}$$

Constraints of multi-satellite mission scheduling are shown as follows:

(1) Each task can only be executed once:

$$x_i \leq 1 (i \in O) \tag{3}$$

(2) the actual starting time of communication rst should be after attitude adjustment:

$$adt + et_i \leq rst_i \tag{4}$$

(3) there can be no overlap between tasks performed:

$$x_i \leq 1 (i \in O) \tag{5}$$

(4) the time when the task actually starts the communication between satellite and earth should be after the earliest satellite visible time stw_j:

$$stw_j \leq rst_i \tag{6}$$

(5) the completion time of the task should be completed before the minimum between the latest task completion time et_i and the latest satellite visible time etw_j:

$$rst_i + d_i \leq \min\{et_i, etw_j\} \tag{9}$$

In summary, the mathematical model of multi-satellite mission scheduling has been basically completed. Solving the abstract and simplified problem model has important guiding significance for practical problems. This problem has been proved to be an NP-Hard problem. Although the intelligent optimization algorithm cannot guarantee the optimal solution, a high-quality solution can be obtained.

3 Method

We propose an improved genetic algorithm with population perturbation and elimination (GA-PS), which can effectively solve the optimization scheduling problem. By adaptively improving the flow of the classical genetic algorithm and the genetic operator for the multi-satellite TT&C scheduling problem, the shortcomings of the low efficiency and slow convergence of the classical genetic algorithm can be effectively overcome. After each update of the population, the GA-PS will get a new task scheduling sequence. The new task scheduling sequence needs to use the TT&C-TW selection algorithm from [10] to determine whether it can be executed. If it can be executed, there will be get the specific start time and end time. The algorithm can also calculate the target function while obtaining the task execution plan through the task sequence. When GA-PS is used to optimize the population, the optimal individual in the population is taken as the final task sequence when the algorithm terminates and the corresponding task execution plan is obtained.

3.1 GA-PS

To improve the slow convergence of the genetic algorithm and easy to repeat search, we analyze and understand the multi-satellite TT&C scheduling problem and the classical genetic algorithm process. By optimizing the coding method, selection operation, fitness function, crossover operation, mutation operation, population perturbation and elimination strategy, termination condition, etc., the optimization effect and optimization efficiency of genetic algorithm are improved.

The flow chart of the GA-PS is shown in Fig. 1. The following parts will introduce the specific process of improving the genetic algorithm, the operation operator in the genetic algorithm and so on.

(1) Population initialization
Before the genetic algorithm population is optimized by genetic operations such as cross mutation, the population needs to be initialized and the population initialization process needs to ensure the individual differences and diversity. The population initialization in GA-PS was generated by a random method, which ensured that there were significant differences in the gene fragments in each individual in the first generation population.
(2) Coding method
The algorithm maps the actual task number to the coding of the genetic algorithm one by one, ensuring that any allele in each individual represents only the unique task in the scheduling sequence. For example, there is a "1 5 7 9" gene fragment, it means that this part of the gene fragment represents the scheduling sequence of task numbers 1, 5, 7, and 9.
(3) Fitness calculation
The fitness value directly affects the probability that an individual in the population is selected to complete the genetic operation [11]. To closely combine the improved genetic algorithm with the deterministic multi-star measurement and control scheduling problem, the objective function proposed in the previous part is used as the fitness function. According to the task execution scheme obtained by the TT&C-TW selection algorithm, the fitness value is obtained by the objective function.
(4) Selection
Individual selection requires both individuals with larger fitness function values to be more likely to be selected, as well as the possibility that individuals with lower fitness function values are selected in a random manner. This article uses the method of roulette to complete the individual selection process.

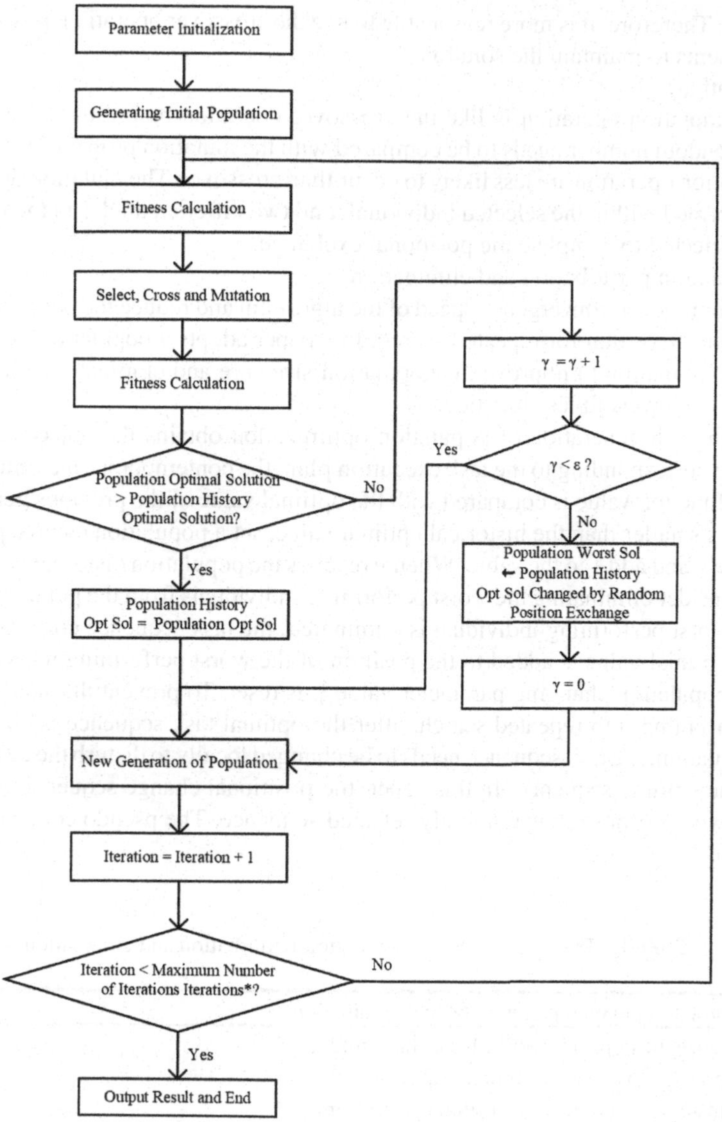

Fig. 1. The flow chart of the GA-PS

(5) Crossover

After selecting one in the population by using the selection method, the random number is compared with the crossover probability p_c to determine whether to perform the cross operation. In GA-PS, cross-operation is performed by means of PMX (partially mapped crossover). Since individual in the population is constructed by real number coding, if two individuals are used for gene segment exchange, a repeated task sequence segment will occur in one individual, and the task sequence at this time violates the constraint that the task is only allowed to be executed once. Therefore, it is more reasonable to use the crossover operation of two internal segments to maintain the solution.

(6) Mutation

The mutation operation is like the crossover operation before the execution that the random number needs to be compared with the mutation probability p_b, but the mutation operation are less likely to occur than crossover. The mutation operation is performed within the selected individual, and two different alleles in the individual are selected to complete the positional exchange.

(7) Population perturbation and elimination

To improve the convergence speed of the algorithm and reduce the possibility of the genetic algorithm for repeated search, this paper adopts a population perturbation and elimination to improve the population structure and eliminate the individuals with the lowest fitness function.

When each generation of population optimization obtains the objective function value corresponding to the task execution plan, the contemporary maximum objective function value is compared with the optimal value of the previous generations. If it is smaller than the historical optimal value, set a population change parameter value γ and add 1 to the value. When γ reaches the population disturbance threshold ε, consider eliminating the worst performing individuals from the population. After the worst performing individual is eliminated, the task sequence corresponding to the optimal value is added to the position of the worst performing individual, and the population changing parameter value γ is reset. To prevent the search process from falling into repeated search, after the optimal task sequence is added to the population, the task sequence needs to be changed locally to disturb the task position in the optimal sequence. In this paper, the positional change sequence is changed by two positions in the randomly selected sequence. The pseudo code is shown in Table 1.

Table 1. The pseudo code of population perturbation and elimination

Algorithm 1 Population perturbation and elimination

```
initialize the population disturbance threshold ε
set population change parameter value γ ← 0
set historically optimal population opt* ← None
for iteration ←1, 2, · · · do
    obtain the optimal of the contemporary Population_iteration opt_iteration
    obtain the worst of the contemporary Population_iteration worst_iteration
    if opt_iteration < opt*
        γ ← γ + 1
        while ε < γ
            worst_iteration ← opt* changed by random position exchange
        end while
    else
        opt* ← opt_iteration
end for
```

(8) Termination
The termination condition of our algorithm setting is that when the number of iterations reaches the preset algebra, the algorithm terminates.

4 Experimental Analysis

4.1 Experimental Design

This part describes experimental examples, experimental environments, comparison algorithms, and evaluation indicators.

Experimental example: The experimental examples are generated based on actual scenarios involving different types of satellites and globally distributed satellite earth stations. The satellite types include low-orbit satellites and high-orbit satellites. The task size ranges from 25–400 tasks.

Experimental environment: The GA-PS and comparison algorithm are completed by MATLAB 2017a on Windows 10 operating system. The hardware support is: Core i7-8750H CPU, 32 GB RAM.

Contrast algorithm: This paper selects three algorithms for comparison experiments. The first algorithm is genetic algorithm (GA) without population perturbation and elimination, the second is adaptive neighborhood local search algorithm (ANLS) from [12], and the third is local search algorithm (LS). Since the proposed algorithm GA-PS in this paper is improved by genetic algorithm, it is regarded as one of our comparison algorithms.

Evaluation indicators: The evaluation indicators are mainly divided into three, the first is the revenue of the task, and the second is the completion rate of the task (TCR). The first indicator focuses on the completion of high-profit tasks, and the second is mainly on the completion of the whole task. The third is the objective function shown in Eq. (1), which weights the completion rate and the profit of the task, and obtains the quality of the overall benefit. In our experiment, we put $w_1 = w_2 = 0.5$.

4.2 Experimental Results

In actual engineering, the multi-star mission planning problem of high-orbit satellites and low-orbit satellites is generally present. So, the experimental process will be carried out in this case. High-orbit satellites have longer timing times and higher profits than low-orbit satellites. At the same time, the difficulty of multi-satellite mission planning for high-orbit satellites has increased correspondingly. During the experiment, we designed seven scenarios of different task sizes. The simulation test was carried out by using the algorithm GA-PS and three comparison algorithms proposed in this paper. The results are shown in Table 2.

As can be seen from Table 2, when the scale of problem is small, the four algorithms all have good performance. But with the increase of the scale, different types of

Table 2. Experiment results for different task sizes

Problems	GA-PS		GA		ALNS		LS	
	Profits	TCR	Profits	TCR	Profits	TCR	Profits	TCR
25	**66**	**1**	66	1	66	1	66	1
50	130	**0.88**	130	0.88	133	0.88	**134**	0.88
75	**237**	**0.8933**	226	0.8933	237	0.8933	216	0.88
100	**291**	**0.90**	291	0.90	291	0.9	290	0.90
125	**305**	**0.848**	300	0.84	298	0.824	298	0.8240
150	**385**	0.8867	380	0.8867	383	**0.88**	371	0.86
175	**456**	**0.8457**	**456**	0.8343	448	0.8229	448	0.8229

algorithms have a decrease in the task completion rate, mainly because of the increased number of tasks caused the conflict between tasks. The algorithm proposed in this paper has advantages in both profits and completion rate.

From a practical point of view, we designed three large-scale task scenarios with 200, 300 and 400 tasks. Through experiment on the profits and completion rate, the result shown in Figs. 2 and 3 is obtained.

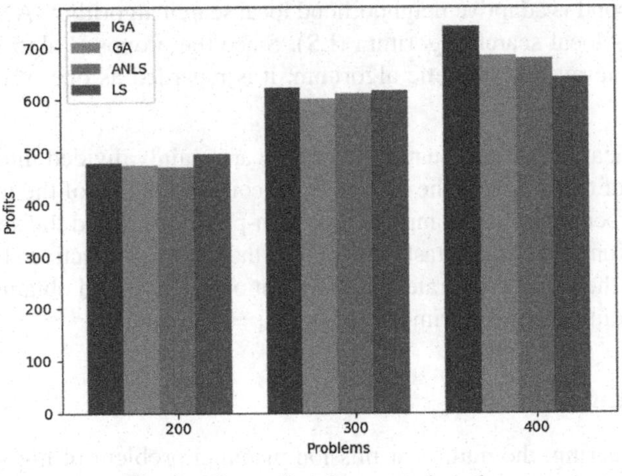

Fig. 2. The profits of large-scale tasks

As can be seen from Fig. 2, in the scenario of large-scale examples, the GA-PS algorithm proposed in this paper obtains the highest benefit compared with other comparison algorithms, and is more suitable for large-scale multi-satellite task scheduling. This situation is especially noticeable when the scale is 400. It is worth noting that as the scale increases, the benefits of the LS algorithm are getting worse than other algorithms. Among these four algorithms, the genetic algorithm has better benefits than the local search algorithm, which reflects the more suitable genetic algorithm for multi-satellite

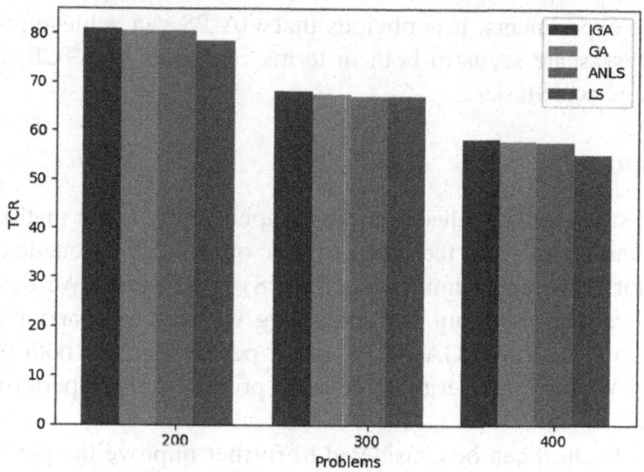

Fig. 3. The TCR of large-scale tasks

task scheduling. Moreover, when compared with the general genetic algorithm GA, GA-PS has achieved good returns. It can be seen that the population perturbation and elimination play an important role in improving the quality of the solution.

It is not difficult to see from Fig. 3 that the four algorithms achieved similar results in the task completion rate, while GA-PS still improved compared with other algorithms. Compared with the results obtained in Fig. 2, GA-PS is more inclined to choose high-yield tasks while ensuring the task completion rate.

Obviously, the objective function value required by the algorithm can be got through formula (1). We applied four algorithms to all the scenes that appeared in the above experiments, and obtained the results shown in Fig. 4. It is easy to see from the figure that the GA-PS has better performance than other algorithms. A performance advantage is especially noticeable in the face of large-scale scenarios.

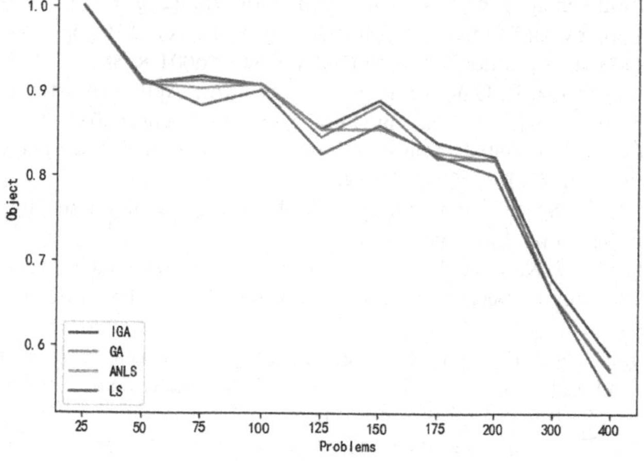

Fig. 4. The result of objective function at different scale tasks

Through the experiments, it is obvious that GA-PS can achieve good scheduling results in any task-scale scenario both in terms of profits and TCR, and it is more prominent in large-scale tasks.

5 Conclusion

Based on multi-satellite task scheduling, this paper establishes a mathematical model by abstracting and simplifying the problem. For our model, a genetic algorithm with population perturbation and elimination (GA-PS) is designed. We designed a series of scenarios to test our algorithm. By comparing with the comparison algorithm GA, ANLS and LS, our algorithm GA-PS has good performance in both profits and task completion rate. When encountering large-scale problems, the superiority of GA-PS is more prominent.

In future research, it can be considered to further improve the performance of the solution by the local search algorithm after the GA-PS solution.

Acknowledgement. This work was supported by the National Natural Science Foundation of China under Grants 71901213, 61473301 and 71690233.

References

1. Li, J., et al.: The TT&C resource scheduling method based on multi-satellite. In: IEEE International Conference on Intelligent Computing & Intelligent Systems. IEEE (2010)
2. Zhang, Z., Zhang, N., Feng, Z.: Multi-satellite control resource scheduling based on ant colony optimization. Expert Syst. Appl. **41**(6), 2816–2823 (2014)
3. Bianchessi, N., et al.: A heuristic for the multi-satellite, multi-orbit and multi-user management of Earth observation satellites. Eur. J. Oper. Res. **177**(2), 750–762 (2007)
4. Gao, K., Wu, G., Zhu, J.: Multi-satellite observation scheduling based on a hybrid ant colony optimization. Adv. Mater. Res. **765–767**, 532–536 (2013)
5. Yang, P., et al.: Heuristic algorithm and conflict-based backjumping algorithm for satellite TT&C resource scheduling. Yuhang Xuebao/J. Astronaut. **6**, 1609–1613 (2007)
6. Chen, Y., Zhang, D., Zhou, M., Zou, H.: Multi-satellite observation scheduling algorithm based on hybrid genetic particle swarm optimization. In: Zeng, D. (ed.) Advances in Information Technology and Industry Applications. LNEE, vol. 136, pp. 441–448. Springer, Heidelberg (2012). https://doi.org/10.1007/978-3-642-26001-8_58
7. Zufferey, N., Amstutz, P., Giaccari, P.: Graph Colouring Approaches for a Satellite Range Scheduling Problem. Kluwer Academic Publishers, Dordrecht (2008)
8. He, R.: Apply constraint satisfaction to optimal allocation of satellite ground station resource. Comput. Eng. Appl. **40**(18), 229–232 (2004)
9. Zhang, T., et al.: Space-ground integrated scheduling based on the hybrid ant colony optimization. Syst. Eng. Electron. (2016)
10. Song, Y., Ma, X., Li, X., Xing, L., Wang, P.: Learning-guided nondominated sorting genetic algorithm II for multi-objective satellite range scheduling problem. Swarm Evol. Comput. **49**, 194–205 (2019)
11. Du, Y., Wang, T., Xin, B., Wang, L., Chen, Y., Xing, L.: A data-driven parallel scheduling approach for multiple agile Earth observation satellites. IEEE Trans. Evol. Comput. (2019). https://doi.org/10.1109/TEVC.2019.2934148
12. Wang, Q., Luo, H., Xiong, J., Song, Y., Zhang, Z.: Evolutionary algorithm for aerospace shell product digital production line scheduling problem. Symmetry **11**, 849 (2019)

A Novel Grey Wolf Optimization Based Combined Feature Selection Method

Haikuan Wang[1], Zhaoyan Hu[1], Zhile Yang[2(✉)], and Yuanjun Guo[2]

[1] Shanghai University, Shanghai 200000, China
[2] Shenzhen Institute of Advanced Technology, Chinese Academy of Sciences, Shenzhen 518055, China
zl.yang@siat.ac.cn

Abstract. In data mining and machine learning area, features targeting and selection are crucial topics in the real world applications. Unfortunately, massive redundant or unrelated features significantly deteriorate the performance of learning algorithm. This paper presents a novel classification model which combined grey wolf optimizer (GWO) and spectral regression discriminant analysis (SRDA) for selecting the most appropriate features. The GWO algorithm is adopted to iteratively update the currently location of the grey wolf population, while the classification algorithm called SRDA is employed to measure the quality of the selected subset of features. The proposed method is compared with genetic algorithm (GA), Jaya, and three recent proposed Rao algorithms also with SRDA as the classifier over a set of UCI machine learning data repository. The experimental results show that the proposed method achieves the lower classification error rate than that of GA and other corresponding methods generally.

Keywords: Grey wolf optimization · Feature selection · Spectral regression discriminant analysis

1 Introduction

Rapid developments of information technology produce a large amount of featured and unfeatured datasets. However, the massive unnecessary data features not only increase the size of the search space, but also significantly reduce classification performance. The process of feature selection aims at selecting the crucial and meaningful features from all available features so as to achieve the purpose of data dimension reduction. The selection is implemented through an optimization process with the specific indicators of the system. It is also a critical part of data preprocessing in pattern recognition. Feature selection effectively eliminates a lot of superfluous and meaningless features so that the model accuracy and the learning algorithm efficiency can be significantly improved [3,5].

The overall process of feature selection can be represented by four parts: generation procedure, evaluation function, stopping criteria, and verification procedure. Generation procedure is the process of searching feature subspace, where

L. Pan et al. (Eds.): BIC-TA 2019, CCIS 1159, pp. 569–580, 2020.
https://doi.org/10.1007/978-981-15-3425-6_45

the search algorithm is divided into three categories: complete search method, heuristic search method, and random search method. Complete search method is divided into exhaustive search [3,9]and non-exhaustive search such as Breadth First Search (BFS). The heuristic search methods [12] consist of sequential forward selection (SFS) [23], sequential backward selection (SBS) [11], bidirectional search(BDS), Plus-L Minus-R Selection (LRS), sequential floating selection and decision tree method (DMT). As for SFS, if a feature has ever been selected, it cannot be deleted throughout the process, even if the features added later make it redundant. The LRS [21] algorithm combines the ideas of SFS and SBS algorithms, which is an improvement of the SFS and SBS algorithms. However, it lacks theoretical guidance on the optimal choice of L and R values. Bidirectional Search (BDS) uses SFS and SBS to start searching from both ends respectively, and both search will stop when they find the same feature subset. Since BDS is a combination of SFS and SBS, as a result, its time complexity is smaller than SFS or SBS, but it also has the disadvantages of both SFS and SBS. SFS is improved by LRS algorithm, which differs from the LRS algorithm in that the L and R values of the sequence float selection are not fixed but changeable. The SFS is divided into Sequential Floating Forward Selection (SFFS) [16], and Sequential Floating Backward Selection (SFBS) [15] according to the search directions. The above search methods are easy to be trapped in the local optimum,in order to effectively solve the aforementioned shortcomings, researchers began to employ the evolutionary algorithm with random search strategy [3,9,10] for solving the feature selection problem.

According to the relationship between evaluation function and classifier, feature selection methods can be currently divided into two categories: filter methods [9] and wrapper methods [3,5]. The main idea of the filter methods is to "score" the features of each dimension, which assigns weights to the features of each dimension, leading the irrelevant features to be quickly eliminated. However, filter methods do not consider the classifier model employed during the process of feature selection, which may result in the selected feature subsets not working very well with subsequent classification algorithms to affect its performance. Therefore, the wrapper methods, combined with classification algorithms, select subsets of features that make the final algorithm to achieve higher performance [3,5,8]. The essence of the wrapper method is a classifier, which evaluates the combination by classification accuracy and compares it with other combinatorial optimization algorithms, regarding the selection of features as an optimization problem. Genetic Algorithm (GA) [4,6,20], particle swarm optimization (PSO) [1], and ant colony optimization (ACO) [19,27] are all frequently-used optimization algorithms for solving feature selection problem.

In this paper, an evaluation function on the basis of classification error rate is proposed by GWO with spectral regression discriminant analysis (SRDA) [2] as the classifier. The Grey Wolf Optimizer (GWO) [13] algorithm is compared against Genetic Algorithm (GA), Jaya algorithm [17], and three RAO algorithms [18] for feature selection on four different datasets. The experimental results show that the GWO obtains the better classification performance compared to the other five algorithms employed in this paper.

This paper incorporates six sections. Section 2 introduces the related work of feature selection. Section 3 describes the detailed information of grey wolf optimizer. Section 4 presents the details of the presented method. Section 4 illustrates the experimental design and results. Eventually, Sect. 5 draws the conclusions.

2 Related Work

2.1 Evolutionary Based Feature Selection Algorithms

Among many feature subset selection methods, swarm intelligence-based optimization algorithms have been extensively developed. Kashef et al. [7] proposed a novel feature selection algorithm based on Ant Colony Optimization (ACO), in which features are used as graph nodes to build graph models. There are two nodes for each feature, one for selecting the feature and the other one for deleting it. Vafaie et al. [22] presented a method that the genetic algorithm (GA) is adopted to identify and select the best subset of features to be used in the traditional rule induction system. Nazir et al. [14] combined PSO and GA to select the most key features to enhance the classification accuracy rate as well as reduce the data size dimension, thereby enhancing the classification performance. Gheyas et al. [4] presented a mixture algorithm called SAGA, combining the simulated annealing and GA with the excellent local search capability of greedy algorithms as well as generalized regression neural networks.

2.2 Feature Selection Problem Formulation

Feature selection can be modeled as a single objective problem to minimize the classification error rate in wrapper methods. The fitness function (See Eqs. 1, 2) is to minimize the classification error rate attained by the selected feature subset in the process [24].

$$\text{Fitness}_1 = \text{ErrorRate} \tag{1}$$

$$\text{ErrorRate} = \frac{A_1 + A_2}{A_1 + A_2 + B_1 + B_2} \tag{2}$$

where A_1, A_2, B_1 and B_2 stand respectively for false positives, false negatives, true positives and true negatives [24,26].

The representation of each individual in the intelligent swarm algorithm is an m–bit binary string, where m denotes the dimension of the search space and the length of each individual in the population. Each binary string represents an individual in a population, where '1' indicates that the feature is selected, '0' indicates that the feature of the location is not selected [24,26]. In some multi-objective optimization problem formulations, the fitness function contains two characteristics: error rate and number of features. Since Eq. (1) does not get the minimum features number, and the selected features by the optimization algorithm may still contain redundant and unrelated features, and we hope that the same classification accuracy can be obtained by fewer selected features. In order to solve this problem, Xue et al. [24] presented two fitness functions so as to maximize classification performance and minimize feature number. One is represented by Eqs. 3 and 4, while another one is represented by Eq. (5).

$$\text{Fitness}_2 = \alpha_t \cdot \underset{\text{\# All Features}}{\text{\# All Features}} + (1 - \alpha_t) \cdot \frac{\text{ErrorRate}}{\text{Error}_0} \tag{3}$$

$$\alpha_t = \alpha_{\max} \cdot \frac{t}{T} \tag{4}$$

$$\text{Fitness}_3 = \begin{cases} \text{ErrorRate, Stage 1} \\ \alpha \cdot \frac{\text{\# Feates}}{\text{\# All Features}} + (1-\alpha) \cdot \frac{\text{ErrorRate}}{\text{Error}_0}, \text{ Stage 2} \end{cases} \tag{5}$$

where α is constant term and $\alpha \in [0, 1]$. $\alpha_t \in [0, 1]$, t denotes the tth iteration in the search process. # Features indicates the number of the selected features. # All Features represents the number of all the original features. Error-rate is the classification error rate achieved by the selected feature subset. Error_0 is the error rate achieved by employing all available features for classification on the training set. $\alpha_{\max}$ is the predefined maximum value of α_t and $\alpha_{\max} \in [0, 1]$. T denotes the maximum number of iterations in evolutionary computation [24,26].

3 Grey Wolf Optimization

3.1 GWO Algorithm Principle

Gray wolves tend to be the most powerful creature in the food chain, normally living in groups of 5 to 12 individuals. In the hunting process, grey wolves have an rigorous social hierarchy, which makes them have a clear and cooperative division of labor and hunt. In the GWO algorithm, the most powerful grey wolf is recorded as α, taking charge of the management of the pack and the decision-making part during the hunting process. The rest of the grey wolves are known in order of social class as β, δ and ω respectively.

The wolf known as α is the leader of the whole grey wolf pack in the hunting process, which is the wisest and most capable individual. The β wolf and the δ wolf are second-best adaptable individuals, which help the α wolf manage the grey wolf pack and make hunting decisions, also being the candidates for the α wolf. The remaining wolves are known as ω wolves, whose primary function is to balance the internal relations of the grey wolf population and to assist the wolves known as α, β and δ to attack preys. Throughout the hunt, the α wolf leads the pack in the progress of searching, tracking and approaching their preys, when the distance from the prey is as small as possible, the wolves of β and δ will attack the prey under the command of the α wolf, calling on the circumambient wolves to attack the prey simultaneously. As the prey moves, the wolves will form a circle around the prey and move with it until they catch the prey.

3.2 GWO Algorithm Description

The whole hunting process in the GWO can be divided into three stages: encirclement, pursuit and attack [6], with the prey to be caught eventually. The specific algorithm is described as follows:

(1) Encircling
After determining the location of the prey, the wolves first need to surround the prey, during which the distance between the prey and the grey wolf can be expressed in Eqs. 6 and 7.

$$\boldsymbol{D} = |\boldsymbol{C} \cdot \boldsymbol{X_{prey}}(\boldsymbol{t}) - \boldsymbol{X_{wolf}}(\boldsymbol{t})| \tag{6}$$

$$\boldsymbol{X_{wolf}}(\boldsymbol{t}+\mathbf{1}) = \boldsymbol{X_{prey}}(\boldsymbol{t}) - \boldsymbol{A} \cdot \boldsymbol{D} \tag{7}$$

where D denotes the distance between the wolf and its prey, t is the number of iterations. $X_{prey}(t)$ shows the preys location after iterations (i.e. the location of the optimal solution). $X_{wolf}(t)$ is the grey wolf location after t iterations (i.e. the location of the potential solution). A and C are coefficient factors, and the calculation formula is shown as Eq. (8):

$$\begin{aligned} \boldsymbol{A} &= 2\boldsymbol{a} \cdot \boldsymbol{r_1} - \boldsymbol{a} \\ \boldsymbol{C} &= 2 \cdot \boldsymbol{r_2} \end{aligned} \tag{8}$$

Where a decreases linearly from 2 to 0 as the number of iterations increases, r_1 and r_2 are random numbers in [0, 1].

(2) Hunting
Once the prey is surrounded, the β and δ wolves led by α wolves will hunt down the prey. In the process of chasing, the wolves individual position of the pack will change as the prey escapes, and the new position of α, β and δ wolves will be used to determine the location of the prey then. The Wolf pack position updating equation is represented in Eqs. (9)–(11).

$$\begin{cases} \boldsymbol{D_{alpha}} = |\boldsymbol{C_1 X_{alpha}}(\boldsymbol{t}) - \boldsymbol{X}(\boldsymbol{t})| \\ \boldsymbol{D_{beta}} = |\boldsymbol{C_2 X_{beta}}(\boldsymbol{t}) - \boldsymbol{X}(\boldsymbol{t})| \\ \boldsymbol{D_{delta}} = |\boldsymbol{C_3 X_{delta}}(\boldsymbol{t}) - \boldsymbol{X}(\boldsymbol{t})| \end{cases} \tag{9}$$

$$\begin{cases} \boldsymbol{X_1} = \boldsymbol{X_{alpha}}(\boldsymbol{t}) - \boldsymbol{A_1 D_{alpha}} \\ \boldsymbol{X_2} = \boldsymbol{X_{beta}}(\boldsymbol{t}) - \boldsymbol{A_2 D_{beta}} \\ \boldsymbol{X_3} = \boldsymbol{X_{delta}}(\boldsymbol{t}) - \boldsymbol{A_3 D_{delta}} \end{cases} \tag{10}$$

$$X(t+1) = \frac{X_1 + X_2 + X_3}{3} \tag{11}$$

Where D_{alpha}, D_{beta}, D_{delta} represents the distance between α, β, δ wolves and ω wolves respectively.

(3) Attacking

Attacking is the last step of the hunting process, during which wolves attack and capture the prey to get the optimal solution, and the value of a is decreased to achieve this implementation in the process seen in Eq. 8. When the value of a falls from 2 to 0 linearly, the corresponding value of A also changes in the interval $[-a, a]$ [25]. Furthermore, when $|A| \leq 1$, it indicates that the wolves next position will be closer to the prey. When $1 < |A| \leq 2$, the wolves will move away from the prey, causing the GWO algorithm to lose the optimal solution position, falling into local optimum.

3.3 The Proposed Method and Counterparts

A novel method on the basis of GWO algorithm combined with SRDA classifier to search for optimal subsets in search space is presented in the paper. Through spectrum analysis, the optimal cutting point on the data point is found, and the optimal solution is attained by putting forward the specific eigen problem. We can obtain the basic equation of SRDA theoretically by two-steps method, and find the most fitting method in the sense of least square. In addition, only some regularized least squares problems need to be solved, which has a great advantage in saving memory space and computing time.

Therefore, it is more advantageous than other learning algorithms in dealing with large-scale data. The overall proposed algorithm flow chart is shown in Fig. 1. The classification error rate is used as the evaluation function to measure the quality of the selected subsets. Through spectral analysis, SRDA transforms discriminant analysis into a regression framework, which is convenient for efficient calculation and the use of regularization technology. The detailed procedure steps is given in Table 1.

The proposed method is compared with GA, Jaya, and three Rao algorithms. The population by GA is initialized at the beginning, of which the each chromosome represents the probable solution set of the problem, which consists of a certain number of chromosomes encoded by genes. Each individual is, in fact, a characteristic entity of chromosomes. Jaya and three RAO algorithms can achieve good concentration and diversity in the search process, of which the details of variable definition is given in Table 2, and the update formulas of the four algorithms are given in Table 3.

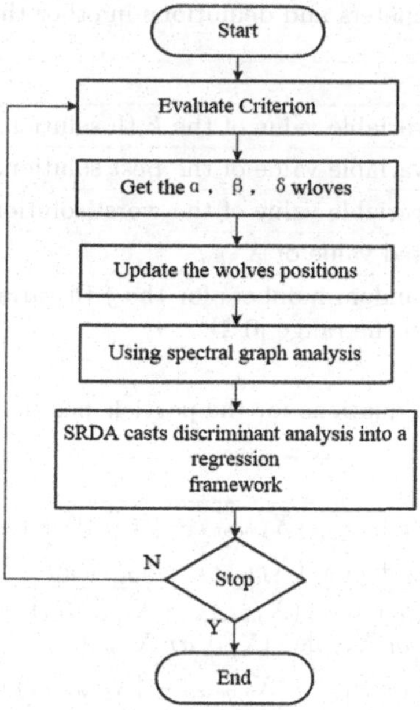

Fig. 1. The overall algorithm flow chart

Table 1. Procedural steps

Step 1	Initialize N, D, t, *pently* and the grey wolf population $X = (X_1, X_2, \ldots, X_N)$, the location of each grey wolf $X_i = (x_{i1}, x_{i2}, \ldots, x_{iD})^T$, $i = 1, 2, ..., N$;
Step 2	Compute the fitness f_i of every candidate individual. The positions of first three grey wolf individuals with fitness values are denoted as X_α, X_β , X_δ, and the minimum fitness value is defined as the best solution
Step 3	Compute the distance between the rest of individuals ω and X_α, X_β, X_δ, according to Eq. (9), and update the positions of α, β, δ wolves and the prey according to Eqs. (10) and (11)
Step 4	Update the values of a, A, C according to formula (8);
Step 5	When the algorithm reaches the maximum number of iterations t, the algorithm ends and the optimal solution X_α is output. Otherwise, return Step 2

4 Experimental Design

4.1 Datasets and Comparison Techniques

The performance of proposed GWO_SRDA method is tested on different datasets, four datasets with different features numbers (from 22 to 256), classes

Table 2. Parameters and definitions in other three algorithms

Parameters	Definitions
$X_{j,k,i}$	The j th variable value of the k th solution in the i th iteration
$X_{j,\text{best},i}$	The j th variable value of the best solution in the i th iteration
$X_{j,\text{worst},i}$	The j th variable value of the worst solution in the i th iteration
$X'_{j,k,i}$	The updated value of $X_{j,k,i}$
$r_{1,j,i}, r_{2,j,i}$	The two random numbers for the j th variable during the i th iteration in the range $[0, 1]$

Table 3. Equations for the particle position updating

Algorithm	Equation
Jaya	$X'_{j,k,i} = X_{j,k,i} + r_{1,j,i}\left(X_{j,best,i} - \lvert X_{j,k,i}\rvert\right) + r_{2,j,i}\left(X_{j,worst,i} - \lvert X_{j,k,i}\rvert\right)$
RAO_1	$X'_{j,k,i} = X_{j,k,i} + r_{1,j,i}\left(X_{j,best,i} - X_{j,worst,i}\right)$
RAO_2	$X'_{j,k,i} = X_{j,k,i} + r_{1,j,i}\left(X_{j,best,i} - X_{j,worst,i}\right) + r_{j,worst,i}\big) + r_{2,j,i}\left(\lvert X_{j,k,i}, \, or \, X_{j,l,i}\rvert - \lvert X_{j,l,i} \, or \, X_{j,k,i}\rvert\right.$
RAO_3	$X'_{j,k,i} = X_{j,k,i} + r_{1,j,i,i}\left(X_{j,\text{best},i} - \lvert X_{j,\text{worst},i}\right) + r_{2,j,i}\left(\lvert X_{j,k,i}, \, or \, X_{j,l,i}\rvert - (X_{j,l,i}, \, or \, X_{j,k,i})\right)$

Table 4. Description of the datasets employed

Datasets	Features	Classes	Instances
USPS	256	10	9298
Parkinsons	22	2	195
CMC	9	3	1473
Semeion	256	80	1593

(from 2 to 26), and instances (from 195 to 9298), named Parkinsons, Semeion, USPS and CMC respectively, are used as the benchmark datasets. All data sets are collected from UCI[1] machine learning repository. Tables 4 summarizes the characteristics of the chosen datasets respectively.

As for the justification of data selection, the main principle is to choose datasets with different instances or features to measure the robustness and practicability of the proposed method. As a result, the experimental results can be compared with other methods based on different evolutionary computation (EC) techniques, the results of which are shown in the next section.

[1] http://archive.ics.uci.edu/ml/index.php.

4.2 Results and Analysis

In programming, we set the parameter number of iterations Gm to be 200 and the population size Np to be 30. The initial population POP consists of 0 or 1 to indicate whether these features are selected or not. The parameter $Tr(Gm, :, j)$ represents the optimal feature selection strategy with the minimum classification error rate. For the experiments on datasets above, Jaya, GA and three algorithms in RAO proposed recently are used as comparison methods. The SRDA is used as a classifier to test their performance by using different EC techniques mentioned above. The feature selection method on the basis of GA is very common in previous work, while Jaya is a recent proposed algorithm and has very limited used for feature selection in recent years. Meanwhile, three algorithms proposed recently named by RAO [18] have been used as comparison methods, between which the difference is the way of population iteration. In the paper, these EC methods all employ the fitness function that only takes the classification error rate into account to choose the most appropriate features numbers so that the classification performance can be improved accurately.

Figure 2 shows the experimental results attained by all the compared techniques. As shown in the experimental curve, GWO gets better performance than other 5 techniques. For USPS dataset, GWO achieves the classification error rate of 10.2% using 61 features, while RAO_3 and RAO_2 all use 134 features to get the error rate of 10.9%. The performance of Jaya and RAO_1 is only slightly worse than that of RAO_2 and RAO_3 but better than that of GA, with Jaya using 122 features and RAO_1 using 145 features. For Parkinsons dataset, the performance of GWO with the classification error rate of 14% is the same as that of RAO_1 but better than that of the other three algorithms, with GWO using 7 features among all 22 features. For Semeion dataset, GWO outperforms other 5 algorithms with respect to the features numbers and the classification performance, with the lowest error rate of 10.5% and the least features of 89 among all 256 features, demonstrating that the GWO achieves the purpose of decreasing the classification error rate without damaging feature reduction. For CMC dataset, the classification error rate of all 6 algorithms are larger than 49%, with GWO using least features and achieving best accuracy among all 6 algorithms. Detailed information about the classification error rate and the number of features for each algorithm is given in Table 5.

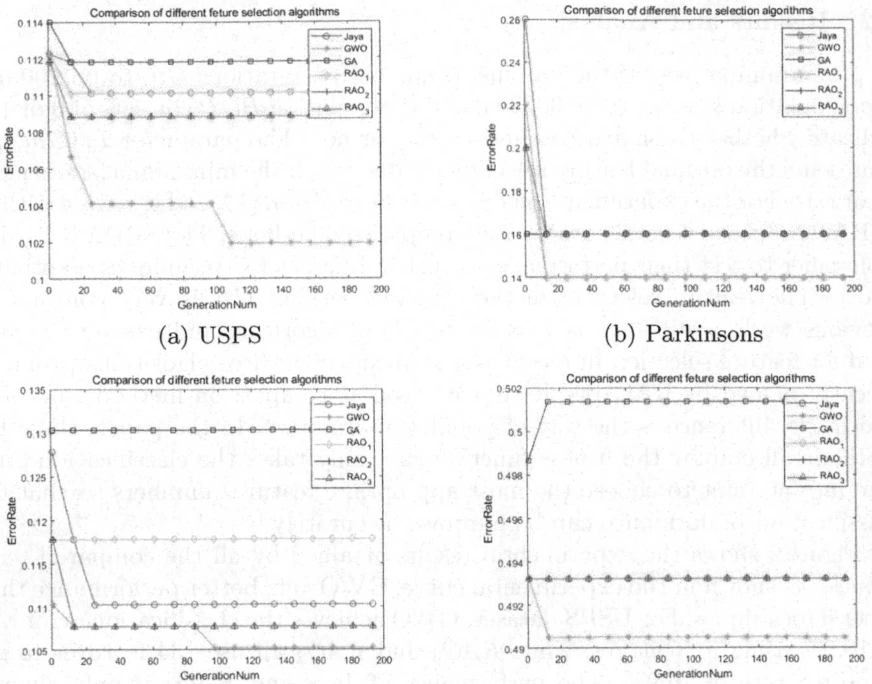

(a) USPS (b) Parkinsons

(c) Semeion (d) CMC

Fig. 2. The experimental results of six algorithms over four datadets

Table 5. Experimental results on four datasets

DataSets	Performance	Method					
		Jaya	GWO	GA	RAO1	RAO2	RAO3
USPS	Error rate	10.92	10.19	11.35	10.92	10.54	10.54
	Num feature	122	61	135	145	134	134
Parkinsons	Error rate	16	14	16	14	16	16
	Num feature	14	7	9	6	8	8
Semeion	Error rate	11.03	10.53	13.03	11.78	10.78	10.78
	Num feature	129	89	129	104	123	123
CMC	Error rate	49.32	49.05	50.14	49.32	49.32	49.32
	Num feature	6	5	5	6	6	6

5 Conclusions

This paper proposed a GWO based combined feature selection algorithm and compared the performance results of several recent proposed swarm intelligence optimization algorithms. The experimental results show that the grey wolf optimization algorithm outperforms other five counterparts in improving the classification accuracy and the feature reduction. It is only a preliminary research for the algorithm application in feature selection. Further work will be addressing more novel optimization algorithms and choose appropriate learning classifiers for each optimization algorithm for feature selection, so as to improve classification accuracy and convergence speed to a large extend.

Acknowledgments. Key Project of Science and Technology Commission of Shanghai Municipality under Grant No. 16010500300, China NSFC under grants 51607177, 61877065, China Postdoctoral Science Foundation (2018M631005) and Natural Science Foundation of Guangdong Province under grants 2018A030310671.

References

1. Agrafiotis, D.K., Cedeno, W.: Feature selection for structure-activity correlation using binary particle swarms. J. Med. Chem. **45**(5), 1098–1107 (2002)
2. Cai, D., He, X., Han, J.: SRDA: an efficient algorithm for large-scale discriminant analysis. IEEE Trans. Knowl. Data Eng. **20**(1), 1–12 (2008)
3. Dash, M., Liu, H.: Feature selection for classification. Intell. Data Anal. **1**(3), 131–156 (1997)
4. Gheyas, I.A., Smith, L.S.: Feature subset selection in large dimensionality domains. Pattern Recogn. **43**(1), 5–13 (2010)
5. Guyon, I., Elisseeff, A.: An introduction to variable and feature selection. J. Mach. Learn. Res. **3**, 1157–1182 (2003)
6. Jeong, Y., Shin, K.S., Jeong, M.K.: An evolutionary algorithm with the partial sequential forward floating search mutation for large-scale feature selection problems. J. Oper. Res. Soc. **66**(4), 529–538 (2015)
7. Kashef, S., Nezamabadipour, H.: A new feature selection algorithm based on binary ant colony optimization, pp. 50–54 (2013)
8. Kohavi, R., John, G.H.: Wrappers for feature subset selection. Artif. Intell. **97**(1), 273–324 (1997)
9. Liu, H., Motoda, H., Setiono, R., Zhao, Z.: Feature selection: an ever evolving frontier in data mining. In: JMLR: Workshop and Conference Proceedings, vol. 10, pp. 4–13 (2010)
10. Liu, H., Yu, L.: Toward integrating feature selection algorithms for classification and clustering. IEEE Trans. Knowl. Data Eng. **17**(4), 491–502 (2005)
11. Marill, T., Green, D.M.: On the effectiveness of receptors in recognition systems. IEEE Trans. Inf. Theory **9**(1), 11–17 (1963)
12. Min, F., Hu, Q., Zhu, W.: Feature selection with test cost constraint. Int. J. Approx. Reason. **55**(1), 167–179 (2014)
13. Mirjalili, S., Mirjalili, S.M., Lewis, A.: Grey wolf optimizer. Adv. Eng. Softw. **69**, 46–61 (2014)
14. Nazir, M., Majidmirza, A., Alikhan, S.: PSO-GA based optimized feature selection using facial and clothing information for gender classification. J. Appl. Res. Technol. **12**(1), 145–152 (2014)
15. Oduntan, I.O., Toulouse, M., Baumgartner, R., Bowman, C.N., Somorjai, R.L., Crainic, T.G.: A multilevel tabu search algorithm for the feature selection problem in biomedical data. Comput. Math. Appl. **55**(5), 1019–1033 (2008)

16. Pudil, P., Novovicova, J., Kittler, J.: Floating search methods in feature selection. Pattern Recogn. Lett. **15**(11), 1119–1125 (1994)
17. Rao, R.: Jaya: a simple and new optimization algorithm for solving constrained and unconstrained optimization problems. Int. J. Ind. Eng. Comput. **7**(1), 19–34 (2016)
18. Rao, R.V.: Rao algorithms: three metaphor-less simple algorithms for solving optimization problems. Int. J. Ind. Eng. Comput. **11**(1), 107–130 (2020)
19. Santana, L.E.A., Silva, L., Canuto, A.M.P., Pintro, F., Vale, K.O.: A comparative analysis of genetic algorithm and ant colony optimization to select attributes for an heterogeneous ensemble of classifiers, pp. 1–8 (2010)
20. Siedlecki, W.W., Sklansky, J.: A note on genetic algorithms for large-scale feature selection. Pattern Recogn. Lett. **10**(5), 335–347 (1989)
21. Sun, J., Zhou, M., Ai, W., Li, H.: Dynamic prediction of relative financial distress based on imbalanced data stream: from the view of one industry. Risk Manage. **21**, 1–28 (2018)
22. Vafaie, H., De Jong, K.: Genetic algorithms as a tool for feature selection in machine learning. In: International Conference on Tools with Artificial Intelligence, pp. 200–203 (1992)
23. Whitney, A.W.: A direct method of nonparametric measurement selection. IEEE Trans. Comput. **20**(9), 1100–1103 (1971)
24. Xue, B., Zhang, M., Browne, W.N.: New fitness functions in binary particle swarm optimisation for feature selection, pp. 1–8 (2012)
25. Xue, B., Zhang, M., Browne, W.N.: Particle swarm optimization for feature selection in classification: a multi-objective approach. IEEE Trans. Syst. Man Cybern. **43**(6), 1656–1671 (2013)
26. Xue, B., Zhang, M., Browne, W.N., Yao, X.: A survey on evolutionary computation approaches to feature selection. IEEE Trans. Evol. Comput. **20**(4), 606–626 (2016)
27. Yan, Z., Yuan, C.: Ant colony optimization for feature selection in face recognition. In: Zhang, D., Jain, A.K. (eds.) ICBA 2004. LNCS, vol. 3072, pp. 221–226. Springer, Heidelberg (2004). https://doi.org/10.1007/978-3-540-25948-0_31

Improved Discrete Artificial Bee Colony Algorithm

Wanying Liang(✉), Shuo Liu, Kang Zhou, Shiji Fan, Xuechun Shang, and Yanzi Yang

School of Math and Computer, Wuhan Polytechnic University, Wuhan 430023, Hubei, China
lwy000208@163.com

Abstract. Grain is an important economic and strategic material of the country. In grain transportation, it is necessary to consider the running time, vehicle number, path length and other factors at the same time, which is a typical multi-objective problem, but also a NP-Hard problem. In this paper, an Artificial Bee Colony algorithm is introduced to solve the routing problem of grain transportation vehicles with multi-objective and time windows. Combined with the practical problems of grain transportation, the standard Artificial Bee Colony algorithm is improved in four aspects: population initialization, domain search mode, bulletin board setting and scout bee search mode, and a Multi-objective Artificial Bee Colony algorithm is proposed by using the strategy of first classification and then sorting and final iteration. The proposed algorithm is compared with other algorithms by using the standard test set in Solomon database. The results show that the Multi-objective Discrete Artificial Bee Colony algorithm has great advantages in solving the routing problem of grain transportation vehicles with time windows.

Keywords: Grain transportation · Multi-objective · Search domain · Discrete artificial bee colony algorithm

1 Foreword

Food is the root of all things, reasonable use of vehicles to transport grain is the top priority. Therefore, the research on the vehicle route problem of grain transport is of great significance. The VRP (Vehicle Routing Problem) was first raised by Dantzig and Ramser [1] in 1959. VRPTW (Vehicle Routing Problems with Time Windows) is an extension of VRP problems, which is more consistent with the actual problems encountered in daily life. Because VRPTW is a NP problem [2], the research of VRPTW is mainly focused on heuristic algorithm, such as Genetic algorithm [3], Ant Colony algorithm [4], Simulated Annealing algorithm [5] and so on. In 1985, Schaffer [11] proposed a vector evaluation genetic algorithm for the first time in combination with multi-objective optimization. In 1989, Goldberg, in his work <Genetic Algorithm for Search, Optimization, and

L. Pan et al. (Eds.): BIC-TA 2019, CCIS 1159, pp. 581–597, 2020.
https://doi.org/10.1007/978-981-15-3425-6_46

Machine Learning> [12], the Pareto theory and the evolutionary algorithm in economics are combined to solve the new idea of multi-objective optimization.

In recent years, the ABC (Artificial Bee Colony) algorithm, as an emerging intelligent algorithm, was first proposed by the Turkish scholar Karboga [6] in 2005, and successfully applied to the problem of solving the function optimization, because it has the advantages of less control parameters, easy control and high convergence speed, good stability and high efficiency are popular by the great scholars. In addition, the ABC algorithm is widely used in the VRP, and Su and others [19] proposed an improved ABC algorithm and applied to solve the traveling salesman problem. At present, the research on artificial bee swarm algorithm has become more and more mature, and there are more and more research on its improved optimization. In 2016, Yu [13] first used ABC algorithm to solve the basic model of VRPTW, and improved it, and introduced a Cluster intelligent algorithm-ABC algorithm as the solution method of the model. In 2018, Jin [7] has good global search ability for ABC algorithm, but there are problems such as easy early maturity, weak local search ability and low precision of optimization solution. Based on the multi-elitist ABC algorithm inspired by particle swarm, a new improved ABC algorithm based on single purity is proposed. In 2018, in order to improve the convergence performance of ABC algorithm, Liang [9] applied the most rapid decrease method to improve the basic ABC algorithm, and proposed the improved ABC algorithm based on the steepest descent method. The algorithm combines the steepest descent method with the modern intelligent optimization ABC algorithm to make full use of the steepest descent method of the fast local search ability and the global optimization ability of ABC algorithm. In 2019, Zhao [8] proposed a hybrid ABC algorithm based on cat swarm, in order to improve the optimization performance of the algorithm through a new search strategy and search process. In 2019, Chao [10] designed an improved multi-objective ABC algorithm based on Knee Points. In the same year, Aslan [11] studies the transition control mechanism of a ABC algorithm. Dervis [12] uses ABC algorithm to find the sequence of DNA. The improvement and optimization of the above ABC algorithm prove the popularity of the algorithm in various fields, and prove that its application is very high, which provides the direction and power for the implementation of ABC algorithm on MOVRPTW. However, the standard ABC algorithm also has some shortcomings, such as easy to fall into local optimization, slow convergence speed, low search accuracy, low population diversity and so on, which greatly affects the practical application of ABC algorithm.

In order to solve the above problems, this paper proposes a multi-objective discrete artificial bee colony(MODABC) algorithm by improving the search method in the field of bee colony and the iterative method of detecting bees, combined with the practical problem of VRPTW, and by the simulation experiment verifies the effectiveness of MODABC algorithm in solving the MOVRPTW (multi-objective vehicle routing problem with time window). Firstly, the basic concept of multi-objective problem and the mathematical model of MOVRPTW problem are introduced. Secondly, based on introducing the standard ABC algorithm, combined with MOVRPTW in grain transportation, and improved DABC algorithm and an improved MODABC algorithm for MOVRPTW is proposed. Finally, many simulation experiments are compared with other algorithms. The experimental results show that the improved MODABC algorithm can get better results in solving the MOVRPTW problem of grain transportation.

2 Fundamental Theory

2.1 Basic Concept of Multi-objective Problem

As the name implies, the multi-objective problem is that there is more than one decision variable, but a variety of cases need to be considered. In order not to lose generality, assume a multi-objective problem that has n decision variables and m target variables [14] is represented as follows.

$$miny = \begin{cases} g_i(x) \leq 0, i = 1, 2, \cdots, u \\ h_j(x) \leq 0, i = 1, 2, \cdots, v \end{cases} \tag{1}$$

Of which, $x = (x_1, \cdots, x_n) \in X \subset R^n$ is the n dimension decision vector. X is the n dimension decision space, $y = (y_1, \cdots, y_n) \in Y \subset R^m$ for the target vector of m dimension. For the target space of dimension. Objective function $F(x)$ defined m mapping functions from decision space to target space. $g_i(x) \leq 0 (i = 1, 2, \cdots, u)$ defined u inequality constraint. $h_j(x) = 0$, $j = 1, 2, \cdots, v$ defined v equality constraints. On the basis, the optimal solution or non-inferior optimal solution in multi-objective optimization problem can be defined as follows [15].

Definition 1 (Pareto Optimal Solution): A solution is $X^* \in X_m$ is referred to as the Pareto optimal solution (or unencumbered solution), only if meet the following conditions:

$$\neg \exists x \in X_m : x > x^* \tag{2}$$

Definition 2 (Pareto optimal solution set): Pareto optimal solution set is the set of all Pareto optimal solutions, defined as follows:

$$P^* \in \{x^* \mid \neg \exists \in X_m : x \succ x^*\} \tag{3}$$

Definition 3 (Pareto frontier): The surface composed of the target vector corresponding to all the Pareto optimal solutions in Pareto optimal solution set P^* is called the Pareto front surface PF^*:

$$PF^* = \{F(x^*) = f_1(x^*), f_{12}(x^*), \cdots, f_{m1}(x^*)^T \mid x^* \in P^*\} \tag{4}$$

For the solution of a multi-objective problem, there are two criteria for evaluation: one is that the solution of the target problem is as close to the Pareto front as possible, and the other is that the distribution of these solutions is as uniform as possible.

2.2 Mathematical Model of MOVRPTW in Grain Transportation

The grain transportation MOVRPTW problem can be described as: the distribution center is provided with m vehicle for distributing grain $n(n \geq k)$ for grain supply points, $V = k$, $k = 1, 2, \cdots, m$, where, m for the number of vehicles to be determined, the load capacity of the vehicle k is Q. In order to serve n grain supply points, the set of the grain supply point is $C = i$, $i = 0, 1, \cdots, n$, and $i = 0$ is the distribution center. The demand for goods at the grain supply point i is q_i, $q_0 = 0$, and the time window for the grain supply point i to allow the service is $[a_i, b_i]$. The distance from the grain supply point i to the grain supply point c_{ij}, the travel time is t_{ij}. Set s_{ik} for the time of the vehicle k to reach the grain supply point that $s_{ik} \in [a_i, b_i]$. How to plan the distribution route so that the number of vehicles assigned m as few as possible, and the total driving distance of the vehicle is as short as possible. Set the definition of x_{ijk} as follows:

$$x_{ijk} = \begin{cases} 1 \ If \text{ vehicle } k \text{ visit } grain \text{ supply } point \text{ i } and \text{ visits } food \text{ supply } point \text{ j} \\ 0 \ Or \end{cases} \tag{5}$$

Then the multi-objective VRPTW objective function is defined as:

$$minZ_1 = \sum_{k \in V} \sum_{i \in C} \sum_{j \in C} C_{ij} x_{ijk} \tag{6}$$

$$minZ_2 = \sum_{k \in V} \sum_{j \in C} x_{0jk} \tag{7}$$

The constraints are:

$$\sum_{k \in V} \sum_{j \in C} x_{ijk} = 1, \forall i \in C \tag{8}$$

$$\sum_{i \in C} q_i \sum_{j \in C} x_{ijk} \leq Q, \forall k \in V \tag{9}$$

$$\sum_{j \in C} x_{0jk} = 1, \forall k \in V \tag{10}$$

$$\sum_{i \in C} x_{ihk} - \sum_{j \in C} x_{hjk} = 0, \forall h \in C, \forall k \in V \tag{11}$$

$$\sum_{i \in C} x_{0jk} = 1, \forall k \in V \tag{12}$$

$$S_{ik} + t_{ij} - K(1 - x_{ijk}) \leq S_{jk}, \forall i, j \in C, \forall k \in V \tag{13}$$

$$a_i \leq S_{ik} \leq b_i, \forall i \in C, \forall k \in V \tag{14}$$

Among, formulas (6) and (7) represent the two objective functions of minimizing the total driving distance of the target vehicle and minimizing the number of vehicles distributed by the target. Formula (8) indicates that each grain supply points need to be served by the vehicle, but only once. Formula (9) indicates that the vehicle does not exceed the maximum load. And formulas (10), (11) and (12) indicate that each vehicles start from the distribution center 0 and delivers services to different customers, and finally returns to the distribution center 0. Formula (13) indicates that if the vehicle k is from the grain supply point i to the grain supply point on the way, it can't ahead time $S_{ik} + t_{ij}$ arrive at grain supply point j (k is a big coefficient). Formula (14) represents the customer's service time window.

2.3 Discrete ABC Algorithm

Setting parameters: the number of foods = the number of lead bees = the number of follow bees (S), the maximum number of domain search is limit, maximum cycles N. The algorithm steps are as follows: generate the initial solution set X_{ij}, $i \in \{1, 2, \cdots, SN\}$, $j \in \{1, 2, \cdots, D\}$. And then, calculating the fitness value of the each solutions X_{ij}, set$i = 1$, $j = 1$, calculate the fitness value of food X_{ij} and calculate the probability value of P_{ij}, the follow bees select food according to P_{ij} and carry on the domain search to produce the new solution V_{ij}, calculate its fitness value, record the best solution so far $j = j + 1$,$i \leq i + 1$, after the limit cycles, it is determined whether there is a solution to be lost or not. If it exists, the detect bees produces a new solution X_{ij} to replace according to the following formula.

$$X_j^i = X_{min}^j + rand(0,1)(X_{max}^j - X_{max}^j) \tag{15}$$

If $j < N$, set $j = 1$. ABC algorithm needs fewer parameters, which is one of the advantages of the algorithm. The home page set that the number of food sources, the number of lead bees and follow bees are the same N, the limited number of food sources is l, MCN is the maximum number of iterations of the whole ABC algorithm, the number of customer points is D, cycle is the variable to determine whether or not to reach the maximum number of iterations, and S is the variable to judge whether or not to reach the limit number of times leading bees to detect bees, limit is the limited number of times that lead bees to detect bees. The flow chart of the standard discrete ABC algorithm for solving VRPTW is as follows (Fig. 1).

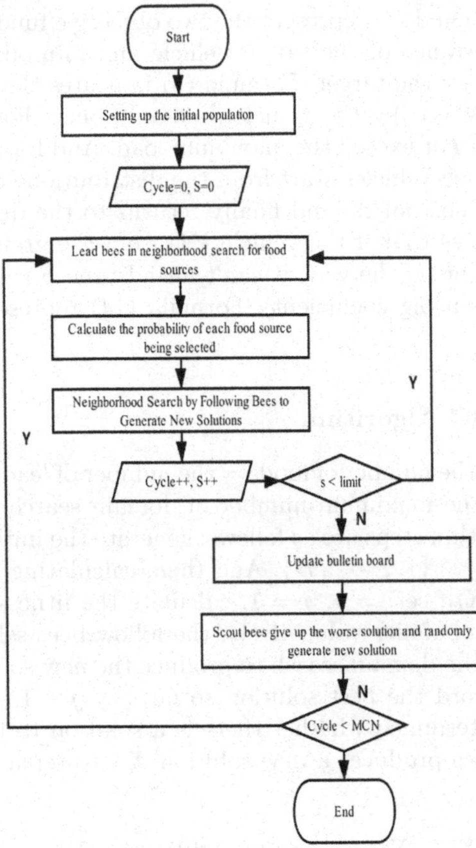

Fig. 1. Flow chart of standard discrete ABC algorithm

3 Improvement of Multi-objective ABC Algorithm Based on Path Optimization Problem with Time Window in Grain Transportation

3.1 Establishment of Initial Populations

The initial population is constructed by random insertion. First, a customer point is randomly selected, the customer point is inserted into the first path, and then the next customer point is randomly selected. If the customer point has been selected, the latter customer point of the customer point is selected, and if all the customer points behind the customer point have been selected, then the first unselected customer point from the starting point is selected to insert into the path, and if the time window and load constraints are not satisfied, adding an additional path and the customer point is inserted into the new path. Thus, the random insertion of the established population can ensure that each solution is a feasible solution and that each solution is a different solution. In the two-dimensional coding method, the number of vehicles is recorded directly, only the path values of the vehicles involved need to be calculated separately, and the whole food source does not need to be decoded, which can save a lot of decoding time.

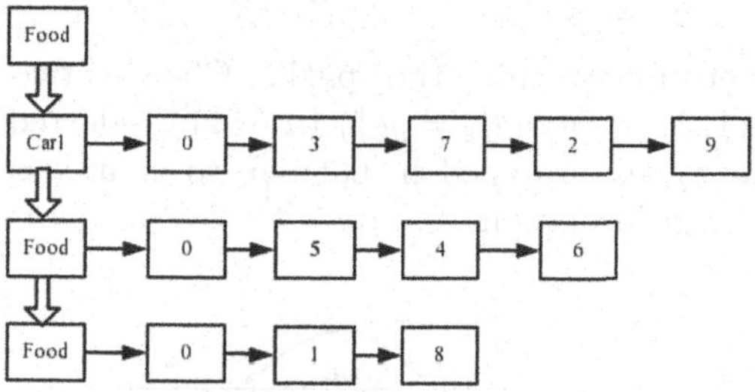

Fig. 2. Coding method

Figure 2 is a specific example of an initial population constructed in a design using a two-dimensional coding method. For example, if the food source (372954618) requires three vehicle services under the condition that the time window and weight constraint are met, the food source structure is shown in Fig. 2, indicating that the vehicle k_1 starts from the storehouse 0, serves the customer 3, 7, 2 and 9 in order, and then returns to the storehouse. The vehicle k_2 starts from the storehouse 0, the vehicle k_2 starts from the customer 5, 4, 6, then returns to the storehouse. The vehicle k_3 starts from the storehouse 0, serves the customer 1, 9, and finally returns to the storehouse 0, thus forming three circuits, the structure is simple and strong extensible.

3.2 Increase in Domain Search Methods

In the standard discrete ABC algorithm, both the lead bee and the follow bee adopt only one domain search method, which will lead to the lack of sufficient depth of the bee's search for the food source so that it can not find a satisfactory solution. The following bee uses gamble on a roulette to select the food source, and the food source with high adaptability is easy to be selected, so the optimization process of follow bee is a kind of deep search. By adding the domain search method of lead bee and follow bee, the search range of bee colony can be increased, and the better solution can be found faster. After discretization, the lead bee and the follow bee use the three domain search methods shown in Figs. 3, 4 and 5 to search the neighborhood to produce a new food source. Before searching the food source in the neighborhood, first, choose one search method from the insert operation and the swap operation, and then choose two paths from all paths. If the two path sequence numbers are the same, select the swap or insert operation within the path. And if the two path sequence numbers are different, select the swap or insert operation between the paths. We can find that the more vehicles needed for the example, the higher the probability that the switching operation between paths will be selected. Under the constraint of the time window, the success rate of insertion and swap operation in path is not high, so we indirectly improve the probability of switching operation between paths, make the path value develop in a better direction, and enhance the diversity of population.

(1) The swap operation within the path. Choose two customer points $j_1, j_2 (j_1, j_2 \in \{1, 2, \cdots, n and j_1 \neq j_2\})$ from any selected path p. The positions of j_1 and j_2 are swapped in path p, so that a new path p' can be obtained after transformation.

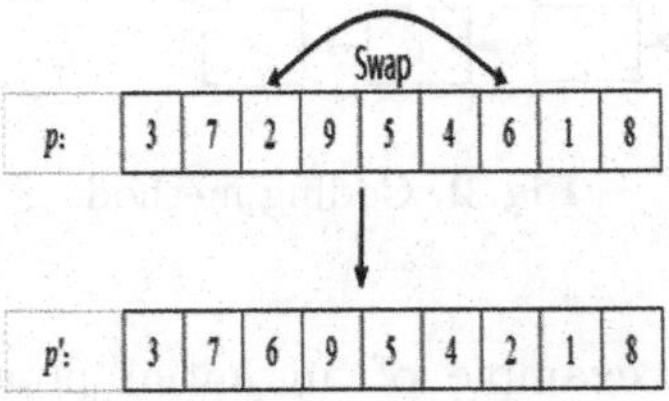

Fig. 3. The swap operation within the path

(2) The swap operation between the path Choose two customer points $j_1, j_2 (j_1, j_2 \in \{1, 2, \cdots, n and j_1 \neq j_2\})$ from any two paths of choice p_1, p_2, the customer point j_1 in path p_1 is swapped with the customer point j_2 in path p_2, so that the new path p_1', p_2' can be obtained after insertion.

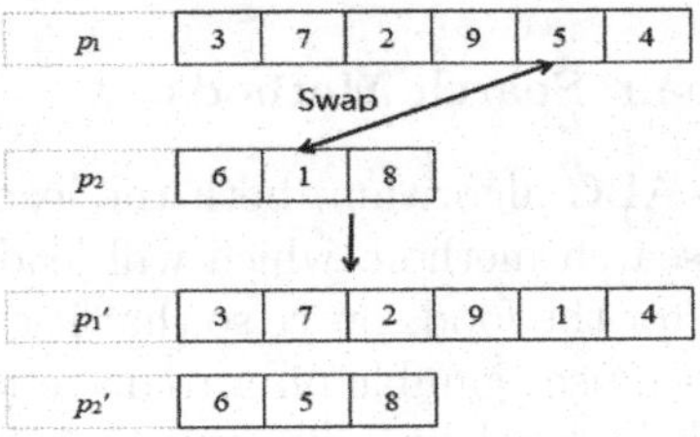

Fig. 4. The swap operation between the path

(3) The insert operation within the path Choose two customer points $j_1, j_2 (j_1, j_2 \in \{1, 2, \cdots, n and j_1 \neq j_2\})$ from any selected path p. After inserting j_2 into j_1 in path p, a new path p' is obtained after transformation.

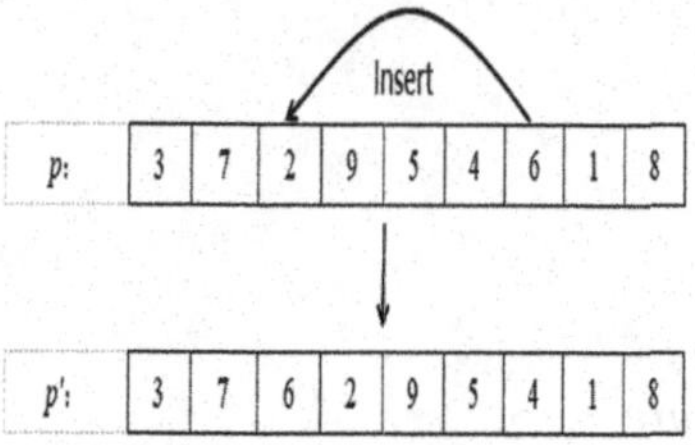

Fig. 5. The insert operation within the path

3.3 Setup of the Bulletin Board

Set up a bulletin board to record the solution value and the number of vehicles corresponding to the optimal food source. After each iteration of limit, the program calls the bulletin board before searching the neighborhood of the detect bee, and compares the best individual in the population with the food source in the bulletin board after this iteration. If the number of vehicles required by the optimal individual is less, the food source in the bulletin board is directly replaced, and if the number of vehicles in the optimal individual and the food source in the bulletin board is the same, the adaptability values of the two food sources are compared, in order to set the appropriate adaptability value, When the number of vehicles is the same, the target path is the shortest, the customer point i and the customer point j are set, the vehicle runs from point i to point j, k is the vehicle, and the c_{ij} is the transportation distance between the customer points i, j, sets the reciprocal of the total driving distance of all vehicles as the fitness value function, which is as follows:

$$f_i = 1/\sum_i \sum_j \sum_k c_{ij} x_{ijk} \tag{16}$$

Food sources with low adaptability are recorded on the bulletin board. If the number of vehicles needed by the optimal individual is more, then give up directly, the food source in the bulletin board will remain unchanged. This idea ensures that the better food source is recorded before it is abandoned, and the optimal solution in the running process of the program is successfully retained. For detailed steps, see the operator $Calculatebest(x_i)$ Prescribed operation:

$f(x_i)$: The corresponding path length of food source i
$G(x_i)$:The corresponding number of vehicles of food source$i$
$FoodNum$:Number of food sources
Input:
x_i :Food source i
Output:
$MinCar$: Minimum number of vehicles
$MinValue$: Optimal path value with number of vehicles as the first target
Begin:

$i = 1$
$while(i \leq FoodNum)$
do
$MinCar = G(x_0)$
$MinValue = f(x_0)$
$ifG(x_i) < MinCar\ then:$
$MinCar = G(x_i)\ continute$
$iff(x_i)MinValue\ then:$
$MinValue = f(x_i)$

return $MinCar$,$MinValue$
end

3.4 Search Mode of Detect Bees

In the standard discrete ABC algorithm, new food sources are generated and then go to the next limit iteration with other food sources that have not been abandoned. However, through the method of randomly generating food sources, these new food sources are a feasible solution, but they are generated randomly, the adaptability value will be poor, and it requires many iterations to narrow the gap between the food sources without being abandoned. In the iterative process, there are some better food sources, which are often eliminated because they do not continue to be updated, and the fitness value of the re-randomly generated food source is not necessarily better than the food source, or even different completely. Among, the better food sources, there may be some path fragments that reach the optimal path.

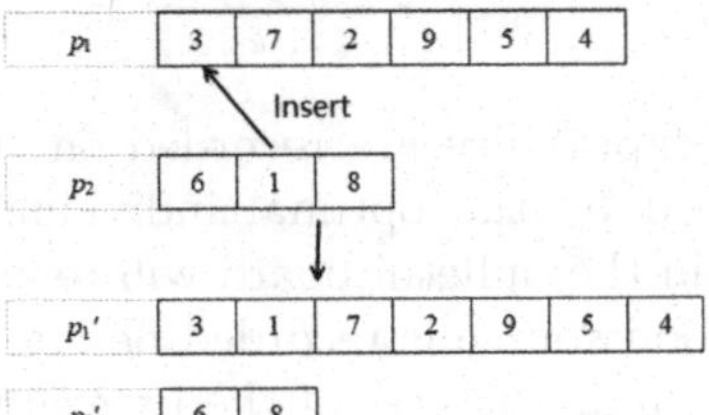

Fig. 6. The insert operation between the path

In order to solve the above two problems, in the detect bee stage, to avoid the re-random food source, to retain the fragments of the optimal path, we use the insert operation between the paths, as shown in Fig. 6, that is, from any choice of two paths p_1, p_2 to choose two customer points $j_1, j_2 (j_1, j_2 \in \{1, 2, \cdots, n\})$ and $j_1 \neq j_2$, insert the customer point j_2 in the path p_2 into the back of the customer point j_1 in the path p_1, so that after inserting, we can get the new path p_1', p_2'.

4 Simulation Experiment and Analysis

4.1 Simulation Experimental Environment and Setting of Experimental Parameters

Experimental Environment. Experimental software: VS 2019. Programming language: C language. Processor: AMD A6 3420M APU with Radeon (tm) HD Graphics 1.50 GHz and Genuine Intel (R) CPU T2080 @ 1.73 GHz. Memory: 8.00 GB. Operating system: Win10 64-bit operating system.

Experimental Analysis and Setting. In this paper, the standard test set in Solomon standard database is used to simulate the MODABC algorithm. In order to verify the applicability and effectiveness of the MODABC algorithm designed in this paper to solve MOVRPTW. Three groups of experiments are carried out in this paper: The standard DABC algorithm is compared with the improved MODABC algorithm. The improved MODABC algorithm is compared with other classical algorithms. The improved MODABC algorithm is used to solve VRPTW. In order to reduce the uncertainty as much as possible, each group of examples were tested 20 times and averaged.

Setting of Experimental Parameters. The number of foraging cycles, the threshold of food source abandonment and the maximum number of iterations all affect the convergence of the algorithm. These factors need to be adjusted according to the size of the problem. Among them, the influence of the maximum number of iterations on the convergence of the algorithm is more complex, and it is difficult to get good results if the value is too small or too large, depending on the size N of the problem. In general, the larger N is, the more iterations are obtained to obtain the optimal solution or the suboptimal solution. The number of foraging cycles and the threshold of food source abandonment should also be adjusted according to N. We take N = 100. In this problem, we choose 2000000 iterations. The number of foraging cycles and the threshold of food source abandonment were tested according to the number of foraging cycles of different sizes, and the most suitable parameter size was finally found. The specific test results are shown in Tables 1 and 2.

It can be seen from Tables 1 and 2 that the number of foraging cycles and the threshold of food source abandonment can not be selected at will, and too large or too small will affect the quality of the solution. And we can know that when the number of foraging cycles is 5000 and the food source abandonment threshold is 4, the best results can be obtained in terms of accuracy. Through the above experiments, the parameters of MODABC algorithm are set as the maximum number of cycles is 200000, the number of foraging cycles is 5000, and the threshold of food source abandonment is 4.

Table 1. Selection of foraging cycles.

Limit	C104.100	C204.100	R104.100	R201.100	RC101.100
100	(10,1020.0)	(3,619.7)	(10,997.0)	(8,1310.4)	(15,1648.9)
200	(10,936.8)	(3,603.4)	(10,1001.2)	(8,1301.2)	(15,1649.5)
500	(10,866.7)	(3,605.9)	(10,990.9))	(8,1246.7)	(15,1655.3)
1000	(10,859.3)	(3,597.5)	(10,988.2)	(8,1251.7)	(15,1644.3)
2000	(10,857.5)	(3,596.7)	(10,983.9)	(8,1227.3)	(15,1643.8)
5000	(10,844.6)	(3,593.9)	(10,982.1)	(10,1178.5)	(15,1639.5)
7000	(10,870.0)	(4,597.4)	(10,981.0)	(10,1211.2)	(15,1642.0)
10000	(10,860.2)	(4,617.8)	(10,983.1)	(10,1244.1)	(15,1649.2)
15000	(10,861.8)	(4,625.1)	(10,990.0)	(10,1237.6)	(15,1647.9)
20000	(10,865.8)	(4,637.9)	(10,987.4)	(11,1249.4)	(16,1663.7)
50000	(10,883.4)	(4,647.3)	(10,997.3)	(11,1261.8)	(16,1669.3)

Table 2. Selection of food source discard threshold.

Value	C104.100	C204.100	R104.100	R201.100	RC101.100
1	(10,852.8)	(4,654.5)	(10,986.1)	(7,1264.6)	(16,1658.7)
2	(10,852.7)	(4,643.6)	(11,987.5)	(7,1267.9)	(15,1652.9)
3	(10,855.6)	(4,638.1)	(11,986.4)	(8,1241.1)	(16,1657.3)
4	(10,851.5)	(3,593.9)	(11,982.1)	(10,1178.5)	(15,1639.5)
5	(10,875.7)	(3,609.0)	(10,985.4)	(10,1233.2)	(16,1650.5)
7	(10, 853.5)	(4, 641.9)	(10, 986.0)	(8,1240.0)	(16,1653.6)
10	(10,862.9)	(4,654.8)	(10,985.7)	(7,1254.4)	(16,1657.6)
15	(10,867.8)	(4,654.3)	(11,989.0)	(7,1253.8)	(15,1658.7)
20	(10,865.2)	(4,656.8)	(10,986.8)	(7,1258.6)	(15,1654.1)
50	(10,852.4)	(4,658.6)	(10,991.2)	(7,1264.5)	(15,1656.5)

4.2 Comparison and Analysis of Standard DABC Algorithm and Improved MODABC Algorithm

Below, we compare and analyze the standard discrete ABC algorithm and the improved DABC algorithm according to different examples, as shown in Table 3.

Through the experimental results of different examples, we can easily find that the improved DABC algorithm is obviously stronger than the standard DABC algorithm in solving the VRPTW problem. The optimal solution and the optimal number of vehicles obtained by the standard DABC algorithm are poor, which shows that the optimization ability of the standard DABC algorithm is not strong, the search depth is not enough, and adding more domain search methods can increase the quality of the solution to a great extent. A single domain search method can easily make the solution fall into local optimization, which will lead

Table 3. Comparison of standard discrete ABC algorithm and improved discrete ABC algorithm for solving VRPTW.

Example	Standard ABC algorithm			Improved ABC algorithm		
	Number of vehicles	Optimal value	Average	Number of vehicles	Optimal value	Average
C101.100	17	1351.6	1415.6	10	828.9	828.9
C102.100	15	1266.2	1341.5	10	828.9	830.7
C103.100	10	1204.0	1278.8	10	828.0	856.6
C201.100	10	1125.6	1286.2	3	591.5	591.5
C202.100	9	1076.3	1188.3	3	591.5	591.5
C203.100	8	1043.0	1113.3	3	591.1	619.7
R101.100	25	1822.6	1894.3	20	1651.6	1658.3
R102.100	22	1679.5	1706.8	18	1474.9	1481.0
R103.100	18	1442.2	1468.5	14	1216.1	1222.0
R201.100	13	1408.2	1466.1	10	1178.5	1215.3
RC101.100	21	1954.9	2011.4	15	1639.5	1660.2
RC102.100	19	1755.3	1822.9	14	1472.0	1484.2
RC103.100	16	1586.4	1641.6	11	1278.1	1297.7
RC201.100	13	1676.3	1741.5	9	1313.3	1358.2
RC202.100	10	1447.4	1573.9	8	1157.5	1191.2

to the fact that the new food source produced by detect bees has no advantage in the competition, and the probability of being selected in the iterative process will be reduced. As a result, the new food source has been delayed in updating. The improved DABC algorithm makes the new food source search many times in the reconnaissance bee stage, which increases the competitiveness of the new food source, makes it easier to jump out of the local optimization, and increases the breadth of the algorithm search. Through comparison, we can draw the following conclusions: the improved DABC algorithm has excellent performance in convergence speed and solution accuracy, which solves the shortcomings that the standard DABC algorithm is easy to fall into local optimization, poor stability and weak search ability, which shows that our improvement direction is correct.

4.3 Comparison of Improved MODABC Algorithm with Other Classical Algorithms

We compare the proposed algorithm with the ABC algorithm introduced in reference [16], the ACO-Tabu algorithm listed in reference [17] and the GA listed in reference [18], as shown in Table 4.

Table 4. Comparison of improved MODABC algorithm with other classical algorithms.

Example	DABC		ABC		ACO-Tabu		GA	
	Number of vehicles	Path value	Number of vehicles	Path value	Number of vehicles	Path value	Number of vehicles	Path value
C102.100	10	828.9	10	828.9	10	828.9	11	868.7
C103.100	10	828.0	10	828.9	10	828.9	11	939.4
C104.100	10	844.6	10	858.9	10	828.0	11	963.7
C202.100	3	591.5	3	591.5	3	591.5	4	683.8
C203.100	3	591.1	3	600.5	3	593.2	4	745.9
C204.100	3	593.9	3	610.0	3	595.5	3	604.9
R102.100	18	1474.9	18	1480.7	18	1491.1	18	1558.5
R103.100	14	1216.1	14	1240.8	14	1243.2	15	1311.8
R104.100	11	981.0	12	1047.0	10	982.0	12	1128.2
R201.100	10	1178.5	8	1185.5	8	1198.1	8	1329.7
RC102.100	14	1472.0	15	1492.8	13	1470.2	17	1603.5
RC103.100	11	1278.1	13	1334.5	12	1196.1	14	1519.8
RC105.100	15	1520.2	15	1546.4	14	1589.9	17	1688.7
RC201.100	9	1313.3	8	1308.6	5	1279.6	10	1565.6
RC202.100	8	1157.5	8	1167.0	5	1157.0	10	1353.2
RC205.100	8	1205.5	7	1210.6	5	1334.5	9	1465.8

From Table 4, we can see that the algorithm in this paper has a good effect in solving the vehicle path value of MOVRPTW example. For the examples of C1, C2, R1 and RC1, most of the algorithms can find a good solution, but the gap is not very large, which shows that it is easy to find a satisfactory solution when solving narrow time window and small load example, while the improved MODABC algorithm has better solution accuracy, which shows that the improved MODABC algorithm has strong deep search ability and has some advantages in solving MOVRPTW problem. Although the MODABC algorithm is used to solve the dataset R2, RC2 No optimal results are obtained, but in terms of the number of vehicles, MOTS algorithm is all superior to GA algorithm, and some of them are superior to other algorithms. Considering the accuracy of MODABC algorithm in solving MOVRPTW, MODABC is a very competitive algorithm.

4.4 Improved MODABC Algorithm for Solving VRPTW

In order to test the effectiveness and efficiency of the improved discrete ABC algorithm in solving VRPTW, we use several different kinds of VRPTW examples to test the data. The test results of the specific data are shown in Table 5.

Table 5. The results of improved discrete ABC algorithm for solving VRPTW.

Example	Improved DABC algorithm				Database optimal solution	
	Number of vehicles	Optimal value	Deviation degree	Average	Number of vehicles	Optimal value
C101.100	10	828.9	0.001	828.9	10	827.3
C102.100	10	828.9	0.001	830.7	10	827.3
C103.100	10	828.0	0.001	856.6	10	826.3
C104.100	10	844.6	0.026	881.8	10	822.9
C105.100	10	828.9	0.001	828.9	10	827.3
C201.100	3	591.5	0.004	591.5	3	589.1
C203.100	3	591.1	0.004	619.7	3	588.7
C204.100	3	593.9	0.009	647.9	3	588.1
C205.100	3	588.8	0.004	588.8	3	586.4
R101.100	20	1651.6	0.008	1658.3	20	1637.7
R102.100	18	1474.9	0.005	1481.0	18	1466.6
R109.100	12	1154.9	0.006	1165.8	13	1146.9
R110.100	11	1084.0	0.014	1099.7	12	1068
R111.100	12	1055.5	0.006	1068.5	12	1048.7
R201.100	10	1178.5	0.030	1215.3	8	1143.2
R202.50	5	702.7	0.005	711.5	5	698.5
R203.50	5	607.6	0.003	614.1	5	605.3
R204.50	2	509.2	0.005	512.0	2	506.4
R205.50	3	712.1	0.031	727.3	4	690.1
R206.50	4	634.9	0.003	648.2	4	632.4
R209.50	4	610.5	0.016	643.1	4	600.6
R210.50	4	652.6	0.011	659.8	4	645.6
RC101.100	15	1639.5	0.012	1660.2	15	1619.8
RC102.100	14	1472.0	0.010	1484.2	14	1457.4
RC103.100	11	1278.1	0.016	1297.7	11	1258.0
RC104.50	5	546.5	0.001	548.4	8	545.8
RC201.100	9	1313.3	0.040	1358.2	9	1261.8
RC202.100	8	1157.5	0.059	1191.2	8	1092.3
RC205.100	8	1205.5	0.044	1243.9	7	1154.0

From Table 5, we can see that for sets C1 and C2, the improved MODABC algorithm can have little difference in the optimal solution in the database, and the number of vehicles is the same as the optimal solution. For the set R1 and RC1, the optimal number of vehicles can be found and the accuracy of the optimal solution value is less than 1% and 2 %. For R2 and RC2, in the databases, only R201, R202, RC201, RC202, RC205 are given at 100 points, so for these two other examples, only 50 points of customer size are selected. When 50 points are selected, the accuracy of the optimal solution obtained by the algorithm is still less than 1%. For R109, R110, R205, RC104 examples, we can find fewer vehicles than the database. These results further show that the improved DABC algorithm is a very competitive algorithm for solving MOVRPTW, and has very good performance in global optimization, convergence speed and solution accuracy. The average value of each example is not different from the optimal solution value, which also shows that the improved MODABC algorithm has good stability and robustness. But for the customer size of the R2 and RC2 collections of 100 points, although there is still a certain gap between the optimal solution value of the database and the optimal solution value of the database, but the error rate is basically less than 5%. The algorithm can find a satisfactory solution, but it does not reach the accuracy of other examples.

5 Conclusion

Based on the analysis of the three factors of grain transportation time, vehicle number and path length, an improved MODABC algorithm with time window is proposed on the basis of the establishment of initial population, the increase of domain search mode, the setting of bulletin board and the search mode of detect bee. By comparing the standard DABC algorithm with the improved MODABC algorithm, the correctness of the improvement direction is proved. In comparison with other classical algorithms, MODABC algorithm is a very competitive algorithm considering the accuracy of MOVRPTW. The improved MODABC algorithm for solving VRPTW, fully shows that the improved MODABC has good stability and robustness in solving the VRPTW problem of grain transportation, and the accuracy of the solution is very high and the depth search ability is strong.

Acknowledgments. The work was supported by National Natural Science Foundation of China (Grant No. 61179032 and 61303116), the Special Scientific Research Fund of Food Public Welfare Profession of China (Grant No. 2015130043), the Research and Practice Project of Graduate Education Teaching Reform of Polytechnic University (YZ2015002), the Scientific research project of Wuhan Polytechnic University (2019), Key Project of Philosophy and Social Science Research Project of Hubei Provincial Department of Education in 2019(19D59), Science and Technology Research Project of Hubei Provincial Department of Education (D20191604).

References

1. Dantzig, G., Ramser, J.: The truck dispatching problem. Manag. Sci. **6**, 80–91 (1959)
2. Lang, M., Hu, S.: Research on solving logistics distribution path optimization problem with hybrid genetic algorithm. Chin. Manag. Sci. **10**(5), 51–56 (2002)

3. Lv, X., Liao, T.: Research on postal vehicle routing problem with time window based on genetic algorithm. J. Shandong Univ. **06**(44), 46–50 (2009)
4. Lin, F., Guo, H.: Simulation research on workshop distribution path optimization based on ant colony algorithm. Mech. Des. Manuf. **10**(10), 13–15 (2007)
5. Tang, Y., Liu, F.: A new genetic simulated annealing algorithm for solving VRPTW problem. Comput. Eng. Appl. **42**(7), 12–14 (2006)
6. Karaboga, D.: An idea based on honey bee swarm for numercial optimization. Technical Report-TR06. Erciyes University, 13–15 (2005)
7. Jin, Y., Sun, Y., Wang, J., Wang, D.: An improved elite artificial bee colony algorithm based on simplex. J. Zhengzhou Univ. **39**(6), 13–15 (2018)
8. Zhao, Y., Xu, X., Huang, W., Ma, Y.: Hybrid artificial bee swarm algorithm based on cat swarm idea. Comput. Technol. Dev. **29**(1), 11–12 (2019)
9. Liang, X., Zhao, X.: An improved artificial bee swarm algorithm based on steepest drop method. J. Beijing Univ. Archit. **34**(3), 49–56 (2018)
10. Chao, X., Li, W.: Feature selection method for artificial bee swarm algorithm optimization. Comput. Sci. Explor. **13**(2), 300–309 (2019)
11. Aslan, S.: A transition control mechanism for artificial bee colony algorithm. Comput. Intell. Neurosci. **4**(6), 1–23 (2019). https://doi.org/10.1155/2019/5012313
12. Dervis, K.: Discovery of conserved regions in DNA sequences by Artificial Bee Colony (ABC) algorithm based methods. Nat. Comput. **15**(6) (2019). https://doi.org/10.1007/s11047-018-9674-1
13. Yu, X.: Research on vehicle routing problem with time window considering carbon emission based on artificial bee swarm algorithm. Master's thesis. Dalian University of Technology, vol. 1, no. 5, pp. 88–89 (2016)
14. Deb, K.: Multi-Objective Optimization Using Evolutionary Algorithms. Wiley, Chichester (2001)
15. Gong, M., Jiao, L., Yang, D.: Research on evolutionary multiobjective optimization algorithm. J. Softw. **20**(2), 271–289 (2009)
16. Alzaqebah, M., Abdullah, S., Jawarneh, S.: Modified artificial bee colony for the vehicle routing problems with time windows. SpringerPlus **5**, 1298 (2016)
17. Tan, K.C., Lee, L.H., Zhu, Q.L., Ou, K.: Heuristic methods for vehicle routing problem with time windows. Artif. Intell. Eng. **15**, 281–295 (2001)
18. Yu, B., Yang, Z.Z., Yao, B.Z.: A hybrid algorithm for vehicle routing problem with time windows. Expert Syst. Appl. **38**, 435–441 (2011)
19. Su, X., Sun, H., Pan, X.: Simulation of traveling Salesman problem based on improved bee swarm algorithm. Comput. Eng. Des. **34**(4), 1420–1424 (2013)

UAV 3D Path Planning Based on Multi-Population Ensemble Differential Evolution

Xuzhao Chai, Junming Xiao, Zhishuai Zheng, Liang Zhang, Boyang Qu(✉), Li Yan, Sumarga Kumar Sah Tyagi, Lu Yang, Chao Feng, and Hang Sun

Zhongyuan University of Technology, Zhengzhou 450007, China
{xzchai, quboyang}@zut.edu.cn

Abstract. The three-dimensional (3D) environmental path planning of unmanned aerial vehicles (UAVs) is studied systemically. The model has been constructed, including 3D environment, threat, path length, yaw angle, climb angle and height. All the objective costs have been converted into a single objective optimization problem by the linear weight method. The multi-population ensemble differential evolution (MPEDE) algorithm has been applied to the 3D UAV path planning, and has shown the best performance compared with the self-adaptive differential evolution (SaDE) and differential evolution (DE) algorithms. The MPEDE algorithm is feasible in solve the UAV path planning problem.

Keywords: 3D path planning · MPEDE algorithm · Linear weighting method · Differential evolution

1 Introduction

Unmanned aerial vehicles (UAVs) have a high potential application in military and civil area. Meanwhile, how to guide UAVs to the destination is a great challenge. The path planning has been becoming a key role in the challenge, and has been a hot topic studied by scholars [1]. The UAV path planning is an optimization problem that gets an optional or near-optional path from a start point to an end point [2, 3]. This work is motivated by the challenge to develop a practical path planner for UAVs since many factors are taken in modeling, such as terrain, height constraint, fuel consumption and so on.

Path planning is typically a non-deterministic polynomial-time (NP) hard problem for UAVs, which has been widely studied by scholars. In the literature, many methods have been proposed to solve this problem, including Mixed-Integer Linear Programming [4], A-star [5], Voronoi Diagram [6, 7], Nonlinear Programming [8], and computational intelligence methods [3, 9, 10]. Most of the methods in these reports are limited to solve 2D scene which can be simplified with maintaining the UAV flight height. In recent years, the path planning in 3D environment has been widely studied, and its difficulties increase exponentially compared with the 2D UAV path planning due to the environment complexity [11]. Intelligent algorithms have been focused in handing

L. Pan et al. (Eds.): BIC-TA 2019, CCIS 1159, pp. 598–610, 2020.
https://doi.org/10.1007/978-981-15-3425-6_47

3D path-planning problems, e.g. the particle swarm optimization (PSO) algorithm [12], the genetic algorithm (GA) [13], the artificial bee colony (ABC) algorithm [14], the ant colony algorithm (ACO) [15], the differential evolution (DE) algorithm [16], etc. DE algorithm has an excellent performance in spite of its simplicity, which has been widely used in many application areas. Recently, many modified DE version has been reported for solving real-world problems [17, 18]. For example, Takahana et al. has designed the ε constrained differential evolution algorithm (εDE) with the α constrained method [19]; Wang et al. has proposed a hybrid DE algorithm with level comparison to solve optimization problems [20]; Wu et al. has proposed a new DE variant named multi-population ensemble DE (MPEDE) with an ensemble of multiple strategies [21]. In this work, the 3D environment is modeled considering the multi-objective problems, such as threats, path length, yaw angle, climbing angle and height. The MPEDE algorithm is adopted tentatively to solve the 3D path planning.

The rest of the article is as follows. Section 2 presents the mathematical models of the UAV cost and related constraints. Section 3 introduces the standard DE algorithm and the MPEDE algorithm. The simulation results and analysis of the MPEDE algorithm are compared with the DE and self-adapted DE (SaDE) algorithms in Sect. 4 and Sect. 5 is the conclusion.

2 Problem Description

UAV path planning is a complex multi-objective optimization problem. The main task is to plan an optimal path from the initial point to the termination point when the corresponding constraints are satisfied.

2.1 Path Representation

The path definition is to create a distance to guide a UAV from the starting point to the goal point. Figure 1 shows the UAV path model. S and G are represented as starting and goal points. The total path is divided into m + 1 parts by the waypoints $(R_2, R_3, \ldots, R_{m+1})$. Each segment between the waypoints i-th and (i + 1)th is equally divided into n + 1 parts, and the corresponding divided points are $p_1, p_2, \ldots, p_n$. The total number of points in a path is calculated as.

$$N = (m+1)(n+1) + 1 \tag{1}$$

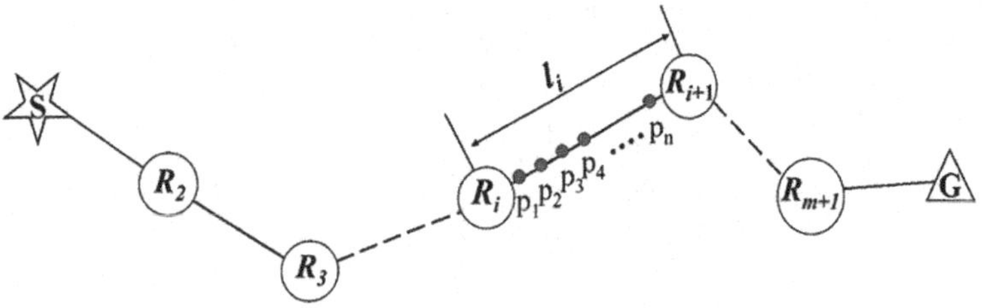

Fig. 1. UAV path diagram

Threat Cost. The threats can decrease effectively the success probability of the UAV mission, such as radars and artilleries. The radar threat and artillery cost model are described in the Eqs. (2) and (3), respectively, which are derived from the Ref. [16]. Here the d_r denotes the distance from a point on the path to the threat center. The d_{t-a} and d_{t-r} are the threat radius of radars and artilleries, respectively. This parameter represents the strength of the threat.

$$T_{p-r} = \begin{cases} 0 & d_r \geq d_{t-r} \\ (1/d_r)^4 & d_r < d_{t-r} \end{cases} \tag{2}$$

$$T_{p-a} = \begin{cases} 0 & d_r \geq d_{t-a} \\ (d_{t-a})^4 \Big/ \left((d_{t-a})^4 + (d_r)^4 \right) & d_r < d_{t-a} \end{cases} \tag{3}$$

Path Length Cost. For a UAV, getting a shortest path is necessary, which can reduce flight time and fuel consumption. The total path length L_p can be obtained by accumulating all the segments [12]. L_{EC} is the Euclidean distance between the starting point $S(x_g, y_g, z_g)$ and the goal point $G(x_s, y_s, z_s)$.

$$L_p = \sum_{i=1}^{N} \left(\sqrt{(x_{i+1} - x_i)^2 + (y_{i+1} - y_i)^2 + (z_{i+1} - z_i)^2} \right) \Big/ L_{EC} \tag{4}$$

$$L_{EC} = \sqrt{(x_g - x_s)^2 + (y_g - y_s)^2 + (z_g - z_s)^2} \tag{5}$$

Yaw Angle Cost. UAVs need to satisfy the turning radius due to the inertia when UAVs adjust courses requires. Here L is parametric equation as shown in Eq. (6), and $X(x_1, x_2, \ldots, x_N)$, $Y(y_1, y_2, \ldots, y_N)$ and $Z(z_1, z_2, \ldots, z_N)$ are the vectors of the N points in a whole path. The curvature radius r_{cv} of the points can be calculated according to the Eq. (7). $\dot{L}$ and $\ddot{L}$ are the first and second derivative of L, respectively. The yaw angle cost θ_p can be described in Eq. (8). $r_{\min}$ is the minimum turning radius. The θ_p is equal to 0, and the UAVs satisfy the flying condition when the $r_{cv} > r_{\min}$; the θ_p is equal to 1, and the UAVs cannot fly when the $r_{cv} < r_{\min}$.

$$\boldsymbol{L} = (\boldsymbol{X}, \boldsymbol{Y}, \boldsymbol{Z}) \tag{6}$$

$$r_{cv} = \frac{\|\dot{\boldsymbol{L}}\|^3}{\|\dot{\boldsymbol{L}} \times \ddot{\boldsymbol{L}}\|} \tag{7}$$

$$\theta_p = \begin{cases} 0 & r_{cv} \geq r_{\min} \\ 1 & r_{cv} \leq r_{\min} \end{cases} \tag{8}$$

Climbing Angle Cost. Climbing angle is another parameter for the UAV flyable condition, which is similar to the yaw angle. Figure 2 shows the climbing angle diagram for the UAVs. A and B are the adjacent two point in the path. The climbing angle ri can be calculated with the Eq. (9). The climbing angle cost is given in Eq. (10) [12]. The γ_p is equal to 0, and the UAVs satisfy the flying condition when the $\gamma_i \leq \gamma_{max}$; the γ_p is equal to 1, and the UAVs cannot fly when the $\gamma_i \geq \gamma_{max}$.

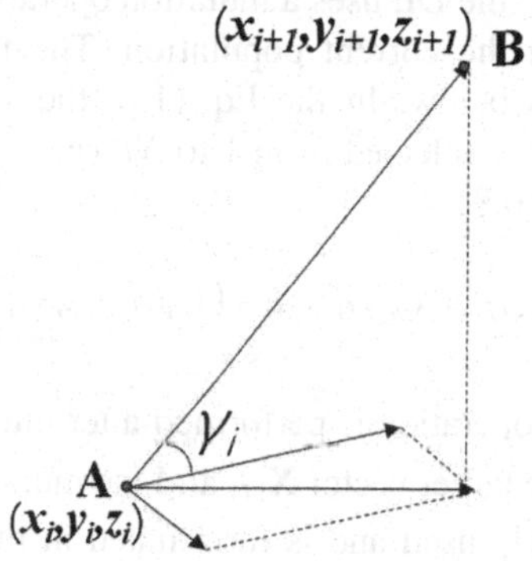

Fig. 2. UAV climbing operation

$$\gamma_i = \arctan\left(\frac{|z_{i+1} - z_i|}{\sqrt{\left|(x_{i+1} - x_i)^2 + (y_{i+1} - y_i)^2\right|}}\right), (i = 1, 2, \cdots, N) \quad (9)$$

$$\gamma_p = \begin{cases} 1 & \gamma_i \geq \gamma_{max} \\ 0 & \gamma_i \leq \gamma_{max} \end{cases}, (i = 1, 2, \cdots, N) \quad (10)$$

Height Cost. The flight height of UAVs is limited to a minimum flight altitude in order to avoid collision with mountain. The Eq. (11) shows the UAV height cost. Here H_i is the UAV flight altitude at the i-th point. H_l is the mountain height at the i-th point.

$$H_p = \begin{cases} 0 & (H_i - H_l) \leq H_{min} \\ (H_i - H_l)/N & (H_i - H_l) \geq H_{min} \end{cases}, (i = 1, 2, \cdots, N) \quad (11)$$

Obviously, this is a multi-objective optimization problem. The sum cost can be combined into a single objective optimization problem by the linear weight method (Eq. (12)). C is the total cost, and $w1$, $w2$, $w3$, $w4$ and $w5$ are corresponding to the weight coefficients.

$$C = min\left(w1 \cdot (T_{p-r} + T_{p-a}) + w2 \cdot L_p + w3 \cdot \theta_p + w4 \cdot \gamma_p + w5 \cdot H_p\right) \quad (12)$$

3 MPEDE Algorithms for the 3D Path Planning

3.1 Standard DE Algorithm

The DE algorithm is a population based stochastic algorithm, which was proposed by Storn and Price in 1997 [23]. The DE evolution process includes three steps, namely mutation, crossover and selection.

Mutation. At generation G, the DE uses a mutation operation to generate a vector $\mathbf{V}_{i,G}$ for each individual $\mathbf{X}_{i,G}$ in the current population. The typical mutation strategy is DE/rand/1, and is listed as below. In the Eq. (13), the subscript *r*1, *r*2, and *r*3 are integer number, and randomly selected from 1 to *NP* ($r1 \neq r2 \neq r3$). F is the mutation factor, ranging from 0.2 to 0.8.

$$v_{i,G} = x_{best,G} + F \cdot \left(x_{r1,G} - x_{r2,G}\right) \tag{13}$$

Crossover. The crossover operation is performed after mutation, and aim to produce a trial vector $\mathbf{U}^{j}_{i,G}$ between the target vector $\mathbf{X}_{i,G}$ and the mutant vector $\mathbf{V}_{i,G}$. The binomial crossover operation is widely used and is formulated in the Eq. (14).

$$u^{j}_{i,G} = \begin{cases} v^{j}_{i,G} & rand(0,1) \leq CR \quad or \quad j = j_{rand}(j = 1, 2, \cdots, D) \\ x^{j}_{i,G} & otherwise \end{cases} \tag{14}$$

In Eq. (14), *CR* is generated in the range of [0,1]. $u^{j}_{i,G}$ is the i-th individual of trial vector $\mathbf{U}_{i,G}$ after crossover on the j-th dimension. $v^{j}_{i,G}$ is the i-th individual of mutant vector $\mathbf{V}_{i,G}$. j_{rand} is a randomly generated integer in the range [1,D].

Selection. After mutation and crossover operation, comparing the objective function value of trial vector $\mathbf{U}_{i,G}$ with its corresponding target vector $\mathbf{V}_{i,G}$, chooses the trial vector enter the next generation as new target vector when its objective function value is smaller than that of the target vector. Otherwise, the target vector will be maintained.

$$\mathbf{X}_{i,G+1} = \begin{cases} \mathbf{U}_{i,G} & f(\mathbf{U}_{i,G}) < f(\mathbf{X}_{i,G}) \\ \mathbf{X}_{i,G} & otherwise \end{cases} \tag{15}$$

The above three operations are repeated in each generation until finding the global optimal value or the termination condition of iteration is satisfied.

3.2 Multi-Population Ensemble DE (MPEDE) Algorithm

The MPEDE algorithm is a modified version based on the standard DE algorithm, which was proposed by Wu in 2016 [21]. In the algorithm, the whole population is divided into indicator subpopulations and the reward population at each generation. Different mutation strategies are assigned to different subpopulation. The mutation strategy that performs outstandingly is assigned to the reward subpopulation.

Mutation Strategy. In general, different real-world optimization problem needs different mutation strategies, and even the different evolution stages in a specific problem needs different mutation strategies. Therefore, the mutation strategy selection determines the efficient of a modified DE algorithm. According to the Ref. [21], three mutation strategies are selected as follows [21].

Strategy 1: current-to-pbest/1

$$V_{i,G} = X_{i,G} + F \cdot \left(X_{pbest,G} - X_{i,G} + X_{r1,G} - \tilde{X}_{r2,G}\right) \tag{16}$$

Strategy 2: current-to-rand/1

$$U_{i,G} = X_{i,G} + K \cdot \left(X_{r1,G} - X_{i,G}\right) + F \cdot \left(X_{r2,G} - X_{r3,G}\right) \tag{17}$$

Strategy 3: rand/1

$$V_{i,G} = X_{r1,G} + F \cdot \left(X_{r2,G} - X_{r3,G}\right) \tag{18}$$

The "current-to-pbest/1" strategy is more effective in solving complex problems. The "current-to-rand/1" strategy is more effective for the algorithm to solve rotated problems. The "DE/rand/1" is the most popular and has good robustness [21, 22].

At the beginning, the whole population is divided into four subpopulations, namely p_1, p_2, p_3 and p_4. Here the subpopulations p_1, p_2, and p_3 have the same population size, and are considered as indicator subpopulation. The subpopulation p_4 are set as reward subpopulation. The subpopulations p_1, p_2, and p_3 are randomly assigned with the above three mutation strategies, and the reward subpopulation is randomly assigned to one of these three mutation strategy. After the ng generations, the performance of the mutation strategy is evaluated by $\Delta f_j/\Delta Fes_j$. Here, Δf_j represents the cumulative variation of the objective function value for the j-th strategy. ΔFes_j represents the number of the generation with the j-th strategy. This process guarantees that most of computing resources are distributed to the most effective mutation strategy.

Parameter Adaptation. The typical DE algorithm needs the setting of a serial of parameter, such as cross factor (CR) and scaling factor (F). The parameter adaptation is also an effective method, and has been proposed in previous studies [24, 25]. In this work, the adaptation method is adopted, and the CR and F are updated by the below formulas [21]:

$$CR_{i,j} = randn_{i,j}\left(\mu CR_j, 0.1\right) \tag{19}$$

$$\mu CR_j = (1 - c) \cdot \mu CR_j + c \cdot mean_A\left(S_{CR,j}\right) \tag{20}$$

$$F_{i,j} = randc_{i,j}\left(\mu F_j, 0.1\right) \tag{21}$$

$$\mu F_j = (1 - c) \cdot \mu F_j + c \cdot mean_L\left(S_{F,j}\right) \tag{22}$$

$$mean_L = \frac{\sum_{F \in S_F} F^2}{\sum_{F \in S_F} F} \tag{23}$$

Here, $CR_{i,j}$ is the crossover probability of the individual X_i with the j-th mutation strategy, and follows the normal distribution. $F_{i,j}$ is the scaling factor of the individual X_i with the j-th mutation strategy, and follows the Cauchy distribution. $mean_A$ and $mean_L$ represents the arithmetic mean and the Lehmer mean, respectively. μCR_j is the mean value, and μF_j is the location parameter. c is a positive constant between 0 and 1. $S_{CR,j}$ is the number of cross factor that generate effective solutions. $S_{F,j}$ is the number of scaling factor that generate effective solutions.

3.3 Algorithm Steps

The algorithm steps are described in detail as below.

Step 1: Parameter setting
Setting specific parameters of the algorithm, such as population size *NP*, ng, etc.
Step 2: Initialize population
The whole population is divided into three indicator subpopulations and one reward subpopulation. and the reward subpopulation is randomly assigned one of the three mutation strategy, and allocated to the corresponding indicator population.
Step 3: Parameter update
The *CR* and *F* for the three mutation strategies are calculated and updated according to the Eqs. (19)–(23).
Step 4: Population redistribution
After the ng generations, the preformation of the three strategies are evaluated by the formula $\Delta f_j / \Delta Fes_j$. According to the evaluation value, the reward population is assigned the most effective mutation strategy, and then is achieved to redistribute.
Step 5: Termination of iteration
The iteration updating is terminated and the optimum value outputs when the number of iterations reaches the setting value. Otherwise, go to the step 3.

4 Experiments and Analysis

4.1 Experimental Parameters

In this work, the UAV mission area is 1000 × 1000 m, and the height range is [0,0.25] km as shown in Fig. 3. The coordinates of start point, goal point and threats as shown in Table 1. Table 2 is the parameters of algorithm. MATLAB R2014b is used as the simulation platform, and the simulation is carried out on a computer with Intel(R) Core (TM) i7-7700 CPU @ 3.60 GHz, 8 GB RAM and windows 10 64-bit operating system.

Table 1. The parameters of the UAV and threats

Parameter	Coordinate	Radius
Start point	(50, 30, 0.005)	
Goal point	(900, 950, 0.005)	
Radar1	(700, 900, 0.15)	90
Radar2	(850, 600, 0.15)	100
Radar3	(200, 900, 0.15)	50
Radar4	(500, 500, 0.15)	50
Artillery1	(500, 700, 0.15)	70
Artillery2	(800, 400, 0.15)	80
Artillery3	(500, 150, 0.15)	60
Artillery4	(350, 500, 0.15)	60

Table 2. The parameters setting of the algorithm

Parameter	Value
The ng generations	20
The population size NP	100
The maximum generation G	200
The mean value of cross factor μCR_j	0.5
The mean value of scaling factor μF_j	0.5

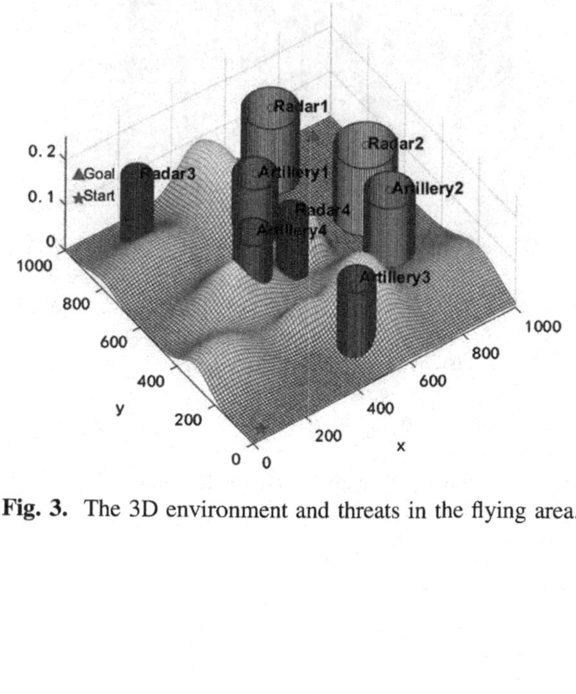

Fig. 3. The 3D environment and threats in the flying area.

4.2 Comparison Analysis with the DE and SaDE Algorithms

In this work, the MPEDE algorithm is compared with the DE and SaDE algorithms, and runs independently 30 times under the same environment condition. The Fig. 4 shows that UAV can get the safe path with these three algorithms, avoiding the height, radar, and artillery threats. Table 3 shows the comparison of the three algorithms when all the weight coefficients ($w1$, $w2$, $w3$, $w4$ and $w5$) are equal to 0.2. As a whole, the MPEDE shows a best performance for the UAV 3D path planning. The MPEDE algorithm is similar to the SaDE algorithm for the best, worst and mean value, but is better for the standard deviation value. The MPEDE algorithm is more effective than the SaDE and DE algorithm in solving the UAV 3D path planning.

Table 3. Path planning results of the MPEDE, SaDE and DE algorithms

Total cost	Best value	Worst value	Mean value	Standard deviation value
DE	0.201012	0.206051	0.202219	1.014×10^{-3}
SaDE	0.200970	0.201151	0.201002	5.500×10^{-5}
MPEDE	**0.200969**	**0.201072**	**0.200976**	$\mathbf{2.000 \times 10^{-5}}$

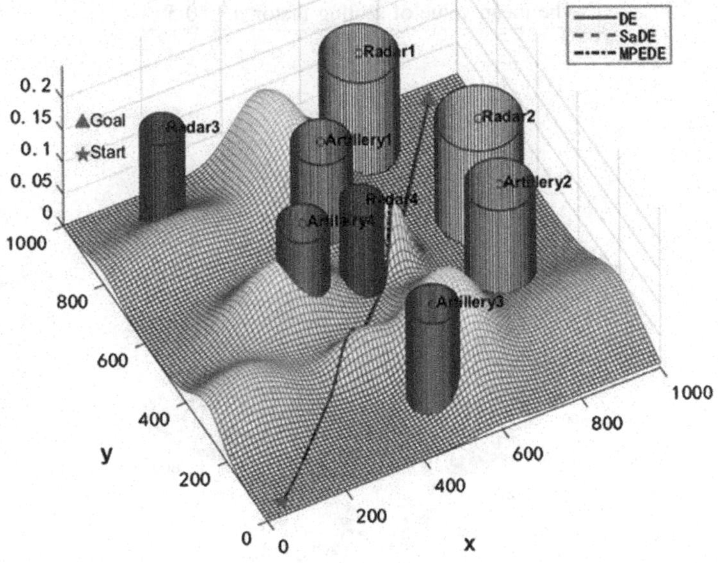

Fig. 4. UAV optimal path with the DE, SaDE and MPEDE algorithms

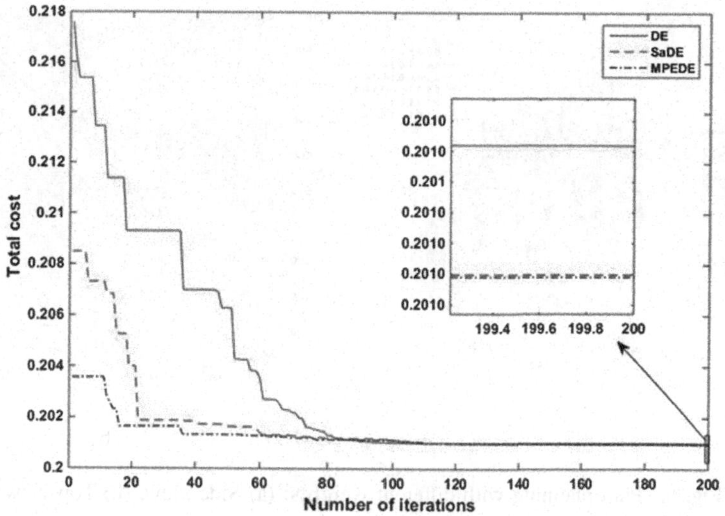

Fig. 5. The convergence curves of the DE, SaDE and MPEDE algorithms

The convergence curves of the DE, SaDE and MPEDE algorithms algorithm are shown in Fig. 5. It is observed that the MPEDE and SaDE algorithms reach to the optimal path at about 80 generations, and are much better than the DE algorithm at the convergence speed. On the screenshot window in Fig. 5, we can see that MPEDE algorithm has a little better than the SaDE algorithm for the optimal solution. This conclusion is the similar to that derived from Table 3.

4.3 The Effect of Cost Weight

The cost weights have a key effect on the UAV 3D path planning. In the following experiments, the MPEDE algorithm is used, and the cost weight coefficients ($w1$, $w2$, $w3$, $w4$ *and* $w5$) are set as follows: (0.6, 0.1, 0.1, 0.1, 0.1), (0.1, 0.6, 0.1, 0.1, 0.1), (0.1, 0.1, 0.6, 0.1, 0.1), (0.1, 0.1, 0.1, 0.6, 0.1) and (0.1, 0.1, 0.1, 0.1, 0.6).

Table 4. Effect of cost weights on the total cost

Weight coefficients	Total cost	$T_{P-r} + T_{P-a}$	L_P	θ_P	γ_P	H_P
(0.6, 0.1, 0.1, 0.1, 0.1)	0.100485	0	0.100485	0	0	0
(0.1, 0.6, 0.1,0.1, 0.1)	0.600792	0.000339	0.600453	0	0	0
(0.1, 0.1, 0.6, 0.1, 0.1)	0.100200	0	0.100200	0	0	0
(0.1, 0.1, 0.1, 0.6, 0.1)	0.100205	0	0.100205	0	0	0
(0.1, 0.1, 0.1, 0.1, 0.6)	0.100484	0	0.100484	0	0	0
(0.2, 0.2, 0.2, 0.2, 0.2)	0.200969	0	0.200969	0	0	0

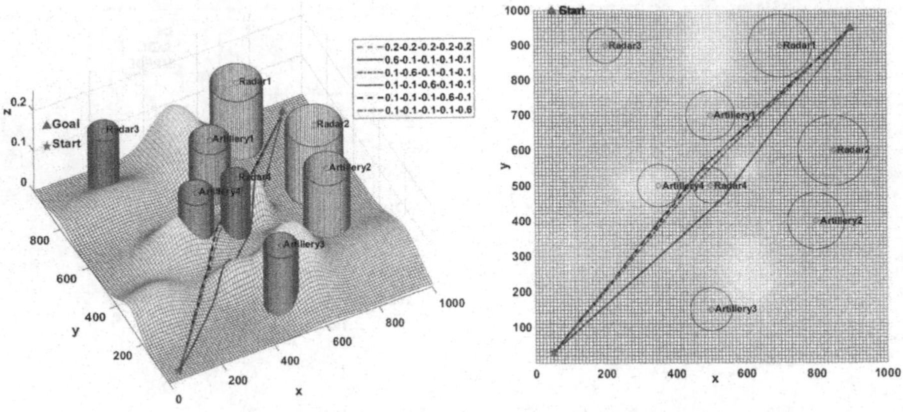

(a) (b)

Fig. 6. Path planning with different weights: (a) Side view, (b) Top view

As it is shown in Table 4, the threat cost $T_{p-r} + T_{p-a}$ increases from 0 to 0.000339 when the weight coefficient of the path length increase up to 0.6. The path length is 0.100076, which is derived from $w1 \times L_p/w2$. The optimized length is smaller than all others. This fact is confirmed in Fig. 6. This indicates that the weight increase of the path length weakens the threat effect. In the UAV path planning problem, the UAV safety is priority to the path length. Therefore, the weight coefficient of the threat cost should be priority to consider.

5 Conclusion

In this work, the 3D environment map of the UAV mission area has been constructed, and multi objectives have been considered, including threats, path length, yaw angle, climbing angle and height. All the objectives have been converted into a single objective optimization problem by the linear weighted method. The DE, SaDE and MPEDE algorithms have been applied to the UAV 3D path planning. The results show that the MPEDE algorithm is more robust, and is feasible in solve the UAV path planning problem.

Acknowledgements. Project supported by the National Natural Science Foundation of China (Grant Nos. 61876169 and 61673404, 61976237), the Key Scientific Research Projects in Colleges and Universities of Henan Province (Grant Nos. 19A120014, 20A120013), and the Science and Technique Project of the China National Textile and Apparel Council (Grant No. 2017054, 2018104), The Research Award Fund for Outstanding Yong Teachers in Zhongyuan University of Technology (Grant No. 2018XQG09, 2019XQG03).

References

1. Zhao, Y.J., Zheng, Z., Liu, Y.: Survey on computational-intelligence-based UAV path planning. Knowl.-Based Syst. **158**(5), 54–64 (2018)
2. Zheng, C., Li, L., Xu, F., Sun, F.: Evolutionary route planner for unmanned air vehicles. IEEE Trans. Robot. **21**(4), 609–620 (2006)
3. Zhang, B., Duan, H.B.: Three-dimensional path planning for uninhabited combat aerial vehicle based on predator-prey pigeon-inspired optimization in dynamic environment. IEEE/ACM Trans. Comput. Biol. Bioinf. **14**(1), 97–107 (2017)
4. Richards, A., How, J.: Aircraft trajectory planning with collision avoidance using mixed integer linear programming. In: Proceedings of the American Control Conference, vol. 3, pp. 1936–1941 (2002)
5. Szczerba, R.J., Galkowski, P., Glicktein, I.S.: Robust algorithm for real-time route planning. IEEE Trans. Aerosp. Electron. Syst. **36**(3), 869–878 (2000)
6. Bhattacharya, P., Gavrilova, M.L.: Voronoi diagram in optimal path planning. In: Proceedings of IEEE International Symposium on Voronoi Diagrams in Science and Engineering, pp. 38–47 (2007)
7. Bhattacharya, P., Gavrilova, M.L.: Roadmap-based path planning - using the voronoi diagram for a clearance-based shortest path. IEEE Robot. Autom. Mag. **15**(2), 58–66 (2008)
8. Raghunathan, A., Gopal, V., Subramanian, D.: Dynamic optimization strategies for 3D conflict resolution of multiple aircraft. AIAA J. Guid. Control Dyn. **27**(4), 586–594 (2004)
9. Abdurrahman, B., Mehmetnder, E.: FPGA based offline 3D UAV local path planner using evolutionary algorithms for unknown environments, In: Proceedings of the Conference of the IEEE Industries Electronics Society, pp. 4778–4783 (2016)
10. Yao, M., Zhao, M.: Unmanned aerial vehicle dynamic path planning in an uncertain environment. Robotica **33**(3), 611–621 (2015)
11. Zhao, Y.J., Zheng, Z., Liu, Y.: Survey on computational-intelligence-based UAV path planning. Knowl.-Based Syst. **158**, 54–64 (2018)
12. Huang, C., Fei, J.Y.: UAV path planning based on particle swarm optimization with global best path competition. Int. J. Pattern Recognit. Artif. Intell. **32**(6), 1859008 (2018)
13. Arantes, J.D., Toledo, F.M., Charleswilliams, B.: Heuristic and genetic algorithm approaches for UAV path planning under critical situation. Int. J. Artif. Intell. Tools **26**(1), 1–30 (2017)
14. Li, B., Gong, L., Yang, W.: An improved artificial bee colony algorithm based on balance-evolution strategy for unmanned combat aerial vehicle path planning. Sci. World J. **2014**(1), 95–104 (2014)
15. Zhang, C., Zhen, Z., Wang, D., Li, M.: UAV path planning method based on ant colony optimization, In: Chinese Control Decision Conference, pp. 3790–3792 (2010)
16. Zhang, X., Duan, H.: An improved constrained differential evolution algorithm for unmanned aerial vehicle global route planning. Appl. Soft Comput. **26**, 270–284 (2015)
17. Neri, F., Tirronen, V.: Recent advances in differential evolution: a review and experimental analysis. Artif. Intell. Rev. **33**(1), 61–106 (2010)
18. Das, S., Suganthan, P.N.: Differential evolution: a survey of the state-of-the-art. IEEE Trans. Evol. Comput. **9**(1), 4–31 (2011)
19. Takahama, T., Sakai, S.: Constrained optimization by applying the alpha constrained method to the nonlinear simplex method with mutations. IEEE Trans. Evol. Comput. **9**(5), 437–451 (2005)
20. Wang, L., Li, L.P.: An effective differential evolution with level comparison for constrained engineering design. Struct. Multidiscip. Optim. **41**(6), 947–963 (2010)

21. Wu, G.H., Mallipeddi, R., Suganthan, P.N.: Differential evolution with multi-population based ensemble of mutation strategies. Inf. Sci. **329**, 329–345 (2016)
22. Das, S., Abraham, A., Chakraborty, U.K.: Differential evolution using a neighborhood-based mutation operator. IEEE Trans. Evol. Comput. **13**, 526–553 (2009)
23. Storn, R., Price, K.V.: Differential evolution-a simple and efficient heuristic for global optimization over continuous spaces. J. Global Optim. **11**(4), 341–359 (1997)
24. Zhang, J., Sanderson, A.C.: JADE: adaptive differential evolution with optional external archive. IEEE Trans. Evol. Comput. **13**, 945–958 (2009)
25. Qin, A.K., Huang, V.L., Sganthan, P.N.: Differential evolution algorithm with strategy adaptation for global numerical optimization. IEEE Trans. Evol. Comput. **13**, 398–417 (2009)

Multi-objective Feature Selection Based on Artificial Bee Colony for Hyperspectral Images

Chun-lin He[1], Yong Zhang[1](✉), Dun-wei Gong[1], and Bin Wu[2]

[1] School of Information and Control Engineering, China University of Mining and Technology, Xuzhou, China
yongzh401@126.com

[2] School of Economics and Management, Nanjing Technology University, Nanjing, China

Abstract. Most of existing band selection methods based on evolutionary optimization emphasis on single indicator, ignoring the whole characteristic of hyperspectral data. This paper establishes a new multi-objective unsupervised band selection model by considering both the information contained in bands and the correlation between bands. And, an improved multi-objective feature selection based on artificial bee colony (MABCBS) is proposed by combining two new operators, i.e., multi-strategy hybrid search mechanism for employ bees and variable space crowing degree based search mechanism for onlooker bees. Appling in two typical data sets, experiment results verify the superiority of the proposed model and the proposed algorithm.

Keywords: Artificial bee colony · Feature selection · Multi-objective · Hyperspectral images

1 Introduction

With the rapid development of imaging spectrometer, the target recognition accuracies in hyperspectral images have been significantly improved, but the data volume and dimension of hyperspectral images have also been obviously increased. Processing high-dimensional data with fewer samples inevitably increases the generalization error of classifier, resulting in the problem "the curse of dimensionality" [1]. Therefore, it is necessary to preprocess hyperspectral images by feature selection approach.

The purpose of feature selection is to select part features from the original feature set, so as to reduce the cost of machine learning and optimize the settled index of performance [2]. Feature selection of hyperspectral image is also called hyperspectral band selection. For hyperspectral band selection problems, the most commonly used methods are sorting [11] and clustering [3–6, 12]. Sorting approach calculates the relevant indicator of each band, and selects bands with better indicators after sorting all the bands based on those indicator values. This approach pays less attention to the correlation between bands, resulting in high correlation between selected bands. Clustering approach usually takes the mutual information between bands as indicators to cluster bands, and then selects

L. Pan et al. (Eds.): BIC-TA 2019, CCIS 1159, pp. 611–621, 2020.
https://doi.org/10.1007/978-981-15-3425-6_48

representative bands from each class. This kind of approach can cover the whole spectral information well and remove redundant bands. However, their performance is greatly affected by the distance formula used.

Due to the ability to find optimal solution of the problem through global search, evolutionary feature selection has gradually become a hot technology in recent years [13–19]. Paoli et al. [20] established a new clustering method for hyperspectral image within the framework of multi-objective particle swarm optimization by considering three criteria, i.e., logarithmic likelihood function, Babbitt statistical distance between classes and minimum description length. Zhu et al. [21] designed a sparse multi-task-based objective function for unsupervised band selection, and proposed an immune cloning algorithm to select the optimal band subset. Zhang et al. [22] proposed an unsupervised band selection method combining fuzzy clustering (FCM) and particle swarm optimization for hyperspectral unsupervised feature selection. Feng et al. [23] designed an objective function based on mutual information and entropy to select key bands, so as to maximize the quantity of information and minimize the redundant information contained in the band subset. Zhang et al. [24] established a multi-objective model based on important information retention and redundant information removal, and designed a multi-objective artificial immune algorithm to optimize the model. As the data volume and dimension of hyperspectral images increase, applying multi-objective evolutionary theory to hyperspectral band selection is still an open problem.

This paper studies the multi-objective evolutionary solving approach for hyperspectral band selection problem. The main contributions are as follows: (1) Establishing an unsupervised multi-objective band selection model of hyperspectral images based on inter-spectral correlation and amount of information; (2) Presenting an improved multi-objective feature selection algorithm based on artificial bee colony. Two new operators, multi-strategy hybrid search mechanism for employ bees and variable space crowing degree based search mechanism for onlooker bees, significantly improve the performance of the proposed algorithm.

2 Related Work

2.1 Multi-objective Optimization

A multi-objective optimization problem is generally described as [7, 8]:

$$\begin{aligned} &\min\ F(x) = (f_1(x), f_2(x), \cdots, f_M(x)) \\ &st.\ x \in S \end{aligned} \tag{1}$$

Where x is a decision variable in the decision space S, and $f_m(x)$ represents the m-th objective function.

Definition 1: Pareto dominance. Let $x_1, x_2 \in S$ is two solutions of the problem (1), if satisfy, $\forall m \in \{1, 2, \cdots, M\}$, $f_m(x_1) \leq f_m(x_2)$, and $\exists i \in \{1, 2, \cdots, M\}$, $f_i(x_1) < f_i(x_2)$, we call that x_1 dominates x_2, denoted as $x_1 \prec x_2$.

Definition 2: Pareto optimal solution. For a solution $x^* \in S$, if there is no other solution x, satisfying $x \prec x^*$, we call that x^* is the optimal solution of the problem (1).

All Pareto optimal solutions compose the Pareto optimal solution set of the problem (1). [9] The map of Pareto optimal solution set in the objective space is called as the Pareto optimal front of the problem (1).

2.2 Artificial Bee Colony Algorithm

Artificial Bee Colony Algorithm (ABC) [10] is a heuristic optimization method mimicking the foraging behavior of bees. It abstracts a food source as a solution of the optimized problem, and all the bees searching for food sources are divided into three categories: employed bees, onlooker bees and scout bees. Employed bees search around food sources first, and a new food source will replace the old one if it has better fitness. When all the employed bees fly back to the hive after searching, the rest unemployed bees in the hive turn into onlookers. After that, the food sources with good fitness are selected with a certain probability, and onlooker bees will search around these food sources. If a food source has not been updated for a long time, its onlooker bee is transformed into a scout to search randomly new food sources.

(1) Phase of employed bee: the update formula of employed bees in this phase is:

$$Cx_{i,d} = x_{i,d} + \varphi(x_{i,d} - x_{j,d}) \tag{2}$$

Where, i is the index of the food source corresponding to the current employed bee, j $(j \neq i)$ is the index of an random food source, d is the dimension of decision variable, φ is a random number within $[-1.1]$, $Cx_{i,d}$ is the new position of the current employed bee.

(2) Phase of onlooker bee: according to the fitness of all the food sources, each onlooker bee continuously searches for a better food source then the selected ones. Taking the i-th food source as an example, its selected probability is:

$$P_i = \frac{fit_i}{\sum_{j=1}^{FN} fit_j} \tag{3}$$

$$fit_i = \begin{cases} \dfrac{1}{f(x_i)+1}, & f(x_i) \geq 0 \\ 1 + |f(x_i)|, & f(x_i) < 0 \end{cases} \tag{4}$$

Where, $f()$ is the objective function, and FN is the number of food sources.

(3) Scout bee phase: In this phase, if a food source cannot be improved after a predetermined number of iterations (*limit*), its associated onlooker bee becomes a scout bee. Then the scout bee will randomly generate a new position for this food source, as follows:

$$x_{i,d} = x_{\min,d} + rand(0,1)(x_{\max,d} - x_{\min,d}) \tag{5}$$

3 Multi-objective Unsupervised Band Selection Model

For a hyperspectral band selection problem, a band subset with the maximum information should be selected in terms of information theory; the correlation between selected bands should be weak in terms of the feature redundancy; and the difference between spectral characteristics of objects should be big by using selected bands. Since there is no label, it is impossible to accurately judge the difference between spectral characteristics of target objects. Therefore, considering the first two criteria, we have the following model:

$$\begin{cases} \min \ f_1(x) = 1/\sum_{i=1}^{K} LC(x_i) \\ \min \ f_2(x) = \frac{2}{K(K-1)} \sum_{i=1}^{K} \sum_{j=i+1}^{K} R(x_i, x_j) \end{cases} \tag{6}$$

Where, K is the number of bands in a subset. f_1 is used to describe the amount of information of the selected bands; f_2 is used to evaluate the correlation between bands.

In Eq. (6), $LC(x_i)$ is coefficient of dispersion, which is used to represent the amount of information contained in the i-th band:

$$LC(x_i) = \delta(x_i)/u_i$$
$$s.t. \ \delta(x_i) = \sqrt{\sum_{z_1=1}^{|z_1|} \sum_{z_2=1}^{|z_2|} (h_i(z_1, z_2) - u_i)^2/|z_1||z_2|} \tag{7}$$

Where, (z_1, z_2) is a pixel point, h_i is the gray matrix of the i-th band, u_i is the gray mean of the i-th band, $|z_1|$ and $|z_2|$ is the length and width of image respectively.

The function f_2 represents the correlation between bands. In this paper we use cross correlation function to evaluate the correlation between bands. Specifically, the cross-correlation function $r(m, n)$ is defined as follows:

$$r(m, n) = \iint \left[h(z_1 + m, z_2 + n) - u_f\right]\left[h'(z_1, z_2) - u_g\right] dz_1 dz_2 \tag{8}$$

Where, $h(z_1, z_2)$ is the gray value of the image at (z_1, z_2), $h'(z_1, z_2)$ is the gray value of the template image at (z_1, z_2); u_h is the gray mean of $h(z_1, z_2)$, and $u_{h'}$ is the gray mean of $h'(z_1, z_2)$. m, n are the change value of the row and column position of pixels respectively. Normalizing Eq. (8), we have:

$$r(m, n) = \frac{\sum_{z_1=1}^{|z_1|} \sum_{z_2=1}^{|z_2|} [h(z_1 + m, z_2 + n) - u_h]\left[h'(z_1, z_2) - u_{h'}\right]}{\sqrt{\left(\sum_{z_1=1}^{|z_1|} \sum_{z_2=1}^{|z_2|} [h(z_1, z_2) - u_h]^2\right)\left(\sum_{z_1=1}^{|z_1|} \sum_{z_2=1}^{|z_2|} [h(z_1, z_2) - u_{h'}]^2\right)}} \tag{9}$$

where, $r(0, 0)$ represents the value of mutual correlation between two images. Suppose the gray matrix of the band x_i is h_i, the gray matrix of the band x_j is h_j, and their gray means are u_i and u_j respectively, the mutual correlation between the two bands is:

$$R(x_i, x_j) = \frac{\sum_{z_1=1}^{|z_1|} \sum_{z_2=1}^{|z_2|} [h_i(z_1, z_2) - u_i][h_j(z_1, z_2) - u_j]}{\sqrt{\left(\sum_{z_1=1}^{|z_1|} \sum_{z_2=1}^{|z_2|} [h_i(z_1, z_2) - u_i]^2\right)\left(\sum_{z_1=1}^{|z_1|} \sum_{z_2=1}^{|z_2|} [h_j(z_1, z_2) - u_j]^2\right)}} \tag{10}$$

Simplify to:

$$\mathrm{R}(x_i, x_j) = \frac{E(h_i - u_i)(h_i - u_i)}{\sqrt{E(h_i - u_i)^2 E(h_i - u_i)^2}} \tag{11}$$

Where, E are the mean function.

4 Multi-objective Feature Selection Algorithm Based on Artificial Bee Colony

In traditional artificial bee colony algorithm, employed and onlooker bees use the same search strategy to search food sources, limiting the search efficiency of the population. By embedding two new operators, i.e., multi-strategy hybrid search mechanism for employed bees and variable space crowing degree based for onlooker bees, this section presents an improved multi-objective artificial bee colony optimization algorithm with variable search strategy.

4.1 Multi-strategy Hybrid Search Mechanism for Employ Bees

In this section, the crowding degree in the objective space $Crow$ is introduced to evaluate the distribution of food sources, and the SC measure is used to evaluate the convergence of bee colony, and different search strategies are selected for employed bees according to the values of $Crow$ and SC. Specifically, when the crowding degree is high, employed bees select a search strategy, which can explore the diversity of food sources. When the crowding degree is low and the convergence becomes bad, employed bees choose a search strategy, which can accelerate the convergence. When the crowding degree is low and the convergence is good, employed bees adopt a crossover strategy with well local exploitation.

The Crowding Measure in the Objective Space

To measure the distribution of food sources in the objective space, a crowding measure based on grid partition is presented. Firstly, we divide the objective space into $L \times L$ grids. After that, the number of food sources contained in each grid is counted, and the non-zero grids are selected. Supposing that the number of food sources contained in the non-zero grids are $e_1, e_2 \cdots, e_k$, respectively, the crowding value of food sources in the objective space is calculated as follows:

$$Crow(t) = \frac{\sum_{i=1}^{k} (e_i - \bar{e})^2}{k} \tag{12}$$

Where, k is the number of non-zero grids, and $\bar{e}$ is the average of food sources contained in the k grids.

The Convergence Measure Based on SC Index

The set covering index (SC) is mainly used to describe the degree of a Pareto front dominated by another [26]. Supposing that there are two Pareto optimal solution sets, A and B, the degree that A dominates B is:

$$SC(A, B) = \frac{|\{b \in B / \exists a \in A : a \succ b\}|}{|B|} \tag{13}$$

Supposing that A is the solution set on the Pareto front obtained in the current generation, and B is the solution set on the Pareto front obtained in the previous generation, SC(A, B) = 1 means that each solution in B is dominated by solutions in A, and SC(A, B) = 0 means that no solution in B is dominated by solutions in A.

Adaptive Updating Strategy for Employed Bees

In the iteration process, this paper determines different strategies for employed bees according to the distribution and convergence of food sources.

(1) $Crow$ is greater than the threshold θ_1. At this time, the bee colony is crowded and its diversity is poor, so we should increase the diversity of the bee colony. Inspired by the literature [25], the following update formula is used:

$$\mathrm{C}x_{i,d} = x_{r,d} + \varphi(x_{i,d} - x_{k,d}), \quad r, k \in \{1, 2, \ldots, N\} \tag{14}$$

(2) $Ocrow \leq \theta_1$, and $SC(A, B) \leq \theta_2$. At this time, the diversity of the bee colony is relatively good, and the optimal solutions are greatly improved compared with the last iteration. Hence, we can continue to accelerate the convergence of the bee colony. Inspired by the idea of particle swarm optimization [28], the global optimal food source is selected to guide the search of employed bees for accelerating the convergence speed of the bee colony. Specifically, the following search formula is used:

$$\mathrm{C}x_{i,d} = x_{best,d} + \varphi(x_{k,d} - x_{r,d}), \quad i \neq k \neq r, r, k \in \{1, 2, \ldots, N\} \tag{15}$$

Where x_{best} is the non-inferior solution randomly selected from the archive.

(3) $Ocrow \leq \theta_1$ and $SC(A, B) \leq \theta_2$. Here, the diversity of the bee colony is relatively good, but the improvement of the optimal solutions is not significant compared to the last iteration. This means that the algorithm is close to the state of convergence to a large extent, and the local exploitation capability of the bee colony should be strengthened. Based on the crossover in genetic algorithm, the following formula is given:

$$Cx_{i,d} = \begin{cases} x_{best,d}, & rand < p_c \\ x_{i,d}, & otherwise \end{cases} \tag{16}$$

4.2 Variable Space Crowing Degree Based Search Mechanism for Onlooker Bees

The traditional update strategy of onlooker bees usually determines a following object (i.e. a food source) for an onlooker bee, according to the distribution and convergence of food sources in the objective space. In order to improve the distribution of those selected following objects in the variable space, a new update strategy based on the variable space crowding degree is proposed.

(1) Variable space crowding degree. Taking the *i*-th food source as an example, and its crowding value $Scrow_i$ in the variable space is:

$$Scrow_i = \sum_j^D \frac{a_{i,j}}{FN} \tag{17}$$

Where, $a_{i,j}$ is the number of elements that have the same value as the *j*-th element of the *i*-th food source in all remaining food sources; *FN* is the number of food sources, and *D* is the dimension of decision variables.

(2) Search strategy of onlooker bees. In the onlooker bee phase, each onlooker bee chooses an object to follow, according to the fitness of food sources. For a multi-objective optimization problem, formulas (3) and (4) are no longer applicable. This paper presents a new selection strategy of following objects by combining ranking and variable space crowding degree. Specifically, the probability that the *i*-th food source is selected by onlooker bees is:

$$\begin{aligned} p_{sel}^i &= newfit_i / \max(newfit_j | j = 1, 2, \cdots, FN) \\ newfit_i &= 1/(rank_i + Scrow_i/D) \end{aligned} \tag{18}$$

Where, $rank_i$ is the rank value obtained by the fast non-dominant sorting in the reference [27]. After determining the food source to follow, each onlooker still uses formula (2) to update its position.

4.3 Steps of the Proposed Algorithm

Based on the improved strategies above, the proposed multi-objective feature selection algorithm based on artificial bee colony is as follows:

Step 1: Preprocess of the hyperspectral data set. Transform the hyperspectral data into grayscale images between 0 and 255.
Step 2: Initialization: Set the size of bee colony N, the numbers of employed bees and food sources $FN = N/2$, the predetermined number of iterations $limit$, the maximum iteration time $T_{\max}$. Initialize *FN* food sources randomly in the search space.
Step 3: The employed bee phase. Implement the search strategy proposed in Sect. 4.1 to update the positions of employed bees. If the new position of an employed bee is superior to its corresponding food source, the new position is used to replace the food source, and is saved in the external archive. This paper updates the reserve set through the method proposed in reference [13].
Step 4: The onlooker bee phase. Implement the onlooker bee search strategy proposed in Sect. 4.2 to update the position of each onlooker bee. If the new position of an onlooker bee is superior to its corresponding food source, the new position is used to replace the food source, and is saved in the external archive.
Step 5: The scout bee phase. If a food source cannot be improved after a predetermined number of iterations (*Limit*), its associated employed bee becomes a scout bee. The scout bee will then generate a new food source randomly, and the old one will be deleted.
Step 6: Determine whether the end condition is satisfied. If yes, stop the algorithm and output the result; Otherwise, return to step 3.

5 Experiment and Analysis

5.1 Compared Algorithms and Experiment Settings

To verify the feasibility of the proposed multi-objective band selection model and the proposed MABCBS algorithm, Maximum Variance Principal Component Analysis (MVPCA) and Waludi are selected as compared algorithms. Set the bee colony size to be 200, the numbers of food sources and employed bees to be 100, the maximum number of iterations to be 300, the size of the archive to be 100, the crowding threshold in the objective space $\theta_1 = 40$, the threshold of convergence $\theta_2 = 0.1$, the crowding threshold in the variable space $\theta_3 = 0.95 \times D$.

We evaluate a feature selection algorithm by measuring the classification performance of the band subset obtained by this algorithm. Two classifier are used, i.e., support vector machine (SVM) and K Nearest Neighbor (KNN). In order to evaluate the performance of classification, overall accuracy (OA) and average accuracy (AA) are used. The overall accuracy represents the ratio of the number of total pixels to the number of pixels classified properly. The average accuracy is the average of the classification accuracy of each category, which reflects the classification effect of each category.

Two real hyperspectral data sets are selected, namely Indian Pines and Pavia University. Indian Pine is obtained by the sensor AVIRIS. It contains 220 bands, whose wavelength ranges from 400 to 2,500 nm, and each band image consists of 145×145 pixels. Pavia University is obtained by the sensor ROSIS. The data is obtained over Pavia University and consists of 610×340 pixels and 115 bands.

5.2 Comparison of Experimental Results

Implementing MVPCA and Waludi once can get an optimal solution, while running the proposed MABCBS algorithm once on a test data set can get a set of Pareto optimal solutions. In the experiment, we select the solution with the best classification accuracy from the obtained Pareto optimal solution set as the final result, and compare it with the results of MVPCA and Waludi. MABCBS is run independently 30 times for each data set to obtain statistical results. Tables 1 and 2 show the OA and AA values obtained by the three algorithms.

Table 1. OA and AA values (mean ± variance) for the three algorithms on Indian Pines

Classifier	Measure	MVPCA	Waludi	MABCBS
SVM	OA	0.7112 ± 0.0702	0.7405 ± 0.0616	**0.7638 ± 0.0325**
	AA	0.6233 ± 0.0932	0.6717 ± 0.1020	**0.6896 ± 0.0804**
KNN	OA	0.6302 ± 0.0154	0.6085 ± 0.0372	**0.6749 ± 0.0297**
	AA	0.5754 ± 0.0181	0.5585 ± 0.0441	**0.6226 ± 0.0504**

Table 2. OA and AA values (mean ± variance) for the three algorithms on Pavia University

Classifier	Measure	MVPCA	Waludi	MABCBS
SVM	OA	**0.8904 ± 0.0394**	0.8646 ± 0.0307	0.8753 ± 0.0325
	AA	0.8055 ± 0.0682	0.7966 ± 0.0722	**0.8145 ± 0.0624**
KNN	OA	0.8132 ± 0.0318	0.8058 ± 0.0188	**0.8434 ± 0.0208**
	AA	0.7949 ± 0.0289	0.7840 ± 0.0185	**0.8201 ± 0.0198**

We can see that, for the dataset Indian Pines, MABCBS achieves better classification accuracy both in terms of OA and AA, compared with MVPCA and Waludi. For the second dataset Pavia University, MABCBS obtains the best OA and AA values for the KNN classifier, and has the best AA value for the SVM classifier. Therefore, compared with MVPCA and Waludi, MABCBS can obtain more competitive optimization results.

5.3 Analyze the Key Operators

This section analyzes the influence of two new strategies, namely, multi-strategy hybrid search mechanism for employ bees and variable space crowing degree based search mechanism for onlooker bees, on the performance of the proposed MABCBS algorithm. For the convenience of comparison, the original artificial bee colony algorithm is called ABCBS. The dataset Indian Pines dataset is selected, and the classification accuracies of bands within [3, 30] are calculated. Figure 1 shows OA and AA values of ABCBS and MABCBS. It can be seen that, with the help of the proposed two new strategy, MABCBS significantly improves the overall and average accuracies of the optimal results.

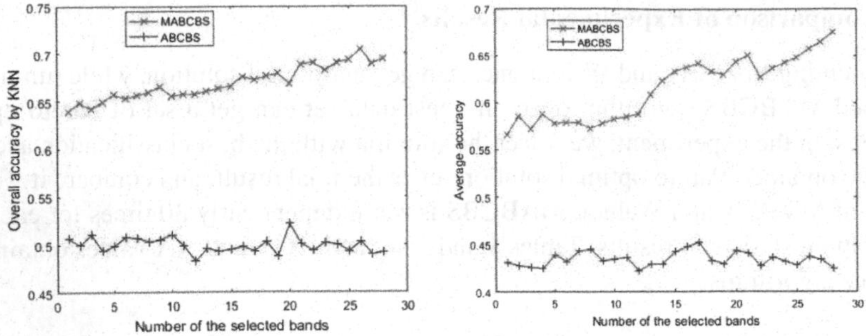

Fig. 1. OA and AA values obtained by MABCBS and ABCBS for Indian Pines

6 Conclusion

In this paper, an unsupervised multi-objective band selection model for hyperspectral images is established by considering the band information and the correlation between bands, and an improved multi-objective artificial bee colony optimization algorithm with variable search strategy is proposed. The proposed two new strategies, multi-strategy hybrid search mechanism for employ bees and variable space crowing degree based search mechanism for onlooker bees, significantly improve the performance of the proposed MABCBS algorithm. Compared with typical band selection algorithms, the experimental results show that the proposed MABCBS algorithm can obtain more competitive optimization results.

References

1. Hughes, G.: On the mean accuracy of statistical pattern recognizers. IEEE Trans. Inf. Theory **14**(1), 55–63 (1968)
2. Li, J., et al.: Feature selection. ACM Comput. Surv. **50**(6), 1–45 (2017)
3. Martinez-Uso, A., Pla, F., Sotoca, J.M.: Clustering-based hyperspectral band selection using information measures. IEEE Trans. Geosci. Remote Sens. **45**(12), 4158–4171 (2008)
4. Cariou, C., Chehdi, K., Moan, S.: BandClust: an unsupervised band reduction method for hyperspectral remote sensing. IEEE Geosci. Remote Sens. Lett. **8**(3), 565–569 (2011)
5. Su, H., Yang, H., Du, Q.: Semi-supervised band clustering for dimensionality reduction of hyperspectral imagery. IEEE Geosci. Remote Sens. Lett. **8**(6), 1135–1139 (2011)
6. Jia, S., Zhen, J., Qian, Y.: Unsupervised band selection for hyperspectral imagery classification without manual band removal. IEEE J. Sel. Top. Appl. Earth Obs. Remote Sens. **5**(2), 531–543 (2012)
7. Cai, X., Mei, Z., Fan, Z., Zhang, Q.: A constrained decomposition approach with grids for evolutionary multi-objective optimization. IEEE Trans. Evol. Comput. **22**(4), 99–104 (2018)
8. Pan, L., He, C., Tian, Y., Wang, H., Zhang, X., Jin, Y.: A classification-based surrogate-assisted evolutionary algorithm for expensive many-objective optimization. IEEE Trans. Evol. Comput. **23**(1), 74–88 (2018)
9. He, C., Tian, Y., Jin, Y., Zhang, X., Pan, L.: A radial space division based evolutionary algorithm for many-objective optimization. Appl. Soft Comput. **61**, 603–621 (2017)
10. Karaboga, D.: An idea based on honey bee swarm for numerical optimization. Erciyes University, Kayseri (2005)

11. Chang, C., Wang, S.: Constrained band selection for hyperspectral imagery. IEEE Trans. Geosci. Remote Sens. **45**(12), 4158–4171 (2007)
12. Martínez-Uso, A., Pla, F., Sotoca, J.M.: Clustering-based hyperspectral band selection using information measures. IEEE Trans. Geosci. Remote Sens. **45**(12), 4158–4171 (2007)
13. Xue, B., Zhang, M., Browne, W., Yao, X.: A survey on evolutionary computation approaches to feature selection. IEEE Trans. Evol. Comput. **20**(4), 606–626 (2016)
14. Zhang, Y., Song, X., Gong, D.: A return-cost-based binary firefly algorithm for feature selection. Inf. Sci. **418**(419), 561–574 (2017)
15. Zhang, Y., Gong, D., Cheng, J.: Multi-objective particle swarm optimization approach for cost-based feature selection in classification. IEEE/ACM Trans. Comput. Biol. Bioinf. **14**(1), 64–75 (2017)
16. Zhang, Y., Cheng, S., Gong, D., Shi, Y., Zhao, X.: Cost-sensitive feature selection using two-archive multi-objective artificial bee colony algorithm. Expert Syst. Appl. **137**, 46–58 (2019)
17. Zhang, Y., Li, H., Wang, Q.: A filter-based bare-bone particle swarm optimization algorithm for unsupervised feature selection. Appl. Intell. **49**(8), 2889–2898 (2019)
18. Xue, B., Zhang, M., Browne, W.: Particle swarm optimization for feature selection in classification: a multi-objective approach. IEEE Trans. Cybern. **43**(6), 1656–1671 (2013)
19. Xue, Y., Xue, B., Zhang, M.: Self-adaptive particle swarm optimization for large-scale feature selection in classification. ACM Trans. Knowl. Discov. Data **13**(5), 1–28 (2019)
20. Paoli, A., Melgani, F., Pasolli, E.: Clustering of hyperspectral images based on multi-objective particle swarm optimization. IEEE Trans. Geosci. Remote Sens. **47**(12), 4175–4188 (2009)
21. Zhu, Y.: Hyperspectral band selection by multitask sparsity pursuit. IEEE Trans. Geosci. Remote Sens. **2**(53), 631–644 (2015)
22. Zhang, M., Ma, J., Gong, M.: Unsupervised hyperspectral band selection by fuzzy clustering with particle swarm optimization. IEEE Geosci. Remote Sens. Lett. **14**(5), 773–777 (2017)
23. Feng, J., Jiao, L., Liu, F.: Unsupervised feature selection based on maximum information and minimum redundancy for hyperspectral images. Pattern Recogn. **51**(C), 295–309 (2016)
24. Zhang, M., Gong, M., Chan, Y.: Hyperspectral band selection based on multi-objective optimization with high information and low redundancy. Appl. Soft Comput. **70**, 604–621 (2018)
25. Kiran, M., Hakli, H., Gunduz, M.: Artificial bee colony algorithm with variable search strategy for continuous optimization. Inf. Sci. **300**, 140–157 (2015)
26. Zhang, X., Tian, Y., Cheng, R.: An efficient approach to nondominated sorting for evolutionary multi-objective optimization. IEEE Trans. Evol. Comput. **19**(2), 201–213 (2015)
27. Deb, K., Pratap, A., Agarwal, S.: A fast and elitist multi-objective genetic algorithm NSGA-II. IEEE Trans. Evol. Comput. **6**(2), 182–197 (2002)
28. Xue, Y., Jiang, J., Zhao, B., Ma, T.: A self-adaptive artificial bee colony algorithm based on global best for global optimization. Soft. Comput. **22**(9), 2935–2952 (2018)

Meta-heuristic Hybrid Algorithmic Approach for Solving Combinatorial Optimization Problem (TSP)

Usman Ashraf[1(✉)], Jing Liang[1(✉)], Aleena Akhtar[1], Kunjie Yu[1], Yi Hu[1], Caitong Yue[1], Abdul Mannan Masood[2], and Muhammad Kashif[3]

[1] School of Electrical Engineering, Zhengzhou University, Zhengzhou 450001, China
usmanjadoon@ymail.com, liangjing@zzu.edu.cn

[2] School of Computer Science and Engineering, Jiangsu University of Science and Technology, Jiangsu, China

[3] School of Computer Science and Technology, Beijing Institute of Technology, Beijing, China

Abstract. Solving and optimizing combinatorial problems require high computational power because of their exponential growth and requirement of high processing power. During this study, a hybrid algorithm (Genetic Ant Colony Optimization Algorithm) is proposed in comparison with standard algorithm (Ant Colony Optimization Algorithm). Further the parameters for Ant Colony Optimization Algorithm are instinctively tuned to different levels of all heuristics to obtain suboptimal level, then multiple crossovers and mutation operators are used alongside those selected parameters while generating results with hybrid algorithm. The main emphasis of the proposed algorithm is the selection and tuning of parameters, which is extremely influential in this case. The algorithm has been tested on six benchmarks of TSPLIB. The results were compared with standard ACO algorithm, the hybrid algorithm outperformed the standard ACO algorithm.

Keywords: Ant Colony Algorithm (ACO) · Genetic Algorithm (GA) · Traveling Salesman Problem (TSP) · Heuristics · Meta-heuristics · Optimization

1 Introduction

Finding optimal solution is the main focus of combinatorial optimization from a finite set of objects. These optimization problems have always been a prime focus in every field of Engineering. With the increasing importance of Optimization problems the main focus is to utilize less resources and make it more efficient [1]. Traveling Salesman Problem, TSP for short, can be best defined as "Given a number of specific cities along with the cost from Point A to all other points, finding the efficient and cheapest way back to initial point after visiting all cities

L. Pan et al. (Eds.): BIC-TA 2019, CCIS 1159, pp. 622–633, 2020.
https://doi.org/10.1007/978-981-15-3425-6_49

once and only once" [2]. One complete cycle through all the points once and only once by a traveling salesman is called a tour.

Traveling Salesman Problem (TSP) is known Non Polynomial(NP) hard problem which is excessively studied for the research work in field of computer science and especially Artificial Intelligence [3]. The general description of the symmetric TSP can be seen in Fig. 1. There are 5 cities which are called nodes and the cost from one node to every other node is also labeled [2,4].

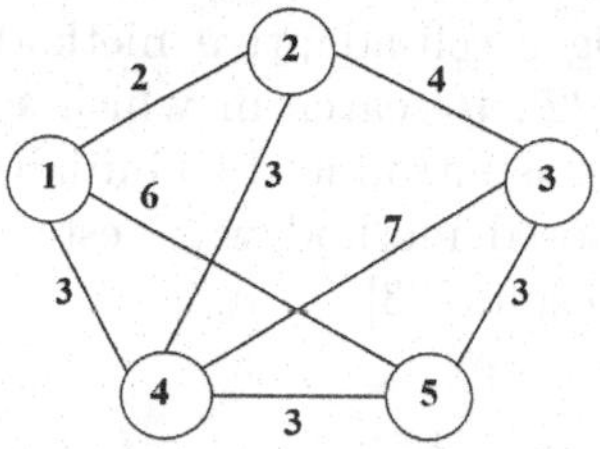

Fig. 1. General TSP representation.

TSP is generally interpreted by the complete edge weighted graph,

$$G = (V, E) \tag{1}$$

Where V is absolute number set, for number of cities or nodes while $E \subseteq V \times V$ which represents the set of edges. Each are $(i, j) \in E$ assign value d_{ij} to the cities from city i to city j where i and j belongs to V. The set of TSP data can either be symmetric or asymmetric. The difference between the symmetric and asymmetric is that the distance between two nodes remains constant from i to j and j to i presented $d_{ij} = d_{ji}$ while in asymmetric the i to j and j to i is never the same which can be interpreted as $d_{ij} \neq d_{ji}$. Optimal solution for TSP is the permutation (π) of the cities indices from $\{1...n\}$ such that $f(\pi)$ is the possible minimum which can be represented mathematically by [5–7];

$$f(\pi) = \sum_{i-1}^{n-1} d\,\pi(i)\;\pi(i+1) + d\,\pi(n)\;\pi(1) \tag{2}$$

In recent decades, researchers have focused on different computational intelligence algorithms to solve the traveling salesman problem, such as neural network, simulated annealing method, genetic algorithm GA, particle swarm optimization and so on [8,9]. An Italian scholar M.Dorigo first proposed it, taking full advantage by introducing similarity between both, travelling salesman problem TSP and ant colony search process [10]. Recently researchers are using ant colony algorithm to solve non-deterministic polynomial (NP) problems, e.g. Traveling Salesman Problem [11], wireless sensor networks [12], Capacitated Location Routing Problem, software project scheduling problem [13], Course Timetabling Problem [14], time series prediction and dynamic topology optimization [15], etc.

In a research for improved ACO for solving TSP, local optimal search strategy and change of parameters were introduced to avoid premature stagnation phenomenon of standard ACO [16]. In another study researchers compared three algorithms to solve TSP Ant colony algorithm, Evolution based and Annealing algorithm and the results concluded that the nature inspired algorithm ACO has outperformed the Evolution based and Annealing algorithm [17].

Multiple literatures have discussed the use of GA for solving TSP, and many literatures generally focused on designing the representation of solution, and initializing an initial population [18,19], designing crossover, mutation and selection operator and designing a self-adaptive method to use multiple mutation or crossover operators [20–22]. Research in which a new method called greedy permuting method (GPM) was introduced to initialize an initial population and then the efficiency of proposed method was tested on some TSP benchmark problems, with promising results [23].

2 Methodology

Generally, the parameter tuning for GA is problem oriented and set as per requirements but in case of TSP it has been tuned as per the work requirement. The selection of parameter which should be tuned are selection, cross over and mutation probability which in general varies from 0.1 to 1. There should be a proper balance between exploitation and exploration. GA always fails because of the not being balanced properly between crossover and mutation probability. Proper allocation of values will help the algorithm not to be trapped in local solution.

2.1 Ant Colony Optimization Algorithm

For the purpose of finding the best parameters for ACO TSPLIB's instance Berlin52 was taken for instance and parameters were tuned to see the effects on the results. The factors which effect the results are discussed as follows. At this point basic parameters used are as shown in Table 1.

Table 1. Basic parameters

α	β	ρ	ϱ	$MaxIt$	$nAnT$
1	2	0.9	10	1000	30

A. Pheromone Rate. First, the other parameters will be kept constant, only the value of α will be changed. Result can be seen as shown below in Table 2.

Table 2. Pheromone rates

α	β	ρ	ϱ	$MaxIt$	$nAnT$	Best solution
1	**2**	**0.9**	**10**	**1000**	**30**	**7548.9927**
3	2	0.9	10	1000	30	7758.7176
5	2	0.9	10	1000	30	8971.4955
7	2	0.9	10	1000	30	9090.2158
9	2	0.9	10	1000	30	8804.8284

B. Heuristic Rate. Now keeping other parameters constant while changing the value of β, and observe the changes in results as shown Table 3.

Table 3. Effects of Heuristic rates

α	β	ρ	ϱ	$MaxIt$	$nAnT$	Best solution
1	1	0.9	10	1000	30	8083.8025
1	**2**	**0.9**	**10**	**1000**	**30**	**7548.9927**
1	3	0.9	10	1000	30	7663.5851
1	5	0.9	10	1000	30	7721.2979
1	7	0.9	10	1000	30	7681.4537
1	9	0.9	10	1000	30	7681.4537

C. Pheromone Evaporation Rate. Now keeping other parameters constant while changing the value of ρ, and observe the change in results as shown in Table 4.

Table 4. Effects of pheromone evaporation

α	β	ρ	ϱ	$MaxIt$	$nAnT$	Best solution
1	2	0.1	10	1000	30	7677.6608
1	2	0.3	10	1000	30	7662.8103
1	2	0.5	10	1000	30	7601.0492
1	2	0.7	10	1000	30	7629.4472
1	**2**	**0.9**	**10**	**1000**	**30**	**7548.9927**

2.2 Genetic Ant Colony Optimization Algorithm

The hybrid algorithm solving TSP can be expressed as follows:

Algorithm 1. Genetic Hybrid Ant Colony Optimization Algorithm

1) $c \leftarrow 0$ (c is iteration number).
2) Generate 100 tours through Ant Colony Optimization algorithm,
   a. $nc \leftarrow 0$ (nc is iteration number).
   b. Choose the next city j according to

$$p_{ij}^{k}(t) = \begin{cases} \frac{\tau_{ij}^{\alpha}(t).\eta_{ij}^{\beta}}{\sum_{s \in allowed_k} \tau_{is}^{\alpha}(t).\eta_{is}^{\beta}} & f\, j \in allowed_k \\ 0 & otherwise \end{cases}$$

   c. Update trail (pheromone) values.

$$\tau_{ij}(t+n) = \rho\tau_{ij}(t) + \Delta\tau_{ij}$$
$$\Delta\tau_{ij} = \sum_{k=1}^{m} \Delta\tau_{ij}^{k}$$

$$\Delta\tau_{ij}^{k} = \begin{cases} \frac{Q}{L_k} & \text{if k} - \text{th ant uses edge(i,j) in its tour} \\ 0 & otherwise \end{cases}$$

   d. $nc \leftarrow nc+1$
   e. If the iteration number nc reaches the maximum iteration number, then go to Step a. Otherwise, go to Step 3.
3) Choose the better 30 tours based on the cost L_k (length of tour done by ant k) from these 100 tours, and pheromone laid on edge of these 30 better tours.
4) Crossover and Mutate the tours and calculate the new evaluated tours,
   a. $mc \leftarrow 0$ (mc is iteration number).
   b. Choose the next city j
   c. $mc \leftarrow mc+1$
   d. If the iteration number mc reaches the maximum iteration number, then go to Step a. Otherwise, go to Step 5.
5) Compute $L_k(k = 1, 2, .., m)$ (L_k is the length of tour done by ant k). Save the current best tour.
6) $c \leftarrow n+1$
7) If the iteration number c reaches the maximum iteration number, then go to Step 8. Otherwise, go to Step 2.
8) Print the current best tour.

In simple ACO Algorithms, in start the value of pheromone matrix is equal. Multiple iterations process is needed by Ants to find the best tour. Trail can be found on each visited edge, after an ant completes a tour. No matter the tour is better or worse, even for the worse tour ants lay the trail on each edge, disturbing the ants which are following them which results in the slower convergence speed of ACO algorithm. At first the length of the tour is calculated and then it is compared with the given value. If the calculated value is lesser than the given value then the trail value is updated, else it remains the same. This results in improvement of ants laying on the better tours, and also affect the ants which are following. A large number of tours can be generated (e.g. 100), out of which some better tours (e.g. 30) are then selected. These selected best tours undergo mutation and crossover operations under Genetic Algorithm, and after these operations are performed then we can calculate the cost of newly mutated tours.

A. Genetic Ant Colony Optimization Parameters. Results from basic Ant Colony Optimization Algorithm shows that the following parameters would be proved to get the better results. Parameters are shown in Table 7.

B. Genetic Algorithm Parameters. In this methodology we only used two fundamental Genetic Algorithm parameters, namely mutation and crossover. Several Mutation methods are used to improve the discussed hybrid algorithm.

Mutation

1. **Mutation A**
   Strategy: Swap a randomly selected city with next visiting city. Detail: Choose one city j_1 randomly and then swap it with the next visiting city in the tour C_0, this will generate a new tour calling C_1.
2. **Mutation B**
   Strategy: swap a randomly selected city with the city after next visiting city. Detail: Choose one city j_1 randomly and then swap it with the city after next visiting city in the tour C_0, this will generate a new tour calling C_1.
3. **Mutation C**
   Strategy: Swap a randomly selected city and the city adjacent to it with the next visiting city.
   Detail: Choose random city j_1, then swap $j_1$and next adjacent city with after coming next city the tour C_0, this will generate a new tour calling C_1.
4. **Mutation D**
   Strategy: Swap randomly chosen 2 adjacent cities with 2 other random adjacent cities.
   Detail: Choose two adjacent cities j_1 and j_2 randomly and then swap them adjacently at a randomly selected place in the tour C_0, this will generate a new tour calling C_1.
5. **Mutation E**
   Strategy: Put randomly chosen 2 adjacent cities at a random place.
   Detail: Choose two adjacent cities j_1 and j_2 randomly and then put them adjacently at a randomly selected place in the tour C_0, this will generate a new tour calling C_1.
6. **Mutation F** Strategy: 2 different randomly selected cities put adjacently at a place adjacently. Detail: Choose two random cities j_1 and j_2, then put them adjacently at a randomly selected place in the tour C_0, this will generate a new tour calling C_1.
7. **Mutation G**
   Strategy: 2 different randomly selected cities put adjacently at 2 different random locations.
   Detail: Choose two random cities j_1 and j_2, then swap them at randomly selected places separately in the tour C_0, this will generate a new tour calling'C_1.

Crossover

Similarly, several Crossover methods were used to improve the currently discussed hybrid algorithm. Let's suppose we have two parent tours given by old1 and old2. A substring from old2 for crossover purpose will be selected and it will be called as donor.

1. **Crossover A**
   Strategy: Replace random length substring at the same place in the old1.
   Detail: We will swap a substring of random length generated from old2. This substring will be then replaced at the exact starting location of the old1.
2. **Crossover B**
   Strategy: Replace random length substring at the beginning at the old1.
   Detail: A substring is selected randomly of random length. Then we insert the substring into the beginning of old1.
3. **Crossover C**
   Strategy: Replace random length substring at the end at the old1.
   Detail: A substring is selected randomly of random length. Then we insert the substring into the end of old1.
4. **Crossover D**
   Strategy: Replace random length substring at the random place at the old1.
   Detail: A substring is selected randomly of random length. Then we insert the substring at a random place of old1.
5. **Crossover E**
   Strategy: Replace random length substring starting in first half and ending in second half of old2, placed at a random location.
   Detail: substring will be selected from random place of first half of the string and ending in second half of the string. The length of this substring will random. Then we insert the substring into random location of old1.
6. **Crossover F**
   Strategy: Replace substring of $1/4^{th}$ length of old2 starting in first half of old2, placed at a random location.
   Detail: this substring will be selected from random place of first half of the string. The length of this substring will be $1/4^{th}$ of the length of full string. Then we insert the substring into random location of old1.
7. **Crossover G**
   Strategy: Replace substring of $1/4^{th}$ length of old2 starting in first $1/3^{rd}$ part of old2, placed at a random location.
   Detail: This substring will be selected from random place of first $1/3^{rd}$ part of the string. The length of this substring will be $1/4^{th}$ of the length of full string. Then we insert the substring into random location of old1.
8. **Crossover H**
   Strategy: Replace substring of $1/4^{th}$ length of old2 starting in middle half part of old2, placed at a random location.
   Detail: this substring will be selected from random place of first middle half of the string. The length of this substring will $1/4^{th}$ of the length of full string. Then we insert the substring into random location of old1.

9. **Crossover I**
   Strategy: Replace substring of 4 adjacent cities from old2 starting in middle half part of old2, placed at a random location.
   Detail: this substring will be selected from random place of first middle half of the string. The length of this substring will be of 4 cities. Then we insert the substring into random location of old1.
10. **Crossover J**
   Strategy: Replace substring of 4 adjacent cities from old2, placed at a random location.
   Detail: this substring will be selected from random place of first middle half of the string. The length of this substring will be of 4 cities. Then we insert the substring into random location of old1.

3 Results

Ant Colony Optimization and Genetic Ant Colony Optimization, which are discussed above have been run for 10 times with the best parameters known as per previous testing. After applying the best parameters, the algorithms have been run for 10 times on the six Benchmarks of TSPLIB varying from 16 cities to 120 cities. The Benchmarks used as the test bed are Ulysses16, Ulysses22, bayg29, att48, berlin52 and g120m. All results are rounded up to two decimal points. The results obtained for the benchmark of TSP are shown below. The Results are compared with basic ACO Algorithm, and it can be clearly noticed that the hybrid Algorithm outperformed the basic ACO algorithm. Environment used to run the above said strategies was suitable enough to obtain good results and has the following characteristics.

Table 5. Minimum result comparison

Algorithms	Ulysses16	Ulysses22	bayg29	att48	berlin52	g120m
Basic ACO algorithm	53.75	56.61	9054.3	35945	8180.36	612.45
Crossover operator A + Mutation operator A	53.65	56.52	8988.2	35676	7835.3	604.61
Crossover operator B + Mutation operator A	51.65	56.76	8497.09	21089.2	8024.58	1245.26
Crossover operator B + Mutation operator D	55.21	65.59	9625.45	35298.62	5970.83	912.95
Crossover operator B + Mutation operator F	54.25	57.76	9871.96	13908.4	8561.05	893.93
Crossover operator E + Mutation operator D	56.92	59.5	9258.99	27373.26	7893.41	852.49
Crossover operator J + Mutation operator F	53.22	65.57	6357.52	37003.63	8540.21	1437.02
Minimum values	51.65	56.52	6357.52	13908.4	5970.83	604.61

- **Processor:** INTEL(R) core(TM) i7-6500U CPU@2.50 GHZ
- **RAM:** 16 Gb
- **Graphic Card:** AMD Radeon R5 M335
- **Operating System:** Microsoft Windows 10
- **Simulation Software:** MATLAB R2015a

All the benchmarks of the TSPLIB varying from 16 cities to 120 cities ran ten times to get the better results. Each of the TSPLIB has been tested with all the combinations of proposed mutations and crossovers that produces 70 different variations of Hybrid Algorithm. To check the difference between different variations of Hybrid Algorithm, 5 minimum values have been selected from each TSPLIB. Further to refine these results only highly productive variations of Hybrid Algorithm were selected that have been shown in Table 5 and then plotted in Fig. 2. Where it can be easily deduced that as the number of cities or the distance between the cities increases hybrid algorithm produces better results than basic ACO.

Further to check the stability of the Hybrid Algorithm, statistical analysis has been performed on the data, and the values of mean and standard deviation are presented in Table 6.

Table 6. Mean (μ) and standard deviation (σ)

Algorithms	Ulysses16	Ulysses22	bayg29	att48	berlin52	g120m
Crossover Operator A+ Mutation Operator A	59.12 ±4.41E+0	61.96 ±4.33E+0	9328.06 ±6.52E+2	37144.06 ±2.63E+3	8076.23 ±5.65E+2	698.14 ±4.88E+1
Crossover Operator B+ Mutation Operator A	72.09 ±5.04E+0	69.67 ±4.87E+0	10968.28 ±7.68E+2	25714.56 ±1.80E+2	9418.77 ±6.59E+2	1428.16 ±9.99E+2
Crossover Operator B+ Mutation Operator D	78.89 ±5.52E+0	91.11 ±6.37E+0	12397.15 ±8.67E+2	52689.81 ±3.69E+3	8698.34 ±6.08E+2	1174.95 ±8.22E+1
Crossover Operator B+ Mutation Operator F	77.26 ±5.40E+0	79.03 ±5.53E+0	11944.33 ±8.36E+3	20122.17 ±1.41E+3	10440.93 ±7.30E+2	1146.93 ±8.02E+1
Crossover Operator E+ Mutation Operator D	83.99 ±5.88E+0	82.77 ±5.79E+0	13042.37 ±9.13E+2	37047.24 ±2.59E+3	9916.16 ±6.94E+2	1248.59 ±8.74E+1
Crossover Operator J+ Mutation Operator F	79.44 ±5.56E+0	84.09 ±5.89E+0	8556.23 ±5.99E+2	48696.73 ±3.41E+3	10385.86 ±7.27E+2	1639.78 ±1.15E+2

The ANT Colony Optimization Algorithm has been run for the above mentioned Benchmarks for 10 times and then those results are used as benchmarks to compare with other results generated from hybrid Genetic Ant Colony Algorithm. Best solution has been taken from the 10 test runs. The parameter set for the algorithms are as follows in Table 7.

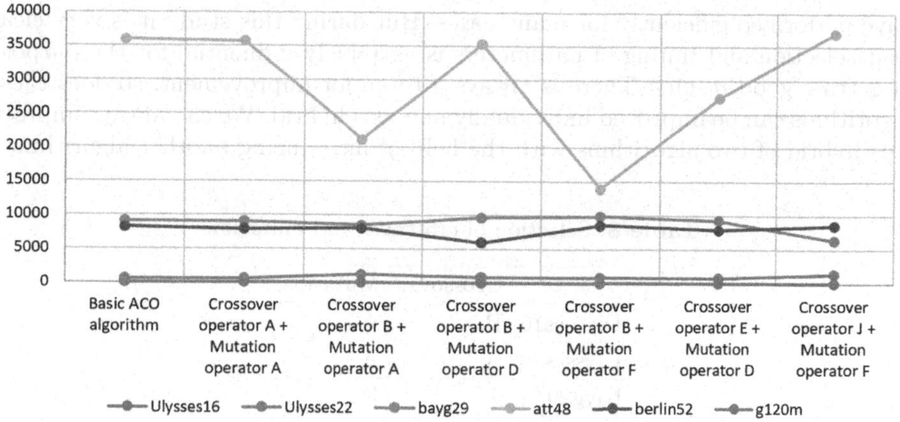

Fig. 2. Results comparison

Table 7. Genetic Ant colony optimization parameters

Parameter	Value
Pheromone exponential weight (α)	1
Heuristic exponential weight (β)	2
Evaporation rate (ρ)	0.9
ϱ	10
Number of ants	30
Selected best tours from ACO	30
Number of tours generated ($MaxIt$)	1000
Number of mutations	1000

4 Conclusion

This research proposed a hybrid algorithm to see the change in results with respect to standard algorithm (Ant Colony Optimization Algorithm) and Hybrid algorithm (Genetic Ant Colony Optimization Algorithm). Further the perimeters for Ant Colony Optimization Algorithm and values are instinctively tuned to different levels of all heuristics to obtain suboptimal level. And then multiple crossovers and mutations are used alongside those selected parameters while generating results with hybrid algorithm. From the results in Sect. 3 the following information can be concluded for the selection of crossover and mutation as shown in Table 8. The results show that if the mutation, crossover and their combination in the proposed hybrid Genetic Ant Colony Algorithm is chosen accordingly, this hybrid algorithm can generate better results. There are other hybrid algorithms that have been presented with the passage of time which have performed efficiently for many cases. But during this study it is very clear that selection and tuning of parameters is extremely influential for the purpose of getting good results. There is always a room for improvement, so does these algorithms can be improved individually and as a hybrid. We can always improve this hybrid of two algorithms with the help of introducing another algorithm.

Table 8. Selection of crossover and mutation

	Crossover	Mutation
Ulysses16	B	A
Ulysses22	A	A
bayg29	J	F
att48	B	F
berlin52	B	D
g120m	A	A

Acknowledgement. This work is supported by the National Natural Science Foundation of China (61922072, 61876169, 61673404, 61976237).

References

1. Ilavarasi, K., Joseph, K.: Variants of travelling salesman problem: a survey. In: International Conference on Information Communication and Embedded Systems, ICICES 2014 (2014)
2. David, A., William, C.: The Traveling Salesman Problem: A Computational Study. Princeton Series in Applied Mathematics, pp. 1–5, 12–17. Princeton University Press, Princeton (2007)
3. Malik, M., Iqbal, A.: Heuristic approaches to solve traveling salesman. TELKOMNIKA Indones. J. Electr. Eng. **15**(2), 390–396 (2015)
4. CO@W: History of the TSP. The Traveling Salesman Problem. Georgia Tech, October 2009. http://www.tsp.gatech.edu/index.html
5. Amanur, R.: The Traveling Salesman problem, Indiana State University USA, 11 April 2012
6. Sureja, N., Chawda, B.: Random travelling salesman problem using SA. Int. J. Emerg. Technol. Adv. Eng. **2**(3), 621–624 (2012)
7. Wang, S., Zhao, A.: An improved hybrid genetic algorithm for traveling salesman problem. In: IEEE Proceedings of the International Conference on Computational Intelligence and Software Engineering (2009)
8. Nagata, Y., Kobayashi, S.: A powerful genetic algorithm using edge assembly crossover for the traveling salesman problem. INFORMS J. Comput. **25**(2), 346–363 (2013)
9. Tian, P., Yang, Z.: An improved simulated annealing algorithm with genetic characteristics and the traveling salesman problem. J. Inf. Optimiz. Sci. **14**(3), 241–255 (2013)

10. Birattari, M., Pellegrini, P., Dorigo, M.: On the invariance of ant colony optimization. IEEE Trans. Evol. Comput. **11**(6), 732–742 (2008)
11. Lizárraga, E., Castillo, O., Soria, J.: A method to solve the traveling salesman problem using ant colony optimization variants with ant set partitioning. Stud. Comput. Intell. **451**(1), 237–246 (2013)
12. Ho, J., Shih, H., Liao, B., Chu, S.: A ladder diffusion algorithm using ant colony optimization for wireless sensor networks. Inf. Sci. **192**(6), 204–212 (2012)
13. Xiao, J., Ao, X., Tang, Y.: Solving software project scheduling problems with ant colony optimization. Comput. Oper. Res. **40**(1), 33–46 (2013)
14. Nothegger, C., Mayer, A., Chwatal, A., Raidl, G.: Solving the post enrolment course timetabling problem by ant colony optimization. Ann. Oper. Res. **194**(1), 325–339 (2012)
15. Yoo, K., Han, S.: A modified ant colony optimization algorithm for dynamic topology optimization. Comput. Struct. **123**(4), 68–78 (2013)
16. Yang, X., Wang, J.: Application of improved ant colony optimization algorithm on traveling salesman problem. In: IEEE 2017 International Conference on Robotics and Automation Sciences (ICRAS) (2017)
17. Muhammad, K., Gao, S.: Comparative analysis of meta-heuristic algorithms for solving optimization problems. In: Advances in Intelligent Systems Research, volume 163, Proceedings of the 2018 8th International on Management, Education and Information, MEICI 2018 (2018)
18. Deng, Y., Liu, Y., Zhou, D.: An improved genetic algorithm with initial population strategy for symmetric TSP. Math. Probl. Eng. **2015**(3), 1–6 (2015)
19. Luo, Y., Lu, B., Liu, F.: Neighbor field method for population initialization of TSP. J. Chongqing Univ. **32**(11), 1311–1315 (2009)
20. Hassanat, A., Alkafaween, E.: On enhancing genetic algorithms using new crossovers. Int. J. Comput. Appl. Technol. **55**(3), 202–212 (2017)
21. Noraini, M., John, G.: Genetic algorithm performance with different selection strategies in solving TSP. In: Proceedings of the World Congress on Engineering, WCE 2011, vol. 2, pp. 1134–1139 (2011)
22. Serpell, M., Smith, J.: Self-adaptation of mutation operator and probability for permutation representations in genetic algorithms. Evol. Comput. **18**(3), 491–514 (2010)
23. Liu, J., Li, W.: Greedy permuting method for genetic algorithm on travelling salesman problem. In: IEEE 8th International Conference on Electronics Information and Emergency Communication (ICEIEC) (2018)

An Effective Two-Stage Optimization Method Based on NSGA-II for Green Multi-objective Integrated Process Planning and Scheduling Problem

Xiaoyu Wen, Kanghong Wang(✉), Haiqiang Sun, Hao Li, and Weiwei Zhang

Henan Key Laboratory of Intelligent Manufacturing of Mechanical Equipment, Zhengzhou University of Light Industry, Zhengzhou 450002, China
wangkanghongwin@foxmail.com

Abstract. At present, green manufacturing plays a vital role in the manufacturing industry. In a sense, Integrated Process Planning and Scheduling (IPPS) itself is a complex NP-Complete problem. However, multi-objective problems in green model are more difficult to address. Consequently, solving IPPS problem under green manufacturing environment is a challenging job. The Green Multi-Objective Integrated Process Planning and Scheduling (GMOIPPS) problem is studied in this paper. A mathematical model for GMOIPPS including efficiency objective and energy consumption objective is established. An effective two-stage optimization method is adopted to deal with GMOIPPS problem. The basic NSGA-II algorithm is employed to optimize the flexible process planning stage and provide the near-optimal process plans for job shop scheduling stage dynamically. An improved NSGA-II algorithm (INSGA-II) with N5 neighborhood structure is designed to find the non-dominated scheduling plans in job shop scheduling stage. Three instances with different scales are constructed to verify the validity of the proposed model and optimization method. The experimental results show that the proposed two-stage optimization method can effectively solve the GMOIPPS problem.

Keywords: Green manufacturing · Integrated Process Planning Scheduling · NSGA-II · N5 neighborhood structure

1 Introduction

As the manufacturing industry is severely affected by the environment, the traditional manufacturing model can't adapt well to the needs of social development [1]. The trend of green manufacturing is increasing on a global scale. Green manufacturing is a sustainable strategy based on economic efficiency and environmental friendliness [2]. The development concept can effectively reduce carbon emissions and reduce environmental pollution. The green manufacturing model is a modern manufacturing model that takes into account environmental impacts and resource efficiency. Process planning and scheduling are two

L. Pan et al. (Eds.): BIC-TA 2019, CCIS 1159, pp. 634–648, 2020.
https://doi.org/10.1007/978-981-15-3425-6_50

significant sub-system in manufacturing system. Process planning provides a near-optimal process plans for the scheduling system, and the scheduling system arranges the start-up time and processing sequence of the jobs on different machines according to these process plans. There is a very close contact between process planning and scheduling. Therefore, for green manufacturing development needs, integrated process planning and scheduling (IPPS) plays a more crucial role in the green manufacturing, which have a pivotal impact on the product processing capability, resource utilization and production efficiency of the manufacturing system [3,4]. The solution of solving IPPS problem based on green manufacturing model can improve production efficiency, shorten manufacturing cycles, eliminate site resource conflicts, and at the same time save energy, reduce costs, and reduce carbon emissions. It has very vital theoretical value and practical significance [5]. Based primarily on the existing research, this paper studies the problem of the Green Multi-Objective IPPS (GMOIPPS), and establishes a mathematical model of GMOIPPS that takes into account both efficiency and green objectives [6]. The optimized model is designed to minimize the maximum completion time, reduce the total carbon emission and shorten the tardiness. An effective two-stage optimization method based on NSGA-II is designed to solve the proposed model. The basic NSGA-II algorithm is employed to optimize the flexible process planning stage and provide the near-optimal process plans for job shop scheduling stage dynamically. An improved NSGA-II algorithm (INSGA-II) with N5 neighborhood structure is designed to find the non-dominated scheduling plans in job shop scheduling stage. Three instances with different scales are constructed to verify the validity of the proposed model and optimization method. The experimental results are very obvious to highlight the feasibility and superiority of the proposed optimization method. The reminder of this research is arranged as follow. Section 2 presents the related works. Section 3 gives the formulation of GMOIPPS. Section 4 elaborates the proposed two-stage optimization method. Experiments and discussions are illustrated in Sect. 5. Conclusion and future works is reported in Sect. 6.

2 Related Works

Nowadays, the developing trend of the manufacturing enterprises are moving towards an efficient, intelligent, green model, the role of IPPS in manufacturing is becoming more and more prominent. Rajemi et al. [7] studied the optimization method of optimal turning conditions under the lowest energy consumption objective. Li et al. [8] studied the optimization problem of high-efficiency low-carbon cutting parameters optimized by CNC turning, and optimized the multi-objective optimization model with minimum processing time (high efficiency) and minimum carbon emission (low carbon). Yin et al. [9] divided the processing energy consumption into two categories: direct energy consumption, indirect energy consumption, and established corresponding processing energy consumption assessment models. Liu et al. [10] studied the energy consumption in the job shop scheduling environment, established a non-processing stage energy consumption and total weighted delay time multi-optimization target model, and used a non-dominated sorting genetic algorithm to obtain the Pareto optimal front end. May et al. [11] analyzed the impact of four workshop scheduling strategies on job shop scheduling completion time and total energy consumption, and proposed a green genetic algorithm. Huang et al. [12] established an energy optimization model for process planning and shop scheduling. Using idle power to calculate the energy consumed by the computer, a hybrid algorithm based on simulated annealing and genetic algorithm was designed to solve the completion time and energy, as well as the Multi-objective optimization of consumption. Min [13] established an energy-saving process planning and shop scheduling integration model, in which the energy consumption index includes the total energy consumption of all working machines in the manufacturing system during the production and non-production phases. A multi-objective integrated optimization model that optimizes the machining process, process route, job machining sequence and machine tool assignment of each process to achieve the minimum carbon emission and maximum completion time of the manufacturing process is proposed. NSGA-II was designed to solve the model. Zhang et al. [14] established the machine output rate as a variable. A job-oriented scheduling model for energy consumption is proposed, and a multi-objective genetic algorithm is proposed for solving the model. Zhang et al. [15] proposed a green IPPS optimization model based on nonlinear programming model. Using the kinetic-based machine tool energy consumption model, the energy consumption calculation model of mechanical parts manufacturing process was established. Liu et al. [16] proposed a multi-objective IPPS model with minimum carbon emission and maximum completion time in the manufacturing process, taking into account carbon emissions caused by power consumption of machine tools during processing, and carbon emissions caused by coolant consumption. There are carbon emissions from the consumption of lubricants and carbon emissions from the energy consumption of handling equipment between jobs. Li et al. [17] established a batch IPPS problem model with the lowest total energy consumption and minimum completion time in the workshop. In terms of energy consumption calculation, the energy consumption of the process, the energy consumption of the tool, and the tooling energy were comprehensively considered. Consumption and machine idle waiting for energy consumption in four parts. Claudia et al. [18] studied the energy consumption, completion time and time in job shop scheduling. The interrelationship between robustness points out the trade-off relationship between energy consumption, robustness and completion time. There is a synergistic relationship between energy consumption and robustness.

There are few researches on the multi-objective IPPS for green manufacturing, which is a lack of a more complete low-carbon optimization model that meets the actual production environment [19]. A mathematical model has been established for the GMOIPPS problem, an effective two-stage optimization method based on NSGA-II to solve GMOIPPS problem has been proposed in this paper. Three objectives (makespan, total carbon emissions and total tardiness) are optimized by the proposed method simultaneously.

3 Problem Formulation

The GMOIPPS problem studied in this paper can be described as follows: There are n jobs to be processed on m machines. Each job has diverse operations and alternative machining resources required for the operations. The operators and the processing sequence of the operators for each job, on which machine each operator is processed and processing time on the machine, and the amount of coolant consumed and the amount of lubricant lost when each operator is processed. The aim of GMOIPPS is to choose the appropriate process routes for each job, which ensures the processing sequence and the start-end time of each operation on each machine by satisfying the precedence constraints among operations and achieves the multiple performance objectives (makespan, the total carbon emission, the total tardiness) of the entire system.

In order to clearly describe the GMOIPPS problem, Table 1 gives processing information of three jobs. Job1 has 2 machining features, Job2 has 3 machining features, Job3 has 3 machining features, and a total of 5 machining machines. Each job has various features, these features can be changed by satisfying the precedence constraints, each feature has alternative operations or sequences of operations, and each operation has alternative machines, which is shown in Table 1. For instance, Job1 is given three operations, considering the amount of coolant consumed and the amount of lubricant lost when each operator is processed, the most optimal process plans and the processing machine of operators are selected, The process plan is $O_1(M_1)$-$O_2(M_3)$-$O_3(M_2)$. To further describe this problem, as shown in Table 1, a feasible process plan of the job2 in the Table 1 is given. This reasonable processing plan is (2, 1, 3, 2.1)-(4, 3, 2, 1.2)-(5, 1, 2, 9.75). 2 represents processing operation, 1 represents the processing machine, 2.1 represents the coolant corresponding to the selected processing machine. The processing sequence of the process plan is O_1-O_2-O_4-O_5.

Table 1. Processing information of three jobs

Jobs	Features	Candidate operations	Candidate machines	Precedence constraints	Process time	Coolant
Job1	F_1	O_1	M_1,M_2	Before F_2	5, 8	2.85/4.05
	F_2	O_2-O_3 O_4	M_1, M_3/M_2 M_3, M_2		4, 6/3 2, 6	1.35, 0.86/1.2 0.75/0.9
Job2	F_1	O_1 O_2-O_3	M_2, M_1 M_2, M_1/M_2		2, 3 1, 3/2	1.2/1.1 3.6, 2.1/3.6
	F_2	O_4	M_3, M_1	Before F_1	2, 4	1.2, 1.2
	F_3	O_5	M_3, M_1		1, 2	9.9, 9.75
Job3	F_1	O_1-O_2	M_2, M_3/M_2	Before F_3	1, 2/5	3.6, 1.2/3.6
	F_2	O_3 O_4-O_5	M_1, M_2 M_1, M_3/M_2, M_3	Before F_3	4, 3 1, 1/3, 2	1.2, 1.2 0.6, 2.1/0.75, 1.1
	F_3	O_6 O_7-O_8	M_3, M_1 M_1, M_3/M_2		2, 3 1, 2/2	2.85, 4.05 1.2, 0.7/0.8

To describe the mathematical model more clearly, the following assumptions should be given in advance:

(1) The manufacturing process of individual jobs.
(2) The coolant used on the cooler on the machine tool is of the same type of the same type.
(3) Jobs and machines are independent among each other. All jobs have the same priorities.
(4) Each machine can only handle one operation at a time.
(5) Various operations of each job can't be processed at the same time.
(6) Each operation can't be stopped when being processed.
(7) All the jobs and machines are available at time zero.
(8) The setup time for the operations is independent of the operation sequence and is included in the processing time.

Based on the above assumptions, the mathematical model of GMOIPPS in this paper is shown as follows [20]. The maximal completion time (Makespan), the total carbon emission (TCE) and the total tardiness (TT) are taken into account for GMOIPPS. The aim of GMOIPPS is to minimize the three objectives simultaneously.

The notations used to explain the model are described below:

C_{max}: the completion time of machines
C_{proc}: the processing time of machines
C_{idle}: the idle time of machines
C_{fit}: the adjustment time of machines
C_{cold}: the time of cooling
C_{oil}: the time of liquid lubrication
CA_{total}: the total carbon emission
Ta_{total}: the total tardiness
C_i: the completion time of job i
d_i: the due date of job i

The following three objectives are considered to be optimized simultaneously:

(1) f_1: Minimizing the maximal completion time of machines (Makespan)

$$Min\ f_1 = Min\ C_{max} = \{maxC,\ i = 1, \ldots, n\} \tag{1}$$

(2) f_2: Minimizing the total carbon emission (TCE)

$$Min\ f_2 = Min\ CA_{total} = C_{proc} + C_{idle} + C_{fit} + C_{cold} + C_{oil} \tag{2}$$

(3) f_3: Minimizing the total tardiness (TT)

$$Min\ f_3 = Min\ Ta_{total} = \sum_{i=1}^{n} max\,(0, C_i - d_i) \tag{3}$$

4 Two-Stage Optimization Method for GMOIPPS

An effective two-stage optimization method based on NSGA-II is designed to solve GMOIPPS. In the process planning stage, the basic NSGA-II algorithm is adopted to generate a near-optimal alternative process plans set for each job, dynamically input different process plans for the scheduling. In the scheduling stage, INSGA-II is constructed to get the optimal scheduling solutions. The flowchart of the proposed method is given in Fig. 1.

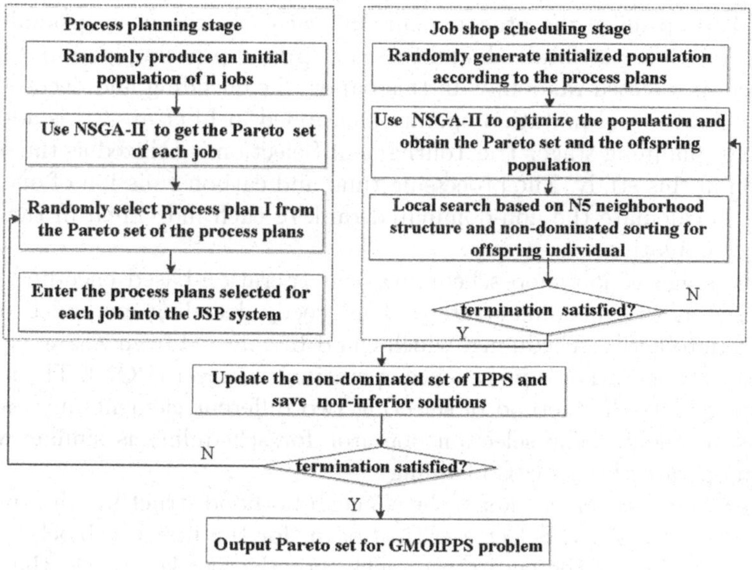

Fig. 1. Framework of the proposed method for GMOIPPS

The main steps of the proposed method for GMOIPPS are described as follows:

Step1: Randomly generate flexible process plan populations of n jobs according to process information of jobs to be processed.
Step 2: Adopt the basic NSGA-II algorithm to optimize the population to obtain the non-dominated solution sets for each job.
Step 3: For each job, randomly select a process plan from the non-dominated solution sets of the process plans.
Step 4: Input the selected process plans selected for each job into the job shop scheduling system.
Step5: Randomly generate an initial population based on the process plans determined by each job.
Step6: Use the NSGA-II algorithm to optimize the population (JSP) to obtain the Pareto solution sets and the offspring population.

Step7: Local search for offspring population based on N5 neighborhood structure described in detail in Fig. 2.
Step8: If the terminate criteria is satisfied, update the non-dominated solution sets of IPPS and save non-inferior solutions generated during the evolution process, otherwise, go to Step 6.
Step9: When the terminate criteria is met, output Pareto solution sets for GMOIPPS problem, otherwise, go to Step1.

The process planning stage mainly provides as many as possible near-optimal process plans for job shop scheduling. In this paper, the basic NSGA-II is employed to optimize the process plans for each job, the near-optimal process plan set of each job is obtained, and then an optimal process plan input scheduling system is selected randomly. In this paper, the encoding and decoding operations, crossover and mutation operations applied in Li et al. [21] is adopted in the process planning stage. The Tournament Selection is utilized as the selection operation in this study. The processing time and carbon emission of one job are adopted to calculate the non-dominated rank of each individual in the process planning population.

In the stage of job shop scheduling, the operation-based encoding method [22] is used as the encoding strategy. The greedy decoding method is designed to decode into active scheduling, which could be referred from Zhang et al. [23]. The crossover operation is precedence operation crossover (POX). The mutation operation adopts the method of selecting two different elements and swapping these two elements. The selection operator for scheduling is similar with the selection operator for process planning.

Figure 2 gives the operation of the N5 neighborhood structure, the main steps are described as follow. Select a critical path, for the first key block, swap the two key processes at the end of the block. For the last key block, the two key processes of the block head are exchanged. For intermediate key blocks, switch the two key processes immediately adjacent to the first and last blocks, and do nothing for other processes. Because the exchange will generate infeasible solution, in order to prevent the generation of infeasible solution, if the key piece of the process consists of a process is not only need to do any action, if two key working procedure of the job is the same, has retained its original position,

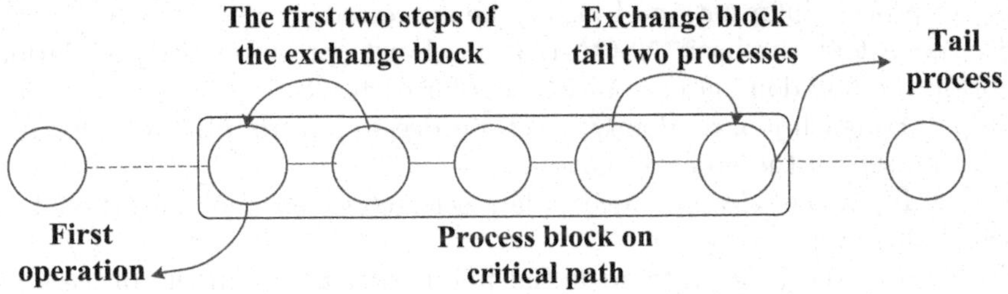

Fig. 2. Mobile operation of the N5 neighborhood structure

among them, the key block is referred to the critical path of continuous procedure set by the same processing machine processing.

The local search strategy based on N5 neighborhood structure is described in Fig. 3. The detailed steps of the local search strategy based on the N5 neighborhood structure are as follows:

Step1: Count the total number of current individuals P_u, let $i = 1$
Step2: If $i < P_u$ is satisfied, go to Step 3, otherwise, go to Step8
Step3: Exchange the individual's first two steps to generate a temporary individual T_i
Step4: If $f(T_i) < f(i)$ is satisfied, Replace T_i with individual i. Otherwise, Retain individual i
Step5: Exchange individual i block and tail processes to generate individual T_i
Step6: If $f(T_i) < f(i))$ is satisfied, Replace T_i with individual i. Otherwise, Retain individual i
Step7: let $i = i + 1$
Step8: Output all P_u individuals

There is at least one corresponding objective function value smaller than the individual i in the objective function value (makespan, carbon emission, total tardiness) of the neighborhood individual i.

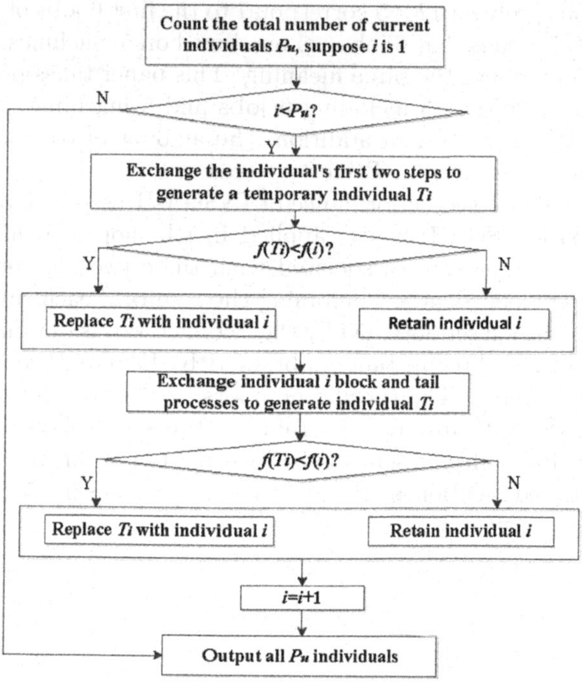

Fig. 3. The local search strategy based on N5 neighborhood structure

5 Experiments and Discussions

Three different instances were adopted to evaluate the performance of the proposed method. C++ programming software is used for optimal solution, the performance of running the computer is intel (R) Core(TM)*i*5-7400 CPU@3.00 GHz. The parameters of the proposed method in Table 2.

Table 2. The parameters in the proposed algorithm

Parameters	Process planning	Scheduling
Population size	100	200
Number of generations	15	100
Probability of mutation operator	0.10	0.10
Probability of crossover operator	0.80	0.80

Three different kinds of scale instances were utilized to evaluate the performance of the proposed optimization method, which include problem 6×5, problem 9×5 and problem 12×5. The problem 6×5 is a case. Based on the problem 6×5, problem 9×5 and problem 12×5 are planned in this paper among them, the last 3 jobs of problem 9×5 correspond to the first 3 jobs of problem 6×5, and the last 6 jobs problem 12×5 correspond to the first 6 jobs of problem 6×5. The problem 6×5 means that 6 jobs are machined on 5 machines, and the latter 9×5 and 12×5 represent the same meaning. This paper takes problem 6×5 as an example shown in Table 3 including 6 jobs and 5 machines, which contains 5 features and 12 operations. In addition, the amount of coolant used in each stage of the machine is given in Table 3.

In order to evaluate the performance of INSGA-II used in job shop scheduling stage, the basic NSGA-II is also applied in job shop scheduling stage. The other operations are the same associated with these two methods. Two methods run 20 times independently. Combining the results of each running, the final non-dominated solution set obtained by the two methods is obtained. The experimental results of the three instances obtained by INSGA-II and NSGA-II are given in Figs. 4, 5 and 6. The specific numbers of Pareto solution obtained by INSGA-II and NSGA-II are given in Table 4. Table 5 displays a detail process planning scheme for a selected solution in Pareto set of problem 9×5. The Gantt chart of the selected solution in Pareto sets for problem 9×5 is illustrated in Fig. 7.

Table 3. The information of problem 6 × 5 with flexibility in process planning

Jobs	Features	Candidate operations	Candidate machines	Precedence constraints	Process time	Coolant
Job1	F_1	O_1	M_1, M_2	Before all	19,27	2.85,4.05
	F_2	O_2-O_3	$M_3, M_4/M_2, M_5$	Before F_4	8, 9/5, 6	1.35, 1.2/4.06, 5.5
	F_3	O_4-O_5 O_6-O_7	$M_3, M_4/M_2, M_5$ $M_1, M_2/M_3, M_4$		9, 9/8, 6 5, 9/8, 8	0.75, 0.9/1.22, 5.4 1.2, 1.2/3.6, 3.6
	F_4	O_8 O_9	M_1, M_2 M_2, M_5		6, 5 8, 7	1.2, 1.2 3.6/3.6
	F_5	O_{10}	M_3, M_4		8, 8	0.6, 0.75
Job2	F_1	O_1-O_2	M_3, M_4	Before F_4	2, 2/4, 5	1.35, 1.2
	F_2	O_3	M_2, M_5	Before all	1, 2	0.75, 0.9
	F_3	O_4-O_5 O_6-O_7	$M_1, M_2/M_3 M_4$ $M_2, M_5/M_1, M_2$		2, 3/5, 3 11, 8/2, 2	5.5, 5.5/1.35, 1.2 0.75, 0.9/3.6, 3.6
	F_4	O_8-O_9 O_{10}	M_3, M_4 M_1		25, 25/10, 11 30	1.2, 1.2 1.05
	F_5	O_{11} O_{12}	M_2 M_1, M_2		12 7, 8	1.25 0.6, 0.7
Job3	F_1	O_1-O_2 O_3	$M_3, M_4/M_1$ M_5		41, 40/5 35	0.15, 0.3/2.45 4.05
	F_2	O_4	M_2, M_5	Before all	40, 41	1.2, 1.2
	F_3	O_5	M_1, M_2	Before F_5	23/23	0.75, 0.9/5.5, 5.5
	F_4	O_6 O_7	M_3, M_4 M_1	Before F_1	11, 12 12	3.6, 3.6 9.9
	F_5	O_8-O_9	$M_5/M_3, M_4$		45/7, 5	1.23/1.27, 1.27
Job4	F_1	O_1	M_1, M_2	Before all	19, 27	0.6, 0.7
	F_2	O_2-O_3	$M_3, M_4/M_2, M_5$	Before F_4	8, 9/5, 6	0.15, 0.3/4.45, 5.5
	F_3	O_4-O_5 O_6-O_7	$M_3, M_4/M_2, M_5$ $M_1, M_2/M_3, M_4$		9, 9/8, 6 5, 9/8, 8	11.7, 11.7/4.15, 5.6 3.6, 3.6/1.23, 1.23
	F_4	O_8 O_9	M_1, M_2 M_2, M_5		6, 5 8, 7	1.8, 1.8 3.8, 5.6
	F_5	O_{10}	M_3, M_4		8, 8	1.5, 1.5
Job5	F_1	O_1-O_2	M_3, M_4	Before F_4	2, 2/4, 5	1.25, 1.24
	F_2	O_3	M_2, M_3	Before all	1, 2	4.05, 1.57
	F_3	O_4-O_5 O_6-O_7	$M_1, M_2/M_3 M_4$ $M_2, M_5/M_1, M_2$		2, 3/5, 3 11, 8/2, 2	0.75, 0.9/7.6, 7.6/1.25, 5.5 2.22, 2.53/3.6, 3.6
	F_4	O_8-O_9 O_{10}	M_3, M_4 M_1		25, 25/10, 11 30	6.15, 6 1.58
	F_5	O_{11} O_{12}	M_2 M_1, M_2		12 7, 8	0.95 6, 6.15
Job6	F_1	O_1-O_2 O_2	$M_3, M_1/M_4$ M_5		41, 40/5 35	2.45, 2.45/1.02 1.65
	F_2	O_3	M_2, M_5	Before all	40, 41	1.2, 1.2
	F_3	O_4	M_1, M_2	Before F_5	23/23	3.8, 3.08
	F_4	O_5-O_6 O_7	M_3, M_4 M_1	Before F_1	11, 12 12	5.5, 5.5 3.6
	F_5	O_8-O_9	$M_5/M_3, M_4$		45/7, 5	9.9/9.75, 1.2

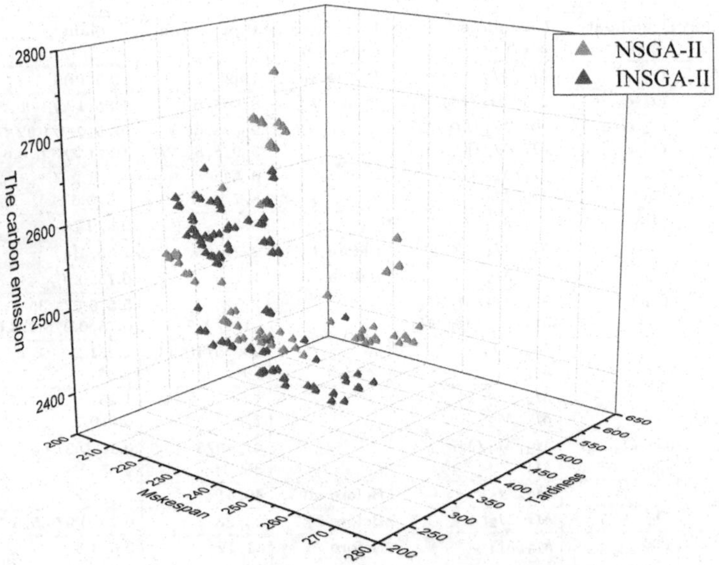

Fig. 4. The Pareto solution sets obtained by the proposed method for problem 6×5

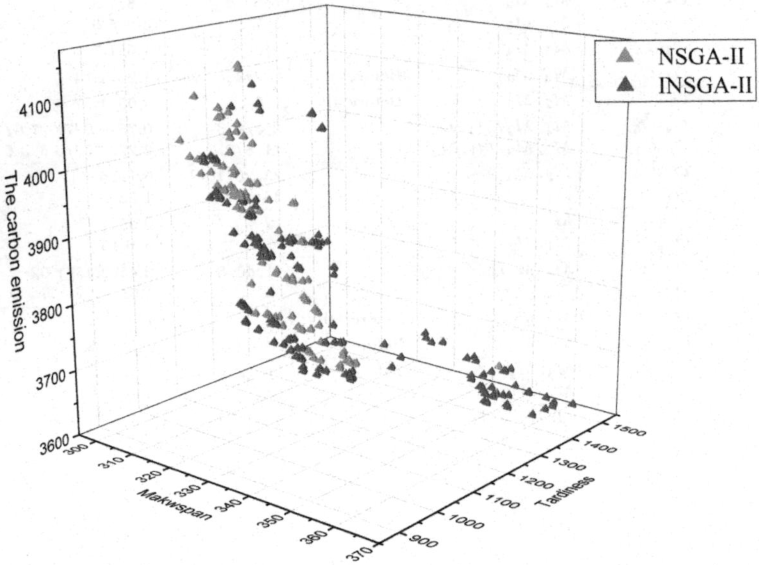

Fig. 5. The Pareto solution sets obtained by the proposed method for problem 9×5

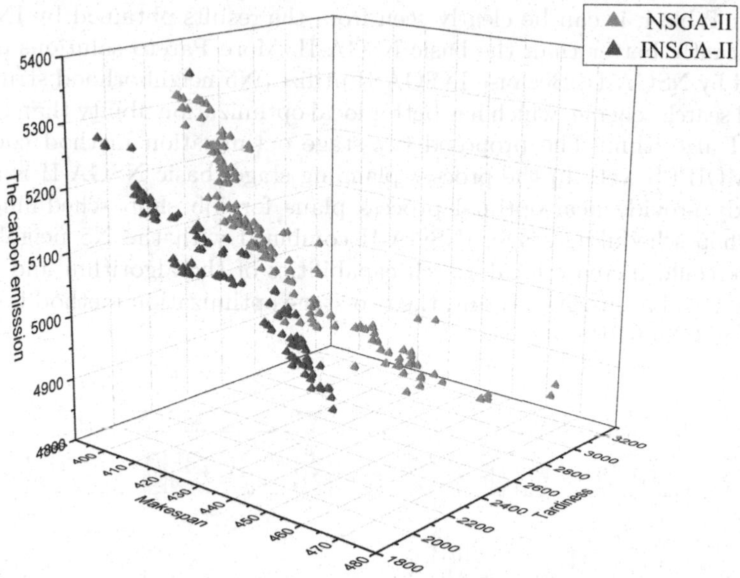

Fig. 6. The Pareto solution sets obtained by the proposed method for problem 12×5

Table 4. The Pareto solutions of three instances compared NSGA-II and INSGA-II

Problem	The numbers of Pareto solutions	NSGA-II	INSGA-II
6×5	69	17	52
9×5	129	44	85
12×5	108	4	104

Table 5. The Process planning for a selected solution in Pareto set for problem 9×5

Job No.	Numbers of operations	Process plans
1	7	$O_1(M_1)$-$O_6(M_1)$-$O_7(M_3)$-$O_2(M_3)$-$O_3(M_2)$-$O_8(M_1)$-$O_{10}(M_3)$
2	7	$O_3(M_2)$-$O_1(M_3)$-$O_2(M_3)$-$O_4(M_1)$-$O_5(M_4)$-$O_{12}(M_1)$-$O_{10}(M_1)$
3	7	$O_4(M_2)$-$O_5(M_1)$-$O_6(M_3)$-$O_1(M_3)$-$O_2(M_1)$-$O_8(M_5)$-$O_9(M_4)$
4	7	$O_1(M_1)$-$O_{10}(M_3)$-$O_2(M_3)$-$O_3(M_2)$-$O_8(M_1)$-$O_6(M_1)$-$O_7(M_3)$
5	7	$O_3(M_2)$-$O_1(M_3)$-$O_2(M_3)$-$O_{10}(M_1)$-$O_{12}(M_1)$-$O_4(M_1)$-$O_5(M_4)$
6	7	$O_4(M_2)$-$O_6(M_3)$-$O_5(M_1)$-$O_8(M_5)$-$O_9(M_4)$-$O_1(M_4)$-$O_2(M_1)$
7	7	$O_1(M_1)$-$O_{10}(M_3)$-$O_2(M_3)$-$O_3(M_2)$-$O_8(M_1)$-$O_6(M_1)$-$O_7(M_3)$
8	7	$O_3(M_2)$-$O_1(M_3)$-$O_2(M_3)$-$O_{12}(M_1)$-$O_4(M_1)$-$O_5(M_4)$-$O_{10}(M_1)$
9	7	$O_4(M_2)$-$O_7(M_1)$-$O_1(M_3)$-$O_2(M_1)$-$O_5(M_1)$-$O_8(M_5)$-$O_9(M_4)$

From Table 4, it can be clearly seen from the results obtained by INSGA-II is significantly better than the basic NSGA-II. More Pareto solutions could be obtained by NSGA-II. Because INSGA-II utilizes N5 neighborhood structure as the local search scheme, which has better local optimization ability than the basic NSGA-II algorithm. The proposed two-stage optimization method could deal with GMOIPPS well. In the process planning stage, basic NSGA-II is adopted to quickly provide near-optimal process plans for job shop scheduling stage. In job shop scheduling stage, INSGA-II combined with the N5 neighborhood structure could increase local search capabilities of the algorithm and find the final Pareto solutions. As a result, the two-stage optimization method is effective for solving GMOIPPS.

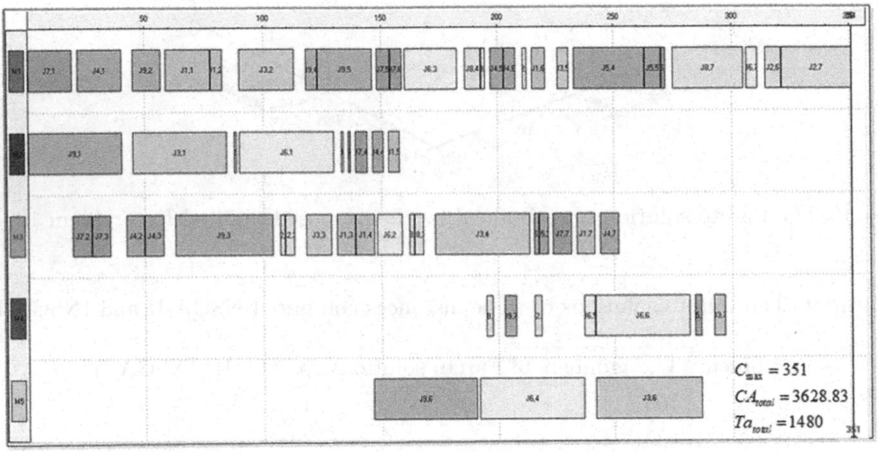

Fig. 7. The Gantt chart of the selected solution in Pareto sets for problem 9×5

6 Conclusion and Future Works

An effective two-stage optimization method algorithm is proposed to solve the GMOIPPS problem in this paper. Three objectives of makespan, total carbon emission and total tardiness are taken into account simultaneously. The INSGA-II used in job shop scheduling stage has better local optimization ability and solution efficiency than basic NSGA-II. The proposed optimization method also has some shortcomings. This paper only considers three objectives. However, there are many objectives to be considered for solving GMOIPPS problem. Therefore, in the future, exploring a more effective method to solve the GMOIPPS problem with more objectives will be a very challenging task.

References

1. Zhu, G.Y., Xu, W.J.: Multi-objective flexible job shop scheduling method for machine tool component production line considering energy consumption and quality. Control Decis. **34**(02), 31–39 (2019)
2. Pan, L., He, C., Tian, Y., Su, Y., Zhang, X.: A region division based diversity maintaining approach for many-objective optimization. Integr. Comput.-Aided Eng. **24**(3), 279–296 (2017)
3. Gao, L., Li, X.Y.: Current research on integrated process planning and scheduling. China Mech. Eng. **22**(8), 1001–1007 (2011)
4. Yang, Y.N., Parsaei, H.R., Leep, H.R.: A prototype of a feature-based multiple-alternative process planning system with scheduling verification. Comput. Ind. Eng. **39**(1–2), 109–124 (2001)
5. He, C., Tian, Y., Jin, Y., Zhang, X., Pan, L.: A radial space division based evolutionary algorithm for many-objective optimization. Appl. Soft Comput. **61**, 603–621 (2017)
6. Pan, L., He, C., Tian, Y., Wang, H., Zhang, X., Jin, Y.: A classification-based surrogate-assisted evolutionary algorithm for expensive many-objective optimization. IEEE Trans. Evol. Comput. **23**(1), 74–88 (2018)
7. Rajemi, M.F., Mativenga, P.T., Aramcharoen, A.: Sustainable machining: selection of optimum turning conditions based on minimum energy considerations. J. Clean. Prod. **18**(10–11), 1059–1065 (2010)
8. Yi, Q., Li, C., Tang, Y., Chen, X.: Multi-objective CNC machining parameters optimization model for high efficiency and low carbon. J. Clean. Prod. **95**, 256–264 (2015)
9. Yin, R.: Energy consumption evaluation model of the machining processes and its application in process planning. Mech. Des. Manuf. **09**, 270–272 (2014)
10. Liu, Y., Dong, H., Lohse, N., Petrovic, S., Gindy, N.: An investigation into minimize total energy consumption and total weighted tardiness in job shops. J. Clean. Prod. **65**(4), 87–96 (2014)
11. May, G., Stahl, B., Taisch, M.: Multi-objective genetic algorithm for energy-efficient job shop scheduling. Int. J. Prod. Res. **53**(23), 1–19 (2015)
12. Huang, Z., Tang, D., Dai, M.: A bi-objective optimization model for integrated process planning and scheduling based on improved algorithm. J. Nanjing Univ. Aeronaut. Astronaut. **47**(1), 88–95 (2015)
13. Min, D., Dunbing, T., Zhi, H., Jun, Y.: Energy-efficient process planning using improved genetic algorithm. Trans. Nanjing Univ. Aeronaut. Astronaut. **33**(05), 602–609 (2016)
14. Zhang, R., Chiong, R.: Solving the energy-efficient job shop scheduling problem: a multi-objective genetic algorithm with enhanced local search for minimizing the total weighted tardiness and total energy consumption. J. Clean. Prod. **112**, 3361–3375 (2016)
15. Zhang, Z., Tang, R., Peng, T., Tao, L., Jia, S.: A method for minimizing the energy consumption of machining system: integration of process planning and scheduling. J. Clean. Prod. **137**, 1647–1662 (2016)
16. Liu, Q., Zhu, M.: Integrated optimization of process planning and shop scheduling for reducing manufacturing carbon emissions. China J. Mech. Eng. **53**(11), 164–174 (2017)
17. Li, C.: A batch splitting flexible job shop scheduling model for energy saving under alternative process plans. J. Mech. Eng. **53**(5), 12–23 (2017)

18. Salido, M.A., Escamilla, J., Barber, F., Giret, A., Tang, D., Dai, M.: Energy efficiency, robustness, and makespan optimality in job-shop scheduling problems. AI EDAM. **30**(03), 300–312 (2016)
19. Pan, L., Li, L., He, C., Tan, K.C.: A subregion division-based evolutionary algorithm with effective mating selection for many-objective optimization. IEEE Trans. Cybern. (2019). https://doi.org/10.1109/TCYB.2019.2906679
20. Li, X., Gao, L., Li, W.: Application of game theory based hybrid algorithm for multi-objective integrated process planning and scheduling. Expert Syst. Appl. **39**(1), 288–297 (2012)
21. Li, X., Gao, L., Wen, X.: Application of an efficient modified particle swarm optimization algorithm for process planning. Int. J. Adv. Manuf. Tech. **67**(5–8), 1355–1369 (2012)
22. Fang, P.R., Corne, D.: A promising genetic algorithm approach to job-shop scheduling, rescheduling, and open-shop scheduling problems. In: Proceedings of the Fifth International Conference genetic Algorithms, pp. 375–382 (1993)
23. Zhang, C., Li, P., Rao, Y., Li, S.: A new hybrid GA/SA algorithm for the job shop scheduling problem. In: Raidl, G.R., Gottlieb, J. (eds.) EvoCOP 2005. LNCS, vol. 3448, pp. 246–259. Springer, Heidelberg (2005). https://doi.org/10.1007/978-3-540-31996-2_23

An Improved Multi-objective Particle Swarm Optimization with Adaptive Penalty Value for Feature Selection

Wentao Chen(✉) and Fei Han

School of Computer Science and Communication Engineering, Jiangsu University, Zhenjiang, China
cwt.7410@163.com

Abstract. Feature selection is an important data-preprocessing technique to eliminate the features with low contributions in classification. Currently, many researches focus their interests on the combination of feature selection and multi-objective particle swarm optimization (PSO). However, these methods exist the problems of large search space and the loss of global search. This paper proposes a multi-objective particle swarm optimization with the method called APPSOFS that the leader archive is updated by an adaptive penalty value mechanism based on PBI parameter. Meanwhile, the random generalized opposition-based learning point (GOBL-R) point is adopted to help jump out of local optima. The proposed method is compared with three multi-objective PSO and MOEA/D on six benchmark datasets. The results have demonstrated that the proposed method has better performance on feature selection.

Keywords: Feature selection · Particle swarm optimization · Multi-objective optimization · Random generalized opposition-based learning

1 Introduction

Feature selection is an important data preprocessing technology of machine learning applications especially classification and clustering. It is aimed at removing irrelevant and redundant features to reduce the dimensionality of the data and increase the performance. Traditional feature selection methods are mainly classified into filter and wrapper methods [10]. The filter method evaluates the relevance of a feature independently from a classifier. Nevertheless, the wrapper method integrates a predetermined learning algorithm with a classifier to group an optimal feature subset according to the prediction accuracy. Although the two types of methods improve classification performance, these methods overemphasize on lower classification error and ignore the number of selected feature subset. Therefore, some researches introduce multi-objective optimization to feature selection, which stresses high classification performance and low number of feature subset simultaneously.

L. Pan et al. (Eds.): BIC-TA 2019, CCIS 1159, pp. 649–661, 2020.
https://doi.org/10.1007/978-981-15-3425-6_51

Nowadays, multi-objective particle swarm optimization (PSO) has attracted a great interest in feature selection for fast convergence on the single objective optimization. On the one hand, a number of methods propose Pareto mechanism to refine archive. NSPSOFS [16] combines multi-objective PSO with NSGAII to obtain Pareto optimal solutions, which adopts crowding distance to distribute solutions uniformly. Meanwhile, CMDPSOFS [16] adopts crowding, mutation and dominance on the external archive to obtain optimal solutions. ISRPSOFS [8] adopts inserting, swapping, and removing on features to obtain better solutions. RFPSOFS [7] adopts a feature elitism mechanism to refine the solution set. Above methods mainly adopt Pareto mechanism to refine archive. Although Pareto mechanism abandons many dominated solutions to accelerate convergence speed, these solutions are good for local search of feature selection.

On the other hand, many methods adopt uniform mutations or non-uniform mutations. However, most existing feature selection methods suffer from stagnation in local optima. HMPSOFS [20] adopts hybrid mutations (uniform mutation and non-uniform mutation) to jump out of local optima and improve the global search. RFPSOFS [7] adopts a comprehensive mutation mechanism to improve global search. However, these mutation mechanisms are randomly adopted to particles and lead some particles to fall into local optima.

Therefore, an adaptive penalty multi-objective PSO feature selection (APPSOFS) is proposed in this paper. Firstly, APPSOFS adopts penalty boundary interaction (PBI) [18] to update the leader archive. Different from Pareto mechanism, PBI may relieve selection pressure and the preserved solutions promote local search. The penalty value is adjusted adaptively according to PBI parameter since fixed penalty value is not beneficial to preserve boundary solutions [17]. Secondly, random generalized opposition-based learning (GOBL-R) points is adopted to whose pbest is not updated for many times in order to help jump out of local optima. The proposed method is compared with MOEA/D [18], MOPSO [1], HMPSOFS [20], RFPSOFS [7] on 6 benchmark datasets. The qualitative analysis of the results is performed by Pareto fronts investigation and quantitative analysis is measured by hypervolume (HV) indicator.

The remainder of this paper is organized as follows. Section 2 introduces related work on the proposed algorithms. Section 3 gives details of APPSOFS and related mechanisms. Section 4 presents the experimental parameters and results. Section 5 presents the experimental conclusion and future work.

2 Preliminaries

2.1 Particle Swarm Optimization

PSO simulates the behavior of bird flocking and is applied to solve single objective problems (SOPs) [2]. In PSO, the particle can be assumed as no weight point and its velocity is controlled by personal best point (*pbest*) and global best point (*gbest*) of the current particle. The velocity and position update equations are listed in the following.

$$v_i(t+1) = wv_i(t) + c_1 r_1 * (pbest_i - x_i(t)) + c_2 r_2 * (gbest_i - x_i(t)), \quad (1)$$

$$x_i(t+1) = x_i(t) + v_i(t+1), i = 1, 2, ..., n \tag{2}$$

where w is the inertia weight; c_1 and c_2 are considered as cognitive and social learning acceleration coefficients respectively; r_1 and r_2 are random values between 0 and 1; t indicates iteration number.

2.2 The PBI Approach

Different from dominance-based mechanism, decomposition approach converts an MOP into a number of scalar subproblems. The decomposition approaches include the weighted sum, weighted Techebycheff and PBI. PBI approach performs better on obtaining a good distribution of solutions. The detailed equations of PBI in the objective space are listed in the following.

$$minimize \quad g(x|\lambda, z^*) = d_1 + \theta d_2, \tag{3}$$

where z^* is the vector including the minimize value of each objective, λ denotes predefined weight vector, θ denotes penalty value and d_1, d_2 are shown as follows:

$$d_1 = \frac{\|(F(x)^T - z^*)\lambda\|}{\|\lambda\|} \quad and \quad d_2 = \|F(x) - (z^* + d_1 \frac{\lambda}{\|\lambda\|})\|, \tag{4}$$

PBI approach is able to approximate a PF very well under an appropriate weight vectors setting [18]. Furthermore, the penalty value θ plays a vital role in the process of solution selection [14]. As Eqs. 3 and 4 show, d_1 represents convergence and d_2 represents diversity [9]. Moreover, a small value of θ accelerates the convergence speed but a large value of θ maintains the diversity of population [17].

2.3 Random Generalized Opposition-Based Learning Points

Opposition-based learning (OBL) mainly sets an opposite number on the objective space or solution space [6]. By setting OBL point, the solution easily jumps out of local optima and promotes global searching. Since OBL was integrated into the DE algorithm, which was called Opposition-based DE [11], some researchers have focused on evolutionary computation.

***Definition* 1 (Opposition-based learning point). Let $x = (x_1, x_2, ..., x(n)) \in R^n$ and $x_i \in [a_i, b_i], \forall i = 1, 2, ..., n$. The opposition point is defined as and listed:**

$$\hat{x}_i = a_i + b_i - x_i, i = 1, ..., n \tag{5}$$

Traditional OBL point may guide the particle to jump out of local optima, but the point bring up the loss of randomness. Random generalized opposition-based learning (GOBL-R) [13] point adds a random number to GOBL for better global searching. The searching range of GOBL-R becomes larger compared with other OBL mechanisms. The detailed equation is listed as follows.

***Definition* 2 (Random generalized opposition-based learning point). Let $x = (x_1, x_2, ..., x_n) \in R^n$ and $x_i \in [a_i, b_i]$, $\forall i = 1, 2, ..., n$.**

$$\hat{x}_i = rand_i \times (a_i + b_i) - x_i, i = 1, ..., n \tag{6}$$

where i denotes the dimension of the current point. In this paper, the algorithm applies the GOBL-R point to help jump out of local optima when pbest particle does not update continuously.

3 The Proposed APPSOFS

3.1 Encoding Particle

For binary PSO, traditional strategy constructs a function to transform a real number into a binary number such as a sigmoid function [3,12]. However, this paper adopts a threshold to control the encoding of 0 and 1. Once the dimension of the position exceeds the threshold, the feature corresponding to the dimension is selected. Take a dataset with D features as an example, the ith particle in the swarm can be encoded by a D-bit real string as follow.

$$X_i = (x_{i,1}, x_{i,2}, ...x_{i,D}), x_{i,j} \in [0, 1], j = 1, 2, ..., D, i = 1, ..., N \tag{7}$$

Equation 7 is applied to continuous optimization. This paper presets a threshold to transform real number into 0 or 1 and the threshold is set to 0.5. Before calculating the fitness of the particle, the particle X_i is encoded to a solution Z_i:

$$z_{i,j} = \begin{cases} 1, & x_{i,j} \geq 0.5 \\ 0, & x_{i,j} < 0.5 \end{cases} \tag{8}$$

3.2 Fitness Evaluation

Many multi-objective feature selection algorithms choose classification accuracy and number of features. However, classification accuracy and number of selected features should be maximization and minimization respectively. Therefore, this paper chooses classification error rate and selected feature rate to minimize. The first objective is classification error rate, which is listed in the following:

$$ErrorRate_i = (FP + FN)/(FP + FN + TP + TN) \tag{9}$$

TP, FP, TN and FN refer to true positive, false positive, true negative and false negative of classifier respectively. Although error rate is not a reliable performance metric [5], this paper ignores this limitation. This paper applies K-fold cross validation to calculate mean value of classification error.

The second objective is the feature rate. Small number of features reduces the computational cost of classification but increases error rate. The definition of the feature rate is the percentage of selected feature in all features, which is listed in the following.

$$FeatureRate_i = f_i/D, f_i = \sum_{j=1}^{D} z_{i,j} \tag{10}$$

where D is the total number of features. In this way, two objectives are normalized in [0, 1].

3.3 Adaptive Penalty Value on Leader Archive

Penalty boundary interaction (PBI) adopts a set of uniformly distributed weight vectors and fixed penalty value mechanism to obtain optimal solutions. For one weight vector, penalty value tends to influence the selection of solution. Figure 1(a) and (b) commonly list different selection results under different penalty values. In Fig. 1(a), the green lines denote the counter lines under $\theta = 5$. Although the point A has lower d_2 than the point B, the point B is selected under $\theta = 5$ and w_1. However, the orange lines of Fig. 1(b) denote the counter lines under $\theta = 10$. The point A is selected under $\theta = 10$ and w_1. The different results show that the fixed penalty value leads to the difficulty of preserving boundary solutions. Meanwhile, penalty value is highly related to d_2, which lower d_2 leads to high penalty value.

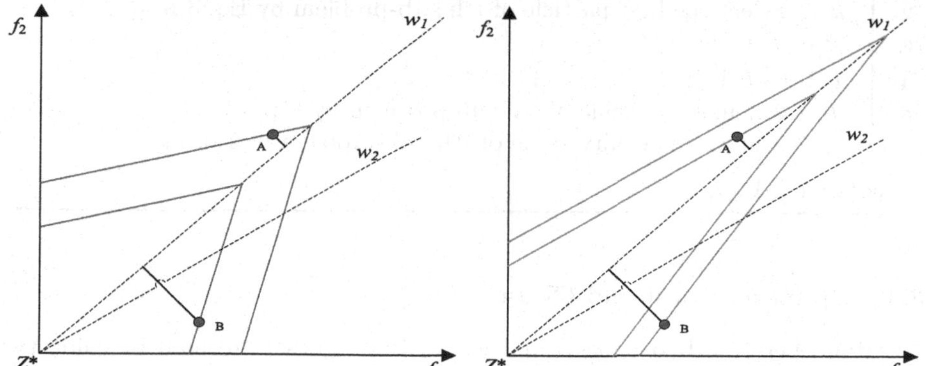

(a) θ=5 (b) θ=10

Fig. 1. The counter line of PBI on different penalty values (Color figure online)

In this paper, a novel adaptive penalty mechanism based on d_2 is applied to update leader archive. In PBI, d_2 denotes the distance between the solution and the weight vector, and indicates the gap between the selected solution and the weight vector. First, for high d_2, decreasing penalty value may ensure that the selected solution gets close to the direction of the weight vector. Second, the penalty value with iteration varying is adopted to avoid high penalty value in the primary stage influencing the speed of convergence. The calculation of penalty value is listed in the following:

$$\theta_i(t) = 1 + (2 + (100 - 2)\frac{t}{maxIteration})e^{-\frac{(d)^2}{2}}, \tag{11}$$

where d denotes d_2 value of the ith solution under ith weight vector in PBI, t denotes the number of current iterations, $maxIteration$ denotes the number of max iteration.

The adaptive penalty value mechanism is applied to leader archive updating. The whole leader archive updating is divided into two parts. The first part is the updating of leader archive for each sub-problem. The leader archive is updated by the improved PBI whose penalty value is provided by Eq. 11. The second part is the updating of penalty value θ. This part includes the calculation of d_2 and updating of θ. The whole mechanism is listed on Algorithm 1.

Algorithm 1. *UpdateLeaderArchive*(LA, LAS, θ)

Input: LA (leader archive), LAS (leader archive size),θ (penalty value of each sub-problem);

Output: LA' (updated leader archive), θ (penalty value of each sub-problem);

for $i = 1$ *to* LAS **do**
- $k \leftarrow$ Select the best particle of ith sub-problem by Eq. 3 from LA;
- $P \leftarrow LA \setminus k$;
- $LA' \leftarrow LA' \cup k$;
- $d \leftarrow$ Calculate d_2 value of ith sub-problem by Eq. 4;
- $\theta_i \leftarrow$ Calculate penalty value of ith sub-problem by Eq. 11;

return LA', θ;

3.4 Updating Pbest and Gbest

For PSO, personal best (pbest) and global best (gbest) are used to guide the search direction [19]. Pbest records the personal best position of the current particle. For each particle, pbest is replaced by the new position if new position is better than current pbest by PBI. Gbest is used as the leader of current particle. For each particle, selecting the solution with lowest aggregate value from neighbour solutions of the leader archive as gbest may enhance convergence of the particle.

3.5 Random Generalized Opposition-Based Learning Mutation

Although a good leader particle helps guide the particle to find optimal solutions, PSO falls into local optima easily due to fast convergence and the loss of exploration. To solve the trouble, many researches propose diverse mutations to enhance diversity of population. HMPSOFS proposes hybrid mutations including re-initialization operator and jumping mutation. RFPSOFS [7] adopts uniform mutation and non-uniform mutation to improve the diversity. However, many mutation are randomly adopted to particles and jump mainly around current position. Thus, incorporating random generalized opposition-based learning point into whose pbest is not updated for many times may help jump out of local optima.

This paper adopts age to record the times of not pbest updating and presets age threshold to control mutation, which is similar to dMOPSO. In one hand, for the particle whose age exceeds age threshold, the particle is updated by the combination of feature rank and GOBL-R. The specific setting of mutation scheme is listed on Eq. 12.

$$x_{ij}(t+1) = rand()_{ij} \times (a_{ij} + b_{ij}) - x_{i,j}(t), i = 1, ..., N, j = 1, ..., D \quad (12)$$

where a denotes lower boundary value of decision variables and b denotes upper boundary value of decision variables. In the equation, $x_{ij}(t)$ stands for the position of ith particle and jth dimension during tth iteration. Also, $rand$ is a random value between 0 and 1. In other hand, for those particles whose ages are under age threshold are updated according to Eqs. 1 and 2.

3.6 The Main Framework of APPSOFS

Algorithm 2 shows the main framework of APPSOFS and integrates above mechanisms. On the process of initialization, the whole dataset is randomly divided into a training set and a test set. A set of N weight vectors are initialized uniformly in the objective space and the positions of N particles are randomly generated. The T neighbour indexes of each weight vector are calculated in advance. The penalty value θ of each sub-problem is set to 5. Also, Algorithm 1 is conducted before the iterations.

On the step of evolution, particles are divided into two parts by age threshold. Particles whose ages are under the age threshold are updated by traditional PSO algorithm. The inertia weight w is set to 0.729, c_1 and c_2 are set to 1.46. The selection of pbest and gbest for each particle are described on Sect. 3.4. However, particles whose ages exceed the age threshold are updated as Sect. 3.5 describes.

On the step of updating leader archive, new penalty values and leader archive are updated by Algorithm 1. Finally, the stopping criteria is set to check whether current iteration is arriving at max iteration.

Algorithm 2. Complete algorithm APPSOFS

Input: N (population size), Ta (a preset age threshold), age (the age of each particle), T (neighbour size);
Output: LA (leader archive);
Divide data into a training set and a test set;
Initialize the population P, a set of weight vectors W, the ideal objective vector z^*, $pbest \leftarrow P$;
for each i=1,...,N, set $B(i) \leftarrow \{i_1,...,i_T\}$ where $w^{i_1},...,w^{i_T}$ are the T closest weight vector to w^i , set $\theta_i \leftarrow 5$ and set $age_i \leftarrow 0$;
$\{LA, \theta\} \leftarrow UpdateLeaderArchive(LA, N, \theta)$;
while *Stopping criterion is not satisfied* **do**
  for $i = 1$ *to* N **do**
    if $age_i < Ta$ **then**
      $U \leftarrow$ Select Bi particles of LA;
      $gbest_i \leftarrow$ Select best particle under ith sub-problem from U by Eq.3;
      Update the velocity v_i and position x_i of P_i by Eqs. (1, 2);
      if P_i *is better than* $pbest_i$ *by Eq. (3)* **then**
        $age_i \leftarrow 0$;
        $pbest_i \leftarrow x_i$;
      else
        $age_i \leftarrow age_i + 1$;
    else
      Update the position x_i of P_i by Eq. (12);
      $age_i \leftarrow 0$;
    Calculate the training error and feature rate of P_i by Eqs. (8, 9, 10);
  $\{LA, \theta\} \leftarrow UpdateLeaderArchive(LA \bigcup P, N, \theta)$;
Calculate test error of solutions in LA;
return LA;

4 Experimental Results and Discussion

4.1 Datasets

The experimental datasets are listed in Table 1. These datasets are chosen from UCI machine learning repository. The experiment selects high number of samples with low features and low number of samples with high features. The dataset is divided into two groups including the first sub-group with two classes (Sonar, Musk1, WBCD) and the second sub-group with more than two classes (Wine, Vehicle and Zoo). The whole experiment is executed on MATLAB R2018b, Intel(R) Core(TM) i5-6500, 3.20 GHz, 4 GB RAM.

Table 1. Six datasets

Datasets	Number of samples	Number of features	Number of classes
Sonar	208	60	2
Musk1	476	166	2
WBCD	569	30	2
Wine	178	13	3
Vehicle	846	18	4
Zoo	101	17	7

4.2 Parameter Setting

The classification of data is performed by K Nearest Neighbored algorithm or KNN for its simplicity. K is set to five, and all datasets are randomly divided into two sets, 70% for training set and 30% for test set. In order to improve the reliability of the algorithms in training set, the 10-fold cross validation technique is used for each subset.

General parameters for each algorithms are: $N = 30$ as the population size, $maxIteration = 100$ as the max number of iterations, $c_1 = c_2 = 1.46$ as acceleration rate and $w = 0.729$ as inertia weight. For MOEA/D, the penalty value θ is 5, the size of neighbour subproblems T is $N/10$, distribution of crossover and mutation are 20. For HMPSOFS [20], the jumping probability is set to 0.01. For MOPSO [1], the number of grids is set to 10; the parameter of selection β is set to 2. For RFPSOFS [7], the related parameters are similar to MOPSO. For the proposed method APPSOFS, the size of neighbour subproblems T is $N/10$ and age threshold Ta is 2. For all real number optimization, the encoding of particles refers to Sect. 3.1. The whole experiment provides the train and test Pareto fronts of comparison algorithms and the proposed algorithm. A total of 30 independent runs of all algorithms are performed.

4.3 Performance Indicators

In order to analyze the performance of the proposed algorithm and comparison algorithms, the hypervolume (HV) [15] is used as the performance indicators. HV not only demonstrates convergence of solutions with true POF but also shows distribution of final solutions [4].

HV is calculated by the volume of region which dominates reference point in the objective space. The setting of reference point is vital to the calculation of HV. The reference point of HV is set to (1.1, 1.1).

$$HV(S) = Leb(\bigcup_{x \in S} [f_1(x), R_1] \times \cdots \times [f_M(x), R_M]), \tag{13}$$

where S is current Pareto front set, f_i denotes ith objective value of S and R_i denotes ith objective value of reference point.

4.4 Results Analysis

The whole datasets are divided into two groups, including two classes and more than two classes. For one dataset, the data is divided randomly into training sets and test sets, thus the final results include training error and test error. Meanwhile, this experiment uses HV metric as main evaluation metric. Firstly, the mean values and the standard deviations of HV in the training set are listed on Table 2. As Table 2 shows, APPSOFS gains better mean value and standard deviations of HV for WBCD, Zoo and Musk, while HMPSOFS gains higher mean value and standard deviations of HV for Wine. For Sonar, HMPSOFS gains best HV mean value, but APPSOFS gets lower standard deviation. APPSOFS is more stable than HMPSOFS and easily find better solution subset. For Musk and WBCD, APPSOFS gains best mean value and standard deviation than other algorithms. For Wine, HMPSOFS performs better than APPSOFS from mean value and standard deviation. However, HMPSOFS may be overfitting for that Wine has low instances. For Vehicle, although HMPSOFS gains better standard deviation for Vehicle, APPSOFS gains best mean value than other algorithms. For Zoo, APPSOFS performs better than other algorithms on mean value and standard deviation.

Table 2. Average and standard deviation of HV indicator values on training sets

Dataset		MOEAD	MOPSO	HMPSOFS	RFPSOFS	APPSOFS
Sonar	Avg.	0.7650	0.8477	0.8644	0.6662	0.8553
	Std.	0.0366	0.0188	0.0206	0.0580	0.0159
Musk	Avg.	0.7287	0.8409	0.8511	0.6464	0.8521
	Std.	0.0271	0.0295	0.0254	0.0511	0.0158
WBCD	Avg.	0.9062	0.9301	0.9405	0.8008	0.9412
	Std.	0.0260	0.0223	0.0181	0.0600	0.0049
Wine	Avg.	0.8676	0.8970	0.9063	0.8242	0.8869
	Std.	0.0157	0.0127	0.0061	0.0308	0.0104
Vehicle	Avg.	0.6525	0.6863	0.6897	0.6098	0.6918
	Std.	0.0219	0.0148	0.0105	0.0310	0.0119
Zoo	Avg.	0.8195	0.8613	0.8719	0.7576	0.8757
	Std.	0.0256	0.0259	0.0132	0.0383	0.0113

Secondly, the mean values and the standard deviations of HV in the test set are listed on Table 3. APPSOFS gains better mean values and standard deviations of HV than other algorithms for Sonar, Musk, WBCD and Zoo. It is obvious to see that APPSOFS performs better than other algorithms for Sonar and Musk. For WBCD, APPSOFS performs a little better than other algorithms.

Table 3. Average and standard deviation of HV indicator values on test sets

Dataset		MOEAD	MOPSO	HMPSOFS	RFPSOFS	APPSOFS
Sonar	Avg.	0.7228	0.7846	0.7951	0.6418	0.8285
	Std.	0.0465	0.0353	0.0327	0.0677	0.0270
Musk	Avg.	0.7029	0.8218	0.8218	0.6413	0.8360
	Std.	0.0338	0.0292	0.0295	0.0508	0.0196
WBCD	Avg.	0.8951	0.9197	0.9307	0.7943	0.9312
	Std.	0.0288	0.0260	0.0227	0.0604	0.0093
Wine	Avg.	0.8650	0.8794	0.8886	0.8157	0.8801
	Std.	0.0202	0.0206	0.0191	0.0337	0.0176
Vehicle	Avg.	0.6392	0.6786	0.6833	0.5930	0.6707
	Std.	0.0212	0.0202	0.0184	0.0375	0.0177
Zoo	Avg.	0.7683	0.8037	0.8132	0.7310	0.8145
	Std.	0.0563	0.0482	0.0531	0.0591	0.0423

Meanwhile, APPSOFS performs more stable than comparison algorithms. For Wine, HMPSOFS performs a little better than APPSOFS, but APPSOFS performs more stable. For Vehicle, HMPSOFS performs much better than APPSOFS, but APPSOFS performs more stable. For Zoo, APPSOFS performs a little better than other algorithms. Meanwhile, APPSOFS performs more stable than comparison algorithms for Zoo.

5 Conclusion and Future Work

Current existing feature selection algorithms mostly adopt single objective optimization. However, feature selection can be seen a multi-objective problem. This paper proposes a novel adaptive penalty value to overcome the problem of not finding boundary solutions. Meanwhile, the archive is updated by PBI approach instead of Pareto mechanism. GOBL-R is applied to generate new particles for some particles whose pbest are updated for many times. The experiment adopts KNN as a classifier to save time of calculating error rate.

Although the experiment has verified the good performance of classification on six datasets, more datasets with high dimensions features are looking forward to verifying the good performance of the proposed method. There is many improvements of the setting of adaptive penalty value mechanism. The feature information of the archive should be of good use.

Acknowledgments. This work was supported by the National Natural Science Foundation of China [Nos. 61976108 and 61572241], the National Key R&D Program of China [No. 2017YFC0806600], the Foundation of the Peak of Six Talents of Jiangsu

Province [No. 2015-DZXX-024] and the Fifth "333 High Level Talented Person Cultivating Project" of Jiangsu Province [No. (2016) III-0845].

References

1. Cagnina, L., Esquivel, S.C., Coello Coello, C.: A particle swarm optimizer for multi-objective optimization. J. Comput. Sci. Technol. **5**(4), 204–210 (2005)
2. Cheng, R., Jin, Y.: A competitive swarm optimizer for large scale optimization. IEEE Trans. Cybern. **45**(2), 191–204 (2014)
3. Chuang, L.Y., Yang, C.H., Li, J.C.: Chaotic maps based on binary particle swarm optimization for feature selection. Appl. Soft Comput. **11**(1), 239–248 (2011)
4. Giagkiozis, I., Fleming, P.J.: Methods for multi-objective optimization: an analysis. Inf. Sci. **293**, 338–350 (2015)
5. Huang, J., Ling, C.X.: Using AUC and accuracy in evaluating learning algorithms. IEEE Trans. Knowl. Data Eng. **17**(3), 299–310 (2005)
6. Mandavi, S., Rahnamayan, S., Deb, K.: Opposition based learning: a literature review. Swarm Evol. Comput. **39**, 1–23 (2018)
7. Maryam, A., Behrouz, M.B.: Optimizing multi-objective PSO based feature selection method using a feature elitism mechanism. Expert Syst. Appl. **113**, 499–514 (2018)
8. Nguyen, B.H., Xue, B., Liu, I., Andreae, P., Zhang, M.: New mechanism for achive maintenance in PSO-based multi-objective feature selection. Soft. Comput. **20**(10), 3927–3946 (2016)
9. Qiao, J., Zhou, H., Yang, C., Yang, S.: A decomposition-based multiobjective evolutionary algorithm with angle-based adaptive penalty. Appl. Soft Comput. **74**(1), 190–205 (2019)
10. Roberto, H.W., George, D.C., Renato, F.C.: A global-ranking local feature selection method for text categorization. Expert Syst. Appl. **39**(17), 12851–12857 (2012)
11. Tang, J., Zhao, X.: On the improvement of opposition-based differential evolution. In: 2010 Sixth International Conference on Natural Computation, pp. 2407–2411 (2010)
12. Unler, A., Murat, A.: A discrete particle swarm optimization method for feature selection in binary classification problems. Eur. J. Oper. Res. **206**(3), 528–539 (2010)
13. Wang, H., Wu, Z., Rahnamayan, S.: Enhanced opposition-based differential evolution for solving high-dimensional continuous optimization problems. Soft. Comput. **15**(11), 2127–2140 (2011)
14. Wang, L., Zhang, Q., Zhou, A., Gong, M., Jiao, L.: Constrained subproblem in a decomposition-based multiobjective evolutionary algorithm. IEEE Trans. Evol. Comput. **20**(3), 475–480 (2016)
15. While, L., Hingston, P., Barone, L., Husband, S.: A faster algorithm for calculating hypervolume. IEEE Trans. Evol. Comput. **10**(1), 29–38 (2006)
16. Xue, B., Zhang, M., Browne, W.N.: Particle swarm optimization for feature selection in classification: a multi-objective approach. IEEE Trans. Cybern. **43**(6), 1656–1671 (2013)
17. Yang, S., Jiang, S., Jiang, Y.: Improving the multiobjective evolutionary algorithm based on decomposition with new penalty schemes. Soft. Comput. **21**(16), 4677–4691 (2016). https://doi.org/10.1007/s00500-016-2076-3
18. Zhang, Q., Li, H.: MOEA/D: a multiobjective evolutionary algorithm based on decomposition. IEEE Trans. Evol. Comput. **11**(6), 712–731 (2008)

19. Zhang, X., Zheng, X., Cheng, R., Qiu, J., Jin, Y.: A competitive mechanism based multi-objective particle swarm optimizer with fast convergence. Inf. Sci. **427**, 63–76 (2018)
20. Zhang, Y., Gong, D., Cheng, J.: Multi-objective particle swarm optimization approach for cost-based feature selection in classification. IEEE Trans. Comput. Biol. Bioinform. **14**(1), 64–75 (2017)

An Adaptive Brain Storm Optimization Algorithm Based on Heuristic Operators for TSP

Yali Wu(✉), Xiaopeng Wang, Jinjin Qi, and Liting Huang

School of Automation and Information Engineering,
Xi'an University of Technology, Xi'an, Shaanxi, China
yliwu@xaut.edu.cn

Abstract. Traveling Salesman Problem (TSP) is a classical combinatorial optimization problem. This paper proposed an MDBSO (Modified Discrete Brain Storm Optimization) algorithm to solve TSP. The convex hull or greedy algorithm was introduced to initialize the population, which can improve the initial population quality. The adaptive inertial selection strategy was proposed to strengthen the global search in the early stage while enhancing the local search in the later stage. The heuristic crossover operator, which is based on the local superior genetic information of the parent to generate new individuals, was designed to improve the search efficiency of the algorithm. The experimental results of different benchmarks show that the MDBSO algorithm proposed in this paper significantly improves the convergence performance in solving TSP.

Keywords: Discrete brain storm algorithm · Prior knowledge · Heuristic crossover operator · Adaptive inertial selection · Traveling Salesman Problems

1 Introduction

TSP is defined as the following: 'A salesman is required to visit once and only once each of n different cities starting from a base city, and returning to this city. What path minimizes the total distance travelled by the salesman?' [1]. It has been proved to be a classical NP combinatorial optimization problem. And the problem has been widely used in engineering and real life, such as calculating of the best distribution route to obtain the most profit, rational planning of driving roads to divert traffic, analysis of crystal structure and so on. With the wider application field, TSP has become a concern of the whole society. How to solve the TSP quickly and effectively is of great significance [2].

TSP is an NP-hard problem, which has remained one of the most challenging problem for a long time in the field of discrete or combinatorial optimization techniques, which are based on linear and non-linear programming. And as the problem scale increases, it cannot be solved easily [3]. The algorithm for solving TSP problem can be divided into two categories. One is the deterministic approach, such as dynamic programming, branch and bound method, and so on. In recent years, researchers found that the intelligent optimization algorithm could search effectively in solving the combinatorial optimization problem. So the intelligent optimization algorithm become the other vital research direction

L. Pan et al. (Eds.): BIC-TA 2019, CCIS 1159, pp. 662–672, 2020.
https://doi.org/10.1007/978-981-15-3425-6_52

for solving the TSP. Bian [4] modifies the PSO algorithm model with the idea of swarm intelligent optimization algorithm to solve TSP. Su [5] proposes the PSO algorithm that uses greedy algorithm to initialize the population to solve the shortcomings of particle swarm optimization algorithm which is easy to fall into local optimum. The results show higher convergence of the algorithm. An improved genetic algorithm based on neighbor choosing strategy was proposed [6], and this paper indicates the effective of the algorithm for solving the problem. MAX-MIN ant colony system (MMAS) based on ant colony optimization algorithm was proposed for the traveling salesman problem [7].

In 2011, Brain Storm Optimization (BSO) Algorithm was proposed by Shi [8], which was inspired by one of human being collective problem solving skills. The algorithm has been successfully applied in many fields, such as the heat and power co-generation scheduling problem [9], job shop scheduling problem [10], multi-UAV formation flight rolling time domain control problem [11], DC brushless motor efficiency problem [12]. In this paper, we proposed the general framework of the BSO algorithm for TSP problem. And then a new versions of discrete BSO algorithm, named MDBSO (Modified Discrete Brain Storm Optimization) algorithm, is designed to solve TSP problems. The experimental results demonstrated that the proposed algorithm improves the convergence performance significantly, and shows its great potential in solving the TSP.

The remaining paper is organized as follows. The related work of TSP problem and BSO algorithm are listed in Sect. 2. The proposed MDBSO method for TSP is described in Sect. 3. And the simulation results and discussion are shown in Sect. 4. In Sect. 5, we conclude the paper with a summarization of results by emphasizing the importance of this study.

2 The Basic Problem Description

2.1 Traveling Salesman Problem

TSP is a classic combinatorial optimization problem. Assume that we know the nodes of the N cities $V = \{V_1, V_2 \ldots\ldots V_N\}$ and the distance between any two cities $d(V_i, V_j)$ under the condition of symmetric TSP, which is means that $d(V_i, V_j) = d(V_j, V_i)$, the problem is that if there is a travel agent who wants to visit N cities, how can he find a shortest access route to complete the process to meet that demand each city to visit and only visit once. The mathematical description is as following,

$$f = d(V_n, V_1) + \min \sum_{i=1}^{n-1} d(V_i, V_{i+1}) \tag{1}$$

Where $d(V_i, V_{i+1})$ represents the distance from city i to city $i + 1$.

2.2 Brain Storm Optimization for TSP

Brain Storm Optimization algorithm is based on the idea of brainstorming of creative thoughts, and it has achieved good results in dealing with continuous and complex optimization problems, it has become the general framework for solving complex optimization problems [13]. The mean procedure of the algorithm as shown in Table 1.

Table 1. Algorithmic Process

The procedure of BSO algorithm:
Population initialization;
while *not terminated* **do**
Evaluating individuals;
Clustering individuals;
Disrupting cluster center;
Updating individuals and population;
Output individuals.

Based on biological and mathematical ecological theory foundation, Discrete Brain Storm Optimization (DBSO) is proposed in the paper to solve TSP. Firstly, the mathematical optimization model should be established according to the target and constraints of the problem. And the appropriate coding method should be determined to map relationship of the mathematical model and the algorithm. Then the algorithm optimizes by means of population iteration, so as to obtain the optimal solution. The key operations of the DBSO algorithm for solving TSP are designed as follows.

2.2.1 Population Initialization

The individual coding method is the primary problem in the initial stage of the population. In this paper, we randomly generate a set of the real value. Gene in each dimension represents the order of the cities to be traveled by salesman. For example, the sequence (8, 5, 1, 4, 10, 7, 6, 3, 2, 9) represents a loop traveled by salesman. The salesman starts from the initial city 8, then visits cities 5, 1,…, in that order, and finally back to the city 8 through city 9. The encoding mode as show in Fig. 1.

8	**5**	**1**	**4**	**10**	**7**	**6**	**3**	**2**	**9**

Fig. 1. Individual coding

2.2.2 Clustering

The clustering algorithm used in the original Brain Storm Optimization algorithm is the k-means clustering method. In the existing literature for BSO, the clustering operation is done in the solution space, while in DBSO algorithm, the clustering operation is done in the objective space. Therefore, the k-means clustering method based on the objective space is adopted in the discrete Brain Storm Optimization algorithm to solve the TSP. In the iteration, we separate all the N ideas of the entire swarm into M different clusters. The basic idea of K-means clustering is to use the average of all data samples in each clustering subset as the representative point of the cluster. The data set is divided into different categories according to the Euclidean distance [14].

2.2.3 The New Individual Generation Strategy

According to the process of generating new individuals, there are two main method, that is, inter-group discussion or intro-group discussion produce new individuals. The details are listed in the following.

(1) Single individual mutation strategy:

- Randomly generate a position and a distance;
- Move the information of the position point to the right with the corresponding distance;
- If the distance exceeds the last digit of the individual, it will go to the first one.

(2) Two individuals merger strategy:

- Randomly select several location points in individual 1 and directly copy the selected location information to the corresponding location of the new individual;
- Delete the information of the selected location point from left to right in the individual 2;
- Fill the remaining of individual 2 in free position of the new location from the left to the right and obtain a complete new individual.

2.2.4 Updating Individuals and Population

During the iterative process, the population size remains the same. Therefore, in process of choosing individuals going to the next iteration, initialized individuals and the generated new individuals will be sorted according to the value of fitness. We choose a certain proportion of optimal individual directly into the next generation and the rest of the individual with the method of roulette wheel selection. In this way, we ensure the population size and population diversity of entering the next iteration.

3 MDBSO Algorithm for TSP

Based on the simple Brain Storm Optimization algorithm, the Modified Discrete Brain Storm Optimization algorithm uses prior knowledge to initialize population in the early phase, and adopts adaptive inertial selection strategy phase in the generation phase to optimize the efficiency of the Brain Storm Optimization algorithm, and uses the heuristic crossover operator to improve the convergence of the algorithm in the new individual generation stage.

3.1 Initialization of the TSP

The basic BSO algorithm generates the initial population by the random method. The primary disadvantages of random method are that it will restrict the convergence speed of the algorithm to a certain extent because of its low fitness. So, a method based on prior knowledge (convex hull and greedy algorithm) is used to generate new populations to enhance convergence [15]. Convex hull algorithm is designed to generate initial populations on a small scale; otherwise a greedy algorithm is used. Some individuals in the population are generated based on prior knowledge and another portion is randomly generated, making the population has a certain target and representativeness [16].

(1) Based on Convex hull algorithm

The implementation process based on the convex hull algorithm. First, the fast convex hull algorithm constructs the convex hull of the city point set, which is a loop that passes through some city points and the rest of the points are inside. Secondly, the remaining city points are sequentially inserted into the loop to form a new loop, so that the length increment of the new loop is minimized until all the city points are on the loop [17].

(2) Based on Greedy algorithm

The implementation process based on the greedy algorithm begins with randomly selection of a city as the currently located city of the traveler agency and adds it to the entity. Then all the cities not included are searched for in the entity and the nearest city to the current city is found. Add it to the entity and make it served as the current city. Keeping to search and adding the next nearest city until all cities are included in the entity so as to achieve the primary optimization of the entity [18].

3.2 Generation of New Individuals

The vital of the Brain Storm Optimization algorithm are the generation and selection of new individuals, which are the same as that in other swarm intelligence algorithms. There are two ways to generate new individuals: (1) intra-group discussion based on individuals in a class; (2) inter-group discussion based on individuals in two classes [10]. Due to the diversity of individual methods, the BSO algorithm can search for the optimal solution faster when solving the TSP. There are two method in order to speed up convergence and increase diversity.

3.2.1 The Adaptive Inertial Selection Strategy

The basic BSO algorithm sets a fix parameters to select the way to generate new individuals. But MDBSO algorithm introduces a dynamic parameter P_{6b} to regulation method in the individual updating process. The adaptive transform of P_{6b} is according to formula (2) in the process of MDBSO algorithm [19]. We can increase the probability P_{6b} of the inter-group discussion in the early stage of the algorithm to locate in the optimal solution area quickly, and improve the probability of intra-group discussion in the later stage to enhance the fine search nearby the optimal solution. The experimental result indicates efficiency of dynamic parameter.

$$P_{6b}(Iter) = P_{max} - (P_{max} - P_{min}) \times \frac{Iter}{MaxIt} \tag{2}$$

Where, P_{max} and P_{min} are the maximum and minimum values of P_{6b}. *Iter* and *MaxIt* respectively represent the current iteration number and the maximum.x

3.2.2 Heuristic Crossover Operator

Considering the complexity characteristics of TSP, this paper uses the heuristic crossover operator to generate offspring by the local excellent genetic information of the parent to accelerate the search speed of the algorithm [20]. Parent1 and Parent2 are selected in the initial population and crossed by a heuristic crossover operator to generate two child individuals Child1 and Child2. The specific process showed in Fig. 3 (Fig. 2).

The procedure of the Heuristic crossover operator:

1 Randomly select a city *m* as the starting point of the individual city. And the city *m* in the parent individual is deleted;
2 **while** *not terminated* **do**
3 Find the next cities *right*1 and *right*2 to the right of the city m of two parent individuals and the leftward cities *left*1 and *left*2 of the city m of the two parent individuals. Then compare the distance between *m* and *right*1, *right*2,and the distance between *m* and *left*1,*left*2;
4 If $d(m,right1)<d(m,right2)$, then select city *right*1 as the second visiting city of child generation Child1 and update Child1. Simultaneously, delete *right*1in Parent1.If $d(m,right1)>d(m,right2)$, then select city *right*2 as the second visiting city of child generation Child1 and update Child1. Simultaneously, delete *right*2 in Parent2until the remaining one of the parent individuals;
5 Similarly, update Child2 until the end, otherwise **1** is executed;
6 Output individuals.

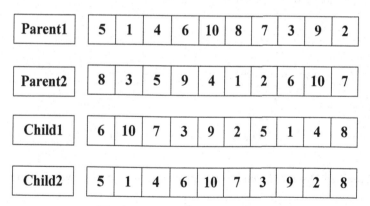

Parent1	5	1	4	6	10	8	7	3	9	2
Parent2	8	3	5	9	4	1	2	6	10	7
Child1	6	10	7	3	9	2	5	1	4	8
Child2	5	1	4	6	10	7	3	9	2	8

Fig. 2. Heuristic crossover operator

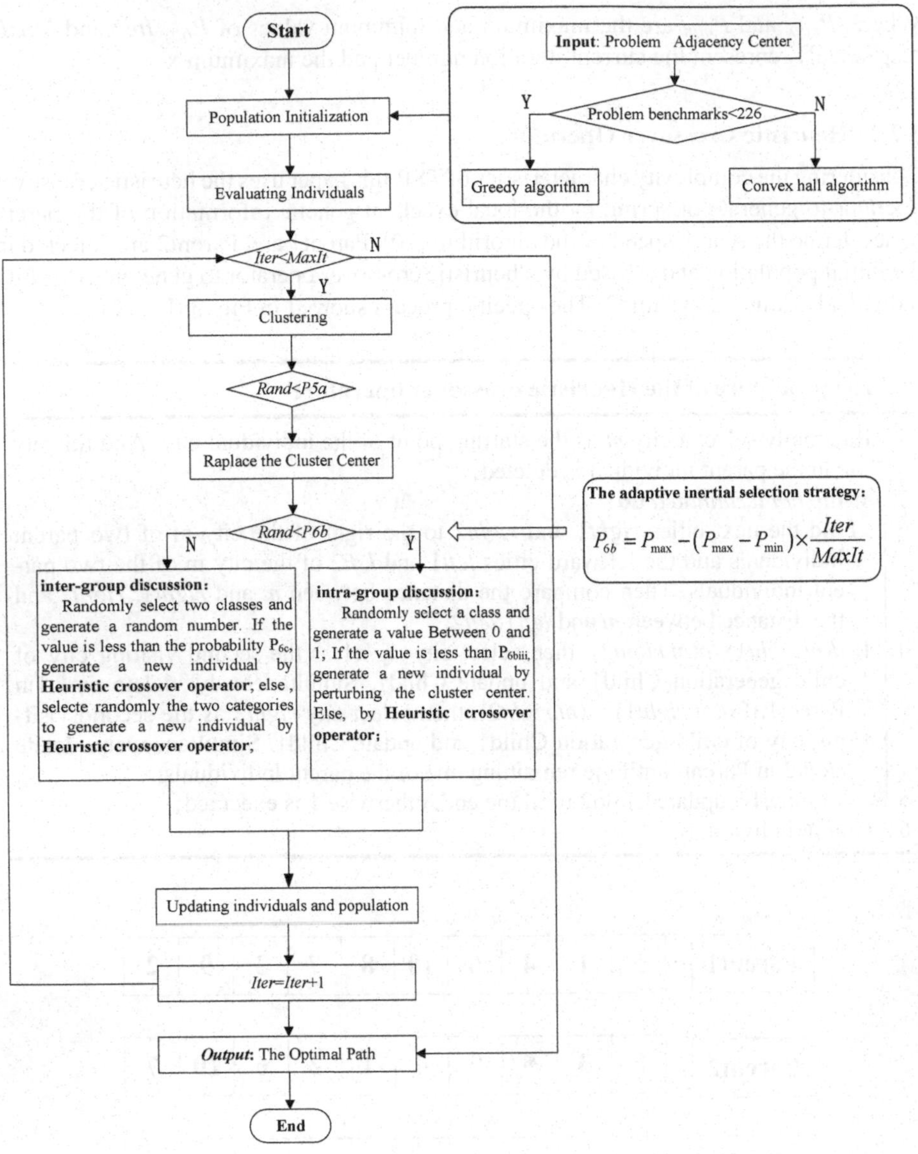

Fig. 3. The MDBSO algorithm

The Modified Discrete Brain Storm Optimization algorithm was shown in Fig. 3.

4 Experimental Results and Analysis

In this section, all numerical studies were carried out on a PC with an Intel(R) Core(TM) i5-4460 CPU 3.20 GHz.

4.1 Testing Problems

In order to verify the effectiveness of MDBSO algorithm, some examples are selected from the TSP instance library (http://eli.zib.de/pub/mptestdata/tsp/tsplib/). All TSP instances can divided into two class according to their problem size. Class 1 of instance are considered as small sized and class2 are considered as medium sized TSP instances. The data sets in TSPLIB adopted in this paper as follow: eil51, st70, kroA100, ch150, pr226, pcb442 (51, 70, 100,150, 226, 442 cities respectively).

4.2 Parameter Setting

Considering the randomness of the algorithm, 30 independent runs were conducted to collect statistical results. In addition, during all the simulation runs, the other parameters of MDBSO are consistent with the parameters of the PKDBSO (Discrete Brain Storm Optimization algorithm based on Prior Knowledge) [3] algorithm. We set the probability parameter of the BSO algorithm as: cluster center selection probability $P_{5a} = 0.5$, mutation individual selection probability $P_{6biii} = 0.4$, $P_{6c} = 0.5$.

This paper uses the error rate to illustrate the effectiveness of the algorithm. The calculation formula of its error rate is as follows:

$$Error = \frac{Average\ \ value\ \ - Optimal\ value}{Optimal\ \ value} \tag{3}$$

Where, *Average value* represents the optimal solution to the algorithm operation, *Optimal value* represents the known optimal solution.

4.3 Simulation Results and Analysis

To compare the search performance of the two algorithms, we collected statistical results. The dynamic average convergence curves with two algorithms for solving different scale problems is shown in Fig. 4. The Fig. 4 indicates the effectiveness and practicability of the DMBSO algorithm. The result of six different scale problem (the standard database has given the optimal solution) verified the effectiveness of the improved algorithm for solving TSP. Experimental result demonstrated that the MDBSO algorithm can avoid being stagnated in the local optima, more effectively and steadily find the better result than PKDBSO algorithm in TSP.

Table 2 shows the simulation results of different scale by different algorithms. It can be seen that MDBSO algorithm has searched better results than PKDBSO in solving TSP in all cities. Obviously, the result of MDBSO solving 51 cities is 2.44%, which is 3.85% less than that of PKDBSO 6.29%. And similarly we can see MDBSO has an error rate of 1.5% for st70 less than before 2.8%. For the city size of kroA100, ch150, pr226, Pcb442, the error rate of MDBSO is reduced by 3.14%, 6.77%, 4.88%, 11.91% successively. All in all, experiment data indicate MDBSO achieved better results. MDBSO can jump out of the local optimal solution faster and improve the convergence performance and speed with the increase of the complexity of the problem. The reason is that some convex global algorithm or greedy algorithm is used to obtain some solutions closed to the global optimal, and the convergence is improved in the initialization stage of the

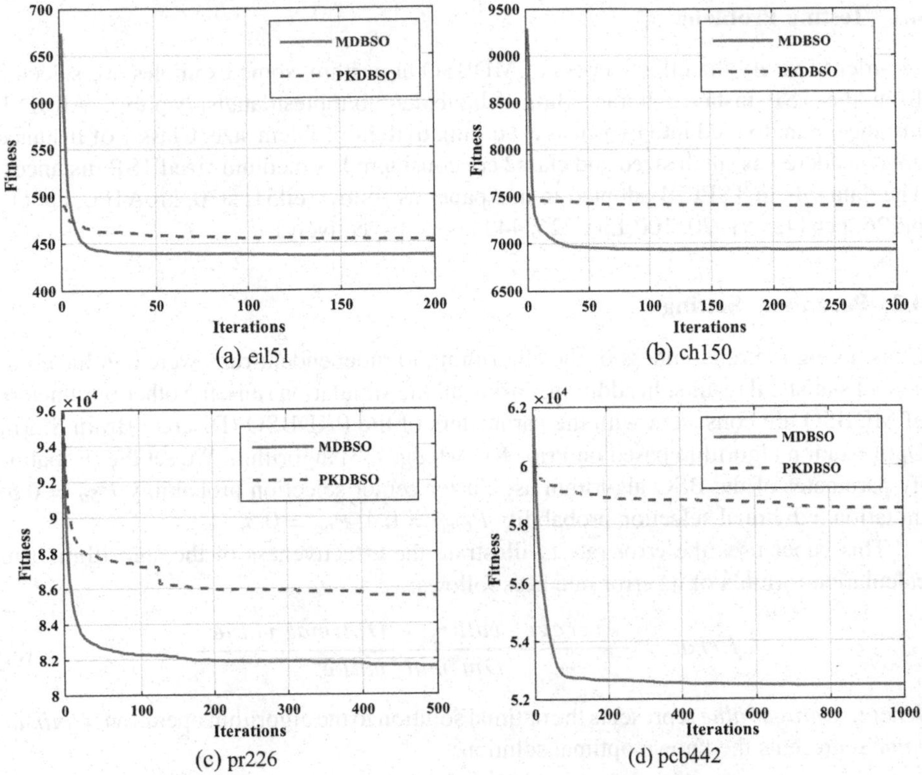

Fig. 4. The dynamic average convergence curves

Table 2. Experimental data

Data	Opt.	Algorithm	Average optimal solution	Error rate (%)
eil51	426	PKDBSO	452.8	6.29
		MDBSO	**436.4**	**2.44**
St70	675	PKDBSO	694.3	2.80
		MDBSO	**685.5**	**1.50**
kroA100	21282	PKDBSO	22023	3.40
		MDBSO	**21339**	**0.26**
ch150	6528	PKDBSO	7381	13.06
		MDBSO	**6939**	**6.29**
pr226	80369	PKDBSO	85526	6.41
		MDBSO	**81606**	1.53
Pcb442	50778	PKDBSO	58516	15.23
		MDBSO	**52467**	**3.32**

algorithm. At the same time, in order to avoid algorithm trapped in the local optimal, the adaptive selection strategy is adopted. In the early stage of the algorithm, the diversity of the individual groups is ensured by increasing the probability of discussion between groups, and the fine search ability is strengthened to ensure the convergence speed of the algorithm. The advantages of MDBSO become more and more obvious in large scale.

5 Conclusion

In this paper, the MDBSO Algorithm for TSP was proposed and was compared with the PKDBSO by testing on different scale cities. The prior knowledge is used to initialize the population to enhance convergence performance of improved algorithm in the initial stage. The MDBSO algorithm controls the search direction and mutation mechanism in the iteration process by linear increment of parameter. We can strength the global search ability in the early stage and enhance the fine search power in the later stage on the basis of ensuring the individual diversity of the population. Taking into account the characteristics of the TSP, the Heuristic Crossover Operator is used to enhance the information interactions capability and to prevent the algorithm from jumping into local optimum. The experimental results demonstrated that the MDBSO generally performs better than the PKDBSO to solution accuracy for TSP and performs well enough for large-scale TSP. But there are still some problems that computation time is longer as the problem scale increases. So how to reduce computation time on solving large-scale is the direction of the next research.

Acknowledgment. This paper was supported by National Natural Science Foundation of China under Grant Number 2018YFB1703004.

References

1. Bellman, R.: Dynamic programming treatment of the travelling salesman problem. J. ACM **9**(1), 61–63 (1962)
2. Sun, Q., Zhang, J., Wang, Y.: Ant colony algorithm optimization strategy review. Inf. Secur. Technol. **5**(2), 22–23 (2014)
3. Xu, Y., Wu, Y., Fu, Y.: Discrete brain storm optimization algorithm based on prior knowledge for traveling salesman problems. In: 2018 13th IEEE Conference on Industrial Electronics and Applications (ICIEA), pp. 2740–2745. IEEE (2018)
4. Bian, F.: An improved quantum particle swarm optimization for solving travelling salesman prom. Comput. Appl. Softw. **26**(11), 218–220 (2009)
5. Su, J., Wang, J.: Improved particle swarm optimization for traveling salesman problem. Comput. Eng. Appl. **46**(4), 52–53 (2010)
6. Tao, L., Guo, J.: Application of solving TSP based on improved genetic algorithm. Comput. Eng. Appl. **45**(33), 45–47 (2009)
7. Gu, W.: Parallel performance of an ant colony optimization algorithm for TSP. In: International Conference on Intelligent Computation Technology & Automation. IEEE (2015)
8. Shi, Y.: Brain storm optimization algorithm. In: Tan, Y., Shi, Y., Chai, Y., Wang, G. (eds.) ICSI 2011. LNCS, vol. 6728, pp. 303–309. Springer, Heidelberg (2011). https://doi.org/10.1007/978-3-642-21515-5_36

9. Ramanand, K.R., Krishnanand, K.R., Panigrahi, B.K., Mallick, M.K.: Brain storming incorporated teaching–learning–based algorithm with application to electric power dispatch. In: Panigrahi, B.K., Das, S., Suganthan, P.N., Nanda, P.K. (eds.) SEMCCO 2012. LNCS, vol. 7677, pp. 476–483. Springer, Heidelberg (2012). https://doi.org/10.1007/978-3-642-35380-2_56
10. Wu, X., Zhang, Z.: A brain storm optimization algorithm integrating diversity and discussion mechanism for solving discrete production scheduling problem. Control Decis. **32**(9), 1583–1590 (2017)
11. Qiu, H., Duan, H.: Receding horizon control for multiple UAV formation flight based on modified brain storm optimization. Nonlinear Dyn. **78**(3), 1973–1988 (2014)
12. Duan, H., Li, S., Shi, Y.: Predator-prey brain storm optimization for DC brushless motor. IEEE Trans. Magn. **49**(10), 5336–5340 (2013)
13. Dinh Thanh, P., Thi Thanh Binh, H., Thu Lam, B.: New mechanism of combination crossover operators in genetic algorithm for solving the traveling salesman problem. In: Nguyen, V.-H., Le, A.-C., Huynh, V.-N. (eds.) Knowledge and Systems Engineering. AISC, vol. 326, pp. 367–379. Springer, Cham (2015). https://doi.org/10.1007/978-3-319-11680-8_29
14. Wu, Y., Jiao, S.: Theory and Application of Brainstorming Optimization Algorithm, pp. 1–185. Science Press, Beijing (2017)
15. Manyawu, A.: An improved genetic algorithm using the convex hull for traveling salesman problem. In: IEEE International Conference on Systems (1998)
16. Meeran, S.: Optimum path planning using convex hull and local search heuristic algorithms. Mechatronics **7**(8), 737–756 (1997)
17. Deĭneko, V.: The convex-hull-and-k-line travelling salesman problem. Inf. Process. Lett. **59**(6), 295–301 (2012)
18. Chen, J.: Hybrid genetic algorithm based on strategy of greedy for TSP. J. Lanzhou Jiaotong Univ. **28**(3), 58–61 (2009)
19. Yang, L., Kong, F.: Self-adaptive selection strategy for artificial bee colony algorithm. J. Guangxi Univ. Technol. **23**(3), 37–44 (2012)
20. Li, K., Xu, F., Ping, H.: A new best-worst ant system with heuristic crossover operator for solving TSP. In: International Conference on Natural Computation. IEEE Computer Society (2009)

A Modified JAYA Algorithm for Optimization in Brushless DC Wheel Motor

Li Yan, Chuang Zhang, Boyang Qu(✉), Fangfang Bian, and Chao Li

School of Electric and Information Engineering, Zhongyuan University of Technology, Zhengzhou 450007, China
quboyang@zut.edu.cn

Abstract. Brushless DC wheel motor has been widely used in various industrial fields and life, which has the advantages of high efficiency, light weight, speed range and so on. The parameters design of BLDC wheel motor is of great significance to its stability, efficiency and accuracy of motor operation. This paper proposes a modified JAYA (MJAYA) algorithm to optimize the parameters of the brushless DC wheel motor. In MJAYA, an experience-based learning strategy is proposed to improve the diversity of the population to avoid falling into the local optima. Experimental results show that the proposed MJAYA is feasible and effective in solving the parameters design problem of brushless DC wheel motor. In addition, MJAYA shows a superior performance compared with other commonly used algorithms.

Keywords: Brushless DC wheel motor · Parameters design · JAYA algorithm · Learning strategy

1 Introduction

Brushless DC wheel motor (BLDC) is a typical mechatronics, which converts direct current into mechanical energy. BLDC wheel motor has the merits of small size, light weight, fast response, high starting torque, high efficiency and wide speed range. It has been successfully applied in automobile, electric vehicle, unmanned aerial vehicle, industrial drive, servo control, military and aerospace fields [1–4]. Thus, it is important to optimize the design of BLDC wheel motor to improve its efficiency. In order to improve the performance of BLDC wheel motor, the parameter design problem is converted to an optimization problem. And different optimization methods based on swarm intelligence have been proposed to solve this problem such as bat-inspired optimization approach [5], cuckoo optimization algorithm [6], multi-objective krill herd algorithm [7], brain storm optimization algorithm [8], and others. Although these methods have achieved satisfied results, the performance of the algorithms are related closely with the settings of the algorithm-specific parameters. It is difficult for users to set the appropriate parameters for a specific or new optimization problem. Moreover, the

L. Pan et al. (Eds.): BIC-TA 2019, CCIS 1159, pp. 673–681, 2020.
https://doi.org/10.1007/978-981-15-3425-6_53

improper tuning of the parameters may increase the computational complexity or obtain a local optimal solution.

JAYA algorithm is a new but powerful heuristic method proposed by Rao for constrained and unconstrained optimization problems in 2016 [9]. The basic idea of JAYA is that the candidate solution obtained for a given problem should move towards the best solution and should avoid the worst solution. The algorithm does not require any algorithm-specific parameters except two common controlling parameters namely population size and number of generations. JAYA has been successfully applied in many fields, such as intelligent facial emotion recognition [10], optimization design of truss structures [11], thermal performance optimization of the underground power cable system [12], and so on. In JAYA, the individual is only guided by the best solution and the worst solution. Although this method makes the algorithm show a good convergence ability, the population diversity may not be maintained and thus it is prone to show a premature convergence. In addition, to our best knowledge, JAYA has not yet been applied in solving the BLDC wheel motor problem.

In this paper, a modified JAYA (MJAYA) algorithm is proposed to solve the parameters design problem of the BLDC wheel motor and to maximize its efficiency. In MJAYA, an experience-based learning strategy is introduced to improve the diversity of population by learning from the experience of other individuals of the current generation. In order to verify the effectiveness of the proposed method, MJAYA is employed to solve the parameters optimization problem, and it is also compared with two well-established algorithms.

The rest of this paper is organized as follows. Section 2 describes the optimization problem of a BLDC wheel motor. In Sect. 3, the basic JAYA and proposed MJAYA algorithms are presented. Experiments and the results analysis are conducted in Sect. 4, Sect. 5 concludes the paper.

2 Problem Formulation

This section presents the model of BLDC wheel motor [13]. The model is composed of 78 nonlinear equations that implemented with five design variables and six inequality constraints [13]. The objective of the design is to optimize the five parameters to maximize the efficiency of the BLDC wheel motor on the condition that the given constraints can be satisfied. Through the optimization of the BLDC wheel motor, the feasibility and the effectiveness of the proposed MJAYA can be verified. As shown in Table 1, the five design variables are given, and the constraints are shown in Table 2.

The parameters design of the BLDC wheel motor can be expressed as the following optimization problem:

$$Maximize \quad \eta = f(D_s, B_d, \delta, B_e, B_{cs}) \tag{1}$$

with

$$150 \le D_s \le 330, 0.9 \le B_d \le 1.8, 2.0 \le \delta \le 5.0,$$
$$0.5 \le B_e \le 0.76, 0.6 \le B_{cs} \le 1.6$$

s.t.

$$M_{tot} \leq 15, D_{ext} \leq 340, D_{int} \leq 76,$$
$$I_{max} \geq 125, discr(D_s, B_d, \delta, B_e) \geq 0, T_a \leq 120.$$

Table 1. Description of the optimization problem.

	Variables	Description	Value
Objective	η (%)	Efficiency	–
Design variables	D_s(mm)	Stator diameter	[150, 330]
	B_d(T)	Magnetic density in the teeth	[0.9, 1.8]
	δ(A/mm^2)	Current density on the windings	[2.0, 5.0]
	B_e(T)	Magnetic density in the air gap	[0.5, 0.76]
	B_{cs}(T)	Magnetic density in the stator back iron	[0.6, 1.6]

Table 2. Description of the problem constraints.

	Variables	Description	Value
Constraints	M_{tot}	Total mass of active parts	$M_{tot} \leq 15$
	D_{ext}(mm)	Outer diameter	$D_{ext} \leq 340$
	D_{int}(mm)	Inner diameter	$D_{int} \geq 76$
	I_{max}(A)	Maximum current in the phases	$I_{max} \geq 125$
	$discr(D_s, B_d, \delta, B_e)$	Determinant used for the calculation of the slot height	$discr(D_s, B_d, \delta, B_e) \geq 0$
	T_a (°C)	Temperature of motor	$T_a \leq 120$

3 Problem Formulation

3.1 JAYA Algorithm

JAYA algorithm is a simple and powerful optimization algorithm for solving constrained and unconstrained optimization problems [9]. The main idea of JAYA algorithm is to approach the best solution and avoid the worst solution. The JAYA performance is not affected by the special parameters of the algorithm, and it only needs two common parameters, namely: population size and iteration number.

For an objective function $f(\mathbf{x})$ with D dimensional variables ($j = 1, 2, \ldots, D$), $\mathbf{x}_i = (x_{i,1}, x_{i,2}, \ldots, x_{i,D})$ is the position of the ith candidate solution, and $x_{i,j}$ is the value of the jth variable. The best candidate solution $\mathbf{x}_{best} = (x_{best,1}, x_{best,2}, \ldots, x_{best,D})$ has the best value of $f(\mathbf{x})$ in the current population, while the worst candidate solution $\mathbf{x}_{worst} = (x_{worst,1}, x_{worst,2}, \ldots, x_{worst,D})$ has the worst value of $f(\mathbf{x})$ in the current population. Then, $x_{i,j}$ is updated using Eq. (2).

$$x'_{i,j} = x_{i,j} + r_1 \cdot (x_{best,j} - |x_{i,j}|) - r_2 \cdot (x_{worst,j} - |x_{i,j}|) \tag{2}$$

where $x_{best,j}$ and $x_{worst,j}$ are the values of the jth variable for the best and worst solutions, respectively. $x'_{i,j}$ is the updated value of $x_{i,j}$, and $|x_{i,j}|$ is the absolute value of $x_{i,j}$. r_1 and r_2 are two uniformly distributed random numbers within [0, 1]. In Eq. (2), the term $r_1 \cdot (x_{best,j} - |x_{i,j}|)$ stand for the tendency of the solution attracted by the best solution and the term $-r_2 \cdot (x_{worst,j} - |x_{i,j}|)$ represents the tendency of the current solution to shun the worst solution. The updated solution $\mathbf{x}_i = (x_{i,1}, x_{i,2}, \ldots, x_{i,D})$ is accepted if it gives a better function value.

3.2 Experience-Based Learning Strategy

In the basic JAYA algorithm, the search directions of the individual depend on the best solution and the worst solution simultaneously. This method is able to enhance the population exploration ability and accelerate the convergence rate of the algorithm. However, it may reduce the diversity of the population with the fast convergence rate, and thus the population is easy to be trapped into the local optimal region. Therefore, a learning strategy based on the experience of other individuals is proposed to increase the population diversity and to enhance the local exploitation ability at the same time. Meantime, the individual is also guided by the best solution to ensure the exploration ability. The learning method can be expressed concretely as follow:

$$x'_{i,j} = \begin{cases} x_{i,j} + r_1 \cdot (x_{best,j} - |x_{i,j}|) + r_2 \cdot (x_{m,j} - x_{n,j}), iff(\mathbf{x}_m) < f(\mathbf{x}_n) \\ x_{i,j} + r_1 \cdot (x_{best,j} - |x_{i,j}|) + r_2 \cdot (x_{n,j} - x_{m,j}), otherwise \end{cases} \tag{3}$$

where $x_{best,j}$ is the values of the jth variable for the best individual, $\mathbf{x}_m$ and $\mathbf{x}_n$ are randomly selected from the current population, $x_{m,j}$ and $x_{n,j}$ are the values of the jth variable for the m and n individuals ($m \neq n \neq i$), respectively. r_1 and r_2 are two random numbers in the range [0,1].

Moreover, during the iteration search process, the update Eqs. (2) and (3) are chosen randomly for each individual to balance the exploration and exploitation abilities of the proposed algorithm.

3.3 Framework of MJAYA

Based on the above description, the pseudo code of MJAYA is summarized in Algorithm 1 and the flow chart of MJAYA is shown in Fig. 1.

Algorithm 1. Pseudo code of MJAYA algorithm

1. Initialize the population size and the maximum number of function evaluations;
2. Initialize the population randomly and evaluate the objective function value for each individual;
3. $FES = NP$;
4. **While** $FES < Max\_FES$ **do**
5. Choose the best individual $\mathbf{X}_{best}$ and the worst individual $\mathbf{X}_{worst}$ from population;
6. **For** $i = 1$ to NP **do**
7. **If** (a<b|a,b U(0,1)) **then**
8. Update the jth ($j = 1, 2, \ldots, D$) variable value of $\mathbf{X}_i$ using Eq. (2); //* Basic JAYA *//
9. **Else**
10. Select the two individuals $\mathbf{X}_m$ and $\mathbf{X}_n$ from population randomly ($\mathbf{X}_m \neq \mathbf{X}_n \neq i$);
11. Update the jth ($j = 1, 2, \ldots, D$) variable value of $\mathbf{X}_i$ using Eq. (3); //* Experience-based learning strategy *//
12. **End**
13. Calculate the function value for the updated individual;
14. $FES = FES + 1$;
15. Accept the new solution if it is better than the old one
16. **End for**
17. **End while**

4 Experimental Results

For comparison, PSO [14], PPBSO [15], and the basic JAYA are also employed to optimize the BLDC wheel motor. The control parameters of all algorithms are given in Table 3, and the parameter settings for all the compared algorithms are based on the advices in the corresponding literatures. For fair comparison, each algorithm is run 30 times independently for the same problem and all of the algorithms use the same maximum number of function evaluations (Max_FES) 4000 in each run.

The statistic results of the simulations are presented in Table 4, and the best solutions of all algorithms among the 30 independent runs are shown in Table 5. In order to make the comparisons clear, the overall best results among all algorithms are highlighted in gray boldface. From Table 4, it is obvious that the proposed MJAYA is superior to other algorithms, and MJAYA achieves the best efficiency and shows a more stable performance than the rest of the algorithms. Although PSO reaches the maximum value 0.95318 also, MJAYA is better than PSO in terms of other three indexes. From Table 5, it is obvious that all the parameters of the solutions obtained by the proposed MJAYA, the basic JAYA and other two compared algorithms satisfy the value limits.

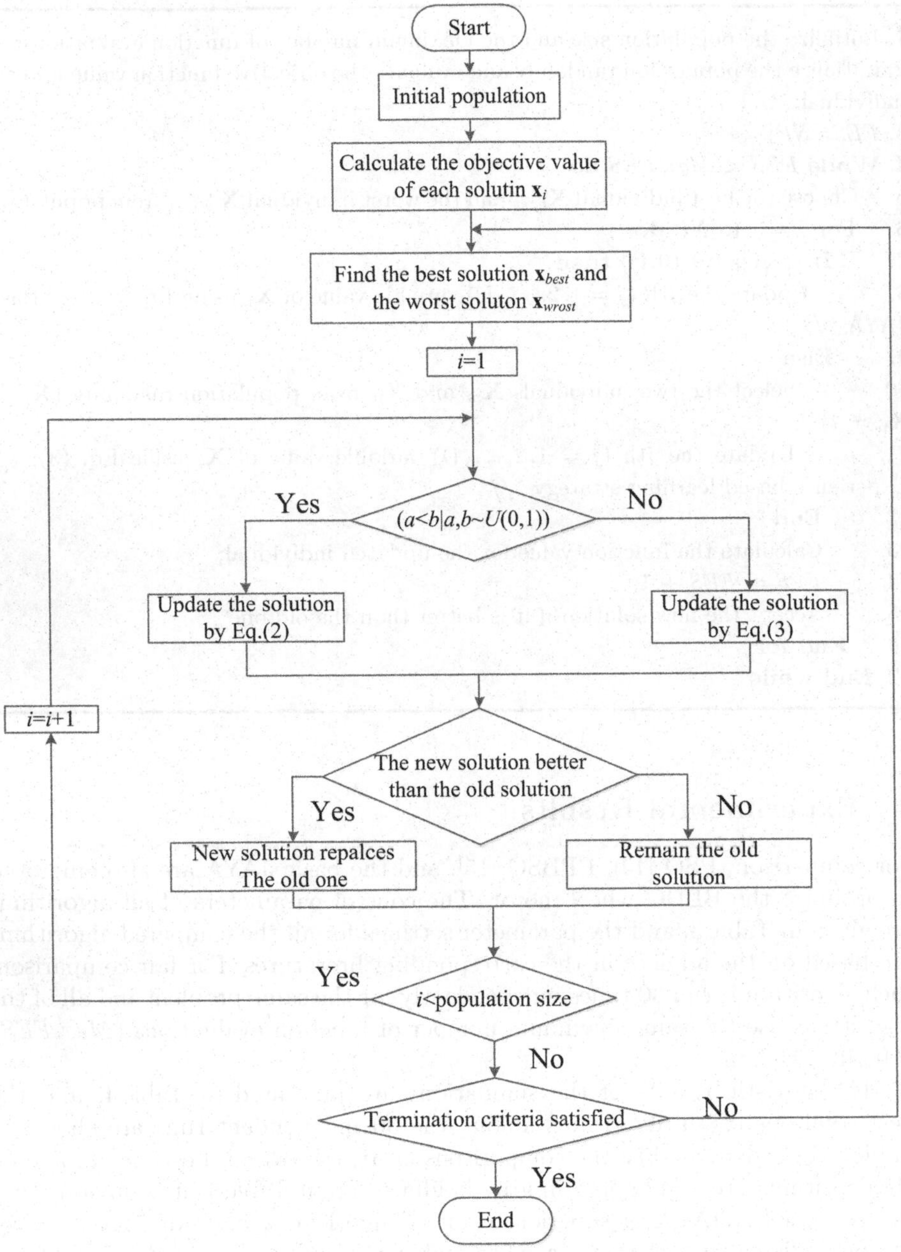

Fig. 1. The flow chart of MJAYA

Table 3. Parameter settings for the involved algorithms

Algorithm	Parameters
MJAYA	NP=40, Nc=100
JAYA	NP=40, Nc=100
PSO [14]	NP=40, Nc=100, $C_1 = [0.4, 1]$, $C_2 = 1.6$, $C_3 = 1.8$
PPBSO [15]	NP=40, Nc=100, K=3, $P_{5a} = 0.2$, $P_{6b} = 0.8$, $P_{6b3} = 0.4$, $P_{6c} = 0.5$, $W_{predator} = 0.05$, $P_{prey} = 0.1$.

Table 4. The statistical results of different algorithms

Algorithm	Min	Max	Mean	SD
MJAYA	**9.5312E-01**	**9.5318E-01**	**9.5317E-01**	**1.2341E-05**
JAYA	9.4595E-01	9.5311E-01	9.5137E-03	1.5946E-03
PSO	9.5128E-01	**9.5318E-01**	9.5303E-04	4.0405E-04
PPBSO	9.5259E-01	9.5315E-01	9.5306E-01	1.0858E-04

Table 5. The final results of the simulation

Algorithm	D_s(mm)	B_d(T)	δ(A/mm^2)	B_e(T)	B_{cs}(T)
MJAYA	201.6	1.8	2.028	0.64	0.89
JAYA	202.1	1.8	2.030	0.64	0.92
PSO	201.6	1.8	2.000	0.64	0.95
PPBSO	202.2	1.8	2.012	0.64	0.96

In order to illustrate the convergence performance of the algorithms, the change curves of the efficiency with the number of iterations are shown in Fig. 2, where the best results of the 30 runs of the four algorithms are displayed. As shown in Fig. 2, the final efficiency value and the convergence rate of the proposed MJAYA are better than the basic JAYA. MJAYA achieves the best efficiency 0.95318 first at the 60th generation. Although the efficiency of the PPBSO efficiency improves as the iteration progresses, the final value is not as good as MJAYA. PSO shows a good convergence rate, but it may struggle with the local optima during the search process and achieves the best value at the 90th generation.

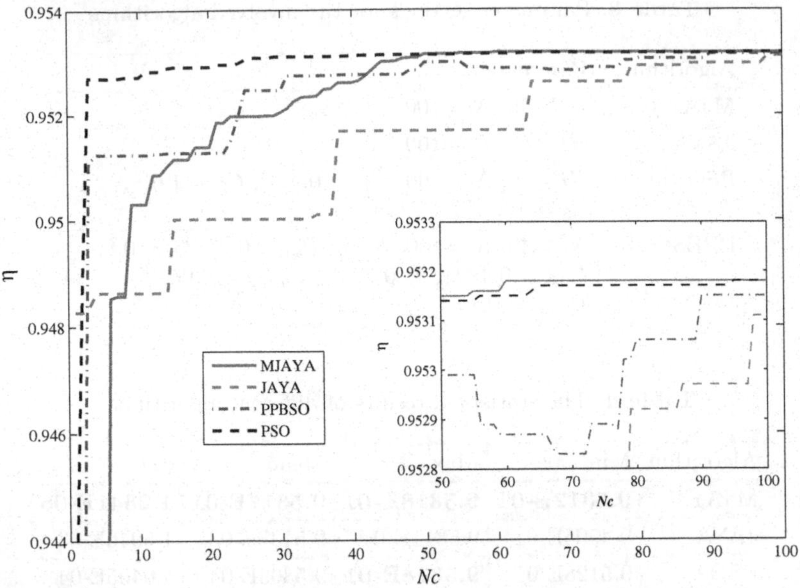

Fig. 2. Comparison of the convergence performance.

5 Conclusion

In this paper, a modified JAYA algorithm(MJAYA) is proposed to solve brushless DC wheel motor design problem. In the proposed algorithm, it makes full use of other individuals experience to learn, where the experience-based learning strategy is developed to improve the population diversity of the MJAYA. The proposed algorithm is compared with three algorithms also. Experiment results demonstrated that the MJAYA shows a better performance and higher efficiency than other three compared algorithms. In conclusion, MJAYA is a promising method to solve the parameters design parameter of brushless DC wheel motor.

Acknowledgement. This work was supported by the National Natural Science Foundation of China (Grant Nos. 61673404, 61876169, 61976237), the Key Scientific Research Projects in Colleges and Universities of Henan Province (Grant No. 19A120014, 20A120013).

References

1. Dadashnialehi, A., Bab-Hadiashar, A., Cao, Z.: Intelligent sensorless antilock braking system for brushless in-wheel electric vehicles. IEEE Trans. Ind. Electron. **62**(3), 1629–1638 (2014)
2. Markovic, M., Perriard, Y.: Optimization design of a segmented Halbach permanent-magnet motor using an analytical model. IEEE Trans. Magn. **45**(7), 2955–2960 (2009)

3. Crevecoeur, G., Sergeant, P., Dupre, L., Van de Walle, R.: A two level genetic algorithm for electromagnetic optimization. IEEE Trans. Magn. **46**(7), 2585–2595 (2010)
4. Liu, X., Hu, H., Zhao, J., Belahcen, A., Tang, L., Yang, L.: Analytical solution of the magnetic field and EMF calculation in ironless BLDC motor. IEEE Trans. Magn. **52**(2), 1–10 (2016)
5. Bora, T.C., Coelho, L.D.S., Lebensztajn, L.: Bat-Insoired optimization approach for the brushless DC wheel motor problem. IEEE Trans. Magn. **48**(2), 947–950 (2012)
6. Azari, M.N., Samami, M., Pahnehkolaei, S.A.: Optimal design of a brushless DC motor, by cuckoo optimization algorithm. Int. J. Eng. **30**(5), 668–677 (2017)
7. Ayala, H.V., Segundo, E.H., Mariani, V.C., Coelho, L.D.S.: Multiobjective krill herd algorithm for electromagnetic optimization. IEEE Trans. Magn. **52**(3), 1–4 (2015)
8. Shi, Y.: Brain storm optimization algorithm. In: Tan, Y., Shi, Y., Chai, Y., Wang, G. (eds.) ICSI 2011. LNCS, vol. 6728, pp. 303–309. Springer, Heidelberg (2011). https://doi.org/10.1007/978-3-642-21515-5_36
9. Rao, R.: Jaya: a simple and new optimization algorithm for solving constrained and unconstrained optimization problems. Int. J. Ind. Eng. Comput. **7**(1), 19–34 (2016)
10. Wang, S.H., Phillips, P., Dong, Z.C., Zhang, Y.D.: Intelligent facial emotion recognition base on stationary wavelet entropy and Jaya algorithm. Neurocomputing **272**(10), 668–676 (2018)
11. Degertekin, S.O., Lamberti, L., Ugur, I.B.: Sizing, layout and topology design optimization of truss structures using the Jaya algorithm. Appl. Soft Comput. **70**, 903–928 (2018)
12. Oclon, P., et al.: Thermal performance optimization of the underground power cable system by using a modified Jaya algorithm. Int. J. Therm. Sci. **123**, 162–180 (2018)
13. Brisset, S., Brochet, P.: Analytical model for the optimal design of brushless DC wheel motor. COMPEL-Int. J. Comput. Math. Electr. Electron. Eng. **24**(3), 829–848 (2005)
14. Moussouni, F., Brisset, S., Brochet, P.: Comparison of two multi-agent algorithms: ACO and PSO for optimization of the brushless DC motor. Intell. Comput. Tech. Appl. Electromagnet. **119**(3), 3–10 (2008)
15. Duan, H., Li, S., Shi, Y.: Predator-Prey brain storm optimization for DC brushless motor. IEEE Trans. Magn. **49**(10), 5336–5340 (2013)

Genetic Action Sequence for Integration of Agent Actions

Man-Je Kim, Jun Suk Kim, Donghyeon Lee, and Chang Wook Ahn(✉)

School of Electrical Engineering and Computer Science,
Gwangju Institute of Science and Technology (GIST),
123 Cheomdangwagi-ro, Buk-gu, Gwangju 61005, Republic of Korea
{jaykim0104,junsuk89,cheetos,cwan}@gist.ac.kr

Abstract. Reinforcement learning has remarkable achievements in areas such as GO, Game, and autonomous vehicles, and is attracting attention as the most promising future technology among neural network-based technique. In spite of its capability, Reinforcement learning suffers from its greedy nature that causes selective behaviors. This is because the agent learns in such a way that each action seeks an optimal reward. However, the pursuit of maximum rewards by each action does not guarantee the maximization of the total reward. This is an obstacle to improving the overall performance of reinforcement learning. This paper introduces a concise illustration of Action Sequence, a method that encourages the trained model to seek grander strategies generating more desirable results. Especially aiming at highly complexed problems where plain actions can hardly handle any tasks. The model was tested under a video game environment to verify its improved proficiency in choosing suitable actions.

Keywords: Artificial intelligence · Evolutionary computing · Genetic algorithms · Video game

1 Introduction

Reinforcement learning differs from traditional machine learning, which has a data-based learning method in which the agent collects data directly from the environment without data. However, the instability of learning and a large amount of computation caused by agents gathering data in new environments has made it difficult to use reinforcement learning. It seems clear that the demanded huge computational cost can only be dealt with by a few rich users. Among the two problems, the problem of a large amount of computation gradually has solved by compact learning, and methods to minimize the exploration [2,3]. However, learning instability in a greedy way that maximizes individual rewards

This work was supported by GIST Research Institute(GRI) grant funded by the GIST in 2019.

L. Pan et al. (Eds.): BIC-TA 2019, CCIS 1159, pp. 682–688, 2020.
https://doi.org/10.1007/978-981-15-3425-6_54

of reinforcement learning and learning instability caused by unestablished data has not yet been resolved.

We propose a new method Action Sequence Method to solve this problem. Integrating genetic algorithms with Monte-Carlo Tree Search (MCTS), a root idea of our Action Sequence (AS) method, could help alleviate pressure from the RL's permanent problem [4,5]. Unlike the method of maximizing the reward for a single action of reinforcement learning, this method defines the reward in Action Sequence which is composed of several Action Groups. This enables linkage behavior that was difficult to occur in order to maximize single behavior, thereby improving the performance of reinforcement learning. In order to prove this result, we selected a fighting game in which reinforcement learning has been conducted for a long time while maximizing rewards through linked actions [1,2,6]. Fighting games have a variety of games, we chose FightingICE, a real-time, dual fighting video game, as an environment to test our method. It is a game played by many potent AI players in Fighting Game AI Competition held by the IEEE Conference on Games. The crucial feature of the in-game setting is that each game status frame lasts only for 16.67 ms, far from sufficient to consider all the available actions. Top FightingICE AIs have thus adapted for discovering optimal actions under the heavy time pressure as if they're exposed to the real-world environment. Under such circumstances, our AS-trained AI player competed against other prominent AI players with the official settings and rules, eventually showing its notable skill to properly choose optimal actions [6].

2 Theory

The greedy nature of RL is optimized for achieving rewards from short operations, but its lack of foresight to see potential, better substitutes makes RL an undesirable choice for real-world problems that contain huge consequential variances. Although, in FightingICE, optimizing the agent's action for rather simple, fragmentary short-term goals is prioritized, making sure that it possesses any form of an integrated and compound plan should also be a significant concern. For example, initiating a combination of actions in a certain, complex series could constantly pressure the opponent into changing its combat policy, bringing confusion to its intended mechanism and eventually ruining it. In a fundamental sense, our method aims to assist the AI in calculating long-term reward expectancy so that it can ultimately accumulate the higher rewards.

Figure 1 shows the overall procedure of the Action Sequence-based method. In FightingICE, there are 56 actions in total that a player character can perform. Among these, we filtered out 40 actions, such as jumping up in the mid-air, which can barely be followed by any subsequent, desirable actions. The rest 16 actions are expected to be capable of yielding higher rewards. Each gene in our structure is an action randomly chosen out of the 16 candidates, and 4 such genes, duplicated or not, form up an Action Sequence (AS) chromosome. An AS chromosome is the basic unit of our method, representing single action combo in a distinct order, which we aimed to optimize. While it is very important to

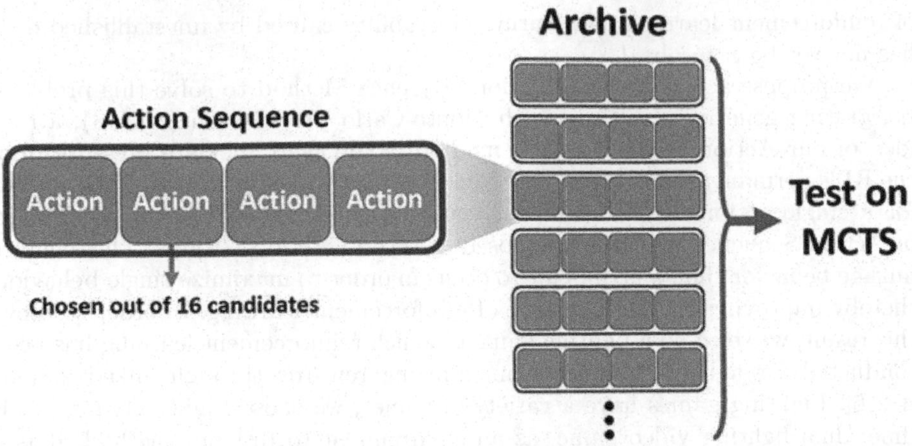

Fig. 1. General Action Sequence structure

find the best action from the first gene, diversifying the population is also a crucial part that adapts the model to as many different situations as possible. Therefore, we divided the implementation of Action Sequence into two parts.

(1) Construct an action sequence that does not duplicate the first gene of the action sequence.
(2) Finding the optimal solutions even if the genes in the action sequence are duplicated.

The first method is to find the best Action Sequence of genes with each filtered action, thereby identifying each action's long-term potential as well as the agent's variety of action choices; the second is to find good actions independently of possible overlap that would occur in any of the first genes. In pursuit of the population diversity, no AS chromosomes share the same first gene, i.e. the same initial action.

Through iterative generations, the AS chromosomes are processed with genetic operators, including selection, crossover, and mutation, and then stored in a genetic archive. The archive then selects chromosomes possessing the first genes with the highest fitness values. This architecture was derived from an intuition that choosing an initial action with positive reward would result in a preferable action combo. Next, unlike the previous step, the second method chooses fitness-wise optimal chromosomes. Finally, the archive is mounted on a MCTS which primarily controls the playing AI. Although the periodic time limit of 16.67 ms is a harsh constrict that would impede any attempt to find the optimal solution, MCTS-based AIs have constantly shown marked excellence in the past competitions [4]. In our model, MCTS chooses the agent's action in each time frame. Initially, it activates the first gene in each chromosome from the archive and, instead of making a new choice for the subsequent action, operates the wholes series of genes of the corresponding chromosome in order. Action Sequence is intended to achieve successful combinations of such chosen actions.

Table 1. Parameter setting

Parameter	Value
Population size	48
Selection probability (K)	0.9
Crossover method	Uniform
Chromosome size	4
Mutation probability	0.01

3 Experiment and Result

We decided to set the population size to 48, which is three times the number of actions expected to yield positive rewards, as stated in the previous section. The fitness value of each AS chromosome, defined as the difference between the health points (HP) of the two playing agents, was obtained via 10 rounds (against a random acting AI) with genetic operations under the parameter setting stated in Table 1. The fitness value of each AS chromosome, defined as the difference between the health points (HP) of the two playing agents, was obtained via 10 rounds (against a random-acting AI) with genetic operations under the parameter setting as follows: population size: 48 (three times the aforementioned 16 actions); selection K: 0.9; 3-point crossover; chromosome size: 4; mutation probability: 0.01. After 50 such generations, 16 chromosomes with the best first genes formed an archive, which ultimately chose 8 particular chromosomes with positive rewards for MCTS. MCTS' greedy nature causes shallow depth of action steps under the real-time pressure of 16.67 ms. Our main idea here is to provide it with 3 optimized, subsequent actions evaluated via genetic operations without spending extra time for manually figuring them out.

To precisely measure how much our method advances from the pure MCTS, [6] we played 100 rounds of our AS-based AI against the basic MCTS-based AI. We first mounted each of the 8 best chromosomes on MCTS and measured their performances as in score separately, which are presented in the of Fig. 2. The graph shows that 4 of them outperforms the MCTS-based AIs. Second, we combined those 4 chromosomes with MCTS at once and recorded its scores against the MCTS-based AI in 100 rounds. As seen in the graph of Fig. 3, our model overwhelms the MCTS AI with 99% winning ratio, proving that integrating top chromosomes reinforces the performance more effectively than mounting them individually. Our approach presents genetic algorithms as a qualified means to accompany RL.

We conducted the experiment based on the top 8 sequences from the second method's archive, combined with MCTS. Figure 4 shows their scores, each earned via 5 rounds, against Machete, the 2015 AI Competition winner. The overall result indicates the high performance of Archive 2. Furthermore, note that the second best action sequence, which would have been eliminated via the first

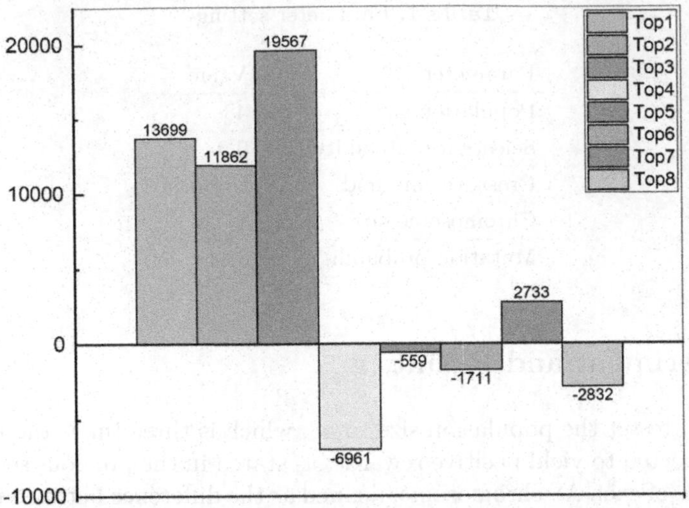

Fig. 2. Scores of AS-based AI with each individual chromosome

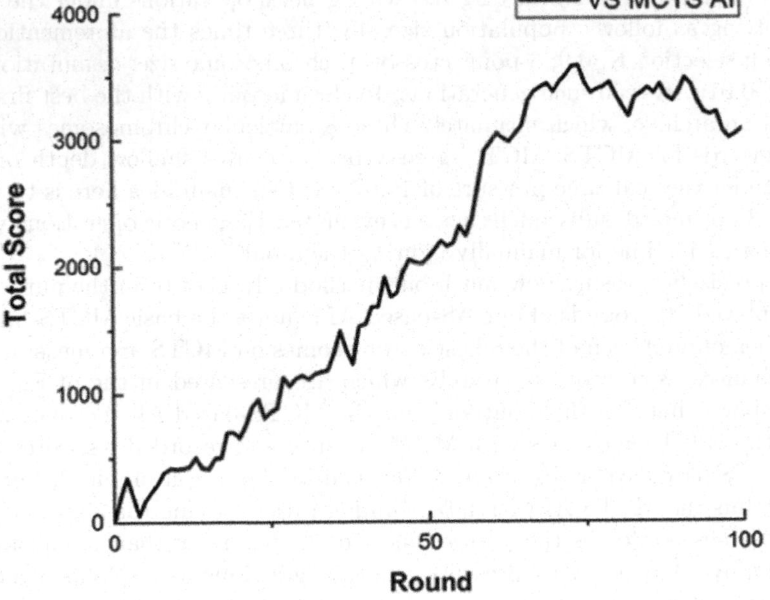

Fig. 3. Accumulated score of AS-based AI with integrated top chromosomes.

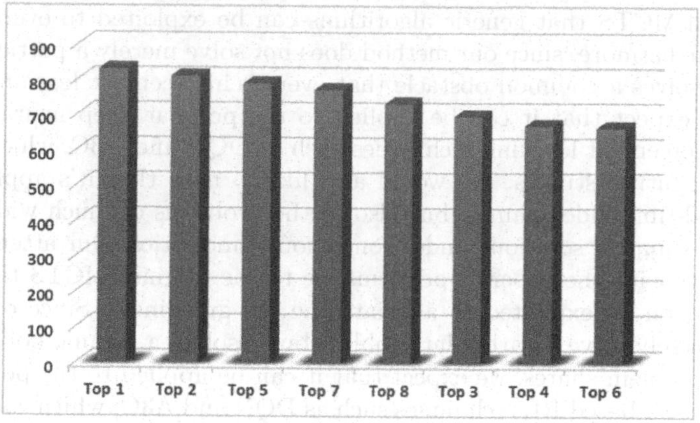

Fig. 4. Score of AS-based AI with Machete (2015's AI Competition Winner)

method, shares the same first gene with the best sequence. It is clearly shown that the two sequences only differ slightly in their performance. Refer to Fig. 5 for the fitness result from the action sequences that utilize the same sequence.

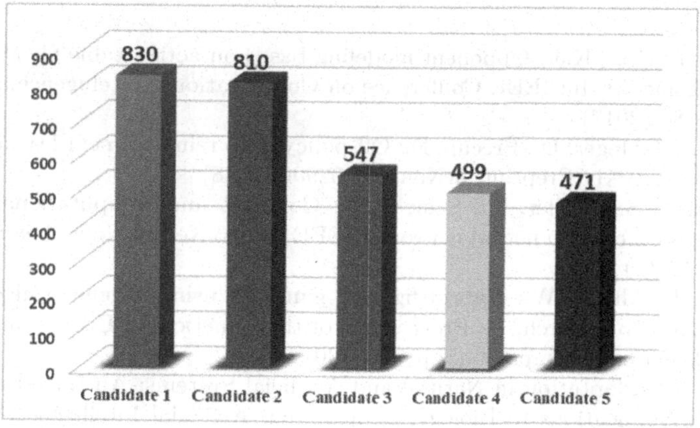

Fig. 5. Score of AS candidate AI with Machete (2015's AI Competition Winner)

4 Conclusion

We suggested and explained the Action Sequence method in order to compensate the reinforcement learning's inherent deficiency. Although reinforcement learning, including MCTS, has recently shown outstanding works in the AI industry, its greedy nature causes lots of difficulties in its actual implementations. Our attempt in this paper, however, shows with the superior performance to the original MCTS that genetic algorithms can be exploited to overcome such defects. Furthermore, since our method does not solve merely a particular problem but resolves a common obstacle that every reinforcement learning problem shares, we expect that it can be applied to the popular deep neural network-based reinforcement learning techniques such as DQN and A3C, which could be covered for future studies. We would also like to note that it's applicable not only to real-time video games, but also to the problems of much wider aspects containing complex solutions under continuous time series. Our attempt in this paper shows with the superior performance to the original MCTS that genetic algorithms can be exploited to alleviate the RL greediness. Since our method does not merely solve a particular problem but resolves a common obstacle that every RL problem shares, we expect that it can be applied to the popular deep neural network-based RL techniques such as DQN and A3C, which could be covered for future studies. We would also like to note that it's applicable not only to real-time video games but also to the problems of much wider aspects containing complex solutions under continuous time series. The most representative example is applicable to Real-time Strategy game (RTS game) including StarCraft II, Aeon of Strife game (AOS Game) including Dota II, and is expected to be applied to various fields besides game.

References

1. Kim, M.-J., Kim, K.J.: Opponent modeling based on action table for MCTS-based fighting game AI. In: IEEE Conference on Computational Intelligence and Games, pp. 178–180 (2017)
2. Fujimoto, S., Meger, D., Precup, D.: Off-policy deep reinforcement learning without exploration. ArXiv Preprint, arXiv:1812.02900 (2018)
3. Wiedemann, S., Müller, K., Samek, W.: Compact and computationally efficient representation of deep neural networks. IEEE Trans. Neural Netw. Learn. Syst. **31**, 772–785 (2019)
4. Kim, M.-J., Ahn, C.W.: Hybrid fighting game AI using a genetic algorithm and Monte Carlo tree search. In: Proceedings of the Genetic and Evolutionary Computation Conference Companion, pp. 129–130 (2018)
5. Holland, J.: Adaptation in Natural and Artificial Systems: An Introductory Analysis with Applications to Biology, Control, and Artificial Intelligence. MIT press, Cambridge (1992)
6. Yoshida, S., Ishihara, M., Miyazaki, T., et al.: Application of Monte-Carlo tree search in a fighting game AI. In: IEEE 5th Global Conference on Consumer Electronics, pp. 1–2 (2016)

Based on Fuzzy Non-dominant and Sparse Individuals to Improve Many-Objective Differential Evolutionary

Yulong Xu[1,2](✉), Xu Pan[1], Xiaomin Jiao[1], Yali Lv[1], and Ting Song[1]

[1] School of Information Technology, Henan University of Chinese Medicine, Zhengzhou 450046, Henan, China
flyxyl@126.com

[2] School of Information Engineering, Zhengzhou University, Zhengzhou 450002, Henan, China

Abstract. In the classical multi-objective algorithm, after sorting all the solutions of population, the non-dominant sorting method selects the better half of the solutions to enter the next generation. In solving the many objective optimization problem, due to the small environmental selection pressure, there will be a lot of redundancy when the selection solution enters the next generation. In order to solve this problem, a fuzzy non-dominant sorting method is proposed to sort the individual population. This method can determine the dominant relationship between the two individuals by comparing to three objectives just, which makes the calculation cost of many-objective approximate to three objectives. Meanwhile, the proposed method increases the selection pressure and improves the computational efficiency. In addition, due to the number of objectives is large, the traditional methods cannot accurately calculate the congestion distance. We design a method based on inflection point and hyperplane to calculate the congestion distance; thus, the selected individuals are evenly distributed by means of sparse individuals. A many-objective evolution algorithm is obtained by using the differential evolution search framework and combining the proposed methods of fuzzy non-dominant sorting and sparse individuals. Finally, the proposed algorithm is used to solve the standard test function DTLZ1-DTLZ6. The simulation results show that the performance of our algorithm on the IGD is similar with the famous comparison algorithms, and that is better than the famous comparison algorithm in running time.

Keywords: Many-objective · Non-dominant ranking · Crowded distance · Differential evolutionary

1 Introduction

Multi-objective optimization problems (MOPs) are common in practical application, especially in biology, engineering and economics [1, 5]. After mathematical modeling, these problems have more than two targets to be optimized. If you have more than three optimization goals, they are called many-objective optimization problems (MaOPs). Due to the non-tunable between the optimization objectives. The contradiction of the

L. Pan et al. (Eds.): BIC-TA 2019, CCIS 1159, pp. 689–702, 2020.
https://doi.org/10.1007/978-981-15-3425-6_55

solution, so it is impossible to use a solution as the global optimal solution, but requires a set of compromise solutions, called Pareto optimal solution set (Pareto-optimal set) [6].

Because it is difficult to solve the high objective problem, the search is very difficult. Because of the complex behavior, high-dimensional multi-objective optimization is a challenging problem in the field of evolutionary computing at home and abroad. Because with the increase of target dimension, it will face the problems of small evolutionary power, ultra-high space-time complexity, difficult to predict the search behavior and so on.

In order to solve these problems, many scholars at home and abroad have put forward relevant solutions, including: KnEA [7] replaces Pareto's non-dominant relationship by defining a new dominant relationship. This solves the loss of selection pressure caused by Pareto's discretionary sequencing; Yang et al. Rotami and Neri [9] improve the selection strategy, propose an adaptive hypervolume mesh algorithm, divide the search space into one grid, and then solve it according to these grids to calculate the super volume, reduce the difficulty of calculating the super volume, and improve the performance of the algorithm. NSGAIII [10] evaluates an individual's fitness by defining a reference point and evaluating the relative relationship between each individual and the reference point, thereby selecting a better one. Although MOEA/ D [11] is a multi-objective optimization algorithm, MOEA/D algorithm is usually an effective algorithm to solve the problem of high-dimensional objective optimization because it is less affected by the number of targets. When there is a great correlation between multiple optimization objectives in the high-dimensional goal, the dimension of the target can be reduced, and only a few of the optimization objectives can be solved. For example, Jaimes et al. [12] divide the target space by calculating the conflict information between the optimization objectives in the current Pareto front end, and transform the high-dimensional multi-objective problem into several multi-objective evolution problems through different target spatial information. Because each multi-objective problem preserves the information of the original problem as much as possible, it can solve the high objective problem well. Tan et al. [13] proposed a uniform design to generate weight direction for each subproblem. The quantitative method makes the algorithm not need to increase the population size according to the problem size and does not consider the weight problem.

In our previous work, we studied the optimization algorithms of two objectives and three objectives [14, 15]. However, when the congestion distance algorithm used in the traditional multi-objective optimization algorithm is used to solve the high-dimensional multi-objective optimization problem, the complexity of the algorithm becomes very large because of the increase of the target dimension, and it is difficult to select the evenly distributed individuals from the search space based on the previous results. Therefore, we use the previous basis to study the high-dimensional multi-objective optimization algorithm. In this paper, by using the generated inflection point and referring to the distance between the individual and the hyperplane, some individuals uniformly distributed in the search space are selected to enter the next generation, so as to improve the diversity of the algorithm.

2 Related Theory

2.1 Differential Evolution Algorithm

Differential evolution algorithm (Differential Evolution, *DE*) [16] is a heuristic algorithm, which will search for targets with good performance in the target space for a limited time. This kind of algorithm has few parameters and good convergence speed, so there are many methods combined with *DE* framework to solve the multi-objective optimization problem. For the framework of the algorithm, see algorithm 1.

```
The algorithm 1
1 Create initial population(N).
2 While (does not meet termination conditions)
3 Variation operation: after the population Pi was mutated, it
became the population Vi.
4 Cross operation: after crossing, the population Vi becomes the
population Ui.
5 Selection operation: the better N individuals were selected from
the combined population (2 N individuals) of Pi ∪Ui to enter the next
generation.
6 End while
7 Evolutionary completion
```

Common *DE* mutation methods are:

1. rand driven mutation strategy (*DE/ rand/ 1*)

$$v_{i,G} = x_{r1,G} + F_i \cdot \left(x_{r2,G} - x_{r3,G}\right) \tag{1}$$

2. best driven mutation strategy (*DE/best/1*)

$$v_{i,G} = x_{best,G} + F_i \cdot \left(x_{r1,G} - x_{r2,G}\right) \tag{2}$$

3. *current-to-best* driven variation strategy (*DE/current-to-best/1*)

$$v_{i,G} = x_{i,G} + F_i \cdot \left(x_{best,G} - x_{i,G}\right) + F_i \cdot \left(x_{r1G} - x_{r2G}\right) \tag{3}$$

The variation strategy of *DE/rand/1* is selected in this paper. Since binomial crossing is simple, with few parameters and better performance, this paper selects binomial crossing as the crossing strategy. See Formula (4) for details.

$$u^j_{i,G} = \begin{cases} v^j_{i,g} & if\ rand_{i,j}[0,1] \le Cr\ or\ j = j_{rand} \\ x^j_{i,g} & otherwise \end{cases} \tag{4}$$

In this paper, the greedy selection strategy is used, which evaluates the children and parents. If the children are better than the parents, the preferred children are selected, otherwise, the parents are selected. See Eq. (5) for details.

$$\mathbf{X}_{i,g+1} = \begin{cases} u^j_{i,g} & if\ f\left(u_{i,g}\right) \le f\left(x_{i,g}\right) \\ x^j_{i,g} & otherwise \end{cases} \tag{5}$$

2.2 High Dimensional Multi-objective Optimization Problems

Because optimization objectives in multi-objective optimization are often in conflict, the optimal solution of all problems is usually a solution set. In single objective optimization, the optimal solution is usually unique. Therefore, multi-objective optimization and single objective optimization have essential differences. For example, see Eq. (6).

$$\begin{cases} Min\ y = F(x) = (f_1(x), f_2(x), \ldots, f_m(x)) \\ x = (x_1, \ldots, x_n) \in X \end{cases} \tag{6}$$

For decision variables are *n*, minimize the number *m* of target variables, where X is the decision space, y is the target vector, $F(x)$ is m fitness function f_i, $i \in 1, 2, \ldots, m$.

For the non-dominant relation of X_A and X_B, if $X_A \succ X_B$, Says that X_A dominate X_B, vice versa. See the formula 7.

$$x_A \succ x_B \equiv \{\forall i = 1, 2, \ldots, m,\ f_i(x_A) \le f_i(x_B) \wedge \exists j = 1, 2, \ldots, m,\ f_j(x_A) < f_j(x_B)\} \tag{7}$$

For any x^* there is no set of solutions that governs Pareto's optimal solution set P^*, be called Pareto leading surface, be written as PF^*, See formula (8) for the definition.

$$\begin{cases} P^* \equiv \{x^* | \neg\exists x \in X_f : x \succ x^*\} \\ PF^* \equiv \left\{F(x^*) = (f_1(x^*), f_2(x^*) \ldots, f_m(x^*))^T x^* \in P^*\right\} \end{cases} \tag{8}$$

This section briefly introduces the minimization of multi-objective optimization problem, and the detailed description is described in literature [16, 17].

3 An Improved High-Dimensional Target Differential Evolutionary Algorithm

3.1 Improved Non-dominant Ranking

The purpose of the non-dominant ranking of the population is to distinguish all the individuals in the population from the front surface, and then select the better front surface (starting from the first front surface) to enter the next generation. In order to select N individuals from 2 N individuals (population size N) to enter the next generation, the traditional non-dominant sorting method is to select the next generation operation after undominated sorting of the whole population. After in-depth analysis of non-dominant sorting, redundant operations are found. Therefore, we add a truncation operation to reduce computational redundancy by about 50%. The proposed algorithm not only selects the next generation of individuals, but also selects the next generation of individuals to reduce the computational complexity. Miscellaneous degrees. In addition, because the traditional multi-objective non-dominated sorting will face the phenomenon of selection pressure loss when solving the high-dimensional multi-objective problem, a non-dominant sorting method based on fuzzy domination is proposed to sort the individual population and distribute the frontier operation. The proposed method can determine the dominant relationship between two bodies by comparing up to three objectives, so

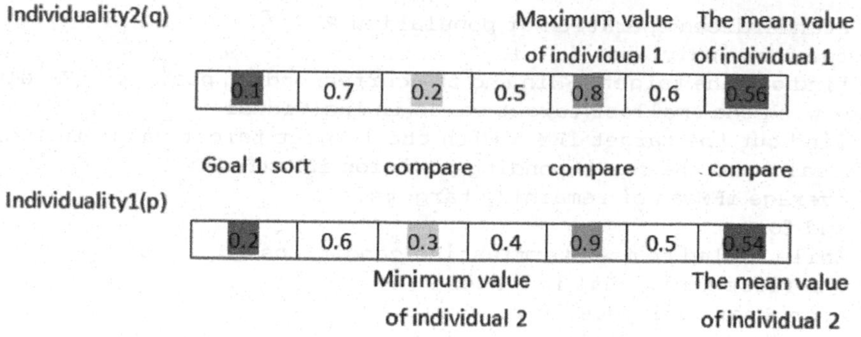

Fig. 1. Examples of fuzzy domination

the computational cost of non-dominant ranking for high-dimensional target problems is not more than three objectives, which greatly increases the complexity of time and space.

As shown in Fig. 1, two six target individuals 1 (named p) = (0.2, 0.6, 0.3, 0.4, 0.9, 0.5) and 2 (named Q) = (0.1, 0.7, 0.2, 0.5, 0.8, 0.6), the algorithm first sorts p and Q in ascending order according to the first function value. Then we calculate the maximum function value qFMax = 0.8, of p except the first function in the minimum function value qFMax = 0.8, except the first optimization objective, and then calculate the maximum function value qFMax = 0.8, of p in addition to the first optimization objective, and then calculate the maximum function value qFMax = 0.8, of p in addition to the first optimization objective, and then calculate the maximum function value qFMax = 0.8, of p in addition to the first optimization objective.

The mean values pFMean = 0.54 and qFMean = 0.56; of the remaining optimization objectives outside the standard are pFMin = 0.3, pFMinIndex = 3, qFMax = 0.8, qFMaxIndex = 5, pFMean = 0.54 and qFMean = 0.56, respectively. The first function value of p is greater than the first function value of Q, so it is only necessary to prove that the other value in Q is greater than the corresponding value in p to prove their nondominant relationship. Next, we compare pFMin with the value of pFMinIndex function in Q. If the relationship is less than, then we can prove that p, Q is in dominant. By comparison, pFMin = 0.3 > qFMin = 0.2, can not prove that they are nondominant, and then compare the value of qFMaxIndex function in qFMax = 0.8 and p, qFMax = 0.8 < 0.9, so we cannot prove their nondominant relationship, and finally we can compare pFMean = 0.54 and qFMean = 0.56, to know that pFMean = 0.5 is non-dominant, so we can't prove that they are nondominant, so we can't prove that they are nondominant. Finally, we can compare pFMean = 0.54 and qFMean = 0.56, to know that pFMean = 0.5 is non-dominant. 4 and qFMean = 0.56, obviously satisfy the non-dominant relationship, so p, Q is the non-dominant relationship. See algorithm 2 for details.

```
Algorithm 2:
Input: population P, the number of optimization objectives m.
Output: the leading surface F to which each individual belongs.
1 sort by the first function value of the population.
```

```
2 standardized operation of population P.
3 for each individual i in P.
4 find out the target iFMin and the corresponding position iFMinIndex with the smallest target value in individual i.
5 find out the target iFMax with the largest target value in individual i and the corresponding position iFMaxIndex.
6 average iFMean of remaining targets.
7 end for.
8 while failed to meet termination conditions.
9 The first individual in P adds F[k].
10 for P per individual i.
11 for F[k] per individual j.
12 if the function value of iFMin less than the corresponding position iFMinIndex of j individual.
13 the non-dominant relationship between two individuals.
14 else if jFMax is greater than the function value of the corresponding position jFM MaxIndex.
15 the non-dominant relationship between two individuals.
16 else if iFMean < jFMean.
17 the non-dominant relationship between two individuals.
18 else
19 two individuals are not non-dominant
20 end if
21 if two individuals are non-dominant
22 add i to F[k]
23 end if
24 end for
25 end while
26 return F
```

3.2 Calculation of Congestion Degree Based on Inflection Point and Hyperplane

After the non-dominant ranking, when all the last frontier individuals enter the next generation, if the population will overflow, it is necessary to rank the last frontier at crowded distance, and then select some individuals with uniform distribution to enter the next generation, so as to ensure the total number of individuals selected to enter the next generation on the basis of population diversity.

Deb [18] proposed congestion distance, as a diversity preservation strategy, is often used to solve multi-objective problems. When there are fewer optimization objectives (2 ≤ 3), this method can effectively calculate the congestion distance, but with the increase of the target dimension, it will consume a lot of resources to calculate the congestion distance of each individual, and it is difficult to select the evenly distributed individuals from the region. Based on the above problems, this paper proposes a hyperplane based on inflection point and hyperplane. First, according to the extreme value of each optimization objective, the hyperplane belonging to the highest front surface is calculated, and then the distance between hyperplane and hyperplane is calculated for each individual in the frontier surface. Secondly, the number of individuals that need to be selected from the maximum frontier Fi to enter the next generation remainSize, is calculated, and then the neighborhood is set according to the target search space *S*. If: find 5 out of 10 individuals, the algorithm first calculates the scope of the whole search space *S* [$1 \leq 9$]. Because the extreme point of the search space is sparse individuals, only 3 individuals need to be selected, so set the neighborhood range $R \leq 3$. If there is no other individual in a neighborhood, the individual is called a sparse individual, as shown in Fig. 2. If the number of sparse individuals selected is greater than or equal to remainSize, from all sparse Among the sparse individuals, those who are farther from the hyperplane are selected to enter the next generation, and if less than remainSize, from the non-sparse individuals, the individuals who are farther from the hyperplane are selected to enter the next generation.

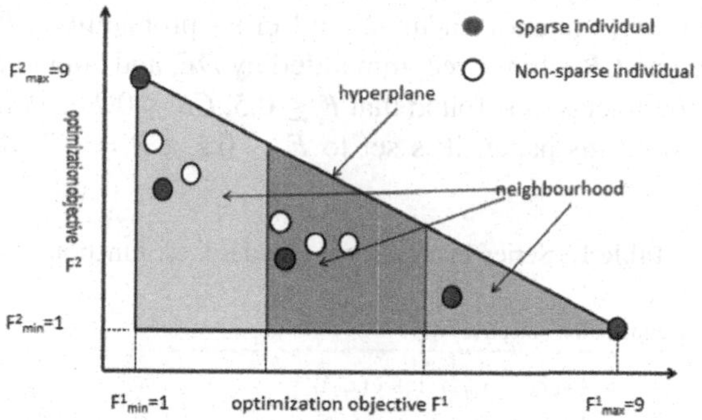

Fig. 2. Examples of uniform selection of sparse individuals

By using the generated sparse individuals, the algorithm can select more uniform individuals to enter the next generation, thus reducing the time complexity and improving the diversity. See algorithm 3 for details.

```
Algorithm 3:
Input: the number of remainSize selected into the next generation
in the front of the Fi, front of the individual
Output: the k point in the front surface individual and the distance
D of the distance hyperplane
1 Calculate the number of individuals in Fi pNum, the extreme value
E of all individuals in Fi and calculate the hyperplane L
2 Calculate the ratio of remainSize-m to pNum r
3 Find out the maximum Fmax, minimum value Fmin on each optimization
goal in Fi
4 Calculate the distance D between each individual and the hyper-
plane in Fi
5 Using the formula (R = (Fmax-Fmin).*r) to calculate R
6 According to the distance D between each individual and the hyper-
plane, the set rank is obtained
7 For each individual k in rank
8 For each individual p in Fi
9 If the distance between k and p functions is less than R, P is not
K point
```

```
10 End For
11 End For
12 Return K point and distance D
```

4 Simulation Experiment

4.1 Experimental Environment and Parameter Setting

All the parameters in the comparison algorithm listed in this section are according to the suggested parameters given in the original text to ensure the fairness of the contrast experiment. The FMODE_MAO proposed in this paper is based on DE framework, so only two parameters, mutation factor *F* and cross probability *CR*, need to be set. According to the *F* and *CR* settings recommended by *DE*, and after comparing different combinations in experiments, it is found that $F \leq 0.5$, $CR \leq 0.2$ can make the algorithm perform the best, so in this paper, it is set to $F \leq 0.5$, $CR = 0.2$. In order to ensure

Table 1. Series international standard test function.

Question	Span	Optimization objective	Dimension
DTLZ1	x ∈ [0, 1]	$f_1(x) = \frac{1}{2}x_1x_2 \ldots x_{m-1}(1 + g(x_m))$ $f_2(x) = \frac{1}{2}x_1x_2 \ldots (1 - x_{m-1})(1 + g(x_m))$ $\vdots$ $f_{m-1}(x) = \frac{1}{2}x_1(1 - x_2)(1 + g(x_m))$ $f_m(x) = \frac{1}{2}(1 - x_1)(1 + g(x_m))$ $g(x_m) = 100\left[|x_m| + \sum_{x_i \in x_m}((x_i - 0.5)^2 - \cos(20\pi(x_i - 0.5)))\right]$	m + 4
DTLZ2	x ∈ [0, 1]	$f_1(x) = (1 + g(x_m))\cos(\frac{x_1\pi}{2})\cos(\frac{x_2\pi}{2})\ldots\cos(\frac{x_{m-2}\pi}{2})\cos(\frac{x_{m-1}\pi}{2})$ $f_2(x) = (1 + g(x_m))\cos(\frac{x_1\pi}{2})\cos(\frac{x_2\pi}{2})\ldots\cos(\frac{x_{m-2}\pi}{2})\sin(\frac{x_{m-1}\pi}{2})$ $f_3(x) = (1 + g(x_m))\cos(\frac{x_1\pi}{2})\cos(\frac{x_2\pi}{2})\ldots\sin(\frac{x_{m-2}\pi}{2})$ $\vdots$ $f_{m-1}(x) = (1 + g(x_m))\cos(\frac{x_1\pi}{2})\sin(\frac{x_2\pi}{2})$ $f_m(x) = (1 + g(x_m))\sin(\frac{x_1\pi}{2})$ $g(x_m) = \sum_{x_i \in x_x}(x_i - 0.5)^2$	m + 9
DTLZ3	x ∈ [0, 1]	$f_1(x) = (1 + g(x_m))\cos(\frac{x_1\pi}{2})\cos(\frac{x_2\pi}{2})\ldots\cos(\frac{x_{m-2}\pi}{2})\cos(\frac{x_{m-1}\pi}{2})$ $f_2(x) = (1 + g(x_m))\cos(\frac{x_1\pi}{2})\cos(\frac{x_2\pi}{2})\cdots\cos(\frac{x_{m-2}\pi}{2})\sin(\frac{x_{m-1}\pi}{2})$ $f_3(x) = (1 + g(x_m))\cos(\frac{x_1\pi}{2})\cos(\frac{x_2\pi}{2})\ldots\sin(\frac{x_{m-2}\pi}{2})$ $\vdots$ $f_{m-1}(x) = (1 + g(x_m))\cos(\frac{x_1\pi}{2})\sin(\frac{x_2\pi}{2})f_m(x) = (1 + g(x_m))\sin(\frac{x_1\pi}{2})$ $g(x_m) = 100\left[|x_m| + \sum_{x_i \in x_m}((x_i - 0.5)^2 - \cos(20\pi(x_i - 0.5)))\right]$	m + 9

(*continued*)

Table 1. (*continued*)

Question	Span	Optimization objective	Dimension
DTLZ4	x ∈ [0, 1]	$f_1(x) = (1+g(x_m))\cos\left(\frac{x_1^a\pi}{2}\right)\cos\left(\frac{x_2^a\pi}{2}\right)\ldots\cos\left(\frac{x_{m-2}^a\pi}{2}\right)\cos\left(\frac{x_{m-1}^a\pi}{2}\right)\Big]$ $f_2(x) = (1+g(x_m))\cos\left(\frac{x_1^a\pi}{2}\right)\cos\left(\frac{x_2^a\pi}{2}\right)\ldots\cos\left(\frac{x_{m-2}^a\pi}{2}\right)\sin\left(\frac{x_{m-1}^a\pi}{2}\right)$ $f_3(x) = (1+g(x_m))\cos\left(\frac{x_1^a\pi}{2}\right)\cos\left(\frac{x_2^a\pi}{2}\right)\ldots\sin\left(\frac{x_{m-2}^a\pi}{2}\right)$ $\vdots$ $f_{m-1}(x) = (1+g(x_m))\cos\left(\frac{x_1^c\pi}{2}\right)\sin\left(\frac{x_2^x\pi}{2}\right)$ $f_m(x) = (1+g(x_m))\sin\left(\frac{x_1^a\pi}{2}\right)$ $g(x_m) = \sum_{x_i\in x_m}(x_i-0.5)^2 \quad ,\alpha = 100$	m + 9
DTLZ5	x ∈ [0, 1]	$f_1(x) = (1+g(x_m))\cos(\frac{x_1\pi}{2})\cos(\frac{x_2\pi}{2})\ldots\cos(\frac{x_{m-2}\pi}{2})\cos(\frac{x_{m-1}\pi}{2})$ $f_2(x) = (1+g(x_m))\cos\left(\frac{\theta_1\pi}{2}\right)\cos\left(\frac{\theta_2\pi}{2}\right)\ldots\cos\left(\frac{\theta_{m-2}\pi}{2}\right)\sin\left(\frac{\theta_{m-1}\pi}{2}\right)$ $f_3(x) = (1+g(x_m))\cos\left(\frac{\theta_1\pi}{2}\right)\cos\left(\frac{\theta_2\pi}{2}\right)\ldots\sin\left(\frac{\theta_{m-2}\pi}{2}\right)$ $\vdots$ $f_{m-1}(x) = (1+g(x_m))\cos\left(\frac{\theta_1\pi}{2}\right)\sin\left(\frac{\theta_2\pi}{2}\right)$ $f_m(x) = (1+g(x_m))\sin\left(\frac{\theta_1\pi}{2}\right)$ $\theta_i = \frac{\pi}{4(1+g(x_m))}(1+2g(x_m)x_i), \forall i = 2, 3, \ldots, (m+1)$ $g(x_m) = \sum_{x_i\in x_m}(x_i-0.5)^2$	m + 9
DTLZ6	x ∈ [0, 1]	$f_1(x) = x_1$ $f_2(x) = x_2$ $\vdots$ $f_{m-1}(x) = x_{m-1}$ $f_m(x) = (1+g(x_m))h(f_1, f_2, \ldots, f_{m-1}, g)$ $g(x_m) = 1 + \frac{9}{\lvert x_m \rvert}\sum_{x_i\in x_m} x_i$ $h(f_1, f_2, \ldots, f_{m-1}, g) = m - \sum_{i=1}^{m}\left[\frac{f_i}{1+g}(1+\sin(3\pi f_i))\right]$	m + 9

fairness, the population size is set to 150. 5%. Evolutionary algebra is based on the choices commonly used in standard test sets, DTLZ1 is 700 generations, DTLZ3 is 1000 generations, and the rest of the problems are set to 250 generations.

4.2 Comparison of Experimental Results

The overall performance of FMODE_MAO is compared with that of some excellent algorithms in this field. All the algorithms run independently 10 times, and then the IGD and time-consuming mean values are obtained. the comparison results are shown in Table 1 $\leq$ 2, and the rough data is the optimal data under the same condition.

Table 1 shows the IGD values of the algorithm in solving each test function on 4, 6, 8 and 10 targets, and the optimal IGD values are shown in bold. At the same time, the IGD values of each algorithm for solving DTLZ series functions with different number of targets are given, as shown in Fig. 3.

Table 2. Performance comparison of algorithm in high dimensional DTLZ series test functions.

Function	Mean error						
	m	GrEA [8]	MOEAD [11]	NSGA-III [10]	KnEA [7]	FMODE_MAO	≈/ ±
DTLZ1	4	4.88E-02	9.29E-02	**4.02E-02**	5.11E-02	4.36E-02	2/2/0
	6	1.05E-01	2.04E-01	8.02E-02	1.62E-01	**7.80E-02**	1/3/0
	8	**1.23E-01**	1.98E-01	1.38E-01	2.65E-01	1.28E-01	3/1/0
	10	1.78E-01	2.25E-01	**1.34E-01**	2.44E-01	1.66E-01	3/1/0
DTLZ2	4	1.25E-01	2.37E-01	**1.16E-01**	1.25E-01	1.35E-01	3/1/0
	6	2.56E-01	4.78E-01	2.59E-01	2.55E-01	**2.47E-01**	3/1/0
	8	3.50E-01	7.65E-01	3.88E-01	3.48E-01	**3.29E-01**	3/1/0
	10	3.44E-01	8.90E-01	4.23E-01	**3.28E-01**	4.62E-01	2/1/1
DTLZ3	4	1.41E-01	2.39E-01	**1.17E-01**	1.92E-01	1.42E-01	3/1/0
	6	3.31E-01	7.46E-01	2.82E-01	5.56E-01	**2.77E-01**	1/3/0
	8	**4.28E-01**	9.58E-01	4.94E-01	8.87E-01	4.36E-01	2/2/0
	10	4.94E-01	1.04E + 00	5.01E-01	8.79E-01	**4.85E-01**	1/3/0
DTLZ4	4	1.45E-01	5.12E-01	**1.33E-01**	1.26E-01	1.64E-01	3/1/0
	6	**2.55E-01**	6.40E-01	2.71E-01	2.54E-01	3.05E-01	0/1/3
	8	**3.47E-01**	7.44E-01	3.93E-01	3.39E-01	4.12E-01	0/1/3
	10	3.47E-01	8.31E-01	4.10E-01	**3.26E-01**	4.77E-01	1/1/2
DTLZ5	4	**1.60E-02**	2.85E-02	4.53E-02	8.39E-02	2.42E-01	1/2/1
	6	**1.31E-01**	7.67E-02	3.16E-01	2.01E-01	1.93E-01	1/3/0
	8	2.24E-01	**6.94E-02**	3.17E-01	2.51E-01	2.92E-01	2/1/1
	10	3.10E-01	**8.11E-02**	4.20E-01	2.51E-01	2.39E-01	1/3/0
DTLZ6	4	2.98E-02	4.69E-02	**1.86E-01**	2.20E-01	1.27E+00	0/0/4
	6	2.30E-01	**1.50E-01**	1.47E+00	3.91E-01	2.50E+00	0/0/4
	8	4.16E-01	**1.33E-01**	2.87E+00	3.81E-01	3.31E+00	1/0/3
	10	4.74E-01	**2.42E-01**	3.77E+00	3.75E-01	3.79E+00	1/0/3

Table 3. Algorithm execution of all DTL test functions run time comparison table.

m	GrEA [8]	MOEAD [11]	NSGA-III [16]	KnEA [7]	FMODE_MAO
4	227	283	**25**	72	31
6	304	330	55	68	**35**
8	338	332	35	68	**32**
10	335	294	36	69	**29**

In the comparison of computational time complexity of the algorithm, all DTLZ functions are executed at one time in different target dimensions, and the sum of time consuming is compared, as shown in Tables 2 and 3.

As can be seen from Tables 1, 2 and Figs. 3 and 4, The NSGA-III and GrEA algorithms perform well in the optimal number of IGD values. There are six IGD values that are optimal. NSGA-III performs well in solving DTLZ series problems when the number

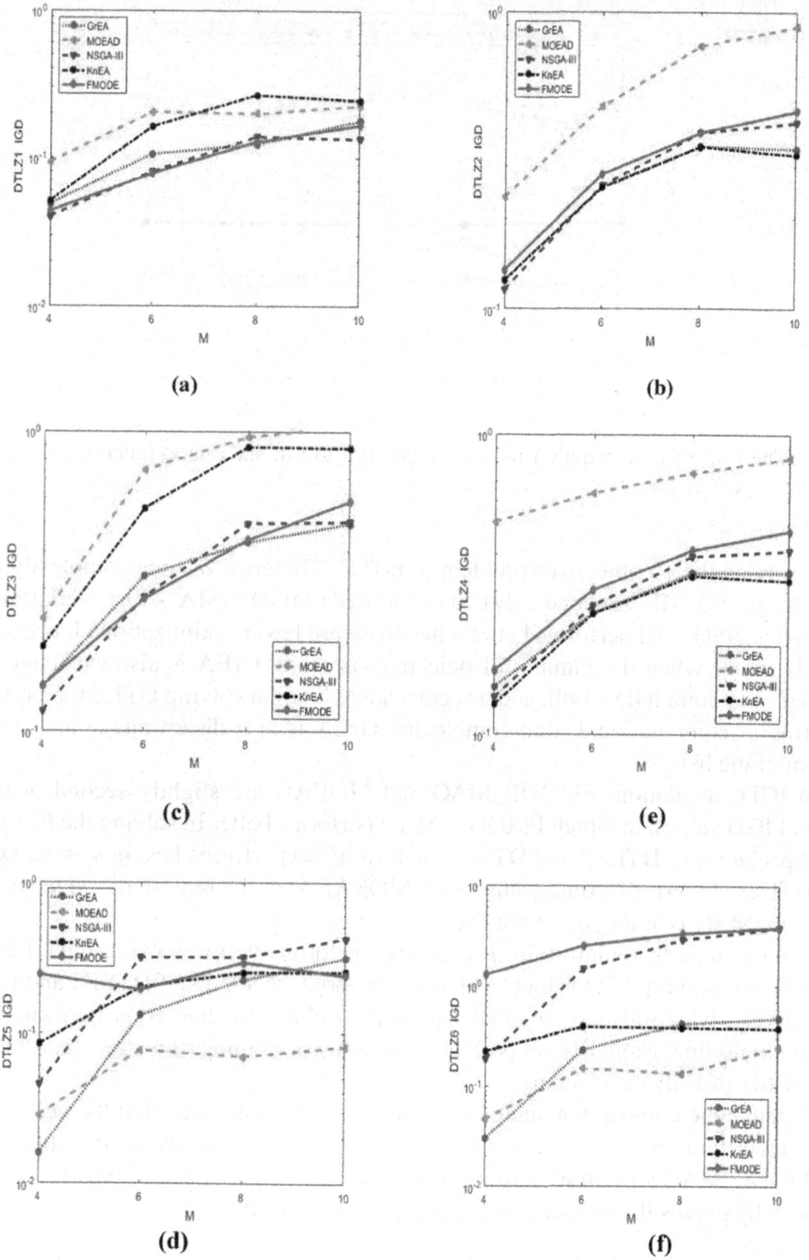

Fig. 3. a–f the comparison of IGD mean values is obtained by each algorithm under different target numbers (M = 4, 6, 8, 10)

of targets is 4, and when solving four objectives of DTLZ1–4 and DTLZ6, IGD value is optimal, However, when the number of targets is 6, 8 and 10, the IGD value is optimal only

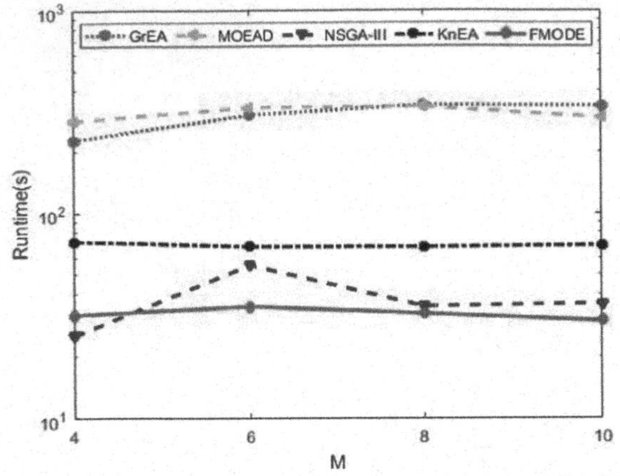

Fig. 4. The sum of time required for each algorithm to run six DTLZ functions at a time at different target numbers

when solving the 10 objectives problem of DTLZ1. In terms of time complexity of the algorithm, NSGA-III is second only to the proposed FMODE_MAO. The overall analysis shows that NSGA-III performs better when there are fewer optimization objectives, but slightly worse when the number of objectives is more. GrEA is also excellent in the number of optimal IGD worth, and has great advantages in solving DTLZ4–5 problems. In terms of computational time complexity, GrEA is at a disadvantage and is at the bottom of the list.

In IGD calculation, FMODE_MAO and MOEAD are slightly second, with five optimal IGD values, in which FMODE_MAO performs better in solving the 6, 8 and 10 target problems of DTLZ2 and DTLZ3, and MOEAD performs best in solving DTLZ5 and DTLZ6. In terms of time complexity, MOEAD is at the bottom of the table, while FMODE_MAO is at the top of the list.

In the comparison algorithm, KnEA also performs well in solving IGD, but there are only two optimal IGD values, which are 10 target problems of DTLZ2 and DTLZ4 respectively, which also shows that KnEA algorithm also has good performance in solving high-dimensional target problems. In terms of running time, the algorithm time complexity of KnEA ranks third.

Through the comparative analysis of the data, it is concluded that the performance of the algorithm on IGD value is close to that of contrast. Because of the optimization of FMODE_MAO algorithm in time complexity, the performance of IGD is also good, which fully proves the advantages of the proposed algorithm.

5 Conclusion

The improved non-dominant sorting algorithm proposed in this paper will first standardize the population when selecting the next generation of individuals to solve the high-dimensional target problem, so that the dominant relationship between individuals can be judged according to the maximum three objectives in the high-dimensional target.

In addition, the algorithm will use the uniform selection algorithm based on inflection point to calculate the congestion degree. Finally, combined with the framework of differential evolution, a differential evolution algorithm based on improved non-dominant sorting is proposed to solve the DTLZ series functions of high-dimensional targets. The proposed algorithm is less than other algorithms, but the IGD value is similar to the contrast algorithm, which fully proves the proposed algorithm. FMODE_MAO algorithm has high efficiency.

References

1. Herrero, J.G., Berlang, A., et al.: Effective evolutionary algorithms for many-specifications attainment: application to air traffic control tracking filters. IEEE Trans. Evol. Comput. **13**(1), 151–168 (2009)
2. Ishibuchi, H., Murata, T.: A multi-objective genetic local search algorithm and its application to flowshop scheduling. IEEE Trans. Syst. Man Cybern. Part C Appl. Rev. **28**(3), 392–403 (1998)
3. Yeung, S.H., Man, K.F., et al.: A trapeizform U-slot folded patch feed antenna design optimized with jumping genes evolutionary algorithm. IEEE Trans. Antennas Propag. **56**(2), 571–577 (2008)
4. Handl, J., Kell, D.B., et al.: Multi-objective optimization in bioinformatics and computational biology. IEEE/ACM Trans. Comput. Biol. Bioinf. **4**(2), 279–292 (2007)
5. Ponsich, A., Jaimes, A.L., et al.: A survey on multi-objective evolutionary algorithms for the solution of the portfolio optimization problem and other finance and economics applications. IEEE Trans. Evol. Comput. **17**(3), 321–344 (2013)
6. Coello, C.C., Gary, B.L., et al.: Evolutionary Algorithms for Solving Multi-Objective Problems, pp. 1–140. Springer, Heidelberg (2007). https://doi.org/10.1007/978-0-387-36797-2
7. Zhang, X., Tian, Y., et al.: Approximate non-dominated sorting for evolutionary many-objective optimization. Inf. Sci. **369**, 14–33 (2016)
8. Yang, S., Li, M., et al.: A grid-based evolutionary algorithm for many-objective optimization. IEEE Trans. Evol. Comput. **17**(5), 721–736 (2013)
9. Gomez, R.H., Coello, C.A.C.: MOMBI: a new metaheuristic for many-objective optimization based on the R2 indicator. In: Proceedings of IEEE Congress on Evolutionary Computation, pp. 2488–2495. IEEE, New York (2013)
10. Deb, K., Jain, H.: An evolutionary many-objective optimization algorithm using reference point based non-dominated sorting approach, part I: solving problems with box constraints. IEEE Trans. Evol. Comput. (2014). https://doi.org/10.1109/tevc.2013.2281535
11. Zhang, Q., Li, H.: MOEA/D: a multi-objective evolutionary algorithm based on decomposition. IEEE Trans. Evol. Comput. **11**(6), 712–731 (2007)
12. Jaimes, A.L., Coello, C., et al.: Objective space partitioning using conflict information for solving many-objective problems. Inf. Sci. **268**, 305–327 (2014)
13. Tan, Y., Jiao, Y., et al.: MOEA/D uniform design: a new version of MOEA/D for optimization problems with many objectives. Comput. Oper. Res. **40**, 1648–1660 (2013)
14. Xu, Y., Pan, X.: A method of efficient three-objective differential Evolution using Pareto non-dominant relation. Res. Comput. Appl. **36**(03), 817–823+828 (2019)
15. Xu, Y., Fang, J.: A fast multi-objective differential evolutionary algorithm based on non-dominant solution ordering. Res. Comput. Appl. **34**(9), 2547–2551 (2014)
16. Das, S., Mullick, S.S., et al.: Recent advances in differential evolution-an updated survey. Swarm Evol. Comput. **4**(27), 1–30 (2016)

17. Xie, T., Chen, H., et al.: An evolutionary algorithm for multi-objective optimization. Comput. J. **26**(8), 173–181 (2003)
18. Deb, K., Pratap, A., et al.: A fast and elitist multi-objective genetic algorithm: NSGA-II. IEEE Trans. Evol. Comput. **6**(2), 182–197 (2002)

KnEA with Ensemble Approach for Parameter Selection for Many-Objective Optimization

Vikas Palakonda and Rammohan Mallipeddi(✉)

School of Electronics Engineering, Kyungpook National University, Daegu 702 701, South Korea
mallipeddi.ram@gmail.com

Abstract. Multi-objective evolutionary algorithms (MOEAs), in the past few decades have received immense recognition in effectively solving the multi-objective problems (MOPs), but these MOEAs encounter difficulties in handling the Many-Objective problems (MaOPs) due to the curse of dimensionality. In the literature, a knee point-driven evolutionary algorithm (KnEA) is proposed for handling the many-objective optimization which employs the concept of knee-points and weighted distance to select the better solutions in the mating and environmental selection. In the existing KnEA, to identify the knee-points, an adaptive strategy is adopted which employs a parameter *T* which is problem-specific and should be specified for each test problem separately. Hence, in our paper, we propose an ensemble approach to the existing KnEA by employing a set of values for the parameter *T*. In other words, instead of employing different value of the parameter *T*, for each test case, we would like to employ a unique set of values for the parameter *T* for all the test problems through the ensemble approach. For analyzing the performance of the ensemble approach, we have conducted experiments on 16 benchmark problems and the experimental results demonstrate that the proposed ensemble performs competitively with the state-of-art algorithms.

Keywords: Multi-objective optimization · Many-objective optimization · Ensemble approach · Knee-points

1 Introduction

Evolutionary algorithms (EAs) were widely employed in the literature to handle the optimization problems that consists of single-, multi- and many-objectives. Recently, the necessity for handling the problem with multiple conflicting objectives has increased drastically due to their wide-range of applications in the real-world scenario [1–3]. While solving the multi-objective problems (MOPs), instead of a unique optimal solution, a set of Pareto-optimal solutions are required. In the literature, various methodologies were proposed to handle MOPs, among them multi-objective evolutionary algorithms (MOEAs) are popular as they are capable of obtaining the set of Pareto-optimal solutions in one run [4, 5]. The main of MOEAs while handling the MOPs is to obtain a trade-off between two conflicting indicators, convergence and diversity. However, the task of

L. Pan et al. (Eds.): BIC-TA 2019, CCIS 1159, pp. 703–713, 2020.
https://doi.org/10.1007/978-981-15-3425-6_56

balancing the convergence and diversity is challenging as the convergence rate tends to get affected if the good performance of diversity is considered [6].

Most of MOEAs proposed in the literature are capable of solving the MOPs effectively but all these MOEAs encounter difficulties in solving the many-objective optimization problems (MaOPs) due to the increasing dimensionality [1–3]. In other words, MaOPs are considered as the special case of MOPs with more than three objectives. While handling the MaOPs, achieving balance between convergence and diversity while tackling the MaOPs is a daunting task [6, 7]. In the past few decades, to converge faster to the optimal Pareto front with better distribution, various selection approaches were proposed. Based on the selection techniques adopted, MOEAs can be classified into different groups. The first group of MOEAs are Pareto-dominance based MOEAs (PDMOEAs) that prioritize solutions based on Pareto rank and the solutions with better Pareto rank are given more importance [6].

The main problem associated with the PDMOEAs is that the effect of Pareto-dominance gradually reduces as the number of objectives increases. In other words, the proportion of nondominated solutions in the population increases due to increase in number of objective and enforces the MOEAs to rely on additional selection operator to distinguish the solutions [1, 2]. In the literature, PDMOEAs with different secondary selection metrics are proposed that prioritize the solutions based on the employed additional selection operator [3–6]. The second group of MOEAs are Decomposition–based MOEAs that decompose the problems with multi- and many-objectives into a single-objective and handles each sub-problem separately [8, 9]. The third group of MOEAs are the Indicator-based MOEAs, in which the solutions are given preferences based on the indicator values [10, 11]. Along with these approaches, many other MOEAs are proposed such as Relaxed dominance-based MOEAs [12], Reference set based MOEAs [13, 14] Preference-based MOEAs [15], Two archive MOEAs [16, 17] and few other approaches like [18–21] have been recently proposed for MaOPs.

In this paper, we consider an existing PDMOEA, a knee point-driven evolutionary algorithm (KnEA) [5], in which to select the elite solutions in mating and environmental selection the concept of knee-points and weighted distance are adopted. In the existing KnEA [5] algorithm, to identify the knee-points the authors have employed an adaptive strategy that consists of a parameter T that is problem-specific and should be mentioned manually. Hence, in this paper, we would like to propose an ensemble approach for the KnEA [5] algorithm by employing different values for the parameter T. In other words, we employ a set of values for the parameter T and stochastically select one value from the set of values assigned for parameter T [5]. We have analyzed the effect of different combination of ensembles in our work.

The rest of this paper is organized as follows. In Sect. 2, we briefly describe the framework of the proposed KnEA and the motivation behind the work. In the Sect. 3, we have presented the proposed work and experimental results and discussion is presented in the Sect. 4. Finally, Sect. 4 concludes the paper.

2 Related Work and Motivation

In this section, we briefly explain the framework of the existing KnEA [5] algorithm which starts with random initialization of the parent population of size *N*. The initialization procedure is then followed by the mating selection procedure, in which a binary tournament selection is employed to select better individuals to generate the offspring population. After obtaining the offspring population, both the parent and offspring populations are combined and then the Pareto-dominance procedure is applied on the combined population. After Pareto-dominance, an adaptive strategy to identify the knee-points is adopted and weighted distance is measured. Finally, the environmental selection procedure is adopted to preserve the elite solutions for the next iterations. [5]. The framework of the existing KnEA is depicted in the Fig. 1.

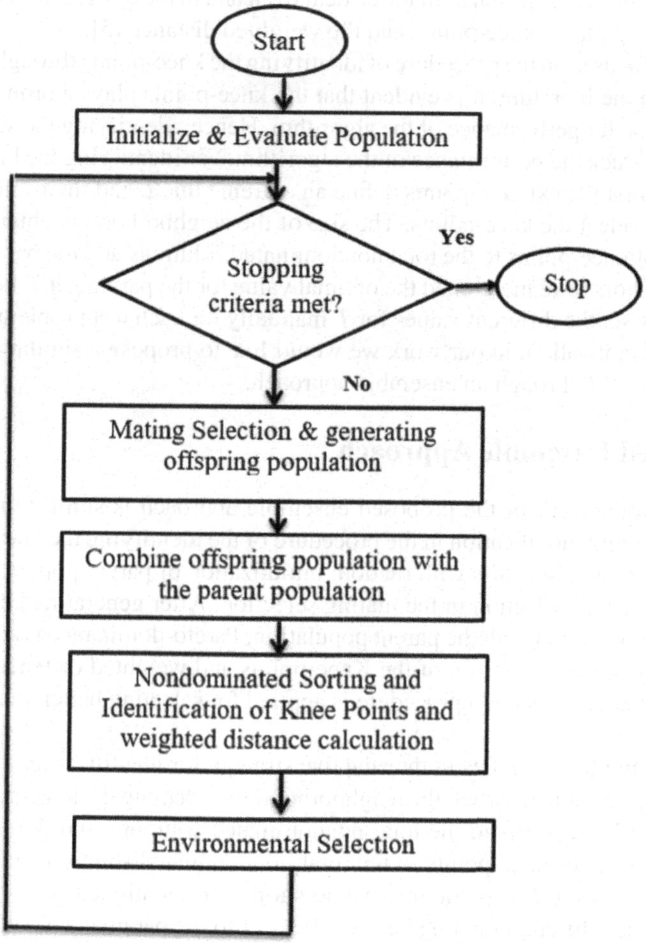

Fig. 1. General framework of the KnEA

In the mating selection, a binary tournament selection is adopted to select the superior individuals to produce promising offspring population. For the tournament selection, three metrics are adopted namely Pareto-dominance, knee-points and weighted distance. Hence, in the mating selection of KnEA, two individuals are selected randomly and if the solution one individual dominates the other individual, then the dominating solution will be chosen. If both the solutions are nondominated with each other then the solution, which is a knee-point, will be chosen. If both the individuals are not knee-points, then the solution with higher weighted distance will be selected. If both the solutions have same weighted distance then the one solution is randomly chosen [5].

In the environmental selection, the concept of Pareto-dominance is used along with the knee-points and the weighted distance. After combining the parent population with the offspring population, Pareto-dominance procedure is adopted which segregates the solutions into different nondominated levels termed as fronts. The solutions in each nondominated front are chosen until the critical front and in the critical front the solutions are chosen based on the knee-points and the weighted distance [5].

Our main focus is on the procedure of identifying the knee-points through an adaptive strategy. From the literature, it is evident that the knee-points plays a prominent role in achieving the better performance of the algorithm. Hence, identifying the suitable knee-points will enhance the performance of the algorithm. For identifying the knee-points in KnEA [5], at first the extreme points define an extreme line L and then a neighborhood is designed to select the knee-points. The size of the neighborhood is obtained with the help of ratio of knee-points to the total nondominated solutions and the parameters T. In the [5], the authors have mentioned the optimal value for the parameter T lies between 0 and 1 and they set the different values for T manually for each test problems separately. Based on this motivation, in our work we would like to propose a similar set of values for the parameter T through an ensemble approach.

3 Proposed Ensemble Approach

The general framework of the proposed ensemble approach is similar to the existing KnEA with a slight modification in the procedure of the identifying the knee-points. The proposed ensemble also starts with random initialization of parent population and with a binary tournament selection in the mating selection. After generating and combining the offspring population with the parent population, Pareto-dominance is adopted which is followed by the identification of the Knee-points and weighted distance calculation. Then, environmental selection procedure is applied for selecting better solutions for the next generations.

Our main modification lies in the adaptive strategy for identification of knee-points procedure. As mentioned earlier, the neighborhood for selecting the knee-points depends on the ratio of knee-points to the total nondominated solutions and the parameters T. However, the ratio of knee-points to the total nondominated solutions will be adapted but for the parameter T, a predefined value should be mentioned for every problem. Hence, we assign different ranging between 0 and 1 to the parameter T, select one value stochastically for each nondominated front in every iteration as shown in the Fig. 2. In the experimental setup, we have analyzed different combinations of ensemble that have different range of values for the parameter T.

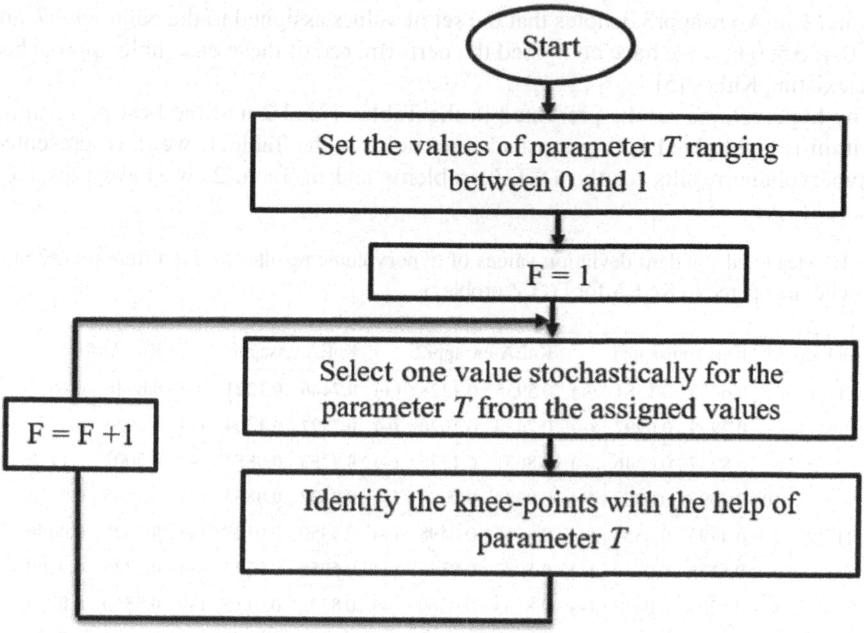

Fig. 2. Flowchart of the proposed Ensemble approach

4 Experiments Results and Discussion

In this section, we have presented the experimental results to analyze the performance of the proposed ensemble approach by conducting the experiments on two popular test problems DTLZ [22] (consists of seven benchmark problems DTLZ1-DTLZ7) and WFG [23] (consists of nine benchmark problems WFG1 to WFG9). We have conducted the experiments in two different experimental setups. In the first experimental setup, we compare the existing KnEA with different combination of the ensemble proposed in this paper. In the second experimental setup, we compare the best ensemble combination among the proposed approaches with the state-of-art algorithms. We have simulated each algorithm for 30 runs and the final population obtained for each run are saved for the comparison of the algorithms. To analyze the performance of the algorithms, we have adopted the hypervolume indicator [24] since the hypervolume indicator considers both the converging and distribution properties of the algorithms. We have conducted Wilcoxon's rank-sum test to obtain the statistical significance and presented the mean and standard deviation results of Hypervolume results in the Tables 1, 2, 3 and 4.

4.1 Experimental Conducted to Analyze Different Ensemble Approaches

In this section, we have conducted the experiments to analyze the performance different ensemble approaches. In this section, the KnEA-ensappr1 denotes that the set of values assigned for the parameter T are {0.1, 0.2, 0.3, 0.4, 0.5, 0.6, 0.7, 0.8, 0.9} and KnEA-ensappr2 refers to the set of values assigned to the parameter *T* are {0.1, 0.2, 0.3, 0.4, 0.5,

0.6,} and KnEA-ensappr3 denotes that the set of values assigned to the parameter T are {0.3, 0.4, 0.5, 0.6,}. We have compared the performance of these ensemble approaches to the existing KnEA [5].

The hypervolume results presented in the Tables 1 and 2 and the best performing algorithm is highlighted with grey shade and bold. In the Table 1, we have presented the hypervolume results for the DTLZ problems and in Table 2; we have presented

Table 1. Mean and standard deviation values of hypervolume results for the different ensemble approaches compared to KNEA for DTLZ problems

Problem	M	KnEA-ensappr1			KnEA-ensappr2			KnEA-ensappr3			KnEA [5]	
DTLZ1	4	0.6403	0.1381	(+)	0.5955	0.1228	(+)	**0.7446**	**0.1221**	(-)	0.6948	0.1667
	6	**0.7999**	**0.0992**	(–)	0.7430	0.1026	(–)	**0.7997**	**0.1304**	(–)	0.6953	0.0970
	8	0.8737	0.0846	(–)	0.8031	0.1575	(–)	**0.9253**	**0.0854**	(–)	0.5001	0.1118
	10	0.9997	0.0009	(=)	0.9983	0.0080	(=)	**0.9432**	**0.0863**	(+)	0.9965	0.0123
DTLZ2	4	0.4793	0.0554	(+)	0.4515	0.0498	(+)	0.5350	0.0127	(+)	**0.5528**	**0.0046**
	6	0.6156	0.0753	(+)	0.5983	0.0565	(+)	0.6956	0.0223	(+)	**0.7220**	**0.0100**
	8	0.8202	0.0337	(+)	0.8133	0.0360	(+)	0.8732	0.0115	(+)	**0.8860**	**0.0099**
	10	**0.9993**	**0.0002**	(–)	**0.9992**	**0.0002**	(–)	0.9970	0.0020	(–)	0.9891	0.0079
DTLZ3	4	0.6744	0.2552	(–)	**0.7080**	**0.2953**	(–)	0.6935	0.2478	(–)	0.6686	0.2314
	6	0.9987	0.0013	(=)	0.9980	0.0013	(=)	**0.9995**	**0.0005**	(–)	0.9985	0.0013
	8	**1.0000**	**0**	(–)	**1.0000**	**0.0000**	(–)	**1.0000**	**0.0000**	(–)	0.9994	0.0016
	10	**1.0000**	**0.0000**	(=)	**1.0000**	**0.0000**	(=)	0.9999	0.0001	(+)	**1.0000**	**0.0000**
DTLZ4	4	0.5152	0.0399	(+)	0.5018	0.0410	(+)	0.5573	0.0100	(+)	**0.5717**	**0.0056**
	6	0.8263	0.0318	(+)	0.8165	0.0263	(+)	0.8454	0.0129	(=)	**0.8552**	**0.0055**
	8	0.9938	0.0013	(+)	0.9928	0.0012	(+)	0.9943	0.0007	(=)	**0.9946**	**0.0003**
	10	0.9999	0.0001	(=)	0.9999	0.0000	(=)	0.9998	0.0001	(=)	**0.9999**	**0.0000**
DTLZ5	4	0.7587	0.0104	(–)	**0.7596**	**0.0070**	(–)	0.7571	0.0070	(=)	0.7575	0.0054
	6	0.8164	0.0086	(–)	0.8097	0.0113	(+)	**0.8180**	**0.0032**	(–)	0.8148	0.0053
	8	0.8063	0.0064	(+)	0.8044	0.0069	(+)	0.8026	0.0080	(+)	**0.8075**	**0.0059**
	10	0.8495	0.0052	(–)	**0.8503**	**0.0064**	(–)	0.8347	0.0098	(+)	0.8465	0.0044
DTLZ6	4	0.9133	0.0041	(–)	**0.9140**	**0.0060**	(–)	0.9129	0.0083	(–)	0.9118	0.0099
	6	**0.9768**	**0.0013**	(–)	0.9745	0.0026	(–)	0.9745	0.0057	(–)	0.9728	0.0059
	8	**0.9791**	**0.0050**	(–)	0.9779	0.0023	(–)	0.9593	0.0160	(+)	0.9761	0.0046
	10	**0.9788**	**0.0041**	(–)	0.9785	0.0018	(–)	0.9165	0.0333	(+)	0.9777	0.0025
DTLZ7	4	0.1823	0.0115	(+)	0.1769	0.0089	(+)	**0.1911**	**0.0064**	(–)	0.1900	0.0094
	6	0.1543	0.0199	(+)	0.1379	0.0159	(+)	0.1599	0.0123	(+)	**0.1691**	**0.0118**
	8	0.4703	0.1109	(+)	0.3923	0.1842	(+)	0.4524	0.1143	(+)	**0.5827**	**0.0312**
	10	0.0583	0.0263	(–)	0.0533	0.0223	(–)	**0.0718**	**0.0204**	(–)	0.0434	0.0129
+/=/–		**11/4/13**			**12/4/12**			**12/4/12**				

M—Number of objectives

Table 2. Mean and standard deviation values of hypervolume results for the different ensemble approaches compared to KNEA for WFG problems

Problem	M	KnEA-ensappr1			KnEA-ensappr2			KnEA-ensappr3			KnEA [5]	
WFG1	4	0.9577	0.0076	(+)	0.9537	0.0083	(+)	**0.9658**	**0.0025**	(–)	0.9641	0.0049
	6	0.9787	0.0032	(+)	0.9745	0.0044	(+)	0.9827	0.0026	(+)	**0.9846**	**0.0027**
	8	**0.9878**	**0.0024**	(–)	0.9855	0.0025	(+)	0.9875	0.0021	(=)	0.9874	0.0028
	10	0.9903	0.0013	(+)	0.9892	0.0014	(+)	0.9918	0.0015	(+)	**0.9927**	**0.0026**
WFG2	4	0.4673	0.0712	(+)	0.4464	0.0910	(+)	0.5155	0.0141	(=)	**0.5257**	**0.0189**
	6	0.4677	0.0347	(+)	0.4633	0.0315	(+)	0.4841	0.0270	(+)	**0.5072**	**0.0223**
	8	0.4113	0.0522	(+)	0.4062	0.0516	(+)	0.4522	0.0446	(+)	**0.4797**	**0.0398**
	10	0.5082	0.0525	(+)	0.4885	0.0496	(+)	**0.6490**	**0.0468**	(–)	0.6401	0.0462
WFG3	4	0.2533	0.0051	(+)	0.2482	0.0091	(+)	0.2588	0.0031	(+)	**0.2600**	**0.0032**
	6	0.1474	0.0092	(=)	0.1439	0.0123	(+)	**0.1487**	**0.0061**	(–)	0.1478	0.0122
	8	0.1070	0.0143	(–)	**0.1089**	**0.0151**	(–)	0.1041	0.0107	(–)	0.1039	0.0114
	10	0.0985	0.0147	(+)	**0.1046**	**0.0143**	(–)	0.0992	0.0112	(+)	0.1020	0.0131
WFG4	4	0.3363	0.0267	(+)	0.3296	0.0262	(+)	0.3659	0.0070	(+)	**0.3746**	**0.0047**
	6	0.3113	0.0297	(+)	0.2756	0.0401	(+)	0.3318	0.0150	(=)	**0.3395**	**0.0145**
	8	0.4238	0.0576	(–)	0.3937	0.0591	(+)	**0.4673**	**0.0203**	(–)	0.4036	0.0200
	10	0.4529	0.0619	(–)	0.4463	0.0505	(–)	**0.4700**	**0.0240**	(–)	0.4241	0.0253
WFG5	4	0.2364	0.0172	(+)	0.2307	0.0198	(+)	0.2516	0.0077	(+)	**0.2605**	**0.0041**
	6	0.1343	0.0300	(+)	0.1235	0.0246	(+)	0.1488	0.0181	(+)	**0.1684**	**0.0198**
	8	0.1440	0.0343	(+)	0.1386	0.0438	(+)	0.1543	0.0218	(+)	**0.1842**	**0.0194**
	10	0.1222	0.0321	(+)	0.1083	0.0346	(+)	0.1518	0.0230	(+)	**0.1779**	**0.0244**
WFG6	4	0.0771	0.0284	(+)	0.0773	0.0370	(+)	0.2052	0.0196	(+)	**0.2221**	**0.0214**
	6	0.0840	0.0312	(+)	0.0695	0.0388	(+)	0.1327	0.0213	(+)	**0.1508**	**0.0289**
	8	0.1003	0.0302	(+)	0.0928	0.0333	(+)	0.1390	0.0238	(+)	**0.1526**	**0.0222**
	10	0.1098	0.0297	(+)	0.1023	0.0238	(+)	0.1295	0.0238	(+)	**0.1335**	**0.0241**
WFG7	4	0.4506	0.0259	(+)	0.4464	0.0263	(+)	0.4856	0.0076	(=)	**0.4919**	**0.0037**
	6	0.5134	0.0329	(+)	0.4874	0.0389	(+)	0.5415	0.0094	(=)	**0.5491**	**0.0057**
	8	0.5229	0.0406	(+)	0.5105	0.0488	(+)	0.5561	0.0183	(+)	**0.5800**	**0.0107**
	10	0.5468	0.0291	(=)	0.5320	0.0383	(+)	**0.5871**	**0.0098**	(–)	0.5476	0.0374
WFG8	4	0.0884	0.0215	(+)	0.0882	0.0249	(+)	0.1177	0.0210	(+)	**0.1497**	**0.0298**
	6	0.0963	0.0231	(+)	0.0967	0.0207	(+)	0.1079	0.0200	(+)	**0.1138**	**0.0212**
	8	0.0898	0.0135	(+)	0.0866	0.0172	(+)	0.1029	0.0134	(+)	**0.1127**	**0.0139**
	10	0.0897	0.0137	(+)	0.0804	0.0169	(+)	0.1053	0.0151	(+)	**0.1156**	**0.0135**
WFG9	4	0.4789	0.0383	(+)	0.4695	0.0341	(+)	0.5013	0.0212	(=)	**0.5067**	**0.0242**
	6	0.5501	0.0387	(+)	0.5161	0.0451	(+)	**0.5708**	**0.0258**	(–)	0.5671	0.0320
	8	0.6471	0.0708	(–)	0.6314	0.0719	(–)	**0.6852**	**0.0634**	(–)	0.6246	0.0583
	10	**0.6890**	**0.0646**	(–)	0.6627	0.0767	(–)	0.6795	0.0744	(–)	0.6239	0.0893
+/=/–		**28/2/6**			**31/0/5**			**20/6/10**				

M—Number of objectives

the hypervolume results for the WFG problems. The From the results, we can observe that KnEA-ensappr1, out of possible 64 test instances when compared with KnEA [5] performs worst in 39 instances and equal in 6 and better in 19 instances. KnEA-ensappr2 performs worse, equal and better in 43, 4, 17 instances respectively when compared to the KnEA. KnEA-ensappr3 performs competitively with the existing KnEA [5] with worse performance in 32 instances and equal performance in 10 instances and better performance in 22 instances. From the experimental results presented in this section, we can witness that the KnEA-ensappr3 performs better in the compared three ensemble comparisons and hence we compare the KnEA-ensappr3 with the state-of-art algorithms in the later section.

Table 3. Mean and standard deviation values of hypervolume results for the ensemble approaches compared to state-of-art algorithms for DTLZ problems

Problem	M	NSGA-II [3]			GrEA [4]			Twoarchive2 [16]			NSGA-III [13]			KnEA-ensemble	
DTLZ1	4	0.7911	0.2413	(–)	0.7052	0.1633	(–)	0.0471	0.1272	(+)	**0.9121**	**0.0005**	(–)	0.6835	0.0935
	6	0.1345	0.2610	(+)	0.5867	0.2761	(+)	0	0	(+)	**0.9783**	**0.0060**	(–)	0.6625	0.1701
	8	0.0177	0.0969	(+)	0.4223	0.2210	(+)	0	0	(+)	**0.9729**	**0.1049**	(–)	0.7469	0.1503
	10	0	0	(+)	0.6298	0.2015	(–)	0	0	(+)	**0.9176**	**0.1989**	(–)	0.0053	0.0268
DTLZ2	4	0.4974	0.0092	(+)	0.6025	0.0022	(–)	0.2779	0.0343	(+)	**0.6026**	**0.0008**	(–)	0.5583	0.0121
	6	0.3917	0.1643	(+)	**0.9702**	**0.0004**	(–)	0.7764	0.0263	(+)	0.9681	0.0070	(=)	0.9615	0.0028
	8	0.5479	0.0862	(+)	0.9995	0.0000	(=)	0.9432	0.0188	(+)	**0.9996**	**0.0002**	(=)	0.9995	0.0001
	10	0.8513	0.0342	(+)	**1.0000**	**0.0000**	(=)	0.9959	0.0014	(+)	**1.0000**	**0.0000**	(=)	**1.0000**	**0.0000**
DTLZ3	4	0.5036	0.0119	(–)	0.1613	0.1675	(+)	0	0	(+)	**0.5941**	**0.0031**	(–)	0.4322	0.0829
	6	0.4838	0.3061	(+)	0.9939	0.0059	(=)	0	0	(+)	**0.9999**	**0.0003**	(=)	0.9930	0.0069
	8	0.3766	0.1660	(+)	**1.0000**	**0**	(=)	0.5874	0.1218	(+)	**1.0000**	**0**	(=)	**1.0000**	**0.0000**
	10	0.4107	0.1109	(+)	**1.0000**	**0**	(–)	0.8416	0.0535	(+)	**1.0000**	**0**	(–)	0.9991	0.0016
DTLZ4	4	0.4964	0.0096	(+)	0.5476	0.0815	(+)	0.3947	0.0561	(+)	0.4793	0.1097	(+)	**0.5573**	**0.0102**
	6	0.7923	0.1056	(+)	**0.9985**	**0.0000**	(=)	0.9891	0.0032	(+)	0.9931	0.0052	(=)	**0.9983**	**0.0002**
	8	0.8970	0.0362	(+)	**1.0000**	**0.0000**	(=)	0.9992	0.0003	(+)	0.9999	0.0001	(+)	**1.0000**	**0.0000**
	10	0.9524	0.0134	(+)	**1.0000**	**0.0000**	(=)	0.9997	0.0002	(+)	**1.0000**	**0.0000**	(=)	**1.0000**	**0.0000**
DTLZ5	4	**0.7793**	**0.0010**	(–)	0.7778	0.0010	(–)	0.7611	0.0046	(=)	0.7718	0.0022	(–)	0.7668	0.0066
	6	0.8408	0.0066	(+)	**0.8739**	**0.0033**	(–)	0.8611	0.0042	(=)	0.8400	0.0065	(+)	0.8697	0.0026
	8	0.8141	0.0146	(+)	**0.8714**	**0.0054**	(–)	0.8662	0.0032	(=)	0.8455	0.0087	(+)	0.8687	0.0061
	10	0.8287	0.0175	(+)	0.8695	0.0071	(=)	**0.8774**	**0.0033**	(–)	0.8766	0.0064	(–)	0.8679	0.0076
DTLZ6	4	0.8997	0.0501	(+)	**0.9359**	**0.0004**	(–)	0.3083	0.0568	(+)	0.9351	0.0007	(–)	0.9293	0.0066
	6	0.5336	0.0500	(+)	0.9851	0.0052	(=)	0.3844	0.0535	(+)	0.9844	0.0032	(=)	**0.9864**	**0.0030**
	8	0.5262	0.0465	(+)	0.9826	0.0059	(=)	0.4993	0.0391	(+)	**0.9877**	**0.0039**	(=)	0.9804	0.0076
	10	0.5545	0.0500	(+)	0.9766	0.0118	(–)	0.4753	0.0351	(+)	**0.9873**	**0.0032**	(–)	0.9585	0.0191
DTLZ7	4	0.1560	0.0060	(+)	0.1860	0.0042	(+)	0.0697	0.0434	(+)	0.1864	0.0023	(+)	**0.1913**	**0.0059**
	6	0.0394	0.0121	(+)	**0.1815**	**0.0065**	(–)	0.0546	0.0306	(+)	0.1419	0.0083	(+)	0.1655	0.0113
	8	0.0007	0.0006	(+)	**0.1374**	**0.0050**	(–)	0.0421	0.0337	(+)	0.1050	0.0199	(+)	0.1177	0.0124
	10	0.0012	0.0015	(+)	0.2597	0.0340	(–)	**0.3057**	**0.0147**	(–)	0.2677	0.0599	(–)	0.2036	0.0793
+/=/–		**25/0/3**			**5/10/13**			**23/3/2**			**7/9/12**				

M—Number of objectives

4.2 Comparison of Ensemble Approach with State-of-Art Algorithms

In this section, we have compared the KnEA-ensappr3 (denoted as the KnEA-ensemble) with the state-of-art algorithms NSGA-II [3], GrEA [4], Twoarchive2 [16] and NSGA-III [13] The parameter settings for test problems and the population sizes as employed

Table 4. Mean and standard deviation values of hypervolume results for the ensemble approaches compared to state-of-art algorithms for WFG problems

Problem	M	NSGA-II [3]			GrEA [4]			Twoarchive2 [16]			NSGA-III [13]			KnEA-ensemble	
WFG1	4	0.9717	0.0026	(=)	0.9324	0.0073	(+)	**0.9763**	**0.0010**	(=)	0.9566	0.0550	(+)	0.9711	0.0026
	6	**0.9955**	**0.0005**	(–)	0.9665	0.0047	(+)	0.9937	0.0004	(–)	0.9229	0.0757	(+)	0.9845	0.0026
	8	**0.9986**	**0.0002**	(–)	0.9657	0.0077	(+)	0.9886	0.0021	(=)	0.8813	0.0861	(+)	0.9879	0.0022
	10	**0.9992**	**0.0001**	(=)	0.9708	0.0056	(+)	0.9920	0.0009	(=)	0.8530	0.1085	(+)	0.9919	0.0015
WFG2	4	0.5576	0.0201	(–)	**0.5819**	**0.0144**	(–)	0.1373	0.1790	(+)	0.5542	0.0620	(–)	0.5346	0.0188
	6	0.5158	0.0298	(–)	**0.5729**	**0.0150**	(–)	0.2674	0.1762	(+)	0.5269	0.0632	(–)	0.4706	0.0282
	8	0.5510	0.0222	(–)	**0.6551**	**0.0052**	(–)	0.3300	0.1524	(+)	0.6321	0.0125	(–)	0.4726	0.0407
	10	0.5159	0.0200	(+)	**0.6029**	**0.0056**	(–)	0.4509	0.0667	(+)	0.5694	0.0139	(–)	0.5551	0.0389
WFG3	4	0.2579	0.0023	(=)	0.2569	0.0019	(=)	0.2454	0.0065	(+)	0.1906	0.0563	(+)	**0.2583**	**0.0030**
	6	0.1672	0.0074	(–)	**0.1828**	**0.0033**	(–)	0.1411	0.0088	(=)	0.0717	0.0255	(+)	0.1456	0.0058
	8	0.1454	0.0060	(–)	**0.1690**	**0.0012**	(–)	0.1044	0.0140	(=)	0.0455	0.0214	(+)	0.1047	0.0099
	10	0.1340	0.0046	(–)	**0.1723**	**0.0022**	(–)	0.1034	0.0116	(–)	0.0038	0.0040	(+)	0.0882	0.0099
WFG4	4	0.3084	0.0118	(+)	**0.3808**	**0.0031**	(–)	0.2477	0.0072	(+)	0.3356	0.0471	(+)	0.3672	0.0071
	6	0.2284	0.0124	(+)	**0.3754**	**0.0054**	(–)	0.2069	0.0062	(+)	0.2312	0.0833	(+)	0.3412	0.0150
	8	0.3396	0.0199	(+)	**0.5353**	**0.0077**	(–)	0.2922	0.0081	(+)	0.4479	0.0509	(+)	0.4839	0.0192
	10	0.3258	0.0142	(+)	**0.5938**	**0.0050**	(–)	0.2828	0.0089	(+)	0.3797	0.1386	(+)	0.4898	0.0225
WFG5	4	0.2323	0.0082	(+)	0.2524	0.0023	(=)	0.2450	0.0049	(+)	**0.2642**	**0.0034**	(–)	0.2528	0.0076
	6	0.1828	0.0126	(–)	0.2401	0.0025	(–)	0.1932	0.0096	(–)	**0.2437**	**0.0093**	(–)	0.1351	0.0174
	8	0.1872	0.0172	(–)	**0.2941**	**0.0046**	(–)	0.2053	0.0109	(–)	0.2656	0.0276	(–)	0.1606	0.0205
	10	0.1877	0.0139	(–)	**0.3188**	**0.0040**	(–)	0.1700	0.0139	(–)	0.2771	0.0201	(–)	0.1496	0.0213
WFG6	4	0.2132	0.0205	(–)	0.2584	0.0113	(–)	0.2335	0.0173	(–)	**0.2709**	**0.0264**	(–)	0.2035	0.0193
	6	0.1550	0.0321	(–)	0.2272	0.0232	(–)	0.1779	0.0203	(–)	**0.2339**	**0.0198**	(–)	0.1009	0.0193
	8	0.1377	0.0243	(–)	**0.2377**	**0.0192**	(–)	0.1517	0.0208	(–)	0.2261	0.0505	(–)	0.1164	0.0207
	10	0.1496	0.0267	(–)	**0.2451**	**0.0201**	(–)	0.1431	0.0187	(–)	0.2181	0.0286	(–)	0.1177	0.0193
WFG7	4	0.4377	0.0108	(+)	**0.4936**	**0.0031**	(–)	0.4514	0.0061	(+)	0.4300	0.0775	(+)	0.4825	0.0077
	6	0.4676	0.0087	(+)	0.5342	0.0051	(=)	0.4277	0.0100	(+)	0.4384	0.0842	(+)	**0.5351**	**0.0093**
	8	0.5141	0.0090	(+)	0.5324	0.0066	(+)	0.4038	0.0093	(+)	0.2922	0.1726	(+)	**0.5558**	**0.0182**
	10	0.5451	0.0098	(+)	0.6006	0.0034	(–)	0.3748	0.0128	(+)	0.3194	0.1033	(+)	**0.5855**	**0.0096**
WFG8	4	0.1164	0.0259	(–)	**0.2247**	**0.0255**	(–)	0.0974	0.0262	(–)	0.2116	0.0200	(–)	0.0248	0.0181
	6	0.0760	0.0269	(–)	**0.2053**	**0.0172**	(–)	0.0624	0.0224	(–)	0.1351	0.0193	(–)	0.0158	0.0101
	8	0.0571	0.0189	(–)	**0.2174**	**0.0098**	(–)	0.0651	0.0127	(–)	0.1837	0.0162	(–)	0.0498	0.0093
	10	0.0437	0.0151	(–)	**0.2252**	**0.0078**	(–)	0.0558	0.0100	(–)	0.1606	0.0217	(–)	0.0349	0.0075
WFG9	4	0.4052	0.0232	(+)	**0.5116**	**0.0224**	(–)	0.3596	0.0136	(+)	0.4757	0.0287	(+)	0.5030	0.0213
	6	0.3849	0.0193	(+)	0.5635	0.0282	(+)	0.4096	0.0187	(+)	0.5242	0.0367	(+)	**0.5832**	**0.0264**
	8	0.4528	0.0219	(+)	**0.7019**	**0.0291**	(–)	0.5137	0.0149	(+)	0.6678	0.1266	(+)	0.6837	0.0630
	10	0.5245	0.0203	(+)	**0.7562**	**0.0320**	(–)	0.5542	0.0148	(+)	0.6962	0.0755	(–)	0.6898	0.0774
+/=/–		**14/3/19**			**6/3/27**			**17/5/14**			**18/0/18**				

M—Number of objectives

as presented in [6]. We have presented the hypervolume results for the DTLZ and WFG problems in Tables 3 and 4 respectively for 4–, 6–, 8–, 10–objectives respectively. From the experimental results, we can observe that the proposed ensemble approach out of possible 64 test instances performs better in 39 instances and equal in three and worse in 22 instances when compared with the NSGA-II algorithm. When compared with the GrEA, the performance of the proposed ensemble approach is deteriorating as the proposed ensemble approach performs better, equal and worse in 11, 13, 40 instances respectively. When compared with Twoarchive2, the proposed ensemble approach performs better in 40 instances and competitive performance in eight instances and worse in 16 instances. The performance of proposed approach is competitive when compared with NSGA-III with better performance in 25 instances and equal performance in nine instances and worse in 30 instances.

5 Conclusion

In this paper, we have proposed an ensemble approach for the existing knee pointdriven evolutionary algorithm for many-objective optimization (KnEA) that emphasizes the importance of the knee-points. In the existing KnEA, to identify the knee-points an adaptive strategy is adopted which has a parameter T in default that cannot be adapted and should be specified for each test instance. Hence, in this paper, we propose an ensemble approach, which assign different values for the parameter T and stochastically select one value for the identification of the knee-points. The experimental results suggest that even though the optimal range for the parameter T lies between zero and one, considering the values between 0.3 and 0.6 gives better performance. We have also compared with the state-of-art algorithms and the proposed ensemble approach exhibited competitive performance. In future, we would like to propose an approach for the choosing the appropriate value for the parameter T.

Acknowledgement. This study was supported by the BK21 Plus project funded by the Ministry of Education, Korea (21A20131600011).

References

1. Zhou, A., Qu, B.-Y., Li, H., Zhao, S.-Z., Suganthan, P.N., Zhang, Q.: Multiobjective evolutionary algorithms: a survey of the state of the art. Swarm Evol. Comput. **1**, 32–49 (2011)
2. Li, B., Li, J., Tang, K., Yao, X.: Many-objective evolutionary algorithms: a survey. ACM Comput. Surv. (CSUR) **48**(13), 1–35 (2015)
3. Deb, K., Pratap, A., Agarwal, S., Meyarivan, T.: A fast and elitist multiobjective genetic algorithm: NSGA-II. IEEE Trans. Evol. Comput. **6**, 182–197 (2002)
4. Yang, S., Li, M., Liu, X., Zheng, J.: A grid-based evolutionary algorithm for many-objective optimization. IEEE Trans. Evol. Comput. **17**, 721–736 (2013)
5. Zhang, X., Tian, Y., Jin, Y.: A knee point-driven evolutionary algorithm for many-objective optimization. IEEE Trans. Evol. Comput. **19**, 761–776 (2015)

6. Palakonda, V., Mallipeddi, R.: Pareto dominance-based algorithms with ranking methods for many-objective optimization. IEEE Access **5**, 11043–11053 (2017)
7. Palakonda, V., Ghorbanpour, S., Mallipeddi, R.: Pareto dominance-based MOEA with multiple ranking methods for many-objective optimization. In: 2018 IEEE Symposium Series on Computational Intelligence (SSCI), pp. 958–964 (2018)
8. Zhang, Q., Li, H.: MOEA/D: a multiobjective evolutionary algorithm based on decomposition. IEEE Trans. Evol. Comput. **11**, 712–731 (2007)
9. Asafuddoula, M., Ray, T., Sarker, R.: A decomposition-based evolutionary algorithm for many objective optimization. IEEE Trans. Evol. Comput. **19**, 445–460 (2015)
10. Zitzler, E., Künzli, S.: Indicator-Based Selection in Multiobjective Search. In: Yao, X., et al. (eds.) PPSN 2004. LNCS, vol. 3242, pp. 832–842. Springer, Heidelberg (2004). https://doi.org/10.1007/978-3-540-30217-9_84
11. Pamulapati, T., Mallipeddi, R., Suganthan, P.N.: ISDE+ - An indicator for multi and many-objective optimization. IEEE Trans. Evol. Comput. **23**, 346–352 (2018)
12. Wang, G., Jiang, H.: Fuzzy-dominance and its application in evolutionary many objective optimization. In: International Conference on Computational Intelligence and Security Workshops, CISW 2007, pp. 195–198 (2007)
13. Deb, K., Jain, H.: An evolutionary many-objective optimization algorithm using reference-point-based nondominated sorting approach, Part I: solving problems with box constraints. IEEE Trans. Evol. Comput. **18**, 577–601 (2014)
14. Cheng, R., Jin, Y., Olhofer, M., Sendhoff, B.: A reference vector guided evolutionary algorithm for many-objective optimization. IEEE Trans. Evol. Comput. **20**, 773–791 (2016)
15. di Pierro, F., Khu, S.-T., Savic, D.A.: An investigation on preference order ranking scheme for multiobjective evolutionary optimization. IEEE Trans. Evol. Comput. **11**, 17–45 (2007)
16. Wang, H., Jiao, L., Yao, X.: Two_Arch2: an improved two-archive algorithm for many-objective optimization. IEEE Trans. Evol. Comput. **19**, 524–541 (2015)
17. Ghorbanpour, S., Palakonda, V., Mallipeddi, R.: Ensemble of Pareto-based selections for many-objective optimization. In: 2018 IEEE Symposium Series on Computational Intelligence (SSCI), pp. 981–988 (2018)
18. Pan, L., He, C., Tian, Y., Wang, H., Zhang, X., Jin, Y.: A classification-based surrogate-assisted evolutionary algorithm for expensive many-objective optimization. IEEE Trans. Evol. Comput. **23**, 74–88 (2018)
19. He, C., Tian, Y., Jin, Y., Zhang, X., Pan, L.: A radial space division based evolutionary algorithm for many-objective optimization. Appl. Soft Comput. **61**, 603–621 (2017)
20. Pan, L., He, C., Tian, Y., Su, Y., Zhang, X.: A region division based diversity maintaining approach for many-objective optimization. Integr. Comput. Aided Eng. **24**, 279–296 (2017)
21. Pan, L., Li, L., He, C., Tan, K.C.: A subregion division-based evolutionary algorithm with effective mating selection for many-objective optimization. IEEE Trans. Cybern. (2019)
22. Deb, K., Thiele, L., Laumanns, M., Zitzler, E.: Scalable test problems for evolutionary multiobjective optimization. In: Abraham, A., Jain, L., Goldberg, R. (eds.) Evolutionary Multiobjective Optimization. AI&KP, pp. 105–145. Springer, London (2005). https://doi.org/10.1007/1-84628-137-7_6
23. Huband, S., Hingston, P., Barone, L., While, L.: A review of multiobjective test problems and a scalable test problem toolkit. IEEE Trans. Evol. Comput. **10**, 477–506 (2006)
24. While, L., Hingston, P., Barone, L., Huband, S.: A faster algorithm for calculating hypervolume. IEEE Trans. Evol. Comput. **10**, 29–38 (2006)

Decomposition Based Differentiate Evolution Algorithm with Niching Strategy for Multimodal Multi-objective Optimization

Weiwei Zhang[1(✉)], Ningjun Zhang[1], Hanwen Wan[2], Daoying Huang[1], Xiaoyu Wen[1], and Yinghui Meng[1]

[1] Zhengzhou University of Light Industry, Zhengzhou, China
anqikeli@163.com

[2] International College of Zhengzhou University, Zhengzhou, China

Abstract. Comparing to the multi-objective optimization, multimodal multi-objective optimization brings greater challenge since it involves both the decision space and objective space. The classic multi-objective optimization method MOEA/D could not locate more than one optimal solution in the decision space corresponding to the same optimal solution in PF due to the shortage of strategies for handing multimodality. Therefore, a modified MOEA/D whth niching strategy is proposed, in which the MOEA/D-DE is modified for balancing the convergence and diversity in the objective space while the niching strategy is adopted for save the multiple solutions in the decision space. Further more, the redundant deletion strategy works for removing the redundancy and saving the computational resource. The proposed algorithm is tested on the 22 newly proposed benchmark functions. Experimental results show the competitive performance of the proposed algorithm.

Keywords: Decomposition · Multimodal · Multi-objective optimization · Niching

1 Introduction

There are many practical optimization problems [17,18,20] equipped with more than one conflicting objective, i.e., when one objective gets better, some of the others may get worse. Therefore, multiple objectives should be handled concurrently. For example, in industrial production, both the cost of fuel and the emission of pollution gas are the objectives to be considered. Finding the lowest cost and emission of pollution gas at the meantime forms a two objectives optimization problem. Generally, a maximal multi-objective optimization problem (MOP) can be stated as follows:

$$Maximize \quad F(x) = (f_1(x), \cdots, f_m(x))^T \tag{1}$$

$$\text{Subject to } x \in \Omega$$

L. Pan et al. (Eds.): BIC-TA 2019, CCIS 1159, pp. 714–726, 2020.
https://doi.org/10.1007/978-981-15-3425-6_57

where $x = (x_1, x_2, \cdots, x_n)$ represents an n-dimensional decision variable in the decision space Ω. $F(x)$ is the objective function set with the objective functions $f_i(x)$, $i = 1, \cdots, m$, where m is the number of objectives. In view of there are more than one objective to be concerned, dominate relationship [5] was proposed to evaluate the quality of solutions. The set of non-dominated solutions is called Pareto optimal Set (PS) in decision space while the set of points corresponding to PS is called Pareto Front (PF) in objective space.

In practice, many MOPs are multimodal inherently. There may be more than one Pareto optimal solutions in PS corresponding to a same Pareto optimal solution in PF, which is called the multimodal multi-objective optimization problem (MMOP) [12]. Taking MMF2 [21] as an example, which is shown as follows:

$$\begin{cases} f_1 = x_1 \\ f_2 = \begin{cases} 1 - \sqrt{x} + 2(4(x_2 - \sqrt{x_1})^2 - 2\cos(\frac{20(x_2 - \sqrt{x_1})\pi}{\sqrt{2}}) + 2) & , 0 \leq x_2 \leq 1 \\ 1 - \sqrt{x_1} + 2(4(x_2 - 1 - \sqrt{x_1})^2 - 2\cos(\frac{20(x_2 - 1 - \sqrt{x_1})\pi}{\sqrt{2}}) + 2) & , 1 < x_2 \leq 2 \end{cases} \end{cases} \tag{2}$$

The problem MMF2 has two objectives. The optimal solution of the problem is shown in Fig. 1, in which the Pareto dominate solutions in PS is shown in the left side while the Pareto dominate solutions in PF is in the right side. It is observed that for a solution in PF shown as red dot in Fig. 1(b), two solutions in the PS shown as red dots in the Fig. 1(a) is correspondent. Multiple solutions not only could provide more options to decision maker in case some of them become infeasible but also present the landscape distribution and characteristics of the solution space.

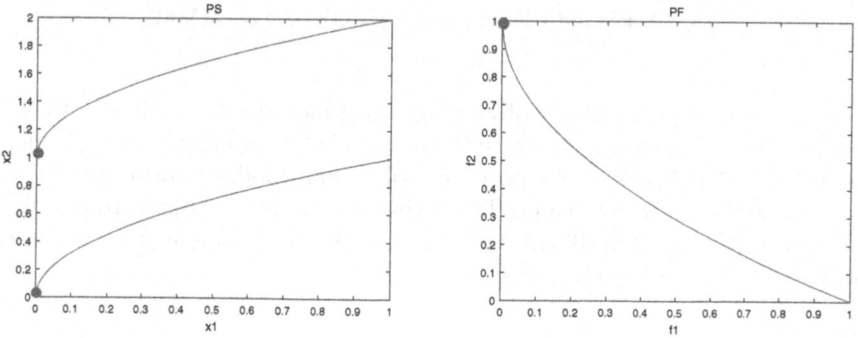

(a) (b)

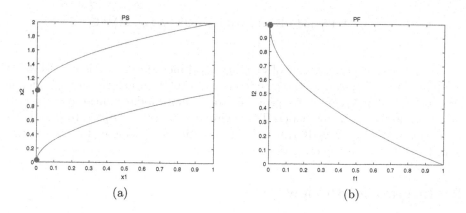

Fig. 1. The true PS and PF of MMF2

Before MMOPs get attentions, a lot of works have been done on solving MOPs, such as NSGA-II [1], SPEA2 [4], MOEA/D [23] and so on [14,15]. Among them, MOEA/D [23] has achieved remarkable results for MOPs and got many success applications. In view of the good performance of MOEA/D in handing the multiobjective optimization problems, a trial of MOEA/D on coping with MMOPs is instinctive. Therefore, MOEA/D was tested on the standard benchmark functions [21] for MMOPs. The experimental results show that MOEA/D performed well in the objective space, but could only locate one solution in PS corresponding one solution in PF. The fragment and discontinuous PS is present in the decision space. Take MMF2 as examples, the PS and PF achieved by MOEA/D on MMF2 in both decision space and objective space are shown in Fig. 2 respectively. It is seen that, even though there are missing fragments on the PF, the results in objective space is acceptable. However, it is observed that there only one solutions could be located in the decision space. Therefore, the strategies of handling the multimodality should be involved when MOEA/D is hired for coping with the MMOP problems.

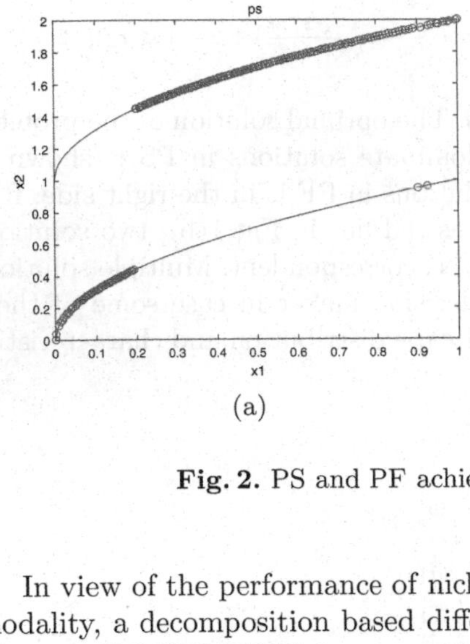

(a)

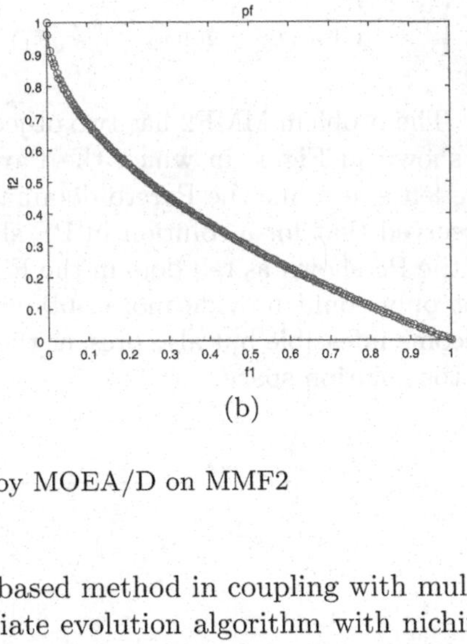

(b)

Fig. 2. PS and PF achieved by MOEA/D on MMF2

In view of the performance of niching based method in coupling with multimodality, a decomposition based differentiate evolution algorithm with niching strategy (MMOEA/DN) is proposed to solve multimodal multi-objective optimization problems in this paper. The niching strategy is adopted to handle the multimodality which the MOEA/D-DE is modified for balancing the convergence and diversity in the objective space.

2 Literatures Review

Compare to the multi-objective optimization, few works has been done on the MMOPs. Liang and Yue et al. [12] proposed a decision space based niching methods multi-objective evolutionary algorithm. Yue and Qu et al. [22] proposed a ring topology (MO_Ring_PSO_SCD) and adopted a special crowding distance concept. Liang and Xu et al. [10] proposed a multimodal multi-objective Differential Evolution optimization algorithm (MMODE). Liu and Gary et al. [13]

proposed a novel multimodal multi-objective evolutionary algorithm using two-archive and recombination strategies (TriMOEA-TA&R) equalized convergence and diversity. Liang and Guo et al. [9] proposed a self-organizing mechanism (SMPSO-MM) to find the multiple solutions in the decision space. Zhang and Li et al. [24] presented a new cluster based PSO with leader updating mechanism and ring-topology (MMO-CLRPSO). Li and Shi et al. [8] proposed a novel Differential Evolution algorithm based on Reinforcement Learning with Fitness Ranking (DE-RLFR). Ryoji and Hisao [19] proposed a niching indicator-based multimodal many-objective optimizer (NIMMO). Li [7] and Qu [22] used a niching method by nearest neighbors particles communicate which demonstrated to performance.

Through literatures review, It is found that the research on MMOP just starts. The decomposition based MOEA haven't been well studied. Few studies based on MOEA/D algorithm have been done on MMOPs [10]. On the other side, niching method has been applied into the MMOP [7,19,22]. Therefore, the decomposition based algorithm is incorporated with niching strategy for MMOP in the paper.

3 Proposed Method

The main framework of the proposed decomposition based differentiate evolution algorithm with niching strategy for MMOP (MMOEA/DN) is shown in Algorithm 1. Firstly, MMOEA/DN decomposes a multi-objective optimization problem into a number of different single objective optimization sub-problems by using weight vectors in objective space. The weight vectors $\lambda_1, \cdots, \lambda_{NA}$ are uniformly distributed in the objective space, where NA is the number of vectors. Each individual of population $s_1, s_2, \cdots, s_N$ is associating with a randomly chosen weight vector. Secondly, set the T closest weight vectors to λ_i as the neighborhood of weigh vector λ_i. In step 5, generate offspring according to the modified MOEA/D-DE by Algorithm 2. Update the neighboring solutions with niching strategy by Algorithm 3. For each vector, the algorithm to ensure well-distributed by control the size of individuals. This is accomplished by non-dominated sorting and special crowding distance method in Algorithm 4. The process is repeated until a termination condition is met.

3.1 Modified MOEA/D-DE

In this section, the way to generate offspring y' according to the modified MOEA/D-DE is described. For each weight vector, randomly choose one individual from $\{s_1, \cdots, s_{nw}\}$ as the parent s_{r1}, where nw is the number of the individuals in the population associating with the weight vector λ_i. Next, all the individuals associate with the neighboring weight vector of λ_i, denoted as $B(i) = \{i_1, \cdots, i_T\}$ are placed to the $popNe_i$, where T is the size of the neighborhood. Then, randomly selected two individuals from $popNe_i$ as s_{r2}, s_{r3}. Finally, new offspring y' is generated based on the chosen s_{r1}, s_{r2}, s_{r3} through DE [6].

Algorithm 1. The framework of proposed MMOEA/DN

Input:
population← $\{s_1, s_2, \cdots, s_N\}$ with N candidate solutions; NA uniform spread of weight vectors $\lambda \leftarrow \{\lambda_1, \cdots, \lambda_{NA}\}$; Reference point $z^* = \{z_1, \cdots, z_m\}^T$,where z_i represent the best f_i found so far; The number of weight vectors in the neighborhood of each weigh vector T; Threshold Th.
Output:
PS and PF.
Step 1) Randomly associate the individuals in the population with the weight vector
Step 2) For each $i = 1, \cdots, NA$, set $B(i) = \{i_1, \cdots, i_T\}$, where $\lambda_{i_1}, \cdots, \lambda_{i_T}$ are the T closest weight vectors to λ_i.
Step 3) While terminate condition is not met
Step 4) For $i = 1, \cdots, NA$
Step 5) Generate offspring y' according to the modified MOEA/D-DE by **Algorithm 2**
Step 6) Update the neighboring solutions with niching strategy by **Algorithm 3.**
Step 7)EndFor
Step 8) Delete the redundant solutions by **Algorithm 4.**
Step 9)Endwhile

Algorithm 2. Modified MOEA/D-DE

Input: weight vector λ_i ; neighboring weight vertor $B(i) = \{i_1, \cdots, i_T\}$, neighboring solutions $popNe = \emptyset$.
Output: offspring y'
Step 1) Find the individuals $\{s_1, \cdots, s_{nw}\}$ associating with weight vector λ_i, where nw is the number of the individuals associating with weight vector λ_i .
Step 2) Randomly choose one individual s_{r1} from $\{s_1, \cdots, s_{nw}\}, r1 \in \{1, \cdots, nw\}$
Step 3) For $j = 1, \cdots, T$
Step 4) Put the individuals associating with weight vector λ_{i_j} into $popNe_i$, where $i_j \in B(i)$
Step 5) EndFor
Step 6) Randomly choose two individuals from $popNe_i$ as s_{r2} and s_{r3}.
Step 7) $y' \leftarrow DE(s_{r1}, s_{r2}, s_{r3})$

3.2 Update the Neighboring Solutions with Niching Strategy

After the new offspring y' is generated, the updating process is implemented as Algorithm 3. The individuals associate with the neighborhood of λ_i are involved. The Euclidean distance between the new generated offspring y' and each individual in the $popNe_i$ are calculated. If the distance is smaller than a user defined threshold Th, they are considered as in the same niching in the decision space. Then, all the individuals in the $popNe_i$ with the g^{te} value worse than y' will be replaced by y'. The reference point z^* represents the best value for each objective at the current search. The g^{te} value is calculated in the same way as in article [23], which is called Tchebycheff Approach. If y' doesn't displace any individual in the population, y' will be assigned to the current weight vector and added into the population.

Algorithm 3. Update the neighboring solutions with niching strategy

Input: y' ; Threshold Th ; weight vector λ_i ; neighboring weight vertor $B(i) = \{i_1, \cdots, i_T\}$ neighboring solutions $popNe_i$ which are the individuals associating with the neighboring weight vectors $B(i) = \{i_1, \cdots, i_T\}$; Reference point z^* .
Output: updated population
Step 1) set FLAG=0
Step 2) For $i = 1, \cdots, T$
Step 3) **If** distance between y' and $popNe_i$ is smaller than Threshold Th
Step 4) **If** $g^{te}(y'|\lambda_i, z^*) < g^{te}(popNe_i|\lambda_i, z^*)$
Step 5) $popNe_i \leftarrow y'$
Step 6) FLAG=1
Step 7) **EndIf**
Step 8) **EndIf**
Step 9) EndFor
Step 10) If FLAG==0
Step 11) associate λ_i with y' and put y' into the population
Step 12) EndIf
Step 13) Output the updated population

3.3 Delete the Redundant Based on Decision Space Method

During the updating process, the number of individuals in the population is increasing continually. It is costly and useless since some of the individuals may be reduplicative. In order to control the number of individuals in the population, the maximal number of the individuals associate with a weight vector is set to a user defined value Nn. If the number of the individuals associate with a weight vector is more than Nn, the individuals with Euclidean distance smaller than Th are deleted at first, and then the individuals with worse g^{te} value will be removed in succession till the limited number Nn.

Algorithm 4. Delete_redundant_vectorpop(vectorpop,i,threshold)

Input: population $\{s_1, s_2, \cdots, s_N\}$; Threshold Th ; weigh verstors $\lambda, \cdots, \lambda_{NA}$;
Output: updated population
Step 1) For $i = 1, \cdots, NA$
Step 2) **While** the number of individuals associating with λ_i is larger than Nn
Step 3) Delete the individual according to Threshold Th and g^{te} value.
Step 4) **EndWhile**
Step 5) EndFor

4 Experiments

In this section, the performance of the proposed algorithm is tested by experiments. A standard benchmark function and the performance indicator about MMOP is elaborated. The algorithm with different threshold Th are tested and analyzed in detail. To test the performance of MMOEA/DN, four representative algorithms are compared.

4.1 Test Functions

There are some test functions have been designed in the previous works, called SYM-PART simple [16], SYM-PART rotated [16], and the Omni-test function [2]. Liang and Yue et al. [12] proposed two test functions namely SS-UF1 and S-UF3 to judge the performance of different algorithms. Yue and Qu et al. [22] proposed six more complicated test functions MMF3-MMF8. On this base, Yue and Qu et al. [21] proposed a novel scalable test problem suite for MMOP. The multimodal multi-objective test functions used in this paper include all of the above test functions mentioned.

4.2 Performance Indicators

To better evaluate the algorithms, it is important to choose an appropriate evaluation criteria. Yue and Qu et al. [22] proposed a new indicator called PSP to reflect the similarity between the obtained PSs and the true PSs. In this paper, two evaluation criteria are adopted: PSP (Pareto Set Proximity) [12], HV (inverted generational distance) [25]. Among them, PSP is the indicator specially for MMOP which could reflect the similarity between acquired PS and real PS for MMOPs. The formula is shown in Eq. (3):

$$PSP = \frac{CR}{IGDX} \tag{3}$$

where CR represents coverage between the obtained PS and the true PS [22], and IGDX represents the Euclidean distance between the Pareto set obtained by the algorithm and the true Pareto set [22].

By computing the volume between the obtained PF and a reference point z*, the diversity and convergence of the algorithm evaluated according to the value of HV, which is shown in Eq. (4):

$$HV(PF, z^*) = volume(\bigcup_{x \in PF} v(x, z^*)) \tag{4}$$

Yue and Qu et al. proposed two indicators 1/PSP and 1/HV instead of PSP and HV, which means the smaller value the better performance [21]. In this section, the experiments use the 1/PSP and 1/HV to test the performance of the algorithms.

4.3 Parameters Analysis and Comparison

This section gives the parameters setting of the test problems and comparison with the other state of the art algorithms. To be fair, the population size N is set to 200 to all the algorithms. The number of weight vector NA is set 200 as also. The size of the weight vectors in the neighborhood T equals to 20. The number of individuals associating with each weight vector λ_i is not more than Nn, which is set to 4. The experiment running on the 22 MMO benchmark functions [22].

Table 1. rPSP value of different threshold

Benchmark functions	5	10	15	20
MMF1	0.069±1.05E−02	0.068±5.24E−03	0.064±6.38E−03	**0.057±3.26E−03**
MMF2	0.028±1.71E−02	0.033±5.92E−03	0.023±3.59E−03	**0.018±2.07E−03**
MMF3	0.034±1.45E−02	0.023±3.04E−03	0.019±2.36E−03	**0.016±1.72E−03**
MMF4	0.053±8.52E−03	0.035±2.71E−03	0.036±2.00E−03	**0.035±2.12E−03**
MMF5	0.137±2.11E−02	0.118±9.07E−03	0.103±5.54E−03	**0.101±7.47E−03**
MMF6	0.122±1.45E−02	0.093±6.15E−03	0.084±5.82E−03	**0.082±5.44E−03**
MMF7	**0.034±9.14E−03**	0.038±6.07E−03	0.045±6.05E−03	0.037±2.68E−03
MMF8	0.216±5.17E−02	0.145±1.83E−02	0.119±1.52E−02	**0.098±9.50E−03**
MMF9	**0.009±2.08E−03**	0.01±8.31E−04	0.015±2.86E−03	0.014±1.30E−03
MMF10	0.186±2.61E−01	0.025±7.33E−03	0.025±7.30E−03	**0.023±3.55E−03**
MMF11	**0.011±2.57E−03**	0.106±3.66E−02	0.096±3.72E−02	0.083±2.94E−02
MMF12	**0.01±6.98E−03**	0.179±2.84E−01	0.06±3.52E−02	0.050±1.16E−02
MMF13	**0.098±9.59E−03**	0.105±1.71E−02	0.103±1.04E−02	0.105±1.06E−02
MMF14	0.05±2.70E−03	0.049±8.77E−04	0.049±1.35E−03	**0.048±9.38E−04**
MMF15	**0.079±1.02E−02**	0.513±9.43E−02	0.222±1.18E−01	0.159±2.38E−02
MMF1_z	0.052±1.19E−02	0.055±5.71E−03	0.051±4.19E−03	**0.044±3.11E−03**
MMF1_e	4.147±5.28E+00	0.486±1.38E−01	0.37±6.13E−02	**0.329±4.96E−02**
MMF14_a	0.072±3.98E−03	0.063±1.87E−03	0.062±1.82E−03	**0.061±1.65E−03**
MMF15_a	0.13±1.64E−02	**0.099±6.30E−03**	0.107±7.66E−03	0.113±8.08E−03
SYM−PART simple	3.597±1.36E+00	0.319±2.74E−01	**0.259±2.74E−02**	0.269±3.13E−02
SYM-PART rotated	3.137±5.30E+00	**0.231±2.61E−02**	0.265±3.12E−02	0.268±2.92E−02
Omni-test	0.644±1.23E−01	0.21±1.60E−02	0.203±7.75E−03	**0.196±8.26E−03**

Each algorithm runs 20 times independently. The maximal number of function evaluations is set to 10,000. The results are mean and standard deviation in the below table.

Experiment Results with Different Threshold *Th*. Algorithms may be sensitive to the parameter. In this section, the effect of threshold Th which is shown in Eq. (5) is discussed.

$$Th = \frac{distance(Xmax - Xmin)}{\sigma} \tag{5}$$

where $Xmax$ and $Xmin$ represent the upper and lower boundaries of decision space respectively. The value of σ is set to be 5, 10, 15, 20 in our experiments. The rPSP and rHV values are shown in Tables 1 and 2, respectively.

It is observed from Table 1 that the threshold Th has big effect on the performance of algorithm. When σ is 20, the algorithm could get the best rPSP value on 13 benchmark functions: MMF1, MMF2, MMF3, MMF4, MMF5, MMF6, MMF8, MMF10, MMF14, MMF1_z, MMF1_e, MMF14_a, and Omni-test. When σ is 5, the algorithm could get the best rPSP value on 6 functions: MMF7, MMF9, MMF11, MMF12, MMF13, and MMF15. When σ equals to 10,

Table 2. rHV value of different threshold

Benchmark functions	5	10	15	20
MMF1	**1.149±3.45E−03**	1.152±1.58E−03	1.154±2.42E−03	1.153±2.00E−03
MMF2	**1.158±8.31E−03**	1.236±3.24E−02	1.315±4.96E+02	1.363±3.24E−02
MMF3	**1.158±8.31E−03**	1.236±3.24E−02	1.315±4.96E+02	1.363±3.24E−02
MMF4	**1.857±2.27E−03**	1.867±2.99E−03	1.869±3.19E−03	1.869±2.75E−03
MMF5	**1.150±3.28E−03**	1.152±1.47E−03	1.152±2.24E−03	1.153±4.69E−03
MMF6	**1.149±2.89E−03**	1.156±1.10E−02	1.157±8.09E−03	1.158±1.15E−02
MMF7	**1.146±8.61E−04**	1.149±7.64E−04	1.154±1.96E−03	1.159±4.49E−03
MMF8	**2.400±8.42E−03**	2.440±9.06E−03	2.458±1.91E−02	2.465±2.75E−02
MMF9	**0.104±2.11E−04**	0.104±2.72E−04	0.106±1.87E−03	0.108±2.11E−03
MMF10	**0.086±8.94E−04**	0.087±1.02E−03	0.087±1.02E−03	0.088±9.93E−04
MMF11	**0.071±7.52E−04**	0.071±8.54E−04	0.772±6.86E−04	0.074±6.53E−04
MMF12	0.772±9.32E−02	**0.686±4.21E−02**	0.759±7.00E−02	0.809±7.10E−02
MMF13	**0.057±8.02E−04**	0.057±3.20E−04	0.057±2.82E−04	0.057±3.41E−04
MMF14	0.335±1.03E−02	**0.317±7.83E−03**	0.320±1.24E−02	0.327±7.85E−03
MMF15	0.239±6.12E−03	**0.217±4.94E−03**	0.223±6.78E−03	0.227±6.77E−03
MMF1_z	**1.147±1.91E−03**	1.152±1.90E−03	1.154±2.29E−03	1.156±4.03E−03
MMF1_e	**1.166±3.34E−02**	1.179±1.17E−02	1.200±1.37E−02	1.226±2.7E−02
MMF14_a	0.323±8.53E−03	**0.312±1.15E−02**	0.321±6.95E−03	0.322±7.97E−03
MMF15_a	0.223±6.64E−03	**0.217±4.17E−03**	0.221±5.49E−03	0.223±5.91E−03
SYM-PART simple	**0.060±3.13E−05**	0.061±3.79E−04	0.062±6.80E−04	0.063±8.43E−04
SYM-PART rotated	**0.060±1.05E−04**	0.061±2.86E−04	0.062±5.05E−04	0.063±9.41E−04
Omni-test	**0.019±5.74E−05**	0.020±1.54E−04	0.020±1.40E−04	0.020±1.29E−04

the algorithm get the best rPSP value for MMF15_a and SYM-PART rotated. It can be seen from the indicator rPSP, the benchmark functions are sensitive to the threshold Th.

As to the HV value which is shown in Table 2, when σ is set to 5, the algorithm get the best rHV value on MMF1, MMF2, MMF3, MMF4, MMF5, MMF6, MMF7, MMF8, MMF9, MMF10, MMF11, MMF13, MMF1_z, MMF1_e, SYM-PART simple, and SYM-PART rotated Omni-test. When σ is 10, the algorithm gets the best rHV value on MMF12, MMF14, MMF15, MMF14_a, MMF15_a. Based on the experimental results, It is found that there are some correlations between the two evaluation criteria. The larger value of the threshold Th has, the better rPSP value the algorithm gets, but the worse rHV value. Therefore, a compromise value as 5 for σ is chosen in the paper.

The performance of the algorithm with σ equals 5 on MMF2 is shown in Fig. 3. Although there are some break points, the algorithm could locate most part of both true PSs in Fig. 3(a). Also, the algorithm also performs well in the objective space in Fig. 3(b).

Comparison with Other Algorithm. In this section, the comparison of the proposed algorithm MMOEA/DN with four state-of-the-art algorithms on the 22 MMO benchmark functions [21] is implemented. The comparing algorithms

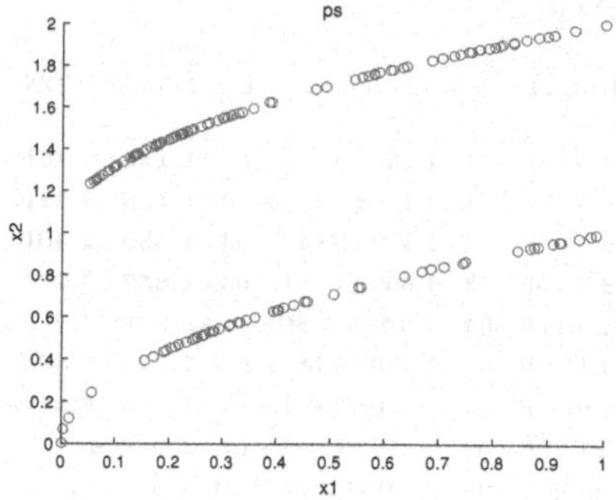

(a)

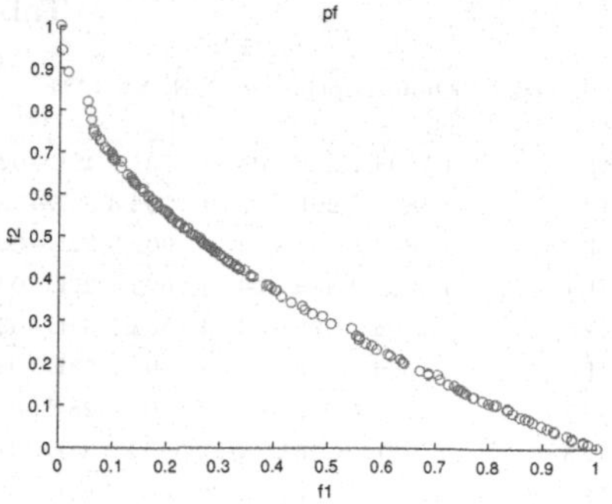

(b)

Fig. 3. Distribution of non-dominated solutions of MMOEA/DN in the decision space and objective space on MMF2

Table 3. rPSP

Benchmark functions	Omini_optimizer [2]	DN_NSGAII [11]	MO_Ring_PSO _SCD [22]	MO_PSO_MM [9]	MMOEA/DN
MMF1	0.096±1.89E−02	0.096±1.59E−02	0.048±1.83E−03	**0.040±1.08E−03**	0.069±1.05E−02
MMF2	0.119±6.54E−02	0.140±8.60E−02	0.045±1.21E−02	0.031±6.82E−03	**0.028±1.71E−02**
MMF3	0.101±4.69E−02	0.143±1.58E−01	0.029±8.21E−03	**0.022±3.53E−03**	0.034±1.45E−02
MMF4	0.086±2.10E−02	0.086±2.34E−02	0.027±1.48E−03	**0.023±9.65E−04**	0.053±8.52E−03
MMF5	0.175±2.61E−02	0.180±1.96E−02	0.086±4.73E−03	**0.073±3.17E−03**	0.137±2.11E−02
MMF6	0.146±1.53E−02	0.147±2.01E−02	0.074±4.56E−03	**0.064±3.43E−03**	0.122±1.45E−02
MMF7	0.049±1.44E−02	0.054±1.07E−02	0.027±1.58E−03	**0.021±9.03E−04**	0.034±9.14E−03
MMF8	0.323±1.60E−01	0.307±1.25E−01	0.068±5.25E−03	**0.058±6.93E−03**	0.216±5.17E−02
MMF9	0.026±1.01E−02	0.024±8.95E−03	0.008±5.66E−04	**0.006±3.54E−04**	0.009±2.08E−03
MMF10	2.580±3.35E+00	1.660±2.86E+00	0.489±1.00E+00	0.771±1.56E+00	**0.186±2.61E−01**
MMF11	1.720±1.44E−01	1.650±1.95E−01	0.480±4.79E−01	1.270±5.13E−01	**0.011±2.57E−03**
MMF12	2.160±2.92E−01	2.220±2.77E−01	0.596±5.74E−01	1.240±5.05E−01	**0.010±6.98E−03**
MMF13	0.606±2.17E−02	0.623±8.12E−02	0.351±1.05E−01	0.485±9.46E−02	**0.098±9.59E−03**
MMF14	0.090±9.79E−03	0.097±9.63E−03	0.053±1.17E−03	0.054±1.68E−03	**0.050±2.70E−03**
MMF15	0.302±1.34E−01	0.247±6.84E−02	0.156±2.09E−02	0.146±1.92E−02	**0.079±1.02E−02**
MMF1_z	0.075±1.53E−02	0.082±1.69E−02	0.035±1.89E−03	**0.029±1.33E−03**	0.052±1.19E−02
MMF1_e	2.380±2.09E+00	1.870±1.43E+00	0.551±1.30E−01	**0.502±1.44E−01**	4.147±5.28E+00
MMF14_a	0.112±1.12E−02	0.120±8.06E−03	0.062±2.35E−03	**0.060±1.78E−03**	0.072±3.98E−03
MMF15_a	0.240±4.26E−02	0.223±3.16E−02	0.166±1.24E−02	0.164±1.51E−02	**0.130±1.64E−02**
SYM-PART simple	5.720±2.78E+00	4.200±9.30E−01	0.176±3.21E−02	0.144±1.83E−02	**3.597±1.36E+00**
SYM-PART rotated	6.99±4.50E+00	6.020±4.28E+00	**0.354±3.91E−01**	0.193±2.90E−02	3.137±5.30E+00
Omni-test	1.980±8.09E−01	1.430±2.44E−01	**0.090±1.83E−03**	0.333±9.57E−02	0.644±1.23E−01

include Omini_optimizer [3], DN_NSGAII [11], MO_Ring_PSO_SCD [22], MO _PSO _MM [9]. For fair competition, every involved algorithms are equipped with the same parameters. The rPSP and rHV values of different algorithms are shown in the Tables 3 and 4 respectively.

Table 4. rHV

Benchmark functions	Omini_optimizer [2]	DN_NSGAII [11]	MO_Ring_PSO _SCD [22]	MO_PSO_MM [9]	MMOEA/DN
MMF1	1.150±1.22E−03	1.150±1.48E−03	1.150±4.98E−04	1.150±3.75E−04	**1.149±3.45E−03**
MMF2	1.180±2.29E−02	1.190±3.05E−02	1.180±6.42E−03	1.170±3.84E−03	**1.158±8.31E−03**
MMF3	1.180±2.69E−02	1.190±3.22E−02	1.170±4.82E−03	1.170±3.91E−03	**1.159±2.40E−02**
MMF4	1.860±9.84E−04	1.860±1.27E−03	1.860±2.30E−03	1.860±1.48E−03	**1.857±2.27E−03**
MMF5	1.150±1.09E−03	1.150±5.64E−03	1.150±6.17E−04	**1.150±3.80E−04**	1.150±3.28E−03
MMF6	1.150±6.51E−04	1.150±1.53E−03	1.150±1.10E−03	1.150±4.58E−04	**1.149±2.89E−03**
MMF7	1.150±5.02E−04	1.150±5.28E−03	1.150±9.25E−04	1.150±3.03E−04	**1.146±8.61E−04**
MMF8	**2.370±8.76E−04**	2.380±3.34E−03	2.410±1.79E−02	2.390±1.42E−02	2.400±8.42E−03
MMF9	0.103±2.19E−05	0.103±2.87E−05	0.103±4.48E−05	**0.103±2.06E−05**	0.104±2.11E−04
MMF10	0.081±2.79E−03	0.082±2.62E−03	0.080±5.05E−04	**0.078±3.22E−04**	0.086±8.94E−04
MMF11	0.069±6.91E−06	**0.069±1.00E−05**	0.069±2.09E−05	0.069±1.28E−05	0.071±7.52E−04
MMF12	**0.636±8.80E−05**	0.636±3.28E−04	0.639±1.16E−03	0.637±1.21E−03	0.772±9.32E−02
MMF13	**0.054±4.17E−06**	0.054±1.94E−05	0.054±2.66E−05	0.054±1.45E−05	0.057±8.02E−04
MMF14	0.336±1.11E−02	0.327±7.19E−03	**0.348±3.45E−02**	0.350±2.32E−02	0.335±1.03E−02
MMF15	0.236±7.45E−03	**0.231±9.60E−03**	0.238±7.88E−03	0.237±1.04E−02	0.239±6.12E−03
MMF1_z	1.150±8.33E−04	1.150±1.23E−03	1.150±5.72E−04	1.150±4.00E−04	**1.147±1.91E−03**
MMF1_e	1.180±2.92E−02	1.530±7.60E−01	1.190±1.75E−02	1.160±5.52E−03	**1.166±3.34E−02**
MMF14_a	0.329±1.07E−02	**0.316±1.19E−02**	0.338±2.54E−02	0.341±1.28E−02	0.323±8.53E−03
MMF15_a	0.230±9.49E−03	0.232±1.35E−02	0.239±8.19E−03	0.238±9.98E−03	**0.223±6.64E−03**
SYM-PART simple	0.060±9.52E−06	**0.060±1.12E−05**	0.060±9.21E−05	0.060±5.18E−05	0.060±3.13E−05
SYM-PART rotated	0.060±9.68E−06	**0.060±1.02E−05**	0.060±9.84E−05	0.060±5.49E−05	0.060±1.05E−04
Omni-test	**0.019±3.35E−07**	0.019±1.32E−06	0.019±1.54E−05	0.019±1.13E−05	0.019±5.74E−05

From the Table 3, it is found that MMOEA/DN obtains the best rPSP value on eight benchmark functions, including MMF2, MMF10, MMF11, MMF12, MMF 13, MMF14, MMF15, and MMF15_a. MO_PSO _MM also obtains competitive results. MO_PSO_MM obtains the best rPSP value on MMF1, MMF3, MMF4, MMF5, MMF6, MMF7, MMF8, MMF9, MMF1_z, MMF1_e, MMF14_a, and SYM-PART simple. The other three algorithms perform a little worse on the rPSP values. MO_Ring_PSO_SCD gets the best rPSP value on SYM-PART rotated and Omni-test.

From the Table 4, it is observed that the rHV values of the algorithms are very closely. MMOEA/DN obtains the best rHV value on nine benchmark functions, including MMF1, MMF2, MMF3, MMF4, MMF6, MMF7, MMF1_z, MMF1_e, and MMF15_a. Omini_optimizer gains the best rHV value on MMF8, MMF12, MMF13, and Omni-test. DN_NSGAII achieves the best value on MMF 11, MMF15, MMF14_a, SYM-PART simple, and SYM-PART rotated. MO_Ring _PSO_SCD only gets the best rHV value on MMF14. MO_PSO_MM obtains the best rHV value on MMF5, MMF9, MMF10. From the rHV value, the proposed algorithm MMOEA/DN achieves the competitive performance among the compared algorithms.

5 Conclusion

Decomposition based differentiate evolution algorithm with niching strategy (MMOEA/DN) for MMOPs is proposed, in which the MOEA/D-DE is modified for balancing the convergence and diversity in the objective space and the niching strategy is adopted to keep multiple solutions in the decision space. The experimental results demonstrate that the proposed MMOEA/DN algorithm is quit competitive.

In the real life, there are a lot of multimodal multi-objective problems. In the proposed MMOEA/DN, the PS distribution of decision space is not well-distributed enough. In future work, we should try our best to fill the blank area in the PS for multimodal multi-objective problems.

Acknowledgements. The work is supported by the National Natural Science Foundation of China (No. 61403349, 61501405), Funding program for key scientific research projects of universities in Henan province (No. 18A210025, 20A520004), science and technology key project of Henan province (No. 182102110399, 192102110203), the training program for key young teachers in henan institutions of higher learning (2019GGJS138).

References

1. Deb, K., Pratap, A., Agarwal, S., Meyarivan, T.: A fast and elitist multiobjective genetic algorithm: NSGA-II. IEEE Trans. Evol. Comput. **6**(2), 182–197 (2002)
2. Deb, K., Tiwari, S.: Omni-optimizer: a procedure for single and multi-objective optimization. In: Coello Coello, C.A., Hernández Aguirre, A., Zitzler, E. (eds.) EMO 2005. LNCS, vol. 3410, pp. 47–61. Springer, Heidelberg (2005). https://doi.org/10.1007/978-3-540-31880-4_4
3. Debab, K.: Omni-optimizer: a generic evolutionary algorithm for single and multi-objective optimization. Eur. J. Oper. Res. **185**(3), 1062–1087 (2008)
4. E. Zitzler, M.L., Thiele, L.: SPEA2: improving the strength pareto evolutionary algorithm. In: Proceedings of the Evolution: Methods Design Optimization Control Application Industrial Problems (2001)
5. Konak, A., Coit, D., Smith, A.: Multi-objective optimization using genetic algorithms: a tutorial. Reliab. Eng. Syst. Saf. **91**(9), 992–1007 (2006)
6. Li, H., Zhang, Q.: Multiobjective optimization problems with complicated pareto sets, MOEA/D and NSGA-II. IEEE Trans. Evol. Comput. **13**(2), 284–302 (2009)
7. Li, X.: Niching without niching parameters: particle swarm optimization using a ring topology. IEEE Trans. Evol. Comput. **14**(1), 150–169 (2010)
8. Li, Z., Shi, L., Yue, C.: Differential evolution based on reinforcement learning with fitness ranking for solving multimodal multiobjective problems. Swarm Evol. Comput. **49**, 234–244 (2019)
9. Liang, J., Guo, Q., Yue, C., Qu, B., Yu, K.: A self-organizing multi-objective particle swarm optimization algorithm for multimodal multi-objective problems. In: Proceedings of International Conference on Swarm Intelligence, pp. 550–560 (2018)
10. Liang, J., Xu, W., Yue, C.: Multimodal multiobjective optimization with differential evolution. Swarm Evol. Comput. **49**, 1028–1059 (2019)

11. Liang, J., Yue, C., Qu, B.: In 2016 IEEE Congress on Evolutionary Computation. CEC 2016 (Institute of Electrical and Electronics Engineers Inc.), pp. 2454–2461 (2016)
12. Liang, J., Yue, C., Qu, B.: Multimodal multi-objective optimization: a preliminary study. In: Evolutionary Computation, pp. 2454–2461 (2016)
13. Liu, Y., Yen, G., Gong, D.: A multi-modal multi-objective evolutionary algorithm using two-archive and recombination strategies. IEEE Trans. Evol. Comput. **23**(4), 660–674 (2019)
14. Pan, L., Cheng, H., Ye, T.: A region division based diversity maintaining approach for many-objective optimization. Integr. Comput. Aided Eng. **24**(3), 279–296 (2017)
15. Pan, L., Li, L., He, C.: A subregion division-based evolutionary algorithm with effective mating selection for many-objective optimization. IEEE Trans. Cybern (2019, in press)
16. Rudolph, G., Naujoks, B., Preuss, M.: Capabilities of EMOA to detect and preserve equivalent pareto subsets. In: Obayashi, S., Deb, K., Poloni, C., Hiroyasu, T., Murata, T. (eds.) EMO 2007. LNCS, vol. 4403, pp. 36–50. Springer, Heidelberg (2007). https://doi.org/10.1007/978-3-540-70928-2_7
17. Santander-Jimnez, S., Vega-Rodrguez, M.A.: Performance evaluation of dominance-based and indicator-based multiobjective approaches for phylogenetic inference. Inf. Sci. **330**, 293–314 (2016)
18. Shelokar, P., Quirin, A., Cordn, S.: A multiobjective evolutionary programming framework for graph-based data mining. Inf. Sci. **237**, 118–136 (2013)
19. Tanabe, R., Ishibuchi, H.: A niching indicator-based multi-modal many-objective optimizer. Swarm Evol. Comput. **49**, 134–146 (2019)
20. Yi, R., Luo, W., Bu, C., Lin, X.: A hybrid genetic algorithm for vehicle routing problems with dynamic requests, pp. 1–8 (2017)
21. Yue, C., Qu, B., Yu, K., Liang, J., Li, X.: A novel scalable test problem suite for multimodal multiobjective optimization. Swarm Evol. Comput. **48**, 62–71 (2019)
22. Yue, C., Qu, B., Liang, J.: A multi-objective particle swarm optimizer using ring topology for solving multimodal multi-objective problems. IEEE Trans. Evol. Comput. **22**(5), 805–817 (2018)
23. Zhang, Q., Hui, L.: MOEA/D: a multiobjective evolutionary algorithm based on decomposition. IEEE Trans. Evol. Comput. **11**(6), 712–731 (2007)
24. Zhang, W., Li, G., Zhang, W.: A cluster based PSO with leader updating mechanism and ring-topology for multimodal multi-objective optimization. Swarm Evol. Comput. **50**, 100569 (2019)
25. Zitzler, E., Thiele, L.: Multiobjective evolutionary algorithms: a comparative case study and the strength pareto approach. IEEE Trans. Evol. Comput. **3**(4), 257–271 (1999)

A Bacterial Foraging Framework for Agent Based Modeling

Mijat Kustudic and Niu Ben(✉)

College of Management, Shenzhen University, Shenzhen 518060, China
Mijat.k.ntc@gmail.com, drniuben@gmail.com

Abstract. Swarm optimization algorithms and agent based modeling (ABM) are two closely related research areas, parts of the multi agent system field, but they are traditionally not combined. Swarm optimization, in this case the bacterial foraging optimization (BFO), searches for an optimal solution while the ABM searches for a conclusion which resembles the real world, and it can be far from optimal. To bridge the gap, the overall goal this paper is to propose a new paradigm in the form of an architecture and operation procedures, thus creating a BFO-ABM hybrid. The other goal is to create a method which enables 3D visualization of the BFO algorithm. Firstly, an environment is created together with bacteria which physically perform all operators of the BFO. Secondly, a way of seamlessly embedding the bacteria from the BFO into the ABM environment is described. The bacteria are then manipulated and motivated with food and toxicity to act in a certain agent-like way. Simulation results prove that the agents can be effectively used as an ABM tool to present agents of all sizes and behaviors resembling numerous things, from companies, vehicles to people.

Keywords: Swarm optimization · Agent based modeling · Evolutionary computation

1 Introduction

1.1 A Subsection Sample

Computational field of Multi agent systems (MAS) is an interesting field of research which encompasses numerous subfields all based on individual and intelligent agents. These agents are autonomous, decentralized and have only their local view. Swarm optimization algorithms and agent based modeling approaches are subfields of MAS but they are not traditionally combined. At first glance they seem to be based on different principles but actually their logic is quite similar. The main difference is the goal which is ought to be made: in optimization there needs to be an optimal solution and in ABM a conclusion that resembles the real world, the result can be far from optimal.

Optimization algorithms have been used to solve different real world optimization and engineering problems. They are usually based on behaviors of certain animals or natural phenomena employed to search for an optimal solution,

L. Pan et al. (Eds.): BIC-TA 2019, CCIS 1159, pp. 727–738, 2020.
https://doi.org/10.1007/978-981-15-3425-6_58

to name a few of them: Particle swarm optimization [1], Hydrologic Cycle Optimization [2], Differential evolution [3], Water Cycle Algorithm [4], Artificial Bee Colony [5], Genetic Algorithm [6], Ant Colony Optimization (ACO) [7], Bacterial Foraging Optimization Algorithm (BFOA) [8] and numerous others which have proven themselves in different areas. These algorithms are always being updated through consideration of new strategies; for example within the BFO change of the chemotactic step length [9], population change [10] and the algorithm being adaptive [11]. Even though these algorithms are based on the real world, only a few tackle the changing habitat where the microorganisms are living [12]. This dynamic relationship makes them more similar to ABMs. They are defined as dynamic optimization tools which are an important focus point in research [13–15]. They consider that surroundings affects organisms but, in turn, they also affect the surroundings making them an important research point [16].

On the other hand, agent based modeling (ABM) is based on analyzing behaviors, emergence and adaption of complex systems with the notion that these systems are built from the bottom up. Their beginnings can be traced to cellular automata [17] and based on those simple principles numerous ABM usages have spawned, ranging from sociology [18], economics [19] and political science [20]. Even though both approaches are a part of MAS, both have intelligent agents and their results are based on their interactive behavior, combining or hybridizing them has not been done.

Bacteria within the BFO move around the environment while searching for food, consume it and move on to other areas. While this phenomena is interesting from the optimization standpoint, ABM is focused on the bacterial behavior and movement patterns. The primary goal of this paper is to create an architecture, mathematical representation and operation procedures in order to bridge the gap between optimization algorithms, namely the BFO, and agent based modeling, we will refer to it as BFO-ABM. Secondary goal is to create a method to enable 3D visualization of the BFO algorithm. To construct it, certain changes to the original are proposed: Creation of a 3D environment where the bacterial agents move and forage. Harnessing the environment (crucial element of the ABM) which the optimizing bacteria can inhabit (crucial element of the BFO). Creation of a methodology which can help in visualizing BFO operators and actions previously described only in writing. Employing the BFO's bacteria, through the ABM prism, as agents representing anything from companies to people.

To show performance and test the modeling capabilities of the BFO-ABM hybrid, we conducted two experiments on an explicit objective function which presents the environment. The first experiment is based on observing the bacterial agents' movement and behavior, disregarding the decline in nutrients and toxicity levels. Second experiment is based on observing bacterial agents' behavior until all the nutrients are completely gone while avoiding the toxic areas. In these two scenarios bacterial health is observed together with the speed of convergence and their effect on the objective function (the environment). This paper is organized as follows: Sect. 2 presents the formation of the environment around which the bacteria move and interact with. Section 3 describes the

BFO-ABM hybrid along with different bacterial actions. Section 4 presents results of two experimental studies together with discussions and potential uses of further modeling, followed by the conclusion in Sect. 5.

2 Environment

Optimization algorithms are usually not visualized, but if we wish to understand their movement we should transfer them to 2 or 3 dimensions. This is environment creation is the crucial step of the BFO-ABM hybrid since it enables the first element to be implemented into the second one. The algorithm envisions a space for activity in the form of a classical grid topology, called the Moore neighborhood, Fig. 1. Bacterial agents are grounded on the grid and are able to follow its topological change in 3D.

The algorithm is consisted out of two basic types of elements: habitat grid and the bacteria. They exist together and effect one another - they are in sync at every moment. As the environment possesses certain positive elements (food, nutrients) and negative ones (toxins and/or no food) the organisms are forced to adapt to it. The change comes from them eating the available food, where it exists and leaving it foodless and uninteresting. As the current is left empty they search for the next best place to forage, which contains the highest quantity of available food.

Grid. The grid covers a 3D space of X × Y × Z through which movement is possible but only across the X × Y grid surface, while their action and implications take effect and are noted on the Z axis. This can be represented as:

$$Grid_t = \{\beta, X, Y, Z \quad (1)$$

Where the $Grid_t$ depicts the habitat over which the agents move at a certain moment t while β is the amount of food which is distributed around it. X And Y are dimensions of the grid which are previously defined and are stationary while Z is a dynamic grid term, which is being changed by the foraging process.

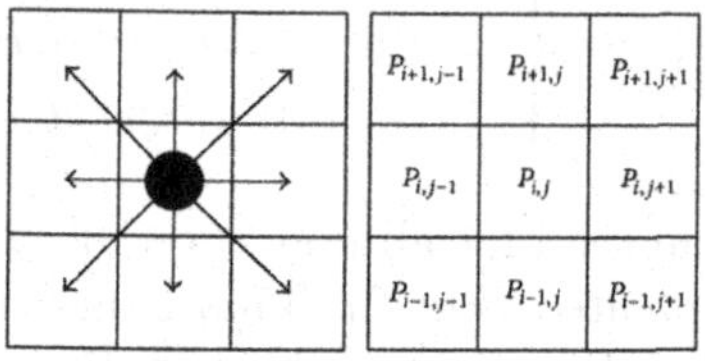

Fig. 1. Moore neighborhood movement possibilities [21]

Bacteria. Each bacteria can be defined as a set of its current characteristics. This means that a bacterium B at a certain time point t is defined and identified by its current position on the grid in the form θ_{xyz} coordinates but also the current iteration step.

$$B = \{\theta_{xyz}, iter \quad (2)$$

All food is randomly distributed across the grid with coordinates x and y and is consumed over t time instances. In both tested versions of the algorithm the food is not created over the course of iterations.

3 Employing the BFO as an ABM

Formation of the Habitat. When constructing the habitat multiple factors need to be taken into account. Analyzing these factors is key for understanding it and extracting some interesting waypoints from it. Surroundings can be presented as a combination of terrain elements, toxic areas, but also areas which are filled with food. We can present the movement space as the Z axis together with its parts:

$$Z = Z_{terrain} + Z_{toxicity} + Z_{food} \quad (3)$$

The bacteria need to consider all of these factors and navigate through them. They will choose to visit and interact with areas which possess the easiest way and the most food – this implies that the minimum of the function is there. To form the Z axis all of the layers are combined into one by addition of their values.

This means that there are certain tradeoffs which need to be considered. Such as: if there is a lot of food at a local area but the terrain is inhospitable, there will be no interest in going there; if there is food but the area is mildly dangerous the overall success needs to be compared with other areas to see if pursuing that area is the best option; and finally if the area is easy to get to and there is a lot of food it will be very interesting to visit. Mathematically we can present it as different points ($A^{Z1}_{x,y}, A^{Z2}_{x,y}, A^{Z3}_{x,y}$) which have the same coordinates but are on different axes, their impact on the Z axis will be calculated as follows:

$$A^{Z1}_{x,y}, A^{Z2}_{x,y}, A^{Z3}_{x,y} = \begin{cases} if sum > 0, interest\ is\ low \\ if sum = 0, interest\ is\ neutral \\ if sum < 0, interest\ is\ high \end{cases} \quad (4)$$

After having this calculation for all points on all axes the summary of them in the form of Z axis is created (Fig. 2). This is also done after each iteration step since the consumption of nutrients reduces their availability in certain areas after each step.

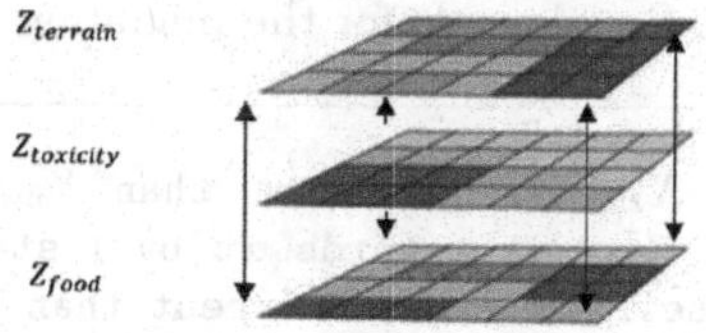

Fig. 2. Z axis formation

Initialization. Bacteria move through and across space which can be created implicitly, using a formula, or explicitly by directly creating it in Excel, for example. Major advantage of the explicit finite method is its relative simplicity and computational ease, on the other hand the implicit environment creates better solutions. In the case of this experiment an explicit one with 20×20 fields is used. Amount of food β is distributed across the environment grid thus forming the Z axis minima. Initial positions of agents at $t = 0$ are randomly generated.

Operators. Bacteria move around their habitat by using their flagella (a whip-like part of their body) with the goal of escaping hazardous and foodless areas and finding ones with food. This process is called chemotaxis but the BFO-ABM considers a Migration term, which will be described further in the text. If a bacterium does not "consume" enough food or is not healthy enough it will die. To keep the population constant the organisms will *reproduce* by dividing into two identical copies which are positioned at the same location. This keeps the population diversified and eliminates some bacterium which are trapped in local minimum. If greater numbers of them are concentrated around a certain food source they will consume it food faster. *Dispersion* is employed to make the population more dynamic and to stimulate exploration and foraging in different areas.

Migration. Migration is defined as moving across the physical and dynamic space whose action requirements are presented by pseudocode in the Table 1. First task is to find the closest minimum to each bacterium, the minimum is defined through its X_{min} and Y_{min} coordinates. After this, the comparison between the current coordinates of the bacterium (X_{Bac},Y_{Bac}) starts. If the coordinates of the food and the bacterium match, then the bacterium is at the right place and the feeding can begin. Important thing to note is that in this algorithm *chemotaxis* operator helps agents move around a local area, migration considers covering of greater - global distances. To conserve energy the bacteria will choose locations where food is plentiful. This implies a greedy selection and helps with focusing on potential global minima.

Agent Waypoints. In the presented case, visualization of nutrient abundance is presented using lower values of the Z axis while the higher values do not possess any food, maybe because it has already been consumed, thus that area

Table 1. Pseudo code for the migration function

```
For I = 1:S
 If X_Bac is different than X_min
    Move it towards it by 1 step
 elseif Y_Bac is different than Y_min
    Move it towards it by 1 step
 End
End
```

is not interesting. That is how the BFO-ABM harnesses the bacterial interest and uses it as an agent modeler. The agents consider these low locations as waypoints or areas of interest to move about. The pseudo code is based on the current location (X_{Bac},Y_{Bac}) and the memorization of the *History* - where the agents have already been. Also important is the $History^{end}$ which memorizes the whole historical path of them but without the last position. This means that the bacterium influences the environment only if it has spent some time at the location, if it has not (meaning that the coordinates of the historical paths and the current position do not match) the bacterium is just passing by and has visited that place for the first time. This comparison and exclusion is made because we want to find out what are the areas of interest and at the same time exclude the path, because we presume it does not possess food. If the coordinates match then a *speed of consumption* value is added to the Z axis point reducing the interest which the bacteria have for that point, in other words the food located at that point is reduced by that amount by a single bacterium foraging activity.

Stopping Criteria. The algorithm stops because of two reasons. First one being the iteration counter reaches its maximum and all the operators have been used. Second reason is that all food in the environment has been consumed (Table 2).

Table 2. Pseudo code for the consumption function

```
For I = 1:S
    History_X(I)=X_Bac && History_Y(I)=Y_Bac
    History_X^end=length(History_X)-1 && History_Y^end=length(History_Y)-1
      If History_X - History_X^end == 0 && History_Y - History_Y^end == 0
        Z(X_Bac,Y_Bac) == Z(X_Bac,Y_Bac) +1
      End
End
```

4 Experimental Studies

To demonstrate the algorithm two scenarios are taken into account. As a control in the experiment the grid considered will be the same size and characteristics, both will have the same number of minima which are corresponding to the amounts of nutrients. It is created using an explicit method based on Excel tables, each corresponding to a specific value of terrain characteristics, toxicity location and food distribution; they are summarized into a single value of the Z axis which size is 20×20 fields. Figure 3 shows the initial shape of the environment grid. The minima are marked as lower values than 7 while values equal to 7 present foodless area which are of no interest. Bacteria will search for food located in the minima. Corresponding values can be found in the Table 3. After the initialization the agents start their foraging activity and their actions take effect.

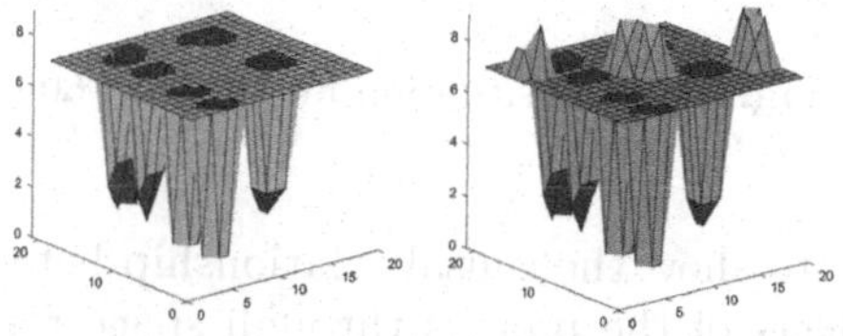

Fig. 3. Initial values of the Z axis for the two experiment scenarios.

4.1 Simulation I

Population for the first experiment is set to 5. Speed of nutrient consumption is set to 0.001. Stopping criterion of the algorithm, relies on the number of iterations and is set to 100. Figure 4 shows a moment at the 30th iteration; bacteria are converging to the absolute minimum which is closest to them. From the moment of entering the minimum they begin foraging for food. This influence and interaction is also visible in the table through the increase of mean, standard deviation and minimum values. Since there are no toxic areas around the environment the bacteria can take the shortest (optimal) route.

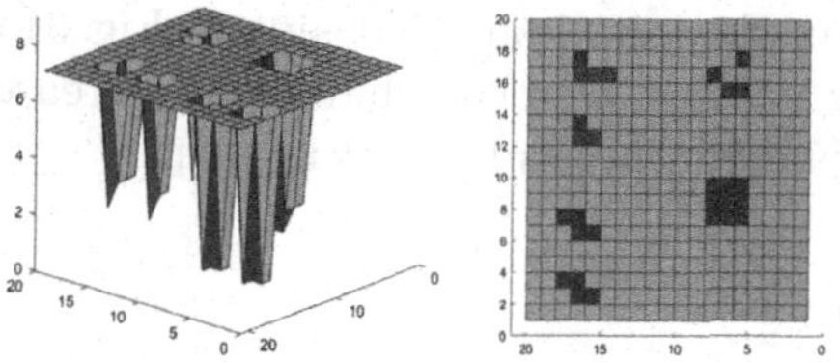

Fig. 4. Z axis situation at iteration 30

On the Fig. 5 we can see gradual warping of the terrain due to the increased foraging. The first minimum has been foraged enough (it is no longer a minimum because of terrain increase) so there is no more food there and other sources are being searched for. Compared to Fig. 4 we can see food sources are getting exhausted and bacteria need to move to other new ones. Note that the all bacteria are positioned on the same coordinates making them a more effective swarm.

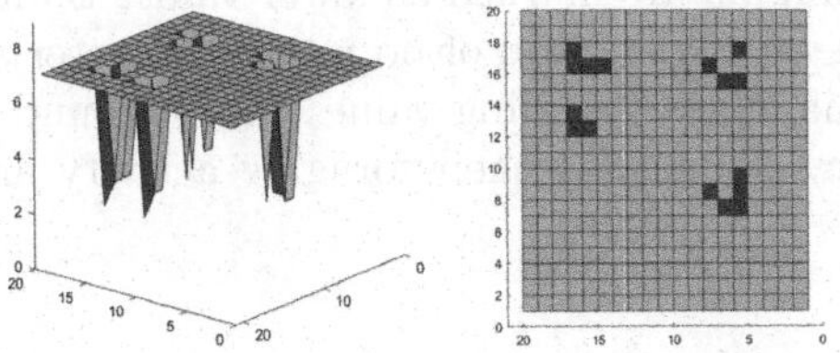

Fig. 5. Z axis situation at iteration 60

Figure 6 continues to show the causal relationship between the bacteria. We see the increasing effects of the process through space warping. Also visible is the gradual change of the minima's color, from dark blue to lighter shades. This is because the lack of food is presented as greater numbers, compared to the global minimum which are darker colors.

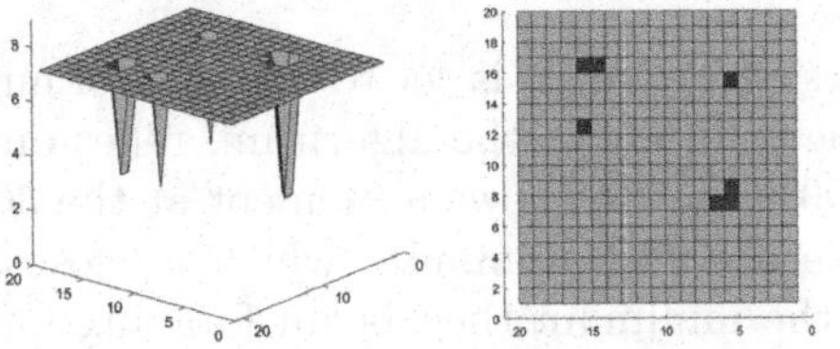

Fig. 6. Z axis situation at iteration 90 (Color figure online)

Final result of the interplay process can be seen on the Fig. 7. We see a very different situation than the initial one (left side of Fig. 3) all minima have been affected because their values have been changed – increased numerically due to the consumption of food which was located in them.

4.2 Simulation II

The second simulation features an increased number of bacteria which is set to 10 together with the incorporation of toxic areas. Initial Z axis shape and values can be seen on the right part of the Fig. 3. Since lower levels of the axis present positive elements, food, negative elements are presented as higher values; bacteria tend to evade them while foraging. The simulation shows how more subjects act faster thus having a greater impact and achieving exploration and exploitation of minima in a shorter time period, regarding the number of iterations. This is due to more competition and the same food level making

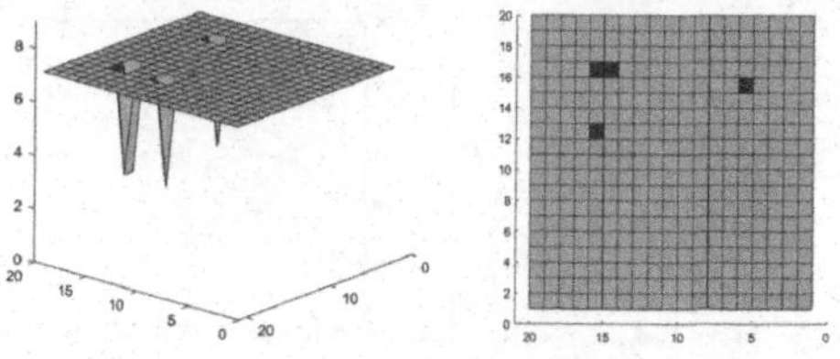

Fig. 7. Z axis situation at the final iteration

dynamics and behaviors different. Toxic levels need to be avoided which changes the path of the bacteria, making it a bit longer.

In the second simulation the stopping criteria was not the iteration limit but the food depletion in the habitat. That is why there are no numbers lower than 7, which exist in the first experiment. As in the final result of the first simulation we can see a great difference and effects which were left by the process (Fig. 8). Also noticeable is the rate and convergence speed which is much greater than it the first experiment; it can be seen on the Z axis. The experiment shows that more agents act quicker and also that their efficiency is increased by swarming and taking effect on a single location.

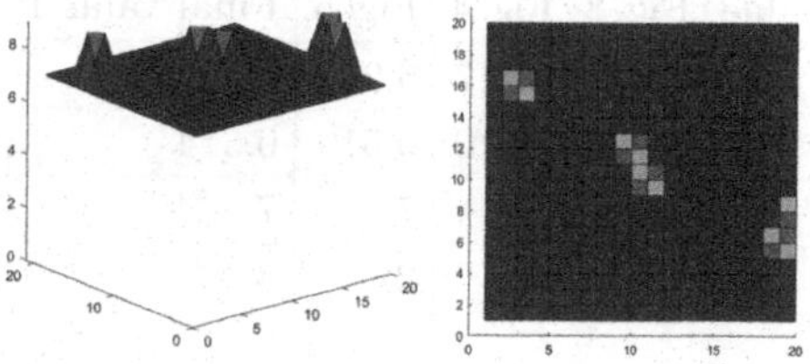

Fig. 8. Z axis situation after all food is consumed

4.3 Discussion

Figure 9 shows the explicit objective function which is being optimized by the interacting bacteria. From it we can observe consumption to movement ratios. Namely, if the line is horizontal it shows the bacteria are moving to the next minimum and during that movement they are not consuming. If the line is steeper than the horizontal one it shows food consumption happening, steeper line means more consumption per unit of time. Left side of the figure shows 5 and the right one 10 bacteria interacting. We can conclude that greater impact is being achieved with 10 bacteria since the conversion rate is faster, even though sometimes they take a longer route to avoid the toxic areas. The line would be also much steeper if food sources were close together or in other words, if the objective function (the environment) was of different shape and value.

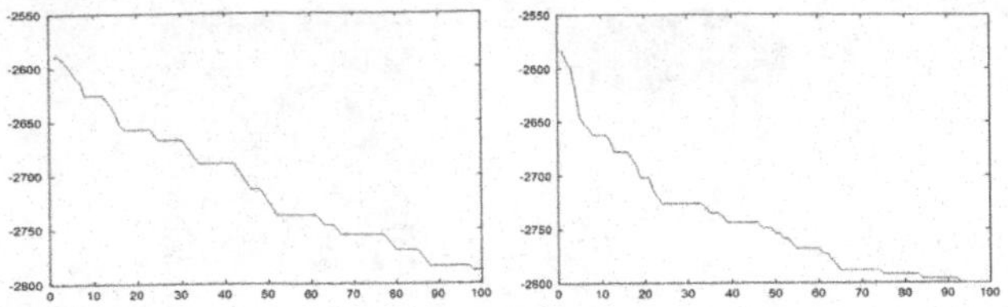

Fig. 9. Objective function from the two simulation scenarios

Table 3 summarizes all values of the Z axis from different experiment steps. In the first experiment the mean and standard deviation are being increased through the foraging process. Mean value has been increased from 6.475 to 6.959. During the course of the experiment standard deviation is being steadily lowered due to the foraging process, 0 will signal that there is no more food available. Final value from the first experiment presents the global minimum which has been increased, from 1 to 3. This is because all minima have been affected by the foraging.

Table 3. Numerical values from the simulation

Result	First value	Fig. 3	Fig. 4	Fig. 5	Final value I	Final value II
Mean	6.475	6.642	6.817	6.922	6.959	7.075
Std	1.008	0.914	0.710	0.519	0.374	0.271
Max	7	7	7	7	7	9
Min	1	1	2	2	3	7

Second simulation has a different stopping criteria than the first one food depletion in the habitat and also toxicity levels are visible. We see that the mean has increased to 7.075 and standard deviation has been lowered to 0.271 signaling deficiency of food but also that the toxicity levels are still active. The overall maximum is 9 which also points to existing dangerous areas for the bacteria. On Fig. 10 we can see how different bacteria have different fitness levels and how it changes over time. It is interesting to note in both experiments that there are several leaders forming, they come to the food source first and forage the most, thus the others do not have the opportunity to eat.

4.4 Bacteria as Agents

It is important to grasp how useful can this BFO-ABM approach be. For example we can consider these agents as companies which need to position themselves on the market to have the best access to the customer base. The first company to set up its business at a market quadrant usually wins, but the classical question

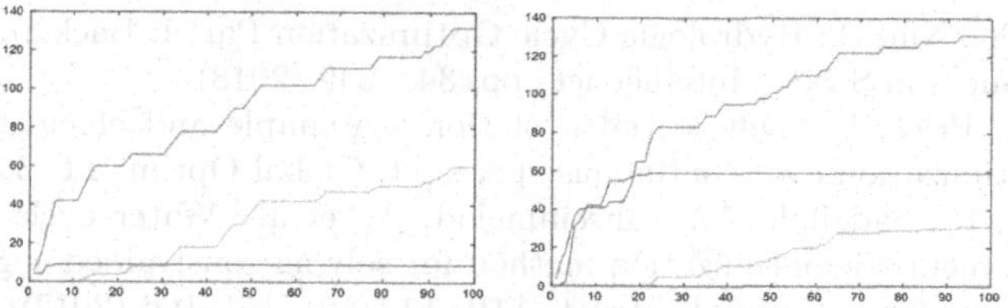

Fig. 10. Individual bacterial fitness from the two simulation scenarios

is whether it is better to be focused only on certain customers or be ready to move about and be more flexible; similarly done in [22].

Other consideration for the bacterial agents can be simulating the movement of people around a certain space, they cannot move through walls and barriers (presented as high toxicity levels); similar simulation is done in [23]. To simulate a fire escape simulation we place the food outside of the room, bounded by "toxicity" serving as walls. Then we can check how bacterial agents act depending on the danger and the environment, as seen in [21]. Apart from these applications the BFO-ABM approach can consider and simulate numerous other agents and scenarios no matter the size or intention.

5 Conclusion

Use of swarm optimization algorithms and agent based modeling is usually kept separate. This is understandable since behind them are different paradigms of operation. Apart from these differences, there are numerous similarities, because they are both from the multi agent systems field. Based on those similarities, this paper has presented a new architecture and framework which is used for modeling bacterial behavior within a 3D environment. It also showed how to seamlessly incorporate bacteria to act according to the ABM rules which makes this approach a hybrid one. The whole environment is presented as an objective function which needs to be optimized by bacterial consumption of food while at the same time avoiding the toxic areas.

Different simulation scenarios showed how bacteria can be manipulated and motivated with food and toxicity to act in a certain agent-like way. Even though the presented environment and scenarios are relatively simple, it shows that they can be used as an effective agent modeling tool. Our future work will be based on using these bacteria in simulating behaviors of companies, vehicles or people and applying them to real world situations.

References

1. Kennedy, J., Eberhart, R.C.: Particle swarm optimization. In: Proceedings of the 1995 IEEE International Conference on Neural Networks, vol. 4, pp. 1942–1948 (1995)

2. Xiaohui, Y., Niu, B.: Hydrologic Cycle Optimization Part I: Background and Theory, Advances in Swarm Intelligence, pp. 341–349 (2018)
3. Storn, R., Price, K.: Differential evolution - a simple and efficient heuristic for global optimization over continuous spaces. J. Global Optim. **11**, 341–359 (1997)
4. Eskandar, H., Sadollah, A., Bahreininejad, A., et al.: Water cycle algorithm - a novel metaheuristic optimization method for solving constrained engineering optimization problems. Comput. Struct. **110–111**(10), 151–166 (2012)
5. Dervis, K., Bahriye, A.: A comparative study of Artificial Bee Colony algorithm. Appl. Math. Comput. **214**(1), 108–132 (2009)
6. Holland, J.: Genetic algorithms. Sci. Am. **267**(1), 66–72 (1992)
7. Dorigo, M., Birattari, M., Stutzle, T.: Ant colony optimization artificial ants as a computational intelligence technique. IEEE Comput. Intell. Mag. **1**(4), 28–39 (2006)
8. Passino, K.M.: Biomimicry of bacterial foraging for distributed optimization and control. IEEE Control Syst. Mag. **22**, 52–67 (2002)
9. Niu, B., Fan, Y., Wang, H., Li, L., Wang, X.: Novel bacterial foraging optimization with time-varying chemotaxis step. Int. J. Artif. Intell. **7**(11), 257–273 (2011)
10. Fernandes, C., Ramos, V., Rosa, A.C.: Varying the population size of artificial foraging swarms on time varying landscapes. In: Duch, W., Kacprzyk, J., Oja, E., Zadrożny, S. (eds.) ICANN 2005. LNCS, vol. 3696, pp. 311–316. Springer, Heidelberg (2005). https://doi.org/10.1007/11550822_49
11. Chen, H.N., Zhu, Y.L., Hu, K.Y.: Adaptive bacterial foraging algorithm. Abstract Appl. Anal. **2011** (2011). Article ID 108269
12. Tang, W.J., Wu, Q.H., Saunders, J.R.: Bacterial foraging algorithm for dynamic environments. In: IEEE Congress on Evolutionary Computation (2006)
13. Morrison, R.W., De Jong, K.A.: A test problem generator for non - stationary environments. In: Proceedings of the 1999 IEEE Congress on Evolutionary Computation, pp. 2047–2053. IEEE Press (1999)
14. Branke, J.: Evolutionary Optimization in Dynamic Environments. Kluwer Academic Publishers, Massachusetts (2002)
15. Tang, W.J., Wu, Q.H., Saunders, J.R.: A novel model for bacterial foraging in varying environments. In: Gavrilova, M., et al. (eds.) ICCSA 2006. LNCS, vol. 3980, pp. 556–565. Springer, Heidelberg (2006). https://doi.org/10.1007/11751540_59
16. Daas, M.S., Batouche, M.: Multi-bacterial foraging optimization for dynamic environments. In: International Conference of Soft Computing and Pattern Recognition (2014)
17. Conway, J.: The game of life. Sci. Am. **223**(4), 4 (1970)
18. Bruch, E., Atwell, J.: Agent-based models in empirical social research. Sociol. Meth. Res. **44**(2), 186–221 (2015)
19. Tesfatsion, L., Judd, K.L. (eds.): Handbook of Computational Economics: Agent-Based Computational Economics, vol. 2. Elsevier, Amsterdam (2006)
20. Cederman, L.E.: Computational models of social forms: advancing generative process theory. Am. J. Sociol. **110**(4), 864–893 (2005)
21. Wang, C., Wang, J.: A modified floor field model combined with risk field for pedestrian simulation. Math. Probl. Eng. **1**(10) (2016)
22. Van, L.E., Lijesen, M.: Agents playing Hotelling's game: an agent-based approach to a game theoretic model. Ann. Reg. Sci. **57**(2–3), 393–411 (2015)
23. Tecchia, F., Loscos, C., Conroy-Dalton, R., Chrysanthou, Y.L.: Agent Behavior Simulator (ABS): a platform for urban behavior development. In: First International Game Technology Conference and Idea Expo (GTEC 2001) (2001)

A Modified Memetic Algorithm for Multi-depot Green Capacitated Arc Routing Problem

Bin Cao[1,2,3](✉), Ruichang Li[1,2,3](✉), and Shuai Chen[1,2,3](✉)

[1] Hebei University of Technology, Tianjin 300401, China
caobin@scse.hebut.edu.cn, {201731704023,201731704047}@stu.hebut.edu.cn
[2] Key Laboratory of Intelligent Perception and Image Understanding of Ministry of Education, Xi'an 710071, China
[3] Hebei Provincial Key Laboratory of Big Data Calculation, Tianjin 300401, China

Abstract. The capacitated arc routing problem (CARP) has important research significance in the area of vehicle routing services. We study a multi-depot green capacitated arc routing problem by minimizing economic cost, makespan and carbon emission cost, and then, present a modified memetic algorithm for multi-depot (MMAMD) that aims to solve the problem efficiently. According to the characteristics of multi-depots in this paper, a specific encoding scheme via the two-dimensional array and the dynamic adjustment strategy of boundary arcs between depots are proposed. By combining ant colony algorithm with local search, we present an ant colony local search strategy with extended neighborhood search for the memetic algorithm. The experimental results show that the proposed algorithm which is used to solve the model is effective, which can provide a reference for waste clearing decision.

Keywords: Carbon emission · Capacitated arc routing problem · Multi-depot · Multi-objective · Memetic algorithm

1 Introduction

In recent years, energy and environment have become hot topics in the world. Carbon dioxide is one of the main sources of greenhouse gases. Transportation accounts for 14% of global carbon emissions, while road carbon accounts for 70% of the entire transport industry [1]. Green transportation has become an inevitable trend to reduce carbon emissions. Besides, due to the rapid growth of urban population and the expansion of urban areas, the amount of municipal solid waste is increasing. How to optimize the collection route of waste vehicles in multi-depots, and how to reduce the pressure of waste collection and management, as well as the carbon emissions brought by waste collection and transportation, are of great significance to environmental governance.

The problem of waste disposal is usually dealt with as a vehicle routing problem (VRP) or capacitated arc routing problem (CARP). Because waste

L. Pan et al. (Eds.): BIC-TA 2019, CCIS 1159, pp. 739–750, 2020.
https://doi.org/10.1007/978-981-15-3425-6_59

cans are distributed on both sides of city streets, the urban waste disposal is an arc routing problem in practical application. CARP is a typical VRP, which is proposed by Golden [2]. On this basis, there are a lot of research achievements. To solve the waste collection problem in Truva, France, Lacomme et al. [3] presented a bi-objective CARP model with the total cost and the makespan. Filippi et al. [4] proposed a cooperative CARP model of two kinds of vehicles, large transport vehicle and small collection vehicle, which used the convenience of small vehicles to clean the streets. Liu et al. [5] proposed the CARP model of the heterogeneous fixed fleet with multi-depots. Krushunsky et al. [6] proposed a two-index MLIP model for asymmetric multi-depot CARP. Some scholars believed that the traditional vehicle transportation problem should consider not only the economic benefits but also consider carbon emissions from the social and environmental aspects. Erfan et al. [7] proposed a single-objective CARP model considering carbon emission costs and economic costs in the process of municipal waste collection for the single depot. Ge et al. [8] studied the open pollution path and put forward the calculation method of carbon emissions.

Capacitated arc routing problem is a typical NP-Hard problem, which is usually solved by the heuristic algorithms. Mei et al. [9] proposed a memetic algorithm based on the decomposition framework (D-MAENS). Shang et al. [10] studied the routing grouping method of bi-objective CARP and proposed an improved IRDG-MAENS algorithm. Amberg et al. [11] proposed a heuristic algorithm to transform the arc routing problem into the minimum spanning tree for multi-depot CARP. Zhu et al. [12] proposed a hybrid genetic algorithm to solve multi-depot CARP.

This paper studies the multi-depot capacitated arc routing optimization problem and constructs a multi-depot multi-objective optimization model including economic costs, makespan and carbon emission costs. Further, a modified memetic algorithm for multi-depot (MMAMD) is proposed. To solve the multi-depot problem, a two-dimensional array solution structure and a dynamic adjustment strategy of boundary arcs between depots are proposed. Besides, an ant colony local search strategy with extended step size is proposed to improve the local search performance of the memetic algorithm.

2 Problem Description and Model Construction

2.1 Problem Description

The capacitated arc routing optimization problem of urban waste collection in the multi-depots considering carbon emissions can be described as follows. The city has multiple waste collection centres, which can send different numbers of waste vehicles to clean and collect the waste on both sides of the road. Each waste vehicle has the same capacity and returns to the original waste depot if it reaches the maximum capacity. The schematic diagram of basic waste collection is shown in Fig. 1.

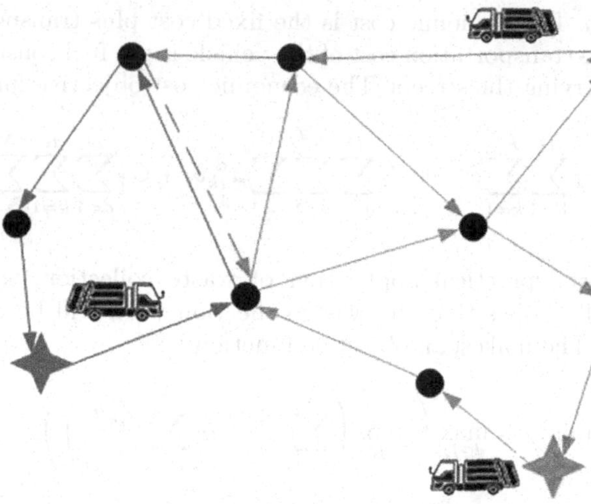

Fig. 1. The schematic diagram of basic waste collection.

The waste collection and transportation problem can be represented by a mixed graph $G = (V, E, A)$, where V is the intersection of the streets, S represents the arc and A stands for the side. During the waste collection process, the vehicle complies with the following assumptions: (1) All streets can be passed by different vehicles, but only one street is served by one vehicle. (2) The amount of waste collected by each vehicle cannot exceed the maximum capacity Q. (3) The vehicle starts from the depot and returns to the original depot after the waste collection is completed.

2.2 Symbol Description

The symbol description is as follows: N: the number of streets in the city; D: the number of waste depots in the city; K^d: the number of vehicles in the depot d; C_1: fixed cost of utilizing the vehicle; C_2: service cost per meter of the vehicle; C_3: the passing cost; q_i: the amount of waste in the street i; b_i: the length of the street i; Q: the maximum capacity of the vehicle; v_1: the speed at which the vehicle serves a street; v_2: the speed at which the vehicle passes through a street; $\rho_1(Q_i)$: the fuel consumption where the vehicle serves street i with load Q_i; $\rho_2(Q_i)$: the fuel consumption when the vehicle passes through street i with load Q_i; C_4: the environmental cost of releasing CO_2 per kilogram; ω: the emission factor of CO_2; Q_i: the weight of the vehicle driving on the street i; $x_{ik}^d = 0$ or 1: the street i is served by the vehicle k of the depot d; $y_{ik}^d = 0$ or 1: the street i is passed by the vehicle k of the depot d;

Economic Cost. The economic cost is the fixed cost plus transportation cost of the vehicle. The transportation cost of the vehicle is the fuel consumption cost for passing and serving the streets. The economic cost objective function is:

$$\min f_1 = C_1 \sum_{d=1}^{D} \sum_{k=1}^{K^d} x_{ik}^d + C_2 \sum_{d=1}^{D} \sum_{k=1}^{K^d} \sum_{i=1}^{N} x_{ik}^d b_i + C_3 \sum_{d=1}^{D} \sum_{k=1}^{K} \sum_{i=1}^{N} y_{ik}^d b_i \quad (1)$$

Makespan. In the practical application of waste collection, the municipal department usually hopes that the waste collection work will be completed as soon as possible. The makespan objective function is:

$$\min f_2 = \max_{d \in D} \left(\max_{k \in K^d} \left(\sum_{i=1}^{N} \frac{x_{ik}^d b_i}{v_1} + \sum_{i=1}^{N} \frac{y_{ik}^d b_i}{v_2} \right) \right) \quad (2)$$

Carbon Emission Cost. The capacitated arc routing problem is different from the vehicle routing problem, and it is necessary to consider not only the carbon emission cost during passing but also the carbon emission cost during servicing. The carbon emission cost objective function is:

$$\min f_3 = C_4 \omega \left(\sum_{d=1}^{D} \sum_{k=1}^{K^d} \sum_{i=1}^{N} x_{ik}^d b_i \rho_1 (Q_i) + \sum_{d=1}^{D} \sum_{k=1}^{K^d} \sum_{i=1}^{N} y_{ik}^d b_i \rho_2 (Q_i) \right) \quad (3)$$

Constraints. The constraints of the objective functions are:

$$\sum_{i=1}^{N} x_{ik}^d q_i \leq Q, \quad \forall k, d \quad (4)$$

$$\sum_{d=1}^{D} \sum_{k=1}^{K^d} x_{ik}^d = 1, \quad \forall i \quad (5)$$

$$\sum_{d=1}^{D} \sum_{k=1}^{K^d} y_{ik}^d \geq 1, \quad \forall i \quad (6)$$

$$y_{0k}^d = y_{Nk}^d, \quad \forall k, d \quad (7)$$

where Eq. (4) defines that the total load of any vehicle cannot exceed the vehicle's largest capacity. Equation (5) means that each street can only be served once. Equation (6) allows the street to be passed many times. Equation (7) expresses that each vehicle starts from the depot, and returns to the original depot after the waste collection is completed.

3 The Design of MMAMD for CARP

The memetic algorithm is an optimization algorithm that combines genetic mechanisms and local search algorithm. The algorithm inherits the global search ability of genetic algorithm, but the local search is relatively weak when solving the routing optimization problem. Ant colony algorithm is robust, and by introducing the ant colony local search strategy, the local search of the memetic algorithm is improved.

3.1 Encoding Design

The numbers of vehicles are as vaired as the depots they are in. Therefore, a two-dimensional array solution structure is proposed in the algorithm. By adopting the solution structure form, the vehicle service route information of different depots can be represented. The encoding design is shown in Fig. 2:

d_1	5	3	12	d_1	1	17	d_1	9	11	d_1
d_2	8	4	13	15	d_2					
d_3	6	10	d_3	2	7	14	16	d_3		

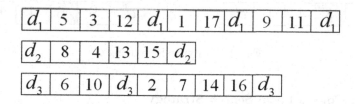

Fig. 2. Two-dimensional array solution structure.

Figure 2 shows a solution with three waste depots. d_1, d_2, and d_3 represent three waste depots. The first row $[d_1,5,3,12,d_1,1,17,d_1,9,11,d_1]$ means that the vehicle starts from the depot d_1, cleans the streets 5, 3 and 12, and returns to the waste depot. The second car cleans the streets 1, 17 and the third car cleans the streets 9, 11. The second row and third row indicate service situations of the vehicles from depot 2 and depot 3.

3.2 Dynamic Adjustment of Boundary Arcs

To deal with this problem, this paper adopts the dynamic adjustment of the boundary arcs, which includes 0-1 dynamic exchange and 1-1 dynamic exchange, as shown in Fig. 3 and Fig. 4.

0-1 Dynamic Exchange. Select randomly arc l_j from the boundary arc table *Tabu*. Record the depot r_1 which l_j belongs to and the adjacent depot r_2. The arc l_j in the task sequence r_1 in the individual x_s is deleted, and the arc l_j is randomly inserted into the task sequence r_2. Thus, a new individual x'_s is generated. If $g(x'_s) < g(x_s)$, update x_s and *Tabu*, i.e. l_j belongs to the depot r_2 and r_1 is the new adjacent depot.

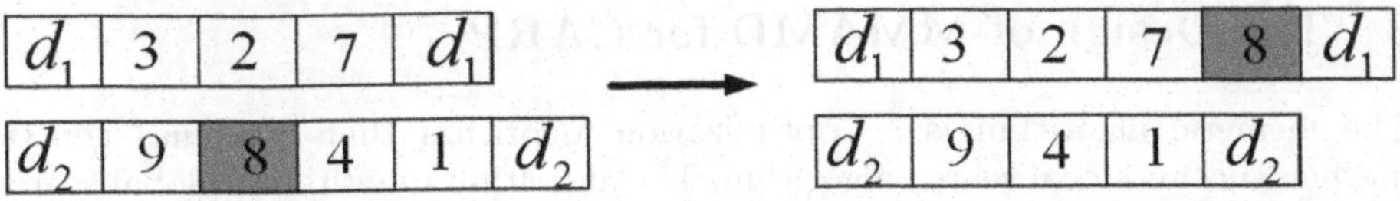

Fig. 3. 0-1 dynamic exchange.

1-1 Dynamic Exchange. From the boundary arc table *Tabu*, select two arcs l_j and l_z that meet the following conditions: the arc l_j belongs to depot r_1 and the adjacent depot is r_2, while l_z belongs to depot r_2 and the adjacent depot is r_1. And then swap the positions of l_j and l_z. Thus, a new individual x_s' is generated. If $g(x_s') < g(x_s)$, update x_s and *Tabu*.

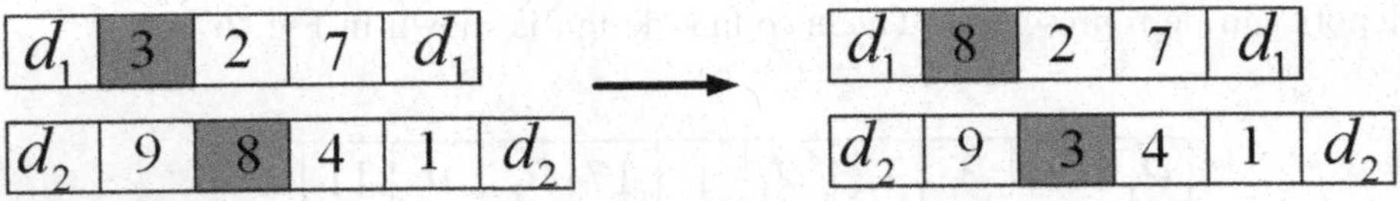

Fig. 4. 1-1 dynamic exchange.

3.3 Ant Colony Local Search Strategy

When an ant colony searches for food in the given map, the ant releases the pheromones on the path it passes, and the next ant will have a higher probability of moving to the places with higher pheromones. The path selection probability is calculated as follows [13]:

$$p_{i,j}^k = \begin{cases} \frac{(\tau_{i,j})^\alpha (\eta_{i,j})^\beta}{\sum_{j\in\Omega_j}(\tau_{i,j})^\alpha(\eta_{i,j})^\beta}, & \text{if } j \in \Omega_l \\ 0, & \text{other} \end{cases} \tag{8}$$

where $\tau_{i,j}$ is the pheromones from street i to street j, $\eta_{i,j}$ is visibility from street i to street$j$, Ω_l defines unserved streets, α and β are pheromone parameter and visibility parameter. The pheromone update formulas is calculated as follows [13]:

$$\tau_{i,j}(t+1) = \rho\tau_{i,j}(t) + (1-\rho)\Delta\tau_{i,j} \tag{9}$$

$$\Delta\tau_{i,j} = \sum_{d=1}^{D}\sum_{k=1}^{K^d}\Delta\tau_{i,j}^k \tag{10}$$

$$\Delta^k\tau_{i,j} = \begin{cases} \frac{W}{L}, & \text{if vehicle } k \text{ moves from street i to } j \\ 0, & \text{other} \end{cases} \tag{11}$$

where ρ indicates the degree of information volatilization and ρ satisfies $0 < \rho < 1$, $\Delta\tau_{i,j}$ is the amount of pheromone from street i and j, W is a constant and L is the arc length.

The ant colony local search strategy is devised as follows:

Step 1: Update pheromones according to formulas (9)–(11).
Step 2: Randomly select a center l from x_s. Then select p loops ($p > 1$). Put the selected street into the taboo table A_{tabu}.
Step 3: An ant starts from the depot l.
Step 4: Select the next street i to be reached from A_{tabu}. If A_{tabu} is empty, go to step 7.
Step 5: Compute the transition probability $p_{i,j}^k$, and use roulette selection strategy to select street j.
Step 6: If A_{tabu} isn't empty, reach the largest vehicle capacity and go to step 4.
Step 7: Record newly generated individuals x'_s. If $g(x'_s) < g(x_s)$, Update x_s.

3.4 Main Steps of MMAMD Algorithm

The main steps of MMAMD are as follows:

Step 1: Determine the size of population N. Set the local search probability pl and the maximum number of iterations G_{max}. Set ant colony related variables $\tau_{i,j}$, W, α, β, ρ. Set $Ite = 1$.
Step 2: Calculate the distances from each arc to each depot using Dijkstra algorithm. Record the distance $D_{i,1}$ from the arc i to the nearest depot and the distance $D_{i,2}$ from the arc i to the neighbor depot. Set a threshold θ. If $\theta < (D_{i,1}/D_{i,2})$, the arc is recorded in the boundary arc table $Tabu$.
Step 3: Initialize the population $X = (x_1, x_2, \ldots, x_N)$.
Step 4: Using a decomposition framework to generate weight coefficients $\lambda^1, \lambda^2, \ldots, \lambda^N$. Decompose the problem into sub-problems and we obtain that $g(x_s) = \lambda_1^s f_1 + \lambda_2^s f_2 + \lambda_3^s f_3$. Define neighborhood vectors B_s.
Step 5: Construct a one-to-one mapping and assign a representative solution to each sub-problems. Set $s = 1$.
Step 6: Randomly choose two parents and use simulated binary crossover (SBX) to generate a new individual x'_s.
Step 7: Randomly generate $rc \in (0, 1)$. If $rc < pc$, conduct dynamic adjustment of boundary arcs. Then randomly generate $bc \in (0, 1)$. If $bc < dc$, conduct $0-1$ dynamic exchange, otherwise conduct $1-1$ dynamic exchange.
Step 8: If $rl < pl$, perform ant colony local search for x'_s.
Step 9: New solution x'_s is saved to Y. $s \leftarrow s + 1$. If $s < N$, go to step 6.
Step 10: $X \leftarrow X \cup Y$. Use fast non-dominated sorting and crowding distance strategy as in NSGA-II.
Step 11: $Ite \leftarrow Ite + 1$. If $Ite < G_{max}$, go to step 4.
Step 12: Output non-dominated solutions.

4 Experimental Results and Comparison

4.1 Instance

To verify the effectiveness of MMAMD, an instance in literature [13] is used. The road network of Chicago is shown in Fig. 5, which has 94 nodes and 195 streets with distance of 44 km and three waste collection centres. The rated weight of vehicle is 9000 kg. No-load constant-speed fuel consumption is 16.5 L/100 km and full-load constant-speed fuel consumption is 37.7 L/100 km. Emission coefficient is 2.63 kg/L.

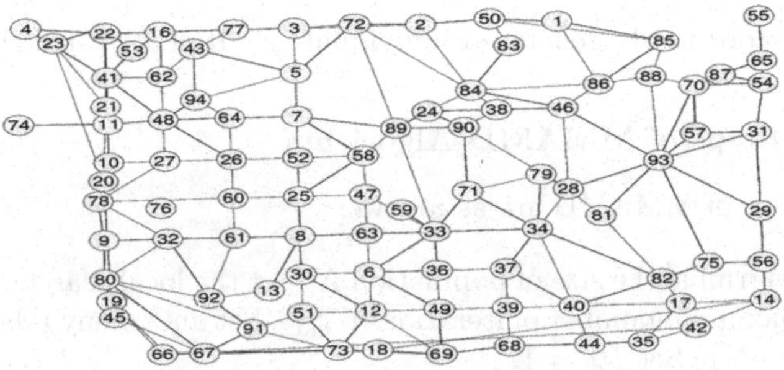

Fig. 5. Chicago city road network.

4.2 Comparison and Results

To make a comparison, all algorithms are implemented in C++. Set the maximum number of iterations to $G_{max} = 500$ and the population size to $N = 60$.

Table 1. Comparison of test results.

Algorithm	f_1^*/Yuan	f_1'/Yuan	f_2^*/minute	f_3'/minute	f_3^*/Yuan	f_3'/Yuan
MMAMD	2435	2471	18	21	688	698
IACO	2488	2501	22	23.2	704	721
MD-NSGA-III	2517	2527	20.1	22	705	719
MD-MAENS	2496	2510	21.6	22.5	702	713

To verify the efficiency of the proposed algorithm, MMAMD is compared with IACO [13], MD-MAENS [9], and NSGA-III [14], all of which run independently ten times. The Pareto front is generated as shown in Fig. 6, and the final results are listed in Table 1. From Fig. 6, it can be seen that MMAMD has more non-dominated solutions, which are more diversified and convergent. It shows that the algorithm has better optimization performance and can search for larger solution space. As can be seen from Table 1, the economic cost f_1' of MMAMD is 1.2%, 2.3% and 16% smaller compared with IACO, NSGA-III, and MD-MAENS. The makespan f_2' of MMAMD is 9.5%, 4.6% and 6.7% smaller. The carbon emission cost f_3' of MMAMD is 3.2%, 2.9% and 2.1% smaller. Therefore, it can be concluded that MMAMD algorithm has a better optimization performance. In Table 1, f_1^* represents the best solution and f_1' represents the average solution.

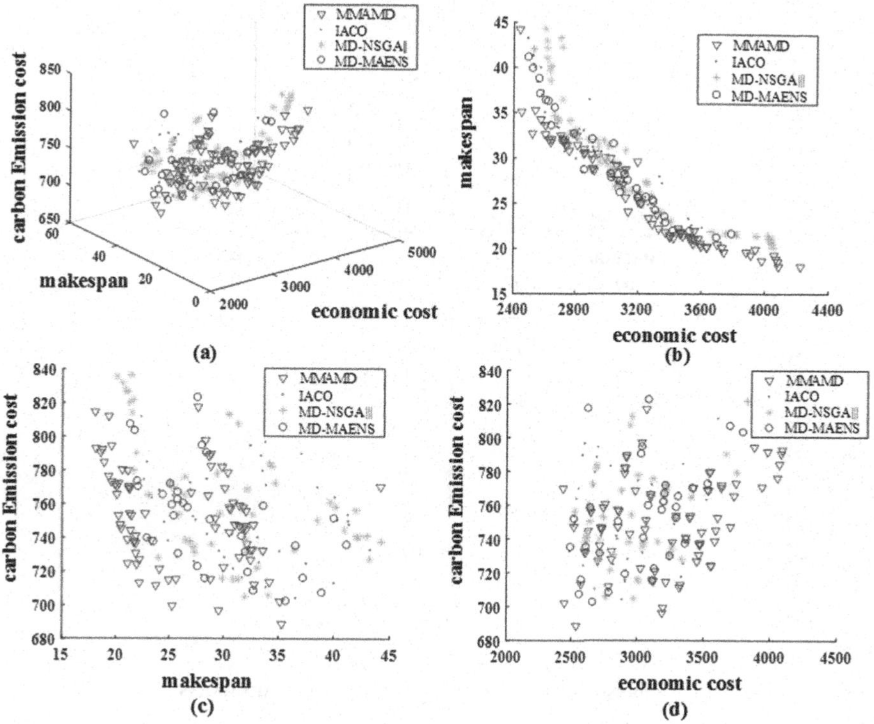

Fig. 6. Spatial distribution of solution.

This paper uses multi-objective visualization indicators [15] for better illustration and comparison. The indicator retains the shape, location and distribution of the non-dominated Pareto solutions. In Fig. 7, we can see that MMAMD algorithm has better performance than the other three algorithms in solving the problem.

To verify the applicability of the algorithm in solving multi-depot problems, instances with two depots, three depots, four depots and five depots are used to carry out experiments. Because the real Pareto solutions of these instances are unknown, this paper chooses HV test index [16], which can evaluate the convergence and diversity of the algorithm, and the bigger the HV value gets, the better the result is. The algorithm runs 20 times independently. We calculates the HV value of each run, and takes the average value as the evaluation criterion of each algorithm. As shown in Table 2, MMAMD has better performance for all cases.

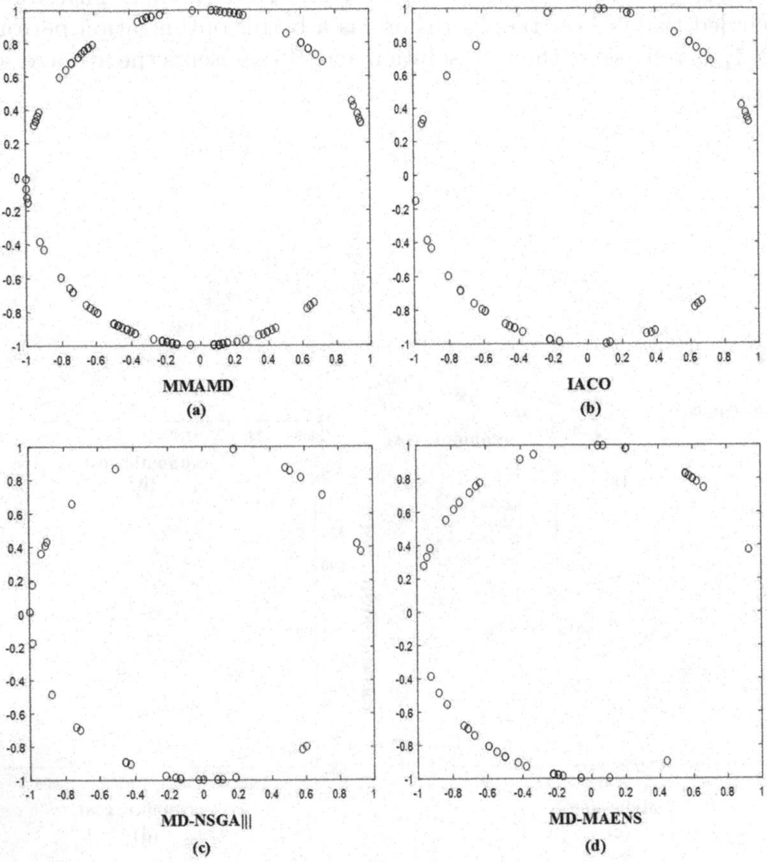

Fig. 7. Visualization of non-dominated solutions.

Table 2. HV value comparison.

Depot number	MMAMD	IACO	MD-NSGA-III	MD-MAENS
2	0.9698	0.7450	0.9050	0.9406
3	0.9622	0.8953	0.9216	0.8852
4	0.9852	0.8964	0.9213	0.8274
5	0.9943	0.7964	0.8650	0.9785

5 Conclusions

In this paper, we study the optimization of the waste collection capacitated arc routing problem with multi-depots by considering carbon emissions. By considering economic cost, makespan, and carbon emissions, a multi-objective capacitated arc routing model is constructed. A modified memetic algorithm for multi-depot (MMAMD) is proposed. To solve the problem of multi-depots, two-dimensional array solution structure and dynamic adjustment strategy of boundary arcs are proposed. The ant colony local search strategy with extended step size is proposed. An instance of Chicago city is used to test the performance of the algorithm. The results show that the algorithm has better convergence and diversity. The model and algorithm can provide methodological support for the optimization of waste collection activities in multi-depots under low carbon environment.

However, this paper only considers the problem of waste collection in multi-depots in low carbon. With the advent of $5G$, intelligent waste facilities will be emerging. In the process of waste collection, how to alert the amount of waste and collect waste by using smart sensors will be an essential research direction in the future.

Acknowledgements. This work was supported in part by the National Natural Science Foundation of China (NSFC) under Grant No. 61976242, in part by the Opening Project of Guangdong Province Key Laboratory of Computational Science at the Sun Yat-sen University under Grant No. 2018002, in part by the Open Fund of Key Laboratory of Intelligent Perception and Image Understanding of Ministry of Education under Grant No. IPIU2019003, and in part by the State Key Program of National Natural Science of China under Grant No. 61836009.

References

1. Tirkolaee, E.B., Alinaghian, M., Hosseinabadi, A.A.R., Sasi, M.B., Sangaiah, A.K.: An improved ant colony optimization for the multi-trip Capacitated Arc Routing Problem. Comput. Electr. Eng. **77**, 457–470 (2019)
2. Golden, B.L., Wong, R.T.: Capacitated arc routing problems. Networks **11**(3), 305–315 (1981)
3. Lacomme, P., Prins, C., Ramdane, C.W.: Competitive memetic algorithms for Arc Routing Problems. Ann. Oper. Res. **131**(4), 159–185 (2004). https://doi.org/10.1023/B:ANOR.0000039517.35989.6d
4. Pia, A.D., Filippi, C.: A variable neighborhood descent algorithm for a real waste collection problem with mobile depots. Intl. Trans. Oper. Res. **13**(2), 125–141 (2006)
5. Liu, T., Jiang, Z., Geng, N.: A genetic local search algorithm for the multi-depot heterogene-ous fleet capacitated arc routing problem. Flex. Serv. Manuf. J. **26**(4), 540–564 (2012). https://doi.org/10.1007/s10696-012-9166-z
6. Krushinsky, D., Woensel, T.V.: An approach to the asymmetric multi-depot capacitated arc routing problem. Eur. J. Oper. Res. **244**(1), 100–109 (2015)

7. Erfan, T., Ali, H., Mehdi, S., Arun, S., Jin, W.: A hybrid genetic algorithm for multi-trip Green Capacitated Arc Routing Problem in the scope of urban services. Sustainability **10**(5), 1366–1378 (2018)
8. Ge, X., Miao, G., Tan, B.: Research on optimization modeling and algorithm of open pollution path problem. Ind. Eng. Manage. **20**(4), 46–53 (2015)
9. Mei, Y., Tang, K., Yao, X.: Decomposition-based memetic algorithm for multi-objective Capacitated Arc Routing Problem. IEEE Trans. Evol. Comput. **15**(2), 151–165 (2011)
10. Shang, R., Dai, K., Jiao, L., Stolkin, R.E.: Improved memetic algorithm based on route distance grouping for multiobjective large scale Capacitated Arc Routing Problems. IEEE Trans. Cybern. **46**(4), 1000–1013 (2017)
11. Amberg, A., Domschke, W., Vob, S.: Multiple center capacitated arc routing problems: a tabu search algorithm using capacitated trees. Eur. J. Oper. Res. **124**(2), 360–376 (2000)
12. Zhu, Z., et al.: A hybrid genetic algorithm for the multiple depot capacitated arc routing problem. In: 2007 IEEE International Conference on Automation and Logistics, pp. 2253–2258. IEEE (2007)
13. Li, Y., Soleimani, H., Zohal, M.: An improved ant colony optimization algorithm for the multi-depot green vehicle routing problem with multiple objectives. J. Clean Prod. **10**(227), 1161–1172 (2019)
14. Bi, Z.S., Zheng, J.B., Cai, G.Y.: Multi-vehicle vehicle routing problem based on high-dimensional multi-objective optimization. Comput. Digit. Eng. **45**(07), 1298–1304 (2017)
15. He, Z., Yen, G.: Visualization and performance metric in many-objective optimization. IEEE Trans. Evol. Comput. **20**(3), 386–402 (2016)
16. Jiang, S., Ong, Y.S., Zhang, J.: Consistencies and contradictions of performance metrics in multiobjective optimization. IEEE Trans. Cybern. **44**(12), 2391–2404 (2014)

Multi-objective Pick-up Point Location Optimization Based on a Modified Genetic Algorithm

Shuai Chen[1,2,3](✉), Bin Cao[1,2,3](✉), and Ruichang Li[1,2,3](✉)

[1] Hebei University of Technology, Tianjin 300401, China
caobin@scse.hebut.edu.cn, {201731704047,201731704023}@stu.hebut.edu.cn
[2] Key Laboratory of Intelligent Perception and Image Understanding of Ministry of Education, Xi'an 710071, China
[3] Hebei Provincial Key Laboratory of Big Data Calculation, Tianjin 300401, China

Abstract. To solve the logistics terminal distribution problem, we construct the multi-objective location allocation model of pick-up point. The model considers the characteristics of express logistics system, enterprise demand and customer demand for delivery distance, and two objectives of minimizing location cost and maximizing customer distance satisfaction are included. We set the segmentation distance function according to the distance of the user from the pick-up point, and calculate the satisfaction function. To solve the model, we modify the non-dominated sorting genetic algorithm with elite strategy algorithm (NSGA-II). Consider the needs of the enterprise or user, we modify the crossover operator, and use the tournament method to evaluate the offspring population generated by genetic operations to reduce the loss of elite individuals. We also verify the effectiveness of the model and the modified algorithm through experiments. Experimental results show that the modified algorithm can obtain a better Pareto optimal solution set for the multi-objective location-allocation problem.

Keywords: Pick-up point · Multi-objective location-allocation model · NSGA-II

1 Introduction

The problem of logistics terminal distribution restricts the development of enterprises. As a solution to the last mile delivery problem, pick-up points can effectively improve the operational efficiency of terminal distribution, increase customer recognition, and expand the market share and competitiveness of enterprises. The location allocation problem (LAP) [1] includes the contents that determining the positions of the pick-up points and assigning the users. The types of pick-up points include convenience stores, smart express boxes, campus pick-up points, etc. Brimberg and Revelle [2] proposed a bi-objective model, which contains two objective functions: minimizing the total cost and minimizing the

L. Pan et al. (Eds.): BIC-TA 2019, CCIS 1159, pp. 751–760, 2020.
https://doi.org/10.1007/978-981-15-3425-6_60

number of demand points. In the research of multi-objective location-allocation model, the objective functions usually include cost, profit, customer satisfaction, etc. For the issue of emergency logistics site selection, the environmental risk and reliability are also considered [3–5]. Liao et al. [6] considered inventory to construct the multi-objective model, and studied the joint inventory-location problem. Konak et al. [7] proposed a new modeling approach to the competitive facility site selection problem in which multiple competitors aim to maximize their market shares. Bilir et al. [8] designed an optimization model for multi-objective supply chain networks with the goal of maximizing profit, maximizing sales and minimizing supply chain risk. Sheu et al. [9] took cost and benefit as goals, and used rough set theory and fuzzy decision theory to study the problem of determining the distribution center area. Grunert et al. [10] studied the optimization problem of direct flight network and applied the mixed tabu search algorithm to solve the construct model. Yang et al. [11] studied the two-stage capacitated facility location problem, and proposed a new solution scheme to solve the problem. Zhao et al. [12] constructed a three-objective decision-making optimization model, considering the spatial relationship of facilities and the potential hazards of environmental accidents based on the analysis of the characteristics of major cities, and they used the multi-objective genetic algorithm to solve the problem. Most of the above research on location-allocation problem considers the factor of enterprises. However, this paper studies the two main bodies of enterprises and customers simultaneously.

Based on the characteristics of pick-up points, we construct a multi-objective location-allocation model from perspectives of both enterprises and consumers, and modify the NSGA-II algorithm to solve this problem.

2 Mathematical Formulation

In the case of the location-allocation model, the courier delivers the goods from the distribution center to each pick-up point, and the user will freely arrange the time to pick up the goods from the pick-up point. Under the premise of understanding the user's needs and geographical location, the position of the pick-up point can be determined and the users served by the pick-up point can be identified, as shown in Fig. 1.

2.1 Assumptions and Notions

To simplify some calculations, we assume the following conditions:

(i) This paper studies multi-objective location-allocation problem, and the impact of competitive enterprises on site selection is not considered.
(ii) There are many types of pick-up points. This section only considers the pick-up point of the unattended type.

Fig. 1. Diagram of pick-up point mode.

(iii) The data contains multiple customer groups, and this paper does not consider the distance between individual customers.
(iv) The alternative pick-up points are all within the delivery range of the enterprise.

The symbol description is listed in Table 1.

Table 1. Notions in the formulation.

Set	Desciption	Set	Desciption
M	Alternative collection of points	q_i	Demand for customer group i,$i \in N$
N	Collection of customer groups	d_{ij}	Distance of arc(i, j), $i \in M$, $j \in N$
f_j	Fixed cost, $j \in M$	u	Maximum distance effect
f_1	Unit operating cost	v	Minimum distance effect
f_2	Open cost of unit capacity	Q_j	Point capacity
x_j	Equal to 1 if the pick-up point is open; 0, otherwise	y_{ij}	Equal to 1 if user i is assigned to the point j; 0, otherwise

2.2 Multi-objective Location-Allocation Model

To measure the degree, we use the user-to-candidate distance as the main factor of service quality satisfaction and construct the satisfaction function [13]. The distance is segmented, wherein the [0, u] distance segment is the maximum satisfaction value, and it is set to 1; the distance between [v, ∞] indicates that the user feels unsatisfied, and the value is set to 0. If distance is between [u, v], the satisfaction value is the difference between u and actual value divided by the difference between u and v. Satisfaction can be expressed by $S(d_{ij})$. For pick-up point j, the total demand of customer group d serviced by pick-up point j is $\mathrm{q}_j^{\mathrm{d}}$. The model can be formulated as follows:

$$\min F_1 = \sum_{j \in M} x_j (f_j + Q_j f_2) + \sum_{i \in N} \sum_{j \in M} q_i f_1 x_j y_{ij} \tag{1}$$

$$\max F_2 = \sum_{i \in N} \sum_{j \in M} S(d_{ij}) y_{ij} \tag{2}$$

subject to:

$$\mathrm{q}_j^{\mathrm{d}} < Q_j \tag{3}$$

$$\sum_{j \in M} f_j x_j \leq C \tag{4}$$

$$\sum_{j \in M} x_j \leq \mathrm{n} \tag{5}$$

$$\sum_{j \in M} y_{ij} = 1, i \in N, j \in M \tag{6}$$

$$x_j \in \{0,1\}, \forall j \in M \tag{7}$$

$$y_{ij} \in \{0,1\}. \forall j \in M, \forall i \in N \tag{8}$$

The objective function (1) is used to minimize the cost of construction and operation. The objective function (2) is used to maximize customer distance satisfaction. The constraint condition (3) indicates that the capacity of the pick-up point should be greater than the total demand of the users served. Constraint (4) indicates that the fixed cost should be lower than the budgeted cost. The constraint condition (5) indicates that the actual number of pick-up points should not be higher than the planned number. The constraint (6) indicates that each user group can only be served by one pick-up point. Constraints (7) and (8) indicate x_j and y_{ij} are two-valued.

3 Algorithm Design

3.1 The Modified Algorithm

To solve the multi-objective location-allocation problem, we modify the NSGA-II algorithm, which has been improved on the basis of NSGA by joining the elite strategy [14]. NSGA-II uses a random method to generate initial populations. When dealing with multi-objective location problems, a large number of duplicate individuals will be generated, and it is difficult to avoid premature phenomenon. Therefore, we consider using the tournament evaluation method to improve the algorithm screening mechanism [15].

NSGA-II uses the SBX (Simulated Binary Crossover) operator, and if the number of genes contained in the population is large, it will lead to the occurrence of super-individual. If the number of genes is small, a good individual may lost, which has a certain impact on the optimization efficiency. Therefore, we consider improving the crossover operator via introducing the Gaussian distribution into the iterative process [16]. The specific steps are as follows:

$$A = \frac{2\sigma}{y_{1,i} - y_{2,i}} \tag{9}$$

$$P_1 = \int_{-1/A}^{1/A} \frac{1}{\sqrt{2\pi}} e^{-x^2/2} dX \tag{10}$$

$$P_2 = 1 - P_1 \tag{11}$$

Compared with SBX, when $P_1 = P_2$, A is calculated. Therefore, the random variable is replaced in SBX with $A\,|N(0,1)|$. We named the modified algorithm MNSGA-II.

3.2 Main Steps of the Modified Algorithm

The main steps of the modified algorithm are as follows:

Step 1: Chromosome encoding and decoding. We separately code the user and the pick-up point. The first layer encodes the users' demand point. The number of demand points is a_1, and the number of pick-up points planned to be established is b. So the length of first layer code is $(a_1 + b - 1)$. The specific encoding and decoding process is shown in Fig. 2. The second layer of the coding object is the position of pick-up point, and the coding length is a. The decoding method uses the roulette rules to select the pick-up piont for the first group of customers, and the probability of each point being selected is equal. Then, remove the selected network points, and each network point is selected with a probability of $1/(a-1)$. We repeat this for m times to complete the site selection of the pick-up point.
Step 2: Randomly initialize the population.
Step 3: The fitness of objective 1 is equal to $F1$, and the fitness function of objective function 2 is $FitF2$, $FitF2 = M - F2$, where M is the maximum satisfaction of the customer group under the ideal state, and the maximum satisfaction value is 1.
Step 4: Sort all individuals in the population.
Step 5: After crossover and mutation, the offspring population is generated.
Step 6: Introduce a binary tournament system for the comparison of the offspring population.
Step 7: Mix the individuals in the parent population with the well-selected offspring populations, and adopt the elite strategy to form a new population.
Step 8: If the evolution termination condition is reached, the loop is stopped and the results are outputted. Otherwise, return to Step 2.

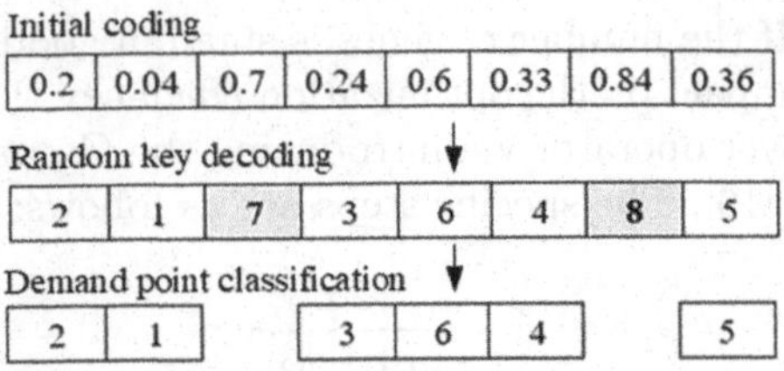

Fig. 2. User allocation coding and decoding diagram.

4 Experiments

4.1 Case Study and Analysis

The experimental data was obtained from the database provided by Augerat and the researchers made the appropriate modifications [17]. The data source is A-n36-k5.vrp, including customer demand and coordinate information. The details of the pick-up point information are listed in Table 2, and user coordinate and demand information is shown in Table 3. The cost of goods circulation is set to 1, and the open cost of the pick-up point capacity is set to 0.3. In real-world application, parameters can change according to actual conditions. The population size is 50, the crossover probability is 0.7, the mutation probability

Table 2. Alternative point information.

Num	x axis	y axis	Capacity	Cost	Num	x axis	y axis	Capacity	Cost
1	15	19	450	640	2	19	75	480	580
3	31	87	500	620	4	71	41	450	400
5	61	83	520	560	6	59	51	480	480

Table 3. User coordinate and demand information.

Num	x	y	q_i	Num	x	y	q_i	Num	x	y	q_i
1	1	49	7	11	19	47	40	21	15	79	37
2	87	25	46	12	57	63	66	22	79	47	22
3	69	65	49	13	5	95	10	23	19	65	43
4	93	91	37	14	65	43	46	24	27	49	61
5	33	31	58	15	69	1	31	25	29	17	43
6	71	61	10	16	3	25	34	26	25	65	28
7	29	9	70	17	19	91	16	27	27	95	58
8	93	7	25	18	21	81	61	28	21	91	37
9	55	47	58	19	67	91	10	29	15	83	40
10	23	13	73	20	41	23	64	30	91	21	37

is 0.4, and the number of iterations is 300. The distance segmentation upper limit is 70 and the lower limit is 30. Algorithm running environment is Matlab R 2016a.

4.2 Optimization Results and Analysis

According to the model and data, after several experiments, the Pareto optimal solution set is generated, as shown in Table 4. We decode the Pareto optimal solutions obtained by the algorithm, and the decoded schemes are shown in Tables 5 and 6, respectively, corresponding to Scheme 2 and Scheme 6, respectively. When making decisions, enterprises should consider cost and satisfaction simultaneously, and choose the most suitable solution according to the actual situation.

Table 4. Pareto solution.

Program	1	2	3	4	5	6
F1	3092	3100	3146	3816	3875	3947
F2	21.9162	25.9823	26.3974	26.1940	27.203	28.2836

Table 5. Pick-up point selection and customer allocation information (Scheme 2).

Point	Customer allocation	User sum	Total demand
2	[1,7,11,13,17,18,21,23,26,27,28,29]	12	447
4	[3,4,8,9,12,14,19,22,30]	9	350
6	[2,5,6,10,15,16,20,24,25]	9	420

Table 6. Pick-up point selection and customer allocation information (Scheme 6).

Point	Customer allocation	User sum	Total demand
1	[5,7,10,11,16,25]	6	318
2	[1,13,17,18,21,23,24,26,27,28,29]	11	398
4	[3,8,9,12,15,20,30]	7	330
6	[2,4,6,14,19,22]	6	171

4.3 Algorithm Comparison

The modified algorithm is compared with the traditional NSGA-II algorithm and the multi-objective particle swarm optimization algorithm (MOPSO) [18]. For MOPSO, the number of particles is 50, the external archive size is 50, and the inertia weight is nonlinearly decremented. The same coding method is adopted in all the three algorithms, the number of iterations is 300, and the algorithms run 10 times. The pick-up point number constraint is set to three. The obtained results are shown in Fig. 3, where a, b and c represent the distribution of Pareto optimal solutions generated by MNSGA-II, NSGA-II and MOPSO, respectively.

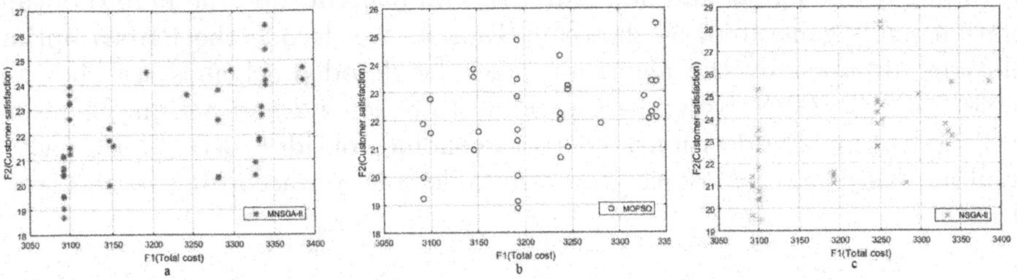

Fig. 3. The distribution of the Pareto optimal solution of the three algorithms on the plane.

To verify the performance of the modified algorithm, we use different numbers of pick-up points for experiments. Since the actual Pareto solutions of this example are unknown, this paper selects the Hypervolume (HV) indicator [19]. The HV index can evaluate the convergence and diversity of the algorithm, and larger values indicate better results. The algorithm runs 10 times. We calculate the HV values and take the average of all runs as the evaluation criteria. As shown in Table 7, MNSGA-II has better optimization performance with respect to all cases.

Table 7. HV value comparison.

Point restriction	MNSGA-II	NSGA-II	MOPSO
3	0.720586736	0.586680535	0.633257646
4	0.842795545	0.721036696	0.671187334
5	0.745143343	0.636704715	0.578207882

5 Conclusions

The application of the pick-up point mode can effectively solve the problem of logistics terminal distribution. For users, they can arrange the pick-up time independently. For logistics companies, they can improve work efficiency and reduce delivery costs. This paper construct a multi-objective location-allocation model, which takes into account the two main participants of the user and the express company, including constraints such as cost and users' expectation distance.

To solve the model, we propose the modified genetic algorithm via introducing the tournament evaluation method and the improved crossover operator, which can further optimize and generate the elite individuals participating in subsequent iterations, and obtain a better Pareto optimal solution set. Through experiments, we obtain a set of Pareto optimal solutions, and give the specific scheme after decoding. Finally, we conduct the comparative experiments, which show that the modified algorithm can gain better Pareto optimal solutions.

Acknowledgements. This work was supported in part by the National Natural Science Foundation of China (NSFC) under Grant No. 61976242, in part by the Opening Project of Guangdong Province Key Laboratory of Computational Science at the Sun Yat-sen University under Grant No. 2018002, in part by the Open Fund of Key Laboratory of Intelligent Perception and Image Understanding of Ministry of Education under Grant No. IPIU2019003, and in part by the State Key Program of National Natural Science of China under Grant No. 61836009.

References

1. Abin, A.A.: Querying beneficial constraints before clustering using facility location analysis. IEEE Trans. Cybern. **48**(1), 312–323 (2018)
2. Brimberg, J., Revelle, C.: A bi-objective plant location problem: cost vs. demand served. Location Sci. **6**(1–4), 121–135 (1998)
3. Akgul, O., Shah, N., Papageorgiou, L.G.: An optimisation framework for a hybrid first/second generation bioethanol supply chain. Comput. Chem. Eng. **42**, 101–114 (2012)
4. Olivares-Benitez, E., Ríos-Mercado, R.Z., González-Velarde, J.L.: A metaheuristic algorithm to solve the selection of transportation channels in supply chain design. Int. J. Prod. Econ. **145**(1), 161–172 (2013)
5. Prakash, A., Chan, F.T.S., Liao, H., Deshmukh, S.G.: Network optimization in supply chain: a KBGA approach. Decis. Support Syst. **52**, 528–538 (2012)
6. Shankar, B.L., Basavarajappa, S., Chen, J.C.H., Kadadevaramath, R.S.: Location and allocation decisions for multi-echelon supply chain network - a multi-objective evolutionary approach. Exp. Syst. Appl. **40**(2), 551–562 (2013)
7. Konak, A., Kulturel-Konak, S., Snyder, L.: A multi-objective approach to the competitive facility location problem. Proc. Comput. Sci. **108**, 1434–1442 (2017)
8. Bilir, C., Ekici, S.O., Ulengin, F.: An integrated multi-objective supply chain network and competitive facility location model. Comput. Ind. Eng. **108**, 136–148 (2017)
9. Sheu, J.B., Lin, A.Y.S.: Hierarchical facility network planning model for global logistics net-work configurations. Appl. Math. Model. **36**(7), 3053–3066 (2012)
10. Büdenbender, K., Grünert, T., Sebastian, H.J.: A hybrid tabu search/ branch-and-bound algorithm for the direct flight network design problem. Transp. Sci. **34**(4), 364–380 (2000)
11. Yang, Z., Chen, H., Chu, F., Wang, N.: An effective hybrid approach to the two-stage capacitated facility location problem. Eur. J. Oper. Res. **275**(2), 467–480 (2019)
12. Zhao, M., Chen, Q.: Risk-based optimization of emergency rescue facilities locations for large-scale environmental accidents to improve urban public safety. Nat. Hazards **75**(1), 163–189 (2014). https://doi.org/10.1007/s11069-014-1313-2

13. Drezner, Z., Wesolowsky, G.O., Drezner, T.: The gradual covering problem. Nav. Res. Log. **51**(6), 841–855 (2004)
14. Deb, K., Pratap, A., Agarwal, S., Meyarivan, T.A.M.T.: A fast and elitist multiobjective genetic algorithm: NSGA-II. IEEE Trans. Evol. Comput. **6**(2), 182–197 (2002)
15. Azad, M.A.K., Fernandes, E.M.G.P.: A modified differential evolution based solution technique for economic dispatch problems. J. Ind. Manage. Optim. **8**(4), 1017–1038 (2012)
16. Jiang, Q., Wang, L., Hei, X., Yu, G.L., Lin, Y.Y., Lu, X.F.: MOEA/D-ARA+SBX: a new multi-objective evolutionary algorithm based on decomposition with artificial raindrop algorithm and simulated binary crossove. Knowl.-Based Syst. **107**, 197–218 (2016)
17. Lu, F., Li, Y.H.: Model and algorithm of inventory path problem for spare parts logistics system. Ind. Eng. Manage. **15**(2), 82–86 (2010)
18. Ding, S., Chen, C., Xin, B., Pardalos, M.P.: A bi-objective load balancing model in a distributed simulation system using NSGA-II and MOPSO approaches. Appl. Soft. Comput. **63**, 249–267 (2018)
19. Jiang, S., Ong, Y.S., Zhang, J., Feng, L.: Consistencies and contradictions of performance metrics in multiobjective optimization. IEEE Trans. Cybern. **44**(12), 2391–2404 (2014)

Efficient Evolutionary Neural Architecture Search (NAS) by Modular Inheritable Crossover

Hao Tan, Cheng He(✉), Dexuan Tang, and Ran Cheng(✉)

Shenzhen Key Laboratory of Computational Intelligence,
University Key Laboratory of Evolving Intelligent Systems of Guangdong Province,
Department of Computer Science and Engineering,
Southern University of Science and Technology,
Shenzhen 518055, China
tanbox@live.com, chenghehust@gmail.com, tdx1997tdx@gmail.com,
ranchengcn@gmail.com

Abstract. The convolution neural network is prominent in image processing, and a large number of excellent deep neural networks have been proposed in recent years. However, the hand-actuated design of a neural network is time-consuming, laborious, and challenging. Thus many neural architecture search (NAS) methods have been proposed, among which the evolutionary NAS methods have achieved encouraging results due to the global search capability of evolutionary algorithms. Nevertheless, most evolutionary NAS methods use only mutation operators for offspring generation, and the generated offspring networks could be quite different from their parent networks. To address this deficiency, we propose an efficient evolutionary NAS method using a tailored crossover operator. Different from existing mutation operators, the proposed crossover operator enables the offspring network to inherit promising modular from their parent networks. Experimental results indicate that our proposed evolutionary NAS method has achieved competitive results in comparison with some state-of-the-art NAS methods. Moreover, the effectiveness of our proposed modular inheritable crossover operator for offspring generation is validated.

Keywords: Neural architecture search · Evolutionary algorithm · Crossover

1 Introduction

Convolutional neural networks (CNNs) have achieved remarkable results in image processing, and many state-of-the-art image classifiers have been designed by experts in the last decade [3,5–7,17]. Nevertheless, the design of deep neural networks manually remains challenging, and it is computationally expensive to design a specific CNN model for solving a specific task. Therefore, people have

L. Pan et al. (Eds.): BIC-TA 2019, CCIS 1159, pp. 761–769, 2020.
https://doi.org/10.1007/978-981-15-3425-6_61

proposed methods which can automatically generate the high-accuracy neural networks for any given task, i.e. automatic neural architecture search (NAS) [14]. Methods such as reinforcement learning (RL), evolutionary algorithm (EA), and gradient-based search have been used for NAS [2,10,18].

With the progress in computing devices, more and more researchers have attempted to conduct NAS by using EAs due to their population-based property and capability in global optimization. The early research in [14] made it possible for evolutionary NAS methods to surpass the human-acted neural network model in terms of both accuracy and computing economization. To be more specific, the population in an EA consists of many individuals (models) and is updated iteratively. At each iteration (or generation in terms of evolution), pairwise neural networks are selected as parent models for offspring generation. Then, the generated offspring models are merged into the population, and the models with poor performance (e.g. in terms of accuracy) are removed from the merged population. Consequently, the accuracy of the best model obtained by EA has surpassed the best human-designed models. Since this method should handle fairly large search space, it requires too much computing resource.

Recently, a number of works have been dedicated efficient evolutionary NAS [19]. For example, some mutation operators introduced function-preserving operations for offspring generation, which reduced the computation time significantly [8]. Some others turn to EAs with efficient environmental selection strategies for accelerating the convergence rate of EAs. For instance, an aging based environmental selection strategy was used in [13], where the oldest model instead of the worst performance one is removed from the population during the evolution. The obtained architectures have high accuracy and outperform the classifiers designed by experts. However, the GPU cost is still too high, since most previous research only uses mutation operators without crossover. Inspire by NEAT [16], in this paper we propose a modular inheritable crossover operator for offspring generation in evolutionary deep NAS. The main new contributions are summarized as follows:

- A modular inheritable crossover operator is proposed in our evolutionary NAS method. To be best of our knowledge, it is one of the very first tailored crossover operators in evolutionary NAS literature, which enables the offspring model to inherit their parent modules.
- The proposed evolutionary NAS method is compared with four state-of-the-art methods. Experimental results have demonstrated the efficiency of our proposed method.

The rest of this paper is organized as follows. The details of the proposed method for NAS are described in Sect. 2. Experimental settings and comparisons of optimization with the state-of-the-art methods on CIFAR-10 classification task are presented in Sect. 3. Conclusions are drawn in Sect. 4.

2 The Proposed Method

In this section, we first describe the definition of search space in our proposed NAS method, followed by the adopted modular inheritable crossover operator and mutation operators. Then the details of the adopted evolutionary algorithm are illustrated.

2.1 Search Space

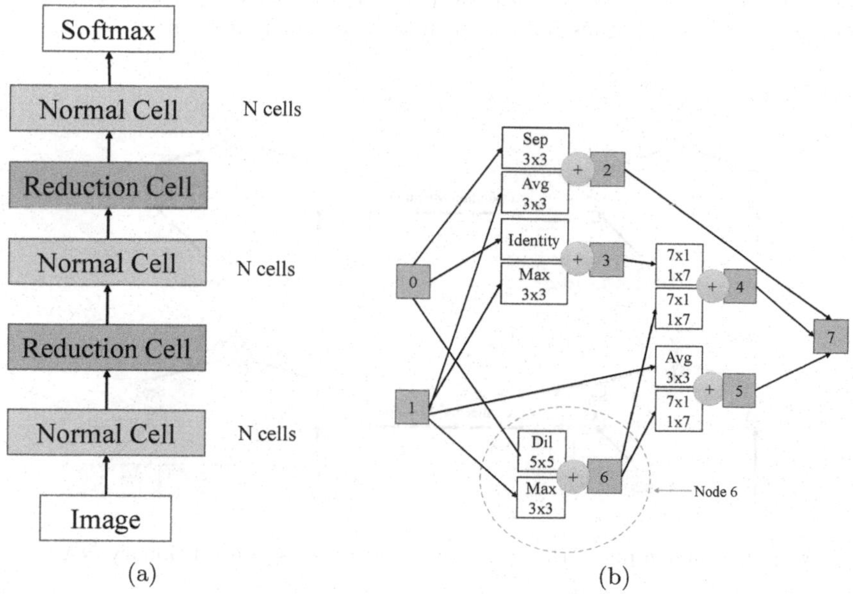

Fig. 1. The illustrative diagram of NASNet. (a) An overview of the cell-based structure in our proposed evolutionary NAS method. (b) An example of the detailed cell structure.

The overall cell-based structure and an example of a cell are presented in Fig. 1, which mainly involves two types of cells [21], i.e., the normal cell and the reduction cell. To be more specific, the normal cell ensures that the dimension of the output is the same as the dimension of the input. As for the reduction cell, it reduces the dimension of the input by a stride with step size of two. In NASNet [21], the reduction cell is connected after a stack of N normal cells. Note that all the normal cells have the same architecture and so as the reduction cells. Figure 1(b) has shown a cell with seven nodes, where node 0 and node 1 are the output of two cells in front of this cell. Each node contains the ID information (denoting its sequential location) of the two input nodes and their corresponding operations. In this work, there are nine different operations, including 3×3 max pooling, 3×3 average pooling, three depth-wise separable convolutions [4] (3×3, 5×5, and 7×7), two dilated separable convolutions (3×3 and 5×5), two norm convolutions (7×1 and 1×7), and identity. Consequently, the search space is restricted in this cell-based structure.

2.2 Mutation and Crossover

There are mainly two types of mutations [13], i.e. the operation mutation and the connection mutation. The first type changes the operation between pairwise nodes, while the second one mutates the input node of a selected node. An example of these two types of mutations are presented in Fig. 2, where the red rectangles with dashed lines indicate the locations of mutation.

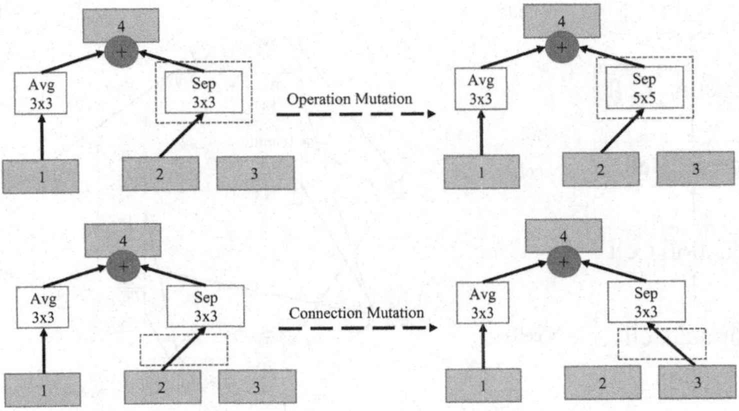

Fig. 2. An example of the two types of mutations in evolutionary NAS.

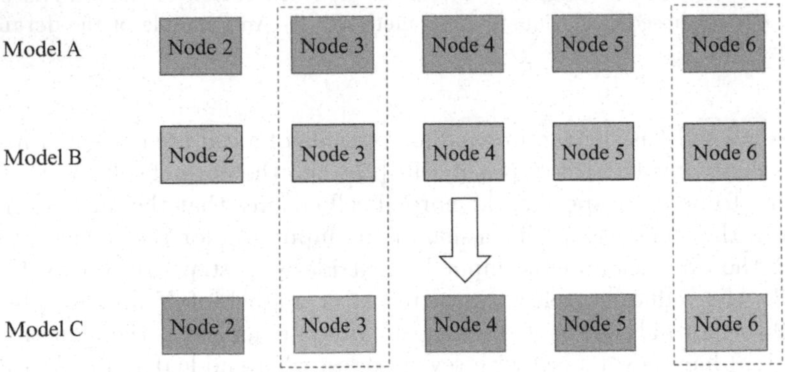

Fig. 3. An example of the modular inheritable crossover operator in our proposed evolutionary NAS method.

Since the mutation operators mutate the model step by step, it takes much time for NAS methods to obtain a promising network. To address this issue, we propose a modular inheritable crossover operator to accelerate the convergence rate of EAs. Note that the nodes of parent models are aligned according to their IDs, and only the nodes with the same IDs can be exchanged. An example of the crossover operator is shown in Fig. 3, where model A and model B are two parent cells and model C is the generated cell. Model C consists of three nodes from model A (i.e., node 2, node 4, and node 5) and two nodes from model B (i.e. node 3 and node 6). It can be observed that model C is quite different from model A and model B, indicating a significant structural variation from its parents while inheriting some modular parts.

2.3 Evolutionary NAS Method

Algorithm 1. Proposed Evolutionary NAS Method

$P \leftarrow \varnothing$
for $i \leftarrow 1 : n$ **do**
  $model(i).dna \leftarrow$ Random Initialization
  $model(i).accuracy \leftarrow$ Fitness($model(i).dna$)
  add $model(i)$ to population P
end for
while termination condition is not satisfied **do**
  $sample \leftarrow$ Randomly select s models from P
  $parents \leftarrow$ Tournament Selection($sample$)
  $child.dna \leftarrow$ Crossover($parents[0].dna, parents[1].dna$)
  $child.dna \leftarrow$ Mutation($child.dna$)
  $child.accuracy \leftarrow$ Fitness($child.dna$)
  remove the worst model in P
end while
return The best model in P

The details of our proposed evolutionary NAS method are presented in Algorithm 1. To begin with, a population P with size n is initialized and their fitness values are the corresponding prediction accuracies, where the initial models are randomly generated. Then we select s parent models randomly, and the tournament selection [1] is applied to select two parent models according to their fitness values. Afterwards, the crossover and mutation operators are applied for offspring generation, and the generated models are merged into P. Note that the mutation operator randomly chooses either operation mutation or connection mutation. Next, the model with the worst performance is removed from P. The above procedures are repeated until the termination condition is satisfied.

3 Experiments

In this part, we first present the detailed settings in our proposed evolutionary NAS method, followed by the parameter settings in the adopted training method. Finally, the ablation study is conducted to validate the effectiveness of our proposed modular inheritable crossover operator.

3.1 Experimental Settings

In our proposed evolutionary NAS method, the population size n is set to 100, the sample size s is set to 10, and the termination condition is set to the maximum number of explored models (it is set to 1000). The probability of operation mutation to identity is set to 0.05, and the probabilities of the rest mutation operations are set to the same as suggested in [13].

During the evolution of population, a small network with 11 cells (i.e. $N = 3$) is trained for 25 epochs, the batch size is set to 384 (for both the training and validation sets), and the initial number of channels is set to 24. Once the search process is finished, we further train the network with 20 cells (i.e. N is expanded to 4) for 600 epochs, the batch size is increased to 560 (for both the training and validation sets), and the initial number of channels is set to 36. The statistic gradient descent (SGD) algorithm [15], is adopted to train the network, where the initial learning rate lr is set to 0.025, the momentum rate is set to 0.9, and the weight decay is set to 3×10^{-4}. In addition, lr is decayed using a cosine annealing schedule [12].

To examine the performance of the proposed method, we compare four state-of-the-art NAS methods on CIFAR-10 classification task [9].

3.2 General Performance

The experimental results on CIFAR-10 achieved by our proposed method and the four compared methods are presented in Table 1. It can be observed that the network obtained by our proposed method is competitive to the four state-of-the-art NAS methods in terms of accuracy. Moreover, our method has obtained a model with low complexity (parameters) and high efficiency (GPU days).

3.3 Detailed Structures

Once the NAS is completed, the model with the highest accuracy in P is selected as the output, and the obtained structures of the normal cell and reduction cell are presented in Fig. 4. It can be observed that the normal cell includes more identity operations, which may be attributed to the fact that the identity operation can reduce the model complexity significantly.

Table 1. Comparison with four state-of-the-art NAS methods on CIFAR-10 classification task.

Model	Test error (%)	Params (M)	Search cost (GPU days)	Search method
NASNet-A [21]	3.3	3.41	2000	RL
AmoebaNet-A [13]	3.2	3.34	3150	EA
Hierarchical evolution [11]	3.75	15.7	300	EA
BlockQNN [20]	3.54	39.8	96	RL
Ours	4.8	3.79	15	EA

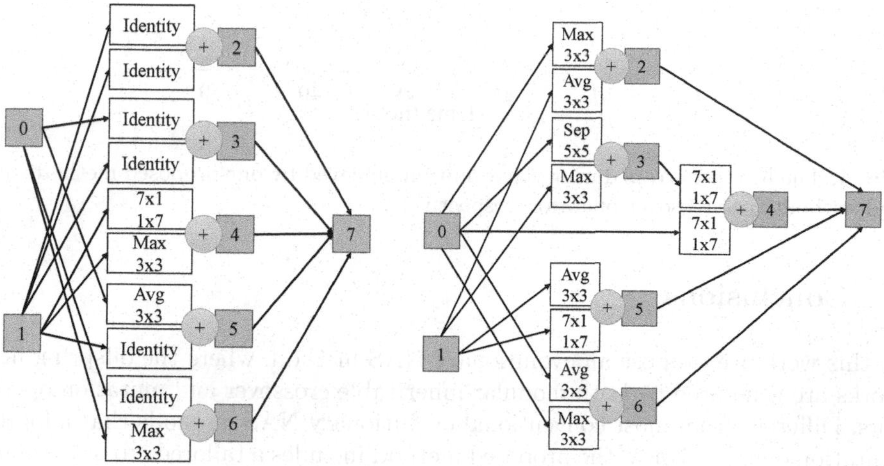

Fig. 4. The obtained structures of the normal cell and reduction cell.

3.4 Ablation Study

To validate the effectiveness of our proposed modular inheritable crossover operator, we compare our proposed method with its modified version whose crossover operator is ablated. The top test errors during the evolution are presented in Fig. 5. It can be observed that our proposed method has outperformed the modified version, indicating the effectiveness of our proposed crossover operator.

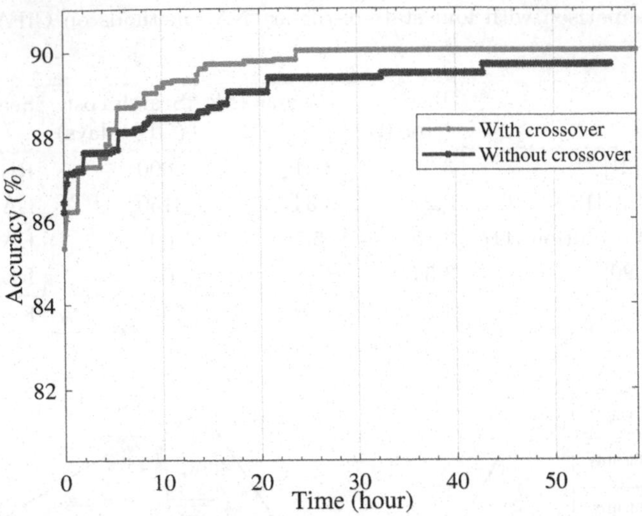

Fig. 5. The top test errors during the evolution achieved by our proposed method with and without the crossover operator, respectively.

4 Conclusions

In this work, we propose an evolutionary NAS method, where the offspring networks are generated by both modular inheritable crossover and mutation operators. Different from most conventional evolutionary NAS methods that adopted mutation operators only, the proposed method includes a tailored crossover operator which is capable of inheriting promising modules from two parent networks. Experimental results on the CIFAR-10 classification task in comparison with four state-of-the-art NAS methods have validated the efficiency of our proposed method. Moreover, the ablation study has demonstrated the effectiveness of the proposed modular inheritable crossover operator.

Acknowledgment. This work was supported in part by the National Natural Science Foundation of China (No. 61903178 and 61906081), in part by the Program for Guangdong Introducing Innovative and Entrepreneurial Teams grant (No. 2017ZT07X386), and in part by the Shenzhen Peacock Plan grant (No. KQTD2016112514355531),

References

1. Blickle, T., Thiele, L.: A mathematical analysis of tournament selection. In: ICGA, vol. 95, pp. 9–15. Citeseer (1995)
2. Cai, H., Chen, T., Zhang, W., Yu, Y., Wang, J.: Reinforcement learning for architecture search by network transformation (2017). arXiv preprint arXiv:1707.04873
3. Chen, Y., Li, J., Xiao, H., Jin, X., Yan, S., Feng, J.: Dual path networks. In: Advances in Neural Information Processing Systems, pp. 4467–4475 (2017)

4. Chollet, F.: Xception: deep learning with depthwise separable convolutions. In: Proceedings of the IEEE Conference on Computer Vision and Pattern Recognition, pp. 1251–1258 (2017)
5. He, K., Zhang, X., Ren, S., Sun, J.: Deep residual learning for image recognition. In: Proceedings of the IEEE Conference on Computer Vision and Pattern Recognition, pp. 770–778 (2016)
6. Hu, J., Shen, L., Sun, G.: Squeeze-and-excitation networks. In: Proceedings of the IEEE Conference on Computer Vision and Pattern Recognition, pp. 7132–7141 (2018)
7. Huang, G., Liu, Z., Van Der Maaten, L., Weinberger, K.Q.: Densely connected convolutional networks. In: Proceedings of the IEEE Conference on Computer Vision and Pattern Recognition, pp. 4700–4708 (2017)
8. Jin, H., Song, Q., Hu, X.: Efficient neural architecture search with network morphism (2018). arXiv preprint arXiv:1806.10282. 9
9. Krizhevsky, A., Hinton, G., et al.: Learning multiple layers of features from tiny images. Technical report Citeseer (2009)
10. Liu, C., et al.: Progressive neural architecture search. In: Proceedings of the European Conference on Computer Vision (ECCV), pp. 19–34 (2018)
11. Liu, H., Simonyan, K., Vinyals, O., Fernando, C., Kavukcuoglu, K.: Hierarchical representations for efficient architecture search (2017). arXiv preprint arXiv:1711.00436
12. Loshchilov, I., Hutter, F.: SGDR: Stochastic gradient descent with warm restarts (2016). arXiv preprint arXiv:1608.03983
13. Real, E., Aggarwal, A., Huang, Y., Le, Q.V.: Regularized evolution for image classifier architecture search. In: Proceedings of the AAAI Conference on Artificial Intelligence, vol. 33, pp. 4780–4789 (2019)
14. Real, E., et al.: Large-scale evolution of image classifiers. In: Proceedings of the 34th International Conference on Machine Learning, vol. 70, pp. 2902–2911. JMLR. org (2017)
15. Ruder, S.: An overview of gradient descent optimization algorithms (2016). arXiv preprint arXiv:1609.04747
16. Stanley, K.O., Miikkulainen, R.: Evolving neural networks through augmenting topologies. Evol. Comput. **10**(2), 99–127 (2002)
17. Szegedy, C., Vanhoucke, V., Ioffe, S., Shlens, J., Wojna, Z.: Rethinking the inception architecture for computer vision. In: Proceedings of the IEEE Conference on Computer Vision and Pattern Recognition, pp. 2818–2826 (2016)
18. Wen, W., Yan, F., Li, H.: Autogrow: automatic layer growing in deep convolutional networks (2019). arXiv preprint arXiv:1906.02909
19. Wistuba, M.: Deep learning architecture search by neuro-cell-based evolution with function-preserving mutations. In: Berlingerio, M., Bonchi, F., Gärtner, T., Hurley, N., Ifrim, G. (eds.) ECML PKDD 2018. LNCS (LNAI), vol. 11052, pp. 243–258. Springer, Cham (2019). https://doi.org/10.1007/978-3-030-10928-8_15
20. Zhong, Z., Yan, J., Wu, W., Shao, J., Liu, C.L.: Practical block-wise neural network architecture generation. In: Proceedings of the IEEE Conference on Computer Vision and Pattern Recognition, pp. 2423–2432 (2018)
21. Zoph, B., Vasudevan, V., Shlens, J., Le, Q.V.: Learning transferable architectures for scalable image recognition. In: Proceedings of the IEEE Conference on Computer Vision and Pattern Recognition, pp. 8697–8710 (2018)

Author Index

Printed in the United States
By Bookmasters